Test Items for
Stewart, Redlin, and Watson's

ALGEBRA AND TRIGONOMETRY

William Steven Harmon
Alex Taler

BROOKS/COLE
THOMSON LEARNING™

Australia • Canada • Mexico • Singapore • Spain • United Kingdom • United States

BROOKS/COLE
THOMSON LEARNING™

For more information about this or any other Brooks/Cole product, contact:
BROOKS/COLE
511 Forest Lodge Road
Pacific Grove, CA 93950 USA
www.brookscole.com
1-800-423-0563 (Thomson Learning Academic Resource Center)

Printed in the United States of America

5 4 3 2 1

ISBN 0-534-38271-1

CONTENTS

CHAPTER ONE

SECTION 1.1

1. The following list of facts illustrates a general principle about numbers. State the principle in algebraic notation.
   $5(2+3) = 5 \cdot 2 + 5 \cdot 3,\ 3(4+6) = 3 \cdot 4 + 3 \cdot 6$
   (a) $a \cdot (b \cdot c) = (a \cdot b) \cdot c$
   (b) $a \cdot (b + c) = a \cdot b + a \cdot c$
   (c) $a + (b + c) = (a + b) + c$
   (d) $(a \cdot b) + c = (a + b) \cdot (a + c)$
   (e) $a + b + c = c + b + a$
   Answer: easy (b)

2. The following list of facts illustrates a general principle about numbers. State the principle in algebraic notation.
   $(4+6)^2 = 4^2 + 2(4 \cdot 6) + 6^2$, $(3+8)^2 = 3^2 + 2(3 \cdot 8) + 8^2$
   (a) $(a+b)(a-b) = a^2 - b^2$
   (b) $a^2 + b^2 = (a+b)^2$
   (c) $(a-b)^2 = a^2 - 2(a \cdot b) + b^2$
   (d) $(a-b)^2 = a^2 - 2(a \cdot b) + b^2$
   (e) $(a+b)^2 = a^2 + 2(a \cdot b) + b^2$
   Answer: easy (e)

3. The following list of facts illustrates a general principle about numbers. State the principle in algebraic notation.
   $(4+6) - 2 = 4 + (6-2)$, $(1+5) - 3 = 1 + (5-3)$
   (a) $(a+b)(a-b) = a^2 - b^2$
   (b) $(a+b) - c = a + (b-c)$
   (c) $(a+b) \cdot c = a \cdot c + b \cdot c$
   (d) $(a-b) + c = a - (b+c)$
   (d) $(a-b) - c = a - (b+c)$
   Answer: easy (b)

4. The following list of facts illustrates a general principle about numbers. State the principle in algebraic notation.
   $(4 \cdot 5)^2 = 4^2 \cdot 5^2$, $(2 \cdot 6)^2 = 2^2 \cdot 6^2$
   (a) $(a \cdot b)^2 = a^2 \cdot b^2$
   (b) $a^2 + b^2 = (a \cdot b)^2$
   (c) $(a+b)^2 = a^2 \cdot b^2$
   (d) $(a \cdot b)^2 = a^2 + 2(a \cdot b) + b^2$
   (e) $(a \cdot b)^2 = a^2 - 2(a \cdot b) + b^2$
   Answer: easy (a)

5. Write an algebraic formula for the area of a rectangle whose length is 6 ft more than its width.
   (a) $A = (w+6)w$, where w is the width of the rectangle
   (b) $A = (w-6)w$, where w is the width of the rectangle
   (c) $A = \ell(\ell+6)$, where ℓ is the length of the rectangle
   (d) $A = (\ell+6)\ell$, where ℓ is the length of the rectangle
   (e) $A = w^2$, where w is the width of the rectangle
   Answer: medium (a)

6. Write an algebraic formula for the number of cents c in d dimes.
   (a) $c = 10d$
   (b) $d = 10c$
   (c) $c = 100d$
   (d) $d = \frac{1}{10}c$
   (e) $c = \frac{1}{100}d$
   Answer: medium (a)

7. Write an algebraic formula for the distance d in meters that a rocket travels in t seconds at a speed of r meters per second.
   (a) $d = r + t$
   (b) $d = r/t$
   (c) $d = rt$
   (d) $d = 100rt$
   (e) $d = 100t$
   Answer: medium (c)
   $r\,\frac{\text{m}}{\text{s}} \times t\,\text{s} = rt\,\text{m}$, so the formula is $d = rt$.

8. Write an algebraic formula for the number of feet f in y yards.
(a) $f = \frac{1}{3}y$ (b) $f = 100y$ (c) $f = 3y$ (d) $y = 3f$ (e) $y = 6f$
Answer: medium (c)

9. Write an algebraic formula for the cost c of a pound of peanuts, if you can buy p pounds for $\$x$.
(a) $c = px$ (b) $c = p/x$ (c) $c = p - x$ (d) $c = x + p$ (e) $c = x/p$
Answer: medium (e)

10. Write an algebraic formula for the product of a number and its cube.
(a) $p = x^3$ (b) $p = 2x^3$ (c) $p = x^2$ (d) $p = x^4$ (e) $p = x + x^3$
Answer: medium (d)

11. Write an algebraic formula for the cost c of sending p postcards, if each stamp costs x cents.
(a) $c = p/x$ (b) $c = px$ (c) $c = 2y$ (d) $c = 2px$ (e) $c = x/p$
Answer: medium (b)

12. Suppose that your scores on three college algebra tests are 76, 72 and 81, and that the only test left to take is the final exam. This exam, however, counts double (this means that your test score on the final exam counts twice in the formula for average).
(a) If you earn a score of 88 on the final exam, what is your average test score for the course?
(b) Write a formula for a student's average test score in terms of the three test scores a, b, and c, and the score f on the final exam.
(c) What score do you need to earn on the final exam in order for your average test score for the course to be 80?
Answer: medium
(a) The average test score is $\frac{76 + 72 + 81 + 2 \cdot 88}{5} = 81$
(b) The average test score is $\frac{a + b + c + 2f}{5}$
(c) $\frac{76 + 72 + 81 + 2f}{5} = 80 \quad \Leftrightarrow \quad 2f = 5 \cdot 80 - (76 + 72 + 81) = 171 \quad \Leftrightarrow \quad f = 85.5$

13. Isabelle can mow a lawn at a rate of 200 square feet per minute.
(a) What area can she mow in half an hour?
(b) What area can she mow in T minutes?
(c) How long would it take her to mow a lawn that is 100 ft wide and 150 ft long?
(d) If she wants to mow this lawn in one hour, what would her rate of mowing have to be?
Answer: medium
(a) $200 \frac{\text{ft}^2}{\text{min}} \times 30 \text{ min} = 6{,}000 \text{ ft}^2$
(b) $200 \frac{\text{ft}^2}{\text{min}} \times T \text{ min} = 200T \text{ ft}^2$
(c) The area of the lawn is $100 \text{ ft} \times 150 \text{ ft} = 15{,}000 \text{ ft}^2$. If T is the time required, then, from part (b), $200T = 15{,}000$, so $T = \frac{15{,}000}{200} = 75 \text{ min}$.
(d) If R is the rate of mowing, then $60R = 15{,}000$, so $R = \frac{15{,}000}{60} = 250 \frac{\text{ft}^2}{\text{min}}$.

14. Write an algebraic formula for the number of n nickels in q quarters.
(a) $n = 4q$ (b) $n = 25q$ (c) $n = \frac{q}{5}$ (d) $n = 5q$ (e) $q = 4n$
Answer: easy (d)

15. Write an algebraic formula for the difference of the cubes of two numbers.
(a) $a^3 - b^3$ (b) $(a - b)^3$ (c) $3(a - b)$ (d) $(a + b)^3$ (e) $(a^3 - b^3)^3$
Answer: easy (a)

16. Write an algebraic formula for the perimeter P of a rectangle whose length l is twice its width w.
(a) $P = 6w + 2l$ (b) $P = 4(w + l)$ (c) $P = 2w + 2l$ (d) $P = 6l$ (e) $P = 6w$

Answer: medium (e)
$P = 2w + 2l$, but $l = 2w$, so $P = 2w + 2(2w) = 6w$.

17. Write an algebraic formula for the volume of a cone whose height h is three times its radius r.
(a) $V = \frac{3\pi hr^2}{2}$ (b) $V = 3\pi hr^3$ (c) $V = \pi r^3$ (d) $V = \frac{\pi r^3}{3}$ (e) $V = \frac{\pi r^3}{h}$

Answer: medium (c)
$V = \frac{1}{3}\pi r^2 h$, but $h = 3r$, so $V = \frac{1}{3}\pi r^2(3r) = \pi r^3$.

18. A cylindrical metal container has a height h that is three times as long as its radius r.
(a) Find the volume of the can if its radius is 1 in.
(b) Find the radius of the container if the volume is 81π in^3.
(c) Write a formula for the volume V in terms of h.
Answer: medium
(a) The volume of a cylinder is $V = \pi r^2 h$, with $h = 3r$ and $r = 1$. Thus $V = \pi r^2(3r) = 3\pi r^3 \Rightarrow V = 3\pi(1)^3 = 3\pi$ in^3.
(b) $81\pi = 3\pi r^3 \Leftrightarrow r^3 = 27 \Leftrightarrow r = 3$ in
(c) $r = \frac{h}{3} \Rightarrow V = \pi r^2 h \Rightarrow V = \pi\left(\frac{h}{3}\right)^2 h = \frac{1}{9}\pi h^3$

SECTION 1.2

1. State the property of real numbers being used:
$4x + 2y = 2y + 4x$
(a) Commutative Property for addition
(b) Commutative Property for multiplication
(c) Associative Property for addition
(d) Associative Property for multiplication
(e) Distributive Property
Answer: easy (a)

2. State the property of real numbers being used:
$3x(4 + 2y) = 12x + 6xy$
(a) Commutative Property for addition
(b) Commutative Property for multiplication
(c) Associative Property for addition
(d) Associative Property for multiplication
(e) Distributive Property
Answer: easy (e)

3. State the property of real numbers being used:
$xyz = zxy$
(a) Commutative Property for addition
(b) Commutative Property for multiplication
(c) Associative Property for addition
(d) Associative Property for multiplication
(e) Distributive Property
Answer: medium (b)

4. State the property of real numbers being used:
$(3x)y = 3(xy)$
(a) Commutative Property for addition
(b) Commutative Property for multiplication
(c) Associative Property for addition
(d) Associative Property for multiplication
(e) Distributive Property
Answer: easy (d)

5. State the property of real numbers being used:
$(55z + 2) + 11y = 55z + (2 + 11y)$
(a) Commutative Property for addition
(b) Commutative Property for multiplication
(c) Associative Property for addition
(d) Associative Property for multiplication
(e) Distributive Property
Answer: easy (c)

6. Use properties of real numbers to write $4(a - b)$ without parentheses.
(a) $4a + 4b$ (b) $4 - 4b$ (c) $4a - 4b$ (d) $4a - b$ (e) $a - 4b$
Answer: easy (c)

7. Use properties of real numbers to write $\frac{1}{3}(12z)$ without parentheses.
(a) $\frac{1}{3}z$ (b) $\frac{1}{4}z$ (c) $36z$ (d) $3z$ (e) $4z$
Answer: easy (e)

8. Use properties of real numbers to write $-2(r + s)$ without parentheses.
(a) $-2r + 2s$ (b) $-2r - 2s$ (c) $2s + 2r$ (d) $-2r + s$ (e) $2r - 2s$
Answer: easy (b)

9. Use properties of real numbers to write $(4a)(b + 2c - d)$ without parentheses.
(a) $4ab + 4ac - 4ad$
(b) $4ab + 8ac - 4ad$
(c) $4(ab + 2ac - d)$
(d) $4ab + 8c - 4d$
(e) $4b + 8c - 4d$
Answer: easy (b)
$(4a)(b + 2c - d) = (4a)b + (4a)2c - (4a)d = 4ab + 8ac - 4ad$

10. Perform the indicated operation: $\dfrac{3}{4} - \dfrac{7}{12}$
(a) 6 (b) $\dfrac{1}{4}$ (c) $\dfrac{1}{12}$ (d) $\dfrac{1}{6}$ (e) $\dfrac{3}{12}$
Answer: easy (d)

11. Perform the indicated operation: $\dfrac{1}{16} \div \dfrac{3}{4}$
(a) $\dfrac{1}{2}$ (b) $\dfrac{1}{3}$ (c) $\dfrac{2}{3}$ (d) 3 (e) $\dfrac{1}{12}$
Answer: easy (e)

12. State whether the inequality $-2 < -4$ is true or false.
Answer: easy
False

13. State whether the inequality $-\sqrt{2} < -1.41$ is true or false.
Answer: medium
True

14. State whether the inequality $-2\pi > -6$ is true or false.
Answer: medium
False

15. State whether the inequality $7 \leq 8$ is true or false.
Answer: easy
True

16. State whether the inequality $2.2 > 2.\overline{2}$ is true or false.
Answer: medium
False

17. Write the statement in terms of inequalities: x is positive.
(a) $x > 0$ (b) $x < 0$ (c) $x \geq 0$ (d) $x \leq 0$ (e) $x = 0$
Answer: easy (a)

18. Write the statement in terms of inequalities: z is greater than -1.
(a) $z \leq -1$ (b) $z \geq -1$ (c) $z < -1$ (d) $z > -1$ (e) $z = 0$
Answer: easy (d)

19. Write the statement in terms of inequalities: c is at most 8.
(a) $c > 8$ (b) $c < 8$ (c) $c \geq 8$ (d) $c \leq 8$ (e) $c = 8$
Answer: easy (d)

20. Write the statement in terms of inequalities: u is positive and is less than or equal to 23.
(a) $0 < u < 23$ (b) $0 \leq u \leq 23$ (c) $u \leq 23$ (d) $0 \leq u < 23$ (e) $0 < u \leq 23$
Answer: easy (e)

21. Write the statement in terms of inequalities: a is at least 4 units from 2π.
(a) $|a - 2\pi| \leq 4$ (b) $|a - 2\pi| \geq 4$ (c) $|a - 2\pi| < 4$ (d) $|a - 2\pi| > 4$ (e) $|a - 4| < 2\pi$
Answer: medium (b)

22. Find the set $A \cap B$ if $A = \{2, 3, 4, 5, 6, 7\}$ and $B = \{3, 5, 7, 9\}$.
(a) $\{2, 4, 6\}$ (b) $\{3, 5, 7, 9\}$ (c) $\{3, 5, 7\}$ (d) $\{5, 7, 9\}$ (e) $\{3, 7, 9\}$
Answer: easy (c)

23. Find the set $B \cup C$ if $B = \{3, 5, 7, 9\}$ and $C = \{8, 9, 10\}$.
(a) $\{3, 5, 7, 8, 9, 10\}$ (b) $\{3, 5, 7, 9, 10\}$ (c) $\{3, 5, 7, 8, 9\}$ (d) $\{3, 8, 9, 10\}$ (e) $\{3, 5, 7, 9\}$
Answer: easy (a)

24. Find the set $A \cap C$ if $A = \{2, 3, 4, 5, 6, 7\}$ and $C = \{8, 9, 10\}$.
(a) $\emptyset$ (b) $\{7, 8\}$ (c) $\{2, 10\}$ (d) $\{2, 4, 6, 8, 10\}$ (e) $\{7, 10\}$
Answer: medium (a)

25. Find $A \cap C$ if $A = \{x \mid x < 3\}$ and $C = \{x \mid -2 < x \leq 5\}$.
(a) $\{x \mid 0 < x < 3\}$ (b) $\{x \mid 3 < x \leq 5\}$ (c) $\{x \mid -2 \leq x \leq 3\}$
(d) $\{x \mid -2 < x \leq 5\}$ (e) $\{x \mid -2 < x < 3\}$
Answer: medium (e)

26. Find $B \cup C$ if $B = \{x \mid x > 4\}$ and $C = \{x \mid -1 < x < 4\}$.
(a) $\{x \mid x > 4 \text{ or } x < -1\}$ (b) $\{x \mid x < 4 \text{ and } x \geq 4\}$ (c) $\{x \mid x > -1\}$
(d) $\{x \mid x \geq -1\}$ (e) $\{x \mid x > -1 \text{ and } x \neq 4\}$
Answer: medium (e)

27. Express the interval $(3, 9]$ in terms of inequalities and graph the corresponding interval.
(a) $(3, 9] = \{x \mid 3 < x < 9\}$ 

(b) $(3, 9] = \{x \mid 3 \leq x \leq 9\}$ 

(c) $(3, 9] = \{x \mid 3 \leq x < 9\}$ 

(d) $(3, 9] = \{x \mid 3 < x \leq 9\}$ 

(e) $(3, 9] = \{x \mid x < 3 \text{ or } x > 9\}$ 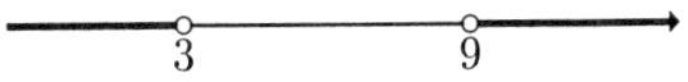

Answer: easy (d)

28. Express the interval $\left[-7, -\frac{3}{2}\right]$ in terms of inequalities and graph the corresponding interval.

(a) $\left[-7, -\frac{3}{2}\right] = \left\{x \mid -7 < x < -\frac{3}{2}\right\}$ 

(b) $\left[-7, -\frac{3}{2}\right] = \left\{x \mid -7 \le x \le -\frac{3}{2}\right\}$ 

(c) $\left[-7, -\frac{3}{2}\right] = \left\{x \mid -7 \le x < -\frac{3}{2}\right\}$ 

(d) $\left[-7, -\frac{3}{2}\right] = \left\{x \mid -7 < x \le -\frac{3}{2}\right\}$ 

(e) $\left[-7, -\frac{3}{2}\right] = \left\{x \mid x < -7 \text{ or } x \ge -\frac{3}{2}\right\}$ 

Answer: medium (b)

29. Express the interval $(-6, \infty)$ in terms of inequalities and graph the corresponding interval.

Answer: easy

$(-6, \infty) = \{x \mid x > -6\}$ 

30. Express the interval $(-\infty, 5)$ in terms of inequalities and graph the corresponding interval.

Answer: easy

$(-\infty, 5) = \{x \mid x < 5\}$ 

31. Express the interval $(0, 2\pi)$ in terms of inequalities and graph the corresponding interval.

Answer: easy

$(0, 2\pi) = \{x \mid 0 < x < 2\pi\}$ 

32. Express the inequality $2 < x < 10$ in interval notation and graph the corresponding interval.

Answer: easy

$2 < x < 10 \quad \Leftrightarrow \quad x \in (2, 10)$ 

33. Express the inequality $x < 4$ in interval notation and graph the corresponding interval.

Answer: easy

$x < 4 \quad \Leftrightarrow \quad x \in (-\infty, 4)$ 

34. Express the inequality $x \ge -8$ in interval notation and graph the corresponding interval.

Answer: easy

$x \ge -8 \quad \Leftrightarrow \quad x \in [-8, \infty)$ 

35. Express the inequality $-3 < x < 4$ in interval notation and graph the corresponding interval.

Answer: easy

$-3 < x < 4 \quad \Leftrightarrow \quad x \in (-3, 4)$ 

36. Graph the set $(-3, -1) \cap (-2, 0)$.

Answer: easy

$(-3, -1) \cap (-2, 0) = (-2, -1)$ 

37. Graph the set $[-6, 4) \cup [-2, 6)$.

Answer: easy

$[-6, 4) \cup [-2, 6) = [-6, 6)$ 

38. Graph the set $(-\infty, 7] \cap (3, 10)$.

Answer: easy

$(-\infty, 7] \cap (3, 10) = (3, 7]$ 

39. Graph the set $(-2,-1] \cup [0,1)$.

Answer: easy

$(-2,-1] \cup [0,1)$ 

40. Evaluate the expression $|-65|$.

Answer: easy

$|-65| = 65$

41. Evaluate the expression $|-6-(-16)|$.

Answer: easy

$|-6-(-16)| = |-6+16| = 10$

42. Evaluate the expression $|2\pi - 8|$.

Answer: medium

$|2\pi - 8| = 8 - 2\pi$ (since $2\pi < 8$)

43. Evaluate the expression $||-3|-|-2||$.

Answer: medium

$||-3|-|-2|| = |3-2| = |1| = 1$

44. Evaluate the expression $|-23-15|$.

Answer: easy

$|-23-15| = |-38| = 38$

45. Evaluate the expression $-2-|2-|-2||$.

Answer: medium

$-2-|2-|-2|| = -2-|2-2| = -2-0 = -2$

46. Find the distance between -6 and 19.

Answer: medium

$|19-(-6)| = |19+6| = 25$

47. Find the distance between 150 and -80.

Answer: medium

$|-80-150| = |-230| = 230$

48. Find the distance between -2.4 and -1.3.

Answer: medium

$|-1.3-(-2.4)| = |-1.3+2.4| = |1.1| = 1.1$

49. (a) Which of the numbers 0.0002337 and 0.002337 is closer to 0?

(b) Give a convincing argument which proves that there is no positive real number closest to zero.

Answer: hard

(a) 0.0002337

(b) Any positive real number less than 1 can be made smaller by putting an extra 0 immediately after the decimal point in its decimal expansion.

50. (a) Is the sum of two irrational numbers always an irrational number?

(b) Is the product of two irrational numbers always an irrational number?

Answer: hard

(a) No. For example, $\sqrt{2} + \left(-\sqrt{2}\right) = 0$, which is rational.

(b) No. For example, $\sqrt{2} \cdot \sqrt{2} = 2$, which is rational.

51. Use the properties of real numbers to write the expression $2x\left(a-b-2c+\frac{d}{2}\right)$ without parentheses.

(a) $xa-xb-2xc+xd$
(b) $2xa-2xb-xc+4xd$
(c) $xa-2xb-4xc+xd$
(d) $2xa-2xb-4xc+xd$
(e) $2xa-xb-2xc+2xd$

Answer: easy (d)

$2x\left(a-b-2c+\frac{d}{2}\right)=2x\left(a\right)+2x\left(-b\right)+2x\left(-2c\right)+2x\left(\frac{d}{2}\right)=2xa-2xb-4xc+xd.$

52. Perform the indicated operation: $\frac{1}{4}+5\left(\frac{1}{4}\cdot\frac{1}{5}\right)-\frac{3}{4}$

(a) $-\frac{3}{4}$
(b) $\frac{3}{4}$
(c) $\frac{1}{25}$
(d) $-\frac{1}{4}$
(e) $\frac{1}{4}$

Answer: easy (d)

$\frac{1}{4}+5\left(\frac{1}{4}\cdot\frac{1}{5}\right)-\frac{3}{4}=\frac{1}{4}+\frac{1}{4}-\frac{3}{4}=-\frac{1}{4}$

53. Perform the indicated operation: $\frac{1}{5}\div\left[\frac{1}{4}\left(\frac{1}{2}+\frac{1}{3}\right)\right]$

(a) $\frac{3}{2}$
(b) $\frac{25}{24}$
(c) 1
(d) $\frac{2}{3}$
(e) $\frac{24}{25}$

Answer: easy (e)

$\frac{1}{5}\div\left[\frac{1}{4}\left(\frac{1}{2}+\frac{1}{3}\right)\right]=\frac{1}{5}\div\left[\frac{1}{4}\left(\frac{3+2}{6}\right)\right]=\frac{1}{5}\div\left(\frac{1}{4}\left(\frac{5}{6}\right)\right)=\frac{1}{5}\div\left(\frac{5}{24}\right)=\frac{24}{25}$

54. State whether the inequality $-\sqrt{2}<-1.41$ is true or false.

Answer: easy

True

55. State whether the inequality $-\frac{1}{10}>-\frac{1}{100}$ is true or false.

Answer: easy

False

56. State whether the inequality $\frac{355}{113}>\pi$ is true or false.

Answer: easy

True

57. Write the statement in terms of inequalities: The distance from x to 3 is at most 6.

(a) $|x-3|\leq 6$
(b) $|x-3|\geq 6$
(c) $|x-3|<6$
(d) $|x-6|\leq 3$
(e) $|x-6|\geq 3$

Answer: medium (a)

$|x-3|\leq 6$

58. Write the statement in terms of inequalities: The distance from r to $3r$ is at least 1 unit.

(a) $|r-3r|\geq 3$
(b) $|2r|\geq 1$
(c) $|r-3r|<3$
(d) $|3r-1|<1$
(e) $|r+3r|\geq 1$

Answer: medium (b)

$|3r-r|\geq 1 \quad\Rightarrow\quad |2r|\geq 1$

59. Find the set $A\cup B$ if $A=\left\{-3,-2,0,\frac{1}{3},\frac{2}{3},6,9\right\}$ and $B=\left\{0,\frac{1}{3},\frac{2}{3}\right\}$.

(a) $\left\{0,-2,-3,6,9,\frac{1}{3},\frac{2}{3}\right\}$
(b) $\left\{0,-2,3,6,9,-\frac{1}{3},\frac{2}{3}\right\}$
(c) $\left\{0,2,-3,6,9,\frac{1}{3},-\frac{2}{3}\right\}$
(d) $\{0,-2,-3,6,-9\}$
(e) $\left\{0,-2,-3,6,9,\frac{2}{3}\right\}$

Answer: easy (a)

60. Find the set $A \cup B \cap C$ if $A = \{0\}$ and $B = \{1, 2\}$ and $C = \{-1, 0, 1, 2\}$.
(a) $\{0, 2\}$ (b) $\{-1, 2\}$ (c) $\{1, 2\}$ (d) $\{-1, 0, 1, 2\}$ (e) $\{0, 1, 2\}$
Answer: medium (e)

61. Find $A \cap C$ if $A = \{x \mid x < 4\}$ and $C = \{x \mid -2 < x \leq 6\}$.
(a) $\{x \mid -2 < x < 4\}$ (b) $\{x \mid -2 < x < 6\}$ (c) $\{x \mid 4 < x < 6\}$ (d) $\{x \mid -2 < x \leq 6\}$ (e) $\emptyset$
Answer: medium (a)

62. Find $A \cup B$ if $A = \{x \mid x > \pi\}$ and $B = \{x \mid -1 < x < \pi\}$.
(a) $\emptyset$
(b) $\{x \mid x > -1 \text{ and } x \neq \pi\}$
(c) $\{x \mid -1 < x < \pi\}$
(d) $\{x \mid -\pi < x \leq 1\}$
(e) $\{x \mid x > -\pi\}$
Answer: medium (b)

63. Evaluate the expression $|2 - |-3| + (-3)|$.
Answer: medium
$|2 - |-3| + (-3)| = |2 - 3 - 3| = |-4| = 4$

64. Evaluate the expression $-|-|(-40)| - |(-2)||$.
Answer: medium
$-|-|(-40)| - |(-2)|| = -|-40 - 2| = -|-42| = -42$

65. Find the distance between π and $\pi - 1$.
Answer: easy
$|(\pi - 1) - \pi| = |-1| = 1$

SECTION 1.3

1. Evaluate -2^5
(a) -32 (b) -16 (c) -2 (d) 16 (e) 32
Answer: medium (a)
$-2^5 = -(2^5) = -32$

2. Evaluate the number 4^{-4}.
(a) $-\frac{1}{256}$ (b) $-\frac{1}{64}$ (c) $\frac{1}{64}$ (d) $\frac{1}{256}$ (e) $\frac{1}{512}$
Answer: medium (d)
$$4^{-4} = \frac{1}{4^4} = \frac{1}{256}$$

3. Evaluate $\left(\frac{1}{3}\right)^3$
(a) $\frac{1}{81}$ (b) $\frac{1}{27}$ (c) $\frac{1}{9}$ (d) 1 (e) 9
Answer: easy (b)
$$\left(\frac{1}{3}\right)^3 = \frac{1}{3^3} = \frac{1}{27}$$

4. Evaluate $\left(\frac{3}{4}\right)^{-1}$
(a) $-\frac{2}{3}$ (b) $-\frac{1}{3}$ (c) $-\frac{1}{4}$ (d) $\frac{3}{2}$ (e) $\frac{4}{3}$
Answer: easy (e)
$\left(\frac{3}{4}\right)^{-1} = \frac{4}{3}$

5. Evaluate $\left(2^2 \cdot 2\right)^3$
(a) 64 (b) 128 (c) 256 (d) 512 (e) 1024
Answer: easy (d)
$(2^2 \cdot 2)^3 = (2^3)^3 = 2^9 = 512$

6. Evaluate $32 \cdot \left(\frac{2}{3}\right)^0 \cdot 2^{-2}$
(a) 64 (b) 32 (c) 24 (d) 16 (e) 8
Answer: medium (e)
$$32 \cdot \left(\frac{2}{3}\right)^0 \cdot 2^{-2} = 2^5 \cdot 1 \cdot \frac{1}{2^2} = 8$$

7. Evaluate $\sqrt{\frac{9}{25}}$.
(a) $\frac{9}{5}$ (b) $\frac{3}{15}$ (c) $\frac{3}{5}$ (d) $\frac{5}{3}$ (e) $\frac{15}{3}$
Answer: easy (c)
$\sqrt{\frac{9}{25}} = \frac{3}{5}$

8. Evaluate $\sqrt[4]{81}$.
(a) 2 (b) 3 (c) 4 (d) 5 (e) 6
Answer: easy (b)
$\sqrt[4]{81} = 3$

9. Evaluate $\sqrt[6]{-64}$.
(a) -4 (b) -2 (c) 2 (d) 4 (e) 8
Answer: medium (b)
$\sqrt[6]{-64} = \sqrt[6]{(-2)^6} = -2$

10. Evaluate $243^{1/5}$.
(a) 2 (b) 3 (c) 5 (d) 7 (e) 9
Answer: medium (b)
$243^{1/5} = \sqrt[5]{243} = 3$

11. Evaluate $(-128)^{5/7}$.
(a) -64 (b) -32 (c) -4 (d) 4 (e) 16
Answer: medium (b)
$(-128)^{5/7} = \left(\sqrt[7]{-128}\right)^5 = (-2)^5 = -32$

12. Evaluate $\left(-\frac{243}{32}\right)^{3/5}$
(a) $-\frac{3}{2}$ (b) -3 (c) $-\frac{27}{8}$ (d) $-\frac{3}{4}$ (e) $-\frac{1}{8}$
Answer: medium (c)
$\left(-\frac{243}{32}\right)^{3/5} = \left(\sqrt[5]{-\frac{243}{32}}\right)^3 = \left(-\frac{3}{2}\right)^3 = -\frac{27}{8}$

13. Evaluate $4^{2/9}4^{7/9}$
(a) 2 (b) 4 (c) 6 (d) 9 (e) 12
Answer: easy (b)
$4^{2/9} \times 4^{7/9} = 4$

14. Evaluate $(0.09)^{-2.5}$

(a) $\frac{1}{0.3}$ (b) $\frac{1}{0.09}$ (c) $\frac{1}{0.027}$ (d) $\frac{1}{0.009}$ (e) $\frac{1}{0.00243}$

Answer: hard (e)

$(0.09)^{-2.5} = \dfrac{1}{(0.09)^{5/2}} = \dfrac{1}{\left(\sqrt{0.09}\right)^5} = \dfrac{1}{0.00243}$

15. Simplify the expression $\sqrt{50} + \sqrt{98}$.

(a) $12\sqrt{2}$ (b) $12\sqrt{3}$ (c) $8\sqrt{2}$ (d) $8\sqrt{3}$ (e) $12\sqrt{5}$

Answer: medium (a)

$\sqrt{50} + \sqrt{98} = 5\sqrt{2} + 7\sqrt{2} = 12\sqrt{2}$

16. Simplify the expression $\sqrt[4]{162} - \sqrt[4]{32}$.

(a) $\sqrt{2}$ (b) $\sqrt[3]{2}$ (c) $\sqrt[4]{2}$ (d) $2\sqrt{2}$ (e) $3\sqrt{2}$

Answer: medium (c)

$\sqrt[4]{162} - \sqrt[4]{32} = 3\sqrt[4]{2} - 2\sqrt[4]{2} = \sqrt[4]{2}$

17. Write the number $3^5 \cdot 9^3$ as a power of 3.

(a) 3^8 (b) 3^9 (c) 3^{10} (d) 3^{11} (e) 3^{12}

Answer: easy (d)

$3^5 \cdot 9^3 = 3^5 \cdot (3^2)^3 = 3^5 \cdot 3^6 = 3^{11}$

18. Write the number $\frac{1}{27}$ as a power of 3.

(a) 3^{-4} (b) 3^{-3} (c) 3^{-1} (d) 3^0 (e) 3^2

Answer: medium (b)

$\frac{1}{27} = \frac{1}{3^3} = 3^{-3}$

19. Write the number 1 as a power of 3.

(a) 3^{-2} (b) 3^{-1} (c) 3^0 (d) 3^1 (e) 3^2

Answer: easy (c)

20. Write $(3^x)^5$ as a power of 3.

(a) 3^{x+5} (b) 3^{x-5} (c) 5^x (d) 3^{x^5} (e) 3^{5x}

Answer: easy (e)

$(3^x)^5 = 3^{5x}$

21. Simplify $x^{-3}x^7$ and eliminate any negative exponents.

(a) x^{-4} (b) x^{-2} (c) x^2 (d) x^4 (e) x^7

Answer: easy (d)

22. Simplify $\left(\dfrac{z^3}{4y}\right)^{-2}$ and eliminate any negative exponents.

(a) $\dfrac{4y}{z^6}$ (b) $\dfrac{16y^2}{z^6}$ (c) $\dfrac{4z^6}{y^4}$ (d) $\dfrac{z^6}{y^2}$ (e) $\dfrac{z}{4y}$

Answer: medium (b)

$\left(\dfrac{z^3}{4y}\right)^{-2} = \left(\dfrac{4y}{z^3}\right)^2 = \dfrac{4^2 \cdot y^2}{z^6} = \dfrac{16y^2}{z^6}$

23. Simplify $a^3(a^{-2})^4$ and eliminate any negative exponents.

(a) $1/a$ (b) $1/a^5$ (c) a (d) a^5 (e) $a^{1/5}$

Answer: medium (b)

$a^3(a^{-2})^4 = a^3 \cdot a^{-8} = a^{-5} = 1/a^5$

24. Simplify the expression $(3x^4)(5x^5)$.

(a) $2x$ (b) $15x^9$ (c) $15x^{20}$ (d) $15x^{22}$ (e) $8x^{20}$

Answer: medium (b)

$(3x^4)(5x^5) = 15x^{4+5} = 15x^9$

25. Eliminate the negative exponents and simplify the expression $\dfrac{a^{-2}b^3}{a^{-6}b^4}$.

(a) ab^4 (b) a^4b^7 (c) a^4/b (d) a^6/b^2 (e) a/b^4

Answer: medium (c)

$\dfrac{a^{-2}b^3}{a^{-6}b^4} = a^{-2+6} \cdot b^{3-4} = a^4 \cdot b^{-1} = a^4/b$

26. Eliminate the negative exponent and simplify the expression $(3s^2t^{-2})\left(\frac{1}{6}s^6\right)(12t^4)$.

(a) $6s^8t^2$ (b) $3s^5t$ (c) $12s^4t^2$ (d) $12s^3t^3$ (e) $18s^4t^6$

Answer: medium (a)

$(3s^2t^{-2})\left(\frac{1}{6}s^6\right)(12t^4) = 6s^{2+6}t^{-2+4} = 6s^8t^2$

27. Eliminate the negative exponent and simplify the expression $(2u^2v^4)^3(3u^2v)^{-1}$.

(a) $\frac{8}{9}u^3v^8$ (b) $\frac{8}{3}u^4v^{11}$ (c) $\frac{3}{4}uv^6$ (d) $\frac{2}{3}u^2v^4$ (e) $\frac{3}{8}u^3v^6$

Answer: medium (b)

$(2u^2v^4)^3(3u^2v)^{-1} = 2^3 \cdot 3^{-1} \cdot u^{6-2} \cdot v^{12-1} = \frac{8}{3}u^4v^{11}$

28. Simplify the expression $\dfrac{(3x^2)^3(2x^2)}{(x^4)^3}$.

Answer: medium

$\dfrac{(3x^2)^3(2x^2)}{(x^4)^3} = 54x^{6+2-12} = \dfrac{54}{x^4}$

29. Simplify the expression $\left(\dfrac{c^3d^4}{c^2d}\right)\left(\dfrac{d^3}{c^5}\right)^2$.

Answer: medium

$\left(\dfrac{c^3d^4}{c^2d}\right)\left(\dfrac{d^3}{c^5}\right)^2 = c^{3-2-10}d^{4-1+6} = \dfrac{d^9}{c^9}$

30. Simplify $\left(\dfrac{x^2y^{-1}z^{-4}}{xy^4z^{-2}}\right)^{-2}$ and eliminate any negative exponents.

Answer: medium

$\left(\dfrac{x^2y^{-1}z^{-4}}{xy^4z^{-2}}\right)^{-2} = \left(\dfrac{x^{2-1}}{y^{4+1}z^{-2+4}}\right)^{-2} = \dfrac{y^{10}z^4}{x^2}$

31. Simplify the expression $\left(-3a^{4/3}\right)\left(6a^{5/9}\right)$.

Answer: medium

$\left(-3a^{4/3}\right)\left(6a^{5/9}\right) = -18a^{17/9}$

32. Eliminate the negative exponent and simplify the expression $(27x^9)^{-4/3}$.

Answer: medium

$(27x^9)^{-4/3} = (3^3)^{-4/3} \cdot x^{-36/3} = 3^{-4} \cdot x^{-12} = \dfrac{1}{81x^{12}}$

33. Simplify the expression $(9x^8y^{12})^{3/2}$.

Answer: medium

$(9x^8y^{12})^{3/2} = (3^2)^{3/2} \cdot (x^8)^{3/2} \cdot (y^{12})^{3/2} = 3^3x^{12}y^{18} = 27x^{12}y^{18}$

34. Eliminate the negative exponent and simplify the expression $\left(a^{3/8}\right)^{-2/5}$.

Answer: medium

$\left(a^{3/8}\right)^{-2/5} = a^{-6/40} = a^{-3/20} = \dfrac{1}{a^{3/20}}$

35. Eliminate the negative exponents and simplify the expression $\left(x^{-3}y^5z^{15}\right)^{-2/5}$.
Answer: medium
$$\left(x^{-3}y^5z^{15}\right)^{-2/5} = \left(x^{-3}\right)^{-2/5} \cdot \left(y^5\right)^{-2/5} \cdot \left(z^{15}\right)^{-2/5} = \frac{x^{6/5}}{y^2z^6}$$

36. Simplify the expression $\left(\frac{-3x^{2/3}}{y^{1/3}z^{1/8}}\right)^3$.
Answer: medium
$$\left(\frac{-3x^{2/3}}{y^{1/3}z^{1/8}}\right)^3 = \frac{(-3)^3x^2}{yz^{3/8}} = \frac{-27x^2}{yz^{3/8}}$$

37. Eliminate the negative exponents and simplify the expression $\frac{\left(y^9z^{-3}\right)^{1/3}}{\left(y^{-4}z^2\right)^{1/4}}$.
Answer: medium
$$\frac{\left(y^9z^{-3}\right)^{1/3}}{\left(y^{-4}z^2\right)^{1/4}} = \frac{y^3z^{-1}}{y^{-1}z^{1/2}} = \frac{y^{3+1}}{z^{1+1/2}} = \frac{y^4}{z^{3/2}}$$

38. Eliminate the negative exponents and simplify the expression $\left(\frac{a^3b^{-2}}{x^3y^{-1}}\right)^2\left(\frac{x^{-3}b^{-2}}{a^{5/2}y^{2/3}}\right)$.
Answer: medium
$$\left(\frac{a^3b^{-2}}{x^3y^{-1}}\right)^2\left(\frac{x^{-3}b^{-2}}{a^{5/2}y^{2/3}}\right) = \frac{a^6b^{-4}}{x^6y^{-2}} \cdot \frac{x^{-3}b^{-2}}{a^{5/2}y^{2/3}} = \frac{a^{6-5/2}y^{2-2/3}}{b^{4+2}x^{6+3}} = \frac{a^{7/2}y^{4/3}}{b^6x^9}$$

39. Simplify the expression $\sqrt[4]{x^4y^{12}}$.
Answer: medium
$$\sqrt[4]{x^4y^{12}} = \left(x^4\right)^{1/4} \cdot \left(y^{12}\right)^{1/4} = \left|xy^3\right|$$

40. Simplify the expression $\sqrt{x^5y^6}$.
Answer: medium
$$\sqrt{x^5y^6} = \left(x^5\right)^{1/2} \cdot \left(y^6\right)^{1/2} = x^{5/2}\left|y^3\right| = x^2\left|y^3\right|\sqrt{x}$$

41. Simplify the expression $\sqrt[3]{ab^2}\sqrt[3]{a^4b^4}$.
Answer: medium
$$\sqrt[3]{ab^2}\sqrt[3]{a^4b^4} = \sqrt[3]{a^5b^6} = a^{5/3}b^2$$

42. Simplify the expression $\sqrt[4]{xy^2z^4}$.
Answer: hard
$$\sqrt[4]{xy^2z^4} = x^{1/4} \cdot \left(y^2\right)^{1/4} \cdot \left(z^4\right)^{1/4} = x^{1/4}\sqrt{|y|}|z|$$

43. Simplify the expression $\sqrt[9]{r^{3n+1}}\sqrt[9]{r^{-1}}$.
Answer: medium
$$\sqrt[9]{r^{3n+1}} \cdot \sqrt[9]{r^{-1}} = r^{(3n+1)/9} \cdot r^{-1/9} = r^{3n/9} = \sqrt[3]{r^n}$$

44. Rationalize the denominator of $\sqrt{\frac{3}{2}}$.
Answer: easy
$$\sqrt{\tfrac{3}{2}} = \tfrac{\sqrt{3}}{\sqrt{2}} = \tfrac{\sqrt{3}}{\sqrt{2}} \cdot \tfrac{\sqrt{2}}{\sqrt{2}} = \tfrac{\sqrt{6}}{2}$$

45. Rationalize the denominator of $\sqrt{x^7/3}$
Answer: medium
$$\sqrt{\frac{x^7}{3}} = \frac{\sqrt{x^7}}{\sqrt{3}} \cdot \frac{\sqrt{3}}{\sqrt{3}} = \frac{\sqrt{3x^7}}{3} = \frac{x^3\sqrt{3x}}{3}$$

46. Rationalize the denominator of $\sqrt{\frac{1}{3x^5y^6}}$ and simplify.

Answer: medium

$$\sqrt{\frac{1}{3x^5y^6}} = \frac{1}{\sqrt{3x^5y^6}} \cdot \frac{\sqrt{3x^5y^6}}{\sqrt{3x^5y^6}} = \frac{x^2y^3\sqrt{3x}}{3x^5y^6} = \frac{\sqrt{3x}}{3x^3y^3}$$

47. Rationalize the denominator of $\frac{1}{\sqrt{x}+\sqrt{y}}$.

Answer: medium

$$\frac{1}{\sqrt{x}+\sqrt{y}} = \frac{1}{\sqrt{x}+\sqrt{y}} \cdot \frac{\sqrt{x}-\sqrt{y}}{\sqrt{x}-\sqrt{y}} = \frac{\sqrt{x}-\sqrt{y}}{x-y}$$

48. Rewrite $\frac{1-\sqrt{3}}{2}$ by rationalizing the numerator; that is, multiply numerator and denominator by the conjugate radical of the numerator and simplify.

Answer: medium

$$\frac{1-\sqrt{3}}{2} = \frac{1-\sqrt{3}}{2} \cdot \frac{1+\sqrt{3}}{1+\sqrt{3}} = \frac{1-3}{2+2\sqrt{3}} = \frac{-2}{2+2\sqrt{3}} = \frac{-1}{1+\sqrt{3}}$$

49. Rewrite $\sqrt{x}-\sqrt{3}$ by rationalizing the numerator; that is, multiply numerator and denominator by the conjugate radical of the numerator and simplify.

Answer: medium

$$\sqrt{x}-\sqrt{3} = \frac{\sqrt{x}-\sqrt{3}}{1} \cdot \frac{\sqrt{x}+\sqrt{3}}{\sqrt{x}+\sqrt{3}} = \frac{x-3}{\sqrt{x}+\sqrt{3}}$$

50. Write 8,221,000,000 in scientific notation.

Answer: easy

$8{,}221{,}000{,}000 = 8.221 \times 10^9$

51. Write 0.000000032 in scientific notation.

Answer: easy

$0.000000032 = 3.2 \times 10^{-8}$

52. Write 0.0004017 in scientific notation.

Answer: easy

$0.0004017 = 4.017 \times 10^{-4}$

53. Write 6.44×10^{-4} in ordinary decimal notation.

Answer: easy

$6.44 \times 10^{-4} = 0.000644$

54. Write 7×10^{14} in ordinary decimal notation.

Answer: easy

$7 \times 10^{14} = 700{,}000{,}000{,}000{,}000$

55. Write 4.116×10^{-11} in ordinary decimal notation.

Answer: easy

$4.116 \times 10^{-11} = 0.00000000004116$

56. Write the number in the statement in scientific notation: The distance from the earth to the sun is about ninety-three million miles.

Answer: easy

$93{,}000{,}000 = 9.3 \times 10^7$ mi

57. Write the number in the statement in scientific notation: The mass of an oxygen molecule is about 0.000000000000000000000053 g.

Answer: easy

$0.000000000000000000000053 = 5.3 \times 10^{-23}$ g

58. Write the number in the statement in scientific notation. The mass of the earth is about 5,970,000,000,000,000,000,000,000 kg.

Answer: easy

$5{,}970{,}000{,}000{,}000{,}000{,}000{,}000{,}000 = 5.97 \times 10^{24}$ kg

59. Use scientific notation, laws of exponents, and a calculator to perform the indicated operations. State your answer correct to the number of significant digits indicated by the data. $(3.022 \times 10^{18})(9.46 \times 10^{11})$

Answer: medium

$(3.022 \times 10^{18})(9.46 \times 10^{11}) = (3.022)(9.46) \times 10^{18+11} \approx 2.86 \times 10^{30}$

60. Use scientific notation, laws of exponents, and a calculator to perform the indicated operations. State your answer correct to the number of significant digits indicated by the data. $\dfrac{(14.2)(6.6366 \times 10^{14})}{0.000000026}$

Answer: medium

$\dfrac{(14.2)(6.6366 \times 10^{14})}{0.000000026} = \dfrac{(1.42)(6.6366)}{2.6} \times 10^{1+14+8} \approx 3.6 \times 10^{23}$

61. Use scientific notation, laws of exponents, and a calculator to perform the indicated operations. State your answer correct to the number of significant digits indicated by the data. $\dfrac{(2.614 \times 10^{-4})^8}{(4.04 \times 10^2)^{11}}$

Answer: medium

$\dfrac{(2.614 \times 10^{-4})^8}{(4.04 \times 10^2)^{11}} = \dfrac{(2.614)^8 \times 10^{-32}}{(4.04)^{11} \times 10^{22}} \approx 4.66 \times 10^{-58}$

62. Which of the following pairs of numbers is closer together?

(i) 2^{23} and 2^{37} (ii) 2^{80} and 2^{81}

Answer: medium

(i), since $|2^{37} - 2^{23}| < 2^{37}$ but $|2^{81} - 2^{80}| = 2^{80}$.

63. Due to the curvature of the earth, the maximum distance D that you can see from the top of a tall building of height h is estimated by the formula $D = \sqrt{2rh + h^2}$, where $r = 6372$ km is the radius of the earth and D and h are also measured in kilometers. How far can you see from the top of New York's World Trade Center, 417 m above the ground?

Answer: medium

$D = \sqrt{2rh + h^2} = \sqrt{2 \cdot 6372 \cdot 0.417 + 0.417^2} \approx 72.9\,\text{km}$

64. (a) Prove $\dfrac{a^m}{a^n} = a^{m-n}$ $(a \neq 0)$ for the case where m and n are positive integers and $m > n$.

(b) Prove $\left(\dfrac{a}{b}\right)^n = \dfrac{a^n}{b^n}$ $(b \neq 0)$ for the case where n is a positive integer.

Answer: hard

(a) $\dfrac{a^m}{a^n} = \dfrac{a \cdot a \cdot \cdots \cdot a \text{ [}m\text{ factors]}}{a \cdot a \cdot \cdots \cdot a \text{ [}n\text{ factors]}} = a \cdot a \cdot \cdots \cdot a \text{ [}m-n\text{ factors]} = a^{m-n}$ $(a \neq 0)$

(b) $\left(\dfrac{a}{b}\right)^n = \dfrac{a}{b} \cdot \dfrac{a}{b} \cdot \cdots \cdot \dfrac{a}{b} \text{ [}n\text{ factors]} = \dfrac{a \cdot a \cdot \cdots \cdot a \text{ [}n\text{ factors]}}{b \cdot b \cdot \cdots \cdot b \text{ [}n\text{ factors]}} = \dfrac{a^n}{b^n}$ $(b \neq 0)$

65. Evaluate $\left(-\frac{3}{2}\right)^0 \left(-\frac{2}{3}\right)^{-2}$.

(a) $\frac{9}{4}$ (b) $\frac{4}{9}$ (c) $-\frac{9}{4}$ (d) $\frac{1}{4}$ (e) 1

Answer: easy (a)

$\left(-\frac{3}{2}\right)^0 \left(-\frac{2}{3}\right)^{-2} = \frac{9}{4}$

66. Evaluate $\left(\sqrt{\frac{36}{9}}\right)^{-2}$.

(a) $\frac{9}{2}$ (b) $\frac{-2}{9}$ (c) $\frac{-3}{4}$ (d) $\frac{1}{4}$ (e) $\frac{1}{12}$

Answer: easy (d)

$\left(\sqrt{\frac{36}{9}}\right)^{-2} = \frac{1}{4}$

67. Evaluate $\left(\frac{1}{4}\right)^{-2}\left(\frac{1}{2}\right)^{-4}$.

(a) $\frac{1}{64}$ (b) $\frac{1}{256}$ (c) 64 (d) $-\frac{1}{4}$ (e) 256

Answer: easy (e)

$\left(\frac{1}{4}\right)^{-2}\left(\frac{1}{2}\right)^{-4} = 256$

68. Evaluate $(512)^{-\frac{1}{9}}$.

(a) $\frac{1}{4}$ (b) $\frac{1}{2}$ (c) 16 (d) $\frac{1}{16}$ (e) 256

Answer: easy (b)

$(512)^{-1/9} = \sqrt[9]{\frac{1}{512}} = \frac{1}{2}$

69. Evaluate $(32)^{-0.2}$.

(a) 2 (b) $\frac{1}{2}$ (c) 16 (d) $\frac{1}{4}$ (e) $\frac{1}{1024}$

Answer: medium (b)

$(32)^{-0.2} = (32)^{-1/5} = \left(\frac{1}{\sqrt[5]{32}}\right) = \frac{1}{2}$

70. Simplify the expression $\sqrt{27} - \sqrt{48}$.

(a) 21 (b) -1 (c) $4\sqrt{3}$ (d) $-3\sqrt{3}$ (e) $-\sqrt{3}$

Answer: easy (e)

$\sqrt{27} - \sqrt{48} = 3\sqrt{3} - 4\sqrt{3} = -\sqrt{3}$

71. Simplify $\dfrac{x^{1/3}y^{1/3}}{(xy)^{-2/3}}$ and eliminate any negative exponents.

(a) xy (b) $(xy)^{-2/3}$ (c) $(xy)^{1/3}$ (d) y (e) $x^{1/3}y$

Answer: easy (a)

$$\frac{x^{1/3}y^{1/3}}{(xy)^{-2/3}} = \frac{(xy)^{1/3}}{(xy)^{-2/3}} = (xy)^{(1/3)-(-2/3)} = xy$$

72. Simplify $\left(x^2y^{-5}z^5\right)^{-2/5}$ and eliminate any negative exponents.

(a) $\dfrac{y^2}{z^2\left(\sqrt{x}\right)^5}$ (b) $\dfrac{y^2}{z^2\left(\sqrt[5]{x}\right)^4}$ (c) $\dfrac{y}{z^2\left(\sqrt[4]{x}\right)^4}$ (d) $\dfrac{y^2}{z\sqrt{x}^2}$ (e) $\dfrac{y^4}{z^2\left(\sqrt[5]{x}\right)}$

Answer: easy (b)

$$\left(x^2y^{-5}z^5\right)^{-2/5} = \left(x^{-4/5}y^2z^{-2}\right) = \frac{y^2}{z^2\left(\sqrt[5]{x}\right)^4}$$

73. Simplify $3ab^3c^2\left(\frac{2a^2b}{c^4}\right)^{-1}$ and eliminate any negative exponents.

(a) $3b^2c^6$ (b) $\frac{3b^6c^2}{2a^2}$ (c) $\frac{2b^2c^6}{3a}$ (d) $\frac{3b^2c^6}{2a}$ (e) b^2c^6

Answer: easy (d)

$$3ab^3c^2\left(\frac{2a^2b}{c^4}\right)^{-1} = 3ab^3c^2\frac{c^4}{2a^2b} = \frac{3b^2c^6}{2a}$$

74. Simplify $\left(\frac{a^2b^2c^2}{ab^3c^4}\right)^3\left(\frac{c^2}{a^2}\right)^2$ and eliminate any negative exponents.

Answer: medium

$$\left(\frac{a^2b^2c^2}{ab^3c^4}\right)^3\left(\frac{c^2}{a^2}\right)^2 = \left(\frac{a}{bc^2}\right)^3\left(\frac{c^4}{a^4}\right) = \left(\frac{a^3}{b^3c^6}\right)\left(\frac{c^4}{a^4}\right) = \frac{1}{ab^3c^2}$$

75. Simplify $\left[\left(a^{-3}b^{-4}c^2\right)^{-2}(ab)^{-3}b^{-1}c\right]^{-2}$ and eliminate any negative exponents.

Answer: medium

$$\left[\left(a^{-3}b^{-4}c^2\right)^{-2}(ab)^{-3}b^{-1}c\right]^{-2} = \left[\frac{a^6b^8}{c^4}\frac{1}{(ab)^3}\frac{c}{b}\right]^{-2} = \left(\frac{a^3b^4}{c^3}\right)^{-2} = \frac{1}{\left(\frac{a^3b^4}{c^3}\right)^2} = \frac{c^6}{a^6b^8}$$

76. Simplify the expression $\sqrt[3]{\frac{y^3z^3}{64}}$.

Answer: easy

$$\sqrt[3]{\frac{y^3z^3}{64}} = \sqrt[3]{\frac{(yz)^3}{64}} = \frac{yz}{4}$$

77. Simplify the expression $\sqrt[3]{x^{n+1}}\,\sqrt[3]{x}$.

Answer: medium

$$\sqrt[3]{x^{n+1}}\,\sqrt[3]{x} = x^{(n+1)/3}x^{1/3} = x^{(n/3)+1/3}x^{1/3} = x^{(n+2)/3} = \sqrt[3]{x^{n+2}}$$

78. Which of the numbers $\frac{10}{0.001}$ and $\frac{10}{0.01}$ is smaller?

Answer: easy

$\frac{10}{0.01}$ is smaller because $\frac{10}{0.01} = \frac{10}{10^{-2}} = 10^3 = 1000$, whereas $\frac{10}{0.001} = \frac{10}{10^{-3}} = 10^4 = 10{,}000$.

79. Which of the numbers $-\left(\frac{10}{0.001}\right)^{-1}$ and $-\left(\frac{10}{0.01}\right)^{-2}$ is larger?

Answer: medium

$-\left(\frac{10}{0.01}\right)^{-2}$ is larger because $-\left(\frac{10}{0.01}\right)^{-2} = -\left(10^{-6}\right) = -\frac{1}{1{,}000{,}000}$, whereas $-\left(\frac{10}{0.001}\right)^{-1} = -\left(10^{-4}\right) = -\frac{1}{10{,}000}$.

SECTION 1.4

1. Perform the indicated operations and simplify: $3(2x+5)-2(2x-7)$.
(a) $4x-13$ (b) $2x-13$ (c) $4x-7$ (d) $2x+13$ (e) $2x+29$
Answer: easy (e)
$3(2x+5)-2(2x-7)=6x+15-4x+14=2x+29$

2. Perform the indicated operations and simplify: $(2x^3+4x^2-x+2)+(2x^2+4x-6)$.
(a) $2x^3+6x^2+2x-4$ (b) $2x^3+6x^2+3x-4$ (c) $2x^3+6x^2+3x-1$
(d) x^3+6x^2+3x-4 (e) x^3+6x^2-3x+4
Answer: easy (b)
$(2x^3+4x^2-x+2)+(2x^2+4x-6)=2x^3+6x^2+3x-4$

3. Perform the indicated operations and simplify: $(xy^2-4xy)+2y(2x+y)$.
(a) $2y^2+x^2y+4xy$ (b) $2y^2+2x^2y$ (c) y^2+2x^2y (d) $2y^2+xy^2$ (e) y^2+2xy
Answer: medium (d)
$(xy^2-4xy)+2y(2x+y)=xy^2-4xy+4xy+2y^2=2y^2+xy^2$

4. Perform the indicated operations and simplify: $3(2t-1)-(2t^2+3)-3t(t-1)$.
(a) $-5t^2+9t-6$ (b) $-5t^2+11t+6$ (c) $5t^2+12t-2$ (d) $-5t^2+3t-2$ (e) $5t^2-9t+2$
Answer: medium (a)
$3(2t-1)-(2t^2+3)-3t(t-1)=6t-3-2t^2-3-3t^2+3t=-5t^2+9t-6$

5. Perform the indicated operations and simplify: $x^{5/2}\left(\sqrt{x}-2/\sqrt{x}\right)$.
(a) x^5-2x (b) x^3-2x^4 (c) x^3-2x^2 (d) x^3+2x^2 (e) x^5-2x^3
Answer: hard (c)
$x^{5/2}\left(\sqrt{x}-2/\sqrt{x}\right)=x^3-2x^2$

6. Perform the indicated operations: $(3x-2)(x+5)$
(a) $3x^2+13x-10$ (b) $3x^2+15x-10$ (c) $x^2+13x-5$ (d) $x^2-13x+10$ (e) $x^2-15x+5$
Answer: easy (a)
$(3x-2)(x+5)=3x^2+15x-2x-10=3x^2+13x-10$

7. Perform the indicated operations and simplify: $(t+3)(t-2)-6(t+1)$.
(a) t^2+6t+4 (b) $t^2-5t-12$ (c) $t^2-2t-10$ (d) t^2-3t-5 (e) t^2+5t+4
Answer: medium (b)
$(t+3)(t-2)-6(t+1)=t^2+t-6-6t-6=t^2-5t-12$

8. Perform the indicated operations and simplify: $(4u-2v)(4u+2v)$.
(a) $8u^2-4v^2$ (b) $8u^2+4v^2$ (c) $16u^2+4v^2$ (d) $16u^2+2v^2$ (e) $16u^2-4v^2$
Answer: easy (e)
$(4u-2v)(4u+2v)=16u^2+8uv-8uv-4v^2=16u^2-4v^2$

9. Perform the indicated operation: $(3x+4)^2$.
(a) $9x^2+12x+16$ (b) $9x^2+12x+8$ (c) $9x^2+24x+16$ (d) $3x^2+24x+16$ (e) $3x^2-12x-8$
Answer: medium (c)
$(3x+4)^2=9x^2+24x+16$

10. Perform the indicated operations and simplify: $(2+3x)(x^2+2x+4)$.
(a) $9x^2+12x+16$ (b) $9x^2+12x+8$ (c) $3x^3+8x^2+16x+8$
(d) $3x^2+24x+16$ (e) $3x^2-12x-8$
Answer: medium (c)
$(2+3x)(x^2+2x+4)=2x^2+4x+8+3x^3+6x^2+12x=3x^3+8x^2+16x+8$

11. Perform the indicated operations and simplify: $(3u^2 + 2u - 1)(u^3 + 2)$.
(a) $3u^5 - 2u^4 + u^3 + 6u^2 - 4u - 2$ (b) $3u^5 + 2u^4 - u^3 + 6u^2 + 4u - 2$ (c) $3u^5 + 2u^4 + 6u^2 + 4u$
(d) $3u^5 + u^3 - 4u$ (e) $3u^5 + 2u^4 + u^3 + 6u^2 + 4u + 2$
Answer: medium (b)
$(3u^2 + 2u - 1)(u^3 + 2) = 3u^5 + 2u^4 - u^3 + 6u^2 + 4u - 2$

12. Perform the indicated operations and simplify: $\left(2\sqrt{r} - \sqrt{s}\right)^2$.
(a) $4r^2 - 4rs + s^2$ (b) $4r^2 - 4\sqrt{rs} + s^2$ (c) $4r + 4\sqrt{rs} + s$ (d) $4r + 4\sqrt{rs} - s$ (e) $4r - 4\sqrt{rs} + s$
Answer: medium (e)
$\left(2\sqrt{r} - \sqrt{s}\right)^2 = 4r - 4\sqrt{r}\sqrt{s} + s = 4r - 4\sqrt{rs} + s$

13. Perform the indicated operations and simplify: $(2x - 3)^3$.
(a) $8x^3 - 36x^2 + 54x - 27$ (b) $4x^3 - 18x^2 + 27x - 27$ (c) $8x^3 + 36x^2 - 54x$
(d) $2x^3 - 18x^2 + 9x - 18$ (e) $4x^3 - 18x^2 - 18$
Answer: medium (a)
$(2x - 3)^3 = (2x - 3)(4x^2 - 12x + 9) = 8x^3 - 36x^2 + 54x - 27$

14. Perform the indicated operations and simplify: $\left(c - \dfrac{1}{c}\right)^2$.
(a) $c^2 - 2 + \dfrac{1}{c^2}$ (b) $c - 2\sqrt{c} + \dfrac{1}{c}$ (c) $c^2 - 2c + 1$ (d) $c^2 - 2c + \dfrac{1}{c}$ (e) $c - 2 + \dfrac{1}{c^2}$
Answer: medium (a)
$\left(c - \dfrac{1}{c}\right)^2 = c^2 - \dfrac{c}{c} - \dfrac{c}{c} + \dfrac{1}{c^2} = c^2 - 2 + \dfrac{1}{c^2}$

15. Perform the indicated operations and simplify: $(1 - x - x^2)(1 - x + x^2)$.
(a) $1 + 2x - x^2 + x^4$ (b) $-2x + x^2 + x^4$ (c) $1 + 2x - x^2 - x^4$
(d) $1 - 2x + x^2 - x^4$ (e) $1 - 2x - x^2 + x^4$
Answer: medium (d)
$(1 - x - x^2)(1 - x + x^2) = 1 - x + x^2 - x + x^2 - x^3 - x^2 + x^3 - x^4 = 1 - 2x + x^2 - x^4$

16. Perform the indicated operations and simplify: $(1 - x + x^2)^2$.
(a) $x^4 - 4x^3 + 6x^2 - 4x + 1$ (b) $x^4 + 2x^3 - 3x^2 + 2x - 1$ (c) $x^3 - 2x^2 + 3x - 1$
(d) $x^4 - x^3 + 3x^2 - x + 1$ (e) $x^4 - 2x^3 + 3x^2 - 2x + 1$
Answer: medium (e)
$(1 - x + x^2)^2 = 1 - 2x + 2x^2 + x^2 - 2x^3 + x^4 = x^4 - 2x^3 + 3x^2 - 2x + 1$

17. Perform the indicated operations and simplify: $(x^3y^2 - y^4)(x^2 - xy - y^2)$
(a) $x^5y^2 - x^4y^3 - x^3y^4 - x^2y^4 + xy^5 + y^6$ (b) $x^4y - x^4y^3 - x^2y^4 - xy^4 + y^5 + y^6$
(c) $x^5y^4 - x^4y^5 - x^3y^6 - x^2y + xy^2 + y^3$ (d) $x^5y^2 + x^4y^3 + x^3y^4 + x^2y^4 - xy^5 - y^6$
(e) $x^5y^2 - 2x^4y^3 - 4x^3y^4 - 4x^2y^4 + 2xy^5 + y^6$
Answer: medium (a)
$(x^3y^2 - y^4)(x^2 - xy - y^2) = x^5y^2 - x^4y^3 - x^3y^4 - x^2y^4 + xy^5 + y^6$

18. Factor $6x^4 + 4x^2$.
(a) $x^2(6x^2 + 2)$ (b) $2x^2(3x^2 + 2)$ (c) $2x^2(x^2 + 2)$ (d) $3x^2(2x^2 + 1)$ (e) $4x^2(2x^2 + 1)$
Answer: easy (b)
$6x^4 + 4x^2 = 2x^2(3x^2 + 2)$

19. Factor the expression $4abc - 7ac$.
(a) $ab(4c - 7)$ (b) $4ac(b - 7)$ (c) $ab(4c - 7)$ (d) $ab(4 - 7c)$ (e) $ac(4b - 7)$
Answer: easy (e)
$4abc - 7ac = ac(4b - 7)$

20. Factor $3x^5y^5 - 15x^2y^3 + 24x^4y^2$
   (a) $3x^2y^2(x^3y^3 - 5y + 8x^2)$
   (b) $3x^3y^3(x^2y^2 - 5y^2 + x^3)$
   (c) $x^2y^2(3x^3y^3 - 15y + 8x^2)$
   (d) $3y^2(x^5y^3 - 5x^2y + 8x^2)$
   (e) $3xy^2(x^3y^3 - 5xy + 8x^3)$
   Answer: medium (a)
   $3x^5y^5 - 15x^2y^3 + 24x^4y^2 = 3x^2y^2(x^3y^3 - 5y + 8x^2)$

21. Factor the expression $x^2 + 2x - 8$.
   (a) $(x-4)(x+2)$
   (b) $(x+4)(x+2)$
   (c) $(-x-4)(-x-2)$
   (d) $(x+8)(x-1)$
   (e) $(x+4)(x-2)$
   Answer: easy (e)
   $x^2 + 2x - 8 = (x+4)(x-2)$

22. Factor $x^2 - 8x + 15$.
   (a) $(x+3)(x+5)$ (b) $(x-3)(x+5)$ (c) $(x-3)(x-5)$ (d) $(x+3)(x-5)$ (e) $(-x-3)(x-5)$
   Answer: easy (c)
   $x^2 - 8x + 15 = (x-3)(x-5)$

23. Factor $z^2 + 3z - 18$
   (a) $(z-3)(z+6)$ (b) $(z-9)(z+2)$ (c) $(z+9)(z-2)$ (d) $(z+3)(z+6)$ (e) $(z+3)(z-6)$
   Answer: easy (a)
   $z^2 + 3z - 18 = (z-3)(z+6)$

24. Factor the expression $2x^2 - 7x + 6$.
   (a) $(2x-2)(x-3)$ (b) $(2x-3)(x-2)$ (c) $(x-3)(2x+2)$ (d) $(2x+3)(x+2)$ (e) $(x+3)(2x-2)$
   Answer: medium (b)
   $2x^2 - 7x + 6 = (2x-3)(x-2)$

25. Factor the expression $12x^2 + 8x - 4$.
   (a) $(4x-1)(3x+4)$ (b) $4(3x-1)(x+1)$ (c) $2(2x+1)(2x+1)$ (d) $(2x+4)(3x+1)$ (e) $(3x+2)(4x-2)$
   Answer: medium (b)
   $12x^2 + 8x - 4 = 4(3x-1)(x+1)$

26. Factor the expression $x^2 - 4x + 4$.
   Answer: easy
   $x^2 - 4x + 4 = (x-2)^2$

27. Factor $12 + 7t - 12t^2$
   Answer: medium
   $12 + 7t - 12t^2 = (4-3t)(3+4t)$

28. Factor $3x^2 - 11xy - 4y^2$.
   Answer: medium
   $3x^2 - 11xy - 4y^2 = (3x+y)(x-4y)$

29. Factor the expression $9s^2 - 16t^2$.
   Answer: medium
   $9s^2 - 16t^2 = (3s-4t)(3s+4t)$

30. Factor the expression $x^3 - 8$.
   Answer: medium
   $x^3 - 8 = (x-2)(x^2 + 2x + 4)$

31. Factor the expression $2x^3 - 8x$.
   Answer: medium
   $2x^3 - 8x = 2x(x^2 - 4) = 2x(x-2)(x+2)$

32. Factor $9r^2 - 12rs + 4s^2$.

Answer: hard

$9r^2 - 12rs + 4s^2 = (3r - 2s)^2$

33. Factor the expression $x^6 + 27$.

Answer: hard

$x^6 + 27 = (x^2)^3 + 3^3 = (x^2 + 3)(x^4 - 3x^2 + 9)$

34. Factor the expression $x^4 + 4x^2 + 4$.

Answer: medium

$x^4 + 4x^2 + 4 = (x^2 + 2)^2$

35. Factor the expression $x^3 + 2x^2 - x - 2$.

Answer: medium

$x^3 + 2x^2 - x - 2 = x^2(x + 2) - (x + 2) = (x^2 - 1)(x + 2) = (x - 1)(x + 1)(x + 2)$

36. Factor the expression $y^3 + y^2 - y - 1$.

Answer: medium

$y^3 + y^2 - y - 1 = y^2(y + 1) - 1(y + 1) = (y - 1)(y + 1)^2$

37. Factor $2x^3 + 7x^2 - 4x - 14$.

Answer: medium

$2x^3 + 7x^2 - 4x - 14 = x^2(2x + 7) - 2(2x + 7) = (2x + 7)(x^2 - 2)$

38. Factor the expression $x^8 - 256$.

Answer: hard

$x^8 - 256 = (x^4)^2 - 16^2 = (x^4 - 16)(x^4 + 16) = (x^2 - 4)(x^2 + 4)(x^4 + 16) = (x - 2)(x + 2)(x^2 + 4)(x^4 + 16)$

39. Factor $2t^6 - 20t^4 + 18t^2$

Answer: medium

$2t^6 - 20t^4 + 18t^2 = 2t^2(t^4 - 10t^2 + 9) = 2t^2(t^2 - 1)(t^2 - 9) = 2t^2(t - 1)(t + 1)(t - 3)(t + 3)$

40. Factor $x^5 - 2x^4 - x^2 + 2x$.

Answer: hard

$x^5 - 2x^4 - x^2 + 2x = x(x^4 - 2x^3 - x + 2) = x[x^3(x - 2) - 1(x - 2)] = x(x - 2)(x^3 - 1) = x(x - 2)(x - 1)(x^2 + x + 1)$

41. Factor the expression $4x^{-1/2} + 5x^{1/2} + x^{3/2}$.

Answer: hard

$4x^{-1/2} + 5x^{1/2} + x^{3/2} = (4 + 5x + x^2)x^{-1/2} = (x + 4)(x + 1)x^{-1/2}$

42. Factor the expression $(x - 2)^{7/2} - 4(x - 2)^{3/2}$.

Answer: hard

$(x - 2)^{7/2} - 4(x - 2)^{3/2} = [(x - 2)^2 - 4](x - 2)^{3/2} = (x^2 - 4x)(x - 2)^{3/2} = x(x - 4)(x - 2)^{3/2}$

43. Factor the expression $x^3(1 - x^3)^{1/2} - 2(1 - x^3)^{3/2}$.

Answer: hard

$x^3(1 - x^3)^{1/2} - 2(1 - x^3)^{3/2} = [x^3 - 2(1 - x^3)](1 - x^3)^{1/2} = (3x^3 - 2)(1 - x^3)^{1/2}$

44. Factor $x^{1/2}(y + 1)^{-1/2} + (y + 1)^{1/2} + 2x^{-1/2}(y + 1)^{3/2}$

Answer: hard

$x^{1/2}(y + 1)^{-1/2} + (y + 1)^{1/2} + 2x^{-1/2}(y + 1)^{3/2} = \left[x + x^{1/2}(y + 1) + 2(y + 1)^2\right]x^{-1/2}(y + 1)^{-1/2}$

45. Verify the sum of cubes formula $a^3 + b^3 = (a + b)(a^2 - ab + b^2)$.

Answer: medium

$\text{RHS} = (a + b)(a^2 - ab + b^2) = a^3 - a^2b + ab^2 + ba^2 - ab^2 + b^3 = a^3 + b^3 = \text{LHS}$

46. Factor $y^4 + 4$.

Answer: hard

$y^4 + 4 = (y^4 + 4 + 4y^2) - 4y^2 = (y^2 + 2)^2 - (2y)^2 = (y^2 + 2 - 2y)(y^2 + 2 + 2y)$

47. (a) Verify the formula $(a - b)^2 = a^2 - 2ab + b^2$.

(b) Verify the formula $(a - b)^3 = a^3 - 3a^2b + 3ab^2 - b^3$.

Answer: medium

(a) $\text{LHS} = (a - b)^2 = (a - b)(a - b) = a(a - b) - b(a - b) = a^2 - ab - ba + b^2 = a^2 - 2ab + b^2 = \text{RHS}$

(b) $\text{LHS} = (a - b)^3 = (a - b)(a - b)^2 = (a - b)(a^2 - 2ab + b^2)$

$= a(a^2 - 2ab + b^2) - b(a^2 - 2ab + b^2) = a^3 - 2a^2b + ab^2 - ba^2 + 2ab^2 - b^3 = a^3 - 3a^2b + 3ab^2 - b^3 = \text{RHS}$

48. Perform the indicated operation and simplify: $\left(ax^2 - y^2\right)\left(ax^2 + y^2\right)$.

(a) $a^2x^2 - y^2$ (b) $a^2x^4 + ax^2y^2 - y^4$ (c) $a^2x^4 - y^4$ (d) $a^2x^4 - ax^2y^2 - y^4$ (e) $a^2x^4 + y^4$

Answer: easy (c)

$\left(ax^2 - y^2\right)\left(ax^2 + y^2\right) = a^2x^4 - y^4$

49. Perform the indicated operation and simplify: $\left(x^2 - b^2y^2\right)\left(x^2 - cxy + ay^2\right)$.

(a) $x^4 - cx^3y + ax^2y^2 - b^2y^2x^2 + b^2cy^3x - ab^2y^4$ (b) $x^4 - x^3cy + x^2ay^2 + b^2y^3cx - b^2y^4a$

(c) $x^2 - x^3y + x^2ay^2 - b^2y^3cx - ab^2y^4$ (d) $a^2x^4 - ax^2y^2 - y^4$

(e) $x^4 - cx^3y + ax^2y^2 - cy^3x - ab^2y^4$

Answer: medium (a)

$\left(x^2 - b^2y^2\right)\left(x^2 - cxy + ay^2\right) = x^4 - cx^3y + ax^2y^2 - b^2y^2x^2 + b^2cy^3x - ab^2y^4$

50. Perform the indicated operation and simplify : $\left(x + \dfrac{1}{y^2}\right)^3$.

(a) $x^3 + \dfrac{3x^2}{y^2} + \dfrac{3x}{y^4} + \dfrac{1}{y^6}$ (b) $x^3 + \dfrac{3x^2}{y^2} + \dfrac{x}{y^2} + \dfrac{1}{y^6}$

(c) $x^3 + \dfrac{3x^2}{y} + \dfrac{3x^2}{y^4} + \dfrac{3}{y^3}$ (d) $x^3 + \dfrac{3x^2}{y} + \dfrac{1}{y^6}$

(e) $x^3 + \dfrac{1}{y^6}$

Answer: medium (a)

$\left(x + \dfrac{1}{y^2}\right)^3 = x^3 + \dfrac{3x^2}{y^2} + \dfrac{3x}{y^4} + \dfrac{1}{y^6}$

51. Perform the indicated operation and simplify : $\left(1 - x^{2/3}\right)\left(2 + x^{1/3}\right)$

(a) $2 + \sqrt[3]{x} - 2\left(\sqrt[3]{x}\right)^3 + x$ (b) $1 + \sqrt[3]{x} - \left(\sqrt[3]{x}\right)^2 - x$

(c) $2 + \sqrt[3]{x} - \sqrt[3]{x} - x^2$ (d) $2 + \sqrt[3]{x} - 2\left(\sqrt[3]{x}\right)^2 - x$

(e) $2 + \sqrt[3]{x} + 2\left(\sqrt[3]{x}\right)^2$

Answer: medium (d)

$\left(1 - x^{2/3}\right)\left(2 + x^{1/3}\right) = 2 + \sqrt[3]{x} - 2\left(\sqrt[3]{x}\right)^2 - x$

52. Factor the expression $y^2 + xy - 2x^2$.

(a) $-(2x + xy)(y - 2x)$

(b) $-(2x + y)(x - y)$

(c) $(2x - y)(x - y)$

(d) $(2x + y)(x + y)$

(e) $-(2x + y)(x - 2y)$

Answer: medium (b)

$y^2 + xy - 2x^2 = -(2x + y)(x - y)$

53. Factor the expression $6x^2 + 13x + 6$.

(a) $(2x + 3)^2$

(b) $(x + 3)(6x + 2)$

(c) $(2x + 1)(3x + 6)$

(d) $(6x + 1)(x + 6)$

(e) $(2x + 3)(3x + 2)$

Answer: medium (e)

$6x^2 + 13x + 6 = (2x + 3)(3x + 2)$

54. Factor the expression $x\left(x - \frac{1}{2}\right)^3 + y\left(x - \frac{1}{2}\right)^2$.

(a) $-\frac{1}{8}(2x - 1)^2\left(2x^2 - x + y\right)$

(b) $\frac{1}{8}(2x - 1)^2\left(2x^2 - x + 2y\right)$

(c) $\frac{1}{4}(2x - 1)^2\left(x^2 + x + y\right)$

(d) $\frac{1}{8}(2x + 1)\left(2x^2 - x + 2y\right)$

(e) $8(2x - 1)\left(2x^2 - x + 2y\right)^2$

Answer: medium (b)

$x\left(x - \frac{1}{2}\right)^3 + y\left(x - \frac{1}{2}\right)^2 = \frac{1}{8}(2x - 1)^2\left(2x^2 - x + 2y\right)$

55. Factor the expression $x^2(x - 5)^3 + x^3(x - 5)^2$.

(a) $x(2x + 5)(x - 5)^2$

(b) $x^2(2x + 5)(x - 5)^2$

(c) $x(2x - 5)^2(x - 5)^2$

(d) $x^2(5x - 5)(2x - 5)$

(e) $x^2(2x - 5)(x - 5)^2$

Answer: medium (e)

$x^2(x - 5)^3 + x^3(x - 5)^2 = x^2(2x - 5)(x - 5)^2$

56. Factor the expression $x^3 + 6x^2 + 12x + 8$.

(a) $(x + 2)^3$

(b) $x^3 + 8$

(c) $(x - 2)^3$

(d) $2(x + 2)\left(x^2 + 2x + 4\right)$

(e) $2(x + 4)^2$

Answer: medium (a)

$x^3 + 6x^2 + 12x + 8 = (x + 2)^3$

57. Factor the expression $64x^3 - 1$.

(a) $(4x - 1)\left(16x^2 - 4x + 1\right)$

(b) $(2x - 1)\left(8x^2 + 4x + 1\right)$

(c) $(4x - 1)\left(16x^2 + 4x + 1\right)$

(d) $(64x - 1)\left(x^2 - 2x + 1\right)$

(e) $(8x - 1)\left(8x^2 + 4x + 1\right)$

Answer: hard (c)

$64x^3 - 1 = (4x - 1)\left(16x^2 + 4x + 1\right)$

58. Factor the expression $2y^2x + xy + 5y + 10y^2$.

(a) $xy\,(2y-1)\,(x+5)$

(b) $y^2\,(y+1)\,(5x+5)$

(c) $y\,(2y+1)\,(x+5)$

(d) $y\,(y+2)\,(5x+1)$

(e) $(y+1)^2\,(x+5)$

Answer: medium (c)

$2y^2x + xy + 5y + 10y^2 = y\,(2y+1)\,(x+5)$

59. Factor the expression $x^2\left(2-x^2\right)^{1/3} - \left(2-x^2\right)^{4/3}$.

(a) $-\sqrt[3]{2+x^2}\,\left(2x^2+2\right)$

(b) $\sqrt[3]{2-2x^2}\,\left(x^2-2\right)$

(c) $\sqrt[3]{2+x^2}\,\left(2x^2-2\right)$

(d) $\left(2-x^2\right)\sqrt[3]{x^2-2}$

(e) $\sqrt[3]{2-x^2}\,\left(2x^2-2\right)$

Answer: hard (e)

$x^2\left(2-x^2\right)^{1/3} - \left(2-x^2\right)^{4/3} = \sqrt[3]{(2-x^2)}\,\left(2x^2-2\right)$

SECTION 1.5

1. Simplify the expression $\dfrac{x^2-x-6}{x^2-4}$.

(a) $\dfrac{x+1}{x+2}$ (b) $\dfrac{x+2}{x+3}$ (c) $\dfrac{x-3}{x-2}$ (d) $\dfrac{x+3}{x+2}$ (e) $\dfrac{x-2}{x-2}$

Answer: medium (c)

$$\frac{x^2-x-6}{x^2-4} = \frac{(x-3)(x+2)}{(x-2)(x+2)} = \frac{x-3}{x-2}$$

2. Simplify the expression $\dfrac{2y^2-15y+18}{y^2-11y+30}$.

(a) $\dfrac{2y-3}{y-5}$ (b) $\dfrac{y-6}{2y+5}$ (c) $\dfrac{y+6}{2y+6}$ (d) $\dfrac{y-5}{2y+6}$ (e) $\dfrac{y-3}{y+6}$

Answer: medium (a)

$$\frac{2y^2-15y+18}{y^2-11y+30} = \frac{(2y-3)(y-6)}{(y-6)(y-5)} = \frac{2y-3}{y-5}$$

3. Simplify the expression $\dfrac{x^2-1}{x^3-1}$.

(a) $\dfrac{-x+1}{x^2-x-1}$ (b) $\dfrac{-1+x}{-x^2-x+1}$ (c) $\dfrac{x-1}{x^2+2x+1}$ (d) $\dfrac{x+1}{x^2+x+1}$ (e) $\dfrac{x+1}{x^2-x-1}$

Answer: medium (d)

$$\frac{x^2-1}{x^3-1} = \frac{(x-1)(x+1)}{(x-1)(x^2+x+1)} = \frac{x+1}{x^2+x+1}$$

4. Simplify the expression $\dfrac{x^2+x-6}{x^2-2x}\cdot\dfrac{x^3-x^2}{x^2+2x-3}$.

(a) $\dfrac{x+2}{x^2}$ (b) $\dfrac{x}{2-x}$ (c) 1 (d) x (e) x^2

Answer: hard (d)

$$\frac{x^2+x-6}{x^2-2x}\cdot\frac{x^3-x^2}{x^2+2x-3} = \frac{(x-2)(x+3)\cdot x^2(x-1)}{x(x-2)(x+3)(x-1)} = x$$

5. Simplify $\dfrac{x^2+2xy+y^2}{x^2-y^2}\cdot\dfrac{x^2+xy-2y^2}{2x^2+xy-y^2}$

(a) $\dfrac{2x+y}{x-2y}$ (b) $\dfrac{x+2y}{2x-y}$ (c) $\dfrac{2x-y}{x+2y}$ (d) $\dfrac{2-xy}{2x+y}$ (e) $\dfrac{x+y+2}{x-2y-2}$

Answer: hard (b)

$$\frac{x^2+2xy+y^2}{x^2-y^2}\cdot\frac{x^2+xy-2y^2}{2x^2+xy-y^2}=\frac{(x+y)^2\cdot(x+2y)(x-y)}{(x-y)(x+y)\cdot(x+y)(2x-y)}=\frac{x+2y}{2x-y}$$

6. Simplify the expression $\dfrac{y^2-9}{2y^2-15y+18}\div\dfrac{2y^2+7y+3}{y^2-5y-6}$.

(a) x (b) $\dfrac{(y-3)(y+3)}{(2y-3)(2y+1)}$ (c) $\dfrac{y-1}{2y+3}$ (d) -1 (e) $\dfrac{2y+3}{2y-3}$

Answer: hard (b)

$$\frac{y^2-9}{2y^2-15y+18}\div\frac{2y^2+7y+3}{y^2-5y-6}=\frac{(y-3)(y+3)}{(2y-3)(y-6)}\cdot\frac{(y-6)(y+1)}{(2y+1)(y+3)}=\frac{(y-3)(y+1)}{(2y-3)(2y+1)}$$

7. Simplify $\dfrac{\dfrac{2x^2-3x-2}{x^2-1}}{\dfrac{2x^2-3x-2}{x^2+3x+2}}$

(a) $\dfrac{x-1}{x+2}$ (b) $\dfrac{x+1}{x-2}$ (c) $\dfrac{x+2}{x-1}$ (d) $\dfrac{x-2}{x+1}$ (e) 2

Answer: hard (c)

$$\frac{\dfrac{2x^2-3x-2}{x^2-1}}{\dfrac{2x^2-3x-2}{x^2+3x+2}}=\frac{(x-2)(2x+1)}{(x-1)(x+1)}\cdot\frac{(x+2)(x+1)}{(2x+1)(x-2)}=\frac{x+2}{x-1}$$

8. Simplify the expression $\dfrac{x}{z/y}$.

(a) $\dfrac{xy}{z}$ (b) $\dfrac{xz}{y}$ (c) $\dfrac{yz}{x}$ (d) xyz (e) $\dfrac{1}{xyz}$

Answer: medium (a)

$$\frac{x}{z/y}=x\cdot\frac{y}{z}=\frac{xy}{z}$$

9. Simplify the expression $\dfrac{1}{x+2}+\dfrac{1}{x-2}$.

(a) $\dfrac{2}{x-2}$ (b) $\dfrac{2x}{2x^2-2}$ (c) $\dfrac{1}{x^2-4}$ (d) $\dfrac{x^2+1}{x^2-2}$ (e) $\dfrac{2x}{x^2-4}$

Answer: medium (e)

$$\frac{1}{x+2}+\frac{1}{x-2}=\frac{x-2+x+2}{(x-2)(x+2)}=\frac{2x}{x^2-4}$$

10. Simplify the expression $\dfrac{x}{x-2}-\dfrac{3}{x+3}$.

(a) $\dfrac{x^2+2x+6}{(x-2)(x+3)}$ (b) $\dfrac{x^2-2x+12}{(x-4)(x-2)}$ (c) $\dfrac{x^2+6}{(x-2)(x+3)}$ (d) $\dfrac{x^2+3}{(x-2)(x+3)}$ (e) $\dfrac{x^2+1}{(x+2)(x-3)}$

Answer: medium (c)

$$\frac{x}{x-2}-\frac{3}{x+3}=\frac{x(x+3)-3(x-2)}{(x-2)(x+3)}=\frac{x^2+6}{(x-2)(x+3)}$$

11. Simplify the expression $\dfrac{5}{2x-1}-\dfrac{2}{(2x-1)^2}$.

(a) $\dfrac{10x-18}{(2x-3)^2}$ (b) $\dfrac{5x-7}{(x-2)^2}$ (c) $\dfrac{10x-7}{(2x-1)^2}$ (d) $\dfrac{10x-2}{(2x+1)^2}$ (e) $\dfrac{2x+9}{(2x-1)^2}$

Answer: medium (c)

$$\frac{5}{2x-1}-\frac{2}{(2x-1)^2}=\frac{5(2x-1)-2}{(2x-1)^2}=\frac{10x-7}{(2x-1)^2}$$

12. Simplify the expression $\dfrac{4}{a^2} - \dfrac{3}{ab} + \dfrac{2}{b^2}$.

(a) $\dfrac{b^2 - 3ab + 4a^2}{2a^2b^2}$ (b) $\dfrac{4b - 3ab + 2a}{2ab}$ (c) $\dfrac{4b^2 - 3ab + 6a^2}{3a^2b^2}$ (d) $\dfrac{4b^2 - 3ab + 2a^2}{a^2b^2}$ (e) $\dfrac{4ab^2 - 3ab + 2a^2b}{a^3b^3}$

Answer: medium (d)

$$\frac{4}{a^2} - \frac{3}{ab} + \frac{2}{b^2} = \frac{4b^2 - 3ab + 2a^2}{a^2b^2}$$

13. Simplify the expression $\dfrac{1}{x} + \dfrac{1}{x^2} + \dfrac{1}{x^3} + \dfrac{1}{x^4}$.

(a) $\dfrac{x^3 + x^2 + x + 1}{x^4}$ (b) $\dfrac{x^3 + x^2 - x - 1}{x^4}$ (c) $\dfrac{x^3 - x^2 + x - 1}{x^3}$ (d) $\dfrac{x^3 - x^2 - x + 1}{x^4}$ (e) $\dfrac{x^3 + x^2 + x - 1}{x^3}$

Answer: medium (a)

$$\frac{1}{x} + \frac{1}{x^2} + \frac{1}{x^3} + \frac{1}{x^4} = \frac{x^3 + x^2 + x + 1}{x^4}$$

14. Simplify the expression $\dfrac{x}{x^2 - 4} + \dfrac{1}{x - 2}$.

(a) $\dfrac{2x + 1}{(x - 2)(x + 1)}$ (b) $\dfrac{2x - 1}{(x - 1)(x + 2)}$ (c) $\dfrac{2(x + 2)}{(x - 2)^2}$ (d) $\dfrac{2(x + 1)}{(x + 2)^2}$ (e) $\dfrac{2(x + 1)}{(x - 2)(x + 2)}$

Answer: medium (e)

$$\frac{x}{x^2 - 4} + \frac{1}{x - 2} = \frac{x + x + 2}{(x - 2)(x + 2)} = \frac{2x + 2}{(x - 2)(x + 2)} = \frac{2(x + 1)}{(x - 2)(x + 2)}$$

15. Simplify the expression $\dfrac{x}{x^2 - x - 6} - \dfrac{2}{x^2 - 7x + 12}$.

(a) $\dfrac{x^2 - 3x - 2}{(x + 1)(x - 2)(x - 3)}$ (b) $\dfrac{x^2 - 2x + 2}{(x - 2)^2(x - 3)}$ (c) $\dfrac{2x^2 - 3x - 6}{(x - 2)(x + 2)(x - 4)}$

(d) $\dfrac{x^2 - 6x - 4}{(x + 2)(x - 4)(x - 3)}$ (e) $\dfrac{x^2 - 6x - 2}{(x + 1)(x + 2)(x - 3)}$

Answer: medium (d)

$$\frac{x}{x^2 - x - 6} - \frac{2}{x^2 - 7x + 12} = \frac{x}{(x + 2)(x - 3)} - \frac{2}{(x - 4)(x - 3)} = \frac{x(x - 4) - 2(x + 2)}{(x + 2)(x - 4)(x - 3)} = \frac{x^2 - 6x - 4}{(x + 2)(x - 4)(x - 3)}$$

16. Simplify the expression $\dfrac{x}{x^2 + x - 6} - \dfrac{1}{x - 2} - \dfrac{2}{x + 3}$.

(a) $\dfrac{-2x + 2}{(x - 2)(x + 2)}$ (b) $\dfrac{2x + 1}{(x - 2)(x - 3)}$ (c) $\dfrac{-2x + 1}{(x - 2)(x + 3)}$ (d) $\dfrac{2x - 2}{(x - 2)(x + 3)}$ (e) $\dfrac{2x - 1}{(x + 2)(x + 3)}$

Answer: medium (c)

$$\frac{x}{x^2 + x - 6} - \frac{1}{x - 2} - \frac{2}{x + 3} = \frac{x - (x + 3) - 2(x - 2)}{(x - 2)(x + 3)} = \frac{-2x + 1}{(x - 2)(x + 3)}$$

17. Simplify the expression $\dfrac{1}{x + 2} - \dfrac{2}{(x + 2)^2} + \dfrac{3}{x^2 - 4}$.

Answer: medium

$$\frac{1}{x + 2} - \frac{2}{(x + 2)^2} + \frac{3}{x^2 - 4} = \frac{(x - 2)(x + 2) - 2(x - 2) + 3(x + 2)}{(x - 2)(x + 2)^2} = \frac{x^2 + x + 6}{(x - 2)(x + 2)^2}$$

18. Simplify the expression $x^2 - \dfrac{y^2}{\dfrac{1}{y^2} + \dfrac{1}{x^2}}$.

Answer: hard

$$x^2 - \frac{y^2}{\dfrac{1}{y^2} + \dfrac{1}{x^2}} = x^2 - \frac{y^2}{\dfrac{x^2 + y^2}{x^2y^2}} = \frac{x^2(x^2 + y^2) - y^2(x^2y^2)}{x^2 + y^2} = \frac{x^2(x^2 + y^2 - y^4)}{x^2 + y^2}$$

19. Simplify the expression $1+\cfrac{1}{1+\cfrac{1}{1+\cfrac{1}{x}}}$.

Answer: hard

$$1+\cfrac{1}{1+\cfrac{1}{1+\cfrac{1}{x}}}=1+\cfrac{1}{1+\cfrac{1}{\cfrac{x+1}{x}}}=1+\cfrac{1}{\cfrac{2x+1}{x+1}}=1+\frac{x+1}{2x+1}=\frac{3x+2}{2x+1}$$

20. Simplify the expression $\dfrac{\dfrac{a+b}{a}+\dfrac{a-b}{b}}{\dfrac{a+b}{b}-\dfrac{a-b}{a}}$.

Answer: hard

$$\frac{\dfrac{a+b}{a}+\dfrac{a-b}{b}}{\dfrac{a+b}{b}-\dfrac{a-b}{a}}=\frac{b(a+b)+a(a-b)}{ab}\div\frac{a(a+b)-b(a-b)}{ab}=1$$

21. Simplify the expression $\dfrac{x^{-1}+y^{-1}}{(x+y)^{-1}}$.

Answer: medium

$$\frac{x^{-1}+y^{-1}}{(x+y)^{-1}}=(x+y)\left(\frac{1}{x}+\frac{1}{y}\right)=\frac{(x+y)(y+x)}{xy}=\frac{(x+y)^2}{xy}$$

22. Simplify the expression $\dfrac{(x+h)^{-2}-x^{-2}}{h}$.

Answer: medium

$$\frac{(x+h)^{-2}-x^{-2}}{h}=\frac{x^2-(x+h)^2}{x^2(x+h)^2}\cdot\frac{1}{h}=\frac{x^2-(x^2+2xh+h^2)}{hx^2(x+h)^2}=-\frac{2x+h}{x^2(x+h)^2}$$

23. Simplify the expression $\dfrac{(x+h)^2-7(x+h)-(x^2-7x)}{h}$.

Answer: medium

$$\frac{(x+h)^2-7(x+h)-(x^2-7x)}{h}=\frac{x^2+2xh+h^2-7x-7h-x^2+7x}{h}=2x+h-7$$

24. Simplify the expression $\sqrt{\left(x^4+\dfrac{1}{4x^4}\right)^2-1}$.

Answer: hard

$$\sqrt{\left(x^4+\frac{1}{4x^4}\right)^2-1}=\sqrt{\left(\frac{4x^8+1}{4x^4}\right)^2-1}=\sqrt{\frac{16x^{16}+8x^8+1-16x^8}{16x^8}}=\frac{\sqrt{16x^{16}-8x^8+1}}{4x^4}=\frac{4x^8-1}{4x^4}$$

25. Rationalize the denominator of the expression $\dfrac{1}{2+\sqrt{3}}$.

Answer: easy

$$\frac{1}{2+\sqrt{3}}=\frac{2-\sqrt{3}}{\left(2+\sqrt{3}\right)\left(2-\sqrt{3}\right)}=\frac{2-\sqrt{3}}{4-3}=2-\sqrt{3}$$

26. Rationalize the denominator of the expression $\dfrac{3}{2-\sqrt{y}}$.

Answer: easy

$$\frac{3}{2-\sqrt{y}}=\frac{3\left(2+\sqrt{y}\right)}{\left(2-\sqrt{y}\right)\left(2+\sqrt{y}\right)}=\frac{6+3\sqrt{y}}{4-y}$$

27. Rationalize the numerator of the expression $\dfrac{\sqrt{2(x+h)+5}-\sqrt{2x+5}}{h}$.

Answer: hard

$$\frac{\sqrt{2(x+h)+5}-\sqrt{2x+5}}{h} = \frac{\left(\sqrt{2(x+h)+5}-\sqrt{2x+5}\right)\left(\sqrt{2(x+h)+5}+\sqrt{2x+5}\right)}{h\left(\sqrt{2(x+h)+5}+\sqrt{2x+5}\right)}$$
$$= \frac{2(x+h)+5-(2x+5)}{h\left(\sqrt{2(x+h)+5}+\sqrt{2x+5}\right)} = \frac{2}{\sqrt{2(x+h)+5}+\sqrt{2x+5}}$$

28. Rationalize the numerator of the expression $\sqrt{x^2+4}-x$.

Answer: medium

$$\sqrt{x^2+4}-x = \frac{\left(\sqrt{x^2+4}-x\right)\left(\sqrt{x^2+4}+x\right)}{\sqrt{x^2+4}+x} = \frac{x^2+4-x^2}{\sqrt{x^2+4}+x} = \frac{4}{\sqrt{x^2+4}+x}$$

29. Rationalize the numerator of the expression $\sqrt{x+2}-\sqrt{x}$.

Answer: medium

$$\sqrt{x+2}-\sqrt{x} = \frac{\left(\sqrt{x+2}-\sqrt{x}\right)\left(\sqrt{x+2}+\sqrt{x}\right)}{\sqrt{x+2}+\sqrt{x}} = \frac{x+2-x}{\sqrt{x+2}+\sqrt{x}} = \frac{2}{\sqrt{x+2}+\sqrt{x}}$$

30. State whether or not the equation $\dfrac{b}{b-c} = 1 - \dfrac{b}{c}$ is true for all values of the variables. (Disregard values that make denominators 0.)

Answer: hard

$\dfrac{b}{b-c} \neq 1 - \dfrac{b}{c}$. For example, take $b = 2$, $c = 1$: LHS $= 2$, RHS $= -1$. So the equation is false.

31. State whether or not the equation $\dfrac{x+1}{y+1} = \dfrac{x}{y}$ is true for all values of the variables. (Disregard values that make denominators 0.)

Answer: hard

$\dfrac{x+1}{y+1} \neq \dfrac{x}{y}$. For example, take $x = 0$, $y = 1$: LHS $= \dfrac{1}{2}$, RHS $= 0$. So the equation is false.

32. State whether or not the equation $2\left(\dfrac{a}{b}\right) = \dfrac{2a}{2b}$ is true for all values of the variables. (Disregard values that make denominators 0.)

Answer: medium

$2\left(\dfrac{a}{b}\right) \neq \dfrac{2a}{2b}$. For example, take $a = 1$, $b = 1$: LHS $= 2$, RHS $= 1$. So the equation is false.

33. State whether or not the equation $\dfrac{1+x+x^2}{x} = \dfrac{1}{x} + 1 + x$ is true for all values of the variables. (Disregard values that make denominators 0.)

Answer: hard

$\dfrac{1+x+x^2}{x} = \dfrac{1}{x} + \dfrac{x}{x} + \dfrac{x^2}{x} = \dfrac{1}{x} + 1 + x$ So, true.

34. State whether or not the equation $\dfrac{x^2-1}{x-1} = x+1$ is true for all values of the variables. (Disregard values that make denominators 0.)

Answer: hard

$\dfrac{x^2-1}{x-1} = \dfrac{(x-1)(x+1)}{x-1} = x+1$ So, true.

35. Simplify the expression $\dfrac{y^2 - \frac{1}{4}}{2y^2 + 4y + \frac{3}{2}}$.

(a) $\dfrac{2y-1}{2y+3}$ (b) $\dfrac{3}{4}\left(\dfrac{y-1}{2y+3}\right)$ (c) $\dfrac{1}{2}\left(\dfrac{2y+3}{2y-1}\right)$ (d) $\dfrac{1}{2}\left(\dfrac{2y-1}{2y+3}\right)$ (e) $2\left(\dfrac{2y+1}{2y-3}\right)$

Answer: medium (d)

$$\frac{y^2 - \frac{1}{4}}{2y^2 + 4y + \frac{3}{2}} = \frac{\frac{1}{4}(2y-1)(2y+1)}{\frac{1}{2}(2y+3)(2y+1)} = \frac{1}{2}\left(\frac{2y-1}{2y+3}\right)$$

36. Simplify the expression $\dfrac{x^2+2x-24}{x^2+3x-10} \cdot \dfrac{x^2-25}{x^2+4x-12}$.

(a) $\dfrac{(x-4)(x-5)}{(x-2)^2}$ (b) $\dfrac{(x-4)(x-5)}{(x-2)}$ (c) $\dfrac{(x-6)(x-5)}{(x+5)^2}$ (d) $\dfrac{(x-4)(x+5)}{(x+2)^2}$ (e) $\dfrac{(x+4)(x-5)}{(x-2)^2}$

Answer: medium (a)

$$\frac{x^2+2x-24}{x^2+3x-10} \cdot \frac{x^2-25}{x^2+4x-12} = \frac{(x+6)(x-4)}{(x+5)(x-2)} \cdot \frac{(x-5)(x+5)}{(x+6)(x-2)} = \frac{(x-4)(x-5)}{(x-2)^2}$$

37. Simplify the expression $\dfrac{\frac{x^4}{x+2}}{\frac{x^3}{x^2+6x+8}}$.

(a) $x(x+2)^2$ (b) $x^2(x-4)$ (c) $x(x+4)$ (d) $-x(x+2)$ (e) $x^3(x+4)$

Answer: medium (c)

$$\frac{\frac{x^4}{x+2}}{\frac{x^3}{x^2+6x+8}} = \frac{x^4}{x+2} \cdot \frac{x^2+6x+8}{x^3} = \frac{x^4}{x+2} \cdot \frac{(x+4)(x+2)}{x^3} = x(x+4)$$

38. Simplify the expression $\dfrac{2y}{y^2-1} + \dfrac{1}{y-1}$.

(a) $\dfrac{3y+1}{(y-1)^2}$

(b) $3\dfrac{y+1}{(3y-1)(3y+1)}$

(c) $\dfrac{3y+1}{(y-1)(y+1)}$

(d) $\dfrac{y+1}{(y-1)(y+1)}$

(e) $\dfrac{(3y+1)(y+1)}{y-1}$

Answer: medium (c)

$$\frac{2y}{y^2-1} + \frac{1}{y-1} = \frac{2y}{y^2-1} + \frac{y-1}{y^2-1} = \frac{3y+1}{(y-1)(y+1)}$$

39. Simplify the expression $\dfrac{\dfrac{x}{x-1}+\dfrac{1}{x+1}}{\dfrac{5}{x-1}-\dfrac{2}{x+1}}$.

(a) $\dfrac{x^2+2x-1}{x-1}$ (b) $\dfrac{3x+7}{x^2+2x-1}$ (c) $\dfrac{3x+7}{x-1}$ (d) $\dfrac{x^2-2x+1}{x-1}$ (e) $\dfrac{x^2+2x-1}{3x+7}$

Answer: medium (e)

$$\frac{\dfrac{x}{x-1}+\dfrac{1}{x+1}}{\dfrac{5}{x-1}-\dfrac{2}{x+1}}=\frac{\dfrac{x^2+2x-1}{(x-1)(x+1)}}{\dfrac{3x+7}{(x-1)(x+1)}}=\frac{x^2+2x-1}{(x-1)(x+1)}\cdot\frac{(x-1)(x+1)}{3x+7}=\frac{x^2+2x-1}{3x+7}$$

40. Simplify the expression $\dfrac{\left(x+\dfrac{1}{y}\right)^r\left(x-\dfrac{1}{y}\right)^r}{\left(y+\dfrac{1}{x}\right)^r\left(y-\dfrac{1}{x}\right)^r}$.

(a) $\dfrac{x^{2r}}{y}$ (b) $\dfrac{x^2}{y^{2r}}$ (c) $\dfrac{y^{2r}}{x^{2r}}$ (d) $\dfrac{x^{2r}}{y^{2r}}$ (e) $\dfrac{x}{y}$

Answer: medium (d)

$$\frac{\left(x+\dfrac{1}{y}\right)^r\left(x-\dfrac{1}{y}\right)^r}{\left(y+\dfrac{1}{x}\right)^r\left(y-\dfrac{1}{x}\right)^r}=\left[\frac{\left(x+\dfrac{1}{y}\right)\left(x-\dfrac{1}{y}\right)}{\left(y+\dfrac{1}{x}\right)\left(y-\dfrac{1}{x}\right)}\right]^r=\left(\frac{\dfrac{xy-1}{y}\dfrac{xy+1}{y}}{\dfrac{xy+1}{x}\dfrac{xy-1}{x}}\right)^r=\left(\frac{\dfrac{x^2y^2-1}{y^2}}{\dfrac{x^2y^2-1}{x^2}}\right)^r$$

$$=\left(\frac{x^2y^2-1}{y^2}\cdot\frac{x^2}{x^2y^2-1}\right)^r=\frac{x^{2r}}{y^{2r}}$$

41. Rationalize the denominator of the expression $\dfrac{5-x}{\sqrt{5}+\sqrt{x}}$.

Answer: easy

$$\frac{5-x}{\sqrt{5}+\sqrt{x}}=\frac{5-x}{\sqrt{5}+\sqrt{x}}\cdot\frac{\sqrt{5}-\sqrt{x}}{\sqrt{5}-\sqrt{x}}=\frac{(5-x)\left(\sqrt{5}-\sqrt{x}\right)}{\left(\sqrt{5}+\sqrt{x}\right)\left(\sqrt{5}-\sqrt{x}\right)}=\frac{(5-x)\left(\sqrt{5}-\sqrt{x}\right)}{5-x}$$

$$=\sqrt{5}-\sqrt{x}$$

42. Rationalize the numerator of the expression $\sqrt{x^2+5x+6}-\sqrt{x}$.

Answer: medium

$$\sqrt{x^2+5x+6}-\sqrt{x}=\left(\sqrt{x^2+5x+6}-\sqrt{x}\right)\cdot\frac{\sqrt{x^2+5x+6}+\sqrt{x}}{\sqrt{x^2+5x+6}+\sqrt{x}}$$

$$=\frac{\left(\sqrt{x^2+5x+6}-\sqrt{x}\right)\left(\sqrt{x^2+5x+6}+\sqrt{x}\right)}{\sqrt{x^2+5x+6}+\sqrt{x}}=\frac{x^2+4x+6}{\sqrt{x^2+5x+6}+\sqrt{x}}$$

SECTION 1.6

1. Determine whether or not the given values of the variables are solutions of the equation
$\left(x-\frac{1}{2}\right)+(2x-1)=2\left(x-\frac{1}{2}\right)-(1-x)+\frac{1}{2}$.
(a) $x=\frac{5}{2}$ (b) $x=\frac{1}{2}$
Answer: easy
(a) LHS $=\left(\frac{5}{2}-\frac{1}{2}\right)+\left(2\left(\frac{5}{2}\right)-1\right)=6$, RHS $=2\left(\frac{5}{2}-\frac{1}{2}\right)-\left(1-\frac{5}{2}\right)+\frac{1}{2}=6=$ LHS. Thus $x=\frac{5}{2}$ is a solution of $\left(x-\frac{1}{2}\right)+(2x-1)=2\left(x-\frac{1}{2}\right)-(1-x)+\frac{1}{2}$.
(b) LHS $=\left(\frac{1}{2}-\frac{1}{2}\right)+\left(2\left(\frac{1}{2}\right)-1\right)=0$, RHS $=2\left(\frac{1}{2}-\frac{1}{2}\right)-\left(1-\frac{1}{2}\right)+\frac{1}{2}=0=$ LHS. Thus $x=\frac{1}{2}$ is a solution of $\left(x-\frac{1}{2}\right)+(2x-1)=2\left(x-\frac{1}{2}\right)-(1-x)+\frac{1}{2}$.

2. Solve the equation $5x-9=22$.
(a) $\frac{13}{5}$ (b) $-\frac{31}{5}$ (c) $\frac{5}{31}$ (d) 6 (e) $\frac{31}{5}$
Answer: easy (e)
$5x-9=22 \quad\Leftrightarrow\quad x=\frac{31}{5}$

3. Solve the equation $\frac{3}{2}\left(t-\frac{2}{3}\right)+7=\frac{t+1}{3}-1$.
(a) $-\frac{40}{7}$ (b) 2 (c) $-\frac{36}{7}$ (d) $\frac{40}{7}$ (e) $\frac{7}{36}$
Answer: medium (a)
$\frac{3}{2}\left(t-\frac{2}{3}\right)+7=\frac{t+1}{3}-1 \quad\Rightarrow\quad \frac{3}{2}t+6=\frac{1}{3}t-\frac{2}{3} \quad\Rightarrow\quad t=-\frac{40}{7}$

4. Solve the equation $\dfrac{3x-2}{1-x}=\dfrac{14}{3}$.
(a) $-\frac{20}{23}$ (b) $\frac{20}{23}$ (c) $\frac{3}{4}$ (d) $\frac{4}{3}$ (e) 23
Answer: medium (b)
$\dfrac{3x-2}{1-x}=\dfrac{14}{3} \quad\Leftrightarrow\quad 3(3x-2)=14(1-x) \quad\Rightarrow\quad 9x-6=14-14x \quad\Leftrightarrow\quad x=\dfrac{20}{23}$.

5. Solve the equation $(y-3)^2=(y+3)^2-6y-9$.
(a) 3 (b) 2 (c) $-\frac{3}{2}$ (d) $\frac{2}{3}$ (e) $\frac{3}{2}$
Answer: medium (e)
$(y-3)^2=(y+3)^2-6y-9 \quad\Rightarrow\quad y^2-6y+9=y^2 \quad\Rightarrow\quad y=\frac{3}{2}$.

6. Solve the equation $\dfrac{1}{y}-\dfrac{1}{5y}+\dfrac{1}{y+1}=\dfrac{5}{2y}$.
(a) $\frac{10}{7}$ (b) 17 (c) 7 (d) $-\frac{7}{17}$ (e) $\frac{7}{17}$
Answer: medium (d)
$\dfrac{1}{y}-\dfrac{1}{5y}+\dfrac{1}{y+1}=\dfrac{5}{2y} \quad\Rightarrow\quad \dfrac{1}{5}\dfrac{9y+4}{y(y+1)}=\dfrac{5}{2y} \quad\Rightarrow\quad 2y(9y+4)=y(y+1) \quad\Rightarrow\quad 18y^2+8y=y^2+y \quad\Rightarrow$
$17y^2=-7y \quad\Rightarrow\quad y=-\frac{7}{17}\ (y\neq 0)$

7. Solve the equation $\dfrac{x}{\dfrac{x}{2}+\dfrac{2(x+3)}{5}}=2$.
(a) -4 (b) -3 (c) 3 (d) 12 (e) 4
Answer: medium (b)
$\dfrac{x}{\dfrac{x}{2}+\dfrac{2(x+3)}{5}}=2 \quad\Rightarrow\quad \dfrac{x}{\frac{9}{10}x+\frac{6}{5}}=2 \quad\Rightarrow\quad x=2\left(\frac{9}{10}x+\frac{6}{5}\right) \quad\Rightarrow\quad x=\frac{9}{5}x+\frac{12}{5} \quad\Rightarrow\quad -\frac{4}{5}x=\frac{12}{5} \quad\Rightarrow\quad x=-3$

8. Solve the equation $\frac{4}{x-7}+\frac{1}{x+7}+\frac{3}{2(x^2-49)}=0$.

(a) $\frac{9}{2}$ (b) $-\frac{2}{9}$ (c) -4 (d) -5 (e) $-\frac{9}{2}$

Answer: hard (e)

$\frac{4}{x-7}+\frac{1}{x+7}+\frac{3}{2(x^2-49)}=0 \Rightarrow 2(x^2-49)\left[\frac{4}{x-7}+\frac{1}{x+7}+\frac{3}{2(x^2-49)}\right]=0 \Rightarrow$

$8(x+7)+2(x-7)+3=0 \Rightarrow 10x+45=0 \Rightarrow x=-\frac{9}{2}$

9. Solve the equation $\frac{x}{4x^2+x}+\frac{2}{x-1}=\frac{3}{4x+1}$.

(a) -1 (b) $-\frac{3}{2}$ (c) 1 (d) $-\frac{2}{3}$ (e) $\frac{2}{3}$

Answer: hard (d)

$\frac{x}{4x^2+x}+\frac{2}{x-1}=\frac{3}{4x+1} \Rightarrow x(4x+1)(x-1)\left(\frac{x}{4x^2+x}+\frac{2}{x-1}\right)=\frac{3}{4x+1}x(4x+1)(x-1) \Rightarrow$

$x(9x+1)=3x(x-1) \Rightarrow 9x^2+x=3x^2-3x \Rightarrow x=-\frac{2}{3}\ (x\neq 0)$

10. Find all real solutions of $x^2=64$.

(a) ± 4 (b) -8 (c) $\{0,8\}$ (d) 8 (e) ± 8

Answer: medium (e)

$x^2=64 \Rightarrow x=\pm 8$

11. Find all real solutions of $7x^2-343=0$.

(a) $\pm\frac{1}{7}$ (b) 7 (c) ± 7 (d) 49 (e) No real solution

Answer: medium (c)

$7x^2-343=0 \Rightarrow x^2=49 \Rightarrow x=\pm 7$

12. Find all real solutions of $5x^2+5=0$.

(a) No real solution (b) ± 1 (c) ± 5 (d) 1 (e) -5

Answer: medium (a)

$5x^2+5=0 \Rightarrow x=\pm i$, so there is no real solution.

13. Find all real solutions of $4(x-3)^2=12$.

(a) $\pm 3\sqrt{3}$ (b) $3\pm\sqrt{3}$ (c) $\sqrt{3}$ (d) -3 (e) $3\pm 2\sqrt{3}$

Answer: medium (b)

$4(x-3)^2=12 \Rightarrow (x-3)^2=3 \Rightarrow x=3\pm\sqrt{3}$

14. Find all real solutions of $x^4-256=0$.

(a) $\sqrt{2}$ (b) No real solution (c) 2 (d) ± 4 (e) 4

Answer: medium (d)

$x^4-256=0 \Rightarrow x=\pm\sqrt[4]{256} \Rightarrow x=\pm 4$

15. Find all real solutions of $7(x-1)^3-2401=0$.

(a) No real solution (b) -8 (c) 8 (d) ± 8 (e) ± 7

Answer: medium (c)

$7(x-1)^3-2401=0 \Rightarrow (x-1)^3=343 \Rightarrow x-1=7 \Rightarrow x=8$

16. Find all real solutions of $8x^{2/3} - 250 = 6 + 4x^{2/3}$.
(a) 64 (b) ± 64 (c) 256 (d) ± 512 (e) No real solution

Answer: medium (d)

$$8x^{2/3} - 250 = 6 + 4x^{2/3} \quad \Rightarrow \quad 4x^{2/3} = 256 \quad \Rightarrow \quad x = \pm\sqrt{(64)^3} = \pm 512$$

17. Find the solution of the equation $3.14\,(t - 2.01) = 2.34t + 1.98$, correct to two decimals.
Answer: easy

$$3.14\,(t - 2.01) = 2.34t + 1.98 \quad \Rightarrow \quad 3.14t - 6.3114 = 2.34t + 1.98 \quad \Rightarrow \quad 0.8t = 8.29 \quad \Rightarrow \quad t = 10.36$$

18. Solve the equation $P = a + art$ for t.
Answer: easy

$$P = a + art \quad \Rightarrow \quad P - a = art \quad \Rightarrow \quad t = \frac{P - a}{ar}$$

19. Solve the equation $S = \dfrac{a}{1 - r}$ for r.
Answer: easy

$$S = \frac{a}{1 - r} \quad \Rightarrow \quad S - Sr = a \quad \Rightarrow \quad r = \frac{S - a}{S}$$

20. Solve the equation $A = 2lw + 2lh + 2w$ for h.
Answer: easy

$$A = 2lw + 2lh + 2w \quad \Rightarrow \quad A - 2lw - 2w = 2lh \quad \Rightarrow \quad h = \frac{A - 2lw - 2w}{2l}$$

21. Solve the equation $A = 2\pi rh + 2\pi r^2$ for h.
Answer: medium

$$A = 2\pi rh + 2\pi r^2 \quad \Rightarrow \quad A - 2\pi r^2 = 2\pi rh \quad \Rightarrow \quad h = \frac{A - 2\pi r^2}{2\pi r}$$

22. Solve the equation $\dfrac{1}{s + a} + \dfrac{2}{s - a} = \dfrac{5}{s + c}$ for c.
Answer: medium

$$\frac{1}{s + a} + \frac{2}{s - a} = \frac{5}{s + c} \quad \Rightarrow \quad (s + a)(s - a)(s + c)\left(\frac{1}{s + a} + \frac{2}{s - a}\right) = (s + a)(s - a)(s + c)\left(\frac{5}{s + c}\right) \quad \Rightarrow$$

$$(s - a)(s + c) + 2(s + a)(s + c) = 5(s + a)(s - a) \quad \Rightarrow \quad (s + c)(3s + a) = 5s^2 - 5a^2 \quad \Rightarrow$$

$$c = \frac{5s^2 - 5a^2}{(3s + a)} - s = \frac{2s^2 - as - 5a^2}{3s + a}$$

23. If $h = -\dfrac{b}{2a}$ and $k = \dfrac{4ac - b^2}{4a}$, show that $a\,(x - h)^2 + k = ax^2 + bx + c$.
Answer: medium

$$a\,(x - h)^2 + k = a\left[x - \left(-\frac{b}{2a}\right)\right]^2 + \frac{4ac - b^2}{4a} = a\left(x^2 + \frac{bx}{a} + \frac{b^2}{4a^2}\right) + \frac{4ac - b^2}{4a}$$

$$= ax^2 + bx + \frac{b^2}{4a} + \frac{4ac - b^2}{4a} = \frac{4a^2x^2 + 4abx + b^2 + 4ac - b^2}{4a}$$

$$= \frac{4a^2x^2 + 4abx + 4ac}{4a} = ax^2 + bx + c$$

CHAPTER 1 REVIEW

1. State the property of real numbers being used: $2(xy) = (2x)y$
   (a) Commutative Property for addition
   (b) Commutative Property for multiplication
   (c) Associative Property for addition
   (d) Associative Property for multiplication
   (e) Distributive Property
   Answer: easy (d)

2. State the property of real numbers being used: $3(x + z) = 3x + 3z$
   (a) Commutative Property for addition
   (b) Commutative Property for multiplication
   (c) Associative Property for addition
   (d) Associative Property for multiplication
   (e) Distributive Property
   Answer: easy (e)

3. Express the interval $(-\infty, -3]$ in terms of inequalities and graph the interval.
   (a) $(-\infty, -3] = \{x \mid x < -3\}$ 

   (b) $(-\infty, -3] = \{x \mid x \le -3\}$ 

   (c) $(-\infty, -3] = \{x \mid x > -3\}$ 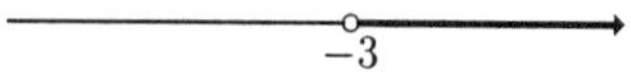

   (d) $(-\infty, -3] = \{x \mid x \ge -3\}$ 

   (e) $(-\infty, -3] = \{x \mid x = -3\}$ 

   Answer: easy (b)
   $(-\infty, -3] = \{x \mid x \le -3\}$ 

4. Express the inequality $-2 < x < 3$ in interval notation and graph the corresponding interval.
   (a) $-2 < x < 3 \quad \Leftrightarrow \quad x \in (-2, 3)$ 

   (b) $-2 < x < 3 \quad \Leftrightarrow \quad x \in (-2, 3]$ 

   (c) $-2 < x < 3 \quad \Leftrightarrow \quad x \in [-2, 3)$ 

   (d) $-2 < x < 3 \quad \Leftrightarrow \quad x \in [-2, 3]$ 

   (e) $-2 < x < 3 \quad \Leftrightarrow \quad x \in (-\infty, -2) \cup (3, \infty)$ 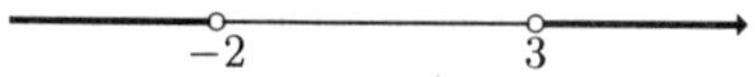

   Answer: easy (a)
   $-2 < x < 3 \quad \Leftrightarrow \quad x \in (-2, 3)$ 

5. Evaluate the number $2 - |1 - |-2||$.
   (a) -2 (b) -1 (c) 0 (d) 1 (e) 2
   Answer: medium (d)
   $2 - |1 - |-2|| = 2 - |1 - 2| = 1$

6. Evaluate the number $\sqrt[3]{-27}$.
   (a) -9 (b) -3 (c) 1 (d) 3 (e) 9
   Answer: easy (b)
   $\sqrt[3]{-27} = \sqrt[3]{(-3)^3} = -3$

7. Evaluate the number $256^{3/4}$.
   (a) 4 (b) 16 (c) 32 (d) 64 (e) 128
   Answer: medium (d)
   $256^{3/4} = (4^4)^{3/4} = 4^3 = 64$

8. Evaluate the number $\sqrt[4]{5}\sqrt[4]{125}$.
(a) $\sqrt[4]{5}$ (b) $\sqrt{5}$ (c) 5 (d) 25 (e) 125
Answer: medium (c)
$\sqrt[4]{5}\cdot\sqrt[4]{125}=\sqrt[4]{5\cdot 125}=\sqrt[4]{625}=\sqrt[4]{5^4}=5$

9. Write the given expression as a power of x: $x^2\sqrt{x}$
(a) $x^{1/2}$ (b) $x^{3/4}$ (c) $x^{5/4}$ (d) $x^{7/4}$ (e) $x^{5/2}$
Answer: medium (e)
$x^2\sqrt{x}=x^2\cdot x^{1/2}=x^{5/2}$

10. Write the given expression as a power of x: $((x^m)^3)^n$
(a) x^{3mn} (b) $x^{mn/3}$ (c) x^{m+n+3} (d) $x^{m/(3n)}$ (e) $x^{3(m+n)}$
Answer: easy (a)
$((x^m)^3)^n=x^{3mn}$

11. Write the given expression as a power of x: $\dfrac{x^p x^q}{x^r}$
(a) x^{pqr} (b) $x^{r/(pq)}$ (c) $x^{pq/r}$ (d) $x^{p/(qr)}$ (e) x^{p+q-r}
Answer: medium (e)
$\dfrac{x^p x^q}{x^r}=x^{p+q-r}$

12. Write the given expression as a power of x: $x^a(x^b)^2(x^c)^3$
(a) x^{6abc} (b) $x^{a+b^2+c^3}$ (c) $x^{a+2b+3c}$ (d) $x^{ab^2c^3}$ (e) $(x^{abc})^6$
Answer: easy (c)
$x^a(x^b)^2(x^c)^3=x^{a+2b+3c}$

13. Simplify the expression $(a^3)^{-2}(a^2b)^3(b^3)^5$.
(a) a^3b^{12} (b) a^6b^{12} (c) a^6b^{15} (d) a^9b^{18} (e) b^{18}
Answer: medium (e)
$(a^3)^{-2}(a^2b)^3(b^3)^5=a^{-6}\cdot a^6b^3\cdot b^{15}=b^{18}$

14. Simplify the expression $\left(\dfrac{r^3s^{5/3}}{r^{2/3}s^2}\right)^6$.
(a) r^2s^{-3} (b) r^6s^{-1} (c) $r^{14}s^{-2}$ (d) r^8s^2 (e) $r^{16}s^{-8}$
Answer: medium (c)
$\left(\dfrac{r^3s^{5/3}}{r^{2/3}s^2}\right)^6=\left(r^{9/3-2/3}\cdot s^{5/3-6/3}\right)^6=r^{14}s^{-2}$

15. Simplify the expression $\sqrt{x^2y^8}$.
(a) xy^2 (b) $|x|y^2$ (c) xy^4 (d) $|x|y^4$ (e) x^2y^4
Answer: hard (d)
$\sqrt{x^2y^8}=\sqrt{(xy^4)^2}=|xy^4|=|x|y^4$

16. Simplify the expression $\dfrac{\sqrt{x}+2}{\sqrt{x}-2}$.
(a) $\dfrac{x+2}{x-4}$ (b) $\dfrac{\sqrt{x+2}}{x-2}$ (c) $\dfrac{(\sqrt{x}+2)^2}{x-4}$ (d) $\dfrac{x^2+2x+4}{x-4}$ (e) $\dfrac{\sqrt{x}-2}{x-2}$
Answer: medium (c)
$\dfrac{\sqrt{x}+2}{\sqrt{x}-2}=\dfrac{(\sqrt{x}+2)(\sqrt{x}+2)}{(\sqrt{x}-2)(\sqrt{x}+2)}=\dfrac{(\sqrt{x}+2)^2}{x-4}$

17. Simplify the expression $\left(\dfrac{ab^{-2}c^3}{3a^{-3}b^4}\right)^{-2}$

(a) $\dfrac{3a^8}{b^{12}c^6}$ (b) $\dfrac{3b^8}{a^{12}c^4}$ (c) $\dfrac{9c^6}{a^{12}b^8}$ (d) $\dfrac{9b^{12}}{a^8c^6}$ (e) $\dfrac{9a^3}{b^4c^8}$

Answer: medium (d)

$\left(\dfrac{ab^{-2}c^3}{3a^{-3}b^4}\right)^{-2} = \left(3^{-1}\, a^{1+3}b^{-2-4}c^3\right)^{-2} = \dfrac{9b^{12}}{a^8c^6}$

18. Write the number 3.01×10^{-9} in ordinary decimal notation.

(a) 0.00000301 (b) 0.000000301 (c) 0.0000000301 (d) 0.00000000301 (e) 0.000000000301

Answer: easy (d)

$3.01 \times 10^{-9} = 0.00000000301$

19. If your heart beats 60 times a minute and you live to be 95 years old, estimate the number of times that your heart beats during your lifetime. State your answer in scientific notation.

(a) 2.61032×10^8 (b) 2.61032×10^9 (c) 2.99592×10^8 (d) 2.99592×10^9 (e) 6.02×10^{23}

Answer: medium (d)

The number of times that the heart beats $= (60)(95)(365)(24)(60) = 2.99592 \times 10^9$

20. Factor $x^2 - 7x + 10$.

(a) $(x-1)(x-3)$ (b) $(x-3)(x-4)$ (c) $(x-3)(x-5)$ (d) $(x-2)(x-4)$ (e) $(x-2)(x-5)$

Answer: easy (e)

$x^2 - 7x + 10 = (x-2)(x-5)$

21. Factor the expression $6x^2 + 8x - 8$.

(a) $2(2x-3)(x+2)$ (b) $2(3x-4)(x+1)$ (c) $2(x+2)(3x-2)$
(d) $2(3x+1)(x-4)$ (e) $2(3x+2)(x+2)$

Answer: medium (c)

$6x^2 + 8x - 8 = 2(3x^2 + 4x - 4) = 2(x+2)(3x-2)$

22. Factor $x^4 - 8x^2 + 16$.

(a) $(x-2)^2(x+2)^2$ (b) $(x-4)^4$ (c) $(x-2)^4(x+2)^4$
(d) $(x-2)^4$ (e) $(x+2)^4$

Answer: medium (a)

$x^4 - 8x^2 + 16 = (x^2-4)^2 = (x-2)^2(x+2)^2$

23. Factor the expression $3y^6 - 3y^2$.

(a) $3y^2(y^2-1)(y+1)^2$ (b) $y^2(3y^2+1)(2y+1)(y-1)$ (c) $y^2(3y^2+1)(y^2-1)$
(d) $3y^2(y^2+1)(y+1)(y-1)$ (e) $3y^2(y+1)^2(y-1)^2$

Answer: medium (d)

$3y^6 - 3y^2 = 3y^2(y^4-1) = 3y^2(y^2-1)(y^2+1) = 3y^2(y^2+1)(y+1)(y-1)$

24. Factor the expression $y^3 - y^2 - 4y + 4$.

(a) $(y-2)(y+2)(y-1)$ (b) $(y-2)^2(y-1)$ (c) $(y+2)^2(y-1)$
(d) $(y-2)(y-1)^2$ (e) $(y-2)(y+1)(y-1)$

Answer: medium (a)

$y^3 - y^2 - 4y + 4 = y^2(y-1) - 4(y-1) = (y^2-4)(y-1) = (y-2)(y+2)(y-1)$

25. Factor the expression $a^5b + a^2b^4$.

(a) $ab(a^2+b^2)(a^2-ab+b^2)$ (b) $ab^2(a+b)(a^2+ab+b^2)$ (c) $a^2b(a+b)(a^2+b^2)$
(d) $a^2b(a+b)(a^2-ab+b^2)$ (e) $ab(a+b^2)(a^2-ab+b^2)$

Answer: medium (d)

$a^5b + a^2b^4 = a^2b(a^3+b^3) = a^2b(a+b)(a^2-ab+b^2)$

26. Factor $x^6 + 8y^3$

(a) $(2x + y^2)(x^4 - x^2y + y^2)$ (b) $(x^2 + 2y)(x^4 - 2x^2y + 4y^2)$ (c) $(x^2 + y)(x^4 - xy^2 + y^2)$

(d) $(2x + y)(x^5 - 2x^2y + 8y^2)$ (e) $(x^2 + 2y)(x^2 - 2x)(y^2 + 4xy)$

Answer: medium (b)

$x^6 + 8y^3 = (x^2)^3 + (2y)^3 = (x^2 + 2y)(x^4 - 2x^2y + 4y^2)$

27. Factor $3x^3 - x^2 + 9x - 3$

Answer: medium

$3x^3 - x^2 + 9x - 3 = x^2(3x - 1) + 3(3x - 1) = (3x - 1)(x^2 + 3)$

28. Perform the indicated operation on $(3y - 5)(3y + 5)$.

Answer: easy

$(3y - 5)(3y + 5) = 9y^2 - 25$

29. Perform the indicated operation on $(2 + x)(1 - x) - (4 - x)(4 + x)$.

Answer: medium

$(2 + x)(1 - x) - (4 - x)(4 + x) = 2 - 2x + x - x^2 - (16 - x^2) = -x - 14$

30. Perform the indicated operation on $(3x + 2)^3$.

Answer: medium

$(3x + 2)^3 = (3x)^3 + 3 \cdot 2(3x)^2 + 3 \cdot 2^2(3x) + 8 = 27x^3 + 54x^2 + 36x + 8$

31. Perform the indicated operation on $\dfrac{2x^3 + 3x^2 + x}{x}$.

Answer: easy

$$\frac{2x^3 + 3x^2 + x}{x} = 2x^2 + 3x + 1$$

32. Perform the indicated operation on $\dfrac{t^3 - 8}{t^2 - 4}$.

Answer: medium

$$\frac{t^3 - 8}{t^2 - 4} = \frac{(t - 2)(t^2 + 2t + 4)}{(t - 2)(t + 2)} = \frac{t^2 + 2t + 4}{t + 2}$$

33. Perform the indicated operations. $\dfrac{x^2/(x^3 - 1)}{x^3/(x - 1)}$

Answer: medium

$$\frac{\frac{x^2}{x^3 - 1}}{\frac{x^3}{x - 1}} = \frac{x^2}{(x - 1)(x^2 + x + 1)} \cdot \frac{x - 1}{x^3} = \frac{1}{x(x^2 + x + 1)}$$

34. Perform the indicated operation on $x^2 - \dfrac{1}{x^2 + 1}$.

Answer: medium

$$x^2 - \frac{1}{x^2 + 1} = \frac{x^2(x^2 + 1) - 1}{x^2 + 1} = \frac{x^4 + x^2 - 1}{x^2 + 1}$$

35. Perform the indicated operation on $\dfrac{1}{x} + \dfrac{2}{x - 1} + \dfrac{3}{(x - 1)^2}$.

Answer: medium

$$\frac{1}{x} + \frac{2}{x - 1} + \frac{3}{(x - 1)^2} = \frac{(x - 1)^2 + 2x(x - 1) + 3x}{x(x - 1)^2} = \frac{x^2 - 2x + 1 + 2x^2 - 2x + 3x}{x(x - 1)^2} = \frac{3x^2 - x + 1}{x(x - 1)^2}$$

36. Perform the indicated operation on $\dfrac{3}{x+1}+\dfrac{2}{x^2-1}-\dfrac{1}{x^2-x-2}$.

Answer: medium

$$\frac{3}{x+1}+\frac{2}{x^2-1}-\frac{1}{x^2-x-2}=\frac{3}{x+1}+\frac{2}{(x-1)(x+1)}-\frac{1}{(x-2)(x+1)}=\frac{3(x-2)(x-1)+2(x-2)-(x-1)}{(x-2)(x-1)(x+1)}$$
$$=\frac{3x^2-9x+6+2x-4-x+1}{(x-2)(x-1)(x+1)}=\frac{3x^2-8x+3}{(x-2)(x-1)(x+1)}$$

37. Perform the indicated operation on $\dfrac{\dfrac{1}{x+1}-\dfrac{1}{x}}{\dfrac{1}{x+1}+\dfrac{1}{x}}$.

Answer: medium

$$\frac{\dfrac{1}{x+1}-\dfrac{1}{x}}{\dfrac{1}{x+1}+\dfrac{1}{x}}=\frac{x-(x+1)}{x(x+1)}\div\frac{x+(x+1)}{x(x+1)}=\frac{-1}{2x+1}$$

38. Rationalize the numerator of $\dfrac{\sqrt{2(x+h)}-\sqrt{2x}}{h}$.

Answer: medium

$$\frac{\sqrt{2(x+h)}-\sqrt{2x}}{h}=\frac{\sqrt{2(x+h)}-\sqrt{2x}}{h}\cdot\frac{\sqrt{2(x+h)}+\sqrt{2x}}{\sqrt{2(x+h)}+\sqrt{2x}}=\frac{2(x+h)-2x}{h(\sqrt{2x+h}+\sqrt{2x})}=\frac{2}{\sqrt{2x+h}+\sqrt{2x}}$$

39. State whether or not the equation $\dfrac{1-a}{1+\sqrt{a}}=1-\sqrt{a}$ is true for all values of the variable. (Disregard values that make denominators 0.)

Answer: hard

$$\text{LHS}=\frac{1-a}{1+\sqrt{a}}=\frac{(1+\sqrt{a})(1-\sqrt{a})}{1+\sqrt{a}}=1-\sqrt{a}=\text{RHS, so the equation is true.}$$

40. State whether or not the equation $\sqrt{a+b}=\sqrt{a}+\sqrt{b}$ is true for all values of the variable.

Answer: medium

Try $a=b=1$: LHS $=\sqrt{a+b}=\sqrt{2}\neq\sqrt{1}+\sqrt{1}=$ RHS and so the equation is false.

41. State whether or not the equation $\dfrac{1}{x+2}=\dfrac{1}{x}+\dfrac{1}{2}$ is true for all values of the variable. (Disregard values that make denominators 0.)

Answer: medium

Try $x=1$: LHS $=\dfrac{1}{x+2}=\dfrac{1}{3}\neq\dfrac{1}{1}+\dfrac{1}{2}=$ RHS, so the equation is false.

42. What is wrong with the following argument?

$$\begin{aligned}a&=b\\a^2&=ab\\a^2-b^2&=ab-b^2\\(a+b)(a-b)&=b(a-b)\\a+b&=b\\\text{Now put } a=b=1:\quad 2&=1\end{aligned}$$

Answer: hard

In going from the fourth line to the fifth line of the argument, the implicit assumption is that $a-b\neq 0$. Otherwise we would be dividing both sides of the equation by zero. However, in the sixth line, we set $a=b=1$, and hence $a-b=0$, which violates our earlier assumption.

CHAPTER TWO

SECTION 2.1

1. Plot the following points in a coordinate plane: $(3,2)$, $(-3,2)$, $(1,5)$, $(1,-5)$, $(-1,5)$, $(-1,-5)$

Answer: easy

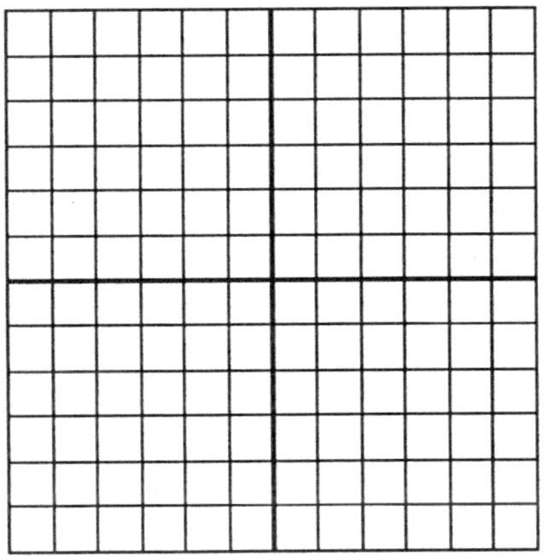

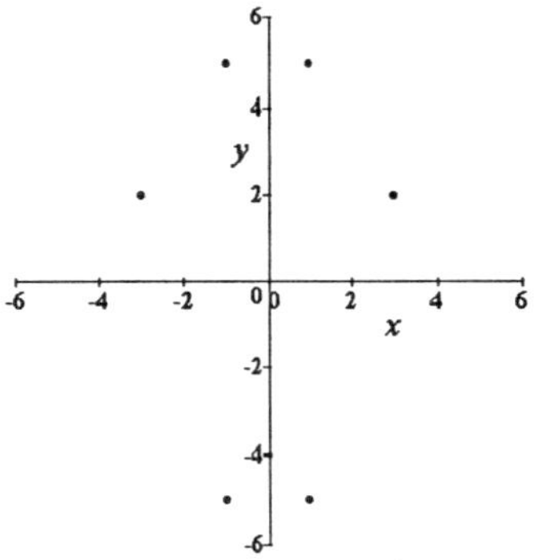

2. Find the coordinates of the points shown in the figure.

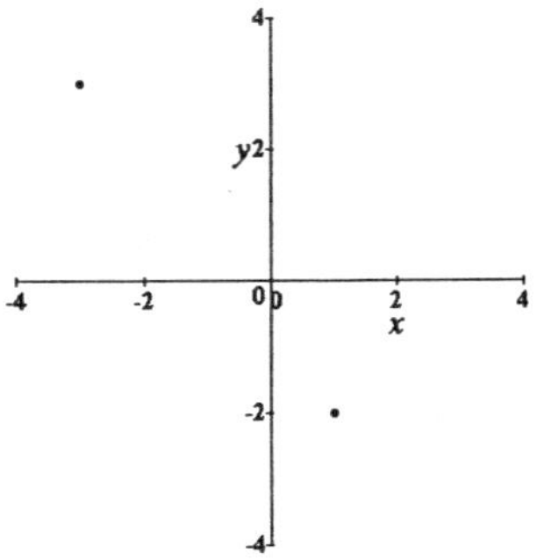

(a) $(1,2)$, $(3,3)$ (b) $(1,-2)$, $(3,3)$ (c) $(-1,2)$, $(3,-3)$ (d) $(1,-2)$, $(3,-3)$ (e) $(1,-2)$, $(-3,3)$

Answer: easy (e)

3. Draw the square with vertices $A(2,4)$, $B(2,8)$, $C(6,8)$, and $D(6,4)$ on a coordinate plane. Find the area of the square.

Answer: medium

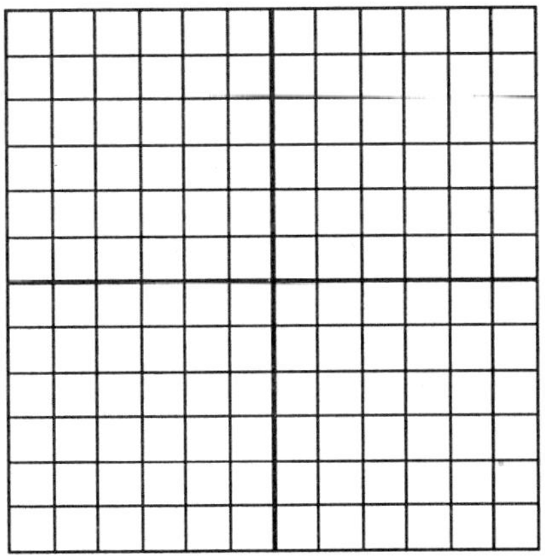

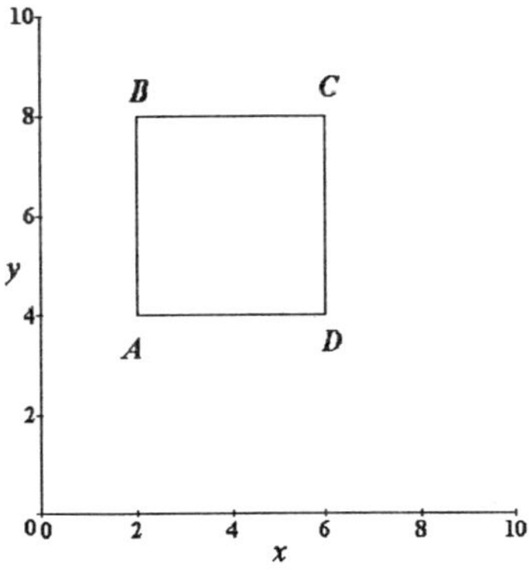

The area of the square is $4^2 = 16$.

4. For the points $(1, 4)$ and $(3, 8)$:
   (a) Plot the points on a coordinate plane.
   (b) Find the distance between them.
   (b) Find the midpoint of the line segment that joins them.

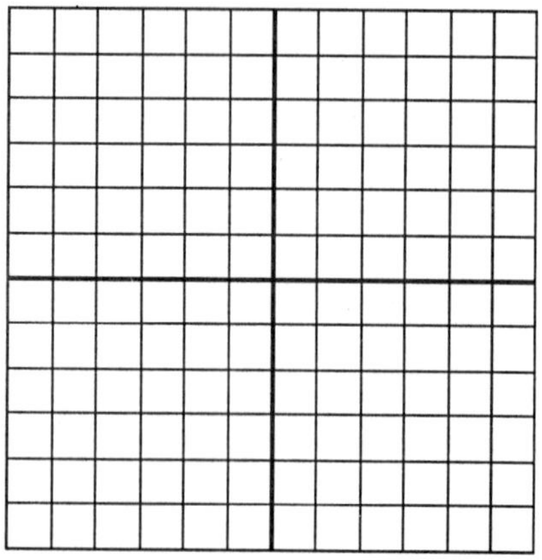

Answer: medium

(a)

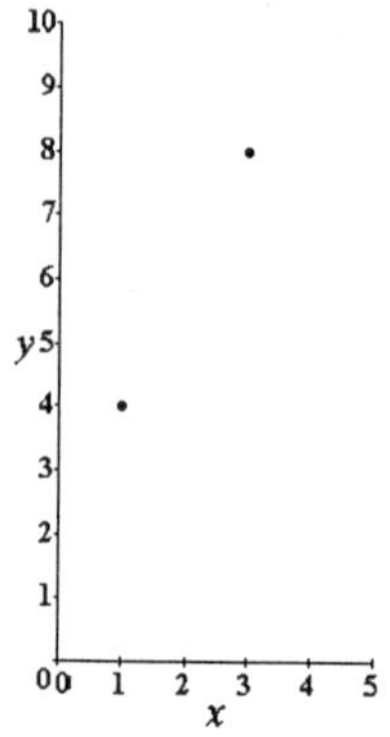

(b) $P_1 = (1, 4)$ and $P_2 = (3, 8)$ so

$$|P_1P_2| = \sqrt{(1-3)^2 + (4-8)^2} = \sqrt{20} = 2\sqrt{5}$$

(c) The midpoint is $\left(\frac{1+3}{2}, \frac{4+8}{2}\right) = (2, 6)$.

5. For the points $(2, -3)$ and $(5, 6)$:
   (a) Plot the points on a coordinate plane.
   (b) Find the distance between them.
   (b) Find the midpoint of the line segment that joins them.

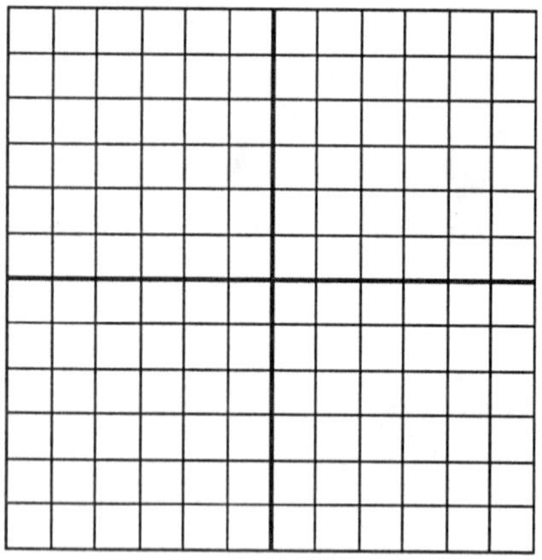

Answer: medium

(a)

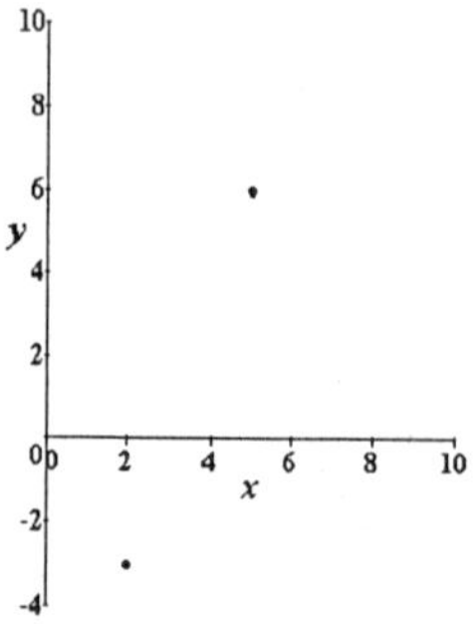

(b) $P_1 = (2, -3)$ and $P_2 = (5, 6)$ so

$$|P_1P_2| = \sqrt{(2-5)^2 + (-3-6)^2} = \sqrt{90} = 3\sqrt{10}$$

(c) The midpoint is $\left(\frac{2+5}{2}, \frac{-3+6}{2}\right) = \left(\frac{7}{2}, \frac{3}{2}\right)$

6. For the points $\left(-2,-\frac{3}{2}\right)$ and $\left(\frac{7}{2},4\right)$:
   (a) Plot the points on a coordinate plane.
   (b) Find the distance between them.
   (b) Find the midpoint of the line segment that joins them.

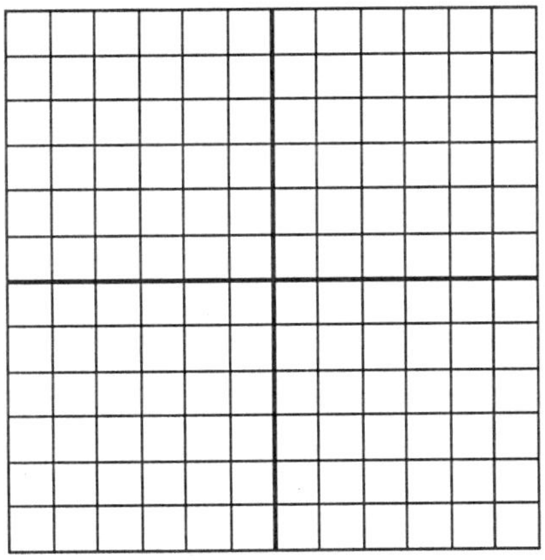

Answer: medium

(a)

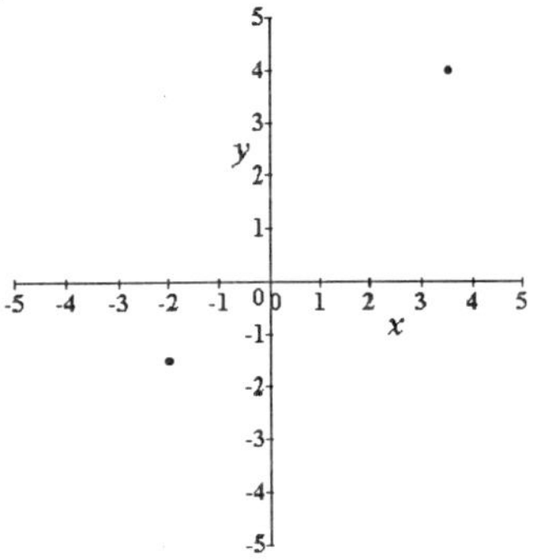

(b) $P_1=\left(-2,-\frac{3}{2}\right)$ and $P_2=\left(\frac{7}{2},4\right)$ so

$$|P_1P_2|=\sqrt{\left(-2-\frac{7}{2}\right)^2+\left(-\frac{3}{2}-4\right)^2}$$
$$=\sqrt{\frac{121}{2}}=\frac{11}{2}\sqrt{2}$$

(c) The midpoint is $\left(\dfrac{-2+\frac{7}{2}}{2},\dfrac{-\frac{3}{2}+4}{2}\right)=\left(\dfrac{3}{4},\dfrac{5}{4}\right)$

7. Plot the points $P(6,2)$, $Q(1,7)$, and $R(-4,2)$ on a coordinate plane. Where must the point S be located so that the quadrilateral $PQRS$ is a square? Find the area of this square.

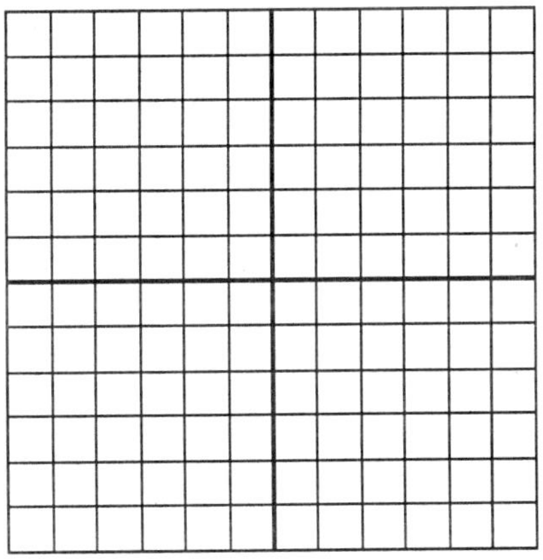

Answer: hard

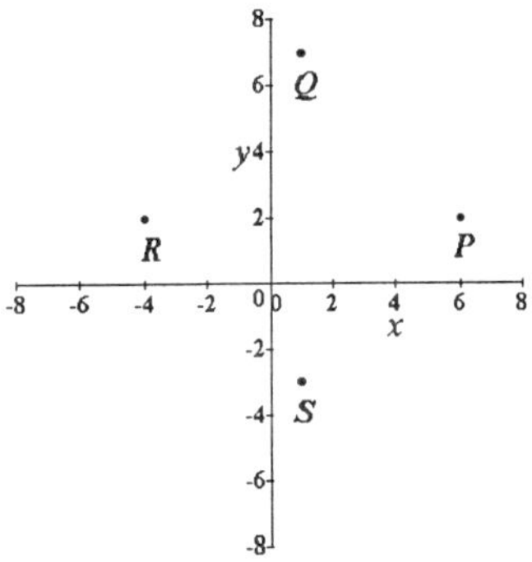

The point S must be located at $(1,-3)$.

Since $|PQ|=\sqrt{(6-1)^2+(2-7)^2}=5\sqrt{2}$,

the area of $PQRS=\left(5\sqrt{2}\right)^2=50$ square units.

8. Sketch the region given by the set $\{(x, y) \mid y > 2\}$.

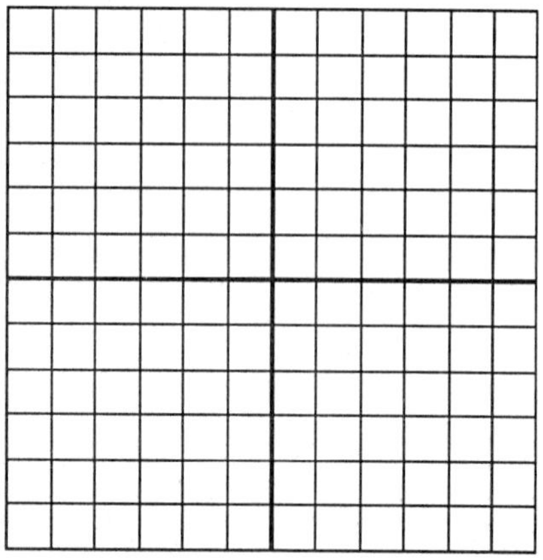

Answer: easy

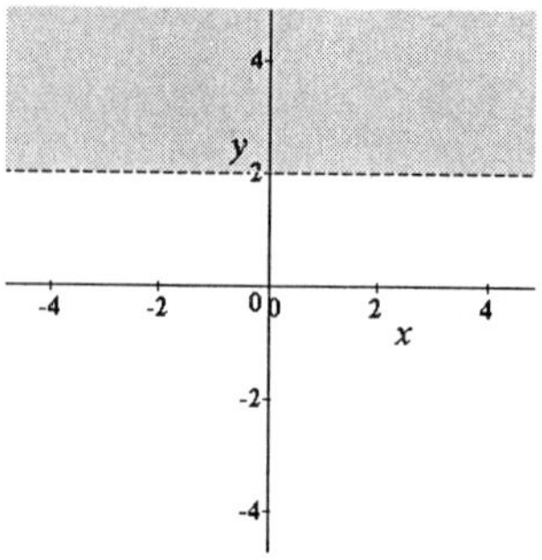

$\{(x, y) \mid y > 2\}$

9. Sketch the region given by the set $\{(x, y) \mid xy < 0\}$.

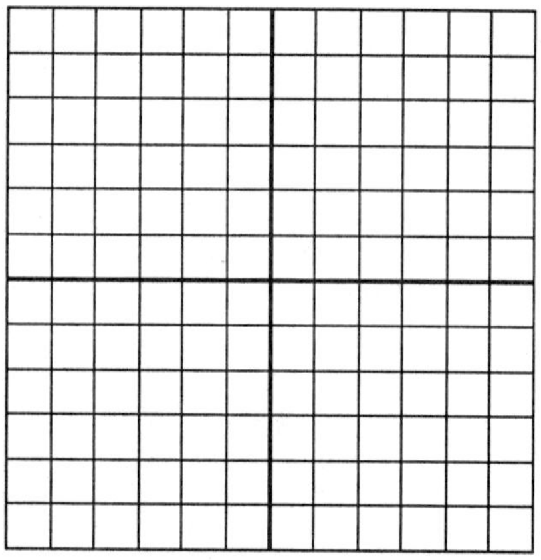

Answer: medium

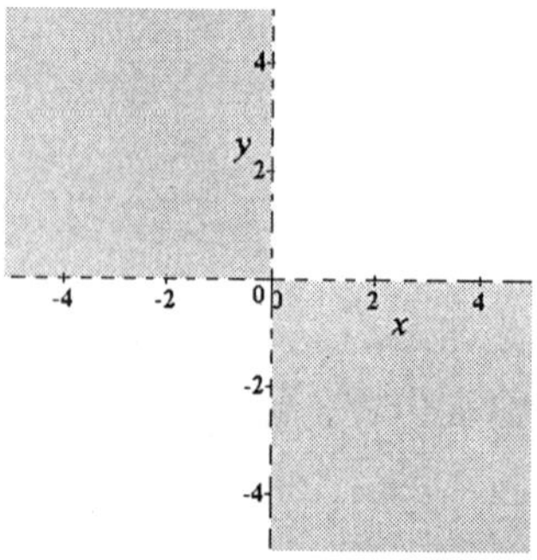

$\{(x, y) \mid xy < 0\}$

10. Sketch the region given by the set $\{(x, y) \mid y = 1\}$.

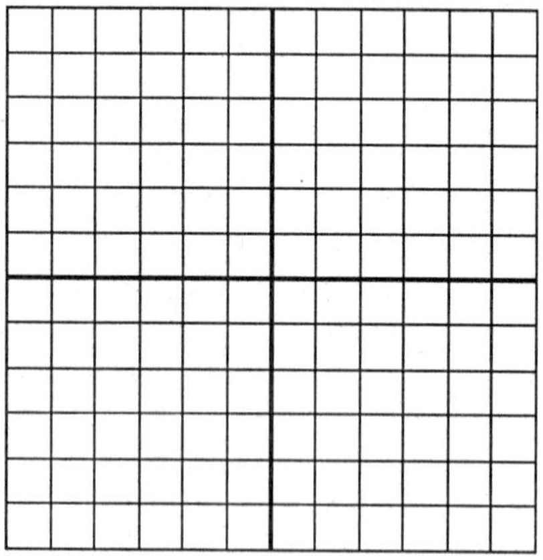

Answer: easy

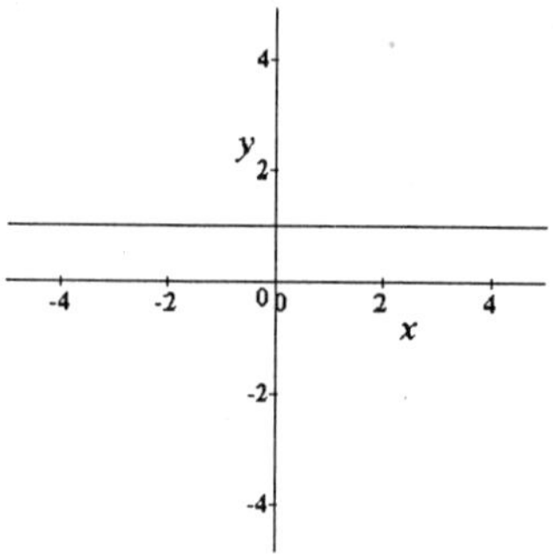

$\{(x, y) \mid y = 1\}$

11. Sketch the region given by the set $\{(x, y) \mid 2 < y \leq 4\}$.

Answer: medium

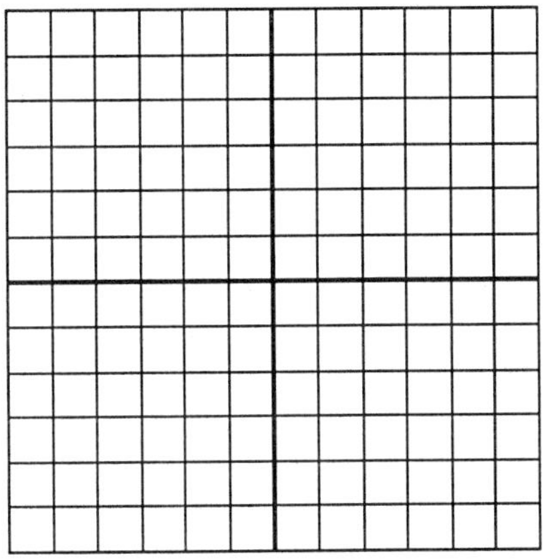

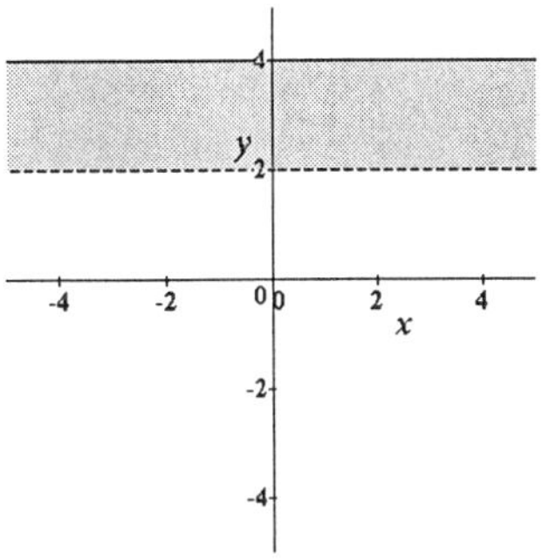

$\{(x, y) \mid 2 < y \leq 4\}$

12. Sketch the region given by the set $\{(x, y) \mid |y| < 2\}$.

Answer: medium

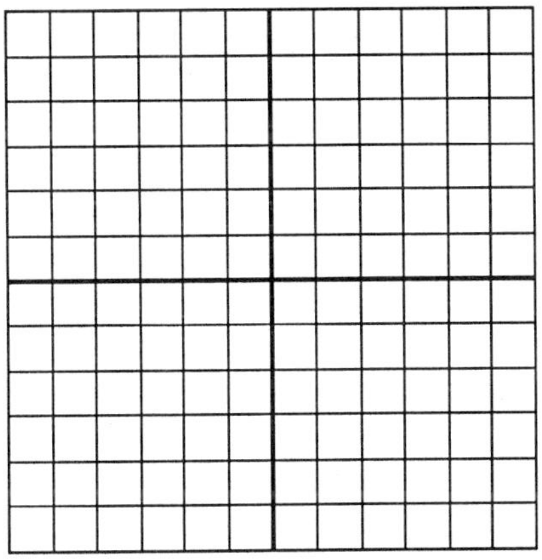

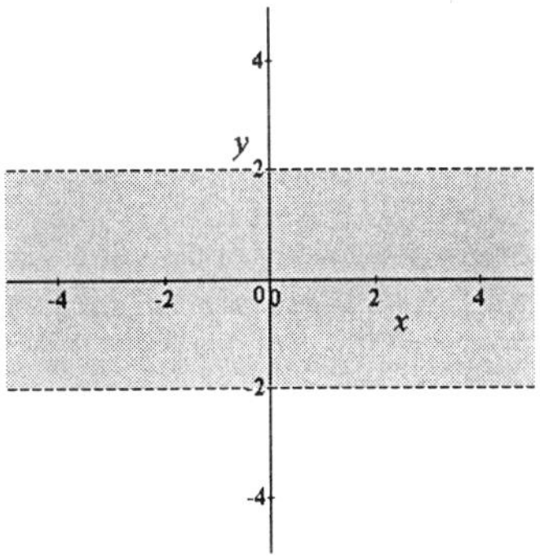

$\{(x, y) \mid |y| < 2\}$

13. Sketch the region given by the set $\{(x, y) \mid x < 4 \text{ and } y > 1\}$.

Answer: medium

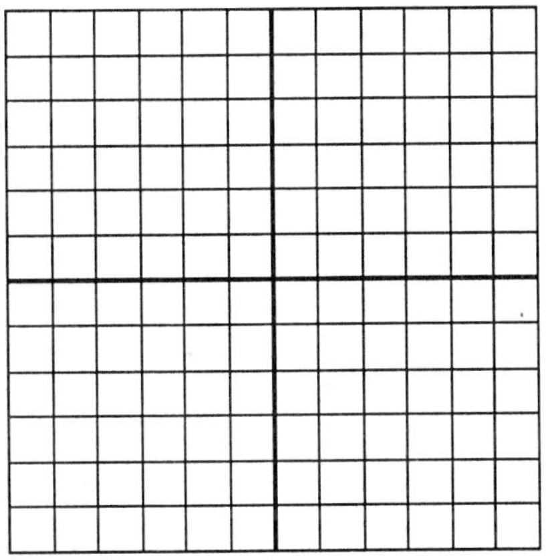

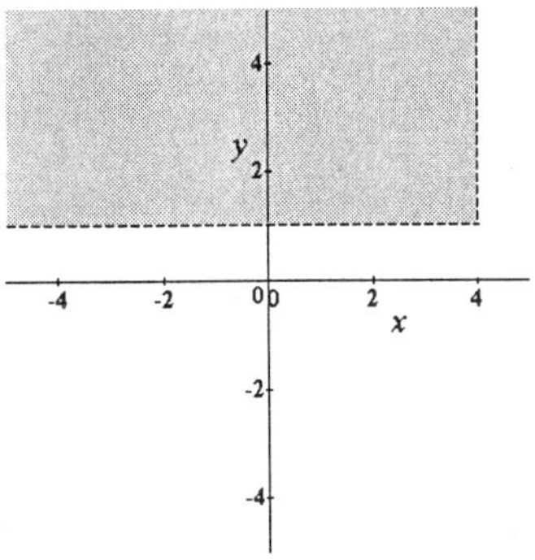

$\{(x, y) \mid x < 4 \text{ and } y > 1\}$

14. Sketch the region given by the set $\{(x, y) \mid |x - 3| < 1 \text{ and } |y + 1| \le 3\}$.

Answer: medium

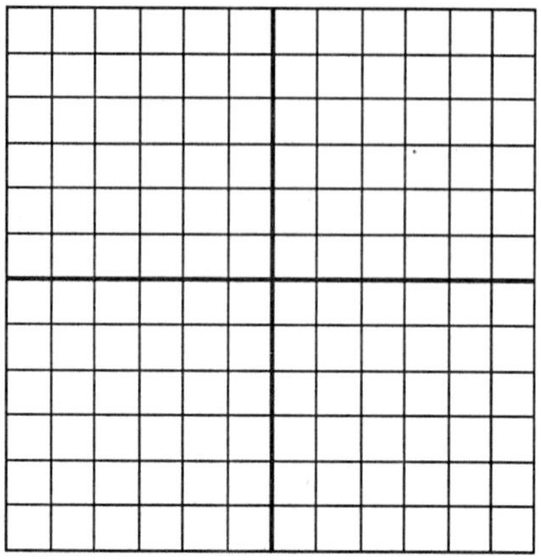

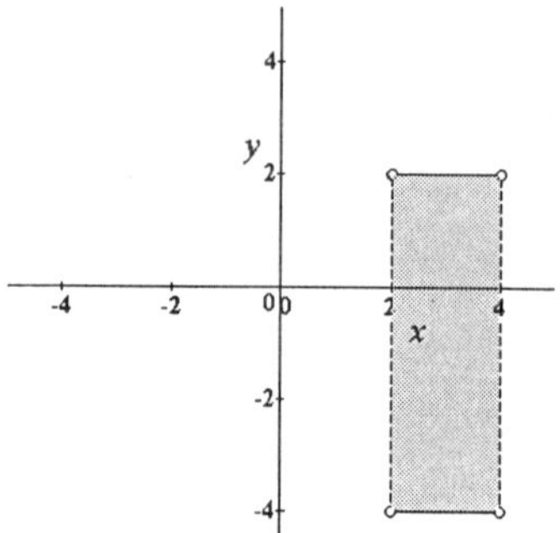

$\{(x, y) \mid |x - 3| < 1 \text{ and } |y + 1| \le 3\}$
$= \{(x, y) \mid 2 < x < 4 \text{ and } -4 \le y \le 2\}$

15. Which of the points $C(-6, 4)$ or $D(3, 0)$ is closer to the point $E(-1, 2)$?

Answer: medium

$C = (-6, 4)$, $D = (3, 0)$, and $E = (-1, 2)$

$|CE| = \sqrt{(-6+1)^2 + (4-2)^2} = \sqrt{5^2 + 2^2} = \sqrt{29}$

$|DE| = \sqrt{(3+1)^2 + (0-2)^2} = \sqrt{4^2 + 2^2} = \sqrt{20} = 2\sqrt{5}$

Because $|CE| > |DE|$, D is closer to E.

16. Show that the triangle with vertices $A(5, -8)$, $B(10, -4)$, and $C(1, -3)$ is a right triangle using the converse of the Pythagorean Theorem. Find the area of the triangle.

Answer: hard

$A = (5, -8)$, $B = (10, -4)$, and $C = (1, -3)$

$|AB| = \sqrt{(5-10)^2 + (-8+4)^2} = \sqrt{41}$, $|AC| = \sqrt{(5-1)^2 + (-8+3)^2} = \sqrt{41}$

$|BC| = \sqrt{(10-1)^2 + (-4+3)^2} = \sqrt{82}$. Because $|AB|^2 + |AC|^2 = 41 + 41 = 82 = |BC|^2$, by the converse of the Pythagorean Theorem, triangle ABC is a right triangle. The area of the triangle is $A = \frac{1}{2} \cdot b \cdot h = \frac{1}{2}\sqrt{41}\sqrt{41} = \frac{41}{2} = 20.5$.

17. Show that the points $A(-3, 4)$, $B(1, 12)$, and $C(3, 16)$ are collinear by showing that $|AB| + |BC| = |AC|$.

Answer: medium

$A = (-3, 4)$, $B = (1, 12)$, and $C = (3, 16)$

$|AB| = \sqrt{(-3-1)^2 + (4-12)^2} = \sqrt{80} = 4\sqrt{5}$

$|BC| = \sqrt{(1-3)^2 + (12-16)^2} = \sqrt{20} = 2\sqrt{5}$

$|AC| = \sqrt{(-3-3)^2 + (4-16)^2} = \sqrt{180} = 6\sqrt{5}$

Because $|AB| + |BC| = 4\sqrt{5} + 2\sqrt{5} = 6\sqrt{5} = |AC|$, A, B, and C are collinear.

18. Find the lengths of the medians of the triangle with vertices $A(3, 1)$, $B(5, 7)$, and $C(1, 5)$. (A median is a line segment from a vertex to the midpoint of the opposite side.)

Answer: medium

$A = (3, 1)$, $B = (5, 7)$, and $C = (1, 5)$

The midpoint of AB is $P_1 = \left(\frac{3+5}{2}, \frac{1+7}{2}\right) = (4, 4)$; the midpoint of AC is $P_2 = \left(\frac{3+1}{2}, \frac{1+5}{2}\right) = (2, 3)$; and the midpoint of BC is $P_3 = \left(\frac{5+1}{2}, \frac{7+5}{2}\right) = (3, 6)$.

So median P_1C has length $\sqrt{(4-1)^2 + (4-5)^2} = \sqrt{10}$; median P_2B has length $\sqrt{(2-5)^2 + (3-7)^2} = 5$, and median P_3A has length $\sqrt{(3-3)^2 + (6-1)^2} = 5$.

19. Plot the points $P(-4,6)$ and $Q(10,4)$. Which (if either) of the points $A(4,12)$ or $B(3,-2)$ lies on the perpendicular bisector of the segment PQ?
Answer: medium
$|PA| = \sqrt{(-4-4)^2 + (6-12)^2} = 10$ and $|QA| = \sqrt{(10-4)^2 + (4-12)^2} = 10$. Since $|PA| = |QA|$, point A lies on the perpendicular bisector of the segment PQ.
$|PB| = \sqrt{(-4-3)^2 + (6+2)^2} = \sqrt{113}$ and $|QB| = \sqrt{(10-3)^2 + (4+2)^2} = \sqrt{85}$. Since $|PB| \neq |QB|$, point B does not lie on the perpendicular bisector of the segment PQ.

20. Suppose that each point in the coordinate plane is shifted 1 unit to the right and 4 units downward.

(a) The point $(1,2)$ is shifted to what new point?

(b) The point (a,b) is shifted to what new point?

(c) Triangle ABC in the figure has been shifted to triangle $A'B'C'$. Find the coordinates of the points A', B', and C'.

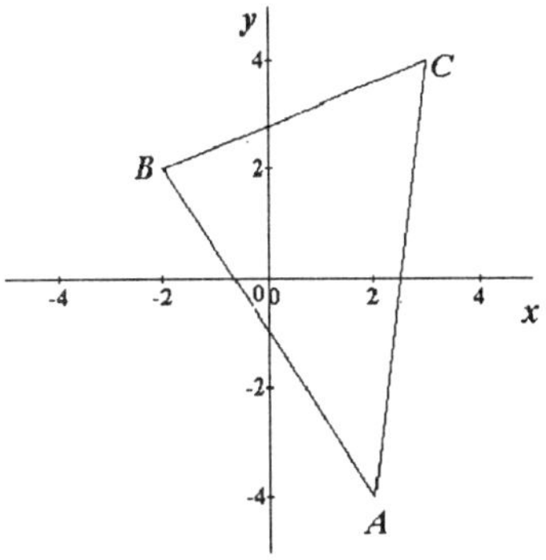

Answer: medium
(a) The point $(1,2)$ is shifted to $(1+1, 2-4) = (2,-2)$.
(b) The point (a,b) is shifted to $(a+1, b-4)$.
(c) $A(2,-4)$ is shifted to $A'(3,-8)$; $B(-2,2)$ is shifted to $B'(-1,-2)$; $C(3,4)$ is shifted to $C'(4,0)$.

21. Triangle A has vertices $P(1,1)$, $Q(5,1)$ and $R(5,5)$ and Triangle B has vertices $S(1,1)$, $T(1,5)$, and $U(5,5)$. Find the sum of the areas of the two triangles.
Answer: easy

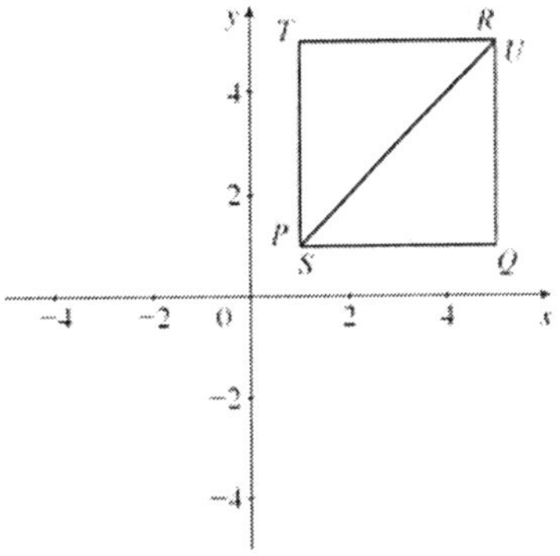

$A_A = \frac{1}{2}(4)(4) = 8$ and $A_B = \frac{1}{2}(4)(4) = 8$, so $A_A + A_B = 16$.

22. For the points $(-2, 5)$ and $(1, -3)$:
(a) Plot the points on a coordinate plane.
(b) Find the distance between them.
(c) Find the distance between the midpoint and the origin.
Answer: easy
(a)

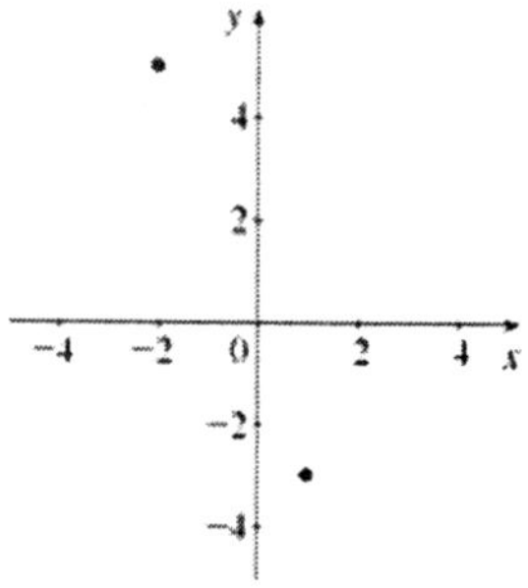

(b) $P_1 = (-2, 5)$ and $P_2 = (1, -3) \Rightarrow |P_1P_2| = \sqrt{(1-(-2))^2 + (-3-5)^2} = \sqrt{73}$.

(c) The midpoint is $\left(\frac{-2+1}{2}, \frac{5+(-3)}{2}\right) = \left(-\frac{1}{2}, 1\right)$, and the distance from $\left(-\frac{1}{2}, 1\right)$ to the origin $(0, 0)$ is $\sqrt{\left[0-\left(-\frac{1}{2}\right)\right]^2 + (0-1)^2} = \frac{\sqrt{5}}{2}$.

23. Find y such that the distance between the points $P(-2, 5)$ and $Q(-3, y)$ is 4 units.
Answer: medium
$D = \sqrt{(x_2 - x_1)^2 + (y_2 - y_1)^2} \Rightarrow 4 = \sqrt{[-3-(-2)]^2 + (y-5)^2} \Rightarrow 16 = 1 + (y-5)^2 \Rightarrow y - 5 = \pm\sqrt{15}$
$\Rightarrow y = 5 \pm \sqrt{15}$

24. If $M(2, 1)$ is the midpoint of the line segment AB, and if A has coordinates $\left(-\frac{1}{2}, 6\right)$, find the coordinates of B.
Answer: medium
$\dfrac{-\frac{1}{2} + x}{2} = 2 \Rightarrow x = \frac{9}{2}, \dfrac{6+y}{2} = 1 \Rightarrow y = -4$. Thus the coordinates of B are $\left(\frac{9}{2}, -4\right)$.

25. Plot the points $P(3, 1)$, $Q(4, 5)$, and $R(8, 7)$ on a coordinate plane. Where should the point S be located so that $PQRS$ is a parallelogram?
Answer: medium
The midpoint of PR is $\left(\frac{3+8}{2}, \frac{1+7}{2}\right) = \left(\frac{11}{2}, 4\right)$. The midpoint of QS is $\left(\frac{4+x}{2}, \frac{5+y}{2}\right)$. The two diagonals must have the same midpoint, therefore $\left(\frac{4+x}{2}, \frac{5+y}{2}\right) = \left(\frac{11}{2}, 4\right) \Rightarrow \frac{4+x}{2} = \frac{11}{2} \Rightarrow x = 7$, $\frac{5+y}{2} = 4 \Rightarrow y = 3$. Thus the point S is $(7, 3)$.

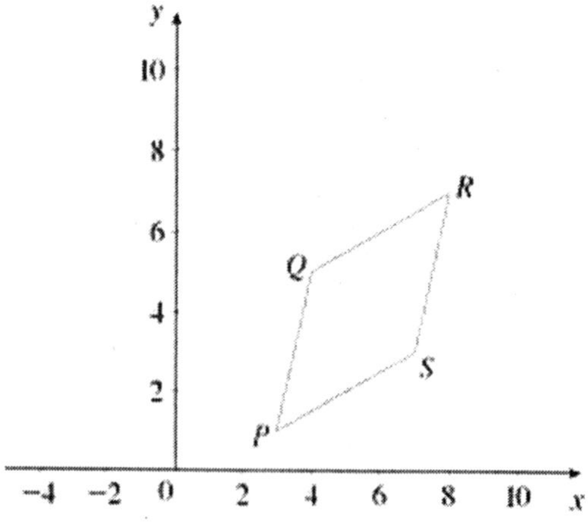

SECTION 2.2

1. Determine which of the points $(1,2)$, $(3,6)$, and $(4,9)$ are on the graph of the equation $y = \frac{1}{2}x^2 + \frac{3}{2}$.
(a) $(1,2)$ (b) $(1,2), (3,6)$ (c) $(3,6)$ (d) $(3,6), (4,9)$ (e) $(1,2), (3,6), (4,9)$
Answer: easy (b)
Substitute to find that $2 = \frac{1}{2}(1^2) + \frac{3}{2}$ and $6 = \frac{1}{2}(3^2) + \frac{3}{2}$, but $9 \neq \frac{1}{2}(4^2) + \frac{3}{2}$.

2. Determine which of the points $(1,-1)$, $(2,-2)$, and $(-3,1)$ are on the graph of the equation $y(1+xy) = 6$.
(a) $(1,-1)$ (b) $(1,-1), (2,-2)$ (c) $(2,-2)$ (d) $(2,-2), (-3,1)$ (e) $(1,-1), (-3,1)$
Answer: easy (c)
Substitute to find that $-2[1+2(-2)] = 6$, but $-1(1+1(-1)) \neq 6$ and $1[1+(-3)1] \neq 6$.

3. Determine which of the points $(1,3)$, $(2,7)$, and $(3,11)$ are on the graph of the equation $y = x^2 + 2x$.
(a) $(1,3)$ (b) $(1,3), (3,11)$ (c) $(1,3), (2,7)$ (d) $(2,7), (3,11)$ (e) $(1,3), (2,7), (3,11)$
Answer: easy (a)
Substitute to find that $3 = 1^2 + 2 \cdot 1$, but $7 \neq 2^2 + 2 \cdot 4$ and $11 \neq 3^2 + 2 \cdot 3$.

4. Find the x- and y-intercepts of the graph of $y = 2x + \frac{3}{2}$.
(a) x-intercept 3, y-intercept $\frac{2}{3}$
(b) x-intercept -3, y-intercept $\frac{3}{4}$
(c) x-intercept $-\frac{3}{4}$, y-intercept $\frac{3}{2}$
(d) x-intercept $-\frac{3}{2}$, y-intercept $\frac{2}{3}$
(e) x-intercept $-\frac{2}{3}$, y-intercept $\frac{3}{2}$
Answer: easy (c)
$y = 0 \quad \Leftrightarrow \quad 2x = -\frac{3}{2} \quad \Leftrightarrow \quad x = -\frac{3}{4}$; $x = 0 \quad \Leftrightarrow \quad y = \frac{3}{2}$

5. Find the x- and y-intercepts of the graph of $y = x^2 - 16$.
(a) x-intercepts ± 3, y-intercept 16
(b) x-intercept 4, y-intercepts ± 16
(c) x-intercept 4, y-intercepts ± 16
(d) x-intercepts ± 4, y-intercept -16
(e) x-intercept -4, y-intercept -16
Answer: medium (d)
$y = 0 \quad \Leftrightarrow \quad x^2 = 16 \quad \Leftrightarrow \quad x = \pm 4$; $x = 0 \quad \Leftrightarrow \quad y = -16$

6. Find the x- and y-intercepts of the graph of $(x+2y)(x-2y) = 1$.
(a) x-intercept 1, y-intercept $1/\sqrt{2}$
(b) x-intercepts ± 1, y-intercepts $\pm 1/\sqrt{2}$
(c) x-intercept -1, no y-intercept
(d) x-intercept 1, no y-intercept
(e) x-intercepts ± 1, no y-intercept
Answer: medium (e)
$y = 0 \quad \Leftrightarrow \quad x^2 = 1 \quad \Leftrightarrow \quad x = \pm 1$; $x = 0 \quad \Leftrightarrow \quad -4y^2 = 1$, which has no real solution

7. Make a table of values and sketch the graph of the equation $y = 3x + 2$. Find x- and y-intercepts and test for symmetry.

Answer: medium

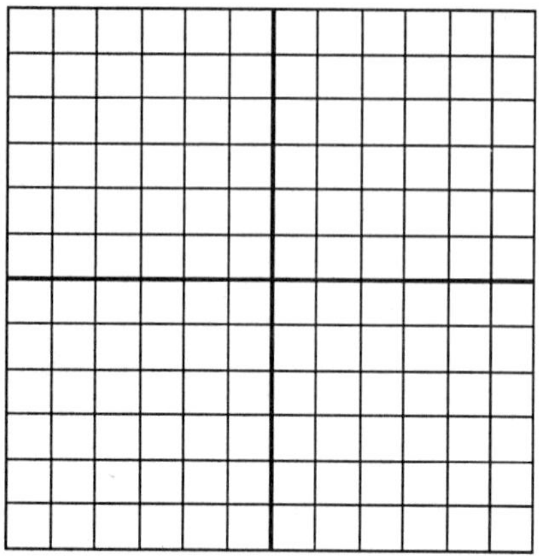

x	$y = 3x + 2$	(x, y)
-3	-7	$(-3, -7)$
-1	-1	$(-1, -1)$
0	2	$(0, 2)$
2	8	$(2, 8)$

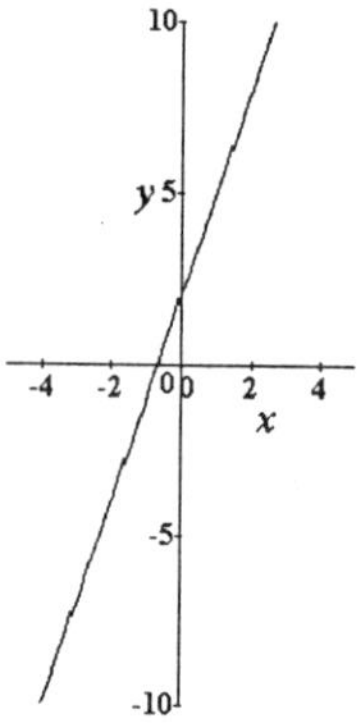

x-intercept: set $y = 0 \Rightarrow x = -\frac{2}{3}$; y-intercept: set $x = 0$ $\Rightarrow y = 2$; symmetry: wrt x-axis $-y = 3x + 2 \Leftrightarrow$ $y = -3x - 2$ which is changed, wrt y-axis $y = -3x + 2$ which is not the original equation, wrt origin $-y = -3x + 2 \Leftrightarrow$ $y = 3x - 2$ which is changed, hence no symmetry.

8. Make a table of values and sketch the graph of the equation $x + y = 2$. Find x- and y-intercepts and test for symmetry.

Answer: medium

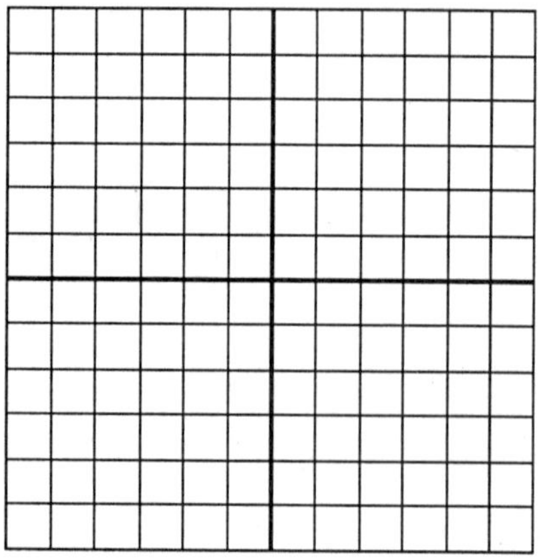

$x + y = 2 \Leftrightarrow y = 2 - x$

x	$y = 2 - x$	(x, y)
-2	4	$(-2, 4)$
0	2	$(0, 2)$
2	0	$(2, 0)$

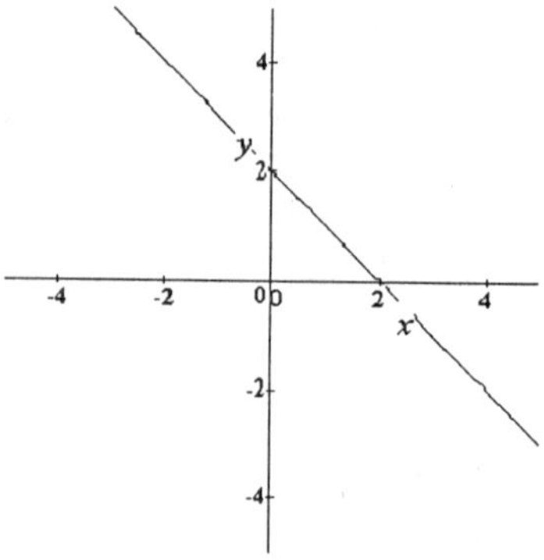

x-intercept: set $y = 0 \Rightarrow x = 2$; y-intercept: set $x = 0 \Rightarrow$ $y = 2$; symmetry: wrt x-axis $-y = 2 - x \Leftrightarrow y = x - 2$ which is changed, wrt y-axis $y = 2 + x$ which is not the original equation, wrt origin $-y = 2 + x \Leftrightarrow$ $y = -2 - x$ which is changed, hence no symmetry.

9. Make a table of values and sketch the graph of the equation $y = x^2 + x$. Find x- and y-intercepts and test for symmetry.

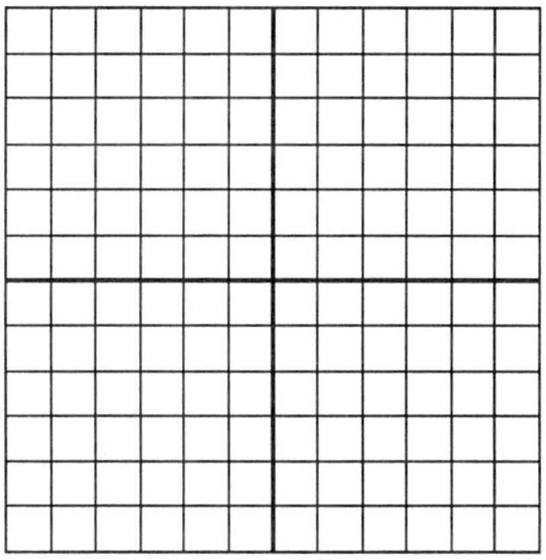

Answer: medium

x	$y = x^2 + x$	(x, y)
-2	2	$(-2, 2)$
-1	0	$(-1, 0)$
0	0	$(0, 0)$
1	2	$(1, 2)$
2	6	$(2, 6)$

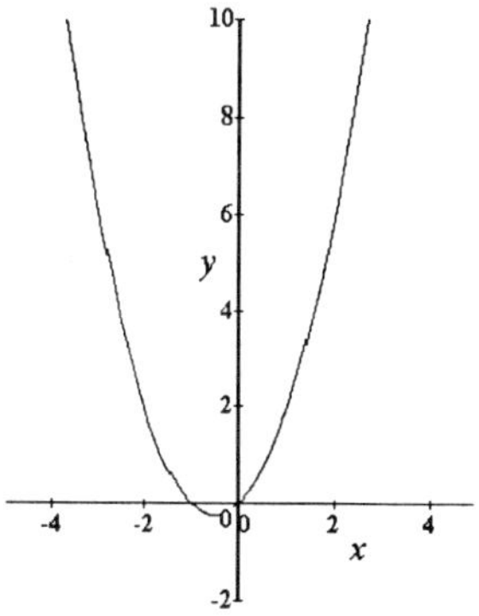

x-intercept: set $y = 0 \Rightarrow x^2 + x = 0 \Leftrightarrow x = 0$ or $x = -1$; y-intercept: set $x = 0 \Rightarrow y = 0$; symmetry: wrt x-axis $-y = x^2 + x \Leftrightarrow y = -x^2 - x$ which is changed; wrt y-axis $y = (-x)^2 - x = x^2 - x$ which is not the original equation, wrt origin $-y = x^2 - x \Leftrightarrow y = -x^2 + x$ which is changed, so no symmetry.

10. Make a table of values and sketch the graph of the equation $4y = x^3$. Find x- and y-intercepts and test for symmetry.

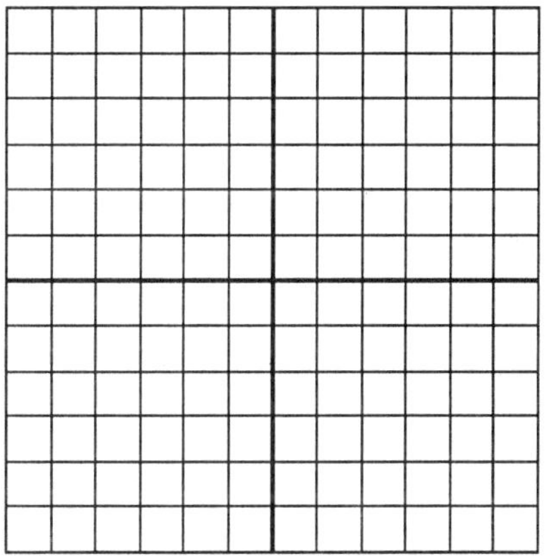

Answer: medium

$4y = x^3 \Leftrightarrow y = x^3/4$

x	$y = x^3/4$	(x, y)
-2	-2	$(-2, -2)$
-1	$-\frac{1}{4}$	$(-1, -\frac{1}{4})$
0	0	$(0, 0)$
2	2	$(2, 2)$
3	$\frac{27}{4}$	$(3, \frac{27}{4})$

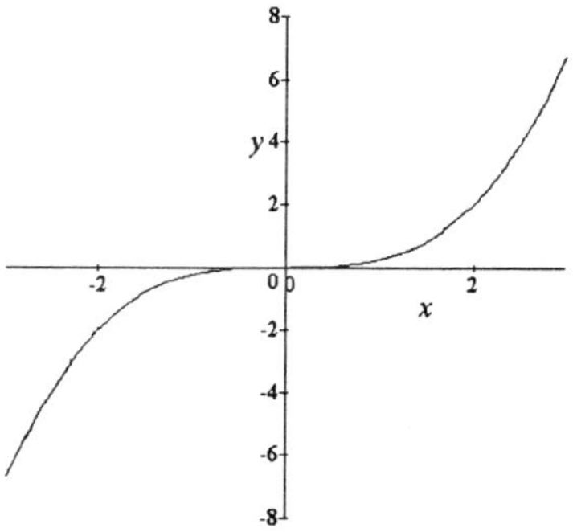

x-intercept: set $y = 0 \Rightarrow x^3/4 = 0 \Leftrightarrow x = 0$; y-intercept: set $x = 0 \Rightarrow y = 0$; symmetry: wrt x-axis $-y = x^3/4 \Leftrightarrow y = -x^3/4$ which is changed, wrt y-axis $y = (-x)^3/4 = -x^3/4$ which is changed, wrt origin $-y = -x^3/4 \Leftrightarrow y = x^3/4$ which is not changed, so symmetric wrt origin.

11. Make a table of values and sketch the graph of the equation $x + y^2 = 9$. Find x- and y-intercepts and test for symmetry.

Answer: medium

$y = \pm\sqrt{9 - x}$

x	$y = \pm\sqrt{9-x}$	(x, y)
-7	± 4	$(-7, \pm 4)$
-3	$\pm 2\sqrt{3}$	$\left(-3, \pm 2\sqrt{3}\right)$
0	± 3	$(0, \pm 3)$
5	± 2	$(5, \pm 2)$
9	0	$(9, 0)$

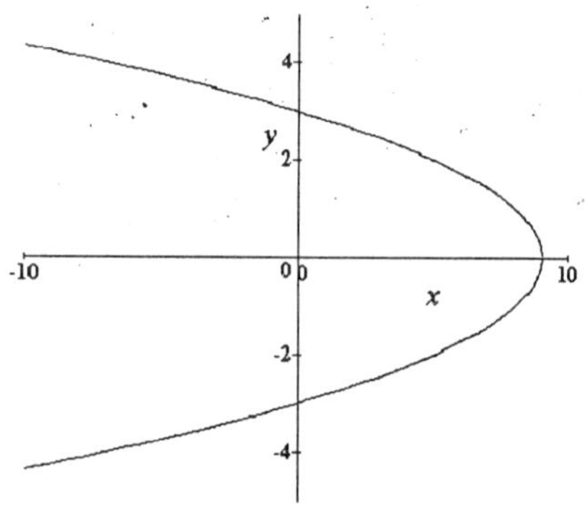

x-intercept: set $y = 0 \Rightarrow \sqrt{9 - x} = 0 \quad \Leftrightarrow \quad x = 9$;
y-intercept: set $x = 0 \Rightarrow y^2 = 9 \quad \Leftrightarrow \quad y = \pm 3$;
symmetry: wrt x-axis $x + (-y)^2 = 9 \quad \Leftrightarrow \quad x + y^2 = 9$ which is not changed, wrt y-axis $-x + y^2 = 9$ which is changed, wrt origin $-x + y^2 = 9$ which is changed, symmetric wrt x-axis.

12. Make a table of values and sketch the graph of the equation $x + \sqrt{y} = 2$. Find x- and y-intercepts and test for symmetry.

Answer: medium

$x + \sqrt{y} = 2 \quad \Leftrightarrow \quad \sqrt{y} = 2 - x \quad \Leftrightarrow \quad y = (2 - x)^2$ for $x \le 2$

x	$y = (2-x)^2$	(x, y)
-1	9	$(-1, 9)$
0	4	$(0, 4)$
1	1	$(1, 1)$
2	0	$(2, 0)$

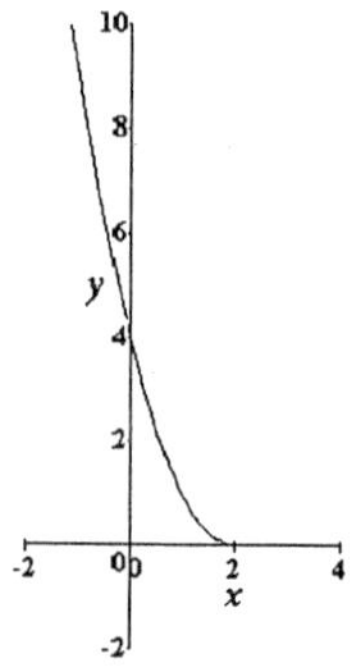

x-intercept: set $y = 0 \Rightarrow (2 - x)^2 = 0 \quad \Leftrightarrow \quad x = 2$;
y-intercept: set $x = 0 \Rightarrow y = 2^2 = 4$; symmetry: wrt x-axis $-y = (2 - x)^2 \quad \Leftrightarrow \quad y = -(2 - x)^2$ which is changed, wrt y-axis $y = (2 + x)^2$ which is changed, wrt origin $-y = (2 + x)^2 \quad \Leftrightarrow \quad y = -(2 + x)^2$ which is changed, so no symmetry.

13. Make a table of values and sketch the graph of the equation $4x^2 + 4y^2 = 81$. Find x- and y-intercepts and test for symmetry.

Answer: medium

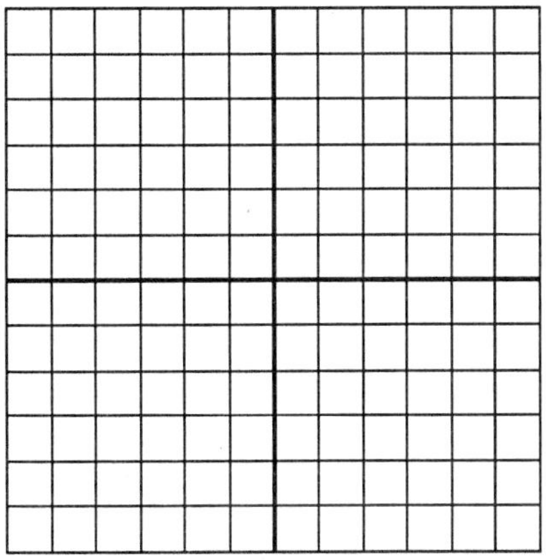

$4x^2 + 4y^2 = 81 \quad \Leftrightarrow \quad x^2 + y^2 = \left(\frac{9}{2}\right)^2$

x	y	(x, y)
$\pm\frac{9}{2}$	0	$\left(\pm\frac{9}{2}, 0\right)$
0	$\pm\frac{9}{2}$	$\left(0, \pm\frac{9}{2}\right)$

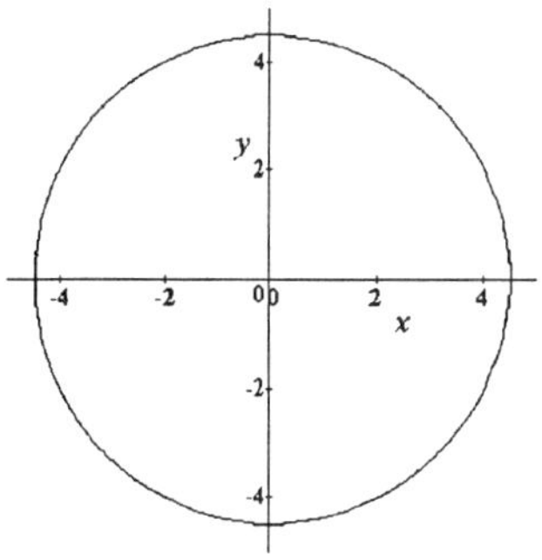

x-intercept: set $y = 0 \Rightarrow x^2 = \left(\frac{9}{2}\right)^2 \quad \Leftrightarrow \quad x = \pm\frac{9}{2}$;

y-intercept: set $x = 0 \Rightarrow y^2 = \left(\frac{9}{2}\right)^2 \quad \Leftrightarrow \quad y = \pm\frac{9}{2}$;

symmetry: wrt x-axis $x^2 + (-y)^2 = \left(\frac{9}{2}\right)^2 \quad \Leftrightarrow \quad x^2 + y^2 = \left(\frac{9}{2}\right)^2$ which is not changed, wrt y-axis $(-x)^2 + y^2 = \left(\frac{9}{2}\right)^2 \quad \Leftrightarrow$ $x^2 + y^2 = \left(\frac{9}{2}\right)^2$ which is not changed, wrt origin $(-x)^2 + (-y)^2 = \left(\frac{9}{2}\right)^2 \quad \Leftrightarrow \quad x^2 + y^2 = \left(\frac{9}{2}\right)^2$ which is not changed, so there is symmetry wrt the x-axis, y-axis, and the origin.

14. Make a table of values and sketch the graph of the equation $x = -\sqrt{16 - y^2}$. Find x- and y-intercepts and test for symmetry.

Answer: hard

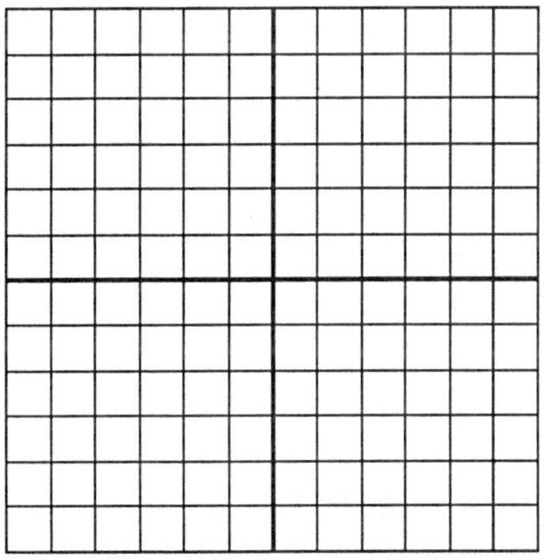

x	y	(x, y)
0	± 4	$(0, \pm 4)$
$-\sqrt{7}$	± 3	$\left(-\sqrt{7}, \pm 3\right)$
$-2\sqrt{3}$	± 2	$\left(-2\sqrt{3}, \pm 2\right)$

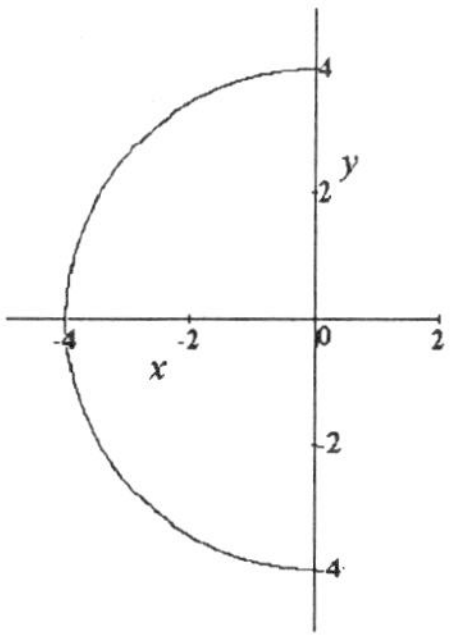

x-intercept: set $y = 0 \Rightarrow x = -\sqrt{16} = -4$;

y-intercept: set $x = 0 \Rightarrow \sqrt{16 - y^2} = 0 \quad \Leftrightarrow \quad y^2 = 16$ $\Leftrightarrow \quad y = \pm 4$; symmetry: wrt x-axis $x = -\sqrt{16 - (-y)^2} = -\sqrt{16 - y^2}$ which is not changed, wrt y-axis $-x = -\sqrt{16 - y^2} \quad \Leftrightarrow \quad x = \sqrt{16 - y^2}$ which is changed, wrt origin $-x = -\sqrt{16 - y^2} \quad \Leftrightarrow$ $x = \sqrt{16 - y^2}$ which is changed, so symmetric wrt x-axis.

15. Make a table of values and sketch the graph of the equation $x = |y| + 2$. Find x- and y-intercepts and test for symmetry.

Answer: medium

x	y	(x,y)
4	± 2	$(4, \pm 2)$
3	± 1	$(3, \pm 1)$
2	0	$(2, 0)$

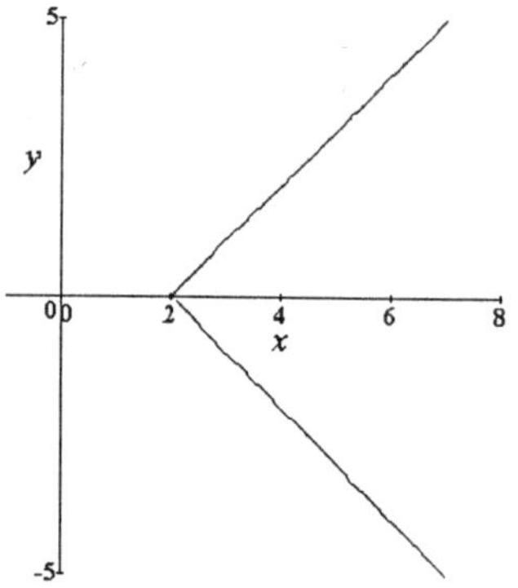

x-intercept: set $y = 0 \Rightarrow x = 2$; y-intercept: set $x = 0 \Rightarrow$ no solution; symmetry: wrt x-axis $x = |-y| + 2 \quad \Leftrightarrow$ $x = |y| + 2$ which is not changed, wrt y-axis $-x = |y| + 2 \quad \Leftrightarrow$ $x = -|y| - 2$ which is changed, wrt origin $-x = |y| + 2 \quad \Leftrightarrow$ $x = -|y| - 2$ which is changed, so symmetric wrt x-axis.

16. Make a table of values and sketch the graph of the equation $y = |2 - x|$. Find x- and y-intercepts and test for symmetry.

Answer: hard

x	$y = \|2 - x\|$	(x,y)
0	2	$(0, 2)$
1	1	$(1, 1)$
2	0	$(2, 0)$
3	1	$(3, 1)$
4	2	$(4, 2)$

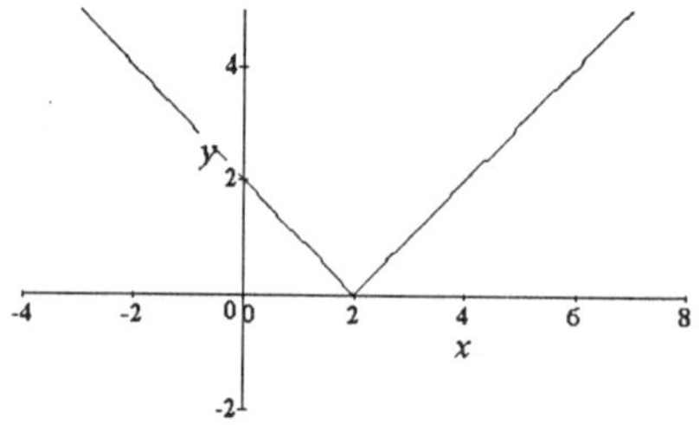

x-intercept: set $y = 0 \Rightarrow |2 - x| = 0 \quad \Leftrightarrow \quad x = 2$;
y-intercept: set $x = 0 \Rightarrow y = 2$; symmetry: wrt x-axis $-y = |2 - x|$ $\Leftrightarrow \quad y = -|2 - x|$ which is changed, wrt y-axis $y = |2 - (-x)| \quad \Leftrightarrow$ $y = |2 + x|$ which is changed, wrt origin $-y = |2 + x| \quad \Leftrightarrow$ $y = -|2 + x|$ which is changed, so no symmetry.

17. Make a table of values and sketch the graph of the equation $y = x^3 - x$. Find x- and y-intercepts and test for symmetry.

Answer: medium

$y = x^3 - x \Leftrightarrow y = x(x-1)(x+1)$

x	$y = x^3 - x$	(x, y)
-2	-6	$(-2, -6)$
-1	0	$(-1, 0)$
0	0	$(0, 0)$
1	0	$(1, 0)$
2	6	$(2, 6)$

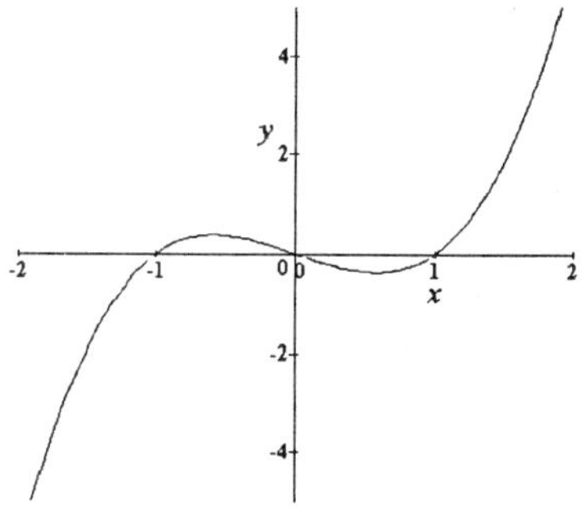

x-intercept: set $y = 0 \Rightarrow x(x-1)(x+1) = 0 \Leftrightarrow x = 0, \pm 1$, y-intercept: set $x = 0 \Rightarrow y = 0$ (wrt x-axis?) $-y = x^3 - x \Leftrightarrow y = -(x^3 - x)$ which is changed, (wrt y-axis?) $y = (-x)^3 - (-x) \Leftrightarrow y = -(x^3 - x)$ which is changed, (wrt origin?) $-y = (-x)^3 - (-x) \Leftrightarrow y = x^3 - x$ which is not changed, so symmetric wrt origin.

18. Make a table of values and sketch the graph of the equation $x = y^3 - y^2$. Find x- and y-intercepts and test for symmetry.

Answer: medium

x	y	(x, y)
-2	-1	$(-2, -1)$
0	0	$(0, 0)$
0	1	$(0, 1)$
4	2	$(4, 2)$

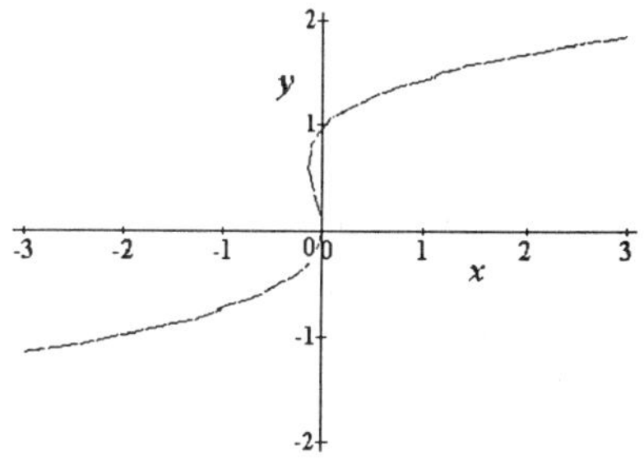

x-intercept: set $y = 0 \Rightarrow x = 0$; y-intercept: set $x = 0 \Rightarrow y^3 - y^2 = 0 \Leftrightarrow y = 0, 1$. Symmetry: (wrt x-axis?) $x = (-y)^3 - (-y)^2 \Leftrightarrow x = -y^3 - y^2$ which is changed, (wrt y-axis?) $-x = y^3 - y^2$ which is changed, (wrt origin?) $-x = (-y)^3 - (-y)^2 \Leftrightarrow -x = -y^3 - y^2$ which is changed, so no symmetry.

19. Sketch a graph of the equation $9x^2 + 9y^2 = 16$.

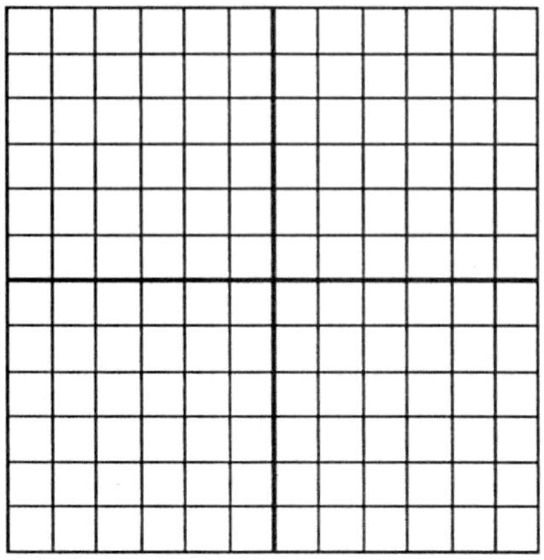

Answer: medium

$9x^2 + 9y^2 = 16$

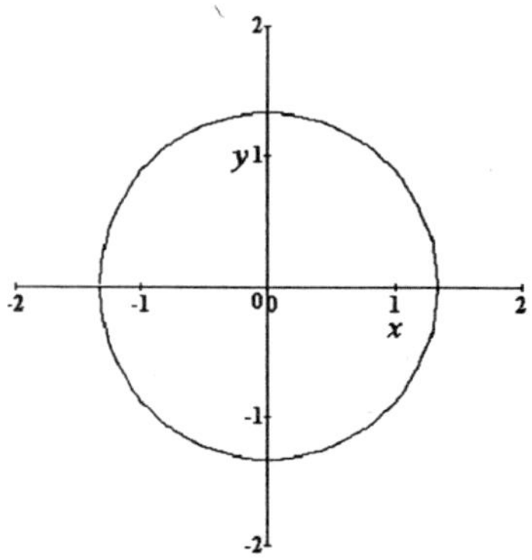

20. Sketch a graph of the equation $(x+1)^2 + (y-1)^2 = 4$.

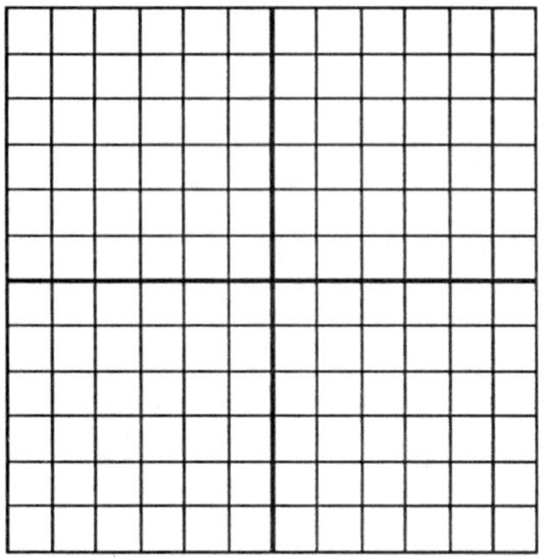

Answer: medium

$(x+1)^2 + (y-1)^2 = 4$

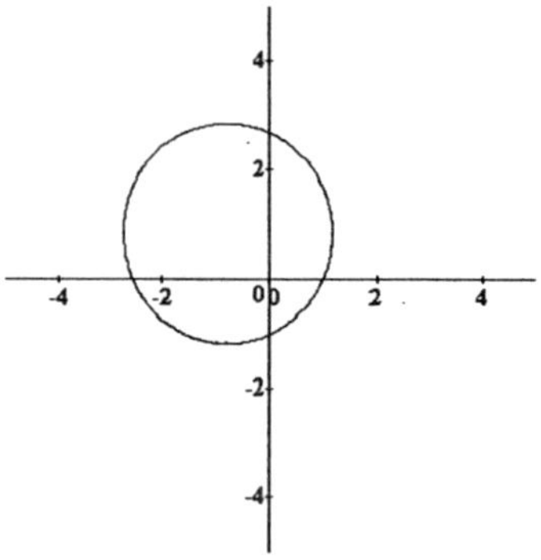

21. Test the equation $y = x^2 - x^3$ for symmetry.

(a) Symmetric about x-axis
(b) Symmetric about y-axis
(c) Symmetric about x- and y-axes
(d) Symmetric about origin
(e) No symmetry

Answer: medium (e)

22. Test the equation $x^2 y + x^4 y^3 = 1$ for symmetry.

(a) Symmetric about x-axis
(b) Symmetric about y-axis
(c) Symmetric about x- and y-axes
(d) Symmetric about origin
(e) No symmetry

Answer: medium (b)

23. Test the equation $y = x^5 + x^3$ for symmetry.

(a) Symmetric about x-axis
(b) Symmetric about y-axis
(c) Symmetric about x- and y-axes
(d) Symmetric about origin
(e) No symmetry

Answer: medium (d)

24. Complete the graph, given that it is symmetric about the y-axis.

Answer: easy

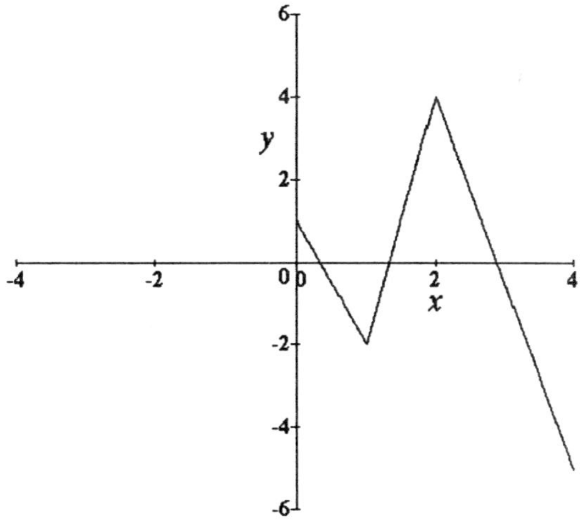

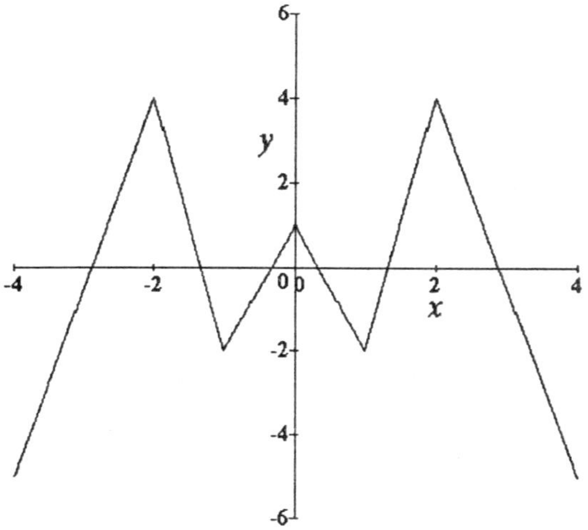

25. Complete the graph, given that it is symmetric about the origin.

Answer: easy

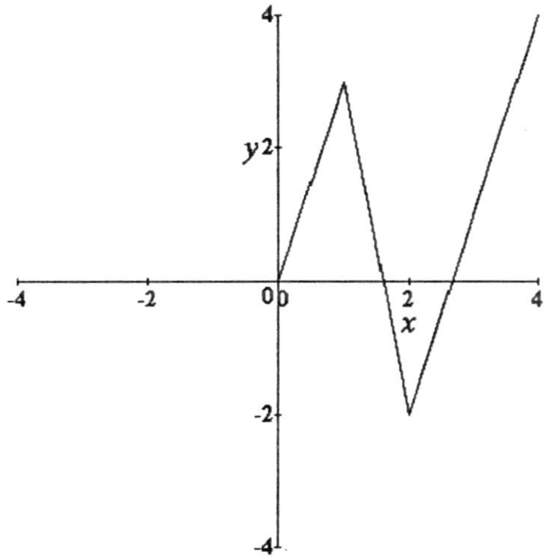

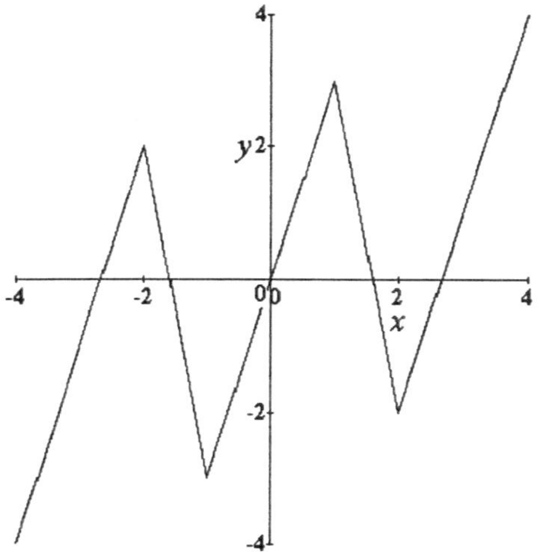

26. Find an equation of the circle with center $(-3, -2)$ and radius 5.

(a) $(x-3)^2 + (y+2)^2 = 25$ (b) $(x+3)^2 + (y-2)^2 = 25$ (c) $(x+3)^2 + y^2 = 25$
(d) $(x+3)^2 + (y+2)^2 = 25$ (e) $x^2 + (y-2)^2 = 25$

Answer: medium (d)

Using the standard notation, $(h, k) = (-3, -2)$ and $r = 5$. So substituting into $(x-h)^2 + (y-k)^2 = r^2$ gives $(x+3)^2 + (y+2)^2 = 25$.

27. Find an equation of the circle with center $(0, 4)$ and radius $\frac{3}{2}$.

(a) $x^2 + (y-4)^2 = \frac{9}{4}$ (b) $x^2 + y^2 = \frac{9}{4}$ (c) $x^2 + (y-2)^2 = 9$
(d) $(x-4)^2 + (y+4)^2 = \frac{3}{2}$ (e) $(x-4)^2 + y^2 = \frac{3}{2}$

Answer: medium (a)

Using the standard notation, $(h, k) = (0, 4)$ and $r = \frac{3}{2}$. So substituting into $(x-h)^2 + (y-k)^2 = r^2$ gives $x^2 + (y-4)^2 = \frac{9}{4}$.

28. Find an equation of the circle that passes through $(-2,-4)$ and has center $(-1,2)$.
(a) $(x-1)^2+(y-2)^2=\sqrt{2}$ (b) $(x+1)^2+(y+2)^2=36$ (c) $(x+1)^2+y^2=36$
(d) $(x+1)^2+(y-2)^2=37$ (e) $x^2+y^2=37$
Answer: medium (d)
$C=(-1,2)$ and $P=(-2,-4)$. Thus $r=|CP|=\sqrt{(-1+2)^2+(2+4)^2}=\sqrt{37}$. Substituting into the standard formula gives $(x+1)^2+(y-2)^2=37$.

29. Find an equation of the circle that satisfies the given conditions: the endpoints of a diameter are $P(-2,5)$ and $Q(4,5)$
(a) $(x-2)^2+(y+1)^2=9$ (b) $(x-1)^2+y^2=34$ (c) $(x+1)^2+y^2=9$
(d) $x^2+y^2=18$ (e) $(x-2)^2+(y-1)^2=18$
Answer: medium (b)
$P=(-2,5)$ and $Q=(4,5)$ are endpoints of a diameter, so the center of the circle is at the midpoint of this diameter, $C=\left(\frac{-2+4}{2},\frac{5-5}{2}\right)=(1,0)$. Therefore the radius is $r=|CP|=\sqrt{(1+2)^2+(0+5)^2}=\sqrt{34}$. The equation of the circle is $(x-1)^2+y^2=34$.

30. Show that the equation $x^2+y^2+4y+3=0$ represents a circle and find the center and radius of the circle.
Answer: medium
$x^2+y^2+4y+3=0 \Leftrightarrow x^2+(y^2+4y)=-3 \Leftrightarrow x^2+(y^2+4y+4)=-3+4 \Leftrightarrow x^2+(y+2)^2=1$. This is a circle with center $(0,-2)$ and radius 1.

31. Show that the equation $x^2+y^2+4x+2y+2=0$ represents a circle and find the center and radius of the circle.
Answer: medium
$x^2+y^2+4x+2y+2=0 \Leftrightarrow (x^2+4x)+(y^2+2y)=-2 \Leftrightarrow (x^2+4x+4)+(y^2+2y+1)=-2+4+1 \Leftrightarrow (x+2)^2+(y+1)^2=3$. This is a circle with center $(-2,-1)$ and radius $\sqrt{3}$.

32. Show that the equation $16x^2+16y^2+8x+24y+4=0$ represents a circle and find the center and radius of the circle.
Answer: hard
$16x^2+16y^2+8x+24y+4=0 \Leftrightarrow (x^2+\frac{1}{2}x)+(y^2+\frac{3}{2}y)=-\frac{1}{4} \Leftrightarrow (x^2+\frac{1}{2}x+\frac{1}{16})+(y^2+\frac{3}{2}y+\frac{9}{16})=-\frac{1}{4}+\frac{1}{16}+\frac{9}{16} \Leftrightarrow (x+\frac{1}{4})^2+(y+\frac{3}{4})^2=\frac{3}{8}$. This is a circle with center $(-\frac{1}{4},-\frac{3}{4})$ and radius $\frac{\sqrt{6}}{4}$.

33. Sketch the graph of the equation $2x^2+2y^2+x=0$.

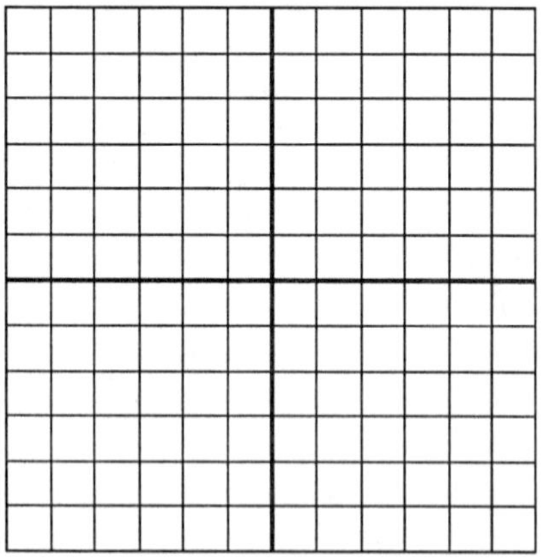

Answer: hard
$2x^2+2y^2+x=0 \Leftrightarrow 2x^2+x+2y^2=0 \Leftrightarrow$
$(x+\frac{1}{4})^2+y^2=\frac{1}{16}$

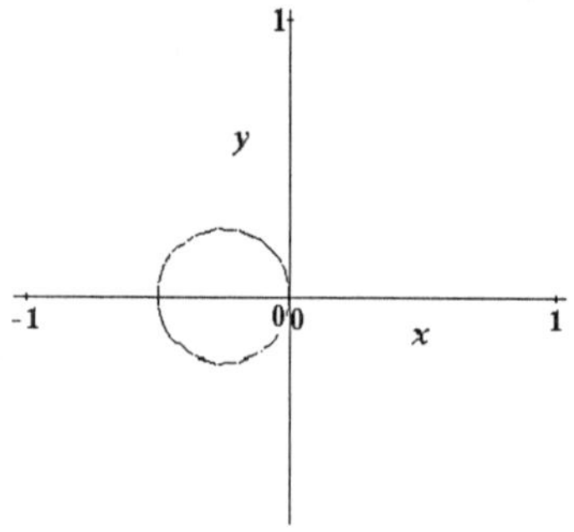

34. Sketch the graph of the equation $x^2 + y^2 - 8x + 4y + 120 = 0$.

Answer: hard

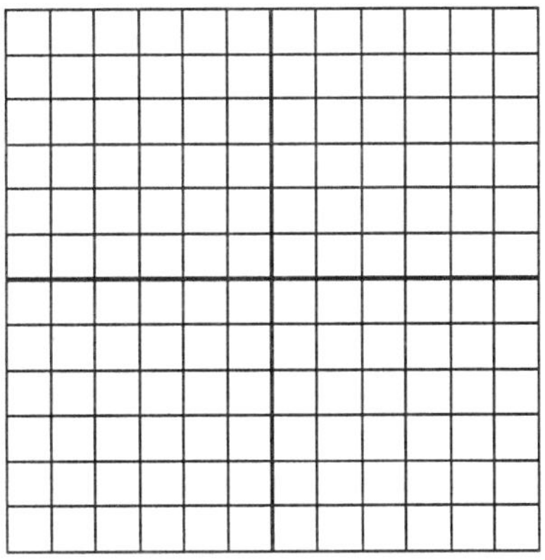

$x^2 + y^2 - 8x + 4y + 120 = 0 \quad \Leftrightarrow \quad (x^2 - 8x) + (y^2 + 4y) = -120$
$\Leftrightarrow \quad (x^2 - 8x + 16) + (y^2 + 4y + 4) = -120 + 16 + 4 \quad \Leftrightarrow$
$(x-4)^2 + (y+2)^2 = -100$. No points satisfy this equation, so the graph is empty.

35. Sketch the region given by the set $\{(x, y) \mid x^2 + y^2 > 8\}$

Answer: medium

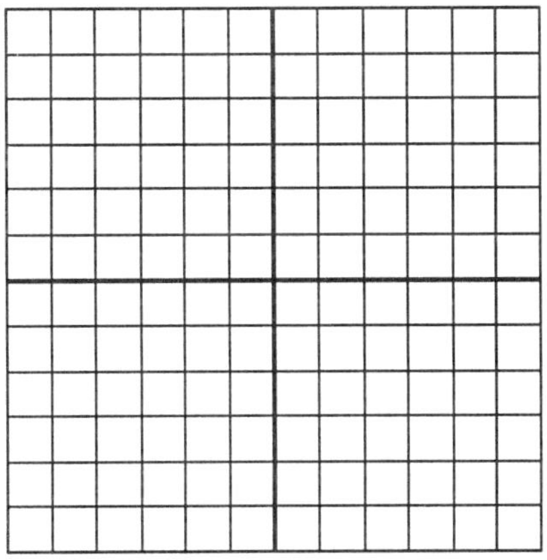

$\{(x, y) \mid x^2 + y^2 > 8\} = \left\{(x, y) \mid \sqrt{x^2 + y^2} > 2\sqrt{2}\right\}$.

The set consists of all points whose distance from the origin is greater than $2\sqrt{2}$, that is, all points outside the circle $x^2 + y^2 = 8$.

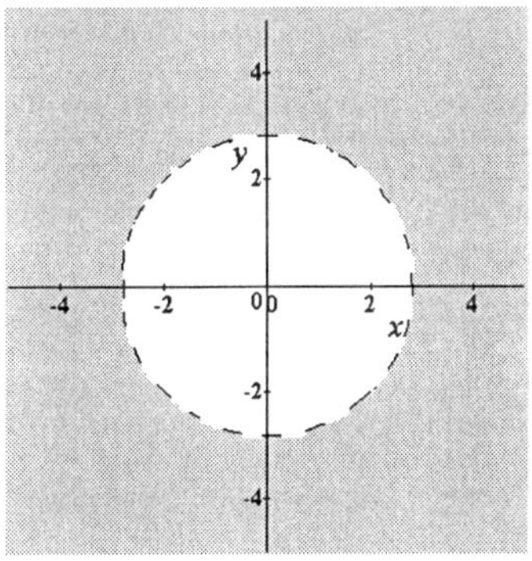

36. Sketch the region given by the set $\{(x, y) \mid 4x \leq x^2 + y^2 < 9\}$

Answer: hard

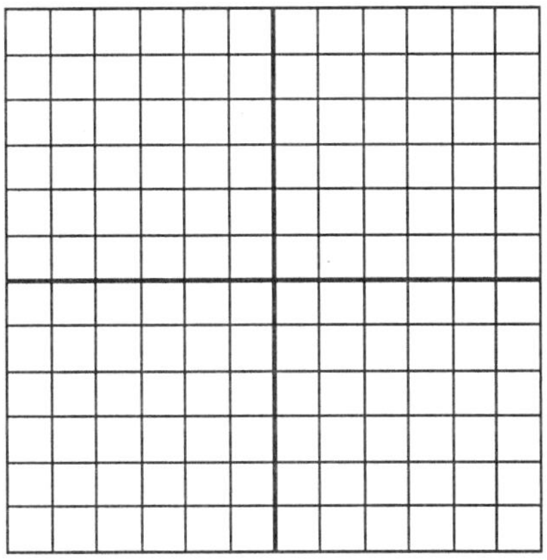

$\{(x, y) \mid 4x \leq x^2 + y^2 < 9\} = \{(x, y) \mid 4x \leq x^2 + y^2 \text{ and } x^2 + y^2 < 9\}$
$= \left\{(x, y) \mid 4 \leq (x-2)^2 + y^2 \text{ and } x^2 + y^2 < 9\right\}$. The set consists of all points that lie inside the circle of radius 3 centered at the origin and on or outside the circle of radius 2 centered at $(2, 0)$.

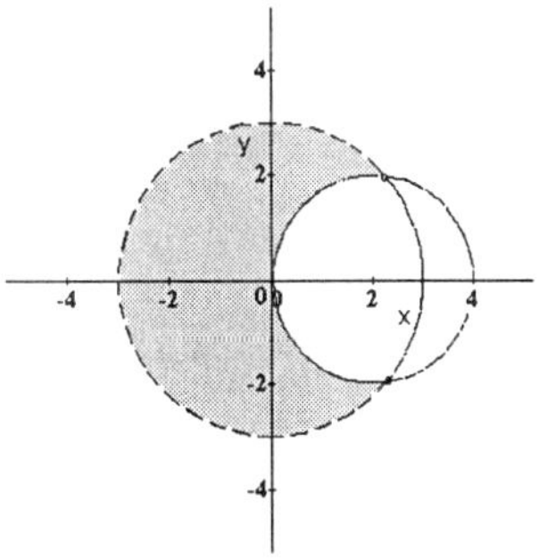

37. Sketch the region in the coordinate plane that satisfies both the inequalities $x^2 + y^2 \le 4$ and $x \ge |y|$. What is the area of this region?

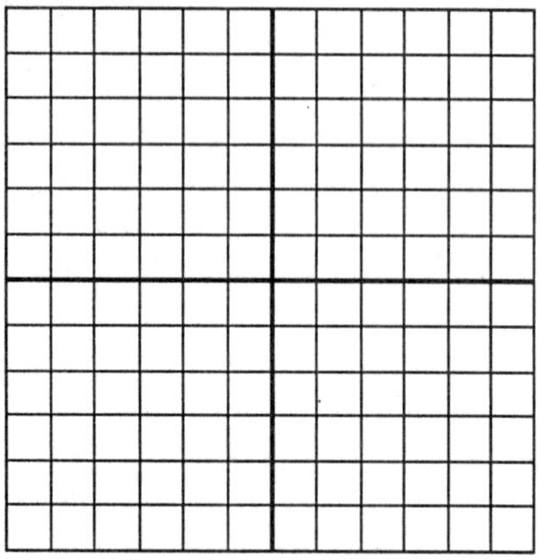

Answer: hard

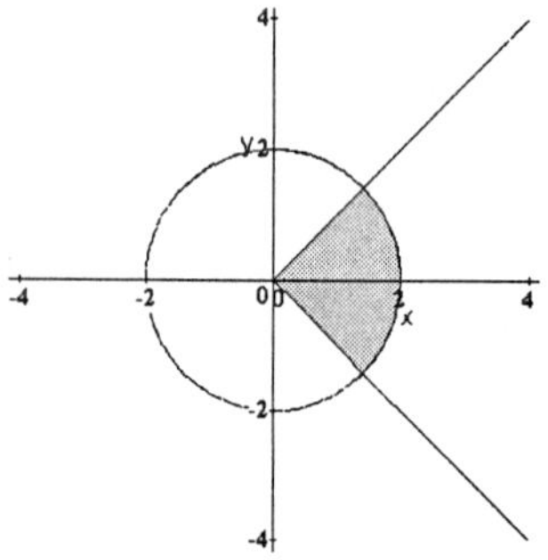

$x^2 + y^2 \le 4$ and $x \ge |y| \quad \Leftrightarrow \quad x^2 + y^2 \le 4$ and $-x \le y \le x$

Since this is one quarter of a circle of radius 2, its area is $\frac{1}{4} \cdot \pi \cdot 2^2 = \pi$.

38. Determine which of the points $(-5, 0)$, $(2, 23)$, $\left(-\sqrt{2}, \sqrt{23}\right)$ are on the graph of the equation $x^2 + y^2 = 25$.

(a) $(2, 23)$, $\left(-\sqrt{2}, \sqrt{23}\right)$

(b) $(-5, 0)$, $\left(-\sqrt{2}, \sqrt{23}\right)$

(c) $(-5, 0)$, $(2, 23)$, $\left(-\sqrt{2}, \sqrt{23}\right)$

(d) $(-5, 0)$, $(2, 23)$

(e) None

Answer: easy (b)

Substitute to find that $(-5)^2 + 0^2 = 25$, $\left(-\sqrt{2}\right)^2 + \left(\sqrt{23}\right)^2 = 25$, but $(2)^2 + (23)^2 \ne 25$.

39. Find the x- and y-intercepts of the graph $x^2 + 2xy + 4y^2 = 4$.

(a) x-intercept 2, y-intercepts ± 1

(b) x-intercepts ± 2, y-intercepts ± 2

(c) x-intercept 2, y-intercept 1

(d) x-intercepts ± 1, y-intercepts $\pm \frac{1}{2}$

(e) x-intercepts ± 2, y-intercepts ± 1

Answer: easy (e)

When $y = 0 \;\Rightarrow\; x^2 = 4 \;\Rightarrow\; x = \pm 2$. When $x = 0 \;\Rightarrow\; 4y^2 = 4 \;\Rightarrow\; y = \pm 1$.

40. Make a table of values and sketch the graph of the equation $x + 3y + 2 = 0$. Find the x- and y-intercepts and test for symmetry.

Answer: easy

$x + 3y + 2 = 0 \;\Leftrightarrow\; y = -\frac{1}{3}x - \frac{2}{3}$.

x	$y = -\frac{1}{3}x - \frac{2}{3}$	(x, y)
-2	0	$(-2, 0)$
-1	$-\frac{1}{3}$	$\left(-1, -\frac{1}{3}\right)$
0	$-\frac{2}{3}$	$\left(0, -\frac{2}{3}\right)$
1	-1	$(1, -1)$

x-intercept: set $y = 0 \;\Rightarrow\; x = -2$. y-intercept: set $x = 0$ $\Rightarrow\; y = -\frac{2}{3}$. Symmetry: with respect to x-axis $-y = -\frac{1}{3}x - \frac{2}{3} \;\Leftrightarrow\; y = \frac{1}{3}x + \frac{2}{3}$ which is changed, with respect to y-axis $y = -\frac{1}{3}(-x) - \frac{2}{3} \;\Leftrightarrow\; y = \frac{1}{3}x - \frac{2}{3}$ which is changed, with respect to origin $-y = -\frac{1}{3}(-x) - \frac{2}{3}$ $\Leftrightarrow\; y = -\frac{1}{3}x + \frac{2}{3}$ which is changed. No symmetry.

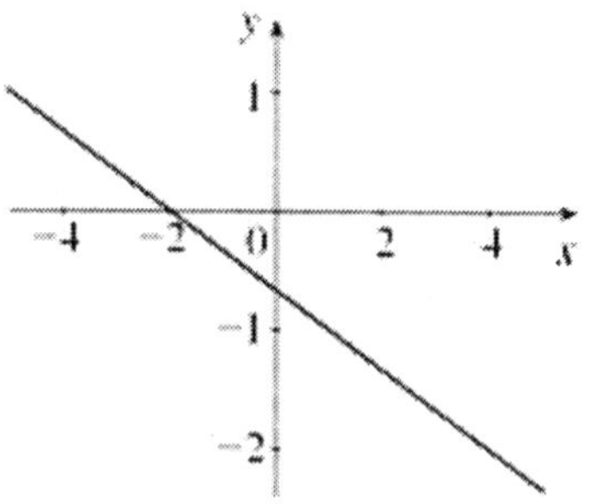

41. Make a table of values and sketch the graph of the equation $y = -|x+2|+3$. Find the x- and y-intercepts and test for symmetry.
Answer: medium
$y = -|x+2|+3$

x	$y = -\|x+2\|+3$	(x,y)
-5	0	$(-5,0)$
-3	2	$(-3,2)$
-2	3	$(-2,3)$
-1	2	$(-1,2)$
0	1	$(0,1)$
1	0	$(1,0)$
2	-1	$(2,-1)$

x-intercept: set $y=0 \Rightarrow x=-5$ or 1. y-intercept: set $x=0 \Rightarrow y=1$. Symmetry: with respect to x-axis $-y=-|x+2|+3 \Leftrightarrow y=|x+2|-3$ which is changed, with respect to y-axis $y=-|(-x)+2|+3 \Leftrightarrow y=-|2-x|+3$ which is changed, with respect to origin $-y=-|(-x)+2|+3 \Leftrightarrow y=|2-x|-3$ which is changed. No symmetry.

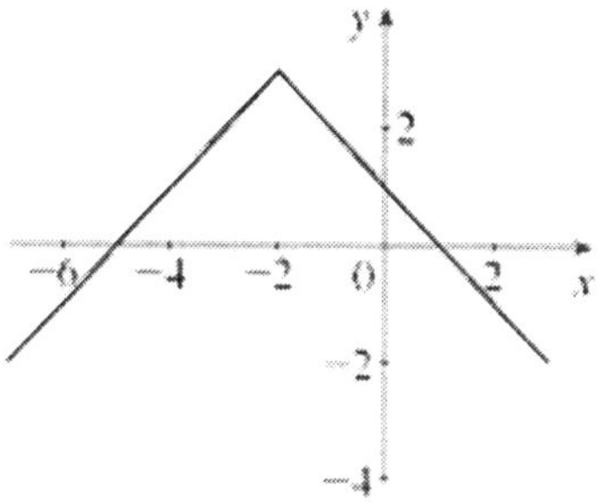

42. Make a table of values and sketch the graph of the equation $y=-\sqrt{25-x^2}$. Find the x- and y-intercepts and test for symmetry.
Answer: medium
$y=-\sqrt{25-x^2}$

x	$y=-\sqrt{25-x^2}$	(x,y)
-5	0	$(-5,0)$
-4	-3	$(-4,-3)$
-3	-4	$(-3,-4)$
0	-5	$(0,-5)$
3	-4	$(3,-4)$
4	-3	$(4,-3)$
5	0	$(5,0)$

x-intercept: set $y=0 \Rightarrow x=-5$ or 5. y-intercept: set $x=0 \Rightarrow y=-5$. Symmetry: with respect to x-axis $-y=-\sqrt{25-x^2} \Leftrightarrow y=\sqrt{25-x^2}$ which is changed, with respect to y-axis $y=-\sqrt{25-(-x)^2} \Leftrightarrow$ $y=-\sqrt{25-x^2}$ which is not changed, with respect to origin $-y=-\sqrt{25-(-x)^2} \Leftrightarrow y=\sqrt{25-x^2}$ which is changed. Symmetry with respect to y-axis.

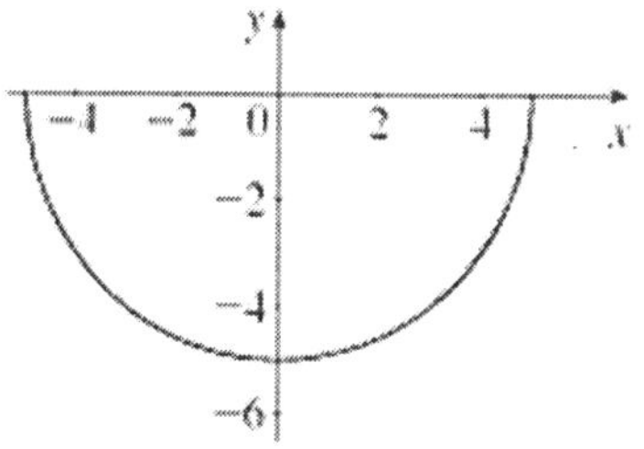

43. Make a table of values and sketch the graph of the equation $y = x^3 - x^2$. Find the x- and y-intercepts and test for symmetry.
Answer: medium

x	$y = x^3 - x^2$	(x, y)
-2	-12	$(-2, -12)$
-1	-2	$(-1, -2)$
0	0	$(0, 0)$
$\frac{1}{2}$	$-\frac{1}{8}$	$(\frac{1}{2}, -\frac{1}{8})$
1	0	$(1, 0)$
2	4	$(2, 4)$
3	18	$(3, 18)$

x-intercept: set $y = 0 \Rightarrow x = 0$ or 1. y-intercept: set $x = 0 \Rightarrow y = 0$. Symmetry: with respect to x-axis $-y = x^3 - x^2 \Leftrightarrow y = x^3 + x^2$ which is changed, with respect to y-axis $y = (-x)^3 - (-x)^2 \Leftrightarrow y = -x^3 - x^2$ which is changed, with respect to origin $-y = (-x)^3 - (-x)^2 \Leftrightarrow y = x^3 + x^2$ which is changed. No symmetry.

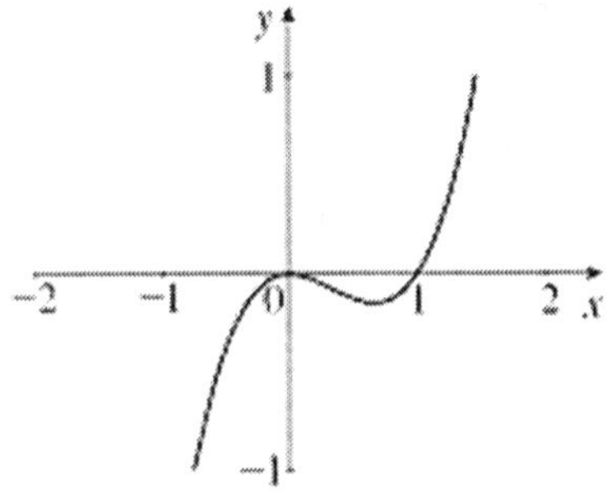

44. Make a table of values and sketch the graph of the equation $y = 3x^4 - 4x^2$. Find the x- and y-intercepts and test for symmetry.
Answer: medium

x	$y = 3x^4 - 4x^2$	(x, y)
-2	32	$(-2, 32)$
-1	-1	$(-1, -1)$
$-\frac{1}{2}$	$-\frac{13}{16}$	$(-\frac{1}{2}, -\frac{13}{16})$
0	0	$(0, 0)$
1	-1	$(1, -1)$
$\frac{1}{2}$	$-\frac{13}{16}$	$(\frac{1}{2}, -\frac{13}{16})$
2	32	$(2, 32)$

x-intercept: set $y = 0 \Rightarrow x = x = \pm\frac{2}{3}\sqrt{3}$. y-intercept: set $x = 0 \Rightarrow y = 0$. Symmetry: with respect to x-axis $-y = 3x^4 - 4x^2 \Leftrightarrow y = -(3x^4 - 4x^2)$ which is changed, with respect to y-axis $y = 3(-x)^4 - 4(-x)^2 \Leftrightarrow y = 3x^4 - 4x^2$ which is not changed, with respect to origin $-y = 3(-x)^4 - 4(-x)^2 \Leftrightarrow y = -(3x^4 - 4x^2)$ which is changed. Symmetry with respect to the y-axis.

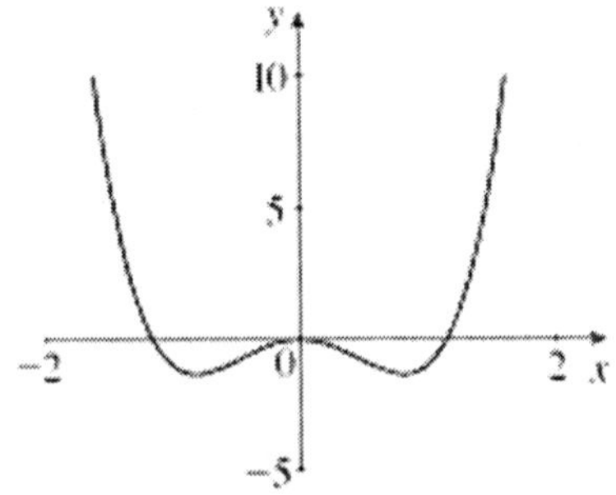

45. Find the equation of the circle with center $(-1,7)$ and radius $\sqrt{2}$.

(a) $x^2+2x+y^2-28y+50=0$

(b) $x^2-2x-y^2-14y-48=0$

(c) $x^2+14x+48+y^2-2y=0$

(d) $x^2-14x+48+y^2-2y=0$

(e) $x^2+2x+y^2-14y+48=0$

Answer: easy (e)

$(h,k)=(-1,7), r=\sqrt{2} \;\Rightarrow\; (x-(-1))^2+(y-7)^2=\left(\sqrt{2}\right)^2 \;\Rightarrow\; (x+1)^2+(y-7)^2=2 \;\Rightarrow$
$x^2+2x+y^2-14y+48=0$

46. Find the equation of the circle with center $(0,-6)$ and tangent to the y-axis.

(a) $(x+6)^2+(y-6)^2=9$

(b) $(x)^2+(y-6)^2=6$

(c) $(x)^2+(y+6)^2=36$

(d) $(x)^2+(y+6)^2=9$

(e) $(x-6)^2+(y)^2=36$

Answer: medium (c)

$(h,k)=(0,-6)$. The radius must be $r=6$, thus $(x)^2+(y+6)^2=36$.

47. Show that $x^2-8x+\frac{89}{4}+y^2-5y=4$ represents a circle, find its center and radius, and sketch its graph.

Answer: medium

$x^2-8x+\frac{89}{4}+y^2-5y=4 \;\Leftrightarrow$

$x^2-8x+y^2-5y=-\frac{73}{4} \;\Leftrightarrow$

$(x-4)^2+\left(y-\frac{5}{2}\right)^2=16+\frac{25}{4}-\frac{73}{4} \;\Leftrightarrow$

$(x-4)^2+\left(y-\frac{5}{2}\right)^2=4$. Center $\left(4,\frac{5}{2}\right)$, $r=2$.

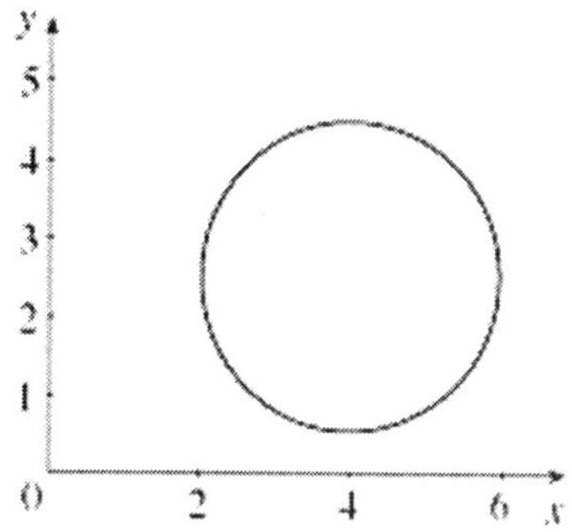

48. Show that $x^2-x+\frac{25}{36}+y^2-\frac{4}{3}y=7$ represents a circle, find its center and radius, and sketch its graph.

Answer: medium

$x^2-x+\frac{25}{36}+y^2-\frac{4}{3}y=7 \;\Leftrightarrow$

$x^2-x+y^2-\frac{4}{3}y=\frac{227}{36} \;\Leftrightarrow$

$\left(x-\frac{1}{2}\right)^2+\left(y-\frac{2}{3}\right)^2=\frac{1}{4}+\frac{4}{9}+\frac{227}{36} \;\Leftrightarrow$

$\left(x-\frac{1}{2}\right)^2+\left(y-\frac{2}{3}\right)^2=7$. Center $\left(\frac{1}{2},\frac{2}{3}\right)$, $r=\sqrt{7}$.

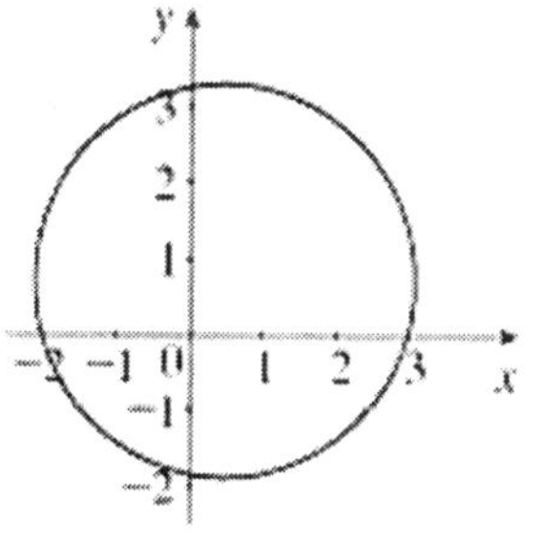

SECTION 2.3

GC 1. Use a graphing calculator or computer to decide which one of the viewing rectangles (a) - (e) produces the most appropriate graph of the equation $y = x^3 - 6$.

(a) $[-2, 2]$ by $[-2, 2]$ (b) $[-2, 4]$ by $[-2, 12]$ (c) $[2, 6]$ by $[-10, 4]$ (d) $[-2, 2]$ by $[-14, 2]$ (e) $[-3, 3]$ by $[-10, -6]$

Answer: easy (d)

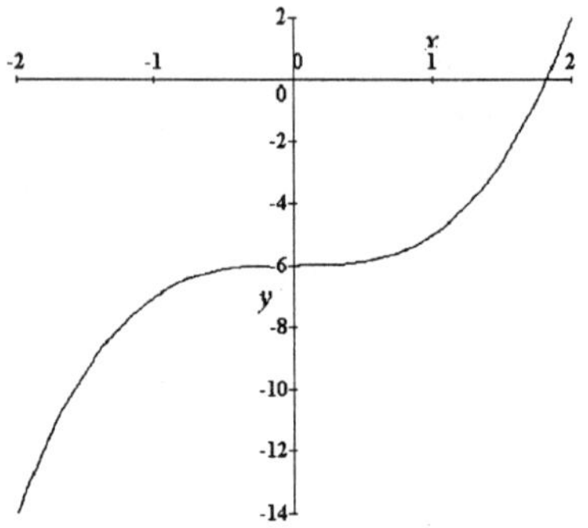

GC 2. Use a graphing calculator or computer to decide which one of the viewing rectangles (a) - (e) produces the most appropriate graph of the equation $y = 2x^2 - 6x + 4$.

(a) $[2, 10]$ by $[-10, 10]$ (b) $[-2, 5]$ by $[-2, 20]$ (c) $[5, 10]$ by $[-10, 50]$ (d) $[-10, 10]$ by $[0, 50]$ (e) $[-10, 10]$ by $[-20, 20]$

Answer: easy (b)

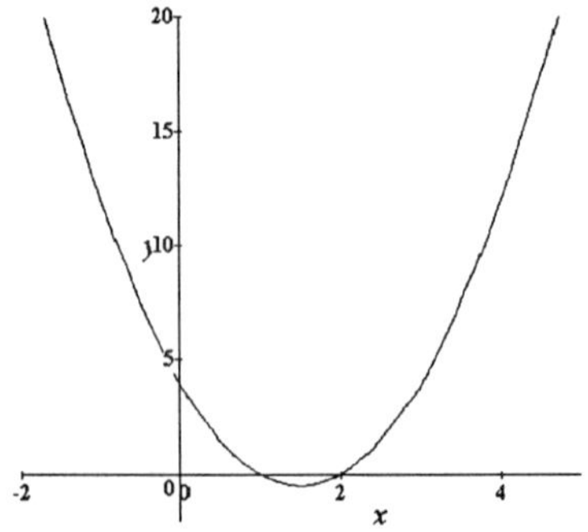

GC 3. Use a graphing calculator or computer to decide which one of the viewing rectangles (a) - (e) produces the most appropriate graph of the equation $y = \sqrt{2x - x^2}$.

(a) $[-1, 3]$ by $[-1, 2]$ (b) $[-4, 4]$ by $[-2, 5]$ (c) $[0, 1]$ by $[0, 2]$ (d) $[-3, 3]$ by $[-1, 3]$ (e) $[-5, 5]$ by $[-10, 10]$

Answer: medium (a)

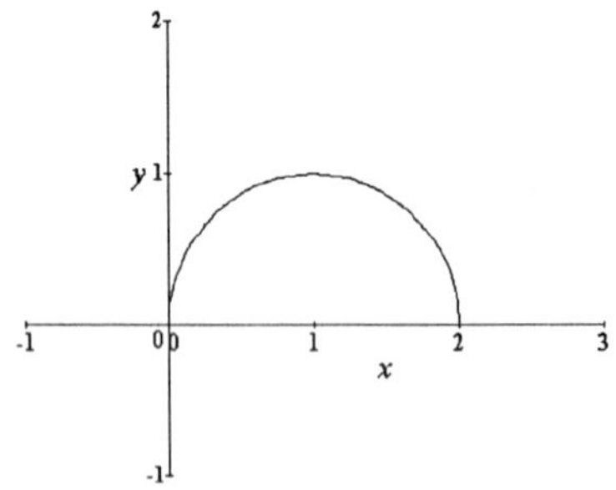

GC 4. Determine an appropriate viewing rectangle for the equation $y = 50x^2 - 200$ and use it to draw the graph.
Answer: medium $[-3, 3]$ by $[-250, 250]$

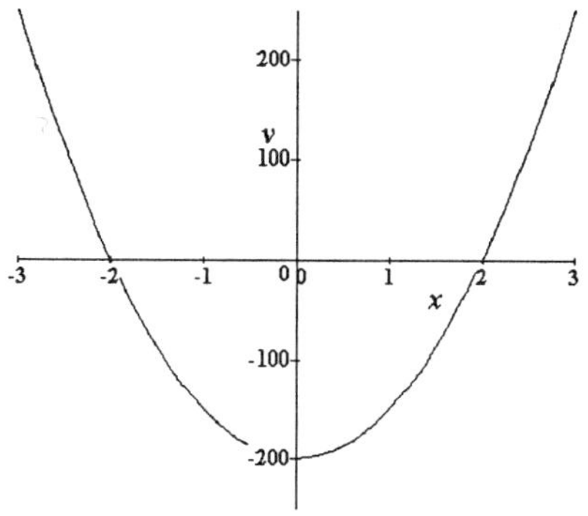

GC 5. Determine an appropriate viewing rectangle for the equation $y = -x^5 + x^4 - x^3 + x^2$ and use it to draw the graph.
Answer: medium $[-1, 2]$ by $[-3, 2]$

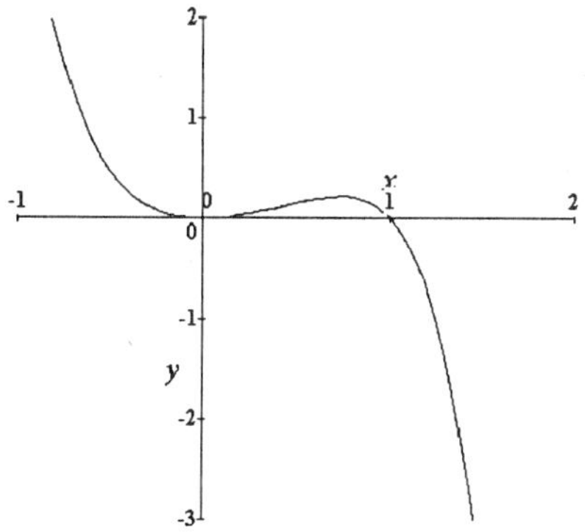

GC 6. Determine an appropriate viewing rectangle for the equation $y = \sqrt{15x - 13}$ and use it to draw the graph.
Answer: medium $[0, 10]$ by $[0, 15]$

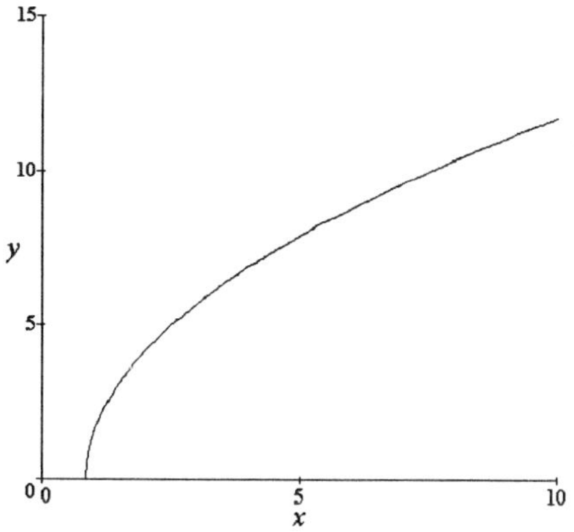

GC 7. Determine an appropriate viewing rectangle for the equation $y = x^2 + \frac{2}{x}$ and use it to draw the graph.

Answer: medium $[-3, 3]$ by $[-10, 10]$

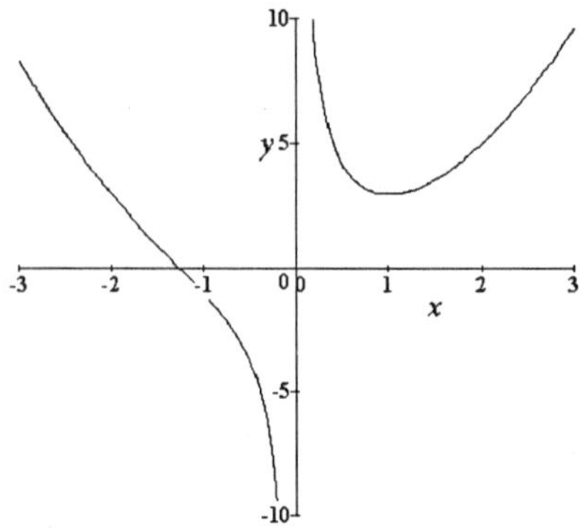

GC 8. Determine an appropriate viewing rectangle for the equation $y = |x^3 + 1| - 3x$ and use it to draw the graph.

Answer: medium $[-2, 3]$ by $[-2, 10]$

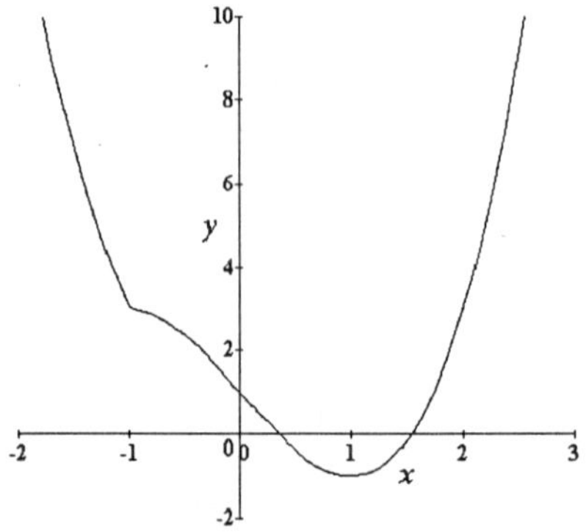

GC 9. Graph the circle $(y-2)^2 + (x+1)^2 = 1$ by solving for y and graphing two equations.

Answer: easy

$(y-2)^2 + (x+1)^2 = 1 \quad \Leftrightarrow \quad (y-2)^2 = 1 - (x+1)^2 \quad \Leftrightarrow \quad y - 2 = \pm\sqrt{1-(x+1)^2} \quad \Leftrightarrow$

$y = \pm\sqrt{1-(x+1)^2} + 2, -2 \le x \le 0$

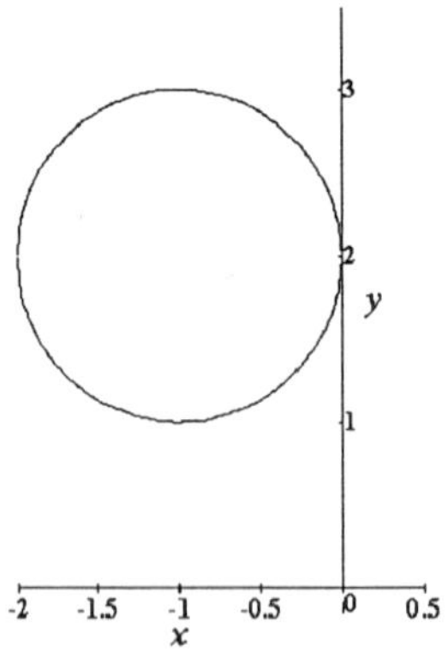

GC 10. Find the approximate solutions of the equation $x^2 - 2x - 4 = 0$ in the interval $[-4, 4]$ by drawing a graph in an appropriate viewing rectangle.

(a) $-2.1, 3.6$ (b) $-2.1, 2.1, 3.6$ (c) $-1.2, 3.2$ (d) $-1.2, 3.4, 3.6$ (e) $-1.4, 3.4$

Answer: medium (c)

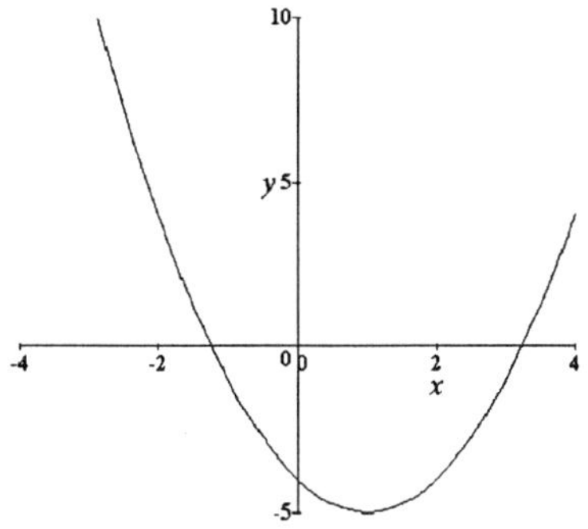

GC 11. Find the approximate solutions of the equation $x^3 - 4x^2 + 4x - 1 = 0$ in the interval $[-1, 4]$ by drawing a graph in an appropriate viewing rectangle.

(a) $1, 2.6$ (b) $0.4, 2.6$ (c) 1 (d) $0.6, 0.8$ (e) $0.4, 1.0, 2.6$

Answer: medium (e)

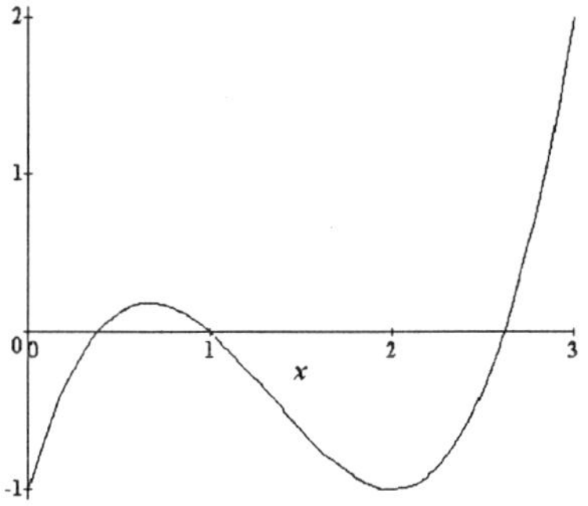

GC 12. Find the approximate solutions of the equation $\sqrt{3 + 7x} = \sqrt{7 - 3x}$ in the interval $[-2, 2]$ by drawing a graph in an appropriate viewing rectangle.

(a) $0.36, 0.63$ (b) 0.40 (c) 0.63 (d) $0.23, 0.37$ (e) 0.26

Answer: medium (b)

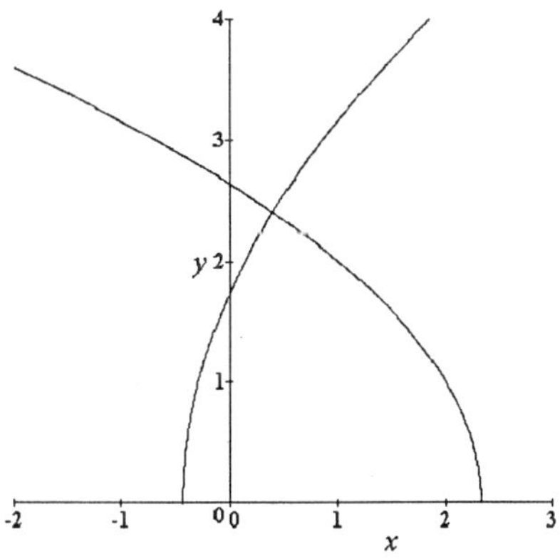

GC 13. Find the approximate solutions of the inequality $0.2x^2 + 0.325x < 0.55$ by drawing appropriate graphs.

(a) $-2.5 < x < 2.0$ (b) $-2.6 < x < 1.0$ (c) $-2.6 < x < 1$ (d) $x < -2.6$ or $x > 1$ (e) $x = 1.0$

Answer: medium (b)

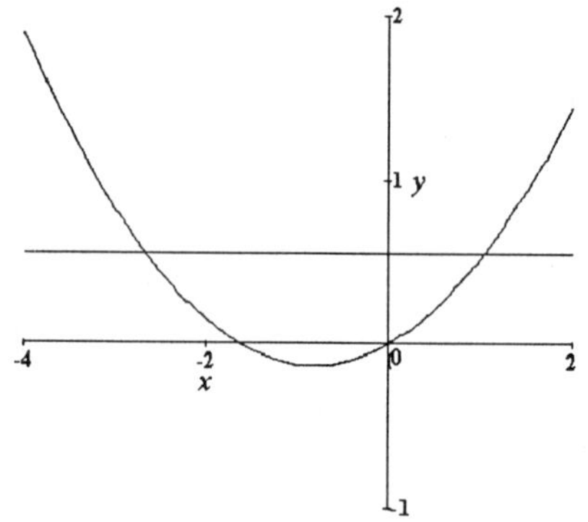

GC 14. Find the approximate solutions of the inequality $3x^2 + 2x < x^3 + 1$ by drawing appropriate graphs.

(a) $-1 < x < 1$ or $2.5 < x$ (b) $-0.8 < x < 0.3$ or $3.5 < x$ (c) $-1 < x < 0.3$

(d) $x < -1.0$ or $2.5 < x$ (e) $x < -0.8$ or $x > 0.3$

Answer: hard (b)

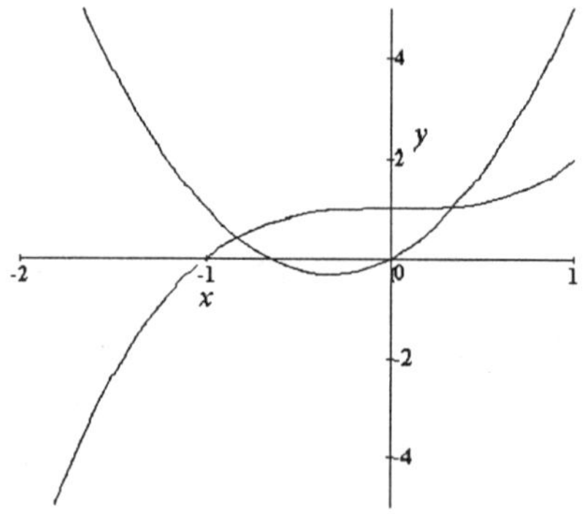

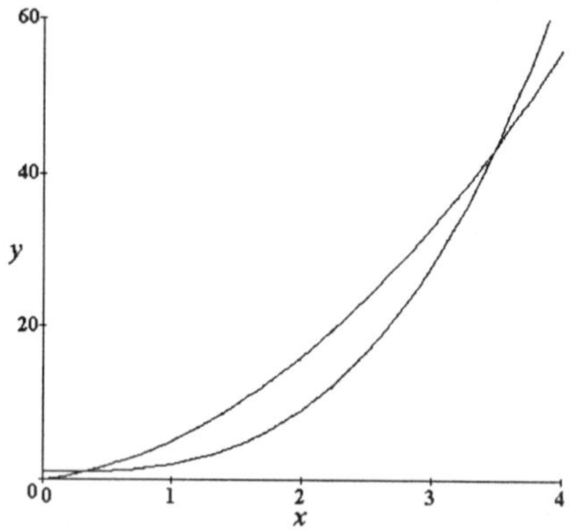

GC 15. Find the approximate solutions of the inequality $\sqrt{x^2 + 3} < 3|x|$ by drawing appropriate graphs.

(a) $-0.8 < x < 0.8$ (b) $x < -0.8$ or $x > 0.8$ (c) $-0.6 < x < 0.6$

(d) $x < -0.6$ or $x > 0.6$ (e) $x = -0.8$ or $x = 0.6$

Answer: medium (d)

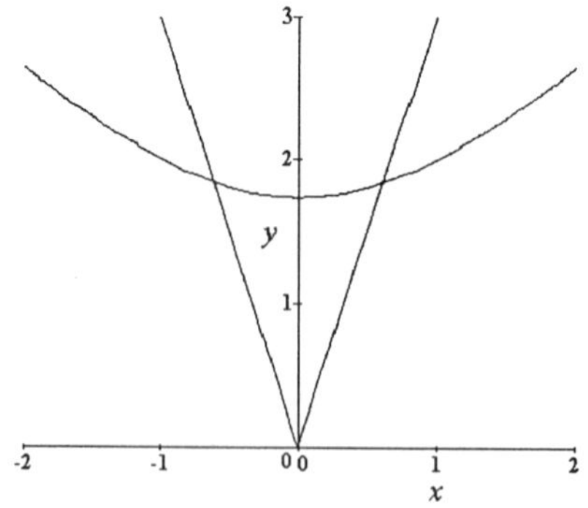

GC 16. Find the approximate solutions of the inequality $5x^3 - 13x^2 < 4x + 1$ by drawing appropriate graphs.

(a) $0 < x < 2.5$ (b) $0 < x < 2.9$ (c) $x > 2.9$ (d) $x < 2.5$ (e) $x < 2.9$

Answer: medium (e)

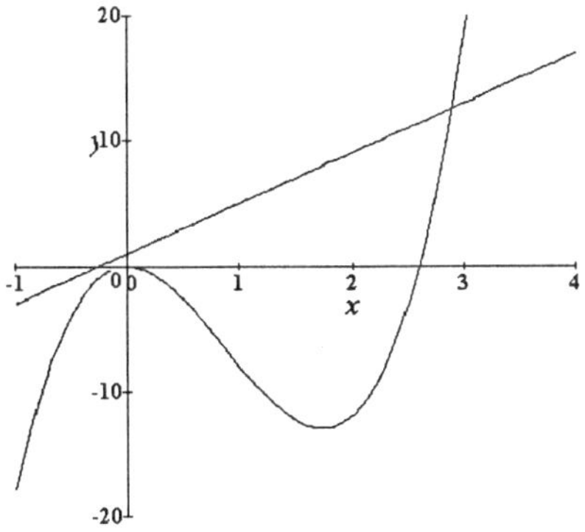

17. Use a graphing calculator or computer to determine which of the viewing rectangles (a)□(e) produces the most appropriate graph of the equation $y = 5x^5 - 4x^4$.

(a) $[-1, 1]$ by $[-1, 1]$ (b) $[-10, 10]$ by $[-10, 10]$

(c) $[-10, 10]$ by $[-1, 1]$ (d) $[-20, 20]$ by $[-1, 1]$

(e) $[-5, 2]$ by $[-1, 1]$

Answer: easy (a)

18. Use a graphing calculator or computer to determine which of the viewing rectangles (a)□(e) produces the most appropriate graph of the equation $y = -|x + 2| - 3$.

(a) $[-1, 1]$ by $[-1, 1]$ (b) $[-6, 1]$ by $[-8, 0]$

(c) $[6, 1]$ by $[-8, 8]$ (d) $[-6, 8]$ by $[8, 0]$

(e) $[-15, 2]$ by $[-1, 1]$

Answer: medium (b)

19. Determine an appropriate viewing rectangle for the equation $y = x^2 + \dfrac{2}{x}$.

Answer: medium
$[-5, 5]$ by $[-10, 10]$

20. In how many distinct points does the graph of $y = \frac{1}{2}x^5 - x^3$ intersect the graph of $y = |x + 1|$ in the viewing rectangle $[-2, 2]$ by $[-5, 5]$?

(a) 1 (b) 2 (c) 3 (d) 4 (e) 5

Answer: easy (c)

21. In how many distinct points does the graph of $y = \frac{3}{2}x^2 - 5x - \frac{1}{4}$ intersect the graph $y = -\frac{1}{2}x^2 + 2x - 7$ in the viewing rectangle $[-10, 10]$ by $[-15, 5]$?

(a) 0 (b) 1 (c) 2 (d) 3 (e) 4

Answer: easy (a)

22. In how many distinct points does the graph of $x^3 - 4x^2 + 5|x| - 2$ intersect the graph of $y = 2x^4 - 4|x^3| + x$ in the viewing rectangle $[-2.5, 2.5]$ by $[-5, 5]$?

(a) 6 (b) 1 (c) 3 (d) 4 (e) 5

Answer: easy (d)

SECTION 2.4

1. Find the slope of the line through $P(-2,4)$ and $Q(4,-2)$.
(a) -2 (b) -1 (c) 0 (d) 1 (e) 2
Answer: easy (b)
$\text{slope} = \dfrac{-2-4}{4-(-2)} = -1$

2. Find the slope of the line through $P(-2,-5)$ and $Q(7,1)$.
(a) $\frac{3}{4}$ (b) $\frac{2}{3}$ (c) $\frac{2}{5}$ (d) $\frac{4}{3}$ (e) $\frac{3}{2}$
Answer: easy (b)
$\text{slope} = \dfrac{1-(-5)}{7-(-2)} = \dfrac{2}{3}$

3. Find the slope of the line through $P(-3,5)$ and $Q(7,7)$.
(a) $\frac{1}{2}$ (b) $\frac{5}{2}$ (c) $\frac{1}{4}$ (d) $\frac{1}{5}$ (e) $\frac{11}{2}$
Answer: easy (d)
$\text{slope} = \dfrac{7-5}{7-(-3)} = \dfrac{1}{5}$

4. Find an equation of the line with slope -2 that passes through $(-3,2)$.
(a) $2x+y+4=0$ (b) $x-2y+4=0$ (c) $x+y-2=0$ (d) $x+y+4=0$ (e) $3x-y-4=0$
Answer: medium (a)
$m=-2$, and $(x_1,y_1)=(-3,2) \Rightarrow y-y_1=m(x-x_1) \Rightarrow y-2=-2(x+3) \quad \Leftrightarrow \quad 2x+y+4=0$

5. Find an equation of the line with slope $-\frac{2}{3}$ that passes through $(-1,2)$.
(a) $2x+y-1=0$ (b) $x+2y-3=0$ (c) $2x+3y-4=0$ (d) $x+3y-2=0$ (e) $4x+2y+4=0$
Answer: medium (c)
$m=-\frac{2}{3}$, and $(x_1,y_1)=(-1,2) \Rightarrow y-y_1=m(x-x_1) \Rightarrow y-2=-\frac{2}{3}(x+1) \quad \Leftrightarrow \quad 2x+3y-4=0$

6. Find an equation of the line that passes through $(-3,-1)$ and $(2,4)$.
(a) $x-2y+1=0$ (b) $x-y+2=0$ (c) $3x-2y-1=0$ (d) $3x-3y+1=0$ (e) $4x+2y=0$
Answer: medium (b)
$m=\dfrac{y_2-y_1}{x_2-x_1}=\dfrac{4-(-1)}{2-(-3)}=1 \Rightarrow y-y_1=m(x-x_1) \Rightarrow y+1=1(x+3) \quad \Leftrightarrow \quad x-y+2=0$

7. Find an equation of the line with slope $\frac{2}{3}$ and y-intercept 6.
(a) $x-y-12=0$ (b) $3x-2y+1=0$ (c) $2x+3y-12=0$ (d) $3x+2y+4=0$ (e) $2x-3y+18=0$
Answer: medium (e)
$m=\frac{2}{3}$, $b=6$. Substituting into $y=mx+b$ gives $y=\frac{2}{3}x+6 \quad \Leftrightarrow \quad 2x-3y+18=0$

8. Find an equation of the line with x-intercept -2 and y-intercept 4.
(a) $3x-2y+1=0$ (b) $2x-y+4=0$ (c) $x+y+2=0$ (d) $2x+2y+3=0$ (e) $x-6y+2=0$
Answer: medium (b)
Because the x-intercept is -2, $(-2,0)$ is a point on the line, and similarly for the y-intercept, $(0,4)$ is a point on the line. Thus $m=\dfrac{4-0}{0+2}=2$ and substituting into $y-y_1=m(x-x_1)$ gives $y-0=2(x+2) \quad \Leftrightarrow \quad 2x-y+4=0$.

9. Find an equation of the line parallel to the y-axis that passes through $(2,3)$.
(a) $x=3$ (b) $x=1$ (c) $y=3$ (d) $x=2$ (e) $y=1$
Answer: medium (d)
The line passes through the point $(2,3)$ and has an undefined slope (since the line is parallel to the y-axis). Hence, the abscissa of the line is a constant 2 and the ordinate is arbitrary. Therefore, the equation of the line is $x=2$.

10. Find an equation of the line with y-intercept 3 and that is parallel to the line $x + 2y + 5 = 0$.

(a) $x + 2y - 6 = 0$ (b) $2x + 2y - 3 = 0$ (c) $3x - y - 2 = 0$ (d) $3x + 4y - 1 = 0$ (e) $x + y - 2 = 0$

Answer: medium (a)

$x + 2y + 5 = 0 \quad \Leftrightarrow \quad 2y = -(x+5) \quad \Leftrightarrow \quad y = -\frac{1}{2}(x+5)$ which is a line with slope $m = -\frac{1}{2}$. Since the unknown line is parallel to this line, it also has slope $m = -\frac{1}{2}$. Substituting for m and $b = 3$ into $y = mx + b$ gives $y = -\frac{1}{2}x + 3 \quad \Leftrightarrow$ $x + 2y - 6 = 0$

11. Find an equation of the line that is perpendicular to the line $2x - 3y = 1$ and that passes through $\left(\frac{1}{4}, -\frac{3}{5}\right)$.

(a) $6x + 4y - 1 = 0$ (b) $6x - 4y - 1 = 0$ (c) $3x + 2y - 9 = 0$ (d) $8x + y - 3 = 0$ (e) $60x + 40y + 9 = 0$

Answer: hard (e)

$2x - 3y = 1 \quad \Leftrightarrow \quad 3y = 2x - 1 \quad \Leftrightarrow \quad y = \frac{2}{3}x - \frac{1}{3}$. Since our unknown line is perpendicular to this line it must have slope $m = -\frac{3}{2}$ and, in addition, it passes through the point $\left(\frac{1}{4}, -\frac{3}{5}\right)$. Substituting into $y - y_1 = m(x - x_1)$ gives $y + \frac{3}{5} = -\frac{3}{2}\left(x - \frac{1}{4}\right) \quad \Leftrightarrow \quad y + \frac{3}{5} = -\frac{3}{2}x + \frac{3}{8} \quad \Leftrightarrow \quad 60x + 40y + 9 = 0.$

12. (a) Sketch the line with slope -3 that passes through the point $(5, -2)$.

(b) Find the equation of this line.

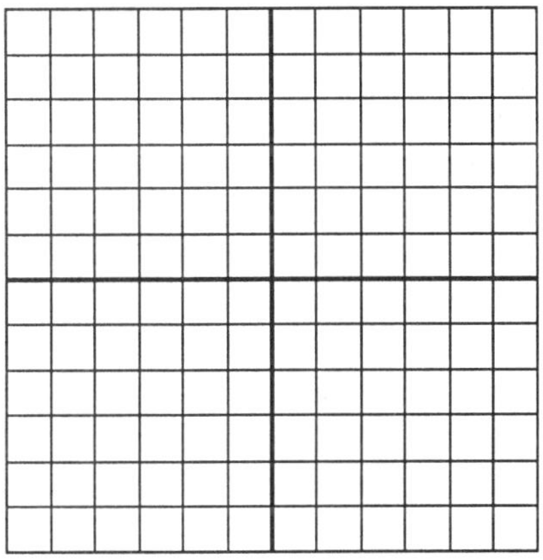

Answer: medium

(a)

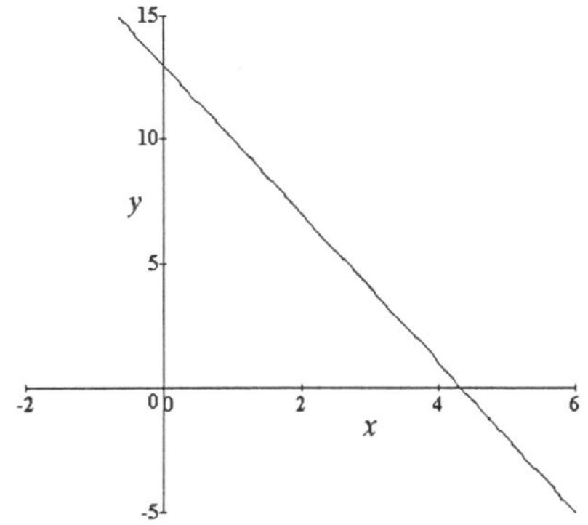

(b) Substituting $(x_1, y_1) = (5, -2)$ and $m = -3$ into $y - y_1 = m(x - x_1)$ gives $y + 2 = -3(x - 5)$ $\Leftrightarrow \quad y + 2 = -3x + 15 \quad \Leftrightarrow \quad 3x + y - 13 = 0$

13. Find the slope and y-intercept of the line $3x - 4y = 5$ and draw its graph.

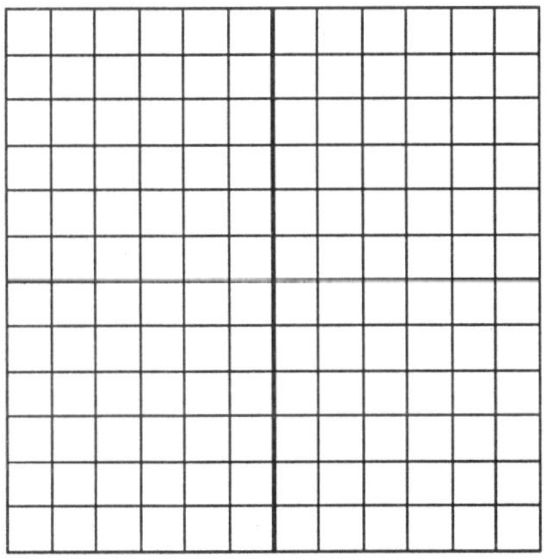

Answer: medium

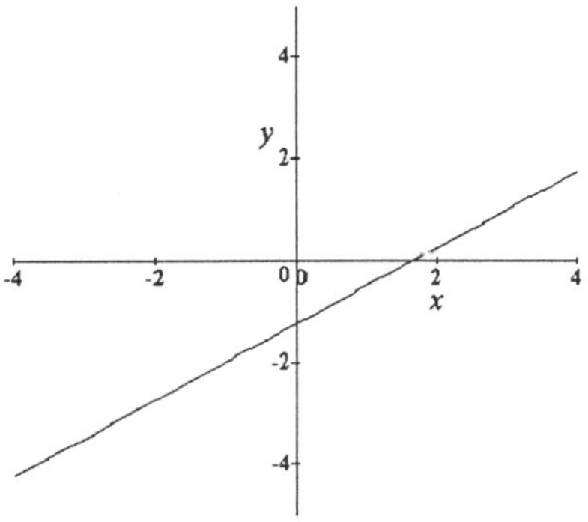

$3x - 4y = 5 \Leftrightarrow y = \frac{3}{4}x - \frac{5}{4} \Rightarrow m = \frac{3}{4}$ and $b = -\frac{5}{4}$

14. Find the slope and y-intercept of the line $3x - 2y = 0$ and draw its graph.

Answer: medium

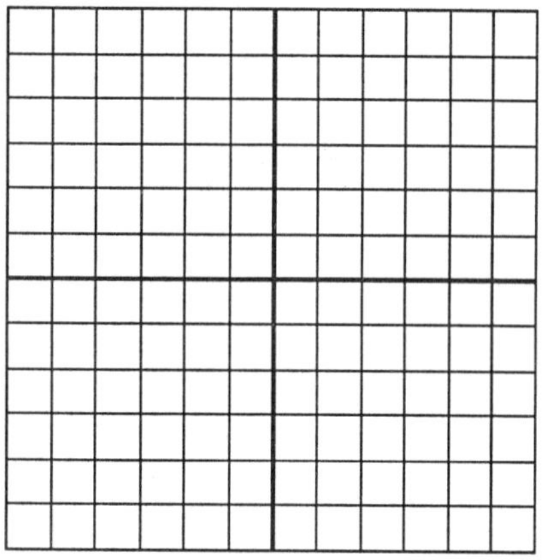

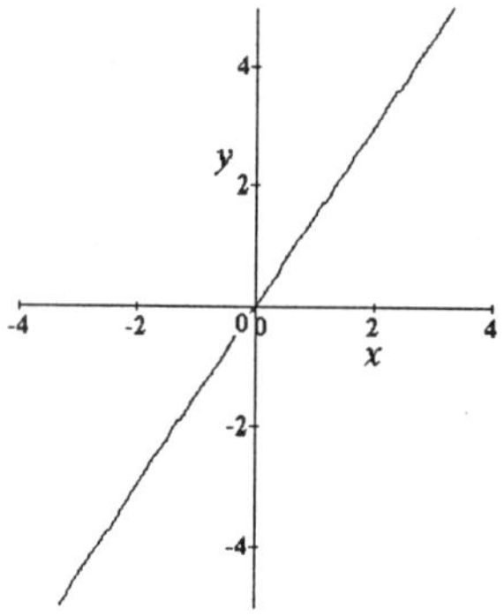

$3x - 2y = 0 \quad \Leftrightarrow \quad y = \frac{3}{2}x \Rightarrow m = \frac{3}{2}$ and $b = 0$

15. Find the slope and y-intercept of the line $-4x - 3y + 25 = 0$ and draw its graph.

Answer: medium

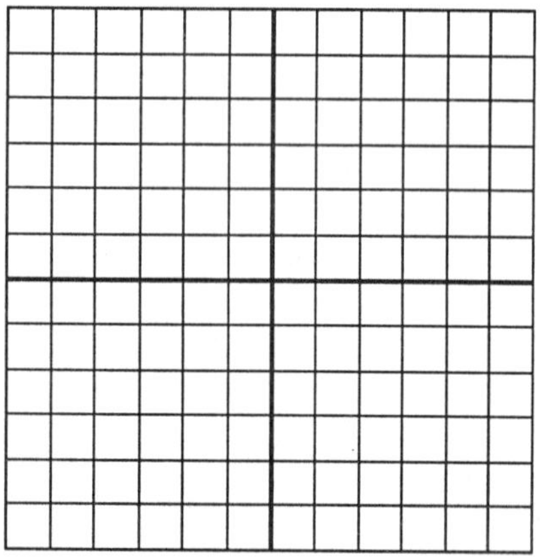

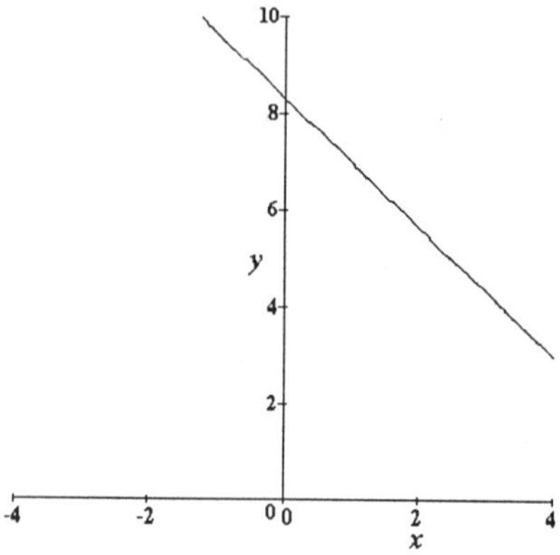

$-4x - 3y + 25 = 0 \Leftrightarrow y = -\frac{4}{3}x + \frac{25}{3} \Rightarrow$

$m = -\frac{4}{3}$ and $b = \frac{25}{3}$

16. Find the slope and y-intercept of the line $3y + 9 = 0$ and draw its graph.

Answer: medium

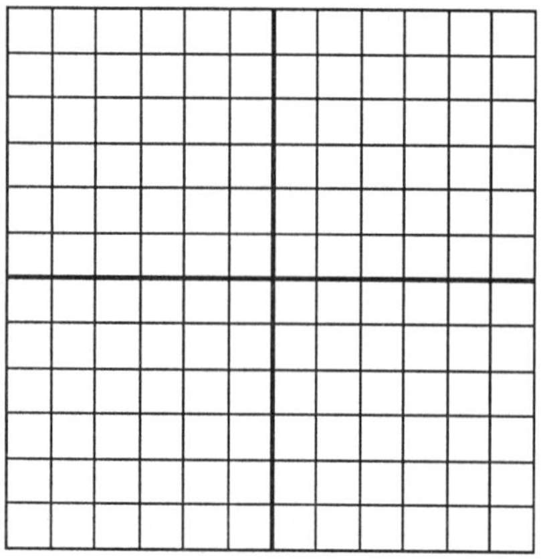

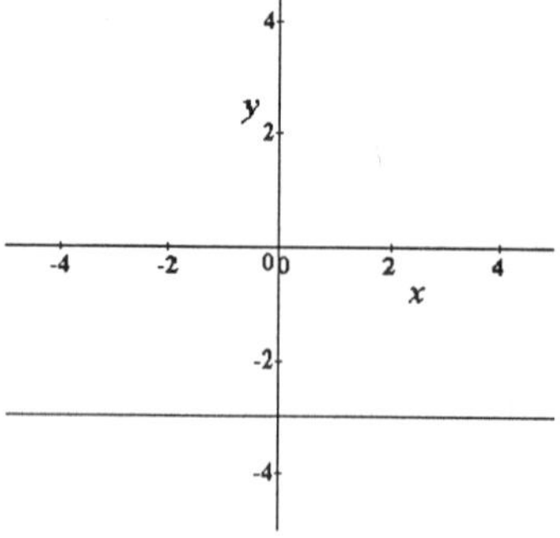

$3y + 9 = 0 \quad \Leftrightarrow \quad y = -3 \Rightarrow m = 0$ and $b = -3$

17. Find the slope and y-intercept of the line $y = 3.5$ and draw its graph.

Answer: medium

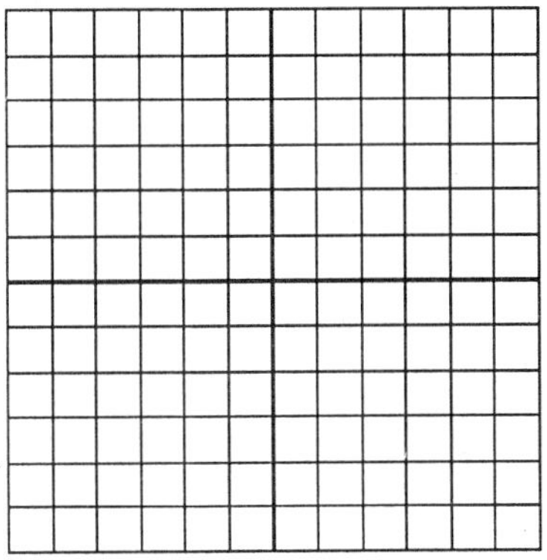

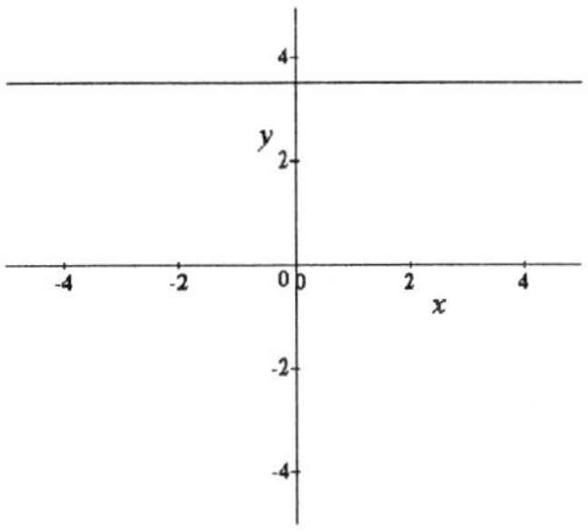

$y = 3.5 \Rightarrow m = 0$ and $b = 3.5$

18. Find the slope and y-intercept of the line $3x + 4y = 8$ and draw its graph.

Answer: medium

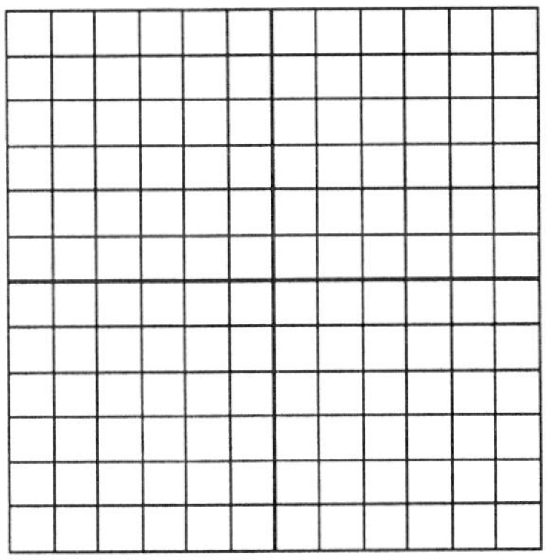

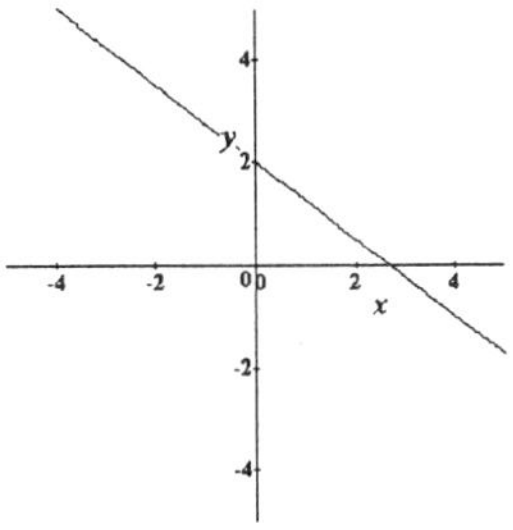

$3x + 4y = 8 \quad \Leftrightarrow \quad 4y = -3x + 8 \quad \Leftrightarrow \quad y = -\frac{3}{4}x + 2 \Rightarrow$

$m = -\frac{3}{4}$ and $b = 2$

19. Show that $A(-2,-2)$, $B(4,2)$, and $C(-8,7)$ are vertices of a right triangle.

Answer: hard

Side AB has slope $m_1 = \dfrac{2+2}{4+2} = \dfrac{2}{3}$. Side BC has slope $m_2 = \dfrac{7-2}{-8-4} = -\dfrac{5}{12}$. Side CA has slope $m_3 = \dfrac{7+2}{-8+2} = -\dfrac{3}{2}$.
Since $m_1 \cdot m_3 = \frac{2}{3}\left(-\frac{3}{2}\right) = -1$, then $AB \perp CA$ and so triangle ABC is a right triangle.

20. Use slopes to determine whether the given points are collinear (lie on a line).

(a) $(5,-2)$, $(7,6)$, and $(10,18)$

(b) $(3,0)$, $(5,4)$, and $(8,12)$

Answer: hard

(a) $P_1 = (5,-2)$ $P_2 = (7,6)$ $P_3 = (10,18)$. Then $m_{12} = \frac{6+2}{7-5} = 4$, $m_{23} = \frac{18-6}{10-7} = 4$, and $m_{13} = \frac{18+2}{10-5} = 4$. Since $m_{12} = m_{23} = m_{13}$, the points are collinear.

(b) $P_1 = (3,0)$ $P_2 = (5,4)$ $P_3 = (8,12)$. Then $m_{12} = \frac{4-0}{5-3} = 2$ and $m_{23} = \frac{12-4}{8-5} = \frac{8}{3}$. Since , $m_{12} \neq m_{23}$, the points are not collinear.

21. (a) Find an equation for the line tangent to the circle $x^2 + y^2 = 9$ at the point $\left(1, 2\sqrt{2}\right)$.

(b) At what other point on the circle will the tangent line be parallel to the line in part (a)?

Answer: medium

(a) Since the circle $x^2 + y^2 = 9$ has center $(0,0)$ and radius 3, then the slope of the line passing through the center $(0,0)$ and the point $\left(1, 2\sqrt{2}\right)$ is $\frac{2\sqrt{2}-0}{1-0} = 2\sqrt{2}$. Hence the slope of the tangent line at $\left(1, 2\sqrt{2}\right)$ is $-\frac{\sqrt{2}}{4} \Rightarrow$
$y - 2\sqrt{2} = -\frac{\sqrt{2}}{4}(x-1) \Leftrightarrow \sqrt{2}x + 4y - 9\sqrt{2} = 0$

(b) Since diametrically opposite points on the circle have parallel tangent lines, the other point is $\left(-1, -2\sqrt{2}\right)$.

22. The manager of a weekend flea market knows from past experience that if he charges x dollars for a space at the flea market, then the number y of spaces he can sell is given by the equation $y = 400 - 6x$.

(a) Sketch a graph of this linear equation. (Remember that the charge per space and the number of spaces sold must both be non-negative quantities).

(b) What do the slope, the y-intercept, and the x-intercept of the graph represent?

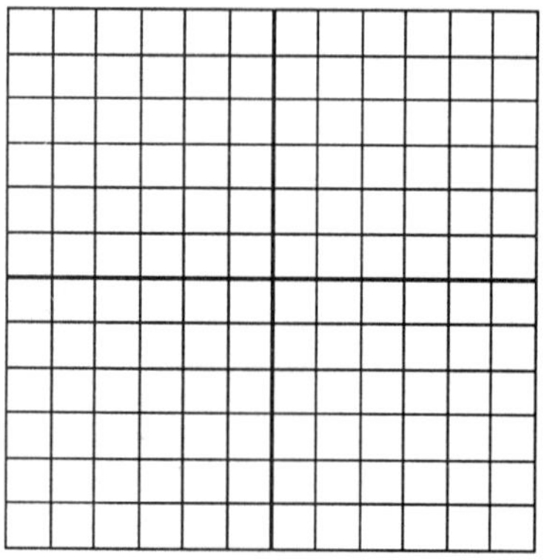

Answer: medium

(a) $y = 400 - 6x$

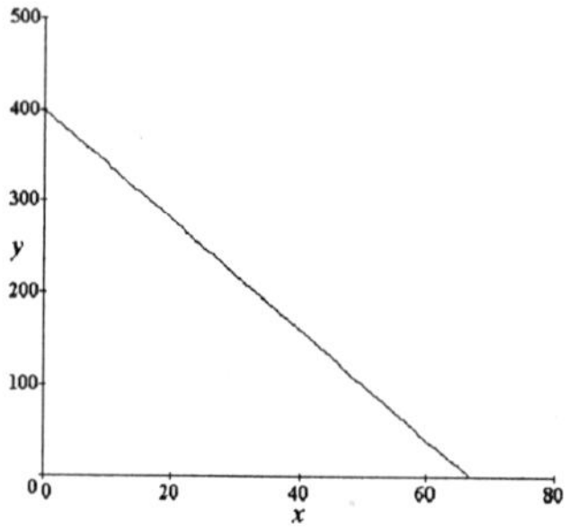

(b) The slope of the line is -6 and it represents that 6 fewer spaces will be sold for each dollar increase in the fee. The y-intercept represents the total number of spaces that the manager has. The x-intercept represents the charge at which he won't sell any spaces.

23. Joe and Chris leave Vancouver at 12:30 P.M. and drive at a constant speed south along the I-5. They pass Seattle, 230 km from Vancouver, at 2:48 P.M.

(a) Express the distance traveled in terms of the time elapsed.

(b) Draw the graph of the equation in part (a).

(c) What is the slope of this line? What does it represent?

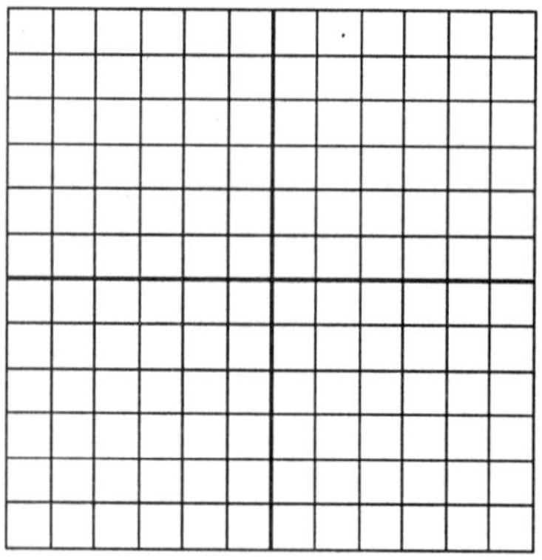

Answer: medium

(a) Let t represent the time in hours and d the distance in kilometers. Given a linear relationship between d and t, then $(0,0)$ and $(2.3,230)$ are two points on the line. Hence the slope of the line is $m = \frac{230-0}{2.3} = \frac{230}{2.3} = 100$ and the equation is $d = 100t$.

(b)

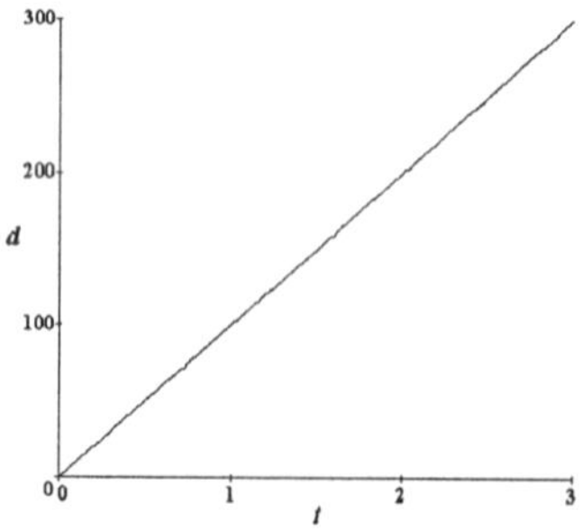

(c) The slope of the line is 100 and it represents the speed in km/h.

24. The manager of a furniture factory finds that it costs \$2000 to manufacture 100 chairs in a day, and \$5400 to produce 300 chairs in a day.

(a) Assuming that the relationship between cost and the number of chairs produced is linear, find an equation that expresses this relationship, and graph the equation.

(b) What is the slope of the line of part (a), and what does it represent?

(c) What is the y-intercept of this line, and what does it represent?

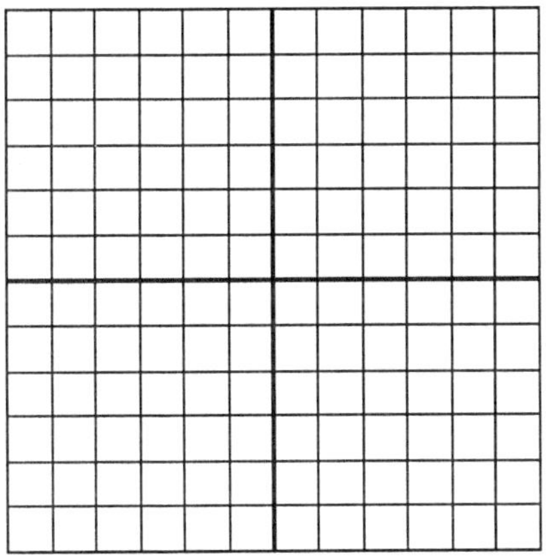

Answer: medium

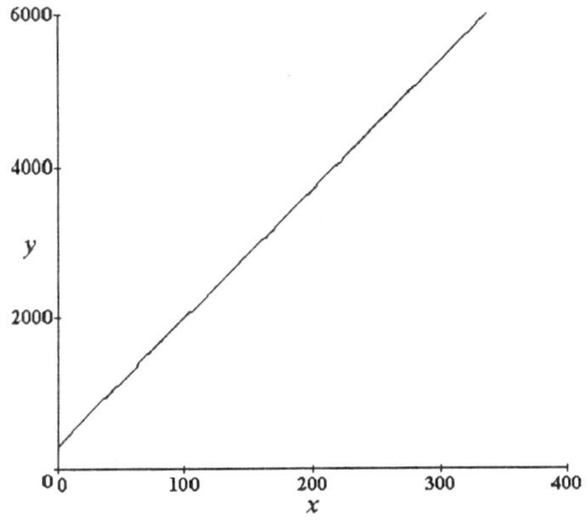

(a) Let y be the cost of producing chairs in dollars and x be the number of chairs. Given the linear relationship between x and y, then $(100, 2000)$ and $(300, 5400)$ are two points on the line. Hence the slope is $\frac{5400-2000}{300-100} = 17$ and the equation is $y - 2000 = 17(x - 100) \quad \Leftrightarrow \quad y = 17x + 300$.

(b) The slope of the line in part (a) is 17 and it represents the increase in cost of producing each additional chair.

(c) The y-intercept of the line is 300 and it represents the fixed daily costs of operating the factory.

25. (a) Draw the graphs of the supply equation $y = 0.3p + 100$ and the demand equation $y = -0.4p + 600$ in an appropriate viewing rectangle.

(b) Find the equilibrium point of the equations in part (a) from the graph.

(c) Find the price and the amount of the commodity produced and sold at equilibrium.

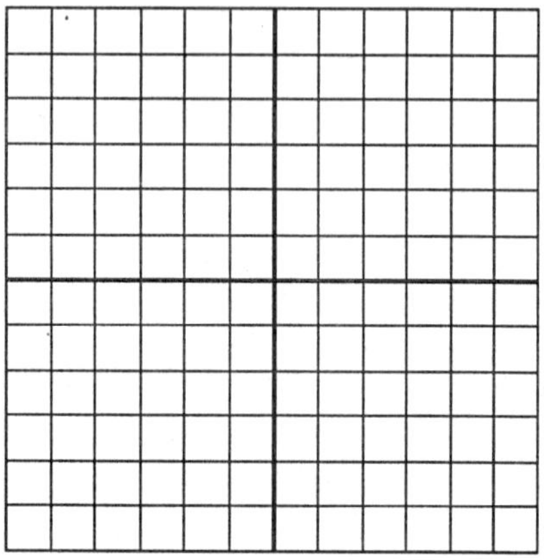

Answer: medium

(a), (b)

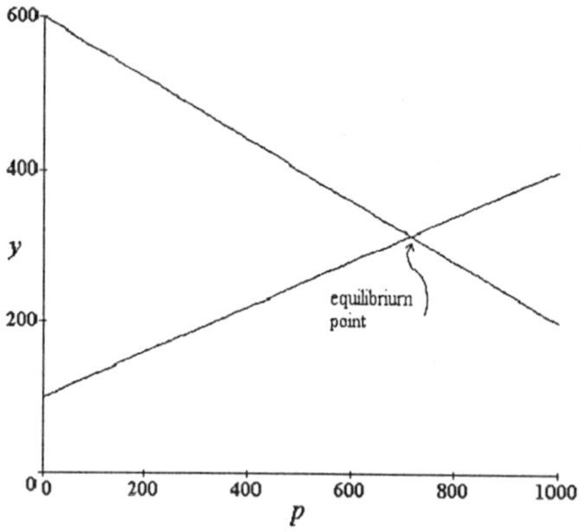

(c) We solve the equation $0.3p + 100 = -0.4p + 600$ $\Leftrightarrow$ $0.7p = 500$ $\Leftrightarrow$ $p \approx 714.29$ and thus $y \approx 0.3(714.29) + 100 \approx 314$.

26. Find the slope of the line through $P(-1, 1)$ and $Q(-2, 10)$.

(a) -9 (b) $-\frac{1}{9}$ (c) Undefined (d) 10 (e) -8

Answer: easy (a)

slope $= \frac{10-1}{-2-(-1)} = -9$

27. Find an equation of a vertical line passing through the point $(-1, -2)$.

(a) $y = -x - 2$ (b) $x = -y$ (c) Undefined (d) $x = -1$ (e) $y = -1$

Answer: easy (d)

The vertical line passing through the point $(-1, y)$ has equation $x = -1$.

28. Find an equation of the line that is parallel to $3x + 11y + 14 = 0$ and passes through the point $\left(-\frac{1}{2}, 6\right)$.

(a) $y = \frac{3}{11}x + 33$ (b) $y = 22 - \frac{3}{11}x$ (c) $y = \frac{64}{11} - x$ (d) $y = \frac{3}{11}x + \frac{129}{22}$ (e) $y = \frac{129}{22} - \frac{3}{11}x$

Answer: easy (e)

$3x + 11y + 14 = 0 \quad \Leftrightarrow \quad y = -\frac{3}{11}x - \frac{14}{11} \quad \Rightarrow \quad m = -\frac{3}{11}$. An equation of the line passing through the point $\left(-\frac{1}{2}, 6\right)$ and parallel to $3x + 11y + 14 = 0$ has the same slope $m = -\frac{3}{11} \quad \Rightarrow \quad y - 6 = -\frac{3}{11}\left(x + \frac{1}{2}\right) \quad \Leftrightarrow \quad y = \frac{129}{22} - \frac{3}{11}x$.

29. Find an equation of the line that is perpendicular to $4x + 12y - 3 = 0$ and passes through the point $(2, 3)$.

(a) $y = 3x - \frac{3}{2}$ (b) $y = 3x - 3$ (c) $y = 2x - 3$ (d) $y = \frac{1}{3}x - 2$ (e) $y = -\frac{1}{3}x + \frac{1}{4}$

Answer: easy (b)

$4x + 12y - 3 = 0 \quad \Leftrightarrow \quad y = -\frac{1}{3}x + \frac{1}{4} \quad \Rightarrow \quad m = -\frac{1}{3}$. An equation of the line passing through the point $(2, 3)$ and perpendicular to $4x + 12y - 3 = 0$ has slope $m_p = -\frac{1}{-1/3} = 3$. Thus $y - 3 = 3(x - 2) \quad \Leftrightarrow \quad y = 3x - 3$.

30. Find an equation of the line that passes through the point $\left(-7, \frac{7}{2}\right)$ and the midpoint of $(-2, 4)$ and $(3, 4)$.

(a) $30y - 2x - 119 = 0$

(b) $2y - 30x = 0$

(c) $40y - 120x - 119 = 0$

(d) $3y - 2x - 19 = 0$

(e) $30y - 40x - 119 = 0$

Answer: medium (a)

The midpoint of $(-2, 4)$ and $(3, 4)$ is $\left(\frac{-2+3}{2}, \frac{4+4}{2}\right) = \left(\frac{1}{2}, 4\right)$. The slope of the line passing through $\left(\frac{1}{2}, 4\right)$ and $\left(-7, \frac{7}{2}\right)$ is $m = \dfrac{\frac{7}{2} - 4}{-7 - \frac{1}{2}} = \frac{1}{15}$. Therefore an equation is $y - \frac{7}{2} = \frac{1}{15}(x + 7) \quad \Leftrightarrow \quad y = \frac{1}{15}x + \frac{119}{30}$ or $30y - 2x - 119 = 0$.

31. Find a line that passes through the midpoint of $(8, -12)$ and $(7, 10)$ and is perpendicular to the line joining these two points.

(a) $4y - 22x + 59 = 0$

(b) $2y - 4x = 59$

(c) $44y - 2x + 59 = 0$

(d) $x - 22y = 164$

(e) $22x - y - 164 = 0$

Answer: medium (c)

The midpoint of $(8, -12)$ and $(7, 10)$ is $\left(\frac{8+7}{2}, \frac{-12+10}{2}\right) = \left(\frac{15}{2}, -1\right)$. The slope of the line that passes through $(8, -12)$ and $(7, 10)$ is $m = \frac{10+12}{7-8} = -22$. Therefore the perpendicular line that passes through $\left(\frac{15}{2}, -1\right)$ has equation $y - (-1) = \frac{1}{22}\left(x - \frac{15}{2}\right) \quad \Leftrightarrow \quad y = \frac{1}{22}x - \frac{59}{44}$ or $44y - 2x + 59 = 0$.

32. Find the area of a triangle formed by the coordinate axes and the line $\frac{1}{2}x + 2y - 5 = 0$.

Answer: medium

Since the base of the triangle is 10 units (the distance from the origin to its x-intercept) and its height is $\frac{5}{2}$ (the distance from the origin to its y-intercept), its area is $A = \frac{1}{2}bh = \frac{1}{2}(10)\left(\frac{5}{2}\right) = \frac{25}{2}$ units^2.

33. A particular mountain highway has a 5% grade. While making your ascent to the top of the mountain, you notice that an elevation sign indicates that you are at an elevation of 2342 ft. (Assume this means a vertical distance from the base where you started and not from sea level). Find the change in your horizontal distance in miles (round to two significant digits).

Answer: hard

Solve for the change in horizontal distance. The slope of the mountain is $\dfrac{\text{change in vertical distance } \Delta y}{\text{change in horizontal distance } \Delta x}$, so $\dfrac{5}{100} = \dfrac{2342}{\Delta x}$

$\Leftrightarrow \quad \Delta x = 46{,}840$ ft. Therefore, the change in horizontal distance is $46{,}840 \text{ ft} \times \dfrac{1 \text{ mile}}{5280 \text{ ft}} \approx 8.8$ miles.

CHAPTER 2 REVIEW

1. For the points $P(1, 1)$ and $Q(-6, 13)$:
   (a) plot the points P and Q on a coordinate plane.
   (b) find the distance from P to Q.
   (c) find the midpoint of the segment PQ.
   (d) Sketch the line determined by P and Q, and find its equation in slope-intercept form.
   (e) Sketch the circle with center P that contains the point Q, and find the equation of this circle.

Answer: medium

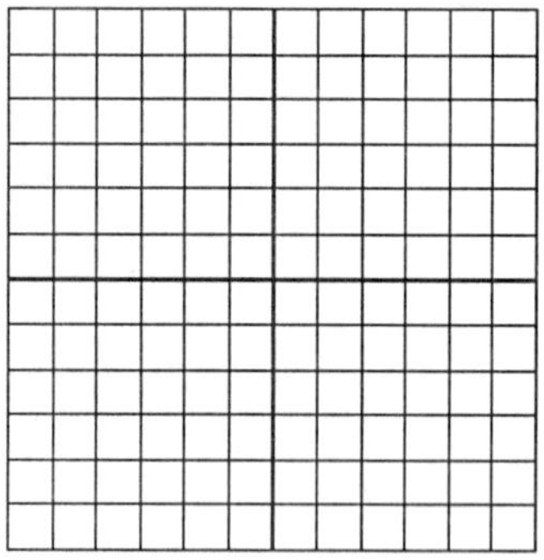

(a), (d), (e)

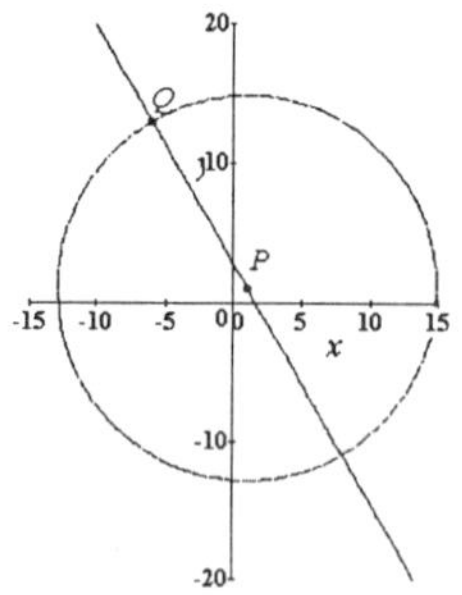

The slope m of the line is $\dfrac{13-1}{-6-1} = -\dfrac{12}{7}$, so $y - 1 = -\frac{12}{7}(x-1) \quad \Leftrightarrow \quad y = -\frac{12}{7}x + \frac{19}{7}$. The equation of the circle is $(x-1)^2 + (y-1)^2 = 193$ [see part (b)]

(b) $P = (1, 1)$ and $Q = (-6, 13)$ so $PQ = \sqrt{(1+6)^2 + (1-13)^2} = \sqrt{193}$

(c) Midpoint $= \left(\dfrac{1-6}{2}, \dfrac{1+13}{2}\right) = \left(-\dfrac{5}{2}, 7\right)$

2. For the points $P(0, 7)$ and $Q(-5, 0)$:
   (a) plot the points P and Q on a coordinate plane.
   (b) find the distance from P to Q.
   (c) find the midpoint of the segment PQ.
   (d) Sketch the line determined by P and Q, and find its equation in slope-intercept form.
   (e) Sketch the circle with center P that contains the point Q, and find the equation of this circle.

Answer: medium

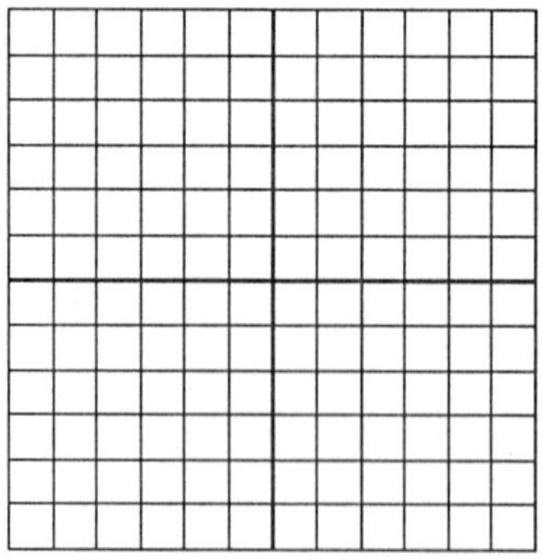

(a), (d), (e)

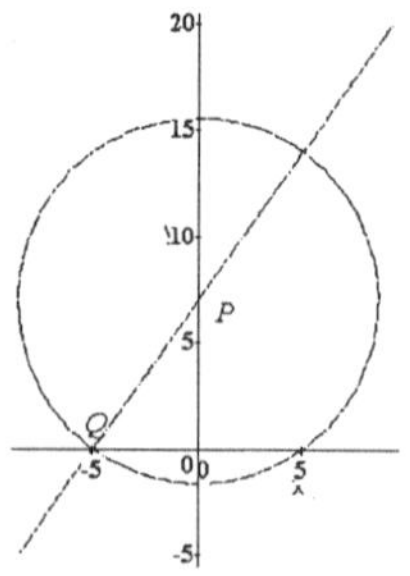

The slope m of the line is $\dfrac{0-7}{-5-0} = \dfrac{7}{5}$, so $y - 7 = \frac{7}{5}(x-0) \quad \Leftrightarrow \quad y = \frac{7}{5}x + 7$. The equation of the circle is $(x-0)^2 + (y-7)^2 = 74 \quad \Leftrightarrow \quad x^2 + (y-7)^2 = 74$ [see part (b)]

(b) $P = (0, 7)$ and $Q = (-5, 0)$ so $PQ = \sqrt{(0+5)^2 + (7-0)^2} = \sqrt{74}$

(c) Midpoint $= \left(\dfrac{0-5}{2}, \dfrac{7+0}{2}\right) = \left(-\dfrac{5}{2}, \dfrac{7}{2}\right)$

3. Sketch the region consisting of all points (x, y) such that $x \geq 4$ and $y \geq 2$.

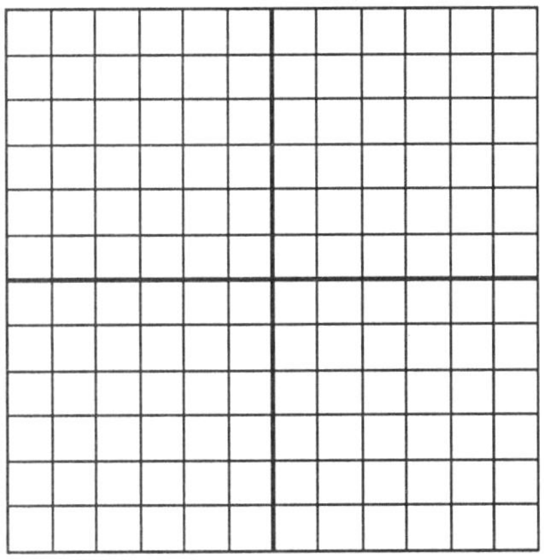

Answer: medium

$\{(x, y) \mid x \geq 4 \text{ and } y \geq 2\}$

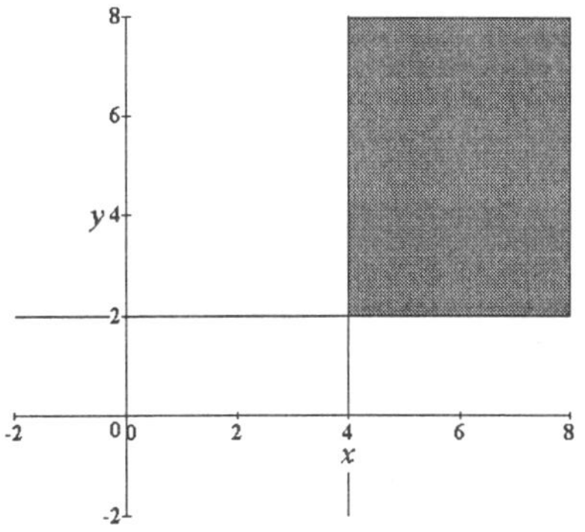

4. (a) Show that the points $A(2,4)$, $B(3,8)$, and $C(10,2)$ are the vertices of a right triangle.
(b) Find the area of the triangle.
Answer: hard
(a) $A = (2,4)$, $B = (3,8)$, and $C = (10,2)$. $|AB| = \sqrt{(3-2)^2 + (8-4)^2} = \sqrt{1^2 + 4^2} = \sqrt{17}$
$|BC| = \sqrt{(10-3)^2 + (2-8)^2} = \sqrt{7^2 + 6^2} = \sqrt{85}$, $|AC| = \sqrt{(10-2)^2 + (2-4)^2} = \sqrt{8^2 + 2^2} = \sqrt{68}$
Since $|AB|^2 + |AC|^2 = 17 + 68 = 85 = |BC|^2$, then by the converse of the Pythagorean Theorem, triangle ABC is a right triangle.
(b) The area of triangle ABC is $\frac{1}{2}bh = \frac{1}{2}|AB||AC| = \frac{1}{2}\sqrt{17} \cdot \sqrt{68} = \frac{1}{2}\sqrt{17} \cdot 2\sqrt{17} = 17$.

5. Find an equation of the circle that has center $(5,1)$ and passes through the origin.
Answer: medium
$C = (5,1)$ and $P = (0,0)$ is a point on the circle. Then the radius of the circle is $r = |PC| = \sqrt{(0-5)^2 + (0-1)^2} = \sqrt{26}$.
Thus the equation of the circle is $(x-5)^2 + (y-1)^2 = 26$.

6. Find an equation for the circle that contains the points $P(-1,3)$ and $Q(1,-2)$ as the endpoints of a diameter.
Answer: medium
$P(-1,3)$ and $Q(1,-2)$, midpoint of $PQ = \left(\dfrac{-1+1}{2}, \dfrac{3-2}{2}\right) = \left(0, \frac{1}{2}\right)$, thus the center is at $\left(0, \frac{1}{2}\right)$.
$|PQ| = \sqrt{(-1-1)^2 + (3+2)^2} = \sqrt{29}$, thus the radius of the circle is $\frac{\sqrt{29}}{2}$. Therefore the equation of the circle is $x^2 + \left(y - \frac{1}{2}\right)^2 = \frac{29}{4}$.

7. Determine whether the equation $4x^2 + 4y^2 - 4x + 16y = 1$ represents a circle, a point, or has no graph. In case the equation represents a circle, find its center and radius.
Answer: medium
$4x^2 + 4y^2 - 4x + 16y = 1 \Leftrightarrow \left(x - \frac{1}{2}\right)^2 + (y+2)^2 = \frac{9}{2} \Rightarrow$ it is a circle whose center is $\left(\frac{1}{2}, -2\right)$ and radius is $\frac{3}{\sqrt{2}}$.

8. Determine whether the equation $2x^2 + 2y^2 - 12x - 20y + 68 = 0$ represents a circle, a point, or has no graph. In case the equation represents a circle, find its center and radius.
Answer: medium
$2x^2 + 2y^2 - 12x - 20y + 68 = 0 \Leftrightarrow (x-3)^2 + (y-5)^2 = 0 \Rightarrow$ it is the point $(3,5)$.

9. Find an equation for the line that has x-intercept 3 and y-intercept 9.
Answer: easy
x-intercept is 3 and y-intercept is 9 $\Rightarrow m = \dfrac{9-0}{0-3} = -3$. Therefore the equation of the line is $y = -3x + 9$

10. Find an equation of the line that passes through the point $(2, 8)$ and is perpendicular to the line $x - 2y + 12 = 0$.

Answer: medium

The line $x - 2y + 12 = 0 \quad \Leftrightarrow \quad 2y = x + 12 \quad \Leftrightarrow \quad y = \frac{1}{2}x + 6$ has slope $\frac{1}{2}$ and hence any perpendicular to this line has slope $m = -2$. Then the line that is perpendicular and passes through the point $(2, 8)$ has equation $y - 8 = -2(x - 2) \quad \Leftrightarrow$ $y = -2x + 12 \quad \Leftrightarrow \quad 2x + y - 12 = 0$

11. Sketch the graph of the equation $x - 2y + 4 = 0$ and test for symmetry.

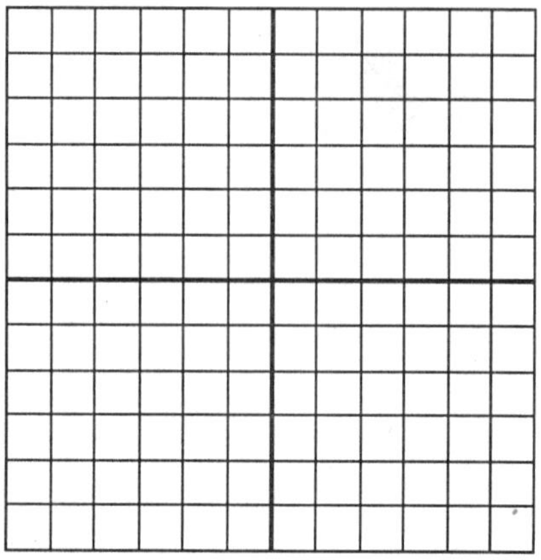

Answer: medium

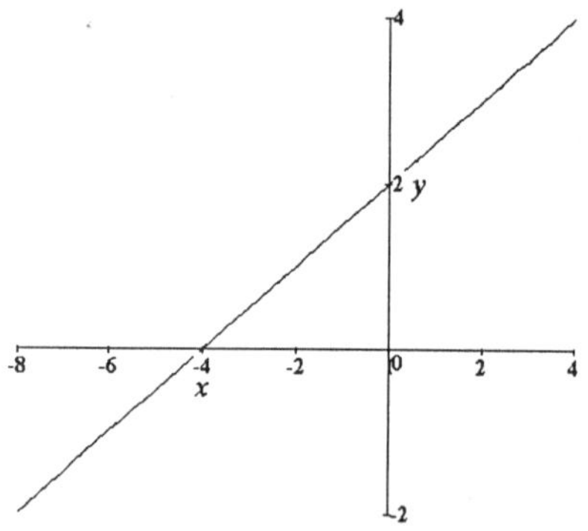

$x - 2y + 4 = 0 \quad \Leftrightarrow \quad y = \frac{1}{2}x + 2$. Symmetry: wrt x-axis $-y = \frac{1}{2}x + 2 \quad \Leftrightarrow \quad y = -\frac{1}{2}x - 2$ which is changed, wrt y-axis $y = \frac{1}{2}(-x) + 2 \quad \Leftrightarrow \quad y = -\frac{1}{2}x + 2$ which is changed, wrt origin $-y = \frac{1}{2}(-x) + 2 \quad \Leftrightarrow \quad y = \frac{1}{2}x - 2$ which is changed; and so no symmetry.

12. Test the equation for symmetry and sketch its graph. $x = 3y + 2$

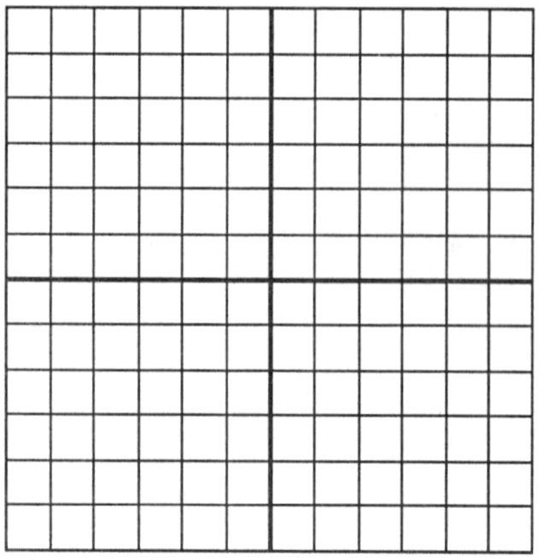

Answer: medium

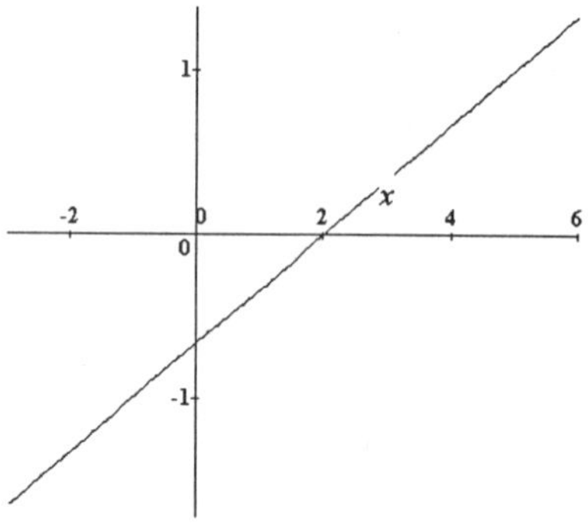

$x = 3y + 2 \quad \Leftrightarrow \quad y = \frac{1}{3}x - \frac{2}{3}$

Symmetry: (wrt x-axis?) $-y = \frac{1}{3}x - \frac{2}{3} \quad \Leftrightarrow \quad y = -\frac{1}{3}x + \frac{2}{3}$, which is changed, (wrt y-axis?) $y = \frac{1}{3}(-x) - \frac{2}{3} \quad \Leftrightarrow$ $y = -\frac{1}{3}x - \frac{2}{3}$, which is changed, (wrt origin?) $-y = -\frac{1}{3}x - \frac{2}{3}$ $\Leftrightarrow \quad y = \frac{1}{3}x + \frac{2}{3}$, which is changed, and so no symmetry.

13. Test the equation for symmetry and sketch its graph. $\frac{x}{3} + \frac{y}{2} = 0$

Answer: medium

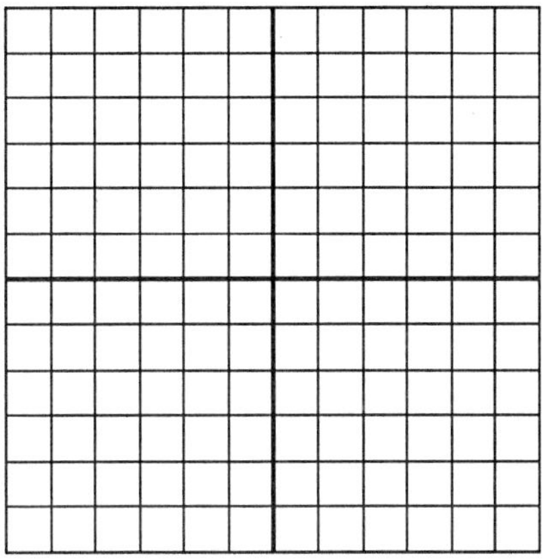

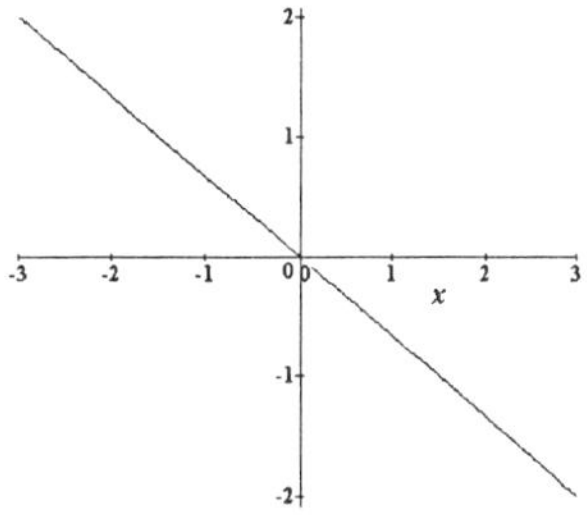

$\frac{x}{3} + \frac{y}{2} = 0 \quad \Leftrightarrow \quad y = -\frac{2}{3}x.$

Symmetry: (wrt x-axis?) $-y = -\frac{2}{3}x \quad \Leftrightarrow \quad y = \frac{2}{3}x,$ which is changed, (wrt y-axis?) $y = -\frac{2}{3}(-x) \quad \Leftrightarrow \quad y = \frac{2}{3}x,$ which is changed, (wrt origin?) $-y = -\frac{2}{3}(-x) \quad \Leftrightarrow \quad y = -\frac{2}{3}x,$ which is not changed, and so symmetry about the origin.

14. Test the equation for symmetry and sketch its graph. $4x - y^2 = 0$

Answer: medium

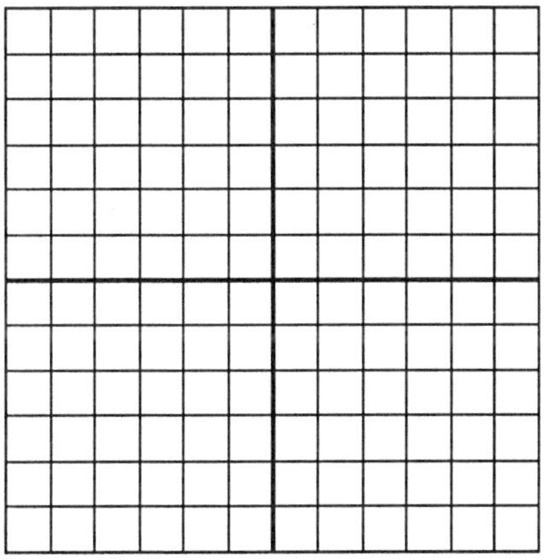

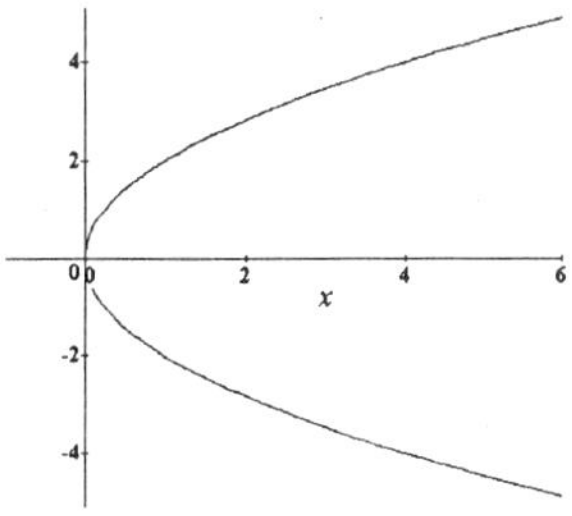

$4x - y^2 = 0$. Symmetry: wrt x-axis $4x - (-y)^2 = 0 \quad \Leftrightarrow$ $4x - y^2 = 0$ which is not changed, wrt y-axis $4(-x) - y^2 = 0$ $\Leftrightarrow \quad y^2 + 4x = 0$ which is changed, wrt origin $4(-x) - (-y)^2 = 0 \quad \Leftrightarrow \quad y^2 + 4x = 0$ which is changed. So, $4x - y^2 = 0$ is symmetric wrt x-axis.

15. Test the equation for symmetry and sketch its graph. $y = -\sqrt{4 - x^2}$

Answer: medium

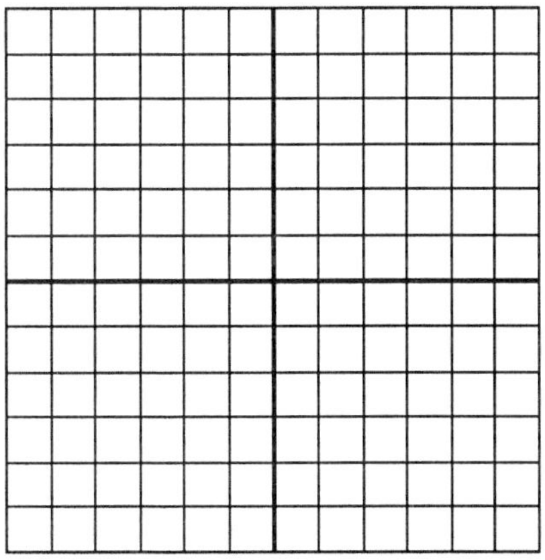

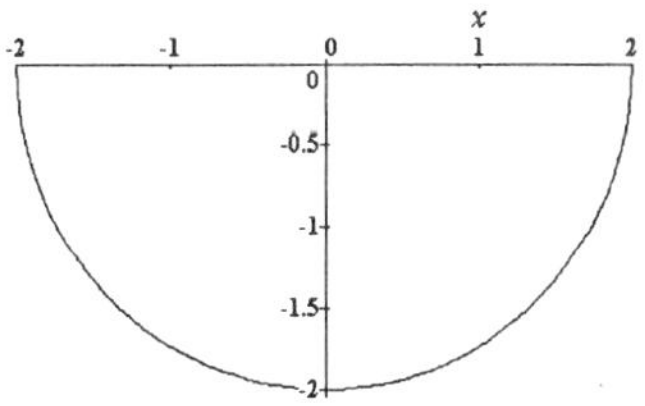

$y = -\sqrt{4 - x^2}$. Symmetry: wrt x-axis $-y = -\sqrt{4 - x^2}$ $\Leftrightarrow \quad y = \sqrt{4 - x^2}$ which is changed, wrt y-axis $y = -\sqrt{4 - (-x)^2} \quad \Leftrightarrow \quad y = -\sqrt{4 - x^2}$ which is not changed, wrt origin $-y = -\sqrt{4 - (-x)^2} \quad \Leftrightarrow \quad y = \sqrt{4 - x^2}$ which is changed, and hence there is symmetry wrt the y-axis.

16. Test the equation for symmetry and sketch its graph. $xy = 1$

Answer: medium

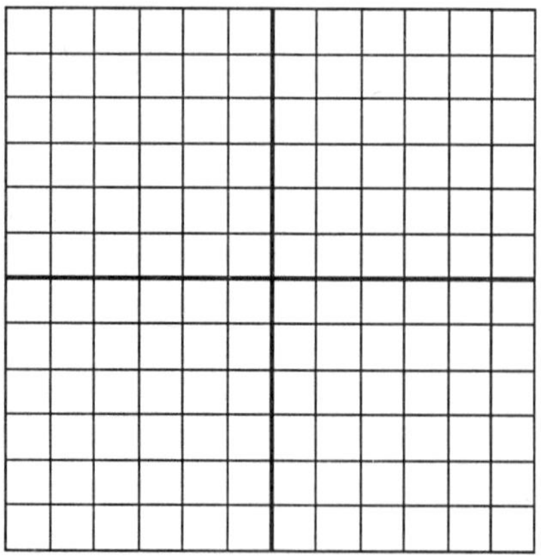

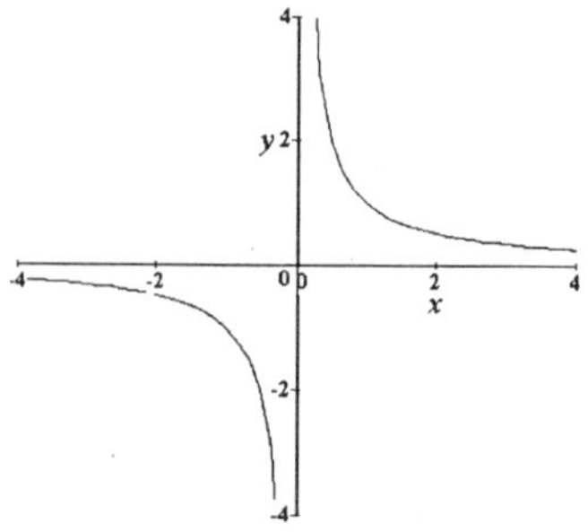

$xy = 1$

Symmetry: (wrt x-axis?) $x(-y) = 1 \quad \Leftrightarrow \quad -xy = 1$, which is changed, (wrt y-axis?) $(-x)y = 1 \quad \Leftrightarrow$ $-xy = 1$, which is changed, (wrt origin?) $(-x)(-y) = 1 \quad \Leftrightarrow$ $xy = 1$, which is not changed, and hence there is symmetry wrt to the origin.

17. Test the equation for symmetry and sketch its graph. $x = y^4 - 6y^2$

Answer: medium

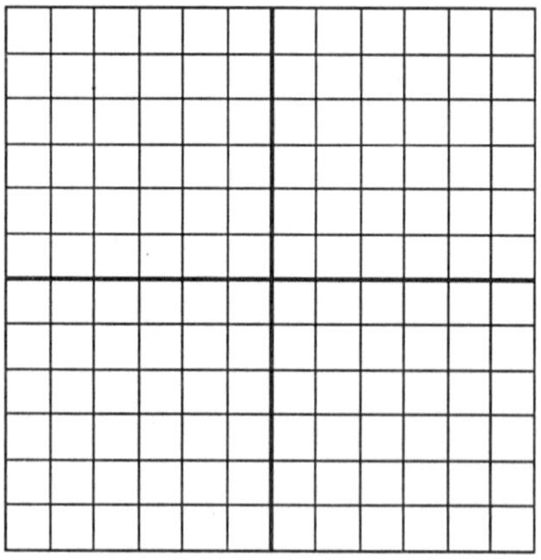

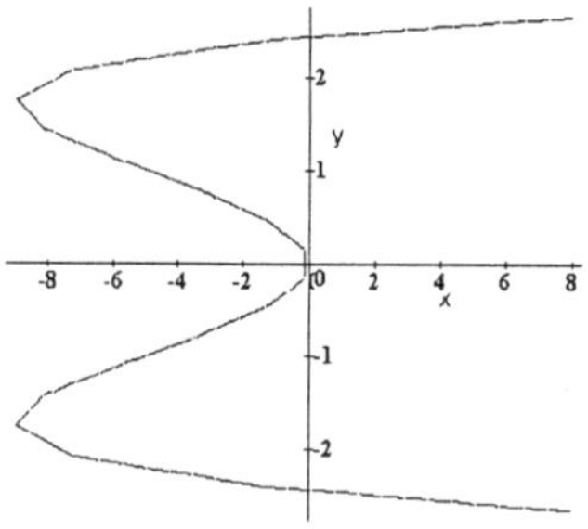

$x = y^4 - 6y^2$

Symmetry: (wrt x-axis?) $x = (-y)^4 - 6(-y)^2 \quad \Leftrightarrow$ $x = y^4 - 6y^2$, which is not changed, (wrt y-axis?) $-x = y^4 - 6y^2 \quad \Leftrightarrow \quad x = 6y^2 - y^4$, which is changed, (wrt origin?) $-x = (-y)^4 - 6(-y)^2 \quad \Leftrightarrow$ $-x = y^4 - 6y^2$, which is changed, and so there is symmetry wrt x-axis.

CHAPTER THREE

SECTION 3.1

1. Solve the equation $3x - 7 = 28$.
   (a) $x = \frac{15}{4}$ (b) $x = 11$ (c) $x = \frac{35}{3}$ (d) $x = \frac{19}{2}$ (e) $x = 13$
   Answer: easy (c)
   $3x - 7 = 28 \quad \Leftrightarrow \quad 3x = 35 \quad \Leftrightarrow \quad x = \frac{35}{3}$

2. Solve the equation $x - 3 = 4x + 33$.
   (a) $x = -12$ (b) $x = 10$ (c) $x = \frac{15}{2}$ (d) $x = -18$ (e) $x = \frac{17}{4}$
   Answer: easy (a)
   $x - 3 = 4x + 33 \quad \Leftrightarrow \quad 3x = -36 \quad \Leftrightarrow \quad x = -12$

3. Solve the equation $\dfrac{z}{14} = \dfrac{2z}{7} + 5$.
   (a) $z = 23$ (b) $z = \frac{47}{2}$ (c) $z = -\frac{49}{2}$ (d) $z = -\frac{70}{3}$ (e) $z = -\frac{47}{2}$
   Answer: easy (d)
   $\dfrac{z}{14} = \dfrac{2z}{7} + 5 \quad \Leftrightarrow \quad z = 4z + 70 \quad \Leftrightarrow \quad 3z = -70 \quad \Leftrightarrow \quad z = -\dfrac{70}{3}$

4. Solve the equation $2(x - 1) - 7 = -2(x - 2) + 3$.
   (a) $x = 4$ (b) $x = -11$ (c) $x = -\frac{67}{2}$ (d) $x = -3$ (e) $x = -1$
   Answer: easy (a)
   $2(x - 1) - 7 = -2(x - 2) + 3 \quad \Leftrightarrow \quad 2x - 9 = -2x + 7 \quad \Leftrightarrow \quad 4x = 16 \quad \Leftrightarrow \quad x = 4$

5. Solve the equation $\dfrac{4}{3}y + \dfrac{3}{2}(y - 1) = \dfrac{y + 1}{18}$.
   (a) $y = 3$ (b) $y = \frac{1}{2}$ (c) $y = \frac{3}{17}$ (d) $y = \frac{14}{25}$ (e) $y = -4$
   Answer: medium (d)
   $\dfrac{4}{3}y + \dfrac{3}{2}(y - 1) = \dfrac{y + 1}{18} \quad \Leftrightarrow \quad \dfrac{24y + 27y - 27}{18} = \dfrac{y + 1}{18} \quad \Leftrightarrow \quad 50y = 28 \quad \Leftrightarrow \quad y = \dfrac{14}{25}$

6. Solve the equation $\dfrac{x - 1}{2x + 7} = \dfrac{1}{3}$.
   (a) $x = -7$ (b) $x = 10$ (c) $x = \frac{19}{3}$ (d) $x = 27$ (e) $x = -21$
   Answer: easy (b)
   $\dfrac{x - 1}{2x + 7} = \dfrac{1}{3} \quad \Leftrightarrow \quad 3(x - 1) = 2x + 7 \quad \Leftrightarrow \quad x - 3 = 7 \quad \Leftrightarrow \quad x = 10$

7. Solve the equation $\dfrac{1}{2t - 4} + \dfrac{t}{4t - 1} = \dfrac{1}{4}$.
   (a) $t = \frac{3}{4}$ (b) $t = \frac{2}{3}$ (c) $t = \frac{4}{9}$ (d) $t = \frac{1}{2}$ (e) $t = -2$
   Answer: medium (c)
   $\dfrac{1}{2t - 4} + \dfrac{t}{4t - 1} = \dfrac{1}{4} \quad \Leftrightarrow \quad 4(4t - 1) + 4t(2t - 4) = (2t - 4)(4t - 1) \quad \Leftrightarrow \quad -4 = -2t - 16t + 4 \quad \Leftrightarrow \quad t = \dfrac{4}{9}$

8. Solve the equation $(t + 5)^2 = (t - 5)^2 + 30$.
   (a) $t = \frac{3}{2}$ (b) $t = 2$ (c) $t = -2$ (d) $t = -5$ (e) $t = -\frac{37}{106}$
   Answer: medium (a)
   $(t + 5)^2 = (t - 5)^2 + 30 \quad \Leftrightarrow \quad t^2 + 10t + 25 = t^2 - 10t + 25 + 30 \quad \Leftrightarrow \quad 20t = 30 \quad \Leftrightarrow \quad t = \frac{3}{2}$

9. Solve the equation $\frac{1}{3}x - \frac{3}{4} = \frac{5}{6}x - \frac{11}{9}$.

(a) $x = \frac{3}{4}$ (b) $x = -\frac{8}{9}$ (c) $x = 2$ (d) $x = -\frac{4}{15}$ (e) $x = \frac{17}{18}$

Answer: medium (e)

$\frac{1}{3}x - \frac{3}{4} = \frac{5}{6}x - \frac{11}{9} \quad \Leftrightarrow \quad 12x - 27 = 30x - 44 \quad \Leftrightarrow \quad 18x = 17 \quad \Leftrightarrow \quad x = \frac{17}{18}$

10. Solve the equation $\dfrac{8}{x-26} = \dfrac{4}{x+13}$.

(a) $x = -13$ (b) $x = -52$ (c) $x = -26$ (d) $x = 16$ (e) $x = 13$

Answer: easy (b)

$\dfrac{8}{x-26} = \dfrac{4}{x+13} \quad \Leftrightarrow \quad 8(x+13) = 4(x-26) \quad \Leftrightarrow \quad 104 + 8x = -104 + 4x \quad \Leftrightarrow \quad x = -52$

11. Solve the equation $\dfrac{5}{x-2} + \dfrac{3}{x+2} = \dfrac{15}{x^2-4}$.

(a) $x = \frac{13}{17}$ (b) $x = 2$ (c) $x = \frac{11}{8}$ (d) $x = \frac{7}{15}$ (e) $x = \frac{5}{4}$

Answer: medium (c)

$\dfrac{5}{x-2} + \dfrac{3}{x+2} = \dfrac{15}{x^2-4} \quad \Leftrightarrow \quad 5(x+2) + 3(x-2) = 15 \quad \Leftrightarrow \quad 8x = 11 \quad \Leftrightarrow \quad x = \dfrac{11}{8}$

12. Solve the equation $\dfrac{2x-3}{12x} = 6 - \dfrac{3}{x}$.

(a) $x = \frac{33}{70}$ (b) $x = \frac{14}{3}$ (c) $x = -\frac{11}{36}$ (d) $x = \frac{5}{6}$ (e) $x = -\frac{2}{3}$

Answer: medium (a)

$\dfrac{2x-3}{12x} = 6 - \dfrac{3}{x} \quad \Leftrightarrow \quad 2x - 3 = 12x\left(6 - \dfrac{3}{x}\right) \quad \Leftrightarrow \quad 2x - 3 = 72x - 36 \quad \Leftrightarrow \quad x = \dfrac{33}{70}$

13. Solve the equation $3x - \dfrac{x}{9} + \dfrac{x-3}{6} = x$.

(a) $x = \frac{1}{17}$ (b) $x = -\frac{4}{23}$ (c) $x = \frac{9}{37}$ (d) $x = -3$ (e) $x = \frac{1}{4}$

Answer: medium (c)

$3x - \dfrac{x}{9} + \dfrac{x-3}{6} = x \quad \Leftrightarrow \quad \dfrac{-2x + 3x - 9}{18} = -2x \Leftrightarrow \quad x - 9 = -36x \quad \Leftrightarrow \quad x = \dfrac{9}{37}$

14. Solve the equation $\dfrac{2}{3 + 1/(1-w)} = 48$.

(a) $w = \frac{3}{48}$ (b) $w = -\frac{2}{93}$ (c) $w = \frac{1}{8}$ (d) $w = \frac{48}{47}$ (e) $w = \frac{95}{71}$

Answer: medium (e)

$\dfrac{2}{3 + 1/(1-w)} = 48 \quad \Leftrightarrow \quad \dfrac{2-2w}{3-3w+1} = 48 \quad \Leftrightarrow \quad 2 - 2w = 48(4 - 3w) \quad \Leftrightarrow \quad 144w - 2w = 192 - 2 \quad \Leftrightarrow$
$w = \frac{190}{142} = \frac{95}{71}$

15. Solve the equation $\dfrac{1}{x} + \dfrac{3}{4-3x} = \dfrac{4}{4x-3x^2}$.

Answer: medium

$\dfrac{1}{x} + \dfrac{3}{4-3x} = \dfrac{4}{4x-3x^2} \quad \Leftrightarrow \quad 4 - 3x + 3x = 4 \quad \Leftrightarrow \quad 4 = 4$, so the equation is an identity. Thus the solution is $\left\{x \in \mathbb{R}\colon x \neq 0, x \neq \frac{4}{3}\right\}$.

16. Find the solution of $7.07 - 3x = 5.77x + 3.00$ correct to two decimals.

Answer: medium

$7.07 - 3x = 5.77x + 3.00 \quad \Leftrightarrow \quad 8.77x = 4.07 \quad \Leftrightarrow \quad x \approx 0.46$

17. Find the solution of $\dfrac{2.37x + 0.84}{1.01 - 3.76x} = 1.32$ correct to two decimals.

Answer: medium

$\dfrac{2.37x + 0.84}{1.01 - 3.76x} = 1.32 \quad \Leftrightarrow \quad 2.37x + 0.84 = 1.3332 - 4.9632x \quad \Leftrightarrow \quad 7.3332x = 0.4932 \quad \Leftrightarrow \quad x \approx 0.067$

18. Solve the equation $F = G\dfrac{mM}{r^2}$ for the variable r.

Answer: medium

$$F = G\frac{mM}{r^2} \quad \Leftrightarrow \quad r^2F = GmM \quad \Leftrightarrow \quad r = \pm\sqrt{\frac{GmM}{F}}$$

19. Solve the equation $a + 3[x - 3(c - b)] = 9$ for the variable x.

Answer: medium

$$a + 3[x - 3(c - b)] = 9 \quad \Leftrightarrow \quad a + 3x - 9c + 9b = 9 \quad \Leftrightarrow \quad x = \frac{9 - a - 9b + 9c}{3}$$

20. Solve the equation $\dfrac{a}{2b} = \dfrac{a}{b+3} + \dfrac{b-1}{2}$ for the variable a.

Answer: hard

$$\frac{a}{2b} = \frac{a}{b+3} + \frac{b-1}{2} \quad \Leftrightarrow \quad \frac{a}{2b} - \frac{a}{b+3} = \frac{b-1}{2} \quad \Leftrightarrow \quad a\left(\frac{1}{2b} - \frac{1}{b+3}\right) = \frac{b-1}{2} \quad \Leftrightarrow \quad a = \frac{b(b+3)}{b+3-2b}(b-1) \quad \Leftrightarrow$$
$$a = \frac{b(b+3)(b-1)}{3-b}, b \neq 3.$$

21. Solve the equation $\dfrac{x-2a}{x-4a} = \dfrac{x+b}{x+2b}$ for the variable x.

Answer: hard

$$\frac{x-2a}{x-4a} = \frac{x+b}{x+2b} \quad \Leftrightarrow \quad (x-2a)(x+2b) = (x+b)(x-4a) \quad \Leftrightarrow \quad x^2 + (2b-2a)x - 4ab = x^2 + (b-4a)x - 4ab$$

$\Leftrightarrow \quad (2b - 2a - b + 4a)x = (4-4)ab \quad \Leftrightarrow \quad (b+2a)x = 0$. Therefore $x = 0$ if $b \neq -2a$. If $b = -2a$, the equation is an identity, with the restriction that $x \neq 4a$ to avoid division by 0.

22. Find a value for k that will make $x = -2$ a solution of the equation $9x - k + 4 = 2kx - k + 4$.

Answer: medium

$9x - k + 4 = 2kx - k + 4\,;\, x = -2 \quad \Leftrightarrow \quad 9(-2) - k + 4 = 2k(-2) - k + 4 \quad \Leftrightarrow \quad -k + 4k + k = 18 - 4 + 4 \quad \Leftrightarrow$
$4k = 18$. Therefore $x = -2$ is a solution when $k = \frac{9}{2}$.

23. The formula for the volume of a right circular cone is $V = \frac{1}{3}\pi r^2 h$, where r is the radius of the base and h is the height. Find the height (correct to two decimals) of a cone with volume 18.00 in^3 and radius 3.50 in.

Answer: medium

$$V = \tfrac{1}{3}\pi r^2 h \quad \Leftrightarrow \quad h = \tfrac{3V}{\pi r^2} \quad \Leftrightarrow \quad h = \frac{3(18)}{\pi(3.5)^2} \approx 1.40 \text{ in.}$$

GC 24. Use a graphing device to find all real solutions of the equation $x^3 - x^2 - 14x + 24 = 0$.

Answer: medium

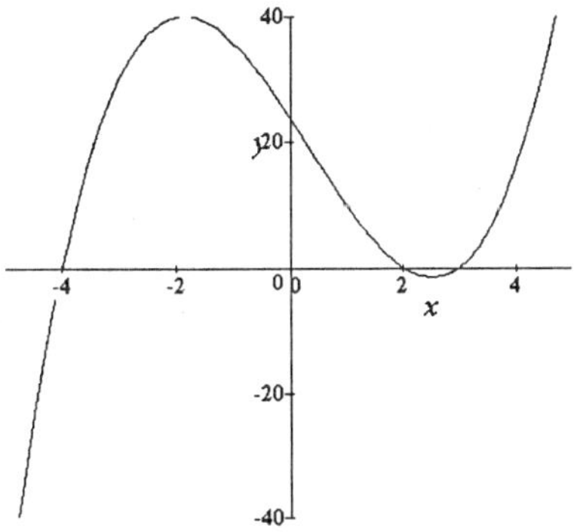

$x^3 - x^2 - 14x + 24 = 0$ By using a graphing device, the real solutions are -4, 2 and 3.

GC 25. Use a graphing device to find all real solutions of the equation $x^5 = x + 4$ correct to two decimals.

Answer: medium

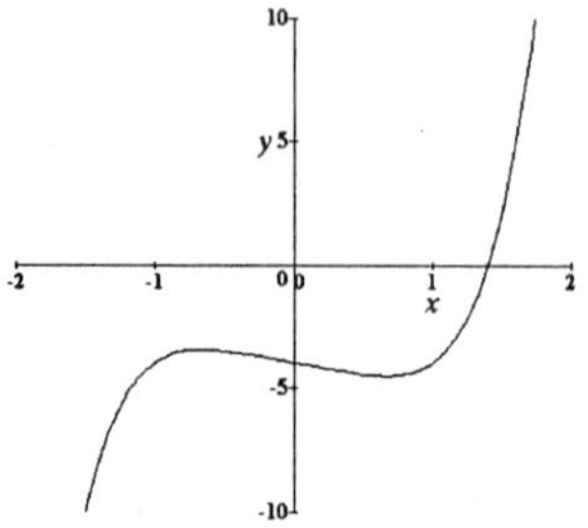

$x^5 = x + 4 \quad \Leftrightarrow \quad x^5 - x - 4 = 0$ By using a graphing device, we find that the real solution is about 1.40.

26. Solve the equation $\frac{1}{3}x - 2 = 4$ both algebraically and graphically.

Answer: easy

$\frac{1}{3}x - 2 = 4 \quad \Leftrightarrow \quad \frac{1}{3}x = 6 \quad \Leftrightarrow \quad x = 18$

27. Solve the equation $-1(x-1) + 3x - 5 = 2(7-x)$ both algebraically and graphically.

Answer: easy

$-1(x-1) + 3x - 5 = 2(7-x) \quad \Leftrightarrow \quad 2x - 4 = 14 - 2x \quad \Leftrightarrow \quad x = \frac{9}{2}$

28. Solve $\frac{1}{x-1} - \frac{2}{x} = \frac{3}{4x-4}$ both algebraically and graphically.

Answer: medium

$\frac{1}{x-1} - \frac{2}{x} = \frac{3}{4x-4} \quad \Leftrightarrow \quad 4x(x-1)\left(\frac{1}{x-1} - \frac{2}{x}\right) = \frac{3}{4x-4}4x(x-1) \quad \Leftrightarrow \quad -4x + 8 = 3x \quad \Leftrightarrow \quad x = \frac{8}{7}$

29. Solve $x^2 - 5x + 6 = 0$ graphically, correct to two decimals, on the interval $[1, 4]$.

Answer: easy

$x^2 - 5x + 6 = 0 \quad \Leftrightarrow \quad x = 2$ or $x = 3$

30. Solve $x^3 - 3x^2 - 2x + 1 = 0$ graphically, correct to two decimals, on the interval $[-2, 4]$.

Answer: medium

$x^3 - 3x^2 - 2x + 1 = 0 \quad \Leftrightarrow \quad x \approx -0.83, x \approx 0.34, x \approx 3.49$

31. Solve the equation $2 - \sqrt{2 + x^2} = -\sqrt{x}$ graphically, correct to two decimals, on the interval $[-1, 5]$.

Answer: medium

$2 - \sqrt{2 + x^2} + \sqrt{x} = 0 \quad \Leftrightarrow \quad x \approx 3.64$

32. Find all real solutions of the equation $2x^3 - x^2 - 2x + 1 = 0$, correct to two decimals.

Answer: medium

$2x^3 - x^2 - 2x + 1 = 0 \quad \Leftrightarrow \quad 2x(x^2-1) - (x^2-1) = 0 \quad \Leftrightarrow \quad (x+1)(2x-1)(x-1) = 0 \quad \Leftrightarrow \quad x = \pm 1, x = \frac{1}{2}$

33. Find all real solutions of the equation $x^4 - 5x^2 + 6 = 0$, correct to two decimals.

Answer: medium

$x^4 - 5x^2 + 6 = 0 \quad \Leftrightarrow \quad (x^2-2)(x^2-3) = 0 \quad \Leftrightarrow \quad x = \pm\sqrt{2}, x = \pm\sqrt{3}$

34. Find all real solutions of the equation $x^4 - 3x^3 = -x^2$, correct to two decimals.

Answer: medium

$x^4 - 3x^3 = -x^2 \quad \Leftrightarrow \quad x^4 - 3x^3 + x^2 = 0 \quad \Leftrightarrow \quad x = 0, x = 0.38, x = 2.61$

35. Find all real solutions of the equation $x^{1/2} + 3x^{1/3} = x$, correct to two decimals.

Answer: medium

$x^{1/2} + 3x^{1/3} = x \quad \Leftrightarrow \quad x^{1/2} + 3x^{1/3} - x = 0 \quad \Leftrightarrow \quad x = 0, x \approx 9.39$

SECTION 3.2

1. Express the given quantity in terms of the indicated variable. The value (in cents) of the change in a purse that contains twice as many dimes as pennies, four fewer nickels than dimes, and as many quarters as dimes and pennies combined; $p =$ number of pennies.
(a) $106p - 20$ (b) $27p - 13$ (c) $100p - 81$ (d) $32p - 14$ (e) $107p - 27$
Answer: medium (a)
$1(p) + 10(2p) + 5(2p - 4) + 25(2p + p) = 106p - 20$, p is the number of pennies.

2. Mozart composed eleven fewer than half the number of symphonies that Haydn composed. Together, they composed 145 symphonies. How many symphonies did Mozart compose?
(a) 9 (b) 10 (c) 36 (d) 41 (e) 91
Answer: medium (d)
Let x be the number of symphonies that Haydn composed. Then Mozart composed $\frac{1}{2}x - 11$ symphonies. We have $x + \frac{1}{2}x - 11 = 145 \quad\Leftrightarrow\quad \frac{3}{2}x = 156 \quad\Leftrightarrow\quad x = 104$. Therefore Mozart composed $\frac{1}{2}(104) - 11$, or 41 symphonies.

3. A change purse contains an equal number of quarters, dimes and nickels. The total value of the coins is \$1.60. How many coins of each type does the purse contain?
(a) 3 (b) 4 (c) 5 (d) 6 (e) 7
Answer: medium (b)
Let x be the number of nickels. Thus $5x + 10x + 25x = 160 \quad\Leftrightarrow\quad x = \frac{160}{40} \quad\Leftrightarrow\quad x = 4$. Therefore the purse contains 4 coins of each type.

4. One-third of Victoria's age last year plus double her age next year is 25 years. How old is Victoria?
(a) 8 (b) 10 (c) 11 (d) 12 (e) 14
Answer: medium (b)
Let x be Victoria's age. Hence next year she will be $x + 1$ and she was $x - 1$ last year. Thus $\frac{1}{3}(x - 1) + 2(x + 1) = 25 \quad\Leftrightarrow\quad (x - 1) + 6(x + 1) = 75 \quad\Leftrightarrow\quad 7x = 75 - 5 \quad\Leftrightarrow\quad x = 10$. Therefore Victoria is 10 years old.

5. In a health food store, a shopper fills his bag with a mixture of plain white rice that costs \$2.00 a pound and long-grain basmati rice that costs \$5.35 a pound, producing a 4.5 lb mixture that costs him \$15.00. Approximately how many pounds of each type of rice did he use in his mixture?
(a) 1.14 lb Basmati, 3.36 lb plain (b) 1.41 lb Basmati, 3.09 lb plain
(c) 1.48 lb Basmati, 3.02 lb plain (d) 1.79 lb Basmati, 2.71 lb plain
(e) 2.02 lb Basmati, 2.48 lb plain
Answer: medium (d)
Let x be the number of pounds of \$5.35 rice. Hence $(4.5 - x)$ is the amount of \$2.00 rice. Thus $5.35x + 2.00(4.5 - x) = 15 \quad\Leftrightarrow\quad 3.35x = 15 - 9 \quad\Leftrightarrow\quad x \approx 1.79$. Therefore he uses approximately 1.79 pounds of Basmati and 2.71 pounds of plain rice.

6. If Rena invests \$7500 at 7% per year, how much additional money must she invest at 12% to ensure that the interest she receives each year is 10% of the total invested?
(a) \$11,250 (b) \$12,375 (c) \$18,000 (d) \$21,000 (e) \$22,500
Answer: medium (a)
Let x be the additional amount of money that she invests at 12%. Hence the total amount invested is $x + 7500$. Thus $0.12x + 0.07(7500) = 0.10(x + 7500) \quad\Leftrightarrow\quad 2x = 22{,}500 \quad\Leftrightarrow\quad x = 11{,}250$. Therefore she must invest \$11,250 more at 12%.

7. A certain engineer's take-home pay was 72% of her gross monthly salary. After receiving a 9% raise in her gross salary, her take-home pay rose by \$83.70 per month, but now her take-home pay was only 68% of her new gross salary. What was her gross monthly salary before her raise?

(a) \$2266.21 (b) \$2784.30 (c) \$3948.11 (d) \$4426.17 (e) \$4872.64

Answer: hard (c)

Let x be her gross monthly salary before the raise. Hence her gross monthly salary after the raise was $(x + 0.09x) = 1.09x$. Her take-home pay before her raise was $0.72x$ and now her take-home pay is $0.68(1.09x)$. Thus $0.68(1.09x) - 0.72x = 83.70$ $\Leftrightarrow$ $0.0212x = 83.70$ $\Leftrightarrow$ $x \approx 3948.11$ Therefore her former gross monthly salary was approximately \$3948.11.

8. After robbing a bank in Dodge City, the robber gallops off at 8 mi/h. Eight minutes later the marshal leaves to pursue him at 18 mi/h. How long does it take the marshal to catch up to the bank robber?

(a) 2.2 minutes (b) 6.4 minutes (c) 8.2 minutes (d) 10 minutes (e) 13.5 minutes

Answer: medium (b)

Let x be the time, in hours, that it takes the marshal to catch up to the bank robber. Hence it takes the robber $x + \frac{8}{60}$ hours to be caught up to. Thus $18x = 8\left(x + \frac{2}{15}\right)$ $\Leftrightarrow$ $10x = \frac{16}{15}$ $\Leftrightarrow$ $x = \frac{16}{150}$ $\Leftrightarrow$ $x = \frac{6.4}{60}$. Therefore it takes the marshal 6.4 minutes to catch up to the robber.

9. An outboard motor boat can pull a water-skier at a cruising speed of 28 mi/h in still water. The boat takes a skier for a ride in a river that flows at 6 mi/h. The boat takes the skier upstream and then returns home. The trip takes one hour. How far upstream does it travel?

(a) 13.36 mi (b) 14.24 mi (c) 15.06 mi (d) 15.65 mi (e) 17.02 mi

Answer: medium (a)

Let x be the distance, in miles, he travels upstream. Thus $\dfrac{x}{28-6} + \dfrac{x}{28+6} = 1$ $\Leftrightarrow$ $\dfrac{x}{22} + \dfrac{x}{34} = 1$ $\Leftrightarrow$ $34x + 22x = (22)(34)$ $\Leftrightarrow$ $x = \frac{748}{56}$. Therefore he travels approximately 13.36 miles upstream.

10. Pete drove from Fresno to Bakersfield at 50 mi/h. Jerome left at the same time but drove at 56 mi/h. When Jerome arrived in Bakersfield, Pete had another 12 mi to drive. How far is it from Fresno to Bakersfield?

Answer: medium

Let x be the distance, in miles, from Fresno to Bakersfield. Thus $\dfrac{x}{56} = \dfrac{x-12}{50}$ $\Leftrightarrow$ $50x = 56x - 672$ $\Leftrightarrow$ $x = \frac{672}{6}$ $\Leftrightarrow$ $x = 112$. Therefore Fresno is 112 miles from Bakersfield.

11. A pilot flew a jet from Toronto to Miami, a distance of 950 mi. On the return trip the speed was increased by 80 mi/h. The round trip took 6 hours and 10 minutes. What was the speed from Toronto to Miami?

Answer: hard

Let v be the speed, in mi/h, from Toronto to Miami. Let t_1 be the flying time, in hours, to Miami, and t_2 be the flying time back to Toronto.

$$\left.\begin{array}{c} vt_1 = 950 \\ (v+80)t_2 = 950 \\ t_1 + t_2 = 6 + 1/6 \end{array}\right\} \Leftrightarrow \left\{\begin{array}{c} t_1 = 950/v \\ (v+80)\left(\frac{37}{6} - t_1\right) = 950 \end{array}\right. \Leftrightarrow (v+80)\left(\tfrac{37}{6} - \tfrac{950}{v}\right) = 950 \Leftrightarrow$$

$$(v+80)(37v - 5700) = 5700v \Leftrightarrow 37v^2 - 8440v - 456{,}000 = 0 \Leftrightarrow v = \frac{8440\sqrt{(-8440)^2 + 4(37)(456{,}000)}}{2(37)} \Leftrightarrow$$

$v = -45.11$ or 273.22, but $v \geq 0$. Therefore the speed from Toronto to Miami was 273.22 mi/h.

12. A woman driving an outboard barge 16 ft. long is passing an 18 ft long Albacore sailboat. The sailboat is traveling at 14 mi/h. How fast must the woman drive the barge so that she can pass the sailboat completely in 12 seconds, from the position where the front of the barge is even with the back of the sailboat to the position where the back of the barge is even with the front of the sailboat? (Hint: use feet and seconds instead of miles and hours.)
Answer: hard
The sailboat is traveling at $14 \text{ mi/h} \cdot \dfrac{1 \text{ h}}{3600 \text{ sec}} \cdot 5280 \text{ ft/mi} = \dfrac{616}{30}$ ft/s. Let x be the speed at which the barge is traveling (in ft/s). Since she wishes to pass the sailboat in 12 s, we have the following information:

	Distance	Rate	Time
Barge	$12x$	x	12
Sailboat	246.4	$\frac{616}{30}$	12

Since she must travel one barge-length plus one sailboat-length farther than the sailboat in this time, we have $12x = 246.4 + 16 + 18 = 280.4 \quad \Rightarrow \quad x \approx 23.37$ ft/s, and since $23.37 \text{ ft/s} \cdot \dfrac{1 \text{ mi}}{5280 \text{ ft}} \cdot 3600 \text{ sec/h} \approx 15.93$ mi/h. The woman must travel at 15.93 mi/h.

13. A marine biologist cultivates salmon in a fish farm. She notices that, on average, each week, the number of new salmon born is always two-thirds of the salmon population at the beginning of the week, but that 84 adult salmon die over the course of a week. If after three weeks she has 230 salmon, how many did she start with (rounded off to the nearest salmon)?
Answer: medium
Let x be the number of salmon that she started with. Hence the number of salmon after the first week is $x + \frac{2}{3}x - 84 = \frac{5}{3}x - 84$. The number of salmon after the second week is $\left(\frac{5}{3}x - 84\right) + \frac{2}{3}\left(\frac{5}{3}x - 84\right) - 84 = \frac{25}{9}x - 224$. The number of salmon after the third week is $\left(\frac{25}{9}x - 224\right) + \frac{2}{3}\left(\frac{25}{9}x - 224\right) - 84 = \frac{125}{27}x - \frac{1372}{3}$. Thus $\frac{125}{27}x - \frac{1372}{3} = 230 \quad \Leftrightarrow \quad x \approx 148$. Therefore she started with 148 salmon.

14. A pot contains 8 liters of brine at a concentration of 145 grams per liter. How much of the water should be boiled off to increase the concentration to 250 g/L?
Answer: hard
The brine has $145 \text{ g/L} \cdot 8 \text{ L} = 1160$ g of salt. Let x be the amount of pure water to be boiled off (in liters). Then there will be $(8 - x)$ L of brine left, and the concentration of salt will be $\dfrac{1160}{8 - x}$. Thus $250 = \dfrac{1160}{8 - x} \quad \Leftrightarrow \quad 1160 = 250(8 - x) \quad \Leftrightarrow \quad 250x = 840 \quad \Leftrightarrow \quad x = 3.36$. Therefore 3.36 L of water should be boiled off.

15. During the winter months, a motorcyclist mixes his gasoline with antifreeze. He has a 5 gal jerry can containing 3% antifreeze mixed with ordinary gasoline. This winter is particularly harsh, and the cyclist wants to use 6% antifreeze in his gasoline. How much of the gasoline mixture should he drain from the jerry can and replace with antifreeze to increase the antifreeze content to the new level?
Answer: medium
Let x be the amount of gasoline mixture to be drained off and replaced with pure antifreeze.
$0.03(5) - 0.03x + 1.00x = 0.06(5) \quad \Leftrightarrow \quad 15 + 97x = 30 \quad \Leftrightarrow \quad x = \frac{15}{97} \approx 0.155$ gal. So 0.155 gal should be drained and replaced.

16. Mrs. Johannson, the director of a boarding house, keeps a 2.5 L tin of Brand A breakfast cereal with raisins, in which the raisins comprise 30% of the cereal by volume. After using up 1 L of the cereal, she then replaces it with an equal amount of a cheaper brand of raisin cereal. If the tin of cereal now contains 21% raisins by volume, what was the concentration in the cheaper brand that Mrs. Johannson added?
Answer: medium
Let x be the concentration of raisins in the cheaper brand that Mrs. Johannson added. Thus $0.3(2.5 - 1) + 1x = 2.5 \cdot 0.21 \quad \Leftrightarrow \quad 1x = 0.075$. Therefore the cheaper brand has a concentration of 7.5% raisins.

17. Gyorgy and Jane can mow the lawn in 35 minutes if they work together. If Jane works three times as fast as Gyorgy, how long would it take him to do the job alone?
Answer: hard
Let x be the time, in minutes, that it takes Gyorgy to do the job alone. Hence Gyorgy can do $1/x$ of the job in a minute. Jane can do $3/x$ of the job in a minute. And both of them can do $1/x + 3/x$ in a minute. Thus $1/x + 3/x = \frac{1}{35} \Leftrightarrow 4/x = \frac{1}{35} \Leftrightarrow x = 140$. Therefore it will take Gyorgy 140 minutes to do the job alone.

18. A running track has straight sides and semicircular ends. If the length of the track is 880 yd and the two straight parts are each 220 yd long, what is the radius of the semicircular part, to the nearest yard?
Answer: medium
Let the radius of the semicircular part be x yd. Thus $2\pi x + 2(220) = 880 \Leftrightarrow x = \dfrac{880 - 440}{2\pi} \Leftrightarrow x \approx 70$. Therefore the radius of the semicircular part is about 70 yd.

19. Burt and Remo, next-door neighbors, use hoses from both houses to fill Burt's swimming pool. They know it takes nine hours using both hoses. They also know that Burt's hose, used alone, takes 25% less time than Remo's hose alone. How much time is required by each hose alone?
Answer: hard
Let x be the time, in hours, required by Remo's hose alone. Hence Burt's hose alone requires $0.75x$ hours. Remo's hose can fill $1/x$ of the swimming pool in an hour. Bob's hose can fill $1/(0.75x)$ of the pool in an hour. Thus $\dfrac{1}{x} + \dfrac{1}{0.75x} = \dfrac{1}{9} \Leftrightarrow \dfrac{1}{x} + \dfrac{100}{75x} = \dfrac{1}{9} \Leftrightarrow 9(75 + 100) = 75x \Leftrightarrow x = 21$. Therefore it takes 21 hours for Remo, and $\frac{3}{4}(21)$, or 15.75 hours for Burt to fill up the pool individually.

20. There are 63 coins in Taylor's piggy bank, all in quarters and dimes. The total value of the contents of his piggy bank is \$9.60. Find the number of each type of coin.
Answer: medium
If x is the number of quarters, then there are $63 - x$ dimes. The total value is $0.25x + 0.10(63 - x) = 9.60 \Leftrightarrow$ there are $x = 22$ quarters and $63 - x = 63 - 22 = 41$ dimes.

21. A money manager invests \$1250 at $4\frac{1}{2}\%$ simple interest and \$1850 at $5\frac{1}{2}\%$ simple interest. How much money must the manager invest at $6\frac{1}{2}\%$ in order to yield his client 6% of the total investment?
Answer: medium
Let x be the amount he needs to invest at $6\frac{1}{2}\%$, and the total investment is $1250 + 1850 + x$. So
$0.065(x) + 0.045(1250) + 0.055(1850) = 0.06(1250 + 1850 + x) \Leftrightarrow 158.0 + 0.065x = 186.0 + 0.06x \Leftrightarrow$ $x = \$5600$. The manager must invest \$5600 at $6\frac{1}{2}\%$ in order to make his client 6% of the total investment.

22. Before delivering a load of goods to Alaska, a driver of a semi-truck wishes to change the mixture of antifreeze in his radiator. The radiator contains 15.5 quarts of 35% antifreeze solution. How much would he need to drain and replace with pure antifreeze in order to get a 55% antifreeze solution?
Answer: medium
Let x be the amount of solution to drained. $0.35(15.5 - x) + x = 0.55(15.5) \Leftrightarrow 5.425 + 0.65x = 8.525 \Leftrightarrow x \approx 4.7$. The driver needs to drain 4.7 quarts from the current system, then replace the same amount with pure antifreeze in order to achieve a 55% antifreeze solution.

23. A running track has the shape of a rectangle with a semicircle on both ends. If the length of the track is 550 yards and the length of the side of the rectangle is 125 yards, what is the radius of the of the semicircular parts?
Answer: medium
If you form a circle from the two semicircles, its circumference would be $P_C = 2\pi r$. The perimeter of the track is therefore $P = 2l + P_C$ or $P = 2l + 2\pi r$. So $P = 2l + 2\pi r \Leftrightarrow 550 = 2(125) + 2\pi r \Leftrightarrow r = \frac{150}{\pi}$ yards.

SECTION 3.3

1. Find all solutions of the equation $4x^2 = 64$

(a) 2 (b) ± 2 (c) 3 (d) -4 (e) ± 4

Answer: easy (e)

$4x^2 = 64 \quad \Leftrightarrow \quad 4x^2 - 64 = 0 \quad \Leftrightarrow \quad 4(x^2 - 16) = 0 \quad \Leftrightarrow \quad x = \pm 4.$

2. Find all solutions of the equation $6x^2 - 216 = 0$

(a) 4 (b) ± 4 (c) 6 (d) ± 6 (e) 8

Answer: easy (d)

$6x^2 - 216 = 0 \quad \Leftrightarrow \quad 6(x^2 - 36) = 0 \quad \Leftrightarrow \quad (x-6)(x+6) = 0.$ So $x = 6$ or $x = -6$.

3. Solve the equation $x^2 + 7x = -12$ by factoring.

(a) $x = -3$ or 4 (b) $x = -4$ or 3 (c) $x = -3$ or -4 (d) $x = 3$ or 4 (e) $x = \pm 3$ or 4

Answer: medium (c)

$x^2 + 7x = -12 \quad \Leftrightarrow \quad x^2 + 7x + 12 = 0 \quad \Leftrightarrow \quad (x+3)(x+4) = 0.$ So $x = -3$ or $x = -4$.

4. Solve the equation $x^2 - 2.7x + 1.8 = 0$ by factoring.

(a) $x = 1.2$ or 1.5 (b) $x = -1.2$ or -1.3 (c) $x = -1.2$ or 1.5 (d) $x = 1.5$ or 2.2 (e) $x = 2.2$ or 4.1

Answer: hard (a)

$x^2 - 2.7x + 1.8 = 0 \quad \Leftrightarrow \quad (x - 1.2)(x - 1.5) = 0.$ So $x = 1.2$ or $x = 1.5$.

5. Solve the equation $10z^2 = z + 3$ by factoring.

(a) $z = \frac{2}{5}$ or $\frac{4}{3}$ (b) $z = \frac{3}{5}$ or $-\frac{1}{5}$ (c) $z = \frac{3}{5}$ or $-\frac{1}{2}$ (d) $z = -\frac{3}{5}$ or $\frac{2}{5}$ (e) $z = \frac{1}{3}$ or $-\frac{2}{3}$

Answer: medium (c)

$10z^2 = z + 3 \quad \Leftrightarrow \quad 10z^2 - z - 3 = 0 \quad \Leftrightarrow \quad (5z-3)(2z+1) = 0.$ So $z = \frac{3}{5}$ or $z = -\frac{1}{2}$.

6. Solve the equation $4a^2x^2 + 3ax - 1 = 0$ $(a > 0)$ by factoring.

(a) $x = \dfrac{1}{2a}$ or $-\dfrac{1}{2a}$ (b) $x = \dfrac{2}{3a}$ or $\dfrac{3a}{2}$ (c) $x = \dfrac{1}{4a}$ or $-\dfrac{1}{a}$ (d) $x = -\dfrac{1}{a}$ or $\dfrac{2}{a}$ (e) $x = \pm\dfrac{1}{2a}$

Answer: hard (c)

$4a^2x^2 + 3ax - 1 = 0 \quad \Leftrightarrow \quad (4ax - 1)(ax + 1) = 0.$ So $x = \dfrac{1}{4a}$ or $x = -\dfrac{1}{a}$.

7. Solve the quadratic equation $x^2 - 10x + 14 = 0$ by completing the square.

(a) $x = 3 \pm \sqrt{13}$ (b) $x = -1 \pm \sqrt{3}$ (c) $x = -5 \pm \sqrt{6}$ (d) $x = 5 \pm \sqrt{11}$ (e) $x = 3 \pm \sqrt{7}$

Answer: medium (d)

$x^2 - 10x + 14 = 0 \quad \Leftrightarrow \quad x^2 - 10x + 25 = 11 \quad \Leftrightarrow \quad (x-5)^2 = 11 \quad \Leftrightarrow \quad x - 5 = \pm\sqrt{11}$ So $x = 5 \pm \sqrt{11}$.

8. Solve the quadratic equation $2x^2 + 12x + 1 = 0$ by completing the square.

(a) $x = -3 \pm \sqrt{\frac{17}{2}}$ (b) $x = -2 \pm \sqrt{\frac{13}{2}}$ (c) $x = 3 \pm \sqrt{\frac{2}{3}}$ (d) $x = -4 \pm \sqrt{\frac{11}{2}}$ (e) $x = -3 \pm \sqrt{\frac{14}{3}}$

Answer: medium (a)

$2x^2 + 12x + 1 = 0 \quad \Leftrightarrow \quad x^2 + 6x + 3^2 = -\frac{1}{2} + 3^2 \quad \Leftrightarrow \quad (x+3)^2 = \frac{17}{2} \quad \Leftrightarrow \quad x + 3 = \pm\sqrt{\frac{17}{2}}$ So $x = -3 \pm \sqrt{\frac{17}{2}}$.

9. Solve the equation $x^2 = \frac{1}{5} - \frac{2}{3}x$ by completing the square.

(a) $\frac{1}{4} \pm \frac{3\sqrt{5}}{7}$ (b) $\frac{2}{3} \pm \frac{\sqrt{3}}{3}$ (c) $\frac{2}{5} \pm \frac{5\sqrt{2}}{2}$ (d) $\frac{1}{3} \pm \frac{\sqrt{70}}{15}$ (e) $\frac{2}{5} \pm \frac{2\sqrt{5}}{5}$

Answer: medium (d)

$x^2 = \frac{1}{5} - \frac{2}{3}x \quad \Leftrightarrow \quad x^2 + \frac{2}{3}x + \left(\frac{1}{3}\right)^2 = \frac{1}{5} + \left(\frac{1}{3}\right)^2 \quad \Leftrightarrow \quad \left(x + \frac{1}{3}\right)^2 = \frac{9+5}{45} \quad \Leftrightarrow \quad x - \frac{1}{3} = \pm\sqrt{\frac{14}{45}}.$ So $x = \frac{1}{3} \pm \frac{\sqrt{70}}{15}$.

10. Find all real solutions of the equation $3x^2 + 4x - 4 = 0$.

(a) $x = -1$ or $\frac{1}{5}$ (b) $x = -2$ or $\frac{2}{3}$ (c) $x = -\frac{3}{2}$ or $\frac{5}{2}$ (d) $x = -\frac{1}{3}$ or $\frac{3}{2}$ (e) $x = -1$ or 2

Answer: medium (b)

$3x^2 + 4x - 4 = 0 \quad \Leftrightarrow \quad (x+2)(3x-2) = 0.$ So $x = -2$ or $x = \frac{2}{3}$.

11. Find all real solutions of the equation $11x^2 + 75x + 81 = 0$.

(a) $x = \dfrac{-75 \pm 3\sqrt{229}}{22}$ (b) $\dfrac{-25 \pm 5\sqrt{19}}{12}$ (c) $\dfrac{-40 \pm 4\sqrt{29}}{6}$

(d) $x = \dfrac{15 \pm \sqrt{109}}{14}$ (e) $x = \dfrac{13 \pm \sqrt{77}}{26}$

Answer: medium (a)

$$11x^2 + 75x + 81 = 0 \quad \Leftrightarrow \quad x = \frac{-75 \pm \sqrt{5625 - 3564}}{22} = \frac{-75 \pm 3\sqrt{229}}{22}.$$

12. Find all real solutions of the equation $2x^2 - 6x + 4 = 0$.

(a) $x = 1$ or 3 (b) $x = 2$ or 3 (c) $x = 2$ or 4 (d) $x = 1$ or 4 (e) $x = 1$ or 2

Answer: medium (e)

$2x^2 - 6x + 4 = 0 \quad \Leftrightarrow \quad x^2 - 3x + 2 = 0 \quad \Leftrightarrow \quad (x-2)(x-1) = 0$. So $x = 2$ or $x = 1$.

13. Find all real solutions of the equation $\theta^2 - \frac{4}{5}\theta + \frac{4}{25} = 0$.

(a) $\theta = \pm\frac{4}{5}$ (b) $\theta = -\frac{1}{5} \pm \frac{\sqrt{13}}{5}$ (c) $\theta = \frac{2}{5}$ (d) $\theta = \pm\frac{2}{5}$ (e) $\theta = -\frac{2}{5}$

Answer: medium (c)

$\theta^2 - \frac{4}{5}\theta + \frac{4}{25} = 0 \quad \Leftrightarrow \quad \left(\theta - \frac{2}{5}\right)^2 = 0 \quad \Leftrightarrow \quad \theta = \frac{2}{5}$.

14. Find all real solutions of the equation $5x = 2x^2 + 1$.

(a) $\dfrac{5 \pm \sqrt{17}}{4}$ (b) $\dfrac{3 \pm \sqrt{14}}{2}$ (c) $\dfrac{7 \pm \sqrt{13}}{6}$ (d) $\dfrac{4 \pm \sqrt{13}}{2}$ (e) $\dfrac{4 \pm \sqrt{17}}{8}$

Answer: medium (a)

$$5x = 2x^2 + 1 \quad \Leftrightarrow \quad 2x^2 - 5x + 1 = 0 \quad \Leftrightarrow \quad x = \frac{5 \pm \sqrt{25 - 8}}{4} \quad \Leftrightarrow \quad x = \frac{5 \pm \sqrt{17}}{4}$$

15. Find all real solutions of the equation $w(w - 2) = 3$.

(a) $w = 2$ or 4 (b) $w = -2$ or -1 (c) $w = -1$ or 3 (d) $w = 0$ or 2 (e) $w = \pm 2$

Answer: medium (c)

$w(w-2) = 3 \quad \Leftrightarrow \quad w^2 - 2w - 3 = 0 \quad \Leftrightarrow \quad (w-3)(w+1) = 0 \quad \Leftrightarrow \quad w = 3$ or $w = -1$.

16. Find all real solutions of the equation $\sqrt{8}x^2 + 2x + \frac{3}{\sqrt{2}} = 0$.

(a) $x = \dfrac{-5 \pm \sqrt{13}}{2}$ (b) $x = \dfrac{5 \pm \sqrt{13}}{4}$ (c) $x = \pm\sqrt{2}$ (d) $x = \dfrac{3 \pm \sqrt{13}}{8}$ (e) No real roots

Answer: hard (e)

$\sqrt{8}x^2 + 2x + \dfrac{3}{\sqrt{2}} = 0 \quad \Leftrightarrow \quad x = \dfrac{-2 \pm \sqrt{4 - 24}}{2\sqrt{8}}$ but $\Delta < 0$, so no real roots.

17. Find all real solutions of the equation $1 + \dfrac{3x}{(x+5)(x-2)} = \dfrac{2}{x+5} + \dfrac{3}{x-2}$.

(a) $x = \dfrac{-3 \pm \sqrt{15}}{6}$ (b) $x = \dfrac{-1 \pm \sqrt{85}}{2}$ (c) $x = \dfrac{2 \pm \sqrt{55}}{4}$ (d) $x = \dfrac{-4 \pm \sqrt{17}}{16}$ (e) $x = \dfrac{3 \pm \sqrt{19}}{4}$

Answer: hard (b)

$$1 + \frac{3x}{(x+5)(x-2)} = \frac{2}{x+5} + \frac{3}{x-2} \quad \Leftrightarrow \quad (x+5)(x-2) + 3x = 2(x-2) + 3(x+5) \quad \Leftrightarrow$$

$$x^2 + 3x - 10 + 3x = 2x + 3x - 4 + 15 \quad \Leftrightarrow \quad x^2 + x - 21 = 0 \quad \Leftrightarrow \quad x = \frac{-1 \pm \sqrt{1 + 84}}{2} = \frac{-1 \pm \sqrt{85}}{2}.$$

18. Find all real solutions of the equation $\dfrac{2x}{2x-5} - \dfrac{x-4}{x+7} = 1$.

(a) $x = \dfrac{3 \pm \sqrt{117}}{4}$ (b) $x = \dfrac{5 \pm \sqrt{241}}{16}$ (c) $x = \dfrac{15 \pm \sqrt{221}}{16}$ (d) $x = \dfrac{1 \pm \sqrt{10}}{4}$ (e) $x = \dfrac{9 \pm \sqrt{111}}{2}$

Answer: hard (e)

$\dfrac{2x}{2x-5} - \dfrac{x-4}{x+7} = 1 \quad \Leftrightarrow \quad 2x(x+7) - (x-4)(2x-5) = (x+7)(2x-5) \quad \Leftrightarrow$

$2x^2 + 14x - 2x^2 + 13x - 20 = 2x^2 + 9x - 35 \quad \Leftrightarrow \quad 2x^2 - 18x - 15 = 0 \quad \Leftrightarrow \quad x = \dfrac{18 \pm \sqrt{324 + 120}}{4}$.

So $x = \dfrac{9 \pm \sqrt{111}}{2}$.

19. Find all real solutions of the equation. $2bx^2 - x - \frac{3}{b} = 0$ $(b \neq 0)$

(a) $x = \dfrac{5}{4b}$ or $-\dfrac{2}{b}$ (b) $x = \dfrac{2b}{5}$ or $-\dfrac{2b}{3}$ (c) $x = \dfrac{3}{2b}$ or $-\dfrac{1}{b}$ (d) $x = \dfrac{5}{6b}$ or $-\dfrac{2}{3b}$ (e) $x = -\dfrac{4}{7b}$ or $\dfrac{6}{5b}$

Answer: medium (c)

$2bx^2 - x - \frac{3}{b} = 0 \quad \Leftrightarrow \quad 2b^2x^2 - bx - 3 = 0 \quad \Leftrightarrow \quad (2bx-3)(bx+1) = 0$. So $x = \dfrac{3}{2b}$ or $x = -\dfrac{1}{b}$.

20. Use a calculator to find all real solutions, correct to three decimals. $3.231x^2 - 6.121x = 5.533$.

Answer: medium

$3.231x^2 - 6.121x = 5.533 \quad \Leftrightarrow \quad 3.231x^2 - 6.121x - 5.533 = 0 \quad \Leftrightarrow$

$x = \dfrac{6.121 \pm \sqrt{(6.121)^2 + 4(3.231)(5.533)}}{2(3.231)} \approx \dfrac{6.121 \pm 10.439}{6.462}$. So $x \approx 2.563$ or $x \approx -0.668$

21. Use a calculator to find all real solutions, correct to three decimals. $2x^2 - 3.998x + 1.894 = 0$

Answer: medium

$2x^2 - 3.998x + 1.894 = 0 \quad \Leftrightarrow \quad x = \dfrac{3.998 \pm \sqrt{(-3.998)^2 - 4(2)(1.894)}}{4} \quad \Leftrightarrow \quad x \approx \dfrac{3.998 \pm 0.912}{4} \quad \Leftrightarrow \quad x \approx 1.228$ or $x \approx 0.771$.

22. Solve the equation $F = G\dfrac{mM}{r^2}$ for the variable G.

Answer: easy

$F = G\dfrac{mM}{r^2} \quad \Leftrightarrow \quad Fr^2 = GmM \quad \Leftrightarrow \quad G = \dfrac{Fr^2}{mM}$

23. Solve the equation $S = \dfrac{(n+1)(n-1)}{2}$ for the variable n.

Answer: medium

$S = \dfrac{(n+1)(n-1)}{2} \quad \Leftrightarrow \quad 2S = n^2 - 1 \quad \Leftrightarrow \quad n^2 = 2S + 1 \quad \Leftrightarrow \quad n = \pm\sqrt{2S+1}$.

24. For the quadratic equation $2x^2 + kx - 4 = 0$ find the value(s) of k that will ensure that $x = -\frac{2}{3}$ and 3 are the solutions of the quadratic equation.

Answer: medium

$2x^2 + kx - 4 = 0$. For $x = -\frac{2}{3}$: $2\left(\frac{2}{3}\right)^2 - \frac{2}{3}k - 4 = 0 \quad \Rightarrow \quad k = -\frac{28}{9}\left(\frac{3}{2}\right) \quad \Rightarrow \quad k = -\frac{14}{3}$. For $x = 3$: $2 \cdot 3^2 + 3k - 4 = 0$ $\Rightarrow \quad k = -\frac{14}{3}$. Thus the value of k is $-\frac{14}{3}$.

25. Using the discriminant, determine how many real solutions the equation $3x^2 = 5 - 6x$ will have, without solving the equations.

Answer: medium

$3x^2 = 5 - 6x \quad \Leftrightarrow \quad 3x^2 + 6x - 5 = 0$, $\Delta = b^2 - 4ac = 36 + 60 > 0 \quad \Rightarrow \quad$ 2 real solutions.

26. Using the discriminant, determine how many real solutions the equation $x^2 + 3.67x + 7.32 = 0$ will have, without solving the equation.

Answer: medium

$x^2 + 3.67x + 7.32 = 0$, $\Delta = b^2 - 4ac = 3.67^2 - 4(7.32) \approx -15.8 < 0 \quad \Rightarrow \quad$ no real solutions.

27. Using the discriminant, determine how many real solutions the equation $x^2 + tx - u = 0$ $(u > 0)$ will have, without solving the equation.
Answer: medium
$x^2 + tx - u = 0$, $\Delta = b^2 - 4ac = t^2 + 4u > 0$ because $u > 0 \quad \Rightarrow \quad$ 2 real solutions.

28. Find all values for k that ensure that the equation $2kx^2 + 18x + k = 0$ has exactly one solution.
Answer: medium
$2kx^2 + 18x + k = 0$, $D = b^2 - 4ac = 18^2 - 8\text{k}^2 = 0$. Thus there is exactly one solution when $18^2 - 8k^2 = 0 \quad \Leftrightarrow$
$k = \pm\frac{18}{\sqrt{8}} \quad \Leftrightarrow \quad k = \pm\frac{9\sqrt{2}}{2}$.

29. The difference of the squares of two consecutive integers is 1707. Find the integers.
Answer: medium
Let the first integer be x. Hence the next consecutive integer is $(x+1)$. we have $(x+1)^2 - (x)^2 = 1707 \quad \Leftrightarrow$
$x^2 - x^2 + 2x + 1 = 1707 \quad \Leftrightarrow \quad 2x + 1 = 1707 \quad \Leftrightarrow \quad x = 853$. Thus the two integers are 853 and 854.

30. A circle has a radius of 7 cm. By how much should the radius be reduced so that the area is reduced by half?
Answer: hard
The area of the original circle is $\pi(7)^2 = 49\pi$. If its radius is reduced by x cm, its area becomes
$\pi(7-x)^2 = 49\pi - 14\pi x + \pi x^2$. We have $49\pi - 14\pi x + \pi x^2 = \dfrac{49\pi}{2} \quad \Leftrightarrow \quad \dfrac{49\pi}{2} - 14\pi x + \pi x^2 = 0 \quad \Leftrightarrow$
$x = \frac{14\pi \pm \sqrt{196\pi^2 - 98\pi^2}}{2\pi} \quad \Leftrightarrow \quad x = \dfrac{14\pi \pm 7\sqrt{2}\pi}{2\pi} \quad \Leftrightarrow \quad x = 7 \pm \frac{7\sqrt{2}}{2}$. But $x < 7$, thus the radius must be reduced by
$\left(7 - \frac{7\sqrt{2}}{2}\right)$ cm.

31. A cylindrical tin can has a volume of 28π cm^3 and a diameter of 3 cm. What is its height?
Answer: medium
Let x be the height, in cm, of the cylindrical tin. The volume V is $\pi r^2 h$, where h is the height of the cylinder. Thus
$28\pi = \pi\left(\frac{3}{2}\right)^2(x) \quad \Leftrightarrow \quad x = \frac{112}{9}$. Therefore the height is $\frac{112}{9}$, or about 12.4 cm.

32. A sailboat's mast is secured on opposite sides by two wires, each of which is 7 feet longer than the pole. The distance between the points where the wires are fixed to the boat's hull is equal to one-third the length of the mast. How tall is the mast (to the nearest inch)?
Answer: hard
Let x be the height, in feet, of the mast. Hence the wires are $x + 7$ ft long. The distance between the wires at the hull is $\frac{1}{3}x$ ft.
Thus $(x+7)^2 = x^2 + \left(\frac{1}{6}x\right)^2 \quad \Leftrightarrow \quad x^2 + 14x + 49 = x^2 + \frac{1}{36}x^2 \quad \Leftrightarrow \quad x^2 - (14)(36)x - (49)(36) \quad \Leftrightarrow$
$x^2 - 504x - 1764 \quad \Leftrightarrow \quad x = \dfrac{504 \pm \sqrt{504^2 + 4(1764)}}{2} \quad \Leftrightarrow \quad x \approx 252 \pm 255.5$. But $x > 0$, so $x \approx 507.5$. Thus the mast is about 507 ft 6 in. tall.

33. How fast would a ball have to be thrown upward to reach a maximum height of 80 ft? (Hint: Use the discriminant on the equation $16t^2 - v_0 t + h = 0$.)
Answer: medium
The discriminant $D = (-v_0)^2 - 4(16)(80) = v_0^2 - 5120 = 0 \quad \Leftrightarrow \quad v_0 \approx 71.55$. Therefore the ball has to be thrown upward at 71.55 ft/s to reach a maximum height of 80 ft.

34. A 240 in long roll of aluminum sheeting with width 48 in is cut into two sheets. One piece is formed into a cylindrical pipe, while the other is formed into a pipe with a square cross-section. If the two pipes have the same volume, what are the lengths of the two pieces of sheeting, to the nearest tenth of an inch?
Answer: hard
Suppose one of the wires is x inches long. Then the other is $(240 - x)$ in long. If the first sheet forms a cylinder, its volume is $48\pi\left(\frac{x}{2\pi}\right)^2$, and if the other forms a square pipe, its volume is $48\left(\frac{240-x}{4}\right)^2$. Thus $48\pi\left(\frac{x}{2\pi}\right)^2 = 48\left(\frac{240-x}{4}\right)^2 \Leftrightarrow$ $\sqrt{\pi}\left(\frac{x}{2\pi}\right) = \frac{240-x}{4}$ (since all lengths are positive) $\Leftrightarrow$ $4x = 2\sqrt{\pi}(240-x)$ $\Leftrightarrow$ $x = \frac{240\sqrt{\pi}}{2+\sqrt{\pi}} \approx 112.8$. Thus one sheet is 112.8 in long and the other is 127.2 in long.

35. Diantha drove from Plattsburgh to Stowe, a distance of 70 mi. She increased her speed by 20 mi/h for the last leg of the trip, the 20 mile drive from Burlington to Stowe. If the total trip took 1.5 hours, what was her speed from Plattsburgh to Burlington?
Answer: medium
Let Diantha's speed from Plattsburgh to Burlington be x mi/hr.

	Distance	Rate	Time
P → B	50	x	$50/x$
B → S	20	$x+20$	$20/(x+20)$

We have $\frac{50}{x} + \frac{20}{x+20} = 1.5$ $\Leftrightarrow$ $50(x+20) + 20x = 1.5x(x+20)$ $\Leftrightarrow$ $1.5x^2 - 40x - 1000 = 0$ $\Leftrightarrow$ $x = \frac{40 \pm \sqrt{40^2 + 4(1.5)(1000)}}{3}$ $\Leftrightarrow$ $x = \frac{40 \pm \sqrt{7600}}{3}$ $\Leftrightarrow$ $x = \frac{20\left(2 \pm \sqrt{19}\right)}{3}$. But $x > 0$, so $x \approx 42.4$. Therefore her speed from Plattsburgh to Burlington was 42.4 mi/h.

36. A certain oriental board game requires a two-toned wood playing surface. The central portion is made of a rectangular piece of pine, measuring 8 in by 11 in. The outer border is made of redwood, and has equal width. The area of the redwood border must be one half of the area of the pine playing surface. What must the width of the redwood border be?
Answer: medium

Let the width of the border be x in. The area of the border is $\frac{1}{2}(8)(11) = 44\text{ in}^2$. The area of the pine surface is 88 in^2. Hence, the board has an area of 132 in^2. We have $(8+2x)(11+2) = 132$ $\Leftrightarrow$ $4x^2 + 38x - 44 = 0$ $\Leftrightarrow$ $2x^2 + 19x - 22 = 0$ $\Leftrightarrow$ $x = \frac{-19 \pm \sqrt{361 + 4(44)}}{4}$ $\Leftrightarrow$ since $x > 0$, $x \approx \frac{-19 + 23.2}{4}$ $\Leftrightarrow$ $x \approx 1.04$ in.

37. Jared, Kitt, and Launa deliver advertising flyers in Baltimore. Working alone, it takes Jared 3 h to deliver all the flyers, and it takes Launa 2 h longer than it takes Kitt. Working together, they can deliver all the flyers in 60% of the time it takes Kitt working alone. How long does it take Kitt to do the delivery alone?
Answer: hard
Let the time that Kitt can deliver all the flyers working alone be x hr. Hence it takes Launa $(x+2)$ hr. Thus $\frac{1}{x+2} + \frac{1}{x} + \frac{1}{3} = \frac{1}{0.6x}$ $\Leftrightarrow$ $6x + 6(x+2) + 2x(x+2) = 10(x+2)$ $\Leftrightarrow$ $2x^2 + 6x - 8 = 0$ $\Leftrightarrow$ $(x+4)(x-1) = 0$ $\Rightarrow$ $x = 1$. Therefore it takes Kitt 1 hr to do the delivery alone.

38. A man wishes to determine how far down the water level is in a deep well. He drops a stone into the well and hears the splash 4 seconds later. If the stone drops $16t^2$ feet after t seconds, and the speed of sound is 1090 ft/s, how far down is the surface of the water (to the nearest foot)?
Answer: hard
Let t be the time, in seconds, that it takes the stone to travel from the ground to the surface of water in the well. Hence it takes $(4-t)$ seconds for the sound to travel back up to the ground. Thus $16t^2 = 1090(4-t)$ $\Leftrightarrow$ $16t^2 + 1090t - 4360 = 0$ $\Leftrightarrow$ $t = \frac{-1090 \pm \sqrt{(1090)^2 + 4(16)(4360)}}{2(16)}$ $\Leftrightarrow$ $t \approx 3.789$, since $t > 0$. Therefore the surface of water is $16(3.789)^2 \approx 230$ ft deep.

39. Solve the equation $10x^2 = 13x - 4$ by factoring.

(a) $x = 1 \pm 5\sqrt{2}$ (b) $x = \frac{5}{2}$ or 6 (c) $x = \frac{1}{2}$ or $\frac{4}{5}$ (d) $x = -\frac{11}{2}$ or $\frac{1}{2}$ (e) $x = \frac{5}{2}$ or 1

Answer: easy (c)

$10x^2 = 13x - 4 \;\Rightarrow\; 10x^2 - 13x + 4 = 0 \;\Rightarrow\; (2x-1)(5x-4) = 0 \;\Rightarrow\; x = \frac{1}{2}, x = \frac{4}{5}$

40. Solve the equation $\frac{1}{2}x^2 - \frac{11}{2}x + 15 = 0$ by completing the square.

(a) $x = -1 \pm 5\sqrt{6}$ (b) $x = 5$ or 6 (c) $x = -5$ or -6 (d) $x = \frac{11}{2}$ or $\frac{1}{2}$ (e) $x = \frac{1}{2}$ or -1

Answer: easy (b)

$\frac{1}{2}x^2 - \frac{11}{2}x + 15 = 0 \;\Rightarrow\; \frac{1}{2}(x^2 - 11x) = -15 \;\Rightarrow\; \frac{1}{2}\left[x^2 - 11x + \left(\frac{11}{2}\right)^2\right] = -15 + \frac{121}{8} \;\Rightarrow\; \frac{1}{2}\left(x - \frac{11}{2}\right)^2 = \frac{1}{8} \;\Rightarrow$
$x = 6, x = 5$

41. Find all real solutions of the equation $12 - 7x^2 + x^4 = 0$.

(a) $x = 1, x = 2, x = \pm\sqrt{3}$ (b) $x = \pm 2, \pm 3$

(c) $x = \pm\sqrt{3}, x = 2 \pm \sqrt{2}$(d) $x = \pm 1, x = \pm\sqrt{3}$

(e) $x = \pm 2, x = \pm\sqrt{3}$

Answer: medium (e)

$12 - 7x^2 + x^4 = (x^2 - 4)(x^2 - 3) = 0 \;\Rightarrow\; (x-2)(x+2)(x^2-3) = 0 \;\Rightarrow\; x = \pm 2, \pm\sqrt{3}$

42. Use a calculator to find all real solutions of the equation $3x^2 - 2.45x - 2.01 = 0$, correct to three decimals.

Answer: easy

$3x^2 - 2.45x - 2.01 = 0 \;\Rightarrow\; x = \frac{2.45 \pm \sqrt{(-2.45)^2 - 4(3)(-2.01)}}{6} \;\Rightarrow\; x \approx -0.506, x \approx 1.321$

43. Solve the equation $\dfrac{1}{t-1} + \dfrac{2}{t} = 3$ for t.

Answer: easy

$\frac{1}{t-1} + \frac{2}{t} - 3 = 0 \;\Rightarrow\; t + 2(t-1) = 3t^2 - 3t \;\Rightarrow\; 3t - 2 = 3t^2 - 3t \;\Rightarrow\; 3t^2 - 6t + 2 = 0 \;\Rightarrow\; t = \frac{6 \pm \sqrt{36 - 4(3)(2)}}{6}$
$\Rightarrow\; t = 1 + \frac{1}{3}\sqrt{3}, t = 1 - \frac{1}{3}\sqrt{3}$

44. Solve the equation $2x^2 - 15x + 2 = 0$ graphically, correct to two decimals.

Answer: easy

$x \approx 0.13, x \approx 7.36$

45. Solve the equation $-1.5x^2 - 2.33x + 4.0 = 0$ graphically, correct to two decimals.

Answer: easy

$x \approx -2.58, x \approx 1.03$

46. The height h of a ball dropped above ground after t seconds is given by the formula $h = -16t^2 + h_0$,where h_0 is the initial height of the ball. Suppose a ball is dropped from the top of a 256-foot building.

(a) After 2 seconds what is the height of ball?

(b) Find the time it takes to fall half the distance to ground level.

(c) Find the time it takes for the ball to reach the ground.

Answer: medium

(a) Since the initial height is $h_0 = 256$ and $t = 2$, the height of the ball is $h = -16(2)^2 + 256 = 192$ feet.

(b) Half the distance is 128 feet, so we have $128 = -16(t)^2 + 256$ or $t \approx 2.8$ s

(c) The ball will hit the ground when $h = 0$ or $0 = -16(t)^2 + 256$, so $t = 4$ s

47. Carolyn and Brian work in an accounting office. If they work together they can complete a corporate tax return in 2.5 hours. Working alone it would take Brian 35 minutes longer than Carolyn to complete the return. How long does it take each person working alone to finish the job? If a third person is used to help complete the task, and works with the same efficiency as Brian, what is the total time it would take to complete the return if all three work together?
Answer: hard
Let t be the time it takes Carolyn to work. Then Brian's time would be $t + 35$ min. Thus $\frac{1}{t} + \frac{1}{t+35} = \frac{1}{150} \Leftrightarrow$ $50(2t + 35) = t^2 + 35t \Leftrightarrow t^2 - 265t - 5250 = 0 \Leftrightarrow t = \pm\frac{\sqrt{(-265)^2 - 4(1)(-5250)}}{2} \approx 283.5$ min or about 4.7 hours. Brian takes 318.5 minutes or 5.3 hours. With the third person helping, the total time is $\frac{1}{283.5} + \frac{2}{318.5} = \frac{1}{T} \Leftrightarrow T \approx 101.1$ min or 1.7 hours.

SECTION 3.4

1. Find the real and imaginary parts of the complex number $i\sqrt{5}$.
(a) real part 0, imaginary part 5
(b) real part 0, imaginary part $\sqrt{5}$
(c) real part $\sqrt{5}$, imaginary part 0
(d) real part 1, imaginary part 5
(e) real part 5, imaginary part $-\sqrt{5}$
Answer: easy (b)
$i\sqrt{5}$ has real part 0 and imaginary part $\sqrt{5}$.

2. Find the real and imaginary parts of the complex number $\dfrac{\sqrt{2} - \sqrt{-6}}{\sqrt{3}}$.
(a) real part $\sqrt{\frac{2}{3}}$, imaginary part $-\sqrt{6}$
(b) real part $\sqrt{\frac{4}{3}}$, imaginary part $\sqrt{6}$
(c) real part $\sqrt{\frac{1}{6}}$, imaginary part $\sqrt{\frac{1}{3}}$
(d) real part $\sqrt{\frac{2}{3}}$, imaginary part $-\sqrt{2}$
(e) real part $\sqrt{\frac{1}{6}}$, imaginary part $\sqrt{6}$
Answer: easy (d)
$\dfrac{\sqrt{2} - \sqrt{-6}}{\sqrt{3}} = \dfrac{\sqrt{2}}{\sqrt{3}} - \dfrac{i\sqrt{6}}{\sqrt{3}} = \sqrt{\dfrac{2}{3}} - i\sqrt{2}$ has real part $\sqrt{\dfrac{2}{3}}$ and imaginary part $-\sqrt{2}$.

3. Find the real and imaginary parts of the complex number $\dfrac{1 - 3i}{\sqrt{-25}}$.
(a) real part $-\frac{3}{5}$, imaginary part $-\frac{1}{5}$
(b) real part $-\frac{1}{5}$, imaginary part $-\frac{3}{25}$
(c) real part $\frac{3}{5}$, imaginary part $-\frac{1}{3}$
(d) real part $\frac{1}{5}$, imaginary part $-\frac{3}{5}$
(e) real part $\frac{2}{5}$, imaginary part $-\frac{2}{5}$
Answer: medium (a)
$\dfrac{1 - 3i}{\sqrt{-25}} = \dfrac{1 - 3i}{5i} = \dfrac{-3i}{5i} + \dfrac{1}{5i} = -\dfrac{3}{5} - \dfrac{1}{5}i$ has real part $-\dfrac{3}{5}$ and imaginary part $\dfrac{1}{5}$.

4. Evaluate the expression $(7 - 3i) + (4 + 9i)$ and write the results in the form $a + bi$.
(a) $3 + 14i$
(b) $6 - 2i$
(c) $12 - 8i$
(d) $14 + 16i$
(e) $11 + 6i$
Answer: easy (e)
$(7 - 3i) + (4 + 9i) = (7 + 4) + (-3i + 9i) = 11 + 6i$

5. Evaluate the expression $(5 - i) - (6 + 3i)$ and write the results in the form $a + bi$.
(a) $-2 + 4i$
(b) $1 - 6i$
(c) $-1 - 4i$
(d) $12 - 4i$
(e) $3 + 4i$
Answer: easy (c)
$(5 - i) - (6 + 3i) = (5 - 6) + (-i - 3i) = -1 - 4i$

6. Evaluate the expression $8i - (2+i)$ and write the results in the form $a+bi$.
(a) $-1+4i$ (b) $-2+7i$ (c) $2-4i$ (d) $5-6i$ (e) $1+5i$
Answer: easy (b)
$8i - (2+i) = -2 + (8i - i) = -2 + 7i$

7. Evaluate the expression $3i\left(\frac{1}{5} - i\right)$ and write the results in the form $a+bi$.
(a) $\frac{1}{5}+4i$ (b) $2+\frac{4}{5}i$ (c) $-2+\frac{1}{5}i$ (d) $\frac{2}{5}+3i$ (e) $3+\frac{3}{5}i$
Answer: easy (e)
$3i \cdot \left(\frac{1}{5} - i\right) = 3i \cdot \frac{1}{5} - 3i \cdot i = \frac{3}{5}i + 3 = 3 + \frac{3}{5}i$

8. Evaluate the expression $(4-7i)(1+3i)$ and write the results in the form $a+bi$.
(a) $5+13i$ (b) $3+2i$ (c) $25+5i$ (d) $6-2i$ (e) $13-5i$
Answer: easy (c)
$(4-7i)(1+3i) = 4 + 12i - 7i - 21i^2 = 4 + 5i + 21 = 25 + 5i$

9. Evaluate the expression $\left(\frac{2}{3}+10i\right)\left(\frac{5}{6}+12i\right)$ and write the results in the form $a+bi$.
(a) $-\frac{947}{9}+\frac{16}{3}i$ (b) $-\frac{1075}{9}+\frac{49}{3}i$ (c) $-32-\frac{17}{3}i$ (d) $\frac{140}{3}+\frac{15}{32}i$ (e) $\frac{1274}{3}+\frac{16}{9}i$
Answer: easy (b)
$\left(\frac{2}{3}+10i\right)\left(\frac{5}{6}+12i\right) = \frac{5}{9} + \frac{25}{3}i + 8i + 120i^2 = -\frac{1075}{9} + \frac{49}{3}i$

10. Evaluate the expression $\dfrac{1}{2+i}$ and write the results in the form $a+bi$.
(a) $\frac{1}{3}+\frac{2}{3}i$ (b) $\frac{1}{5}+\frac{2}{5}i$ (c) $\frac{4}{3}-\frac{2}{3}i$ (d) $\frac{1}{3}-\frac{2}{3}i$ (e) $\frac{2}{5}-\frac{1}{5}i$
Answer: medium (e)
$$\frac{1}{2+i} = \frac{1}{2+i} \cdot \frac{2-i}{2-i} = \frac{2-i}{4-i^2} = \frac{2-i}{5} = \frac{2}{5} - \frac{1}{5}i$$

11. Evaluate the expression $\dfrac{3-i}{2+3i}$ and write the results in the form $a+bi$.
(a) $\frac{5}{11}+\frac{7}{11}i$ (b) $\frac{2}{11}-\frac{2}{15}i$ (c) $\frac{3}{13}-\frac{11}{13}i$ (d) $\frac{5}{17}-\frac{2}{15}i$ (e) $\frac{7}{17}-\frac{9}{11}i$
Answer: medium (c)
$$\frac{3-i}{2+3i} = \frac{(3-i)(2-3i)}{(2+3i)(2-3i)} = \frac{6-9i-2i+3i^2}{4-9i^2} = \frac{3-11i}{13} = \frac{3}{13} - \frac{11}{13}i$$

12. Evaluate the expression $\dfrac{5}{5-2i}$ and write the results in the form $a+bi$.
(a) $\frac{3}{23}+\frac{11}{17}i$ (b) $\frac{25}{29}+\frac{10}{29}i$ (c) $\frac{6}{13}+\frac{11}{43}i$ (d) $\frac{17}{19}+\frac{67}{75}i$ (e) $\frac{12}{13}+\frac{2}{27}i$
Answer: medium (b)
$$\frac{5}{5-2i} = \frac{5(5+2i)}{(5-2i)(5+2i)} = \frac{25+10i}{25-4i^2} = \frac{25+10i}{29} = \frac{25}{29} + \frac{10}{29}i$$

13. Evaluate the expression $(2-6i)^{-1}$ and write the results in the form $a+bi$.
(a) $\frac{1}{18}+\frac{1}{13}i$ (b) $\frac{1}{12}+\frac{5}{18}i$ (c) $\frac{3}{16}+\frac{2}{23}i$ (d) $\frac{3}{40}+\frac{28}{31}i$ (e) $\frac{1}{20}+\frac{3}{20}i$
Answer: medium (e)
$$(2-6i)^{-1} = \frac{1}{2-6i} = \frac{2+6i}{(2-6i)(2+6i)} = \frac{2+6i}{4-36i^2} = \frac{1}{20} + \frac{3}{20}i$$

14. Evaluate the expression $(3i)^4$ and write the results in the form $a+bi$.
(a) -81 (b) -16 (c) 4 (d) 16 (e) 81
Answer: medium (e)
$(3i)^4 = (3i)^2(3i)^2 = 9i^2 \cdot 9i^2 = (-9)(-9) = 81$

15. Evaluate the expression i^{1828} and write the results in the form $a+bi$.
(a) $-i$ (b) i (c) -1 (d) 1 (e) 956
Answer: medium (d)
$i^{1828} = (i^2)^{914} = (-1)^{914} = 1$

16. Evaluate the expression $\sqrt{-\frac{16}{9}}$ and write the results in the form $a+bi$.

(a) $\frac{2}{3}$ (b) $\frac{3}{4}i$ (c) $\frac{4}{3}i$ (d) $12i$ (e) $-i$

Answer: easy (c)

$\sqrt{-\frac{16}{9}} = i\sqrt{\frac{16}{9}} = \frac{4}{3}i$

17. Evaluate the expression $\sqrt{\frac{1}{4}}\sqrt{-64}$ and write the results in the form $a+bi$.

Answer: medium

$\sqrt{\frac{1}{4}}\sqrt{-64} = \frac{1}{2}\cdot 8i = 4i$

18. Evaluate the expression $\dfrac{-1-\sqrt{-1}}{-1+\sqrt{-1}}$ and write the results in the form $a+bi$.

Answer: medium

$$\frac{-1-\sqrt{-1}}{-1+\sqrt{-1}} = \frac{\left(-1-\sqrt{-1}\right)\left(-1-\sqrt{-1}\right)}{\left(-1+\sqrt{-1}\right)\left(-1-\sqrt{-1}\right)} = \frac{1+2\sqrt{-1}+\sqrt{-1}\sqrt{-1}}{2} = \frac{1+2i+i^2}{2} = \frac{2i}{2} = i$$

19. Evaluate the expression $\left(\sqrt{3}-\sqrt{-4}\right)\left(\sqrt{6}-\sqrt{-5}\right)$ and write the results in the form $a+bi$.

Answer: medium

$$\left(\sqrt{3}-\sqrt{-4}\right)\left(\sqrt{6}-\sqrt{-5}\right) = 3\sqrt{2}-i\sqrt{15}-i\sqrt{24}+i^2\sqrt{20} = 3\sqrt{2}-2\sqrt{5}+\left(-\sqrt{15}-2\sqrt{6}\right)i$$

20. Evaluate the expression $\dfrac{\sqrt{-7}\sqrt{-49}}{\sqrt{28}}$ and write the results in the form $a+bi$.

Answer: medium

$$\frac{\sqrt{-7}\,\sqrt{-49}}{\sqrt{28}} = \frac{i\sqrt{7}\cdot 7i}{2\sqrt{7}} = \frac{7i^2}{2} = -\frac{7}{2}$$

21. Find all solutions of the equation $25x^2+9=0$ and express them in the standard form $a+bi$.

Answer: medium

$$25x^2+9=0 \quad\Leftrightarrow\quad x^2=-\tfrac{9}{25} \quad\Leftrightarrow\quad x=\pm\sqrt{-\tfrac{9}{25}} \quad\Leftrightarrow\quad x=\pm\tfrac{3}{5}i$$

22. Find all solutions of the equation $x^2+\frac{1}{2}x+\frac{1}{4}=0$ and express them in the standard form $a+bi$.

Answer: medium

$$x^2+\tfrac{1}{2}x+\tfrac{1}{4}=0 \quad\Leftrightarrow\quad x=\frac{-\frac{1}{2}\pm\sqrt{\left(\frac{1}{2}\right)^2-4(1)\left(\frac{1}{4}\right)}}{2(1)} = \frac{-\frac{1}{2}\pm\sqrt{-\frac{3}{4}}}{2} = \frac{-\frac{1}{2}\pm\frac{\sqrt{3}}{2}i}{2} = -\tfrac{1}{4}\pm\tfrac{\sqrt{3}}{4}i$$

23. Find all solutions of the equation $10x^2+5x+2=0$ and express them in the standard form $a+bi$.

Answer: medium

$$10x^2+5x+2=0 \quad\Leftrightarrow\quad x=\frac{-5\pm\sqrt{5^2-4(10)(2)}}{2(10)} = \frac{-5\pm\sqrt{-55}}{20} = -\tfrac{1}{4}\pm\tfrac{\sqrt{55}}{20}i$$

24. Find all solutions of the equation $2x^2+3=2x$ and express them in the standard form $a+bi$.

Answer: medium

$$2x^2+3=2x \quad\Leftrightarrow\quad 2x^2-2x+3=0 \quad\Leftrightarrow\quad x=\frac{2\pm\sqrt{(-2)^2-4(2)(3)}}{2(2)} = \frac{2\pm\sqrt{-20}}{4} = \tfrac{1}{2}\pm\tfrac{\sqrt{5}}{2}i$$

25. Find all solutions of the equation $z^2-iz=0$ and express them in the standard form $a+bi$.

Answer: easy

$z^2-iz=0 \quad\Leftrightarrow\quad z(z-i)=0 \quad\Leftrightarrow\quad z=0$ or $z=i$

26. Find all solutions of the equation $2x-5i=6+i$ and express them in the standard form $a+bi$.

Answer: easy

$2x-5i=6+i \quad\Leftrightarrow\quad 2x=6i+6 \quad\Leftrightarrow\quad x=3+3i$

27. Find all solutions of the equation $2ix^2 + (4+i)x + 1 = 0$ and express them in the standard form $a + bi$.

Answer: medium

$2ix^2 + (4+i)\,x + 1 = 0 \quad \Leftrightarrow$

$$x = \frac{-(4+i) \pm \sqrt{(4+i)^2 - 4(2i)(1)}}{2(2i)} = \frac{-(4+i) \pm \sqrt{16+8i-1-8i}}{4i} = \frac{-(4+i) \pm \sqrt{15}}{4i} = \frac{-4i - i^2 \pm i\sqrt{15}}{4i^2}$$

$$= -\tfrac{1}{4} + \tfrac{4\pm\sqrt{15}}{4}i$$

28. Find all solutions of the equation $x^2/i - x + i - 1 = 0$ and express them in the standard form $a + bi$.

Answer: medium

$\dfrac{x^2}{i} - x + i - 1 = 0 \quad \Leftrightarrow \quad x^2 - ix - (1+i) = 0 \quad \Leftrightarrow$

$$x = \frac{i \pm \sqrt{(-i)^2 + 4(1)(1+i)}}{2(1)} = \frac{i \pm \sqrt{3+4i}}{2} = \frac{i \pm (2+i)}{2} = -1 \text{ or } 1+i$$

29. Prove the statement $\overline{zw} = \overline{z} \cdot \overline{w}$ for the complex numbers z and w. (Hint: Express z as $a + bi$ and w as $c + di$.)

Answer: medium

$$\overline{zw} = \overline{(a+bi)(c+di)} = \overline{ac + bdi^2 + (bc+ad)i} = ac + bdi^2 - (bc+ad)i = c(a-bi) - di(a-bi) = (a-bi)(c-di)$$
$$= \overline{a+bi} \cdot \overline{c+di} = \overline{z} \cdot \overline{w}$$

30. Recall that the symbol $\overline{z}$ represents the complex conjugate of z. If $z = a + bi$, prove the statement $\overline{\overline{z}} = z$.

Answer: medium

$\overline{\overline{z}} = z$. LHS: $\overline{\overline{z}} = \overline{\overline{(a+bi)}} = \overline{(a-bi)} = a + bi = z =$ RHS.

31. Recall that the symbol $\overline{z}$ represents the complex conjugate of z. If $z = a + bi$, prove the statement $z = \overline{z}$ if and only if z is real.

Answer: hard

If z is real, then $z = a$. Therefore $\overline{z} = a$. This implies $z = \overline{z}$. Conversely if $z = \overline{z}$, then $a + bi = a - bi \quad \Leftrightarrow \quad 2bi = 0 \quad \Leftrightarrow$ $b = 0$. Hence $z = a$, a real number.

32. Prove the statement $\dfrac{z}{\overline{z}} - \dfrac{\overline{z}}{z} = \dfrac{4xyi}{x^2+y^2}$, assuming $z = x + yi$ and $|z| \neq 0$.

Answer: hard

$$\frac{z}{\overline{z}} - \frac{\overline{z}}{z} = \frac{x+yi}{x-yi} - \frac{x-yi}{x+yi} = \frac{(x+yi)(x+yi) - (x-yi)(x-yi)}{(x-yi)(x+yi)} = \frac{x^2 + 2xyi - y^2 - x^2 + 2xyi + y^2}{x^2+y^2} = \frac{4xyi}{x^2+y^2}$$

33. Evaluate the expression $\dfrac{1}{\sqrt{-1}}$ and write the result in the form $a + bi$.

(a) $0 + i$ (b) $1 - i$ (c) i (d) $0 - i$ (e) $1 - i^2$

Answer: easy (d)

$$\tfrac{1}{\sqrt{-1}} = \tfrac{1}{i} \cdot \tfrac{-i}{-i} = \tfrac{-i}{-i^2} = -i = 0 - i$$

34. Evaluate the expression $-\left(1 - \sqrt{-2}\right) + \left[1 - \left(\sqrt{-1} - \sqrt{-2}\right)\right]$ and write the result in the form $a + bi$.

(a) $i\left(2\sqrt{2} - 1\right)$ (b) $0 + i\left(2\sqrt{2} - 1\right)$ (c) $2\sqrt{2} - 1$ (d) $2\sqrt{2} - i$ (e) $1 - \sqrt{2}i$

Answer: easy (b)

$$-\left(1 - \sqrt{-2}\right) + \left[1 - \left(\sqrt{-1} - \sqrt{-2}\right)\right] = 2i\sqrt{2} - i = 0 + i\left(2\sqrt{2} - 1\right)$$

35. Evaluate the expression $\dfrac{4}{3 - \sqrt{-4}} + \dfrac{i}{3 + \sqrt{-4}}$ and write the result in the form $a + bi$.

(a) $\frac{11}{5} - \frac{14}{5}i$ (b) $\frac{14}{5} + \frac{11}{5}i$ (c) $\frac{11}{13} - \frac{5}{13}i$ (d) $\frac{11}{13} + \frac{14}{13}i$ (e) $\frac{14}{13} + \frac{11}{13}i$

Answer: medium (e)

$$\frac{4}{3 - \sqrt{-4}} + \frac{i}{3 + \sqrt{-4}} = \frac{4}{3 - 2i} + \frac{i}{3 + 2i} = \frac{12 + 8i + 3i - 2i^2}{9 - 4i^2} = \tfrac{14}{13} + \tfrac{11}{13}i$$

36. Evaluate the expression $1+\dfrac{\sqrt{-36}\sqrt{-49}i}{\sqrt{21}}$ and write the result in the form $a+bi$.

(a) $\sqrt{21}-42i$ (b) $21-42i$ (c) $1-2i\sqrt{21}$ (d) $-\sqrt{21}+42i$ (e) $-\sqrt{21}-42i$

Answer: medium (c)

$$1+\frac{\sqrt{-36}\sqrt{-49}i}{\sqrt{21}}=\frac{\sqrt{21}+42i^3}{\sqrt{21}}=-\tfrac{\sqrt{21}}{21}\left(-\sqrt{21}+42i\right)=1-2i\sqrt{21}$$

37. Find all solutions of the equation $x^2+5x+25=0$ and express them in standard form $a+bi$.
Answer: easy

$$x^2+5x+25=0 \quad\Leftrightarrow\quad x=\frac{-5\pm\sqrt{25-4(1)(25)}}{2}=-\tfrac{5}{2}\pm\tfrac{5}{2}i\sqrt{3}$$

38. Find all solutions of the equation $6x^2+5x=-24$ and express them in standard form $a+bi$.
Answer: easy

$$6x^2+5x+24=0 \quad\Leftrightarrow\quad x=\frac{-5\pm\sqrt{25-4(6)(24)}}{2(6)}=-\tfrac{5}{12}\pm\tfrac{1}{12}i\sqrt{551}$$

39. Find all solutions of the equation $4x^2-7x=-x-22$ and express them in standard form $a+bi$.
Answer: easy

$$4x^2-7x=-x-22 \quad\Leftrightarrow\quad 4x^2-6x+22=0 \quad\Leftrightarrow\quad x=\frac{6\pm\sqrt{36-4(4)(22)}}{8}=\tfrac{3}{4}\pm\tfrac{1}{4}i\sqrt{79}$$

40. Find all solutions of the equation $y+4+\dfrac{5}{y}=0$ and express them in standard form $a+bi$.

Answer: easy

$$y+4+\frac{5}{y}=0 \quad\Leftrightarrow\quad \frac{y^2+4y+5}{y}=0 \quad\Leftrightarrow\quad y^2+4y+5 \quad\Leftrightarrow\quad y=\frac{-4\pm\sqrt{16-4(1)(5)}}{2}=-2\pm i$$

41. Find all solutions of the equation $\dfrac{t-1}{3}+3+\dfrac{5}{t+1}=0$ and express them in standard form $a+bi$.

Answer: easy

$$\frac{t-1}{3}+3+\frac{5}{t+1}=0 \quad\Leftrightarrow\quad \frac{1}{3}\frac{t^2+9t+23}{t+1}=0 \quad\Leftrightarrow\quad t^2+9t+23=0 \quad\Leftrightarrow$$
$$t=\frac{-9\pm\sqrt{81-4(1)(23)}}{2}=-\tfrac{9}{2}\pm\tfrac{1}{2}i\sqrt{11}$$

SECTION 3.5

1. Find all real solutions of the equation $x^5-32=0$.
(a) $-2\sqrt{2}$ (b) -2 (c) $\sqrt{2}$ (d) 2 (e) 4
Answer: easy (d)
$$x^5-32=0 \quad\Leftrightarrow\quad x=(32)^{1/5}=\left(2^5\right)^{1/5}=2$$

2. Find all real solutions of the equation $8x^6=27$.
(a) $-\sqrt{\frac{3}{2}}$ (b) $\pm\sqrt{\frac{3}{2}}$ (c) $\sqrt{\frac{3}{2}}$ (d) $\frac{9}{4}$ (e) $\pm\frac{9}{2}$
Answer: easy (b)
$$8x^6=27 \quad\Leftrightarrow\quad x^6=\frac{27}{8} \quad\Leftrightarrow\quad x=\pm\left(\frac{3^3}{2^3}\right)^{1/6}=\pm\left(\frac{3}{2}\right)^{1/2}=\pm\sqrt{\frac{3}{2}}$$

3. Find all real solutions of the equation $(x+1)^5 - 4(x+1)^3 = 0$.
(a) $x = \pm 1$ or -3 (b) $x = \pm 1$ or 3 (c) $x = \pm 2$ or -1 (d) $x = -3, -1$ or 2 (e) $x = \pm 3$ or -1
Answer: medium (a)
$(x+1)^5 - 4(x+1)^3 = 0 \Leftrightarrow (x+1)^3\left((x+1)^2 - 4\right) = 0 \Leftrightarrow (x+1)^3(x+1-2)(x+1+2) = 0 \Leftrightarrow$
$(x+1)^3(x-1)(x+3) = 0 \Leftrightarrow x = -1, 1, -3$

4. Find all real solutions of the equation $2x^3 + x^2 - 4x - 2 = 0$.
(a) $x = -2$ or $1 \pm \sqrt{5}$ (b) $x = -\frac{1}{3}, \frac{1}{2}$, or $\sqrt{3}$ (c) $x = -\frac{1}{2}$ or $\pm\sqrt{2}$ (d) $x = \frac{2}{3}, \frac{4}{3}$ or 2 (e) $x = \pm 2$ or $\frac{2}{3}$
Answer: medium (c)
$2x^3 + x^2 - 4x - 2 = 0 \Leftrightarrow x^2(2x+1) - 2(2x+1) = 0 \Leftrightarrow (2x+1)(x^2-2) = 0 \Leftrightarrow x = -\frac{1}{2}, \pm\sqrt{2}$

5. Find all real solutions of the equation $10x^3 - 4x + 2 = x^3 + 6x^2 - x$.
(a) $x = \frac{1}{3}$ or $\pm\frac{2}{\sqrt{3}}$ (b) $x = \frac{2}{3}$ or $\pm\frac{1}{\sqrt{3}}$ (c) $x = \frac{2}{5}$ or $\pm\frac{3}{\sqrt{5}}$ (d) $x = 2$ or $\pm\sqrt{2}$ (e) $x = -\frac{1}{9}$ or $\pm\frac{3}{\sqrt{2}}$
Answer: medium (b)
$10x^3 - 4x + 2 = x^3 + 6x^2 - x \Leftrightarrow 9x^3 - 6x^2 - 3x + 2 = 0 \Leftrightarrow 3x^2(3x-2) - (3x-2) = 0 \Leftrightarrow$
$(3x^2-1)(3x-2) = 0 \Leftrightarrow x = \frac{2}{3}$ or $\pm\frac{1}{\sqrt{3}}$

6. Find all real solutions of the equation $x - \sqrt{16-4x} = 0$.
(a) $x = \pm 2 + 2\sqrt{5}$ (b) $x = -2 + 2\sqrt{5}$ (c) $x = -2 - 2\sqrt{5}$ (d) $3 \pm 2\sqrt{2}$ (e) $-4 \pm 2\sqrt{2}$
Answer: hard (b)
$x - \sqrt{16-4x} = 0 \Leftrightarrow x = \sqrt{16-4x} \Rightarrow x^2 = 16 - 4x \Leftrightarrow x^2 + 4x - 16 = 0 \Leftrightarrow$
$x = \dfrac{-4 \pm \sqrt{16+64}}{2} = \dfrac{-4 \pm 4\sqrt{5}}{2} = -2 \pm 2\sqrt{5}$. Now $x = -2 + 2\sqrt{5}$ is a solution but not $-2 - 2\sqrt{5}$.

7. Find all real solutions of the equation $\sqrt{2x+1} + \sqrt{x+1} = 2$.
(a) $x = -1$ or 2 (b) $x = 0$ or $\sqrt{3}$ (c) -1 or 0 (d) $\sqrt{3}$ (e) 0
Answer: medium (e)
$\sqrt{2x+1} + \sqrt{x+1} = 2 \Leftrightarrow \sqrt{2x+1} = 2 - \sqrt{x+1} \Leftrightarrow 2x + 1 = 4 - 4\sqrt{x+1} + x + 1 \Leftrightarrow$
$4\sqrt{x+1} = 4 - x \Leftrightarrow 16(x+1) = 16 - 8x + x^2 \Leftrightarrow x^2 - 24x = 0 \Leftrightarrow x(x-24) = 0$. Now $x = 0$ is a solution but $x = 24$ is not.

8. Find all real solutions of the equation $x^4 - 5x^2 + 4 = 0$.
(a) $x = \pm 1$ or ± 2 (b) $x = -1$ or 2 (c) $x = -2$ or 1 (d) $x = \pm 1$ or 2 (e) $x = \pm 2$ or 1
Answer: medium (a)
$x^4 - 5x^2 + 4 = 0 \Leftrightarrow (x^2-4)(x^2-1) = 0$. So $x = \pm 2$ or $x = \pm 1$

9. Find all real solutions of the equation $x^6 - 2x^3 - 3 = 0$.
(a) $x = -\sqrt[3]{3}$ or 1 (b) $x = \pm 2$ or -1 (c) $x = \sqrt[3]{3}$ or -1 (d) $x = \pm\sqrt{3}$ or ± 1 (e) $x = \pm\sqrt[3]{2}$ or ± 1
Answer: medium (c)
$x^6 - 2x^3 - 3 = 0 \Leftrightarrow (x^3-3)(x^3+1) = 0$. So $x^3 = 3$ or $x^3 = -1$. Hence $x = \sqrt[3]{3}$ or -1.

10. Find all real solutions of the equation $x^{1/2} + 5x^{-1/2} = 6x^{-3/2}$.
(a) $x = 1$ or $\pm\sqrt{3}$ (b) $x = \pm 1$ (c) $x = -1$ (d) $x = 1$ (e) $x = -1$ or -6
Answer: medium (d)
$x^{1/2} + 5x^{-1/2} = 6x^{-3/2} \Leftrightarrow x^{-3/2}(x^2 + 5x - 6) = 0 \Leftrightarrow x^{-3/2}(x+6)(x-1) = 0$. So $x = 1$ is a solution but $x = 0$ and -6 are not.

11. Find all real solutions of the equation $\sqrt{x} + 4\sqrt[4]{x} - 6 = 0$.
(a) $x = \pm 1$ (b) $x = \pm\sqrt[4]{3}$ (c) $x = \pm\sqrt{3}$ (d) $x = -1$ or 4 (e) $x = 16$
Answer: medium (e)
$\sqrt{x} + \sqrt[4]{x} - 6 = 0 \Leftrightarrow \left(\sqrt[4]{x} - 2\right)\left(\sqrt[4]{x} + 3\right) = 0 \Leftrightarrow \sqrt[4]{x} = 2$ or -3. Since $\sqrt[4]{x} \geq 0$ for real x, we have $x = 2^4 = 16$.

12. Find all real solutions of the equation $x - 3\sqrt{x} + 2 = 0$.
(a) $x = 1$ or 2 (b) $x = 1$ or 4 (c) $x = 2$ or 3 (d) $x = \pm 1$ (e) $x = \pm 2$
Answer: medium (b)
$x - 3\sqrt{x} + 2 = 0 \quad \Leftrightarrow \quad (\sqrt{x} - 1)(\sqrt{x} - 2) = 0 \quad \Leftrightarrow \quad \sqrt{x} = 1$ or $2 \quad \Leftrightarrow \quad x = 1$ or 4.

13. Find all real solutions of the equation $4x^{-4} - 20x^{-2} + 8 = 0$.
(a) $x = \pm\sqrt{\frac{4\pm\sqrt{13}}{2}}$ (b) $x = \pm\sqrt{\frac{3\pm\sqrt{11}}{4}}$ (c) $x = \sqrt{\frac{3\pm\sqrt{11}}{2}}$ (d) $x = \pm\sqrt{\frac{5\pm\sqrt{13}}{2}}$ (e) $x = \pm\sqrt{\frac{5\pm\sqrt{17}}{4}}$
Answer: medium (e)
$4x^{-4} - 20x^{-2} + 8 = 0 \quad \Leftrightarrow \quad 4 - 20x^2 + 8x^4 = 0 \quad \Leftrightarrow \quad 4(2x^4 - 5x^2 + 1) = 0 \quad \Leftrightarrow \quad x^2 = \frac{5\pm\sqrt{17}}{4} \quad \Leftrightarrow$
$x = \pm\frac{\sqrt{5\pm\sqrt{17}}}{2}$.

14. Find all real solutions of the equation $\sqrt{9 - x^2} - \dfrac{2}{\sqrt{9 - x^2}} = 1$.

Answer: hard
$\sqrt{9 - x^2} - \dfrac{2}{\sqrt{9 - x^2}} = 1 \quad \Leftrightarrow \quad 9 - x^2 - 2 = \sqrt{9 - x^2} \quad \Leftrightarrow \quad 7 - x^2 = \sqrt{9 - x^2} \quad \Rightarrow \quad 49 - 14x^2 + x^4 = 9 - x^2$
$\Leftrightarrow \quad x^4 - 13x^2 + 40 = 0 \quad \Leftrightarrow \quad (x^2 - 5)(x^2 - 8) = 0 \quad \Leftrightarrow \quad x = \pm\sqrt{5}, \pm\sqrt{8}$. Now $\pm\sqrt{5}$ are solutions, but $\pm\sqrt{8}$ are not.

15. Find all real solutions of the equation $\sqrt{2 + \sqrt{x + \sqrt{3x + 1}}} = \sqrt{4 + \sqrt{x}}$.

Answer: hard
$\sqrt{2 + \sqrt{x + \sqrt{3x + 1}}} = \sqrt{4 + \sqrt{x}} \quad \Rightarrow \quad 2 + \sqrt{x + \sqrt{3x + 1}} = 4 + \sqrt{x} \quad \Leftrightarrow \quad \sqrt{x + \sqrt{3x + 1}} = 2 + \sqrt{x} \quad \Rightarrow$
$x + \sqrt{3x + 1} = 4 + 4\sqrt{x} + x \quad \Rightarrow \quad 3x + 1 = 16 + 32\sqrt{x} + 16x \quad \Leftrightarrow \quad 13x + 32\sqrt{x} + 15 = 0 \quad \Leftrightarrow$
$\sqrt{x} = \dfrac{-32 \pm \sqrt{32^2 - 4(13)(15)}}{2(13)} = \dfrac{-32 \pm \sqrt{244}}{26}$. Since $32 > \sqrt{244}$, there is no solution.

16. Find all complex number solutions of the equation $x^4 = \frac{256}{81}$.
Answer: medium
$81x^4 = 256 \quad \Leftrightarrow \quad 81x^4 - 256 = 0 \quad \Leftrightarrow \quad (9x^2 - 16)(9x^2 + 16) = 0 \quad \Leftrightarrow \quad (3x - 4)(3x + 4)(9x^2 + 16) = 0 \quad \Leftrightarrow$
$x = \pm\frac{4}{3}, \pm\frac{4}{3}i$

17. Find all complex number solutions of the equation $x^4 - 81 = 0$.
Answer: medium
$x^4 - 81 = 0 \quad \Leftrightarrow \quad (x^2 - 9)(x^2 + 9) = 0 \quad \Leftrightarrow \quad x = \pm 3, \pm 3i$

18. Find all complex number solutions of the equation $x^4 - x^3 + x^2 - x = 0$.
Answer: medium
$x^4 - x^3 + x^2 - x = 0 \quad \Leftrightarrow \quad x^2(x^2 - x) + 1(x^2 - x) = 0 \quad \Leftrightarrow \quad x(x - 1)(x^2 + 1) = 0 \quad \Leftrightarrow \quad x = 0, 1, \pm i$

19. Find all complex number solutions of the equation $x^3 + 4x^2 + 16x + 64 = 0$.
Answer: medium
$x^3 + 4x^2 + 16x + 64 = 0 \quad \Leftrightarrow \quad x^2(x + 4) + 16(x + 4) = 0 \quad \Leftrightarrow \quad (x + 4)(x^2 + 16) = 0 \quad \Leftrightarrow \quad x = -4, \pm 4i$

20. Find all complex number solutions of the equation $x^6 + 16x^4 - 3x^2 - 48 = 0$.
Answer: medium
$x^6 + 16x^4 - 3x^2 - 48 = 0 \quad \Leftrightarrow \quad x^4(x^2 + 16) - 3(x^2 + 16) = 0 \quad \Leftrightarrow \quad (x^2 + 16)(x^4 - 3) = 0 \quad \Leftrightarrow$
$(x^2 + 16)\left(x^2 - \sqrt{3}\right)\left(x^2 + \sqrt{3}\right) = 0 \quad \Leftrightarrow \quad x = \pm 4i, \pm\sqrt[4]{3}, \pm i\sqrt[4]{3}$

21. Find all complex number solutions of the equation $\sqrt{1-\sqrt{3x^2-11}}=\sqrt{6-x^2}$

Answer: hard

$\sqrt{1-\sqrt{3x^2-11}}=\sqrt{6-x^2} \quad \Rightarrow \quad 1-\sqrt{3x^2-11}=6-x^2 \quad \Leftrightarrow \quad \sqrt{3x^2-11}=x^2-5 \quad \Rightarrow$ $3x^2-11=x^4-10x^2+25 \quad \Leftrightarrow \quad x^4-13x^2+36=0 \quad \Leftrightarrow \quad (x^2-4)(x^2-9)=0 \quad \Leftrightarrow \quad x=\pm 2, \pm 3$. $x=\pm 3$ are solutions but not ± 2.

22. A jeweler has three small solid spheres made of gold, of radius 3 mm, 4 mm, and 5 mm. He wishes to melt these down and make just one sphere out of them. What will the radius of this larger sphere be?

Answer: medium

Let the radius of the larger sphere be x mm. Thus $\frac{4}{3}\pi x^3=\frac{4}{3}\pi(3^3+4^3+5^3) \quad \Leftrightarrow \quad x^3=3^3+4^3+5^3 \quad \Leftrightarrow \quad x^3=216$ $\Leftrightarrow \quad x=6$. Therefore the radius of the larger sphere is 6 mm.

23. A boardwalk is parallel to and 265 ft inland from a straight shoreline. There is a sandy beach between the boardwalk and the shoreline. A man is standing on the boardwalk, exactly 689 ft across the sand from his beach umbrella, which is right at the waterline. The man walks 3 ft/s on the boardwalk and 2 ft/s on the sand. How far should he go on the boardwalk before veering off onto the sand if he wishes to reach his umbrella in exactly 4 minutes and 24 seconds?

Answer: hard

Let the distance that he goes on the boardwalk before veering off onto the sand be x ft. The distance along the boardwalk from where he started to where the umbrella lies perpendicular to it is $\sqrt{689^2-265^2}=636$ ft. Hence it takes $\frac{1}{3}x$ s before he veers off onto the sand and $\frac{1}{2}\sqrt{265^2+(636-x^2)}$ s from the boardwalk to the umbrella. Thus $\frac{1}{3}x+\frac{1}{2}\sqrt{265^2+(636-x^2)}=4\cdot 60+24 \quad \Leftrightarrow \quad \sqrt{265^2+(636-x^2)}=528+\frac{2}{3}x \quad \Leftrightarrow$ $70{,}225+404{,}496-1272x+x^2=278{,}784+704x+\frac{4}{9}x^2 \quad \Leftrightarrow \quad x\approx 102.1$ or 3454.7 (using the quadratic formula). The first answer is obviously the one we want, so the distance that he goes on the boardwalk before veering off onto the sand is approximately 102 ft.

24. Two television monitors sitting beside each other on a shelf in an appliance store have the same screen height. One has a conventional screen that is 6 in. wider than it is high. The other has a wider, high-definition screen that is 2.4 times as wide as it is high. The diagonal measure of the wider screen is 16 in. more that the diagonal measure of the smaller. What is the height of the screens? (You may assume that the height is greater than 10 in.)

Answer: medium

Let the height of the screens be x in. Hence the width of the smaller screen is $x+6$ in, and of the bigger screen is $2.4x$ in. Thus $\sqrt{x^2+(2.4x)^2}=\sqrt{x^2+(x+6)^2}+16 \quad \Leftrightarrow \quad 2.60x-16=\sqrt{2x^2+12x+36} \quad \Leftrightarrow$ $6.76x^2-83.2x+256=2x^2+12x+36 \quad \Leftrightarrow \quad 4.76x^2-95.2x+220=0 \quad \Leftrightarrow$ $x=\dfrac{95.2\pm\sqrt{(95.2)^2-4(4.76)(220)}}{2(4.76)}\approx 17.33$ or 2.67. Therefore the height of the screens is about 17.33 in.

25. Find all complex solutions x in the equation $a^3x^3-b^3=0$. The constants a and b are positive real numbers.

Answer: medium

$a^3x^3-b^3=0 \quad \Leftrightarrow \quad (ax-b)(a^2x^2+abx+b^2)=0 \quad \Leftrightarrow \quad x=\dfrac{b}{a}$ or

$$x=\frac{-ab\pm\sqrt{(ab)^2-4(a^2)(b^2)}}{2a^2}=\frac{-ab\pm abi\sqrt{3}}{2a^2}=-\frac{b}{2a}\pm\frac{bi\sqrt{3}}{2a}.$$

26. Solve the equation $\sqrt{x}+a\sqrt[3]{x}-b\sqrt[6]{x}-ab=0$ for the variable x. The constants a and b are positive real numbers.

Answer: hard

Let $w=x^{1/6}$. Then $x^{1/3}=w^2$ and $x^{1/2}=w^3$, and so $w^3+aw^2-bw-ab=0 \quad \Leftrightarrow \quad (w^2-b)(w+a)=0 \quad \Rightarrow \quad w=-a$ or $w=\pm\sqrt{b}$ are the only possible real solutions. But since $x^{1/6}\geq 0$ for nonnegative x, neither $w=-a$ nor $w=-\sqrt{b}$ is real, and so $w=\sqrt{b}$ is the only real solution. Thus $x=b^3$.

27. Find all real solutions of the equation $x^4 = 5x^3 - 6x^2$.
Answer: easy
$x^4 = 5x^3 - 6x^2 \quad \Leftrightarrow \quad x^4 - 5x^3 + 6x^2 = 0 \quad \Leftrightarrow \quad x^2\left(x^2 - 5x + 6\right) = 0 \quad \Leftrightarrow \quad x = 0, x = 2, x = 3$

28. Find all real solutions of the equation $x^3 - 9x^2 - 4x + 36 = 0$.
Answer: easy
$x^3 - 9x^2 - 4x + 36 = 0 \quad \Leftrightarrow \quad x^2(x-9) - 4(x-9) = 0 \quad \Leftrightarrow \quad \left(x^2 - 4\right)(x-9) = 0 \quad \Leftrightarrow \quad x = \pm 2, x = 9$

29. Find all real solutions of the equation $\dfrac{1 + \dfrac{1+2y}{y^2}}{1 - \dfrac{1-2y}{y^2}} = 2$.
Answer: medium
$\dfrac{1 + \dfrac{1+2y}{y^2}}{1 - \dfrac{1-2y}{y^2}} = 2 \quad \Leftrightarrow \quad \dfrac{y^2 + 2y + 1}{y^2 + 2y - 1} = 2 \quad \Leftrightarrow \quad y^2 + 2y + 1 = 2\left(y^2 + 2y - 1\right) \quad \Leftrightarrow \quad y^2 + 2y - 3 = 0 \quad \Leftrightarrow$
$(y-1)(y+3) = 0 \quad \Leftrightarrow \quad y = -3, y = 1$

30. Find all real solutions of the equation $x^4 - 17x^2 + 72 = 0$.
Answer: medium
Let $u = x^2$ so $u^2 = x^4$. $x^4 - 17x^2 + 72 = 0 \quad \Leftrightarrow \quad u^2 - 17u + 72 = 0 \quad \Leftrightarrow \quad (u-8)(u-9) = 0 \quad \Leftrightarrow \quad u = 8, u = 9$ but $u = x^2$ so $x = \pm\sqrt{8}, x = \pm\sqrt{9} \quad \Leftrightarrow \quad x = \pm 3, x = \pm 2\sqrt{2}$

31. Find all real solutions of the equation $14x + 3\sqrt{x} - 2 = 0$.
Answer: medium
Let $w = \sqrt{x}$, $w^2 = x$. $14x + 3\sqrt{x} - 2 = 0 \quad \Leftrightarrow \quad 14w^2 + 3w - 2 = 0 \quad \Leftrightarrow \quad (7w-2)(2w+1) = 0, w = -\frac{1}{2}, w = \frac{2}{7} \quad \Leftrightarrow$ $x = \frac{1}{4}$ or $x = \frac{4}{49}$. Substituting, we find that $x = \frac{1}{4}$ does not work, so $x = \frac{4}{49}$ is the only solution.

32. Find all real solutions of the equation $\sqrt[3]{4x^2 + 16x} = x$.
Answer: hard
$\sqrt[3]{4x^2 + 16x} = x \quad \Leftrightarrow \quad x^3 - 4x^2 - 16x = 0 \quad \Leftrightarrow \quad x\left(x^2 - 4x - 16\right) = 0 \quad \Leftrightarrow \quad x = 0,$
$x = \dfrac{4 \pm \sqrt{16 + 4(1)(16)}}{2} = 2 \pm 2\sqrt{5}$. $x = 0$ and $2 + 2\sqrt{5}$ are the only solutions.

33. Members of a computer club wish to split the cost to purchase a personal computer that sells for \$1400. To further lower the cost by \$25 per user, the group asks one additional member to share in the expense. How many members originally intended to purchase the computer?
Answer: medium
Let x be the number of original members. $\dfrac{1400}{x} - 25 = \dfrac{1400}{x+1} \quad \Leftrightarrow \quad x(x+1)\left(\dfrac{1400}{x} - 25\right) = \dfrac{1400}{x+1}x(x+1) \quad \Leftrightarrow$
$1400x + 1400 - 25x^2 - 25x = 1400x \quad \Leftrightarrow \quad 25x^2 + 25x - 1400 = 0 \quad \Leftrightarrow$
$x = \dfrac{-25 \pm \sqrt{25^2 + 4(25)(1400)}}{2(25)} = 7$ members

34. A cardboard box that is used to store computer software has a volume of 96 in^3. Its length is 8 in. greater than its height, and its width is 2 in. less than its height. What are the dimensions of the box?
Answer: medium
Let the height be x. $V = lwh \quad \Leftrightarrow \quad 96 = x(x-2)(x+8) \quad \Leftrightarrow \quad x^3 + 6x^2 - 16x - 96 = 0 \quad \Leftrightarrow \quad \left(x^2 - 16\right)(x+6) = 0$ $\Leftrightarrow \quad x = 4$. The dimensions are 12 in. by 2 in by 4 in.

35. Before rappelling off the side of a cliff, a mountain climber needs to estimate its height. If she drops a stone off the side of the cliff she knows that it takes $\frac{\sqrt{h}}{4}$ seconds for the stone to strike the ground. She also knows that it takes $\frac{h}{1090}$ seconds for the sound to travel back up. The total elapsed time between dropping the stone and hearing the stone strike the ground is 4.5 seconds. What is the height of the cliff? How much time elapsed before the stone struck the ground?
Answer: medium

(a) Time stone falls + Time sound rises = Total elapsed time $\Leftrightarrow$ $\dfrac{\sqrt{h}}{4} + \dfrac{h}{1090} = 4.5$ $\Leftrightarrow$ $1090\sqrt{h} + 4h = 19620$ $\Rightarrow$ If $x^2 = h$, then $\sqrt{h} = x$, so $4x^2 + 1090x - 19620 = 0$ $\Leftrightarrow$ $x = \dfrac{-1090 + \sqrt{(1090)^2 - 4(4)(-19620)}}{2(4)} \approx 16.95$ $\Leftrightarrow$ $h \approx (16.95)^2 = 287.3$ ft.

(b) $t = \frac{\sqrt{h}}{4} \Rightarrow \frac{\sqrt{287.3}}{4} \approx 4.2$ s

SECTION 3.6

1. Let $S = \left\{-1, 0, \frac{1}{2}, \sqrt{2}, 2\right\}$. Use substitution to determine which of the elements of S satisfy the inequality $x - 1 > 0$.
(a) $\{-1, 0\}$ (b) $\{-1, 0, \frac{1}{2}\}$ (c) $\left\{0, \frac{1}{2}, \sqrt{2}\right\}$ (d) $\left\{\frac{1}{2}, \sqrt{2}, 2\right\}$ (e) $\left\{\sqrt{2}, 2\right\}$
Answer: easy (e)
$\left\{\sqrt{2}, 2\right\}$ satisfy $x - 1 > 0$.

2. Let $S = \left\{-1, 0, \frac{1}{2}, \sqrt{2}, 2\right\}$. Use substitution to determine which of the elements of S satisfy the inequality $4x + 1 \geq 2x$.
(a) $\{-1\}$ (b) $\{-1, 0, \frac{1}{2}\}$ (c) $\left\{0, \frac{1}{2}, \sqrt{2}\right\}$ (d) $\left\{0, \frac{1}{2}, \sqrt{2}, 2\right\}$ (e) $\left\{\frac{1}{2}, \sqrt{2}, 2\right\}$
Answer: easy (d)
$\left\{0, \frac{1}{2}, \sqrt{2}, 2\right\}$ satisfy $4x + 1 \geq 2x$.

3. Let $S = \left\{-1, 0, \frac{1}{2}, \sqrt{2}, 2\right\}$. Use substitution to determine which of the elements of S satisfy the inequality $10 \leq \dfrac{5}{2x}$.
(a) $\emptyset$ (b) $\{-1, 0\}$ (c) $\{-1\}$ (d) $\{0\}$ (e) S
Answer: medium (a)
No element of S satisfies $10 \leq \dfrac{5}{2x}$.

4. Let $S = \left\{-1, 0, \frac{1}{2}, \sqrt{2}, 2\right\}$. Use substitution to determine which of the elements of S satisfy the inequality $\dfrac{1}{1 - x^2} < 0$.
(a) $\left\{\sqrt{2}, 2\right\}$ (b) $\left\{\frac{1}{2}, \sqrt{2}, 2\right\}$ (c) $\left\{0, \frac{1}{2}, \sqrt{2}\right\}$ (d) $\left\{-1, 0, \sqrt{2}, 2\right\}$ (e) $\{-1, 0, 2\}$
Answer: easy (a)
$\left\{\sqrt{2}, 2\right\}$ satisfy $\dfrac{1}{1 - x^2} < 0$.

5. Solve the inequality $3x - 11 < 4$ in terms of intervals and illustrate the solution sets on the real number line.
Answer: easy
$3x - 11 < 4 \quad \Leftrightarrow \quad 3x < 15 \quad \Leftrightarrow \quad x < 5 \quad \Leftrightarrow \quad x \in (-\infty, 5)$
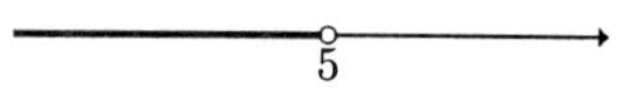

6. Solve the inequality $4 - 3x \geq 6$ in terms of intervals and illustrate the solution sets on the real number line.
Answer: easy
$4 - 3x \geq 6 \quad \Leftrightarrow \quad -3x \geq 2 \quad \Leftrightarrow \quad x \leq -\frac{2}{3} \quad \Leftrightarrow \quad x \in \left(-\infty, -\frac{2}{3}\right]$
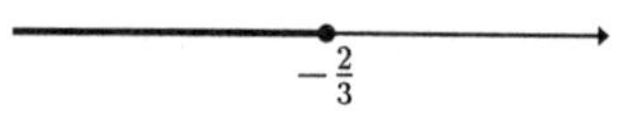

7. Solve the inequality $1 + 5x > 5 - 3x$ in terms of intervals and illustrate the solution sets on the real number line.
Answer: easy
$1 + 5x > 5 - 3x \quad \Leftrightarrow \quad 8x > 4 \quad \Leftrightarrow \quad x > \frac{1}{2} \quad \Leftrightarrow \quad x \in \left(\frac{1}{2}, \infty\right)$

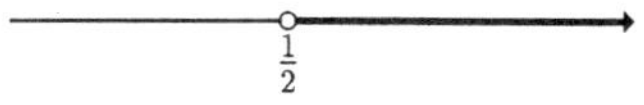

8. Solve the inequality $2(5x - 3) \leq 14x + 16$. Express your solution in the form of intervals and illustrate the solution sets on the real number line.
Answer: easy
$2(5x - 3) \leq 14x + 16 \quad \Leftrightarrow \quad 10x - 6 \leq 14x + 16 \quad \Leftrightarrow \quad 4x \geq -22 \quad \Leftrightarrow \quad x \geq -\frac{11}{2} \quad \Leftrightarrow \quad x \in \left[\frac{11}{2}, \infty\right)$

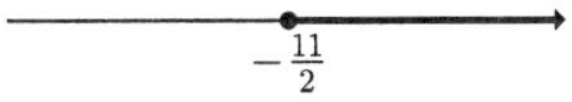

9. Solve the inequality $1 < 2x + 3 \leq 12$ in terms of intervals and illustrate the solution sets on the real number line.
Answer: easy
$1 < 2x + 3 \leq 12 \quad \Leftrightarrow \quad -2 < 2x \leq 9 \quad \Leftrightarrow \quad -1 < x \leq \frac{9}{2} \quad \Leftrightarrow \quad x \in \left(-1, \frac{9}{2}\right]$

10. Solve the inequality $-4 \leq 5 - 3x \leq 10$ in terms of intervals and illustrate the solution sets on the real number line.
Answer: easy
$-4 \leq 5 - 3x \leq 10 \quad \Leftrightarrow \quad -9 \leq -3x \leq 5 \quad \Leftrightarrow \quad 3 \geq x \geq -\frac{5}{3} \quad \Leftrightarrow \quad x \in \left[-\frac{5}{3}, 3\right]$

11. Solve the inequality $3x + 4 < x + 5 < 4x - 2$ in terms of intervals and illustrate the solution sets on the real number line.
Answer: easy
$3x + 4 < x + 5 < 4x - 2$. So $3x + 4 < x + 5 \quad \Leftrightarrow \quad x < \frac{1}{2}$, and $x + 5 < 4x - 2 \quad \Leftrightarrow \quad -3x < -7 \quad \Leftrightarrow \quad x > \frac{7}{3}$. Since $x > \frac{7}{3}$ and $x < \frac{1}{2}$, the solution set is empty.

12. Solve the inequality $x > 2 - x \geq 4 + 3x$ in terms of intervals and illustrate the solution sets on the real number line.
Answer: easy
$x > 2 - x \geq 4 + 3x$. So $x > 2 - x \quad \Leftrightarrow \quad 2x > 2 \quad \Leftrightarrow \quad x > 1$, and $2 - x \geq 4 + 3x \quad \Leftrightarrow \quad -4x \geq 2 \quad \Leftrightarrow \quad x \leq -\frac{1}{2}$. Since $x > 1$ and $x \leq -\frac{1}{2}$, the solution set is empty.

13. Solve the inequality $-\dfrac{1}{3} < \dfrac{3 - 2x}{4} \leq \dfrac{1}{5}$. Express your solution in the form of intervals and illustrate the solution sets on the real number line.
Answer: easy
$-\dfrac{1}{3} < \dfrac{3 - 2x}{4} \leq \dfrac{1}{5} \quad \Leftrightarrow \quad -20 < 15(3 - 2x) \leq 12 \quad \Leftrightarrow \quad -20 < 45 - 30x \leq 12 \quad \Leftrightarrow \quad -65 < -30x \leq -33 \quad \Leftrightarrow$
$\frac{13}{6} > x \geq \frac{11}{10} \quad \Leftrightarrow \quad x \in \left[\frac{11}{10}, \frac{13}{6}\right)$

14. Solve the inequality $3x < \dfrac{2 - x}{12} < \dfrac{3x + 1}{4}$. Express your solution in the form of intervals and illustrate the solution sets on the real number line.
Answer: medium
$3x < \dfrac{2 - x}{12} < \dfrac{3x + 1}{4} \quad \Leftrightarrow \quad 36x < 2 - x < 9x + 3$. So $37x < 2 \quad \Leftrightarrow \quad x < \frac{2}{37}$, and $2 - x < 9x + 3 \quad \Leftrightarrow \quad -1 < 10x$
$\Leftrightarrow \quad -\frac{1}{10} < x$. Therefore $-\frac{1}{10} < x < \frac{2}{37} \quad \Leftrightarrow \quad x \in \left(-\frac{1}{10}, \frac{2}{37}\right)$

15. Solve the inequality $-4 < 1/x \leq 2$ in terms of intervals and illustrate the solution sets on the real number line.
Answer: medium
$-4 < 1/x \leq 2$. So $-4 < 1/x$ is true for all $x > 0$ and implies $x < -\frac{1}{4}$ for $x < 0 \quad \Leftrightarrow \quad x < -\frac{1}{4}$ giving $x < -\frac{1}{4}$ or $x > 0$. Also $1/x \leq 2$ is true for all $x < 0$ and implies $x \geq \frac{1}{2}$ for $x > 0 \quad \Leftrightarrow \quad x \geq \frac{1}{2}$ giving $x < 0$ or $x \geq \frac{1}{2}$. The intersection of these is $x < -\frac{1}{4}$ or $x \geq \frac{1}{2} \quad \Leftrightarrow \quad x \in \left(-\infty, -\frac{1}{4}\right) \cup \left[\frac{1}{2}, \infty\right)$

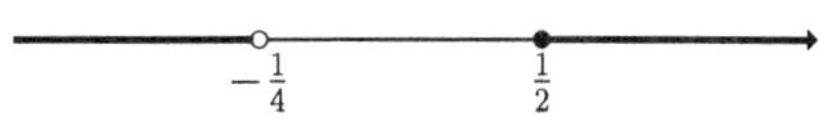

16. Solve the inequality $\frac{1}{x} > 3 \geq \frac{1}{x+3}$. Express your solution in the form of intervals and illustrate the solution sets on the real number line.

Answer: medium

$\frac{1}{x} > 3 \quad \Leftrightarrow \quad 0 < x < \frac{1}{3}$. Also, $3 \geq \frac{1}{x+3} \quad \Leftrightarrow \quad x+3 < 0$ or $x + 3 \geq \frac{1}{3} \quad \Leftrightarrow \quad x < -3$ or $x \geq -\frac{8}{3}$. Taking the intersection, we see that $0 < x < \frac{1}{3} \quad \Leftrightarrow \quad x \in \left(0, \frac{1}{3}\right)$

17. What interval on the Celsius scale corresponds to the temperature range $41 \leq F \leq 86$?

(a) $2 \leq C \leq 15$ (b) $5 \leq C \leq 60$ (c) $5 \leq C \leq 30$ (d) $7 \leq C \leq 35$ (e) $12 \leq C \leq 25$

Answer: medium (c)

$C = \frac{5}{9}(F - 32)$. So $41 \leq F \leq 86$ corresponds to $41 \leq \frac{9}{5}C + 32 \leq 86 \quad \Leftrightarrow \quad 9 \leq \frac{9}{5}C \leq 54 \quad \Leftrightarrow \quad 5 \leq C \leq 30$

18. As dry air moves upward, it expands and in so doing cools at a rate of about 1°C for each 100 m rise, up to about 12 km.

(a) If the ground temperature is 24°C, write a formula for the temperature at height h.

(b) What range of temperatures can be expected if a plane takes off and reaches a maximum height of 6 km?

Answer: medium

(a) $T = 24 - \frac{1}{100}h$, where T is the temperature in °C and h is the height in m.

(b) From (a), we have $h = 100(24 - T)$. So $0 \leq h \leq 6000 \quad \Leftrightarrow \quad 0 \leq 100(24 - T) \leq 6000 \quad \Leftrightarrow \quad 0 \leq 24 - T \leq 60 \quad \Leftrightarrow$ $-24 \leq -T \leq 36 \quad \Leftrightarrow \quad 24 \geq T \geq -36$. So the range of temperatures is from $-36°$ C to $24°$ C.

19. Prove the statement: if $a \leq b$ and $c > 0$ then $ac \leq bc$.

[Hint: Remember that $a < b$ means that $b - a > 0$ and that the product of two positive numbers is positive.]

Answer: hard

If $a \leq b$ then $b - a \geq 0$. Since $c > 0$, then $c(b - a) \geq 0$ because the product of two positive numbers is also positive. Expanding we get $cb - ca \geq 0 \quad \Leftrightarrow \quad cb \geq ca \quad \Leftrightarrow \quad ac \leq bc$.

20. Prove that if $0 < a < b$, then $1/a > 1/b$.

Answer: hard

If $0 < a < b$ then $a < b$ and $a > 0 \quad \Rightarrow \quad 1/a > 0$. Using Rule 3 with $c = 1/a$, we get $\frac{1}{a} \cdot a < \frac{1}{a} \cdot b \quad \Leftrightarrow \quad 1 < \frac{b}{a}$. Since $b > 0 \quad \Leftrightarrow \quad \frac{1}{b} > 0$ and we can multiply both sides of the inequality by this positive value, $\frac{1}{b}$. Thus $\frac{1}{b} \cdot 1 < \frac{1}{b} \cdot \frac{b}{a} \quad \Leftrightarrow \quad \frac{1}{b} < \frac{1}{a}$.

21. Solve the inequality $a \leq bx + c < 3a$ for x assuming that a, b, and c are positive constants.

(a) $\frac{a-c}{b} \leq x < \frac{3a-c}{b}$ (b) $\frac{a-b}{c} \leq x < \frac{2a-3c}{b}$ (c) $\frac{b-c}{a} \leq x < \frac{2c-a}{2b}$

(d) $\frac{b-c}{b} \leq x < \frac{ac-2a}{b}$ (e) $\frac{a-b}{c} \leq x < \frac{2a-3b}{c}$

Answer: medium (a)

$a \leq bx + c < 3a$ where $a, b, c > 0 \quad \Leftrightarrow \quad a - c \leq bx < 3a - c \quad \Leftrightarrow \quad \frac{a-c}{b} \leq x < \frac{3a-c}{b}$

22. Solve the inequality $\frac{ax+b}{c} \geq b$ for x, assuming that a, b, and c are negative constants.

(a) $x \leq \frac{a}{b}(1-c)$ (b) $x \geq \frac{a}{b}(c-1)$ (c) $x \geq \frac{b}{a}(c-1)$ (d) $x \geq \frac{c}{b}(a-1)$ (e) $x \geq \frac{c}{a}(b-1)$

Answer: hard (c)

$\frac{ax+b}{c} \geq b$ where $a, b, c < 0 \quad \Leftrightarrow \quad ax + b \leq bc \quad \Leftrightarrow \quad ax \leq b(c-1) \quad \Leftrightarrow \quad x \geq \frac{b}{a}(c-1)$

23. Show that if $0 < a < b$, then $a^2 < b^2$.

Answer: hard

$0 < a < b \quad \Leftrightarrow \quad 0 < a$ and $a < b \quad \Leftrightarrow \quad a^2 < ab$. Similarly, multiplying both sides by $b > 0$, gives $ab < b^2$, and combining, $a^2 < ab < b^2 \quad \Rightarrow \quad a^2 < b^2$.

24. The solution set for the linear inequality $11 - 5x < -2$ is
(a) $\left(\frac{18}{5}, \infty\right)$ (b) $\left(\frac{13}{5}, \infty\right)$ (c) $\left(-\infty, -\frac{5}{13}\right)$ (d) $\left(-\infty, \frac{13}{5}\right)$ (e) No solution

Answer: easy (b)
$11 - 5x < -2 \quad\Leftrightarrow\quad -5x < -13 \quad\Leftrightarrow\quad x > \frac{13}{5} \quad\Leftrightarrow\quad x \in \left(\frac{13}{5}, \infty\right)$

25. The solution set for the linear inequality $\frac{22}{3}x - 5 > \frac{2}{3}x$ is
(a) $\left[-\frac{4}{3}, \infty\right)$ (b) $\left(-\frac{3}{4}, \frac{3}{4}\right]$ (c) $\left(\frac{3}{4}, \infty\right)$ (d) $(-\infty, 5)$ (e) $\left\{\frac{3}{4}\right\}$

Answer: easy (c)
$\frac{22}{3}x - 5 > \frac{2}{3}x \quad\Leftrightarrow\quad \frac{20}{3}x > 5 \quad\Leftrightarrow\quad x > \frac{3}{4} \quad\Leftrightarrow\quad x \in \left(\frac{3}{4}, \infty\right)$

26. The solution set for the linear inequality $4\left(2 - \frac{1}{2}x\right) \le 6 - \frac{3}{4}x$ is
(a) $\left[\frac{4}{5}, \infty\right)$ (b) $\left[-\frac{5}{8}, \infty\right)$ (c) $\left(-\infty, -\frac{8}{5}\right)$ (d) $\left(-\frac{8}{5}, \infty\right)$ (e) $\left[\frac{8}{5}, \infty\right)$

Answer: easy (e)
$4\left(2 - \frac{1}{2}x\right) \le 6 - \frac{3}{4}x \quad\Leftrightarrow\quad 8 - 2x \le 6 - \frac{3}{4}x \quad\Leftrightarrow\quad x \ge \frac{8}{5} \quad\Leftrightarrow\quad x \in \left[\frac{8}{5}, \infty\right)$

27. The solution set for the linear inequality $3 \le x - 4 < 14$ is
(a) $[7, 18)$ (b) $[7, 18]$ (c) $[-7, 18)$ (d) $\left(-\frac{18}{7}, \infty\right)$ (e) $\left[-\frac{18}{7}, \frac{18}{7}\right)$

Answer: easy (a)
$3 \le x - 4 < 14 \quad\Leftrightarrow\quad 7 \le x < 18 \quad\Leftrightarrow\quad x \in [7, 18)$

28. The solution set for the linear inequality $\frac{2}{3}x - 2 \le 2x - 3 \le 6 + \frac{2}{3}x$ is
(a) $\frac{3}{4}$ (b) $\left[-\frac{27}{4}, \frac{3}{4}\right]$ (c) $\left(\frac{4}{3}, \frac{27}{4}\right]$ (d) $\left[\frac{3}{4}, \frac{27}{4}\right]$ (e) $\left[\frac{-3}{4}, \frac{9}{4}\right]$

Answer: easy (d)
$\frac{2}{3}x - 2 \le 2x - 3 \le 6 + \frac{2}{3}x \quad\Leftrightarrow\quad 1 \le 2x - \frac{2}{3}x \le 9 \quad\Leftrightarrow\quad 1 \le \frac{4}{3}x \le 9 \quad\Leftrightarrow\quad \frac{3}{4} \le x \le \frac{27}{4} \quad\Rightarrow\quad x \in \left[\frac{3}{4}, \frac{27}{4}\right]$

29. The solution set for the linear inequality $\frac{2}{3}x - 2 \le 2x - 3 \le 6 + \frac{2}{3}x$ is
(a) $\frac{3}{4}$ (b) $\left[-\frac{27}{4}, \frac{3}{4}\right]$ (c) $\left(\frac{4}{3}, \frac{27}{4}\right]$ (d) $\left[\frac{3}{4}, \frac{27}{4}\right]$ (e) $\left[\frac{-3}{4}, \frac{9}{4}\right]$

Answer: easy (d)
$\frac{2}{3}x - 2 \le 2x - 3 \le 6 + \frac{2}{3}x \quad\Leftrightarrow\quad 1 \le 2x - \frac{2}{3}x \le 9 \quad\Leftrightarrow\quad 1 \le \frac{4}{3}x \le 9 \quad\Leftrightarrow\quad \frac{3}{4} \le x \le \frac{27}{4} \quad\Rightarrow\quad x \in \left[\frac{3}{4}, \frac{27}{4}\right]$

30. The solution set for the linear inequality $-\frac{3}{10} < \frac{2x-3}{5} \le \frac{6}{5}$ is
(a) $\left(-\frac{9}{2}, \frac{3}{4}\right)$ (b) $\left[\frac{3}{2}, \frac{9}{4}\right]$ (c) $\left(\frac{3}{2}, \frac{9}{4}\right]$ (d) $\left(\frac{-3}{2}, 9\right]$ (e) $\left(\frac{3}{4}, \frac{9}{2}\right]$

Answer: easy (e)
$-\dfrac{3}{10} < \dfrac{2x-3}{5} \le \dfrac{6}{5} \quad\Leftrightarrow\quad -\frac{15}{10} < 2x - 3 \le 6 \quad\Leftrightarrow\quad \frac{3}{2} < 2x \le 9 \quad\Leftrightarrow\quad \frac{3}{4} < x \le \frac{9}{2} \quad\Rightarrow\quad x \in \left(\frac{3}{4}, \frac{9}{2}\right]$

31. A cellular phone company offers two monthly phone packages. The company charges \$19.99 per month and 10 cents per minute for package one and \$39.99 per month and 5 cents per minute for package two. How many minutes per month would one need to use in order that package two be more economical than package one?

Answer: medium
$39.99 + 0.05x < 19.99 + 0.10x \quad\Leftrightarrow\quad 39.99 + 0.05x < 19.99 + 0.10x \quad\Leftrightarrow\quad x > 400$. If one uses more than 400 minutes per month, package two is more economical.

32. Suppose that Norma scores 67, 87, 93, and 95 on her algebra tests, and the only test left to take is the final exam. Her final exam however, accounts for one-half of her overall course average. What range of scores on the final exam would she need in order to achieve an overall course grade of between 80 and 85?
Answer: medium
Since there are four tests, the final exam counts 4 times as much in the formula for course average. Let x be her score on the final exam. Then $80 < \frac{67+87+93+95+4x}{8} < 85 \Leftrightarrow 80 < \frac{342+4x}{8} < 85 \Leftrightarrow 640 < 342+4x < 680 \Leftrightarrow$ $298 < 4x < 338 \Leftrightarrow 74.5 < x < 84.5$. She needs to achieve between 74.5 and 84.5 on her final exam in order to have an overall course grade of between 80 and 85.

SECTION 3.7

1. Solve the inequality $(2x+3)(x-1) \geq 0$ in terms of intervals and illustrate the solution set on the real number line.
Answer: medium
$(2x+3)(x-1) \geq 0$.
Case (i): $2x+3 \geq 0 \Leftrightarrow 2x \geq -3 \Leftrightarrow x \geq -\frac{3}{2}$, and $x-1 \geq 0 \Leftrightarrow x \geq 1$; so $x \geq 1$.
Case (ii): $2x+3 \leq 0 \Leftrightarrow 2x \leq -3 \Leftrightarrow x \leq -\frac{3}{2}$, and $x-1 \leq 0 \Leftrightarrow x \leq 1$; so $x \leq -\frac{3}{2}$. Therefore, $x \leq -\frac{3}{2}$ or $x \geq 1 \Leftrightarrow x \in \left(-\infty, -\frac{3}{2}\right] \cup [1, \infty)$

2. Solve the inequality $x^2 - 5x + 6 > 0$. Express your solution in the form of intervals and illustrate the solution set on the real number line.
Answer: medium
$x^2 - 5x + 6 > 0 \Leftrightarrow (x-2)(x-3) > 0$.
Case (i): $x-2 > 0 \Leftrightarrow x > 2$, and $x-3 > 0 \Leftrightarrow x > 3$; so $x > 3$.
Case (ii): $x-2 < 0 \Leftrightarrow x < 2$, and $x-3 < 0 \Leftrightarrow x < 3$; so $x < 2$. Therefore $x > 3$ or $x < 2 \Leftrightarrow$ $x \in (-\infty, 2) \cup (3, \infty)$
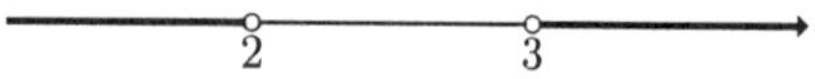

3. Solve the inequality $x^2 < 2x + 8$ in terms of intervals and illustrate the solution set on the real number line.
Answer: medium
$x^2 < 2x + 8 \Leftrightarrow x^2 - 2x - 8 < 0 \Leftrightarrow (x-4)(x+2) < 0$.
Case (i): $x-4 < 0 \Leftrightarrow x < 4$, and $x+2 > 0 \Leftrightarrow x > -2$; so $-2 < x < 4$.
Case (ii): $x-4 > 0 \Leftrightarrow x > 4$, and $x+2 < 0 \Leftrightarrow x < -2$; which is impossible. So, the solution is $-2 < x < 4$ $\Leftrightarrow x \in (-2, 4)$

4. Solve the inequality $5x^2 + 2x \geq 4x^2 + 3$. Express your solution in the form of intervals and illustrate the solution set on the real number line.
Answer: medium
$5x^2 + 2x \geq 4x^2 + 3 \Leftrightarrow x^2 + 2x - 3 \geq 0 \Leftrightarrow (x-1)(x+3) \geq 0$.
Case (i): $x-1 \geq 0 \Leftrightarrow x > 1$, and $x+3 \geq 0 \Leftrightarrow x \geq -3$; so $x \geq 1$.
Case (ii): $x-1 \leq 0 \Leftrightarrow x \leq 1$, and $x+3 \leq 0 \Leftrightarrow x \leq -3$; so $x \leq -3$. Therefore $x \leq -3$ or $x \geq 1 \Leftrightarrow$ $x \in (-\infty, -3] \cup [1, \infty)$

5. Solve the inequality $3x^2 + 2x > 1$ in terms of intervals and illustrate the solution set on the real number line.
Answer: medium
$3x^2 + 2x > 1 \Leftrightarrow 3x^2 + 2x - 1 > 0 \Leftrightarrow (3x-1)(x+1) > 0$.
Case (i): $3x - 1 > 0 \Leftrightarrow 3x > 1 \Leftrightarrow x > \frac{1}{3}$, and $x+1 > 0 \Leftrightarrow x > -1$; so $x > \frac{1}{3}$.
Case (ii): $3x - 1 < 0 \Leftrightarrow x < \frac{1}{3}$, and $x+1 < 0 \Leftrightarrow x < -1$; so $x < -1$.
Therefore, the solution is $x < -1$ or $x > \frac{1}{3} \Leftrightarrow x \in (-\infty, -1) \cup \left(\frac{1}{3}, \infty\right)$

6. Solve the inequality $x^2 \geq 5$ in terms of intervals and illustrate the solution set on the real number line.
Answer: medium
$x^2 \geq 5 \quad \Leftrightarrow \quad x^2 - 5 \geq 0 \quad \Leftrightarrow \quad \left(x - \sqrt{5}\right)\left(x + \sqrt{5}\right) \geq 0.$
Case (i): $x - \sqrt{5} \geq 0 \quad \Leftrightarrow \quad x \geq \sqrt{5}$, and $x + \sqrt{5} \geq 0 \quad \Leftrightarrow \quad x \geq -\sqrt{5}$; so $x \geq \sqrt{5}$.
Case (ii): $x - \sqrt{5} \leq 0 \quad \Leftrightarrow \quad x \leq \sqrt{5}$, and $x + \sqrt{5} \leq 0 \quad \Leftrightarrow \quad x \leq -\sqrt{5}$; so $x \leq -\sqrt{5}$. Therefore, the solution is $x \leq -\sqrt{5}$ or $x \geq \sqrt{5} \quad \Leftrightarrow \quad x \in \left(-\infty, -\sqrt{5}\right] \cup \left[\sqrt{5}, \infty\right)$

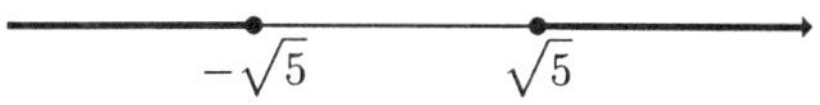

7. Solve the inequality $-x^3 > x^2$. Express your solution in the form of intervals and illustrate the solution set on the real number line.
Answer: hard
$-x^3 > x^2 \quad \Leftrightarrow \quad x^3 + x^2 < 0 \quad \Leftrightarrow \quad x^2(x+1) < 0$. Since x^2 is always nonnegative, this is equivalent to $x + 1 < 0 \quad \Leftrightarrow$ $x < -1$. Therefore $x < -1 \quad \Leftrightarrow \quad x \in (-\infty, -1)$

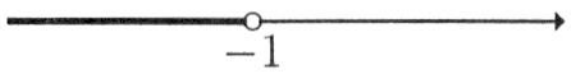

8. Solve the inequality $\dfrac{2x+6}{x-6} < 0$. Express your solution in the form of intervals and illustrate the solution set on the real number line.
Answer: medium
$\dfrac{2x+6}{x-6} < 0.$
Case (i): $2x + 6 < 0 \quad \Leftrightarrow \quad x < -3$, and $x - 6 > 0 \quad \Leftrightarrow \quad x > 6$; which is impossible.
Case (ii): $2x + 6 > 0$ $x > -3$, and $x - 6 < 0 \quad \Leftrightarrow \quad x < 6$; so $-3 < x < 6$. Therefore $-3 < x < 6 \quad \Leftrightarrow \quad x \in (-3, 6)$

9. Solve the inequality $-3 < \dfrac{x+1}{x-3}$. Express your solution in the form of intervals and illustrate the solution set on the real number line.
Answer: medium
$-3 < \dfrac{x+1}{x-3} \quad \Leftrightarrow \quad 0 < \dfrac{x+1}{x-3} + 3 \quad \Leftrightarrow \quad 0 < \dfrac{4x-8}{x-3} = 4\left(\dfrac{x-2}{x-3}\right).$

Interval	$x-2$	$x-3$	$\dfrac{x-2}{x-3}$
$x < 2$	$-$	$-$	$+$
$2 < x < 3$	$+$	$-$	$-$
$x > 3$	$+$	$+$	$+$

From the chart, $x < 2$ or $x > 3 \quad \Leftrightarrow \quad x \in (-\infty, 2) \cup (3, \infty)$

10. Solve the inequality $\dfrac{2+x}{3-x} \leq 1$ in terms of intervals and illustrate the solution set on the real number line.
Answer: medium
$\dfrac{2+x}{3-x} \leq 1.$
Case (i): if $3 - x > 0$ (that is, $x < 3$) then $2 + x \leq 3 - x \quad \Leftrightarrow \quad 2x \leq 1 \quad \Leftrightarrow \quad x \leq \frac{1}{2}$; so $x \leq \frac{1}{2}$.
Case (ii): if $3 - x < 0$ (that is, $x > 3$) then $2 + x \geq 3 - x \quad \Leftrightarrow \quad 2x \geq 1 \quad \Leftrightarrow \quad x \geq \frac{1}{2}$; so $x > 3$. So the solution is $x \leq \frac{1}{2}$ or $x > 3 \quad \Leftrightarrow \quad x \in \left(-\infty, \frac{1}{2}\right] \cup (3, \infty)$

11. Solve the inequality $\dfrac{x}{x+1} > 3$ in terms of intervals and illustrate the solution set on the real number line.
Answer: medium
$\dfrac{x}{x+1} > 3.$
Case (i): if $x + 1 > 0$ (that is, $x > -1$) then $x > 3(x+1) \quad \Leftrightarrow \quad -2x > 3 \quad \Leftrightarrow \quad x < -\frac{3}{2}$; which is impossible.
Case (ii): if $x + 1 < 0$ (that is, $x < -1$) then $x < 3(x+1) \quad \Leftrightarrow \quad -2x < 3 \quad \Leftrightarrow \quad x > -\frac{3}{2}$; so $-\frac{3}{2} < x < -1$. Combining the two cases, the solution is $-\frac{3}{2} < x < -1 \quad \Leftrightarrow \quad x \in \left(-\frac{3}{2}, -1\right)$

12. Solve the inequality $x^3 > x$ in terms of intervals and illustrate the solution set on the real number line.

Answer: hard

$x^3 > x \quad \Leftrightarrow \quad x^3 - x > 0 \quad \Leftrightarrow \quad x(x^2-1) > 0 \quad \Leftrightarrow \quad x(x-1)(x+1) > 0.$

Interval	x	$x-1$	$x+1$	$x(x-1)(x+1)$
$x < -1$	$-$	$-$	$-$	$-$
$-1 < x < 0$	$-$	$-$	$+$	$+$
$0 < x < 1$	$+$	$-$	$+$	$-$
$x > 1$	$+$	$+$	$+$	$+$

From the chart, $-1 < x < 0$ or $x > 1 \quad \Leftrightarrow \quad x \in (-1, 0) \cup (1, \infty)$

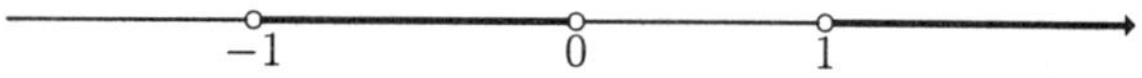

13. Solve the inequality $\dfrac{3}{x-1} - \dfrac{x}{x+1} \geq 1$. Express your solution in the form of intervals and illustrate the solution set on the real number line.

Answer: medium

$$\frac{3}{x-1} - \frac{x}{x+1} \geq 1 \quad \Leftrightarrow \quad \frac{3}{x-1} - \frac{x}{x+1} - 1 \geq 0 \quad \Leftrightarrow \quad \frac{x^2-2x-2}{(x-1)(x+1)} \leq 0 \quad \Leftrightarrow \quad \frac{\left[x-\left(1-\sqrt{3}\right)\right]\left[x-\left(1+\sqrt{3}\right)\right]}{(x-1)(x+1)} \leq 0$$

Interval	x^2-2x-2	$x-1$	$x+1$	$\dfrac{x^2-2x-2}{(x-1)(x+1)}$
$x < -1$	$+$	$-$	$-$	$+$
$-1 < x < 1-\sqrt{3}$	$+$	$-$	$+$	$-$
$1-\sqrt{3} < x < 1$	$-$	$-$	$+$	$+$
$1 < x < 1+\sqrt{3}$	$-$	$+$	$+$	$-$
$x > 1+\sqrt{3}$	$+$	$+$	$+$	$+$

From the chart, $-1 < x \leq 1-\sqrt{3}$ or $1 < x \leq 1+\sqrt{3} \quad \Leftrightarrow \quad x \in \left(-1, 1-\sqrt{3}\right] \cup \left(1, 1+\sqrt{3}\right]$

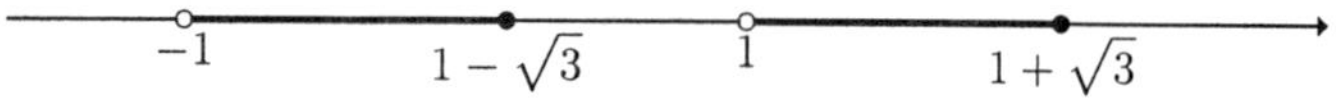

14. Solve the inequality $\dfrac{x^2-16}{x^4-16} > 0$. Express your solution in the form of intervals and illustrate the solution set on the real number line.

Answer: medium

$\dfrac{x^2-16}{x^4-16} > 0 \quad \Leftrightarrow \quad \dfrac{(x-4)(x+4)}{(x-2)(x+2)(x^2+4)} > 0$ has cut points that occur at $x = \pm 4$ and $x = \pm 2$, and $x^2 + 4 > 0$ for all x.

Interval	$x-4$	$x+4$	$x-2$	$x+2$	$\dfrac{x^2-16}{x^4-16}$
$x < -4$	$-$	$-$	$-$	$-$	$+$
$-4 < x < -2$	$-$	$+$	$-$	$-$	$-$
$-2 < x < 2$	$-$	$+$	$-$	$+$	$+$
$2 < x < 4$	$-$	$+$	$+$	$+$	$-$
$x > 4$	$+$	$+$	$+$	$+$	$+$

Since the endpoints are excluded, the solution is $(-\infty, -4) \cup (-2, 2) \cup (4, \infty)$.

15. Solve the inequality $\frac{1}{x} + \frac{1}{x+1} > \frac{2}{x+2}$. Express your solution in the form of intervals and illustrate the solution set on the real number line.

Answer: hard

$\frac{1}{x} + \frac{1}{x+1} > \frac{2}{x+2} \quad \Leftrightarrow \quad \frac{(x+1)(x+2) + x(x+2) - 2x(x+1)}{x(x+1)(x+2)} > 0 \quad \Leftrightarrow$

$\frac{x^2 + 3x + 2 + x^2 + 2x - 2x^2 - 2x}{x(x+1)(x+2)} = \frac{3x+2}{x(x+1)(x+2)} > 0$ has cut points $x = -\frac{2}{3}$, -2, -1, and 0.

Interval	$3x+2$	x	$x+1$	$x+2$	$\frac{3x+2}{x(x+1)(x+2)}$
$x < -2$	$-$	$-$	$-$	$-$	$+$
$-2 < x < -1$	$-$	$-$	$-$	$+$	$-$
$-1 < x < -\frac{2}{3}$	$-$	$-$	$+$	$+$	$+$
$-\frac{2}{3} < x < 0$	$+$	$-$	$+$	$+$	$-$
$x > 0$	$+$	$+$	$+$	$+$	$+$

Since the endpoints are excluded, the solution is $(-\infty, -2) \cup \left(-1, -\frac{2}{3}\right) \cup (0, \infty)$.

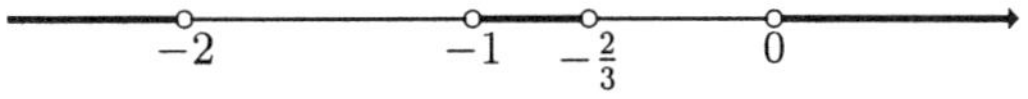

16. Solve the inequality $\frac{(x-1)^2}{(x+1)(x+2)} < 0$. Express your solution in the form of intervals and illustrate the solution set on the real number line.

Answer: medium

$\frac{(x-1)^2}{(x+1)(x+2)} < 0$. Since $(x-1)^2$ is always positive except when $x = 1$, so this is equivalent (for $x \neq 1$) to:

Case (i): $x + 1 > 0 \quad \Leftrightarrow \quad x > -1$, and $x + 2 < 0 \quad \Leftrightarrow \quad x < -2$, which is impossible.

Case (ii): $x + 1 < 0 \quad \Leftrightarrow \quad x < -1$, and $x + 2 > 0 \quad \Leftrightarrow \quad x > -2$. Therefore $-2 < x < -1$.

So the solution is $(-2, -1)$ 

17. Solve the inequality $\frac{(x+2)(x-5)}{2(x+1)} \geq 0$. Express your solution in the form of intervals and illustrate the solution set on the real number line.

Answer: medium

$\frac{(x+2)(x-5)}{2(x+1)} \geq 0$

Interval	$x+2$	$x-5$	$x+1$	$\frac{(x+2)(x-5)}{2(x+1)}$
$x < -2$	$-$	$-$	$-$	$-$
$-2 < x < -1$	$+$	$-$	$-$	$+$
$-1 < x < 5$	$+$	$-$	$+$	$-$
$x > 5$	$+$	$+$	$+$	$+$

From the chart, $-2 \leq x \leq -1$ or $x \geq 5 \quad \Leftrightarrow \quad x \in [-2, -1) \cup [9, \infty)$ 

18. Solve the inequality $\dfrac{1}{x+1}+\dfrac{1}{x+2}\le\dfrac{1}{x+3}$. Express your solution in the form of intervals and illustrate the solution set on the real number line.

Answer: hard

$\dfrac{1}{x+1}+\dfrac{1}{x+2}\le\dfrac{1}{x+3} \quad\Leftrightarrow\quad \dfrac{(x+2)(x+3)+(x+1)(x+3)-(x+1)(x+2)}{(x+1)(x+2)(x+3)}\le 0 \quad\Leftrightarrow$

$\dfrac{x^2+5x+6+x^2+4x+3-x^2-3x-2}{(x+1)(x+2)(x+3)}\le 0 \quad\Leftrightarrow\quad \dfrac{x^2+6x+7}{(x+1)(x+2)(x+3)}\le 0$ has cut points $x=-3,-2,-1$, and

$x=\dfrac{-6\pm\sqrt{36-4\cdot1\cdot7}}{2}=\dfrac{-6\pm2\sqrt{2}}{2}=-3\pm\sqrt{2}$. Denote $F(x)=\dfrac{\left[x-\left(-3-\sqrt{2}\right)\right]\left[x-\left(-3+\sqrt{2}\right)\right]}{(x+1)(x+2)(x+3)}$,

$a=\left(-3-\sqrt{2}\right)$, and $b=\left(-3+\sqrt{2}\right)$.

Interval	$x-a$	$x-b$	$x+1$	$x+2$	$x+3$	$F(x)$
$x<a$	$-$	$-$	$-$	$-$	$-$	$-$
$a<x<-3$	$+$	$-$	$-$	$-$	$-$	$+$
$-3<x<-2$	$+$	$-$	$-$	$-$	$+$	$-$
$-2<x<b$	$+$	$-$	$-$	$+$	$+$	$+$
$b<x<-1$	$+$	$+$	$-$	$+$	$+$	$-$
$x>-1$	$+$	$+$	$+$	$+$	$+$	$+$

Since the function is undefined at $x=-1,-2$, and -3, the solution is $\left(-\infty,-3-\sqrt{2}\right]\cup(-3,-2)\cup\left[-3+\sqrt{2},-1\right)$.

19. Solve the inequality $\frac{2}{3}x^{-1/3}(x+2)^{1/2}+\frac{1}{2}x^{2/3}(x+2)^{-1/2}<0$. Express your solution in the form of intervals and illustrate the solution set on the real number line.

Answer: hard

$\frac{2}{3}x^{-1/3}(x+2)^{1/2}+\frac{1}{2}x^{2/3}(x+2)^{-1/2}<0$. For $(x+2)^{-1/2}$ to be defined, we must have $x+2>0 \quad\Leftrightarrow\quad x>-2$. Now

$\frac{2}{3}x^{-1/3}(x+2)^{1/2}+\frac{1}{2}x^{2/3}(x+2)^{-1/2}<0 \quad\Leftrightarrow\quad \dfrac{2}{3}\dfrac{(x+2)^{1/2}}{x^{1/3}}+\dfrac{1}{2}\dfrac{x^{2/3}}{(x+2)^{1/2}}<0 \quad\Leftrightarrow\quad \dfrac{\frac{2}{3}(x+2)+\frac{1}{2}x}{x^{1/3}(x+2)^{1/2}}<0 \quad\Leftrightarrow$

$\dfrac{7x+8}{6x^{1/3}(x+2)^{1/2}}<0$. Thus the cut points are $x=-2,-\frac{8}{7}$, and 0.

Interval	$7x+8$	$x^{1/3}$	$(x+2)^{1/2}$	$\dfrac{7x+8}{6x^{1/3}(x+2)^{1/2}}$
$-2<x<-\frac{8}{7}$	$-$	$-$	$+$	$+$
$-\frac{8}{7}<x<0$	$+$	$-$	$+$	$-$
$x>0$	$+$	$+$	$+$	$+$

Since the endpoints are excluded, the solution is $\left(-\frac{8}{7},0\right)$.

20. The gravitational force F exerted by the earth on an object having a mass of 50 kg is given by the equation $F=\dfrac{2{,}000{,}000}{d^2}$ where d is the distance (in km) of the object from the center of the earth, and the force F is measured in newtons. For what distances will the gravitational force exerted by the earth on this object be between 0.0002 N and 0.005 N?

Answer: easy

$F=\dfrac{2{,}000{,}000}{d^2}$. Hence $0.0002<\dfrac{2{,}000{,}000}{d^2}<0.005$. Hence $\dfrac{1}{0.0002}>\dfrac{d^2}{2{,}000{,}000}>\dfrac{1}{0.005} \quad\Leftrightarrow\quad 10^{10}>d^2>4\times10^8$

$\Leftrightarrow\quad 10^5>d>2\times10^4$ (since $d>0$). Thus the range of distances is between 20,000 km and 100,000 km.

21. Determine the values of the variable for which $\sqrt{3x^2+x-2}$ is defined as a real number.

Answer: medium

$3x^2+x-2\ge0 \quad\Leftrightarrow\quad (3x-2)(x+1)\ge0$.

Case (i): $3x-2\ge0 \quad\Leftrightarrow\quad x\ge\frac{2}{3}$, and $x+1\ge0 \quad\Leftrightarrow\quad x\ge-1$; so $x\ge\frac{2}{3}$.

Case (ii): $3x-2\le0 \quad\Leftrightarrow\quad x\le\frac{2}{3}$, and $x+1\le0 \quad\Leftrightarrow\quad x\le-1$; so $x\le-1$. Therefore $x\le-1$ or $x\ge\frac{2}{3} \quad\Leftrightarrow$

$x\in(-\infty,-1]\cup\left[\frac{2}{3},\infty\right)$

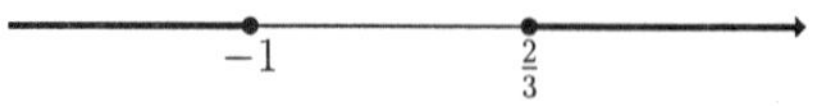

22. Determine the values of the variable for which $\sqrt[4]{\dfrac{2-x}{1+x}}$ is defined as a real number.

Answer: medium

$\dfrac{2-x}{1+x} \geq 0.$

Case (i): $2 - x \geq 0 \quad \Leftrightarrow \quad x \leq 2$, and $1 + x > 0 \quad \Leftrightarrow \quad x > -1$; so $-1 < x \leq 2$.

Case (ii): $2 - x \leq 0 \quad \Leftrightarrow \quad x \geq 2$, and $1 + x < 0 \quad \Leftrightarrow \quad x < -1$; which is impossible. Therefore $-1 < x \leq 2 \quad \Leftrightarrow$

$x \in (-1, 2]$ 

23. Find the solutions of the inequality $x^3 + 14x \leq 5x^2 + 6$ correct to two decimals by drawing appropriate graphs.

Answer: easy

$x^3 + 14x \leq 5x^2 + 6 \quad \Leftrightarrow \quad x \in (-\infty, 0.51]$

24. Find the solutions of the inequality $\left(x + \frac{1}{2}\right)^2 < \left(x - \frac{1}{2}\right)^2$ correct to two decimals by drawing appropriate graphs.

Answer: easy

$\left(x + \frac{1}{2}\right)^2 < \left(x - \frac{1}{2}\right)^2 \quad \Leftrightarrow \quad x \in (-\infty, 0)$

25. Find the solutions of the inequality $4x^3 + 3x^2 \geq x + 0.5$ correct to two decimals by drawing appropriate graphs.

Answer: easy

$4x^3 + 3x^2 \geq x + 0.5 \quad \Leftrightarrow \quad x \in [-0.87, -0.32]$ or $x \in [0.44, \infty)$

SECTION 3.8

1. Use the definition of absolute value to write the expression $|4x - 7|$ without the absolute value symbol.

(a) $|4x-7| = \begin{cases} 4x - 7 \text{ if } x \geq 4 \\ 7 - 4x \text{ if } x < 7 \end{cases}$ (b) $|4x-7| = \begin{cases} 4x - 7 \text{ if } x \geq 0 \\ 7 - 4x \text{ if } x < 0 \end{cases}$ (c) $|4x-7| = \begin{cases} 4x - 7 \text{ if } x \geq \frac{7}{4} \\ 7 - 4x \text{ if } x < \frac{7}{4} \end{cases}$

(d) $|4x-7| = \begin{cases} 4x - 7 \text{ if } x \geq \frac{4}{7} \\ 7 - 4x \text{ if } x < \frac{4}{7} \end{cases}$ (e) $|4x-7| = \begin{cases} 4x - 7 \text{ if } x > \frac{2}{3} \\ 7 - 4x \text{ if } x < \frac{2}{3} \end{cases}$

Answer: medium (c)

$$|4x-7| = \left\{ \begin{array}{c} 4x - 7 \text{ if } 4x - 7 \geq 0 \\ -(4x-7) \text{ if } 4x - 7 < 0 \end{array} \right\} = \begin{cases} 4x - 7 \text{ if } x \geq \frac{7}{4} \\ 7 - 4x \text{ if } x < \frac{7}{4} \end{cases}$$

2. Use the definition of absolute value to write the expression $|8 - 5x|$ without the absolute value symbol.

(a) $|8-5x| = \begin{cases} 8 - 5x \text{ if } x \leq \frac{8}{5} \\ 5x - 8 \text{ if } x > \frac{8}{5} \end{cases}$ (b) $|8-5x| = \begin{cases} 8x - 5 \text{ if } x \leq \frac{5}{8} \\ 8 - 5x \text{ if } x > \frac{5}{8} \end{cases}$ (c) $|8-5x| = \begin{cases} 4 - x \text{ if } x \leq \frac{2}{3} \\ x - 4 \text{ if } x > \frac{2}{3} \end{cases}$

(d) $|8-5x| = \begin{cases} 8 - 5x \text{ if } x \leq \frac{8}{5} \\ 16 - 10x - 8 \text{ if } x > \frac{8}{5} \end{cases}$ (e) $|8-5x| = \begin{cases} 8 - 5x \text{ if } x \leq 0 \\ 5x - 8 \text{ if } x > 0 \end{cases}$

Answer: medium (a)

$$|8-5x| = \left\{ \begin{array}{c} 8 - 5x \text{ if } 8 - 5x \geq 0 \\ -(8-5x) \text{ if } 8 - 5x < 0 \end{array} \right\} = \begin{cases} 8 - 5x \text{ if } x \leq \frac{8}{5} \\ 5x - 8 \text{ if } x > \frac{8}{5} \end{cases}$$

3. Use the definition of absolute value to write the expression $|x^2-1|$ without the absolute value symbol.

(a) $|x^2-1| = \begin{cases} x^2-1 \text{ if } x \le -2 \text{ or } x \ge 2 \\ 1-x^2 \text{ if } -2 \le x \le 2 \end{cases}$
(b) $|x^2-1| = \begin{cases} x-1 \text{ if } x \le -1 \text{ or } x \ge 0 \\ 1-x \text{ if } -1 \le x \le 0 \end{cases}$
(c) $|x^2-1| = \begin{cases} x^2-1 \text{ if } -1 \le x \le 1 \\ 1-x^2 \text{ if } x \le -1 \text{ or } x \ge 1 \end{cases}$
(d) $|x^2-1| = \begin{cases} x^2-1 \text{ if } x \le -1 \text{ or } x \ge 1 \\ 1-x^2 \text{ if } -1 \le x \le 1 \end{cases}$
(e) $|x^2-1| = \begin{cases} x^2-1 \text{ if } x \le -1 \\ x^2+1 \text{ if } x \ge -1 \end{cases}$

Answer: medium (d)

$$|x^2-1| = \left\{ \begin{matrix} x^2-1 \text{ if } x^2-1 \ge 0 \\ -(x^2-1) \text{ if } x^2-1 < 0 \end{matrix} \right\} = \begin{cases} x^2-1 \text{ if } x \le -1 \text{ or } x \ge 1 \\ 1-x^2 \text{ if } -1 \le x \le 1 \end{cases}$$

4. Simplify the expression $|2x-16|$.

(a) $16|x+2|$ (b) $16|x-2|$ (c) $8|x|-2$ (d) $2|x+16|$ (e) $2|x-8|$

Answer: medium (e)

$|2x-16| = |2(x-8)| = |2| \cdot |x-8| = 2|x-8|$

5. Simplify the expression $|-2x-12|$.

(a) $2|x|+6$ (b) $2|x+6|$ (c) $-2|x-6|$ (d) $2|x-6|$ (e) $2|x|-6$

Answer: medium (b)

$|-2x-12| = |-2(x+6)| = |-2||x+6| = 2|x+6|$

6. Simplify the expression $\left|\dfrac{1-x}{x-1}\right|$.

(a) -1 (b) $-|x|$ (c) 0 (d) $|x|$ (e) 1

Answer: medium (e)

$$\left|\frac{1-x}{x-1}\right| = \frac{|1-x|}{|x-1|} = \frac{|-1| \cdot |x-1|}{|x-1|} = |-1| = 1$$

7. Solve the equation $|3x+5| = 1$.

(a) $x = -\frac{4}{3}$ or -2 (b) $x = -\frac{5}{3}$ or 1 (c) $x = -\frac{5}{3}$ or -1 (d) $x = \frac{4}{3}$ or 1 (e) $x = \pm\frac{4}{3}$

Answer: medium (a)

$|3x+5| = 1$. So $3x+5=1 \;\Leftrightarrow\; 3x=-4 \;\Leftrightarrow\; x=-\frac{4}{3}$, or $3x+5=-1 \;\Leftrightarrow\; 3x=-6 \;\Leftrightarrow\; x=-2$. Thus $x=-\frac{4}{3}$ or $x=-2$.

8. Solve the equation $|x-6| = -1$.

(a) $x = \pm 6$ (b) $x=-5$ (c) $x=5$ (d) $x=-6$ (e) No solution

Answer: medium (e)

$|x-6| = -1$. This has no solution for x since $|x-6| \ge 0$ for all x.

9. Solve the equation $4+2|x+5| = 5$

(a) $x = -\frac{13}{2}$ or 11 (b) $x=-7$ or -4 (c) $x = -\frac{2}{3}$ or $-\frac{7}{3}$ (d) $x = -\frac{11}{2}$ or $-\frac{9}{2}$ (e) $x = \pm 5$

Answer: medium (d)

$4+2|x+5| = 5 \;\Leftrightarrow\; |x+5| = \frac{1}{2} \;\Leftrightarrow\; x = -\frac{11}{2}$ or $-\frac{9}{2}$.

10. Solve the equation $|x+2| = |3x+1|$.

(a) $x=-\frac{2}{3}$ or $\frac{1}{3}$ (b) $x=-\frac{1}{2}$ or $\frac{5}{2}$ (c) $x=-\frac{3}{4}$ or $\frac{1}{2}$ (d) $x=-\frac{1}{3}$ or $\frac{2}{3}$ (e) $x=-\frac{1}{4}$ or $\frac{3}{4}$

Answer: medium (c)

$|x+2| = |3x+1|$. So $x+2 = 3x+1 \;\Leftrightarrow\; x=\frac{1}{2}$, or $x+2 = -(3x+1) \;\Leftrightarrow\; x+2 = -3x-1 \;\Leftrightarrow\; x = -\frac{3}{4}$. Therefore the solutions are $x=-\frac{3}{4}$ or $\frac{1}{2}$.

11. Solve the inequality $|x| \ge 3$.

(a) $\{x \mid x \le -3 \text{ or } x \ge 3\}$ (b) $\{x \mid x \le 3 \text{ or } x \ge 6\}$ (c) $\{x \mid x \le -1 \text{ or } x \ge 1\}$
(d) $\{x \mid x \le 0 \text{ or } x \ge 3\}$ (e) $\{x \mid x \ne -3 \text{ or } 3\}$

Answer: easy (a)

$|x| \ge 3 \;\Leftrightarrow\; x \ge 3$ or $x \le -3$. So SS $= \{x \mid x \le -3 \text{ or } x \ge 3\}$

12. Solve the inequality $|x-6|<0.1$.

Answer: easy

$|x-6|<0.1 \quad\Leftrightarrow\quad -0.1<x-6<0.1 \quad\Leftrightarrow\quad 5.9<x<6.1$. So SS $=\{x \mid 5.9<x<6.1\}$

13. Solve the inequality $|x+1|\geq 2$.

Answer: easy

$|x+1|\geq 2$. So $x+1\geq 2 \quad\Leftrightarrow\quad x\geq 1$, or $x+1\leq -2 \quad\Leftrightarrow\quad x\leq -3$. So SS $=\{x \mid x\leq -3 \text{ or } x\geq 1\}$

14. Solve the inequality $|3x-4|<5$.

Answer: easy

$|3x-4|<5 \quad\Leftrightarrow\quad -5<3x-4<5 \quad\Leftrightarrow\quad -1<3x<9 \quad\Leftrightarrow\quad -\frac{1}{3}<x<3$. So SS $=\left\{x \mid -\frac{1}{3}<x<3\right\}$

15. Solve the inequality $\left|\dfrac{x+1}{2}\right|\geq 6$.

Answer: medium

$\left|\dfrac{x+1}{2}\right|\geq 6$. So $\dfrac{x+1}{2}\geq 6 \quad\Leftrightarrow\quad x+1\geq 12 \quad\Leftrightarrow\quad x\geq 11$, or $\dfrac{x+1}{2}\leq -6 \quad\Leftrightarrow\quad x+1\leq -12 \quad\Leftrightarrow\quad x\leq -13$.

Thus, SS $=\{x \mid x\leq -13 \text{ or } x\geq 11\}$

16. Solve the inequality $|x-a|<d$.

Answer: medium

$|x-a|<d \quad\Leftrightarrow\quad -d<x-a<d \quad\Leftrightarrow\quad a-d<x<a+d$. So SS $=\{x \mid a-d<x<a+d\}$

17. Solve the inequality $0<|x-3|<\frac{1}{4}$.

Answer: medium

$0<|x-3|<\frac{1}{4}$. So $|x-3|>0$ which is all $x\neq 3$, and $|x-3|<\frac{1}{4} \quad\Leftrightarrow\quad -\frac{1}{4}<x-3<\frac{1}{4} \quad\Leftrightarrow\quad \frac{11}{4}<x<\frac{13}{4}$. So, SS $=\left\{x \mid \frac{11}{4}<x<3, 3<x<\frac{13}{4}\right\}$

18. Solve the inequality $\left|1-\dfrac{1}{x}\right|\leq 3$. Express your answer using interval notation.

Answer: medium

$\left|1-\dfrac{1}{x}\right|\leq 3 \quad\Leftrightarrow\quad -3\leq \dfrac{x-1}{x}\leq 3$.

For $\dfrac{x-1}{x}\geq -3$; if $x>0$ then $x-1\geq -3x \quad\Leftrightarrow\quad x\geq \frac{1}{4}$. If $x<0$ then $x-1\leq -3x \quad\Leftrightarrow\quad x\leq\frac{1}{4}$; so $x<0$.

For $\dfrac{x-1}{x}\leq 3$; if $x>0$ then $x-1\leq 3x \quad\Leftrightarrow\quad x\geq -\frac{1}{2}$, so $x>0$: if $x<0$ then $x-1\geq 3x \quad\Leftrightarrow\quad x\leq -\frac{1}{2}$, so $x\leq -\frac{1}{2}$.

Combining these cases, $\left(x\geq\frac{1}{4} \text{ or } x<0\right)$ and $\left(x>0 \text{ or } x\leq -\frac{1}{2}\right) \quad\Leftrightarrow\quad x\leq -\frac{1}{2}$ or $x\geq\frac{1}{4} \quad\Leftrightarrow\quad x\in\left(-\infty,-\frac{1}{2}\right]\cup\left[\frac{1}{4},\infty\right)$

19. Solve the inequality $|2x-3|\leq|x+5|$.

Answer: hard

$|2x-3|\leq|x+5| \quad\Leftrightarrow\quad (2x-3)^2\leq(x+5)^2 \quad\Leftrightarrow\quad 4x^2-12x+9\leq x^2+10x+25 \quad\Leftrightarrow\quad 3x^2-22x-16\leq 0 \quad\Leftrightarrow$ $(3x+2)(x-8)\leq 0$.

Case (i): $3x-2\leq 0 \quad\Leftrightarrow\quad x\leq\frac{2}{3}$, and $x-8\geq 0 \quad\Leftrightarrow\quad x\geq 8$; which is impossible.

Case (ii): $3x-2\geq 0 \quad\Leftrightarrow\quad x\geq\frac{2}{3}$, and $x-8\leq 0 \quad\Leftrightarrow\quad x\leq 8$; so $\frac{2}{3}\leq x\leq 8$. Therefore, SS $=\left\{x \mid \frac{2}{3}\leq x\leq 8\right\}$

20. Solve the inequality $\left|\dfrac{1-2x}{2+3x}\right|\leq 4$.

Answer: hard

$\left|\dfrac{1-2x}{2+3x}\right|\leq 4 \quad\Leftrightarrow\quad -4\leq\dfrac{1-2x}{2+3x}\leq 4$.

For $\dfrac{1-2x}{2+3x}\geq -4$: if $2+3x>0$ (that is, $x>-\frac{2}{3}$) then $1-2x\geq -8-12x \quad\Leftrightarrow\quad x\geq -\frac{9}{10}$; so $x>-\frac{2}{3}$. If $2+3x<0$ (that is, $x<-\frac{2}{3}$) then $1-2x\leq -8-12x \quad\Leftrightarrow\quad x\leq -\frac{9}{10}$; so $x\leq -\frac{9}{10}$.

For $\dfrac{1-2x}{2+3x}\leq 4$: if $x>-\frac{2}{3}$ then $1-2x\leq 8+12x \quad\Leftrightarrow\quad x\geq -\frac{1}{2}$; so $x\geq -\frac{1}{2}$: if $x<-\frac{2}{3}$ then $1-2x\geq 8+12x \quad\Leftrightarrow$ $x\leq -\frac{1}{2}$; so $x<-\frac{2}{3}$. Combining the cases, $\left(x>-\frac{2}{3} \text{ or } x\leq -\frac{9}{10}\right)$ and $\left(x\geq -\frac{1}{2} \text{ or } x<-\frac{2}{3}\right)$. So SS $=\left\{x \mid x\leq -\frac{9}{10} \text{ or } x\geq -\frac{1}{2}\right\}$.

21. Prove that $\left|\frac{a}{b}\right| = \frac{|a|}{|b|}$.

Answer: hard

$\left|\frac{a}{b}\right| = \sqrt{\left(\frac{a}{b}\right)^2} = \sqrt{\frac{a^2}{b^2}} = \frac{\sqrt{a^2}}{\sqrt{b^2}} = \frac{|a|}{|b|}$

22. If $|a-2| < 4$ and $|b-3| < 5$, show that $|a+b-5| < 9$.

Answer: hard

$|a-2| < 4$ and $|b-3| < 5$. Using the triangle inequality gives
$|a+b-5| = |(a-2)+(b-3)| \le |a-2| + |b-3| < 4+5 = 9$. So $|a+b-5| < 9$.

23. Show that $|x-y| \ge |x| - |y|$ for all real numbers x and y. (Hint: Use the Triangle Inequality with $a = x-y$ and $b = y$.)

Answer: hard

From the triangle inequality, $|a+b| \le |a| + |b|$. Substituting $a = x-y$ and $b = y$ gives $|x-y+y| = |x| \le |x-y| + |y|$
$\Leftrightarrow \quad |x-y| \ge |x| - |y|$

24. Solve the equation $|2x| = 10$.

(a) $x = -\frac{1}{5}, \frac{1}{5}$ (b) $x = -5, 5$ (c) $x = -2, 2$ (d) $x = -\frac{1}{2}, \frac{1}{5}$ (e) $x = -1, 1$

Answer: easy (b)

$|2x| = 10 \quad \Leftrightarrow \quad 2x = 10$ or $2x = -10 \quad \Leftrightarrow \quad x = 5$ or $x = -5$

25. Solve the equation $|6x - \frac{1}{3}| = 12$.

(a) $x = -\frac{1}{2}, \frac{1}{2}$ (b) $x = \frac{37}{3}, -\frac{35}{3}$ (c) $x = -6, 6$ (d) $x = \frac{37}{18}, -\frac{35}{18}$ (e) $x = -\frac{18}{37}, \frac{18}{35}$

Answer: easy (d)

$\left|6x - \frac{1}{3}\right| = 12 \quad \Leftrightarrow \quad 6x - \frac{1}{3} = 12$ or $6x - \frac{1}{3} = -12 \quad \Leftrightarrow \quad x = \frac{37}{18}$ or $x = -\frac{35}{18}$

26. Solve the equation $6 - 5\left|\frac{1}{3}x - 6\right| = 32$.

(a) $x = \frac{12}{5}$ or $\frac{168}{5}$ (b) $x = \frac{12}{26}$ or $-\frac{12}{5}$ (c) $x = 0$ or 5 (d) No solution (e) $x = \frac{12}{5}$ or $-\frac{168}{5}$

Answer: easy (d)

$6 - 5\left|\frac{1}{3}x - 6\right| = 32 \quad \Leftrightarrow \quad \left|\frac{1}{3}x - 6\right| = -\frac{26}{5}$. This has no solution since $\left|\frac{1}{3}x - 6\right| \ge 0$ for all x.

27. Solve the inequality $\left|t + \frac{1}{2}\right| \ge 2$.

(a) $(-\infty, -2] \cup [3, \infty)$ (b) $\left(-\infty, -\frac{1}{2}\right] \cup \left[\frac{3}{2}, \infty\right)$

(c) $\left(-\infty, -\frac{3}{2}\right] \cup \left[\frac{3}{2}, \infty\right)$ (d) $\left(-\infty, -\frac{5}{2}\right] \cap \left[\frac{3}{2}, \infty\right)$

(e) $\left(-\infty, -\frac{5}{2}\right] \cup \left[\frac{3}{2}, \infty\right)$

Answer: easy (e)

$\left|t + \frac{1}{2}\right| \ge 2 \quad \Leftrightarrow \quad t + \frac{1}{2} \ge 2$ or $t + \frac{1}{2} \le -2 \quad \Leftrightarrow \quad t \ge \frac{3}{2}$ or $t \le -\frac{5}{2}$

28. Solve the inequality $\left|\frac{2x - 0.1}{0.01}\right| \le 2$.

(a) $[0.4, 0.6] \cup [3, \infty)$ (b) $(-\infty, -0.06] \cup (0.04, \infty)$

(c) $[0.04, 0.06]$ (d) $[0.4, 0.6]$

(e) $(-\infty, -0.02) \cup [0.04, 0.06]$

Answer: easy (c)

$\left|\frac{2x - 0.1}{0.01}\right| \le 2 \quad \Leftrightarrow \quad -2 \le \frac{2x - 0.1}{0.01} \le 2 \quad \Leftrightarrow \quad -0.02 \le 2x - 0.1 \le 0.02 \quad \Leftrightarrow \quad 0.04 \le x \le 0.06$

29. Solve the inequality $-2\left|7x-\frac{1}{2}\right|+4\leq -17$.

(a) $\left(-\infty,-\frac{10}{7}\right]\cup\left[\frac{11}{7},\infty\right)$
(b) $\left(-\infty,\frac{10}{7}\right)\cup\left[\frac{11}{7},\infty\right)$
(c) $\left(-\infty,-\frac{10}{7}\right]$
(d) $\left(\frac{11}{7},\infty\right)$
(e) $\left[-\frac{10}{7},\frac{11}{7}\right]$

Answer: easy (a)

$-2\left|7x-\frac{1}{2}\right|+4\leq -17 \quad\Leftrightarrow\quad \left|7x-\frac{1}{2}\right|\geq\frac{21}{2} \quad\Leftrightarrow\quad 7x-\frac{1}{2}\geq\frac{21}{2}$ or $7x-\frac{1}{2}\leq-\frac{21}{2} \quad\Leftrightarrow\quad x\leq-\frac{10}{7}$ or $x\geq\frac{11}{7} \quad\Leftrightarrow$ $x\in\left(-\infty,-\frac{10}{7}\right]\cup\left[\frac{11}{7},\infty\right)$.

30. Solve the equation $|5x|+2x-12=0$ graphically.

Answer: easy

$x\approx 1.71$, $x=-4$

31. Solve the equation $\left|x^2-48\right|=16$ graphically.

Answer: easy

$x=\pm 8$, $x\approx\pm 5.66$

32. Solve the equation $x^2-|5x|=-6$ graphically.

Answer: easy

$x=\pm 2$, $x=\pm 3$

33. Solve the equation $||x-1|-9|=6$ graphically.

Answer: easy

$x=16$, $x=-14$, $x=4$, $x=-2$

CHAPTER 3 REVIEW

1. Find all real solutions of the equation $3x+5=50$.

(a) $x=-12$ (b) $x=15$ (c) $x=3$ (d) $x=-2$ (e) $x=40$

Answer: easy (b)

$3x+5=50 \quad\Leftrightarrow\quad 3x=45 \quad\Leftrightarrow\quad x=15$

2. Find all real solutions of the equation $6x+10=9-4x$.

(a) $x=-\frac{3}{5}$ (b) $x=\frac{1}{5}$ (c) $x=-\frac{2}{5}$ (d) $x=-\frac{1}{10}$ (e) $x=\frac{3}{10}$

Answer: easy (d)

$6x+10=9-4x \quad\Leftrightarrow\quad 10x=-1 \quad\Leftrightarrow\quad x=-\frac{1}{10}$

3. Find all real solutions of the equation $\frac{2}{5}x+2=2x-\frac{1}{10}$.

(a) $x=\frac{3}{4}$ (b) $x=\frac{21}{16}$ (c) $x=\frac{31}{24}$ (d) $x=\frac{9}{10}$ (e) $x=\frac{4}{7}$

Answer: easy (b)

$\frac{2}{5}x+2=2x-\frac{1}{10} \quad\Leftrightarrow\quad 2x-\frac{2}{5}x=2+\frac{1}{10} \quad\Leftrightarrow\quad \frac{8}{5}x=\frac{21}{10} \quad\Leftrightarrow\quad 16x=21 \quad\Leftrightarrow\quad x=\frac{21}{16}$

4. Find all real solutions of the equation $\dfrac{x+3}{2}-\dfrac{2x+3}{6}=\dfrac{5}{12}$.

(a) $x=-\frac{7}{2}$ (b) $x=\frac{2}{5}$ (c) $x=\frac{1}{4}$ (d) $x=-\frac{2}{3}$ (e) $x=\frac{9}{7}$

Answer: easy (a)

$\dfrac{x+3}{2}-\dfrac{2x+3}{6}=\dfrac{5}{12} \quad\Leftrightarrow\quad 6(x+3)-2(2x+3)=5 \quad\Leftrightarrow\quad 6x+18-4x-6=5 \quad\Leftrightarrow\quad 2x=-7 \quad\Leftrightarrow\quad x=-\frac{7}{2}$

5. Find all real solutions of the equation $\dfrac{x}{x+1} - 2 = \dfrac{1}{x+1}$.

(a) $x = -4$ (b) $x = -\frac{7}{2}$ (c) $x = -3$ (d) $x = -\frac{5}{2}$ (e) $x = -2$

Answer: medium (c)

$\dfrac{x}{x+1} - 2 = \dfrac{1}{x+1} \quad \Leftrightarrow \quad x - 2(x+1) = 1 \quad \Leftrightarrow \quad -x - 2 = 1 \quad \Leftrightarrow \quad -x = 3 \quad \Leftrightarrow \quad x = -3$

6. Find all real solutions of the equation $x^2 + 22x + 121 = 0$.

(a) -13 (b) ± 13 (c) 121 (d) -14 (e) -11

Answer: medium (e)

$x^2 + 22x + 121 = 0 \quad \Leftrightarrow \quad (x+11)^2 = 0 \quad \Leftrightarrow \quad x = -11$

7. Find all real solutions of the equation $3x^2 + 8x - 3 = 0$.

(a) $x = -3$ or 2 (b) $x = \frac{2}{3}$ or $\frac{4}{3}$ (c) $x = \frac{8}{3}$ or 4 (d) $x = \pm 2$ (e) $x = \frac{1}{3}$ or -3

Answer: medium (e)

$3x^2 + 8x - 3 = 0 \quad \Leftrightarrow \quad (3x-1)(x+3) = 0 \quad \Leftrightarrow \quad x = \frac{1}{3}$ or -3

8. Find all real solutions of the equation $x^3 - 2x^2 - 4x + 8 = 0$.

(a) $x = \pm\sqrt{5}$ (b) $x = \pm 2$ (c) $x = 2$ or 4 (d) $x = \pm 4$ (e) No solution

Answer: medium (b)

$x^3 - 2x^2 - 4x + 8 = 0 \quad \Leftrightarrow \quad x^2(x-2) - 4(x-2) = 0 \quad \Leftrightarrow \quad (x-2)(x^2-4) = 0 \quad \Leftrightarrow \quad x = 2, -2$

9. Find all real solutions of the equation $x^2 - 3x + 11 = 0$.

(a) $x = \frac{3}{2}$ (b) $x = \pm\frac{3}{2}$ (c) $x = -2 \pm \sqrt{3}$ (d) $x = -3 \pm \sqrt{2}$ (e) No solution

Answer: medium (e)

$x^2 - 3x + 11 = 0 \quad \Leftrightarrow \quad x = \dfrac{3 \pm \sqrt{(-3)^2 - 4(1)(11)}}{2(1)} = \dfrac{3 \pm \sqrt{-35}}{2}$. Therefore there are no real solutions.

10. Find all real solutions of the equation $\dfrac{x}{x-3} + \dfrac{1}{x+3} = \dfrac{9}{x^2-9}$.

(a) $x = -6$ or 2 (b) $x = \pm 3$ (c) $x = -3$ (d) $x = -2$ or 3 (e) $x = -2$ or 4

Answer: medium (a)

$\dfrac{x}{x-3} + \dfrac{1}{x+3} = \dfrac{9}{x^2-9} \quad \Leftrightarrow \quad x(x+3) + (x-3) = 9 \quad \Leftrightarrow \quad x^2 + 4x - 12 = 0 \quad \Leftrightarrow \quad (x+6)(x-2) = 0 \quad \Leftrightarrow$
$x = -6, 2$.

11. Find all real solutions of the equation $x - 3\sqrt{x} = 18$.

(a) $x = 3$ (b) $x = 16$ (c) $x = 36$ (d) $x = -3$ (e) No solution

Answer: medium (c)

$x - 3\sqrt{x} = 18 \quad \Leftrightarrow \quad x - 3\sqrt{x} - 18 = 0 \quad \Leftrightarrow \quad \left(\sqrt{x} - 6\right)\left(\sqrt{x} + 3\right) = 0 \quad \Leftrightarrow \quad \sqrt{x} = 6, -3 \quad \Leftrightarrow \quad x = 36$ (since $\sqrt{x}$ cannot be negative).

12. Find all real solutions of the equation $|3x - 5| = 6$

(a) $x = -2$ or $\frac{5}{3}$ (b) $x = -3$ or 1 (c) $x = \frac{12}{5}$ or $-\frac{7}{5}$ (d) $x = \frac{11}{3}$ or $-\frac{1}{3}$ (e) $x = \frac{2}{3}$ or 5

Answer: easy (d)

$|3x - 5| = 6$.

Case (i): $3x - 5 = 6 \quad \Leftrightarrow \quad 3x = 11 \quad \Leftrightarrow \quad x = \frac{11}{3}$.

Case (ii): $-(3x - 5) = 6 \quad \Leftrightarrow \quad -3x + 5 = 6 \quad \Leftrightarrow \quad 3x = -1 \quad \Leftrightarrow \quad x = -\frac{1}{3}$. So $x = \frac{11}{3}$ or $x = -\frac{1}{3}$.

13. Anthony leaves Kingstown at 2:00 P.M. and drives to Queensville, 180 mi distant, at 55 mi/h. At 2:15 P.M. Hortense leaves Queensville and drives to Kingstown at 40 mi/h. At what time do they pass each other on the road?

(a) 2:35 P.M. (b) 2:42 P.M. (c) 3:30 P.M. (d) 4:00 P.M. (e) 7:30 P.M.

Answer: medium (d)

Let x be the time, in hours, that it takes them to pass each other on the road. Hence they passed each other when Anthony is $55x$ mi away from Kingstown and Hortense is $40\left(x - \frac{15}{60}\right)$ mi away from Queensville. Thus $55x + 40\left(x - \frac{1}{4}\right) = 180 \quad\Leftrightarrow\quad 55x + 40x - 10 = 180 \quad\Leftrightarrow\quad 95x = 190 \quad\Leftrightarrow\quad x = \frac{190}{95} = 2$. Therefore they passed each other on the road at 4:00 P.M.

14. The approximate distance d in feet that a driver travels after noticing that he or she must come to a sudden stop is given by the following formula (where x is the speed of the car in miles per hour): $d = x + \frac{1}{20}x^2$. If a car travels 120 feet before stopping, what must its speed have been before the brakes were applied?

(a) 32 mi/h (b) 36 mi/h (c) 38 mi/h (d) 40 mi/h (e) 44 mi/h

Answer: medium (d)

$d = x + \frac{1}{20}x^2 \quad\Rightarrow\quad 120 = x + \frac{1}{20}x^2 \quad\Leftrightarrow\quad x^2 + 20x - 2400 = 0 \quad\Leftrightarrow\quad (x+60)(x-40) = 0 \quad\Leftrightarrow\quad x = -60, 40.$
Therefore the speed must have been 40 mi/hr before the brakes were applied.

15. Abbie paints twice as fast as Beth and three times as fast as Cathy. If it takes 90 minutes to paint a living room if all three work together, how long would it take Abbie if she works alone?

(a) 125 min (b) 145 min (c) 150 min (d) 155 min (e) 165 min

Answer: hard (e)

Suppose it takes Abbie x min working alone. Hence it takes Beth $2x$ min and $3x$ min for Cathy to work alone. Thus $\dfrac{1}{x} + \dfrac{1}{2x} + \dfrac{1}{3x} = \dfrac{1}{90} \quad\Leftrightarrow\quad 90 + 45 + 30 = x \quad\Leftrightarrow\quad x = 165$. Therefore it takes Abbie 165 min working alone.

16. Evaluate the complex number $(-1+2i) + \left(\frac{1}{3} - i\right)$ and write the result in the form $a + bi$.

(a) $-\frac{2}{3} + i$ (b) $-\frac{1}{3} + 2i$ (c) $\frac{1}{3} - 2i$ (d) $\frac{1}{4} - 6i$ (e) $\frac{2}{3} - i$

Answer: easy (a)

$(-1+2i) + \left(\frac{1}{3} - i\right) = -\frac{2}{3} + i$

17. Evaluate the complex number $2(4-3i)\dfrac{i}{5}$ and write the result in the form $a + bi$.

(a) $\frac{1}{5} - \frac{2}{5}i$ (b) $\frac{6}{5} + \frac{8}{5}i$ (c) $\frac{3}{5} - 4i$ (d) $\frac{7}{5} + \frac{2}{5}i$ (e) $\frac{3}{5} - \frac{6}{5}i$

Answer: medium (b)

$2(4-3i)\dfrac{i}{5} = \frac{8}{5}i - \frac{6}{5}i^2 = \frac{6}{5} + \frac{8}{5}i$

18. Evaluate the complex number $\dfrac{3+i}{2-4i}$ and write the result in the form $a + bi$.

Answer: medium

$\dfrac{3+i}{2-4i} = \dfrac{(3+i)(2+4i)}{(2-4i)(2+4i)} = \dfrac{6+14i+4i^2}{4+16} = \dfrac{2+14i}{20} = \frac{1}{10} + \frac{7}{10}i$

19. Evaluate the complex number $(4-i)^3$ and write the result in the form $a + bi$.

Answer: medium

$(4-i)^3 = 4^3 - 3\cdot 4^2 i + 3\cdot 4 i^2 - i^3 = 64 - 48i - 12 + i = 52 - 47i$

20. Evaluate the complex number $\sqrt{-50}\cdot\sqrt{-2}$ and write the result in the form $a + bi$.

Answer: medium

$\sqrt{-50}\cdot\sqrt{-2} = i\sqrt{50}\cdot i\sqrt{2} = i^2\sqrt{100} = -10$

21. Find all real and imaginary solutions of the equation $x^2 = -18$

Answer: medium

$x^2 = -18 \quad\Leftrightarrow\quad x = \pm\sqrt{-18} = \pm 3i\sqrt{2}$

22. Find all real and imaginary solutions of the equation $2x^2 - 4x + 3 = 0$

Answer: medium

$$2x^2 - 4x + 3 = 0 \quad \Leftrightarrow \quad x = \frac{4 \pm \sqrt{(-4)^2 - 4(2)(3)}}{2(2)} = \frac{4 \pm \sqrt{-8}}{4} = 1 \pm \tfrac{\sqrt{2}}{2} i$$

23. Find all real and imaginary solutions of the equation $ix^2 - 2x - 3i = 0$

Answer: medium

$$ix^2 - 2x - 3i = 0 \quad \Leftrightarrow \quad x = \frac{2 \pm \sqrt{(-2)^2 - 4(i)(-3i)}}{2(i)} = \frac{2 \pm 2\sqrt{2}\,i}{2i} = \pm\sqrt{2} - i$$

24. Find all real and imaginary solutions of the equation $x^3 - 3x^2 + 9x - 27 = 0$

Answer: medium

$x^3 - 3x^2 + 9x - 27 = 0 \quad \Leftrightarrow \quad x^2(x-3) + 9(x-3) = 0 \quad \Leftrightarrow \quad (x-3)(x^2+9) = 0 \quad \Leftrightarrow \quad x = 3, \pm 3i$

25. Find all real and imaginary solutions of the equation $x^3 = 64$

Answer: medium

$x^3 = 64 \quad \Leftrightarrow \quad x^3 - 64 = 0 \quad \Leftrightarrow \quad (x-4)(x^2 + 4x + 16) = 0 \quad \Leftrightarrow \quad x = 4$ or

$\dfrac{-4 \pm \sqrt{4^2 - 4(1)(16)}}{2(1)} = \dfrac{-4 \pm \sqrt{-48}}{2} = \dfrac{-4 \pm 4i\sqrt{3}}{2}$. So $x = 4, -2 \pm 2i\sqrt{3}$

26. Solve the inequality $10 - x \geq 5x$. Express the solutions in terms of intervals and graph the solution set on the real number line.

Answer: medium

$10 - x \geq 5x \quad \Leftrightarrow \quad 10 \geq 6x \quad \Leftrightarrow \quad \frac{5}{3} \geq x \quad \Leftrightarrow \quad x \in \left(-\infty, \frac{5}{3}\right]$

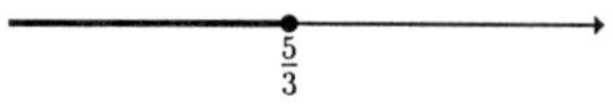

27. Solve the inequality $2 - x \leq 3x - 5$ in terms of intervals and illustrate the solution set on the real number line.

Answer: medium

$2 - x \leq 3x - 5 \quad \Leftrightarrow \quad 7 \leq 4x \quad \Leftrightarrow \quad x \geq \frac{7}{4}$. So the solution is $\left[\frac{7}{4}, \infty\right)$.

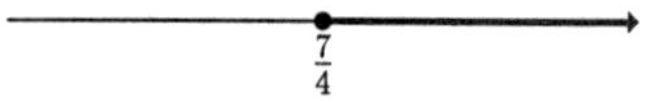

28. Solve the inequality $x^2 \leq 4$ in terms of intervals and illustrate the solution set on the real number line.

Answer: medium

$x^2 \leq 4 \quad \Leftrightarrow \quad -2 \leq x \leq 2$. Therefore the solution is $[-2, 2]$.

29. Solve the inequality $2x^2 \geq 6 - x$. Express the solutions in terms of intervals and graph the solution set on the real number line.

Answer: medium

$2x^2 \geq 6 - x \quad \Leftrightarrow \quad 2x^2 + x - 6 \geq 0 \quad \Leftrightarrow \quad (2x-3)(x+2) \geq 0.$

Case (i): $2x - 3 \geq 0 \quad \Leftrightarrow \quad x \geq \frac{3}{2}$, and $x + 2 \geq 0 \quad \Leftrightarrow \quad x \geq -2$; so $x \geq \frac{3}{2}$.

Case (ii): $2x - 3 \leq 0 \quad \Leftrightarrow \quad x \leq \frac{3}{2}$, and $x + 2 \leq 0 \quad \Leftrightarrow \quad x \leq -2$; so $x \leq -2$. Therefore $x \leq -2$ or $x \geq \frac{3}{2} \quad \Leftrightarrow$

$x \in (-\infty, -2] \cup \left[\frac{3}{2}, \infty\right)$

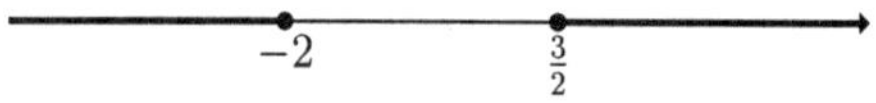

30. Solve the inequality $\dfrac{5}{x^3 - 3x^2 - 4x + 12} < 0$. Express the solutions in terms of intervals and graph the solution set on the real number line.

Answer: hard

$\dfrac{5}{x^3 - 3x^2 - 4x + 12} < 0 \quad \Leftrightarrow \quad \dfrac{5}{x^2(x-3) - 4(x-3)} < 0 \quad \Leftrightarrow \quad \dfrac{5}{(x-3)(x^2-4)} < 0 \quad \Leftrightarrow$
$\frac{5}{(x-3)(x-2)(x+2)} < 0$

Interval	$x-3$	$x-2$	$x+2$	$\dfrac{5}{(x-3)(x-2)(x+2)}$
$x < -2$	$-$	$-$	$-$	$-$
$-2 < x < 2$	$-$	$-$	$+$	$+$
$2 < x < 3$	$-$	$+$	$+$	$-$
$x > 3$	$+$	$+$	$+$	$+$

From the chart, $x < -2$ or $2 < x < 3 \quad \Leftrightarrow \quad x \in (-\infty, -2) \cup (2, 3)$

31. Solve the inequality $|x - 3| < 0.04$. Express the solutions in terms of intervals and graph the solution set on the real number line.

Answer: easy

$|x-3| < 0.04 \quad \Leftrightarrow \quad -0.04 < x - 3 < 0.04 \quad \Leftrightarrow \quad 2.96 < x < 3.04 \quad \Leftrightarrow \quad x \in (2.96, 3.04)$

32. Solve the inequality $|x - 2| < |x - 4|$. Express the solutions in terms of intervals and graph the solution set on the real number line. (Hint: Try interpreting the quantities as distances.)

Answer: hard

$|x-2|$ is the distance between x and 2 on the number line, and $|x-4|$ is the distance between x and 4. We want those points that are closer to 2 than to 4. Since 3 is midway between 2 and 4, we get $x \in (-\infty, 3)$ as the solution.

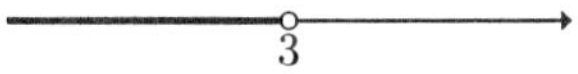

33. The volume of a sphere is given by $V = \frac{4}{3}\pi r^3$, where r is the radius. Find the interval of values of the radius if the volume is to be between 7 ft^3 and 14 ft^3, inclusive.

Answer: hard

$V = \frac{4}{3}\pi r^3$. Then $7 \le V \le 14 \quad \Leftrightarrow \quad 7 \le \frac{4}{3}\pi r^3 \le 14 \quad \Leftrightarrow \quad \dfrac{21}{4\pi} \le r^3 \le \dfrac{21}{2\pi} \quad \Leftrightarrow \quad \sqrt[3]{\dfrac{21}{4\pi}} \le r \le \sqrt[3]{\dfrac{21}{2\pi}}$ which is the interval of values of the radius.

CHAPTER FOUR

SECTION 4.1

1. Express the rule "Multiply by 6, then subtract 3" in function notation.
(a) $f(x) = 3x + 6$ (b) $f(x) = -18x$ (c) $f(x) = 6x - 3$ (d) $f(x) = 6(x + 3)$ (e) $f(x) = 6(x - 3)$
Answer: easy (c)

2. Express the rule "Square, subtract 3, then take the cube root" in function notation.
(a) $\sqrt[3]{x^2 - 3}$ (b) $x^{2/3} - \sqrt[3]{3}$ (c) $x^2 - 3$ (d) $\sqrt[3]{3x^2}$ (e) $-3x^3$
Answer: medium (a)

3. Express the function (or rule) $f(x) = \dfrac{x}{6} - 4$ in words.
(a) Subtract 4, then divide by 6 (b) Subtract 6, then divide by 4 (c) Divide by 6
(d) Divide by 6, then subtract 4 (e) Divide by 6, then add 4
Answer: easy (d)

4. Express the function (or rule) $f(x) = \sqrt{\frac{1}{3}x - 2}$ in words.
(a) Subtract 2, then take the square root (b) Divide by 3, then subtract 2 (c) Divide by 3, then square
(d) Divide by 3, subtract 2, then take the square root (e) Divide by 3, take the square root, then subtract 2
Answer: medium (d)

5. For the function $f(x) = x^2 - 4$ with $1 \le x \le 3$, draw a machine diagram and an arrow diagram.
Answer: medium

Machine Diagram

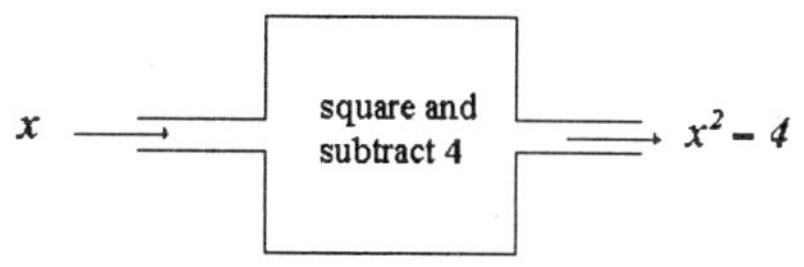

Arrow Diagram

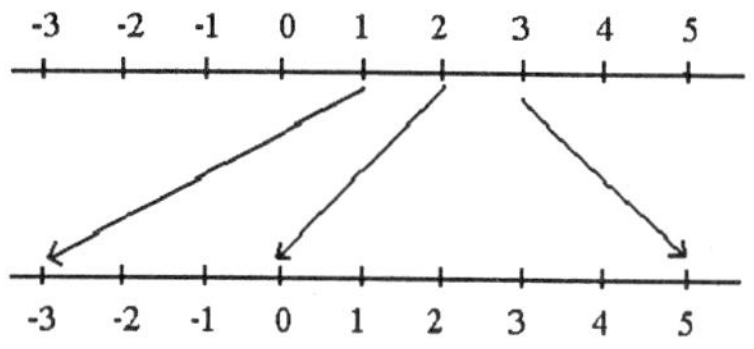

6. If $f(x) = x^3 + 2x - 1$, find $f(0)$, $f(3)$, $f(-3)$, $f(-x)$, and $f(1/a)$.
Answer: easy
$f(x) = x^3 + 2x - 1$, $f(0) = 0^3 + 2 \cdot 0 - 1 = -1$, $f(3) = 3^3 + 2 \cdot 3 - 1 = 27 + 6 - 1 = 32$,
$f(-3) = (-3)^3 + 2(-3) - 1 = -27 - 6 - 1 = -34$, $f(-x) = (-x)^3 + 2(-x) - 1 = -x^3 - 2x - 1$,
$f(1/a) = (1/a)^3 + 2(1/a) - 1 = 1/a^3 + 2/a - 1$

7. If $h(t) = t - 1/t$, find $h(1)$, $h(\pi)$, $h(t+1)$, $h(t) + h(1)$, and $h(x)$.
Answer: easy
$h(t) = t - 1/t$, $h(1) = 1 - 1/1 = 0$, $h(\pi) = \pi - 1/\pi$, $h(t+1) = t + 1 - 1/(t+1)$, $h(t) + h(1) = t - 1/t$,
$h(x) = x - 1/x$

8. If $f(x) = 4x + 1$, find $f(1)$, $f(-1)$, $f\left(\frac{1}{3}\right)$, $f\left(\frac{1}{3}x\right)$, $f(3x)$, $f(x^2)$, and $[f(x)]^2$.
Answer: medium
$f(x) = 4x + 1$, $f(1) = 4 \cdot 1 + 1 = 5$, $f(-1) = 4 \cdot (-1) + 1 = -3$, $f\left(\frac{1}{3}\right) = 4 \cdot \frac{1}{3} + 1 = \frac{7}{3}$, $f\left(\frac{1}{3}x\right) = 4\left(\frac{1}{3}x\right) + 1 = \frac{4}{3}x + 1$,
$f(3x) = 4(3x) + 1 = 12x + 1$, $f(x^2) = 4x^2 + 1$, $[f(x)]^2 = (4x + 1)^2 = 16x^2 + 8x + 1$

9. For the function $f(x) = 2x^2 - x + 1$ find $f(a)$, $f(a) + f(h)$, $f(a+h)$, and $\dfrac{f(a+h) - f(a)}{h}$, where a and h are real numbers and $h \neq 0$.

Answer: medium

$f(x) = 2x^2 - x + 1,\ f(a) = 2a^2 - a + 1, f(a) + f(h) = 2a^2 - a + 1 + 2h^2 - h + 1 = 2a^2 - a + 2h^2 - h + 2,$
$f(a+h) = 2(a+h)^2 - (a+h) + 1 = 2a^2 + 4ah + 2h^2 - a - h + 1,$
$\dfrac{f(a+h) - f(a)}{h} = \dfrac{2a^2 + 4ah + 2h^2 - a - h + 1 - (2a^2 - a + 1)}{h} = 4a + 2h - 1$

10. For the function $f(x) = x^3 + x^2 + 2$ find $f(a)$, $f(a) + f(h)$, $f(a+h)$, and $\dfrac{f(a+h) - f(a)}{h}$, where a and h are real numbers and $h \neq 0$.

Answer: medium

$f(x) = x^3 + x^2 + 2,\ f(a) = a^3 + a^2 + 2,\ f(a) + f(h) = a^3 + a^2 + 2 + h^3 + h^2 + 2 = a^3 + a^2 + h^3 + h^2 + 4,$
$f(a+h) = (a+h)^3 + (a+h)^2 + 2 = a^3 + 3a^2h + 3ah^2 + h^3 + a^2 + 2ah + h^2 + 2,$
$\dfrac{f(a+h) - f(a)}{h} = \dfrac{a^3 + 3a^2h + 3ah^2 + h^3 + a^2 + 2ah + h^2 + 2 - (a^3 + a^2 + 2)}{h} = 3a^2 + 3ah + h^2 + 2a + h$

11. For the function $f(x) = 2x - x^2$ find $f(2+h)$, $f(x+h)$, and $\dfrac{f(x+h) - f(x)}{h}$, where $h \neq 0$.

Answer: medium

$f(x) = 2x - x^2,\ f(2+h) = 2(2+h) - (2+h)^2 = 4 + 2h - (4 + 4h + h^2) = -2h - h^2,$
$f(x+h) = 2(x+h) - (x+h)^2 = 2x + 2h - x^2 - 2xh - h^2,$
$\dfrac{f(x+h) - f(x)}{h} = \dfrac{2x + 2h - x^2 - 2xh - h^2 - (2x - x^2)}{h} = 2 - 2x - h$

12. For the function $f(x) = \dfrac{x+1}{x}$ find $f(2+h)$, $f(x+h)$, and $\dfrac{f(x+h) - f(x)}{h}$, where $h \neq 0$.

Answer: medium

$f(x) = \dfrac{x+1}{x},\ f(2+h) = \dfrac{2+h+1}{2+h} = \dfrac{3+h}{2+h},\ f(x+h) = \dfrac{x+h+1}{x+h},$

$$\frac{f(x+h) - f(x)}{h} = \frac{\frac{x+h+1}{x+h} - \frac{x+1}{x}}{h} = \frac{x(x+h+1) - (x+h)(x+1)}{x(x+h)h} = \frac{x^2 + xh + x - x^2 - xh - x - h}{x(x+h)h} = -\frac{1}{x(x+h)}$$

13. Find the domain and range of the function $f(x) = 2x - 2,\ -1 \leq x \leq 3$.

(a) Domain $[-1, 3]$, Range $[-4, 4]$
(b) Domain $[-1, 3]$, Range $[1, 3]$
(c) Domain $[-3, 1]$, Range $[-2, 2]$
(d) Domain $[-4, 4]$, Range $[-4, 4]$
(e) Domain $[-1, 3]$, Range $[-2, 3]$

Answer: medium (a)

$f(x) = 2x - 2, -1 \leq x \leq 3 \quad \Leftrightarrow \quad -2 \leq 2x \leq 6 \quad \Leftrightarrow \quad -4 \leq 2x - 2 \leq 4$. Then the domain is $[-1, 3]$ and the range is $[-4, 4]$.

14. Find the domain and range of the function $f(x) = 2x^2 + 1,\ -1 \leq x \leq 2$.

(a) Domain $[-2, 2]$, Range $[1, 9]$
(b) Domain $[-1, 1]$, Range $[2, 9]$
(c) Domain $[-1, 2]$, Range $[4, 9]$
(d) Domain $[-1, 2]$, Range $[2, \infty]$
(e) Domain $[-1, 2]$, Range $[1, 9]$

Answer: medium (e)

$f(x) = 2x^2 + 1, -1 \leq x \leq 2 \quad \Leftrightarrow \quad 0 \leq x^2 \leq 4 \quad \Leftrightarrow \quad 1 \leq 2x^2 + 1 \leq 9$. Then the domain is $[-1, 2]$ and the range is $[1, 9]$.

15. Find the domain and range of the function $g(x) = \sqrt{6 - 4x}$.

(a) Domain $[-4, 4]$, Range $[2, \infty)$
(b) Domain $(-\infty, 3]$, Range $[3, \infty)$
(c) Domain $\left(-\infty, \frac{3}{2}\right]$, Range $[0, \infty)$
(d) Domain $(-\infty, 3]$, Range $[0, \infty)$
(e) Domain $[4, 6]$, Range $[0, \infty)$

Answer: medium (c)

$g(x) = \sqrt{6 - 4x}$. Since $6 - 4x \geq 0 \quad \Leftrightarrow \quad 4x \leq 6 \quad \Leftrightarrow \quad x \leq \frac{3}{2}$, the domain is $\left(-\infty, \frac{3}{2}\right]$ and the range is $[0, \infty)$.

16. Find the domain and range of the function $h(x) = 1 - \sqrt{x-1}$.
(a) Domain $[1, \infty)$, Range $[-1, \infty)$ (b) Domain $[-1, \infty)$, Range $(-\infty, 1]$ (c) Domain $(-\infty, -1]$, Range $[1, \infty)$
(d) Domain $[1, \infty)$, Range $(-\infty, 1]$ (e) Domain $[1, \infty)$, Range $(-\infty, -1]$
Answer: medium (d)
$h(x) = 1 - \sqrt{x-1}$. Then $x - 1 \geq 0 \Leftrightarrow x \geq 1$ and $h(x) \leq 1$ for all x. Then the domain is $[1, \infty)$ and the range is $(-\infty, 1]$.

17. Find the domain and range of the function $G(x) = \sqrt{x^2 - 16}$.
(a) Domain $(-\infty, -4] \cup [4, \infty)$, Range $[4, \infty)$ (b) Domain $(-\infty, -4] \cup [4, \infty)$, Range $[-4, \infty)$
(c) Domain $(-\infty, -4] \cup [4, \infty)$, Range $[0, \infty)$ (d) Domain $(-\infty, -2] \cup [2, \infty)$, Range $[2, \infty)$
(e) Domain $(-\infty, \infty)$, Range $[0, \infty)$
Answer: medium (c)
$G(x) = \sqrt{x^2 - 16}$. Since $x^2 - 16 \geq 0 \Leftrightarrow x^2 \geq 16 \Leftrightarrow x \geq 4$ or $x \leq -4$. Then the domain is $(-\infty, -4] \cup [4, \infty)$ and the range is $[0, \infty)$.

18. Find the domain of the function $f(x) = \dfrac{4}{3x-7}$.
Answer: easy
$f(x) = \dfrac{4}{3x-7}$. Since $3x - 7 \neq 0 \Leftrightarrow 3x \neq 7 \Leftrightarrow x \neq \frac{7}{3}$, we have $D = \left\{x \mid x \neq \frac{7}{3}\right\}$

19. Find the domain of the function $f(x) = \dfrac{x^3}{x^2 + 2x - 3}$.
Answer: medium
$f(x) = \dfrac{x^3}{x^2 + 2x - 3} = \dfrac{x^3}{(x+3)(x-1)}$. Since $(x+3)(x-1) \neq 0$ then $x \neq -3$ and $x \neq 1$. Therefore $D = \{x \mid x \neq -3, x \neq 1\}$.

20. Find the domain of the function $f(x) = \dfrac{x^3 + 1}{x^4 - x}$.
Answer: medium
$f(x) = \dfrac{x^3+1}{x^4 - x} = \dfrac{x^3+1}{x(x-1)(x^2+x+1)}$. Since $x(x-1)\left(x^2+x+1\right) \neq 0$ then $x \neq 0$, $x \neq 1$. So $D = \{x \mid x \neq 0, x \neq 1\}$.

21. Find the domain of the function $f(x) = \sqrt[6]{x-9}$.
Answer: medium
$f(x) = \sqrt[6]{x-9}$. Then $x - 9 \geq 0 \Leftrightarrow x \geq 9$ and thus the domain is $[9, \infty)$.

22. Find the domain of the function $f(t) = \sqrt{t^2 + 2}$.
Answer: medium
$f(t) = \sqrt{t^2+2}$. Since $t^2 \geq 0$ for all t, then $t^2 + 2 \geq 2 > 0$, and so the domain is $(-\infty, \infty)$.

23. Find the domain of the function $g(x) = \dfrac{\sqrt{x}}{2x^2 - x - 1}$.
Answer: medium
$g(x) = \dfrac{\sqrt{x}}{2x^2 - x - 1} = \dfrac{\sqrt{x}}{(2x+1)(x-1)}$. Then $x \geq 0$ and $2x + 1 \neq 0$ and $x - 1 \neq 0 \Leftrightarrow x \geq 0$ and $x \neq -\frac{1}{2}$ and $x \neq 1$.
So the domain is $[0, 1) \cup (1, \infty)$.

24. Find the domain of the function $F(x) = x^2 - \dfrac{x}{\sqrt{7-2x}}$.
Answer: medium
$F(x) = x^2 - \dfrac{x}{\sqrt{7-2x}}$. Then $7 - 2x > 0 \Leftrightarrow 7 > 2x \Leftrightarrow x < \frac{7}{2}$. Therefore the domain is $\left(-\infty, \frac{7}{2}\right)$.

25. Find the domain of the function $G(x) = \sqrt{x+1} - 3\sqrt{x-4}$.
Answer: medium
$G(x) = \sqrt{x+1} - 3\sqrt{x-4}$. Then $x+1 \geq 0$ and $x-4 \geq 0 \quad \Leftrightarrow \quad x \geq -1$ and $x \geq 4 \quad \Leftrightarrow \quad x \geq 4$. Then the domain is $[4, \infty)$.

26. Find the domain of the function $f(x) = \sqrt{3-5x^2}$.
Answer: medium
$f(x) = \sqrt{3-5x^2}$. Then $3-5x^2 \geq 0 \quad \Leftrightarrow \quad 3 \geq 5x^2 \quad \Leftrightarrow \quad \frac{3}{5} \geq x^2 \quad \Leftrightarrow \quad -\sqrt{\frac{3}{5}} \leq x \leq \sqrt{\frac{3}{5}}$. Therefore the domain is $\left[-\sqrt{\frac{3}{5}}, \sqrt{\frac{3}{5}}\right]$.

27. Find the domain of the function $g(x) = \sqrt{x^2-3x-4}$.
Answer: hard
$g(x) = \sqrt{x^2-3x-4} = \sqrt{(x-4)(x+1)}$. Then $(x-4)(x+1) \geq 0$.
Case (i): $x-4 \geq 0$ and $x+1 \geq 0 \quad \Leftrightarrow \quad x \geq 4$ and $x \geq -1 \quad \Leftrightarrow \quad x \geq 4$.
Case (ii): $x-4 \leq 0$ and $x+1 \leq 0 \quad \Leftrightarrow \quad x \leq 4$ and $x \leq -1 \quad \Leftrightarrow \quad x \leq -1$. Therefore the domain is $(-\infty, -1] \cup [4, \infty)$.

28. Find the domain of the function $\phi(x) = \sqrt{\dfrac{x^2-3x}{x-2}}$.
Answer: hard
$\phi(x) = \sqrt{\dfrac{x^2-3x}{x-2}} = \sqrt{\dfrac{x(x-3)}{x-2}}$. Then $\dfrac{x(x-3)}{x-2} \geq 0$ and $x-2 \neq 0$.
Case (i): $x \geq 0$, $x-3 \geq 0$ and $x-2 > 0 \quad \Leftrightarrow \quad x \geq 0$ and $x \geq 3$ and $x > 2 \quad \Leftrightarrow \quad x \geq 3$.
Case (ii): $x \leq 0$, $x-3 \leq 0$ and $x-2 > 0 \quad \Leftrightarrow \quad x \leq 0$ and $x \leq 3$ and $x > 2$ which is impossible.
Case (iii): $x \geq 0$ and $x-3 \leq 0$ and $x-2 < 0 \quad \Leftrightarrow \quad 0 \leq x < 2$.
Case (iv): $x \leq 0$ and $x-3 \geq 0$ and $x-2 < 0$ which is impossible.
Combining the four cases gives a domain of $[0, 2) \cup [3, \infty)$.

29. If $f(x) = 3x-1$, find $f(0)$, $f(-1)$, $f(2)$, $f(a)$, $f\left(\frac{1}{a}\right)$, and $f\left(\frac{1}{a+1}\right)$.
Answer: easy
$f(0) = 3(0) - 1 = -1$, $f(-1) = 3(-1) - 1 = -4$, $f(2) = 3(2) - 1 = 5$, $f(a) = 3(a) - 1 = 3a - 1$,
$f\left(\frac{1}{a}\right) = 3\left(\frac{1}{a}\right) - 1 = \frac{3}{a} - 1 = \frac{3-a}{a}$, $f\left(\frac{1}{a+1}\right) = 3\left(\frac{1}{a+1}\right) - 1 = \frac{3}{a+1} - 1 = \frac{2-a}{a+1}$

30. If $f(x) = (x-2)^2 + 5$, find $f(1)$, $f(2)$, $f\left(\frac{5}{2}\right)$, $f(a)$, $f\left(\frac{1}{a}\right)$, and $f(a+h)$.
Answer: easy
$f(1) = (1-2)^2 + 5 = 6$, $f(2) = (2-2)^2 + 5 = 5$, $f\left(\frac{5}{2}\right) = \left(\frac{5}{2} - 2\right)^2 + 5 = \frac{21}{4}$, $f(a) = (a-2)^2 + 5 = 9 - 4a + a^2$,
$f\left(\frac{1}{a}\right) = \left(\frac{1}{a} - 2\right)^2 + 5 = \frac{1-4a+9a^2}{a^2}$, $f(a+h) = ((a+h)-2)^2 + 5 = a^2 + 2ah - 4a + h^2 - 4h + 9$

31. If $g(x) = \frac{1}{x} + x^2$, find $g(-3)$, $g(-1)$, $g(1)$, $g(3)$, $g(a+1)$, and $g(a^2)$.
Answer: easy
$g(-3) = \frac{1}{-3} + (-3)^2 = \frac{26}{3}$, $g(-1) = \frac{1}{-1} + (-1)^2 = 0$, $g(1) = \frac{1}{1} + 1^2 = 2$, $g(3) = \frac{1}{3} + 3^2 = \frac{28}{3}$,
$g(a+1) = \frac{1}{a+1} + (a+1)^2 = \frac{a^3+3a^2+3a+2}{a+1}$, $g(a^2) = \frac{1}{a^2} + (a^2)^2 = \frac{1+a^6}{a^2}$

32. If $h(t) = \dfrac{|t-1|}{t-1}$, find $h(-2)$, $h(-1)$, $h(0)$, $h(2)$, and $h(a+1)$.
Answer: medium
$h(-2) = \frac{|-2-1|}{-2-1} = -1$, $h(-1) = \frac{|-1-1|}{-1-1} = -1$, $h(0) = \frac{|0-1|}{0-1} = -1$, $h(2) = \frac{|2-1|}{2-1} = 1$,
$h(a+1) = \frac{|a+1-1|}{a+1-1} = \frac{|a|}{a} = 1$ if $a > 0$, $\frac{|a|}{a} = -1$ if $a < 0$, $a \neq 0$

33. If $f(x) = \begin{cases} 3x^2 & \text{if } x < 0 \\ 2x+1 & \text{if } x \geq 0 \end{cases}$ evaluate $f(-2)$, $f(-1)$, $f(0)$, $f(1)$,and $f(5)$.

Answer: easy

$f(-2) = 12$, $f(-1) = 3$, $f(0) = 1$, $f(1) = 3$, $f(5) = 11$

34. If $f(x) = \begin{cases} 4x+1 & \text{if } x < 0 \\ -2 & \text{if } 0 \leq x \leq 1 \\ \dfrac{2}{x+1} & \text{if } x > 1 \end{cases}$ evaluate $f(-2)$, $f\left(-\frac{1}{2}\right)$, $f(0)$, $f(1)$, and $f(6)$.

Answer: easy

$f(-2) = -7$, $f\left(-\frac{1}{2}\right) = -1$, $f(0) = -2$, $f(1) = -2$, $f(6) = \frac{2}{7}$

35. If $h(t) = \begin{cases} t^2 & \text{if } t < 0 \\ 0 & \text{if } 0 \leq t < 1 \\ 2 & \text{if } 1 \leq t < 2 \\ 3 & \text{if } 2 \leq t < 3 \\ 3 - \dfrac{t}{3} & \text{if } t \geq 3 \end{cases}$ evaluate $h(-2)$, $h\left(-\frac{1}{2}\right)$, $h(0)$, $h\left(\frac{2}{3}\right)$, $h(1)$, $h(2)$, $h\left(\frac{10}{3}\right)$.

Answer: easy

$h(-2) = 4$, $h\left(-\frac{1}{2}\right) = \frac{1}{4}$, $h(0) = 0$, $h\left(\frac{2}{3}\right) = 0$, $h(1) = 2$, $h(2) = 3$, $h\left(\frac{10}{3}\right) = \frac{17}{9}$

36. Find the domain of the function $f(x) = -\frac{2}{3}x$.

Answer: easy

$(-\infty, \infty)$

37. Find the domain of the function $f(x) = 5x^2 - 5$, $0 \leq x \leq 5$.

Answer: easy

$[0, 5]$

38. Find the domain of the function $f(x) = \dfrac{x+1}{x^2+1}$.

Answer: medium

There is no value of x that would make the denominator 0, so the domain is all real numbers $(-\infty, \infty)$.

39. Find the domain of the function $h(x) = \sqrt{2x^2 + 3x + 1}$.

Answer: medium

For $h(x) = \sqrt{2x^2 + 3x + 1} \Rightarrow 2x^2 + 3x + 1 \geq 0 \Leftrightarrow (x+1)(2x+1) \geq 0$. Case 1: $x + 1 \geq 0$ and $(2x+1) \geq 0 \Leftrightarrow x \geq -1$ and $x \geq -\frac{1}{2} \Rightarrow x \geq -\frac{1}{2}$. Case 2: $x + 1 \leq 0$ and $(2x+1) \leq 0 \Leftrightarrow x \leq -1$ and $x \leq -\frac{1}{2} \Rightarrow x \leq -1$. The domain of the function is $(-\infty, -1] \cup \left[-\frac{1}{2}, \infty\right)$.

SECTION 4.2

1. Sketch the graph of the piecewise-defined function $f(x) = \begin{cases} -2 & \text{if } x \leq -2 \\ 2 & \text{if } x > -2 \end{cases}$

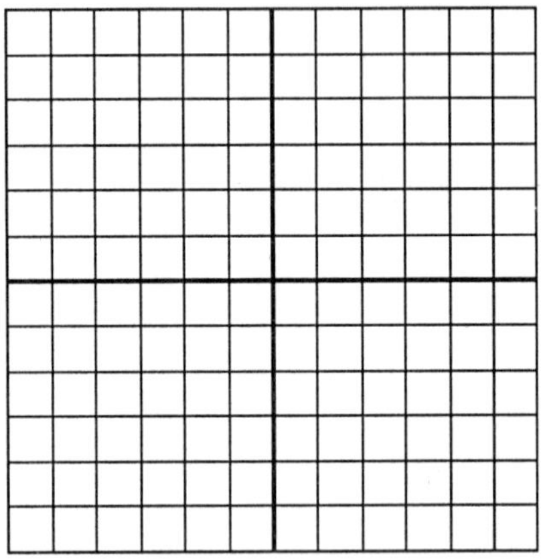

Answer: easy

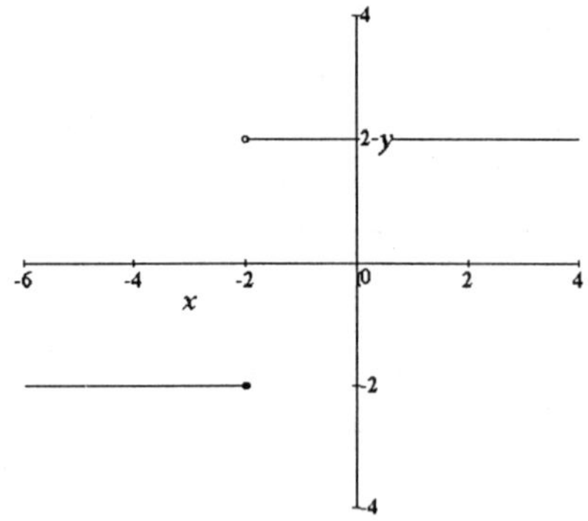

2. Sketch the graph of the piecewise-defined function $f(x) = \begin{cases} -1 & \text{if } x < 1 \\ x+1 & \text{if } x \geq 1 \end{cases}$

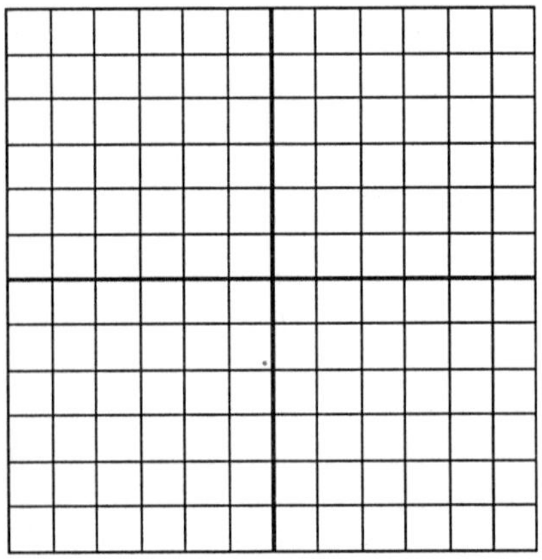

Answer: medium

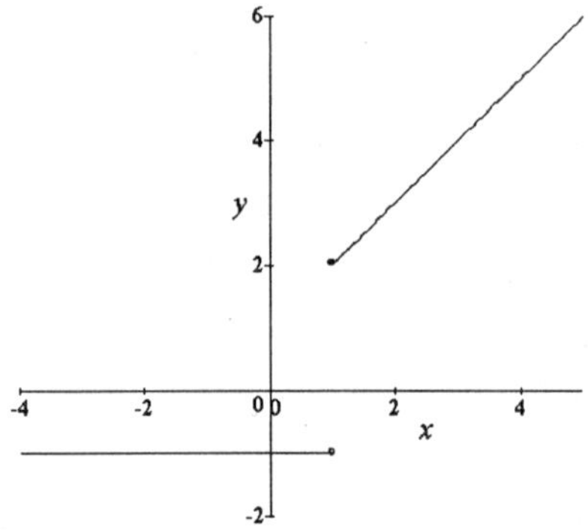

3. Sketch the graph of the piecewise-defined function $f(x) = \begin{cases} 2x+2 & \text{if } x \leq 0 \\ 4x-1 & \text{if } x > 0 \end{cases}$

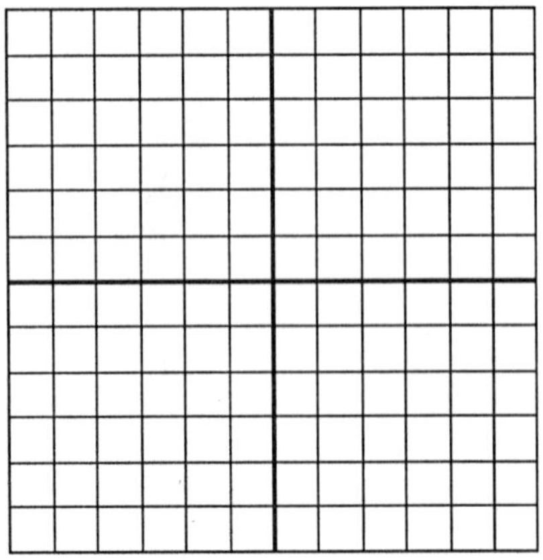

Answer: medium

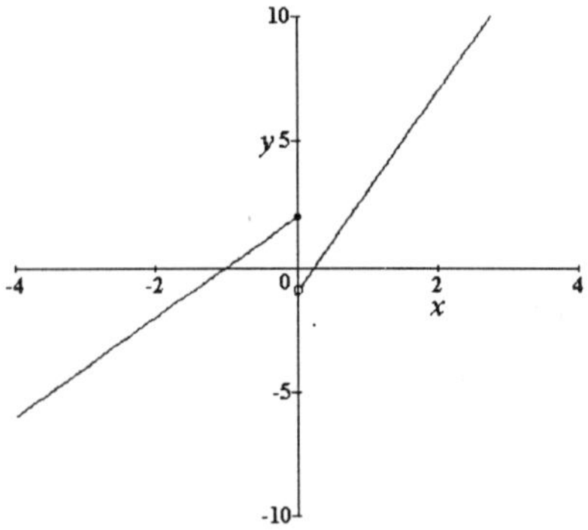

4. Sketch the graph of the piecewise-defined function $f(x) = \begin{cases} -2 & \text{if } x \leq -2 \\ x & \text{if } -2 < x < 2 \\ 4 - x & \text{if } x \geq 2 \end{cases}$

Answer: medium

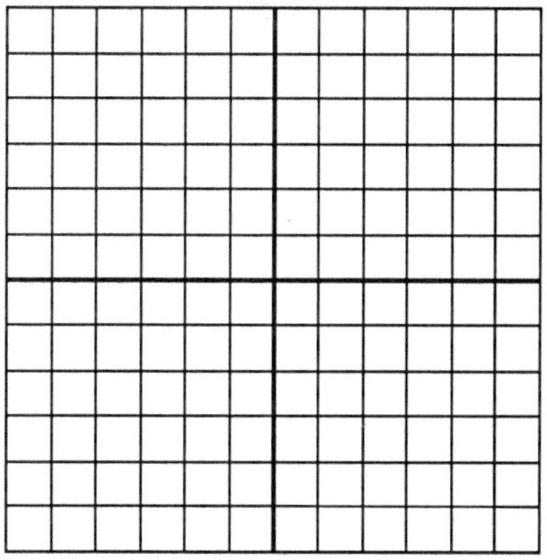

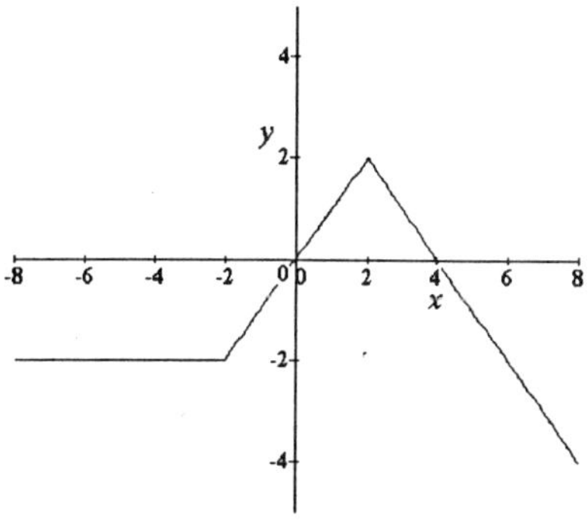

5. Sketch the graph of the piecewise-defined function $f(x) = \begin{cases} x^2 + 1 & \text{if } |x| \leq 1 \\ 2x & \text{if } |x| > 1 \end{cases}$

Answer: hard

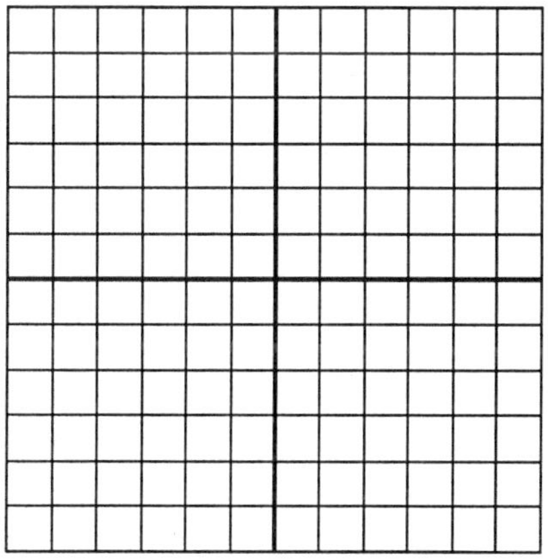

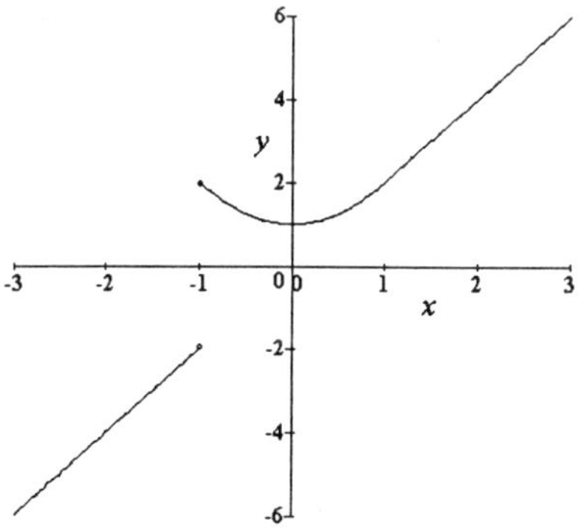

GC 6. Use a graphing device to draw the graph of the piecewise-defined function $f(x) = \begin{cases} x - 2x^2 & \text{if } x \leq -1 \\ x^3 - 2x - 4 & \text{if } x > -1 \end{cases}$

Answer: medium

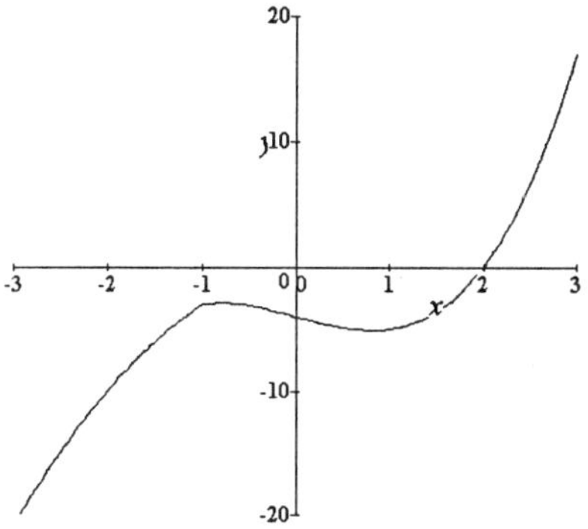

GC 7. Use a graphing device to draw the graph of the piecewise-defined function $f(x) = \begin{cases} \sqrt{-2x+2} & \text{if } x \le 0 \\ \sqrt{4-x^2} & \text{if } 0 < x < 2 \\ \sqrt{x-2} & \text{if } x \ge 2 \end{cases}$

Answer: medium

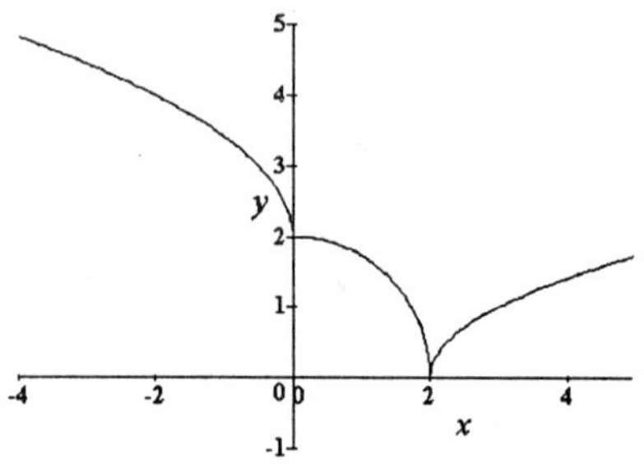

8. Find a function whose graph is the line segment joining the points $(-2, 4)$ and $(3, 1)$.
Answer: medium The slope of the line is $m = \dfrac{1-4}{3-(-2)} = \dfrac{-3}{5}$, so its equation is $y - 1 = -\frac{3}{5}(x-3) \quad \Leftrightarrow$
$y = -\frac{3}{5}x + \frac{14}{5}$, $-2 \le x \le 3$.

9. Find a function whose graph is the top half of the circle $x^2 + (y+1)^2 = 4$.
Answer: medium $x^2 + (y+1)^2 = 4 \quad \Leftrightarrow \quad y + 1 = \pm\sqrt{4-x^2}$. Because we want the top half of the circle, we take the positive root, and so $y = \sqrt{4-x^2} - 1$, $-2 \le x \le 2$.

10. Determine whether the given curve is the graph of a function of x.

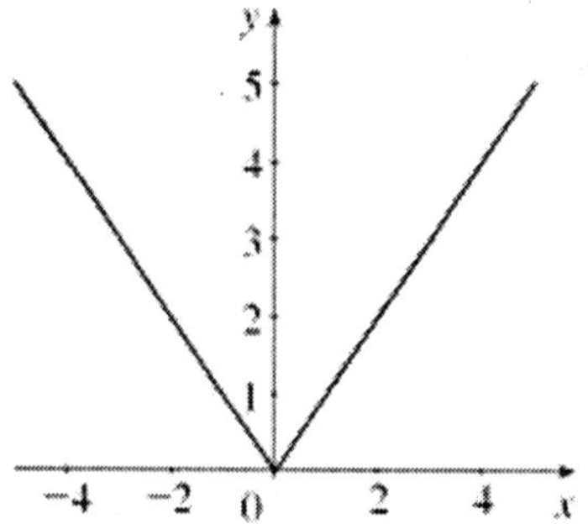

Answer: easy
Yes. Domain: $(-\infty, \infty)$, range: $[0, \infty)$

11. Determine whether the given curve is the graph of a function of x.

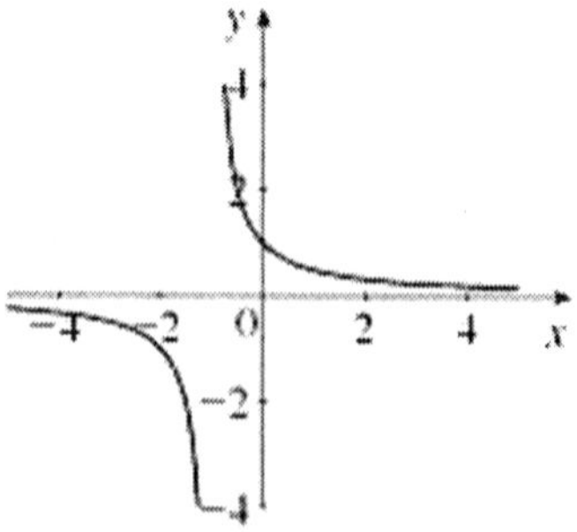

Answer: easy
Yes. Domain: $(-\infty, -1) \cup (-1, \infty)$

12. Determine whether the given curve is the graph of a function of x.

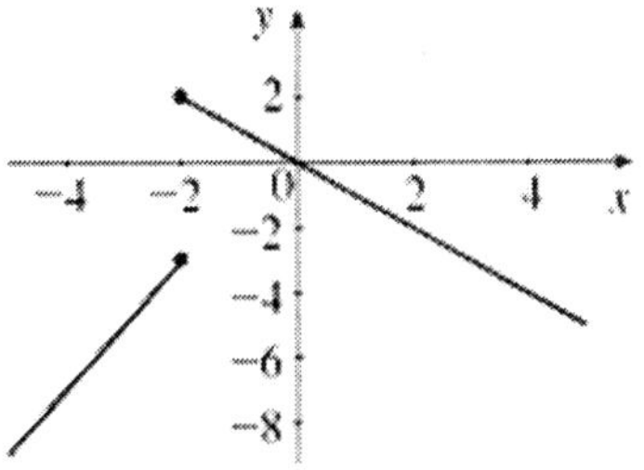

Answer: easy

The graph is not a function of x since there are two different values of y at $x = -2$.

13. Let $f(x) = \sqrt{25 - x^2}$

(a) Sketch the graph of f.

(b) Find the domain and range of f.

Answer: medium

(a) $f(x) = \sqrt{25 - x^2}$

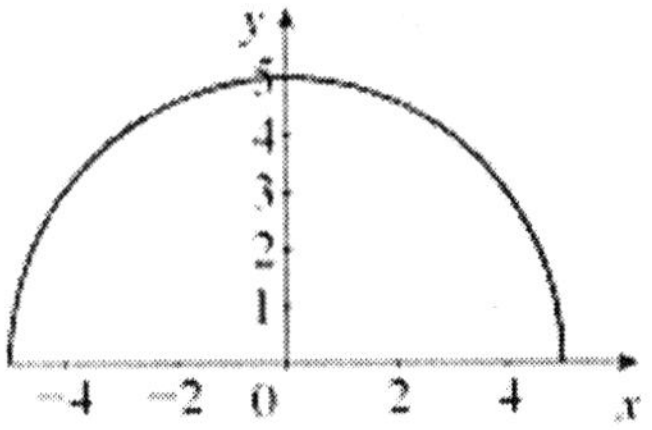

(b) Domain: $[-5, 5]$, range: $[0, 5]$.

14. Let $f(x) = \dfrac{|x-1|}{1-x}$.

(a) Sketch the graph of f.

(b) Find the domain of f.

Answer: medium

(a) $f(x) = \dfrac{|x-1|}{1-x}$

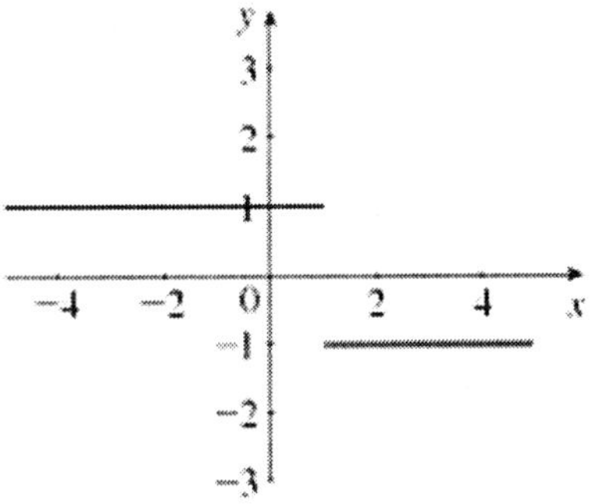

(b) Domain: $(-\infty, 1) \cup (1, \infty)$

15. Let $f(x) = -\frac{|x|}{x^2}$.
(a) Sketch the graph of f.
(b) Find the domain of f.
Answer: medium
(a) $f(x) = -\frac{|x|}{x^2}$

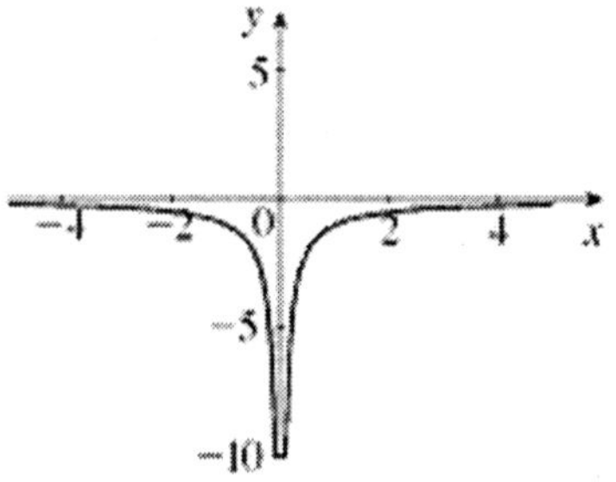

(b) Domain: $(-\infty, 0) \cup (0, \infty)$

16. Determine if the equation $2x^2 + 3y = 4x + 5$ defines y as a function of x. Explain your answer.
Answer: medium
$2x^2 + 3y = 4x + 5 \quad \Leftrightarrow \quad 3y = 5 + 4x - 2x^2 \quad \Leftrightarrow \quad y = \frac{5 + 4x - 2x^2}{3}$. The last equation is a rule that gives one value of y for each value of x, so the original equation does define y as a function of x.

17. Determine if the equation $x^2 + y^2 - 25 = 0$ defines y as a function of x. Explain your answer.
Answer: medium
$x^2 + y^2 - 25 = 0 \quad \Leftrightarrow \quad y^2 = 25 - x^2 \quad \Leftrightarrow \quad y = \pm\sqrt{25 - x^2}$. This equation gives 2 values of y for a given value of x, so the original equation does not define y as a function of x.

18. Determine if the equation $2x^2y - y^2 - 5 = 0$ defines y as a function of x. Explain your answer.
Answer: medium
$2x^2y - y - 5 = 0 \quad \Leftrightarrow \quad y\left(2x^2 - 1\right) = 5 \quad \Leftrightarrow \quad y = \frac{5}{2x^2 - 1}$. The last equation is a rule that gives one value of y for each value of x, so the original equation does define y as a function of x.

19. Determine if the equation $\sqrt{y} + x = 1$ defines y as a function of x. Explain your answer.
Answer: medium
$\sqrt{y} + x = 1 \quad \Leftrightarrow \quad \sqrt{y} = 1 - x \quad \Leftrightarrow \quad y = (1 - x)^2$. The last equation is a rule that gives one value of y for each value of x, so the original equation does define y as a function of x.

20. Determine if the equation $3x - |y| = 0$ defines y as a function of x. Explain your answer.
Answer: medium
$3x - |y| = 0 \quad \Leftrightarrow \quad |y| = 3x \quad \Leftrightarrow \quad y = \pm 3x$. Since this equation gives two values of y for a given value of x, the given equation does not define a function of x.

GC 21. Let $f(x) = x + c$. Graph the family of functions with $c = -1, 0$, and -1 in the viewing rectangle $[-3, 3]$ by $[-3, 3]$. How does the value of c affect the graph?

Answer : medium

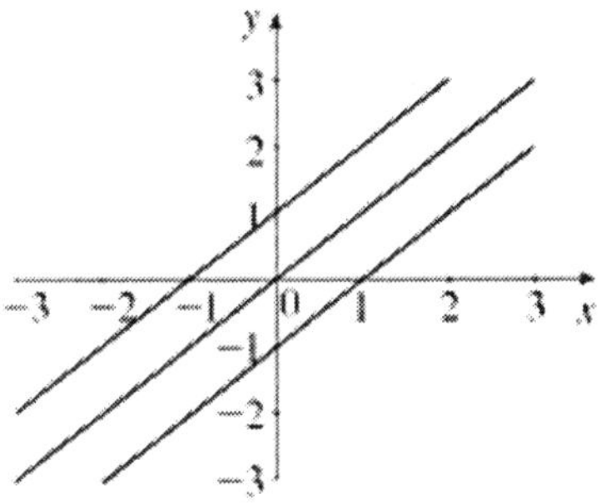

For values of $c < 0$ the graph of the function $f(x) = x$ is moved c units down. For values of $c > 0$, the graph of the function $f(x) = x$ is moved up c units.

22. Let $f(x) = x^3 + c$. Graph the family of functions with $c = -4, -2, 0, 2$, and 4 in the viewing rectangle $[-5, 5]$ by $[-5, 5]$. How does the value of c affect the graph?

Answer: medium

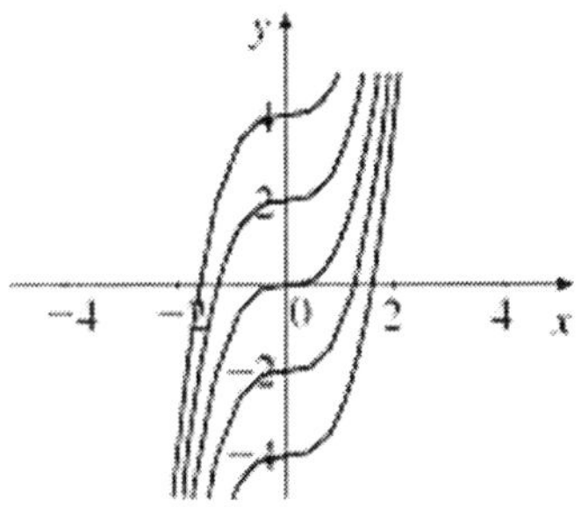

For values of $c < 0$ the graph of the function x^3 is moved c units down. For values of $c > 0$, the graph of the function x^3 is moved up c units.

23. Let $f(x) = cx^3$. Graph the family of functions with $c = \frac{1}{8}, \frac{1}{2}, 1, 2$, and 8 in the viewing rectangle $[-6, 6]$ by$[-6, 6]$. How does the value of c affect the graph?

Answer: medium

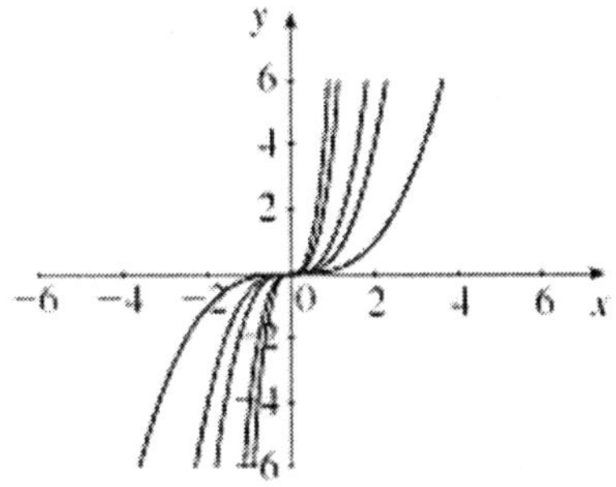

For values $0 < c < 1$, the graph of x^3 is not as steep as for values of $c > 1$.

24. Let $f(x) = cx^4$. Graph the family of functions with $c = \frac{1}{4}, \frac{1}{2}, 1, 2,$ and 4 in the viewing rectangle $[-2, 2]$ by$[-1, 8]$. How does the value of c affect the graph?

Answer: medium

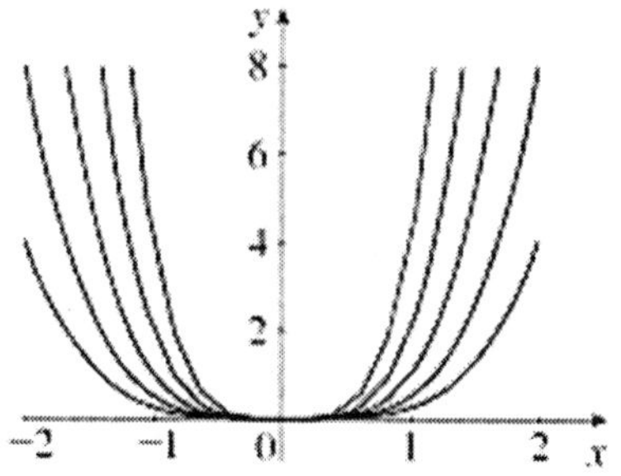

If the values of c are $0 < c < 1$, the graph of x^4 opens □out□ more than if $c > 1$.

25. Sketch the graph of the piecewise-defined function $f(x) = \begin{cases} x^2 & \text{if } |x| \le 1 \\ -x^2 & \text{if } |x| > 1 \end{cases}$.

Answer: medium

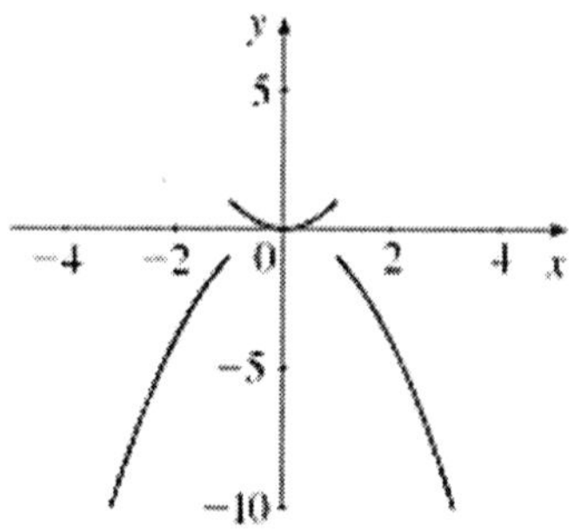

SECTION 4.3

1. In the theory of relativity the mass of an object changes as its speed changes. If m_0 is the rest mass of the object then its mass m is a function of its speed v given by $m(v) = \dfrac{m_0}{\sqrt{1 - v^2/c^2}}$ where $c = 3.0 \times 10^5$ km/s is the speed of light. Sketch the graph of mass m as a function of the speed v for an object with mass 100 kg.

Answer: hard

$$m(v) = \frac{m_0}{\sqrt{1 - v^2/c^2}} = \frac{100}{\sqrt{1 - v^2/(9 \times 10^{10})}}$$

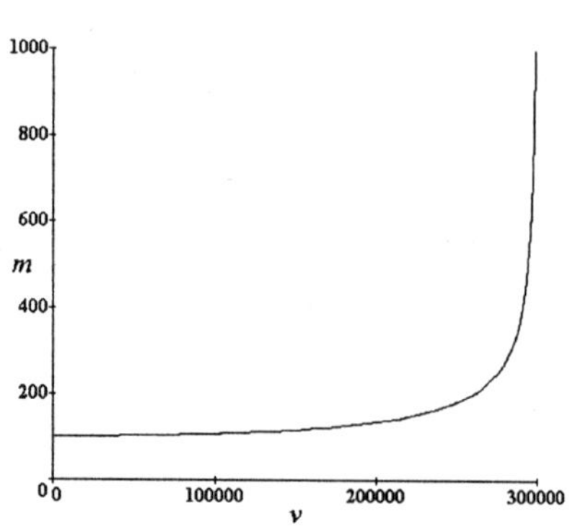

2. Write an equation that expresses the statement that P is directly proportional to q.
Answer: medium
$P = kq$, where k is constant.

3. Write an equation that expresses the statement that w is jointly proportional to m and n.
Answer: medium
$w = kmn$, where k is constant.

4. Write an equation that expresses the statement that P varies inversely as T.
Answer: medium
$P = k/T$, where k is constant.

5. Write an equation that expresses the statement that A is directly proportional to the square of t and inversely proportional to the cube of x.
Answer: medium
$A = kt^2/x^3$, where k is constant.

6. Express the statement as a formula and use the given information to find the constant of proportionality. z varies inversely as t. If $t = 2$, then $z = 7$.
Answer: medium
$z = k/t \quad \Rightarrow \quad 7 = k/2 \quad \Rightarrow \quad k = 14$, so $z = 14/t$.

7. Express the statement as a formula and use the given information to find the constant of proportionality. S varies jointly as p and q. If $p = 3$ and $q = 4$, then $S = 120$.
Answer: medium
$S = kpq \quad \Rightarrow \quad 120 = k \cdot 3 \cdot 4 \quad \Rightarrow \quad k = 10$, so $S = 10pq$.

8. Express the statement as a formula and use the given information to find the constant of proportionality. t is directly proportional to x and s and is inversely proportional to r. If $x = 3$, $s = 4$, and $r = 8$, then $t = 30$.
Answer: medium
$t = k\dfrac{xs}{r} \quad \Rightarrow \quad 30 = k\dfrac{3 \cdot 4}{8} \quad \Rightarrow \quad k = 20$, so $t = 20\dfrac{xs}{r}$.

9. The period of a pendulum (the time it takes for one complete swing of the pendulum) varies directly with the square root of the length of the pendulum.
(a) Express this fact by writing an equation.
(b) In order to triple the period, how would we have to change the length?
Answer: hard
(a) Let T and ℓ be the period and the length of the pendulum, respectively. Hence $T = k\sqrt{\ell}$.
(b) $T = k\sqrt{\ell} \quad \Rightarrow \quad T^2 = k^2\ell \quad \Rightarrow \quad \ell = \dfrac{T^2}{k^2}$. If the period is tripled, the new length is $\dfrac{(3T)^2}{k^2} = 9\dfrac{T^2}{k^2} = 9\ell$.

10. The pressure of a sample of gas is directly proportional to the temperature and inversely proportional to the volume.
(a) Write an equation that expresses this fact if 200 L of gas exerts a pressure of 34.5 kPa at a temperature of 500°K (absolute temperature measured on the Kelvin scale).
(b) If the temperature is increased to 600°K and the volume is decreased to 120 L, what is the pressure of the gas?
Answer: hard
Let P, T and V be the pressure, temperature and volume of the sample of gas, respectively.
(a) $P = \dfrac{kT}{V} \quad \Rightarrow \quad 34.5 = \dfrac{k \cdot 500}{200} \quad \Rightarrow \quad k = 13.8 \quad \Rightarrow \quad P = \dfrac{13.8T}{V}$
(b) $P = \dfrac{13.8 \cdot 600}{120} = 69$ kPa. Hence the pressure of the sample of gas is 69 kPa.

11. Kepler's Third Law of Planetary Motion states that the square of the period T of a planet (the time it takes for the planet to make a complete revolution about the sun) is directly proportional to the cube of its average distance d from the sun.

(a) Express Kepler's Third Law by writing an equation.

(b) Find the constant of proportionality by using the fact that for the planet Neptune, the period is about 164 years and the average distance from the sun is about 2.79×10^9 miles. (Use days and miles in your calculation.)

(c) Find the distance from the Earth to the Sun.

Answer: hard

(a) $T^2 = kd^3$

(b) $(164 \cdot 365)^2 = k \cdot (2.79 \times 10^9)^3 \quad \Rightarrow \quad k \approx 1.65 \times 10^{-19}$

(c) $365^2 = T^2 = kd^3 \approx 1.65 \times 10^{-19} d^3 \quad \Rightarrow \quad d \approx 9.3 \times 10^7$. Hence the distance from the Earth to the Sun is approximately 9.3×10^7 mi.

12. Write an equation that expresses the statement that V is directly proportional to the square root x and inversely proportional to the square of y.
Answer: medium
$V = \dfrac{k\sqrt{x}}{y^2}$, Where k is a constant

13. Write an equation that expresses the statement that T varies jointly as s and the square of r and inversely as the cube root of c.
Answer: medium
$T = \dfrac{kr^2 s}{\sqrt[3]{c}}$, where k is a constant.

14. Express the statement as a formula. Use the information to find the constant of proportionality. P is jointly proportional to the squares of s and t. If $s = 5$ and $t = 6$ then $P = 1800$.
Answer: medium
$P = ks^2t^2 \quad \Rightarrow \quad 1800 = k\,(5)^2\,(6)^2 \quad \Rightarrow \quad k = 2$, so $P = 2s^2t^2$.

15. Express the statement as a formula. Use the information to find the constant of proportionality. Z is jointly proportional to a, b, and c and inversely proportional to d. If c and d have the same value, and if a and b are both 6, then $Z = 360$.
Answer: medium
$Z = k\dfrac{abc}{d}$. Since $c = d$, $Z = kab \quad \Rightarrow \quad 360 = k\,(6)\,(6) \quad \Rightarrow \quad k = 10$, so $Z = 10ab$.

16. In a certain city , the property tax collected for a home is directly proportional to the valuation of the property. The tax on a \$93,000 home is \$2165 per year. What is the value of a home if the tax collected is \$1735?
Answer: medium
$T = kV \quad \Rightarrow \quad 2165 = k93000 \quad \Rightarrow \quad k = 0.023 = 2.3\%$, so $T = 0.023V$. Now $1735 = 0.023V \quad \Rightarrow$
$V = \frac{1735}{0.023} \approx \$75{,}435$.

17. The gravitational force F between two objects is inversely proportional to the square of the distance between the objects. If the force F is 1.62×10^{-6} N when the two objects are 25 cm apart, what is the gravitational force when the objects are 35 cm apart?
Answer: medium
$F = \dfrac{k}{r^2} \quad \Rightarrow \quad 1.62 \times 10^{-6} = \dfrac{k}{25^2} \quad \Rightarrow \quad k = 1.0125 \times 10^{-3} \quad \Rightarrow \quad F = \dfrac{1.0125 \times 10^{-3}}{35^2} = 8.2653 \times 10^{-7}$ N

18. The resistance R of a wire varies directly as its length L and inversely as the square of twice its radius.

(a) Write an equation that expresses this direct variation.

(b) Find the constant of proportionality if a wire 2.4 m long and 0.004 m in diameter has a resistance of 150 ohms.

(c) Using the same constant of proportionality in part (b), find the diameter of a wire that has a resistance of 90 ohms and is 1.2 m long.

Answer: medium

(a) $R = \dfrac{kL}{(2r)^2} = \dfrac{kL}{4r^2}$

(b) $R = \dfrac{kL}{4r^2}$. Since $2r = d$, $R = \dfrac{kL}{d^2} \Rightarrow 150 = \dfrac{k2.4}{0.004} \Rightarrow k = \dfrac{1}{4}$

(c) $R = \dfrac{kL}{d^2} \Rightarrow d^2 = \dfrac{1}{4}\dfrac{L}{R} \Rightarrow d^2 = \dfrac{1}{4}\left(\dfrac{1.2}{90}\right) \Rightarrow d \approx 5.78 \times 10^{-2}$ m

SECTION 4.4

1. The graph of a function f is given.

(a) State the values of $f(-1)$, $f(0)$, $f(1)$, and $f(3)$.

(b) State the domain and range of f.

(c) State the intervals on which f is increasing and on which f is decreasing.

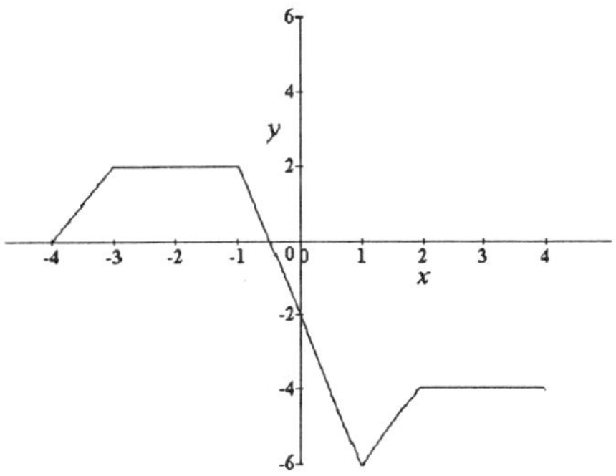

Answer: medium

(a) $f(-1) = 2$, $f(0) = -2$, $f(1) = -6$, $f(3) = -4$

(b) Domain $= [-4, 4]$, Range $= [-6, 2]$

(c) f is increasing on $[-4, -3]$ and $[1, 2]$, and decreasing on $[-1, 1]$.

2. The graph of a function h is given.

(a) State the values of $h(-2)$, $h(-1)$, $h(0)$, $h(2)$, and $h(4)$.

(b) State the domain and range of h.

(c) State the intervals on which h is increasing and on which h is decreasing.

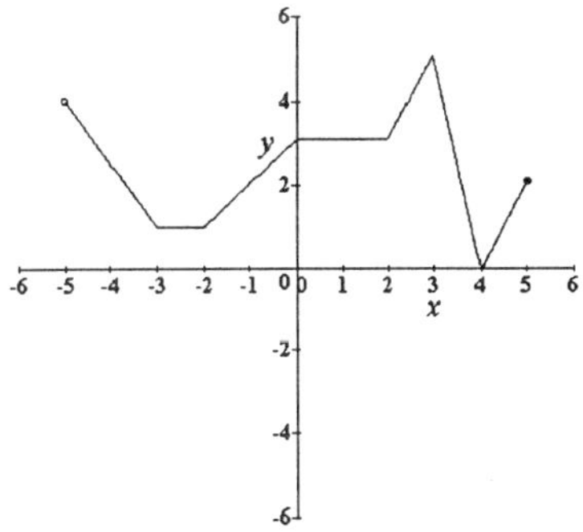

Answer: medium

(a) $h(-2) = 1$, $h(-1) = 2$, $h(0) = 3$, $h(2) = 3$, $h(4) = 0$

(b) Domain $= (-5, 5]$, Range $= [0, 5]$

(c) h is increasing on $[-2, 0]$, $[2, 3]$, and $[4, 5]$, and decreasing on $(-5, -3$ and $[3, 4]$.

3. Let $f(x) = 2 + x$.
   (a) Sketch the graph of f.
   (b) Find the domain of f.
   (c) State the intervals on which f is increasing and on which f is decreasing.

   Answer: medium

   (a)

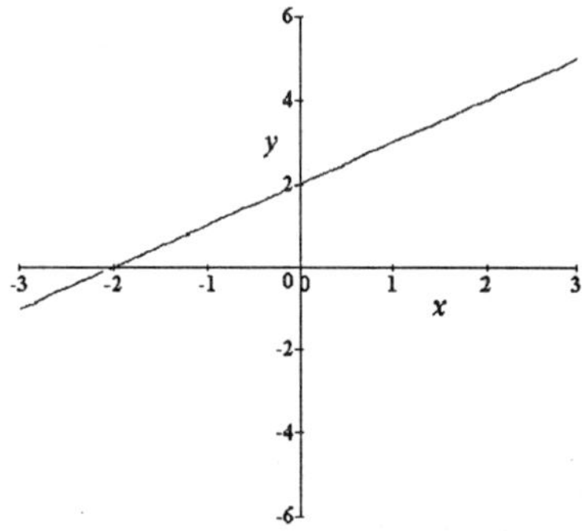

   (b) Domain $= \mathbb{R}$
   (c) f is increasing on $\mathbb{R}$.

4. Let $f(x) = x^2 - 2x$.
   (a) Sketch the graph of f.
   (b) Find the domain of f.
   (c) State the intervals on which f is increasing and on which f is decreasing.

   Answer: medium

   (a)

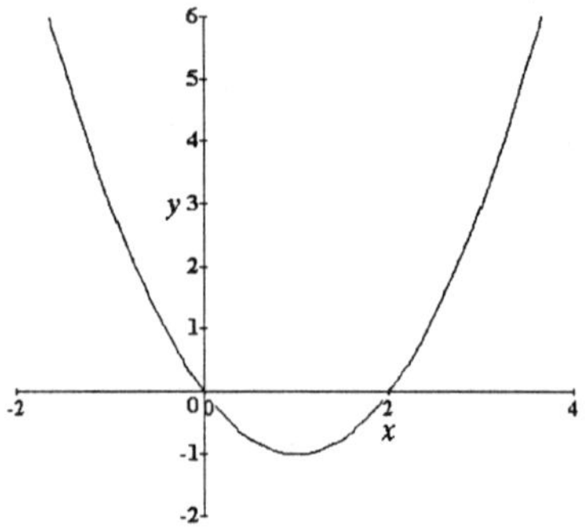

   (b) Domain $= \mathbb{R}$
   (c) f is decreasing on $(-\infty, 1]$ and increasing on $[1, \infty)$.

5. Let $f(x) = \sqrt{16 - x}$.
   (a) Sketch the graph of f.
   (b) Find the domain of f.
   (c) State the intervals on which f is increasing and on which f is decreasing.

   Answer: medium

   (a)

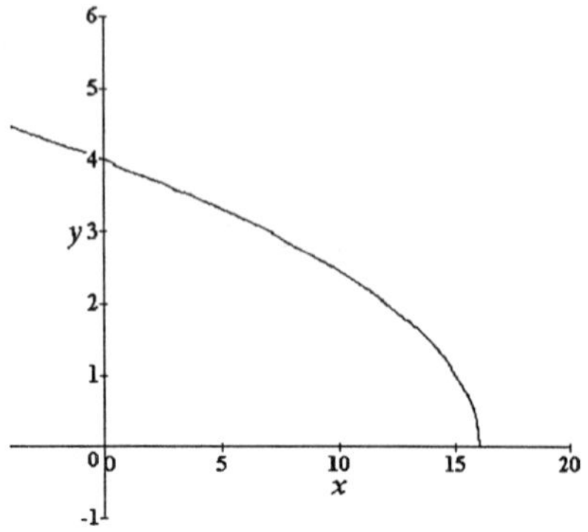

   (b) Domain $= (-\infty, 16]$
   (c) f is decreasing on $(-\infty, 16]$.

6. Let $f(x) = -\sqrt{16 - x^2}$.
   (a) Sketch the graph of f.
   (b) Find the domain of f.
   (c) State the intervals on which f is increasing and on which f is decreasing.
   Answer: medium
   (a)

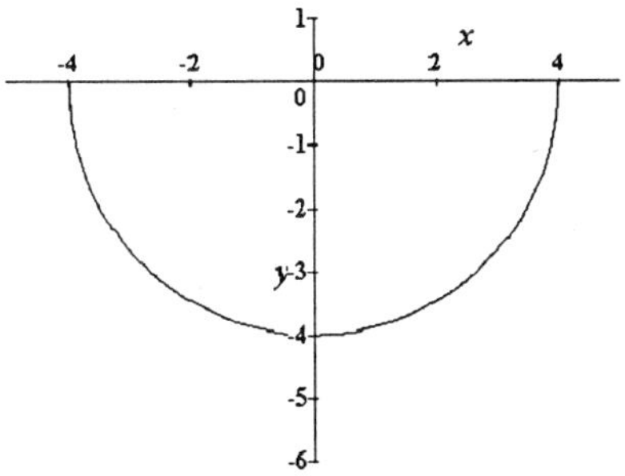

   (b) Domain $= [-4, 4]$
   (c) f is decreasing on $[-4, 0]$ and increasing on $[0, 4]$.

GC 7. Let $f(x) = 2 - x^{4/3}$.
   (a) Use a graphing device to draw the graph of f.
   (b) State the intervals on which f is increasing and on which f is decreasing.
   Answer: medium
   (a)

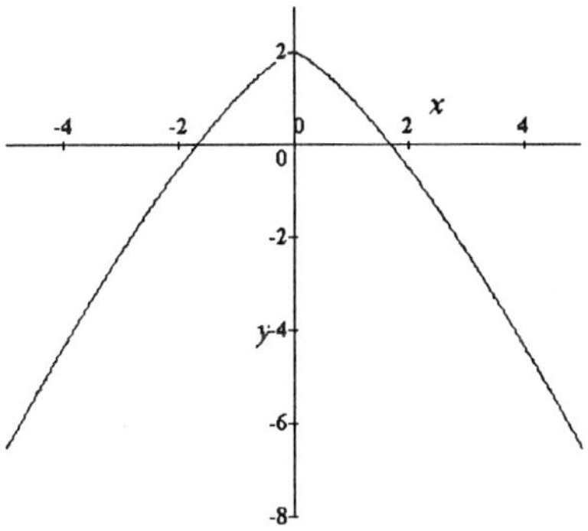

   (b) f is increasing on $(-\infty, 0]$ and decreasing on $[0, \infty)$.

GC 8. Let $f(x) = x^4 - 3x^3 + 2x - 2$.
   (a) Use a graphing device to draw the graph of f.
   (b) State the intervals on which f is increasing and on which f is decreasing.
   Answer: medium
   (a)

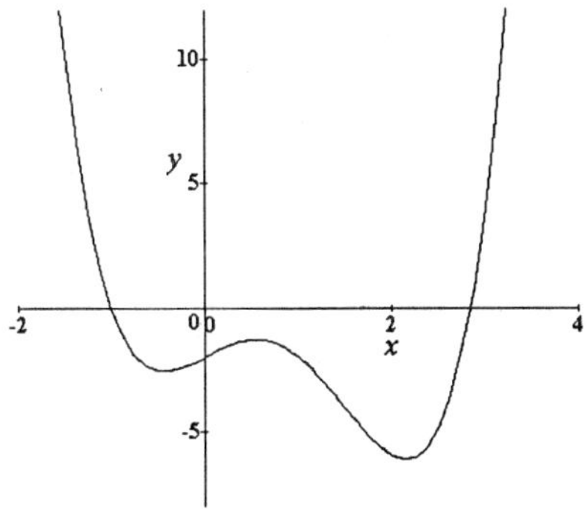

   (b) f is increasing on approximately $[-0.4, 0.5]$ and $[2.1, \infty)$ and decreasing on approximately $(-\infty, -0.4]$ and $[0.5, 2.1]$.

9. Determine the average rate of change of the function $f(x) = 2x + 1$ between $x = 5$ and $x = 6$.
Answer: easy

Average rate of change $= \dfrac{\text{Change in } y}{\text{Change in } x} = \dfrac{f(b) - f(a)}{b - a}$. In this case, average rate of change $= \dfrac{f(6) - f(5)}{6 - 5} = \dfrac{13 - 11}{6 - 5} = 2$.

10. Determine the average rate of change of the function $f(x) = 2x^2 - x$ between $x = -1$ and $x = 0$.
Answer: easy

Average rate of change $= \dfrac{f(0) - f(-1)}{0 - (-1)} = -3$

11. Determine the average rate of change of the function $f(x) = 4x^2 - x^3 + 5$ between $x = 0$ and $x = 10$.
Answer: easy

Average rate of change $= \dfrac{f(10) - f(0)}{10 - 0} = \dfrac{4(10)^2 - (10)^3 + 5 - 5}{10} = -60$

12. Determine the average rate of change of the function $f(x) = \frac{2}{3}x^4 - \frac{1}{3}x^5 - x + 3$ between $x = -1$ and $x = 1$.
Answer: medium

Average rate of change $= \dfrac{f(1) - f(-1)}{1 - (-1)} = \dfrac{\frac{2}{3}(1)^4 - \frac{1}{3}(1)^5 - (1) + 3 - \left[\frac{2}{3}(-1)^4 - \frac{1}{3}(-1)^5 - (-1) + 3\right]}{2} = -\dfrac{4}{3}$

13. Determine the average rate of change of the function $f(x) = 3x^2 - 2$ between $x = 2$ and $x = 2 + h$ $(h \neq 0)$.
Answer: medium

Average rate of change $= \dfrac{f(2+h) - f(2)}{2 + h - 2} = \dfrac{3(2+h)^2 - 12}{h} = \dfrac{12h + 3h^2}{h} = 12 + 3h$

14. Determine the average rate of change of the function $g(t) = \dfrac{1}{3t - 2}$ between $t = 0$ and $t = a + 1$.

Answer: medium

Average rate of change $= \dfrac{g(a+1) - g(0)}{a + 1 - 0} = \dfrac{\dfrac{1}{3a+1} + \dfrac{1}{2}}{a+1} = \dfrac{\dfrac{3}{2}\dfrac{a+1}{3a+1}}{a+1} = \dfrac{3}{2(3a+1)}$

15. Determine the average rate of change of the function $f(t) = \dfrac{2}{t} + 5$ between $t = a$ and $t = a + h$.

Answer: medium

Average rate of change $= \dfrac{f(a+h) - f(a)}{a + h - a} = \dfrac{\dfrac{2}{a+h} - \dfrac{2}{a}}{h} = \dfrac{-2\dfrac{h}{(a+h)a}}{h} = -\dfrac{2}{(a+h)a}$

16. Determine the average rate of change of the function $f(x) = \sqrt{4x}$ between $x = a$ and $x = a + h$ $(h \neq 0)$.
Answer: medium

Average rate of change $= \dfrac{f(a+h) - f(a)}{a + h - a} = \dfrac{\sqrt{4(a+h)} - \sqrt{4a}}{h} = \dfrac{2\sqrt{a+h} - 2\sqrt{a}}{h}$.

17. For the linear function $f(x) = -\frac{1}{2}x + 1$, find the average rate of change between $x = a$ and $x = a + h$. Show that the average rate of change is the same as the slope of the line.
Answer: medium
The slope of the line $y = -\frac{1}{2}x + 1$ is $m = -\frac{1}{2}$. The average rate of change for the function is
$\dfrac{f(a+h) - f(a)}{a + h - a} = \dfrac{-\frac{1}{2}(a+h) + 1 - \left(-\frac{1}{2}a + 1\right)}{h} = \dfrac{-\frac{1}{2}a - \frac{1}{2}h + 1 + \left(\frac{1}{2}a - 1\right)}{h} = \dfrac{-\frac{1}{2}h}{h} = -\dfrac{1}{2}$, equal to the slope of the line.

18. For the linear function $f(x) = 5 - 5x$, find the average rate of change between $x = a$ and $x = a + h$. Show that the average rate of change is the same as the slope of the line.

Answer: medium

The slope of the line $y = 5 - 5x$ is $m = -5$. The average rate of change for the function is

$\dfrac{f(a+h) - f(a)}{a+h-a} = \dfrac{5 - 5(a+h) - (5 - 5a)}{h} = \dfrac{5 - 5a - 5h - 5 + 5a}{h} = \dfrac{-5h}{h} = -5$, equal to the slope of the line.

19. For the given graph of a function, determine the average rate of change between $x = -2$ and $x = 2$.

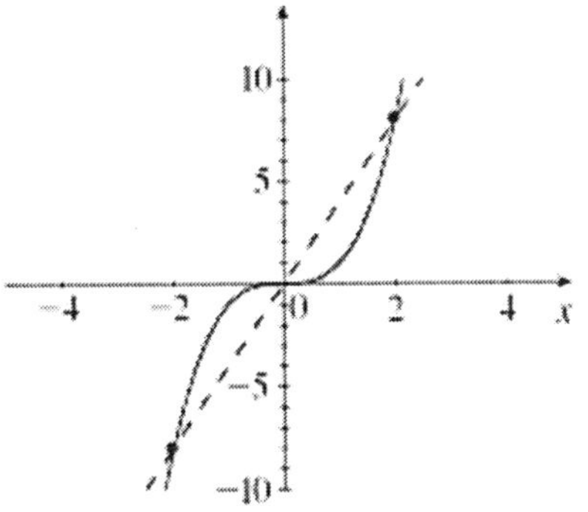

Answer: medium

The average rate of change for the function between the points $(-2, -8)$ and $(2, 8)$ is $\dfrac{8 - (-8)}{2 - (-2)} = 4$.

20. For the given graph of a function, determine the average rate of change between $x = -1$ and $x = 2$.

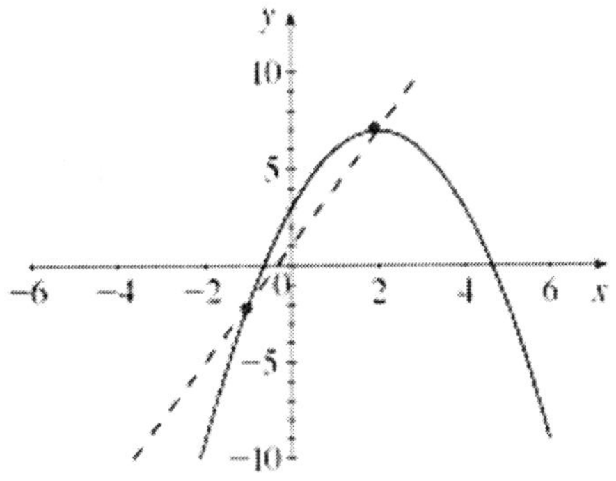

Answer: medium

The average rate of change for the function between the points $(-1, -2)$ and $(2, 7)$ is $\dfrac{7 - (-2)}{2 - (-1)} = 3$.

21. For the given graph of a function, determine the average rate of change between $x = -1$ and $x = 3$.

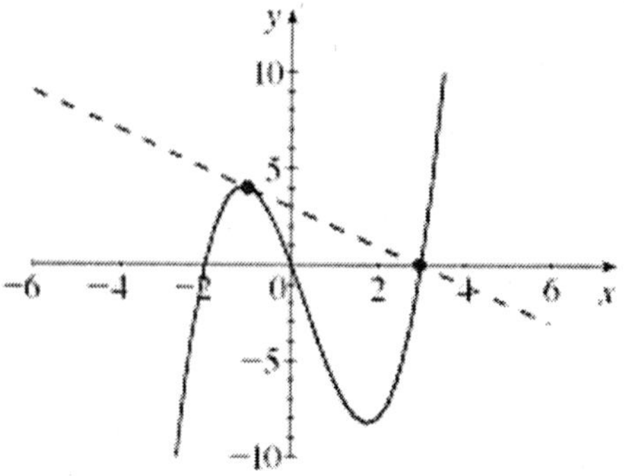

Answer: medium

The average rate of change for the function between the points $(-1, 4)$ and $(3, 0)$ is $\dfrac{0-4}{3-(-1)} = -1$.

22. Use a graphing device to help draw the graph of the function $f(x) = -9 - 3x^2$. State approximately the intervals on which the function is increasing and on which the function is decreasing.
Answer: medium

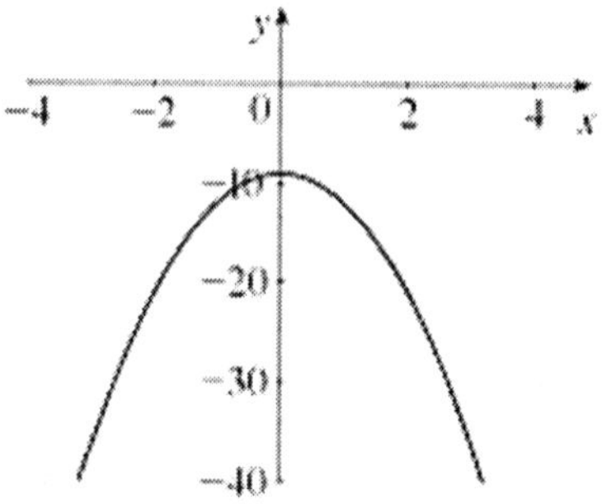

f is increasing on $(-\infty, 0)$, and decreasing on $(0, \infty)$.

23. Use a graphing device to help draw the graph of the function $f(x) = 144x^3 - 144x^2 + 36x$. State approximately the intervals on which the function is increasing and on which the function is decreasing.
Answer: medium

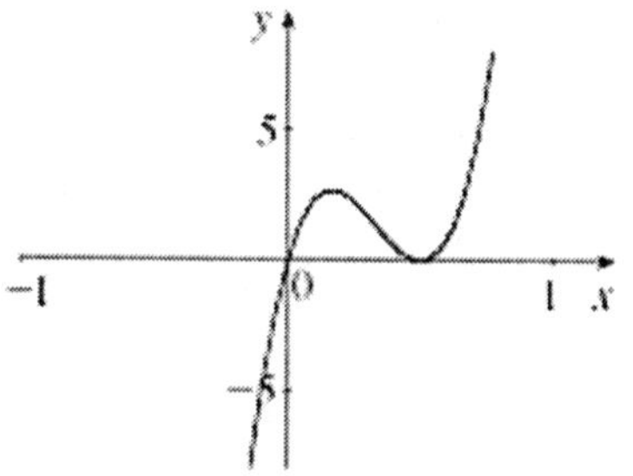

f is increasing on $\left(-\infty, \frac{1}{6}\right)$ and $\left(\frac{1}{2}, \infty\right)$, and decreasing on $\left(\frac{1}{6}, \frac{1}{2}\right)$.

24. Use a graphing device to help draw the graph of the function $f(x) = x^4 + 2x^3 - 3x^2 - 4x + 4$. State approximately the intervals on which the function is increasing and on which the function is decreasing.
Answer: medium

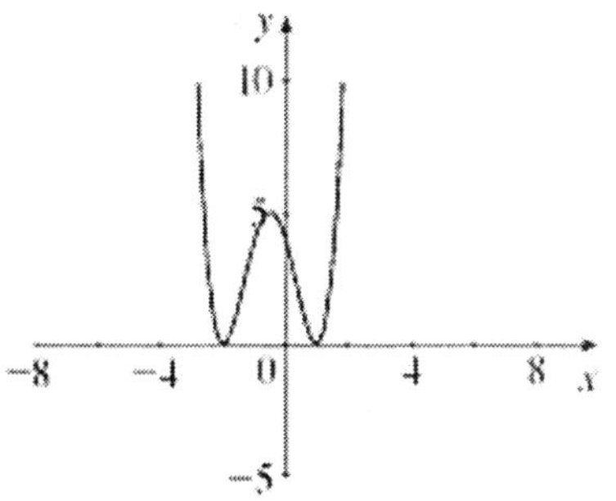

f is increasing on $\left(-2, -\frac{1}{2}\right)$ and $(1, \infty)$,and decreasing on $(-\infty, -2)$ and $\left(-\frac{1}{2}, -1\right)$.

25. Use a graphing device to help draw the graph of the function $f(x) = x^5 - 6x^4 + 11x^3 - 6x^2$. State approximately the intervals on which the function is increasing and on which the function is decreasing.
Answer: medium

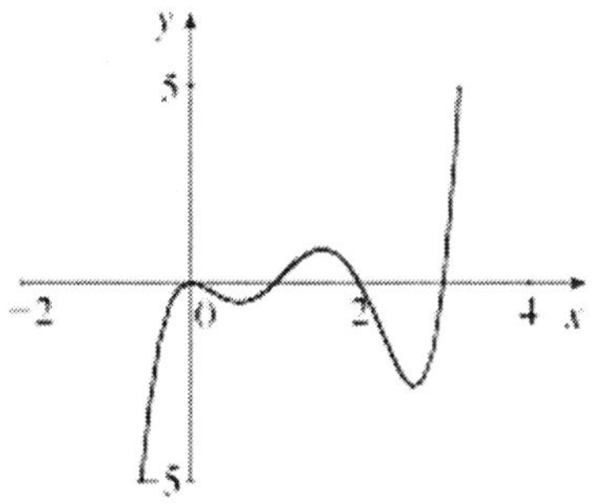

f is increasing on $(-\infty, 0)$, $(0.5, 1.5)$, and $(2.5, \infty)$, and decreasing on $(0, 0.5)$ and $(1.5, 2.5)$.

26. A man standing on a bridge drops a coin into a water fountain from a height of 105 ft. Its distance above the fountain after t seconds is given by $h(t) = 105 - 16t^2$. Find the coin's average speed for the first 2 seconds. What is the average speed of the coin between 2 seconds and the time it takes for the coin to hit the water?
Answer: hard

For the first 2 seconds the coin's average speed is $\dfrac{h(2) - h(0)}{2 - 0} = \dfrac{105 - 16(2)^2 - \left[105 - 16(0)^2\right]}{2} = -32$ ft/s. The coin hits the water when $H(t) = 0 = 105 - 16t^2 \Leftrightarrow t \approx 2.56$ s, and its average speed between $t = 2$ and $t = 2.56$ is $\dfrac{h(2.56) - h(2)}{2.56 - 2} = \dfrac{105 - 16(2.56)^2 - \left[105 - 16(2)^2\right]}{0.56} \approx -72.96$ ft/s.

SECTION 4.5

1. Suppose that the graph of f is given. Describe how the graph of $y = f(x-4)$ can be obtained from the graph of f.
   (a) By shifting the graph of f 4 units to the left
   (b) By shifting the graph of f 4 units down
   (c) By shifting the graph of f 4 units up
   (d) By shifting the graph of f 4 units to the right
   (e) By reflecting the graph of f in the x-axis
   Answer: easy (d)

2. Suppose that the graph of f is given. Describe how the graph of $y = f(x) + 2$ can be obtained from the graph of f.
   (a) By shifting the graph of f 2 units up
   (b) By shifting the graph of f 2 units down
   (c) By shifting the graph of f 2 units to the left
   (d) By shifting the graph of f 2 units to the right
   (e) By reflecting the graph of f in the y-axis
   Answer: easy (a)

3. Suppose that the graph of f is given. Describe how the graph of $y = -f(x)$ can be obtained from the graph of f.
   (a) By reflecting the graph of f in the x-axis
   (b) By reflecting the graph of f in the y-axis
   (c) By reflecting the graph of f in the x- and y-axes
   (d) By shifting the graph of f 1 unit down
   (e) By shifting the graph of f 1 unit to the left
   Answer: easy (a)

4. Suppose that the graph of f is given. Describe how the graph of $y = -\frac{1}{3}f(x)$ can be obtained from the graph of f.
   (a) By shrinking the graph of f vertically by a factor of 3 and then reflecting the result in the y-axis
   (b) By stretching the graph of f vertically by a factor of 3 and then reflecting the result in the y-axis
   (c) By shrinking the graph of f vertically by a factor of 3 and then reflecting the result in the x-axis
   (d) By stretching the graph of f vertically by a factor of 3 and then reflecting the result in the x-axis
   (e) By reflecting the graph of f in the x-axis and then shifting the result 3 units down
   Answer: medium (c)

5. Suppose that the graph of f is given. Describe how the graph of $y = -3f(x-2)$ can be obtained from the graph of f.
   (a) By shifting the graph of f 2 units to the left, then shrinking vertically by a factor of 3, then reflecting in the y-axis
   (b) By shifting the graph of f 2 units to the right, then shrinking vertically by a factor of 3, then reflecting in the x-axis
   (c) By shifting the graph of f 2 units to the left, then shrinking vertically by a factor of 3, then reflecting in the x-axis
   (d) By shifting the graph of f 2 units to the right, then stretching vertically by a factor of 3, then reflecting in the y-axis
   (e) By shifting the graph of f 2 units to the right, then stretching vertically by a factor of 3, then reflecting in the x-axis
   Answer: medium (e)

6. (a) Draw the graph of $g(x) = \sqrt[3]{x}$ by plotting points.

(b) Use the graph of f in part (a) to draw the graphs of the following functions.

(i) $y = \sqrt[3]{x-1}$ (ii) $y = \sqrt[3]{x+1} - 2$ (iii) $y = 3 - \sqrt[3]{x}$

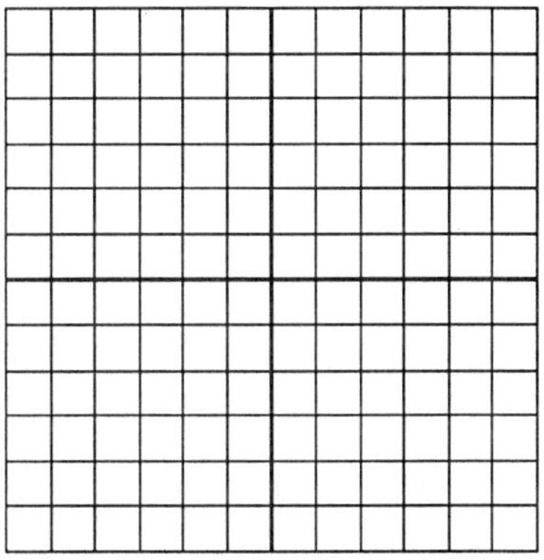

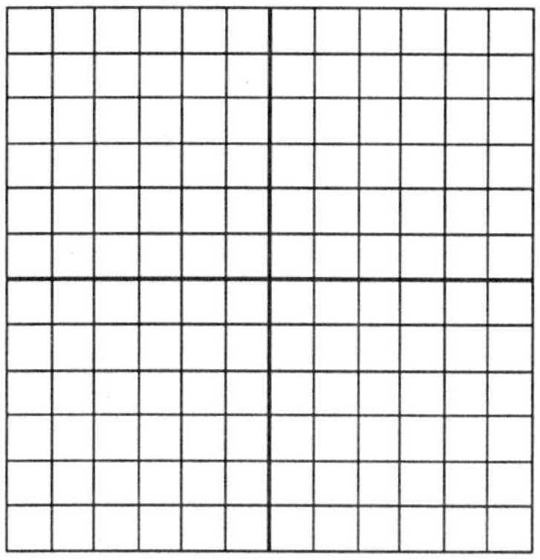

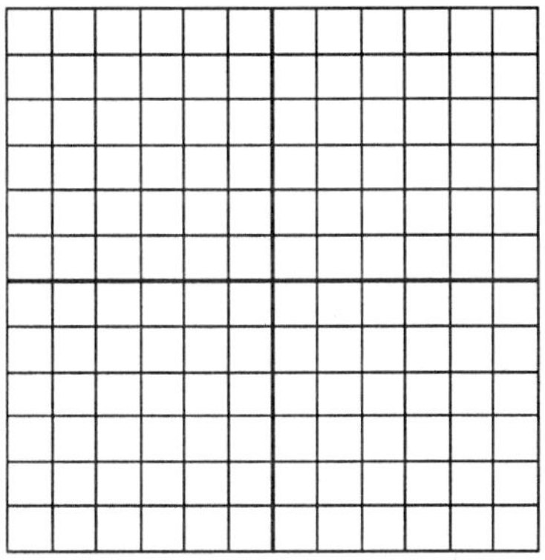

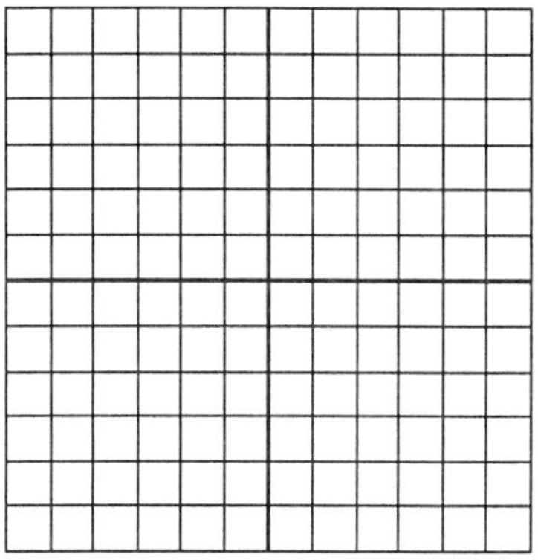

Answer: hard

(a) $g(x) = \sqrt[3]{x}$

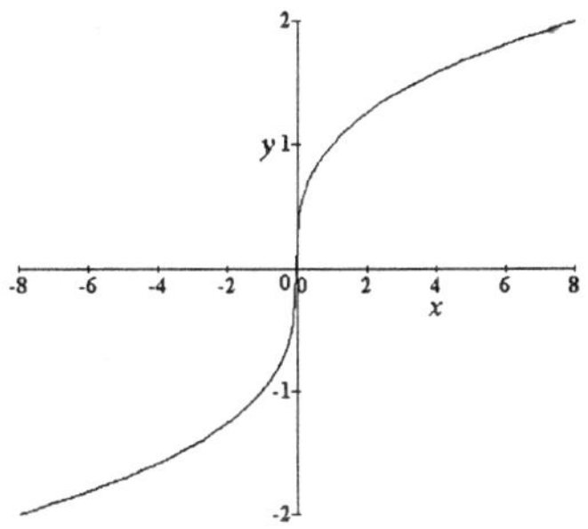

(b) (i) $y = \sqrt[3]{x-1}$

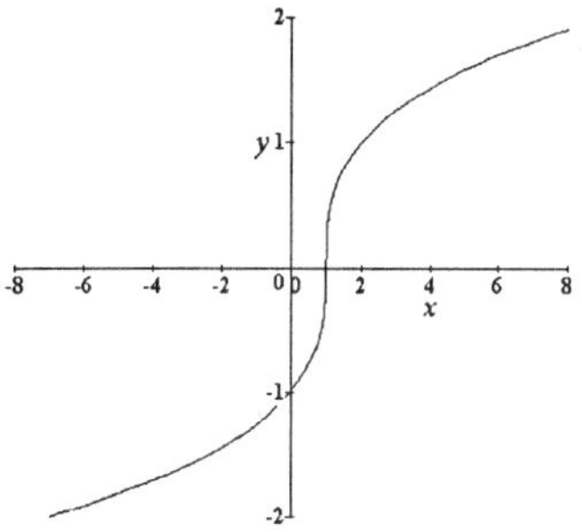

(b) (ii) $y = \sqrt[3]{x+1} - 2$

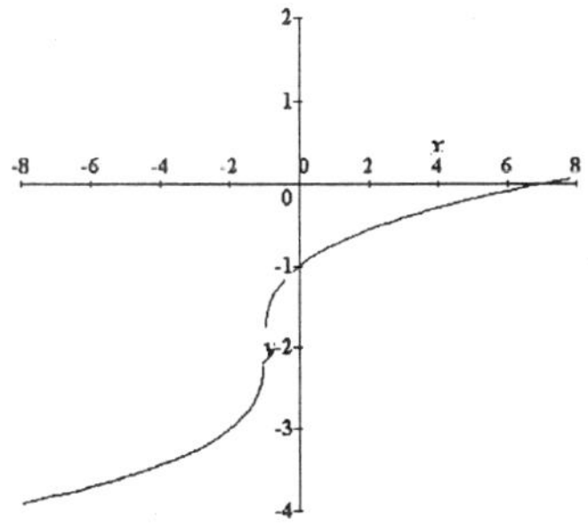

(b) (iii) $y = 3 - \sqrt[3]{x}$

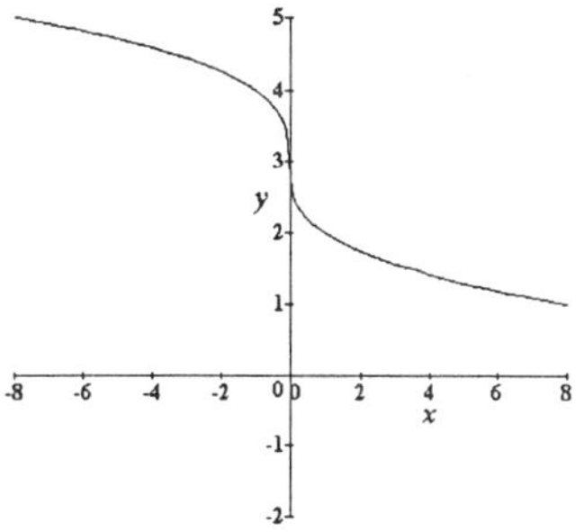

7. Sketch the graph of the function $f(x) = (x+3)^2$, not by plotting points, but by starting with the graphs of standard functions and applying transformations.

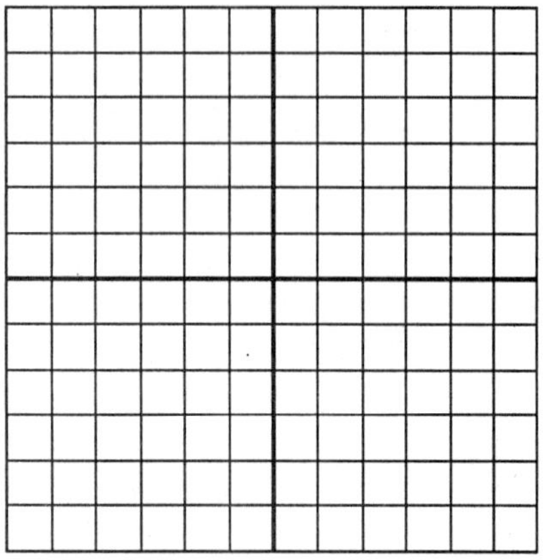

Answer: medium

$f(x) = (x+3)^2$. Shift $y = x^2$ 3 units to the left.

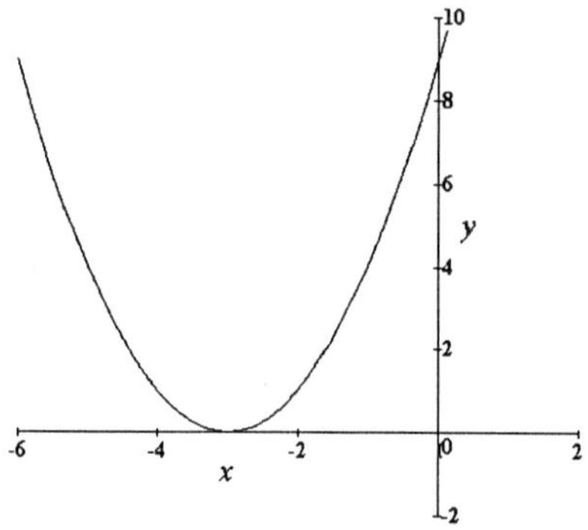

8. Sketch the graph of the function $f(x) = 1 - x^2$, not by plotting points, but by starting with the graphs of standard functions and applying transformations.

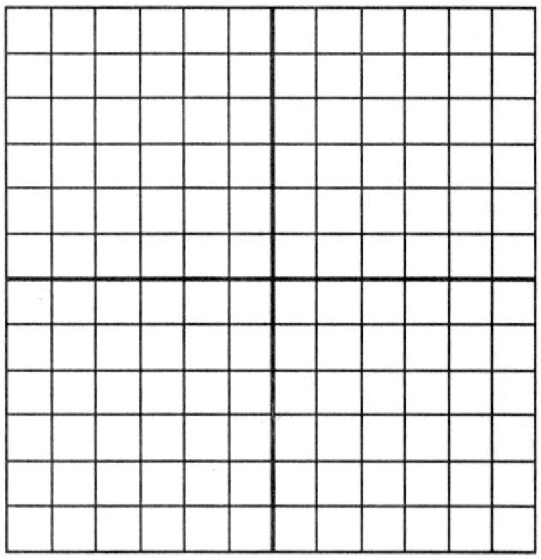

Answer: medium

$f(x) = 1 - x^2$. Reflect $y = x^2$ in the x-axis and then shift 1 unit upwards.

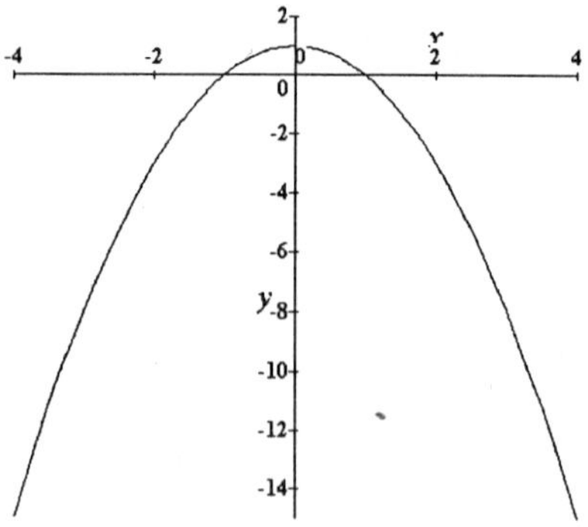

9. Sketch the graph of the function $f(x) = -2x^3$, not by plotting points, but by starting with the graphs of standard functions and applying transformations.

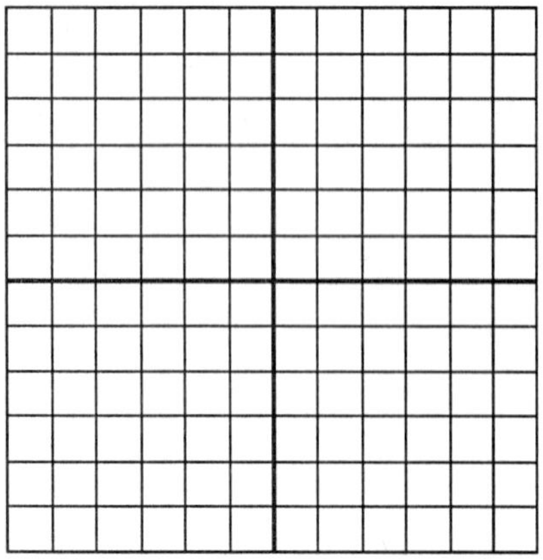

Answer: medium

$f(x) = -2x^3$. Reflect $y = 2x^3$ in the x-axis and stretch vertically by a factor of 2.

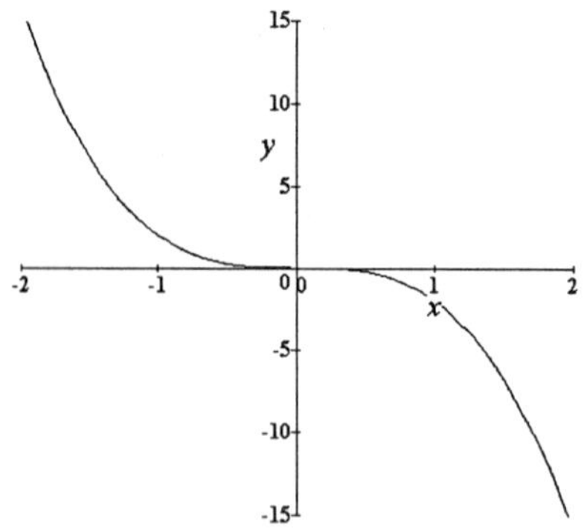

10. Sketch the graph of the function $y = 3 - \sqrt{x-1}$, not by plotting points, but by starting with the graphs of standard functions and applying transformations.

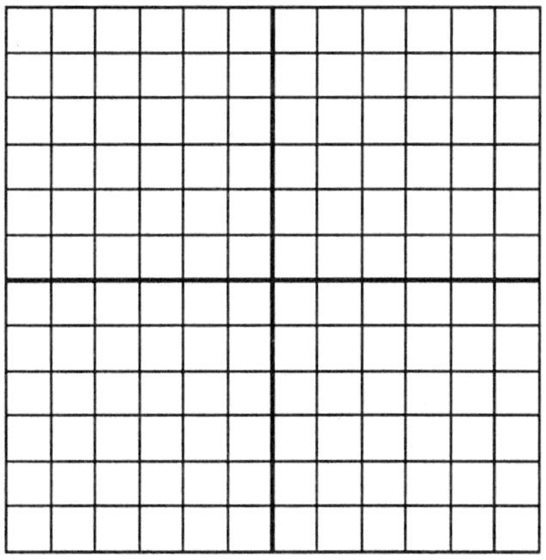

Answer: medium

$y = 3 - \sqrt{x-1}$. Shift $f(x) = \sqrt{x}$ 1 unit to the right, reflect in the x-axis, and then shift 3 units upwards.

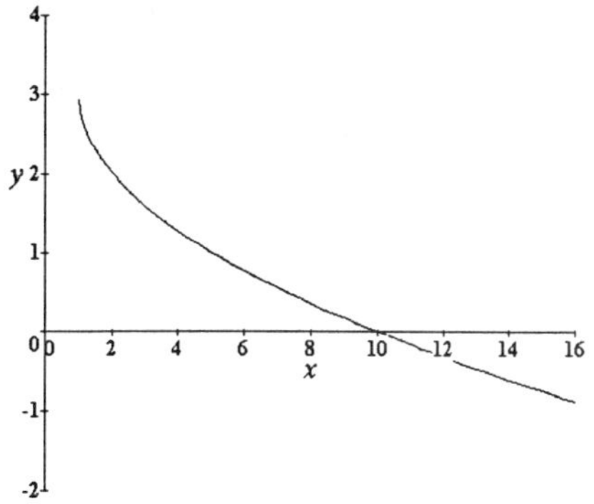

11. Sketch the graph of the function $y = 2 - 3(x+1)^2$, not by plotting points, but by starting with the graphs of standard functions and applying transformations.

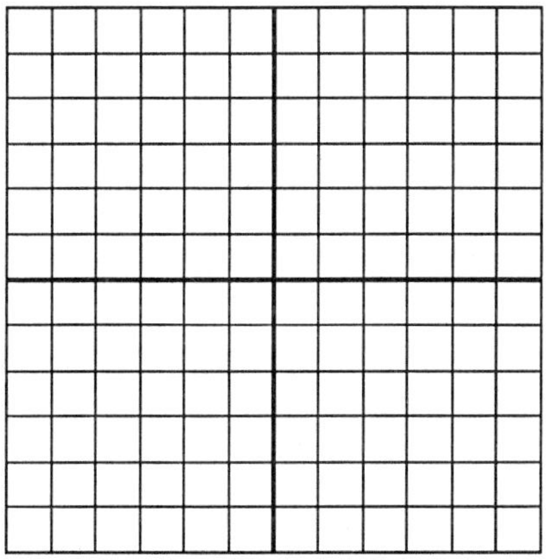

Answer: hard

$y = 2 - 3(x+1)^2$. Shift $f(x) = x^2$ 1 unit to the left, stretch vertically by a factor of 3, reflect in the x-axis, and then shift 2 units upwards.

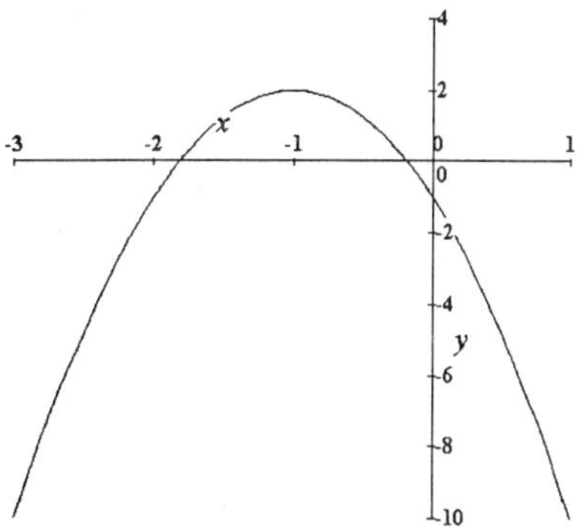

12. Sketch the graph of the function $y = \frac{1}{4}x^3 - 2$, not by plotting points, but by starting with the graphs of standard functions and applying transformations.

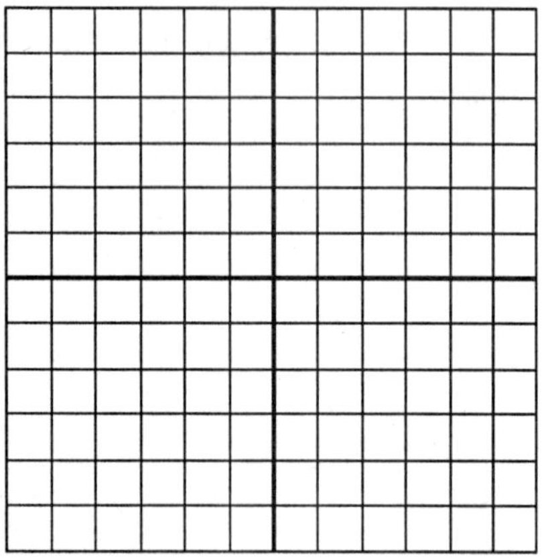

Answer: medium

$y = \frac{1}{4}x^3 - 2$. Shrink $f(x) = x^3$ vertically by a factor of 4 and then shift 2 units down.

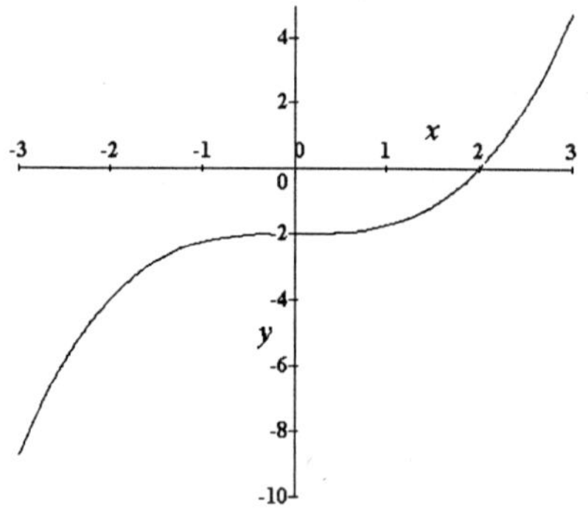

13. Sketch the graph of the function $y = |x + 1|$, not by plotting points, but by starting with the graphs of standard functions and applying transformations.

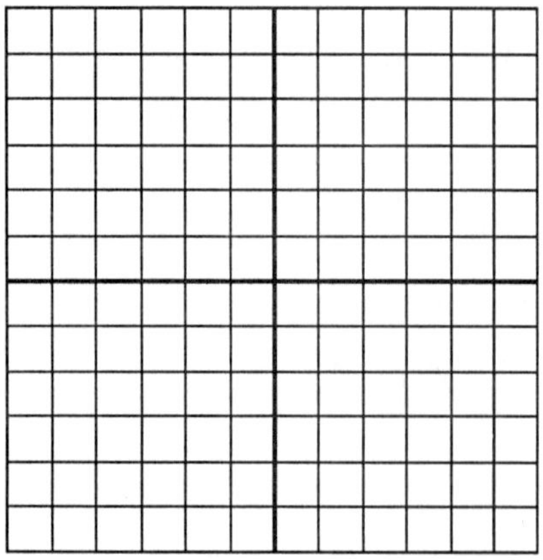

Answer: medium

$y = |x + 1|$. Shift $f(x) = |x|$ 1 unit to the left.

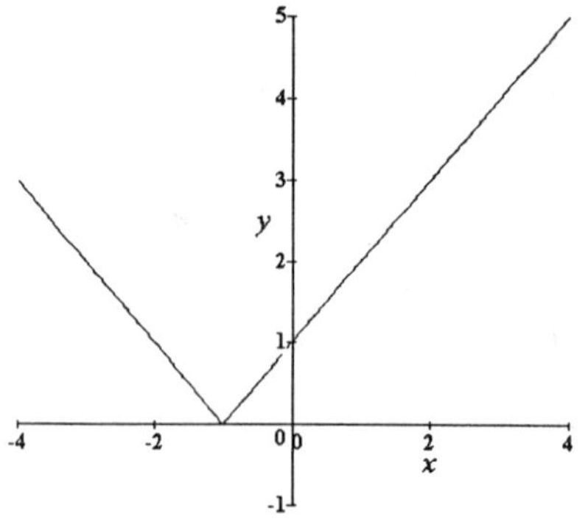

14. Sketch the graph of the function $y = 3 - |x|$, not by plotting points, but by starting with the graphs of standard functions and applying transformations.

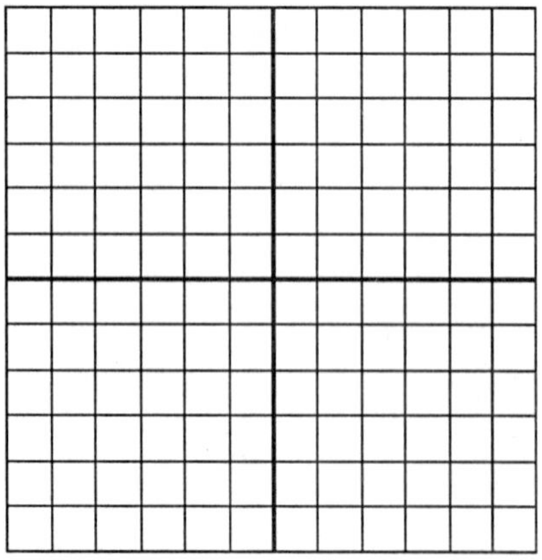

Answer: medium

$y = 3 - |x|$. Reflect $f(x) = |x|$ in the x-axis and then shift 3 units upwards.

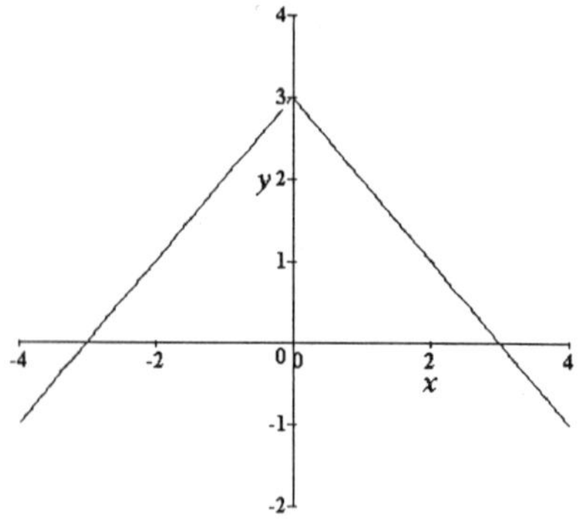

GC 15. Graph the given functions on the same screen using the viewing rectangle $[-8,8]$ by $[-6,6]$.

(a) $y = |x|$ (b) $y = -|x+1|$ (c) $y = 4 - 2|x|$ (d) $y = -3|x-4|$

Answer: medium

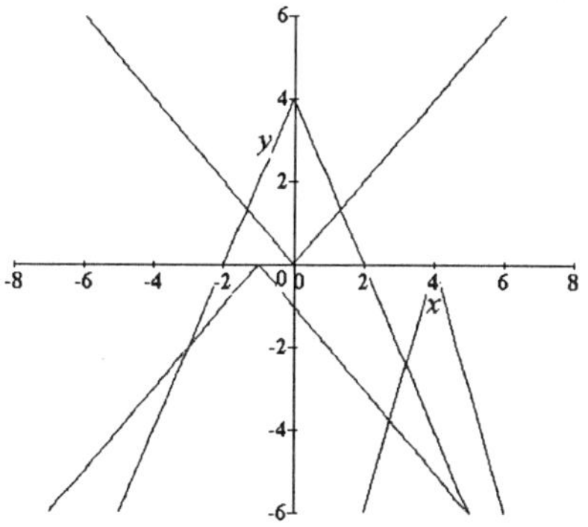

GC 16. Graph the given functions on the same screen using the given viewing rectangle. Viewing rectangle $[-4, 5]$ by $[-6,6]$

(a) $y = x^3$ (b) $\frac{1}{4}x^3$ (c) $-\frac{1}{4}x^3$ (d) $-\frac{1}{4}(x-2)^3$

Answer: medium

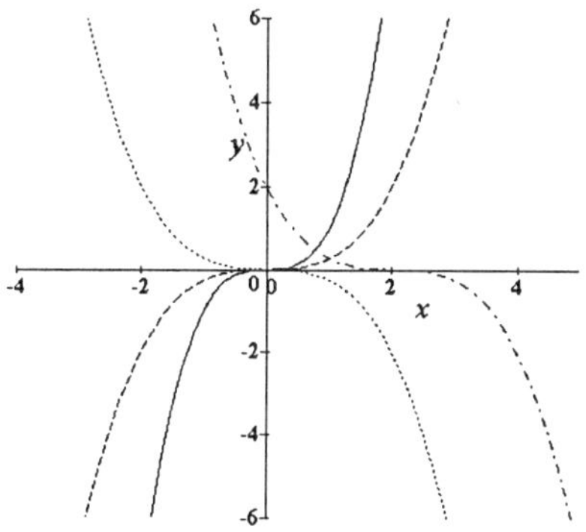

GC 17. If $f(x) = \sqrt{3x - x^2}$, graph the following functions in the viewing rectangle $[-4, 4]$ by $[-4, 4]$. How is each graph related to the graph in part (a)?

(a) $y = f(x)$ (b) $y = f(-x)$ (c) $y = -f(-x)$ (d) $y = f(-2x)$ (e) $y = f(-\frac{3}{4}x)$

Answer: medium

(a) $y = f(x) = \sqrt{3x - x^2}$

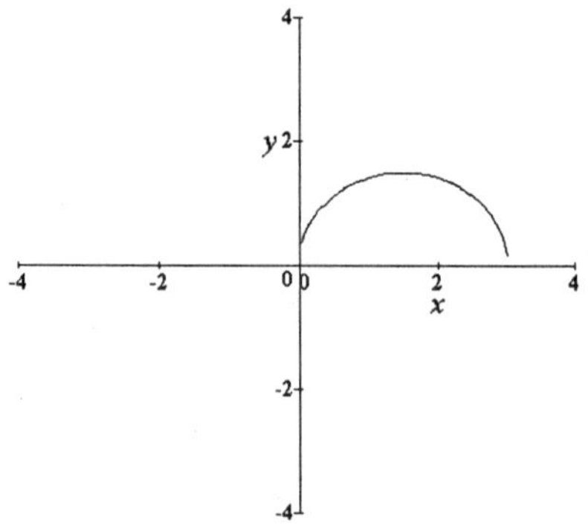

(b) $y = f(-x)$

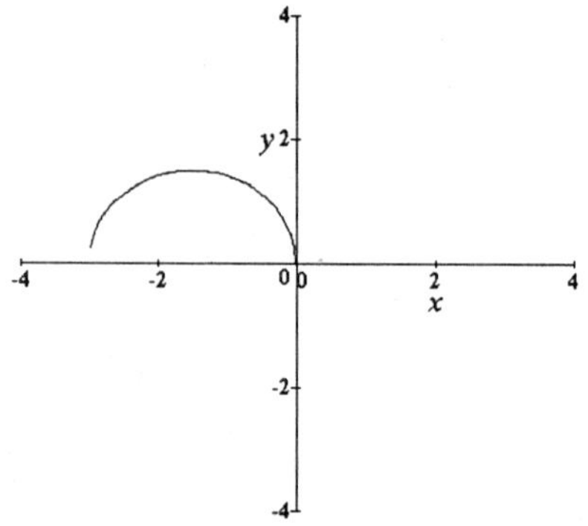

(c) $y = -f(-x)$

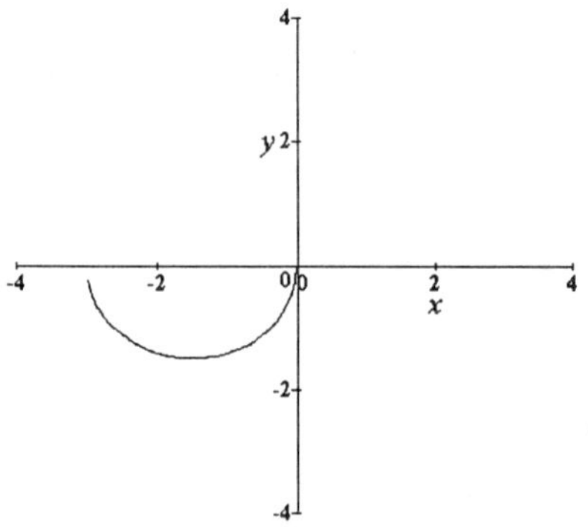

(d) $y = f(-2x)$

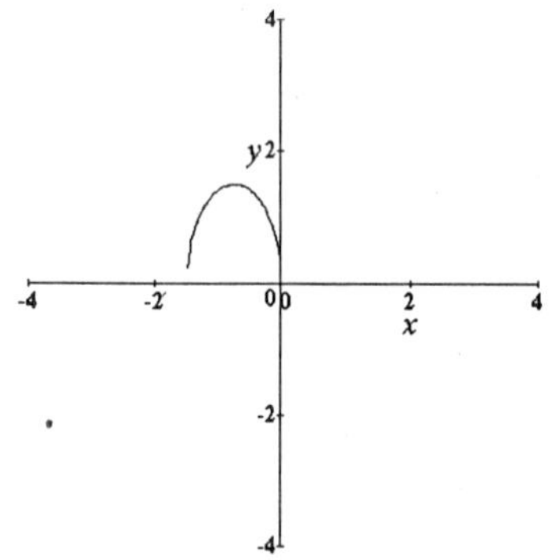

(e) $y = f(-\frac{3}{4}x)$

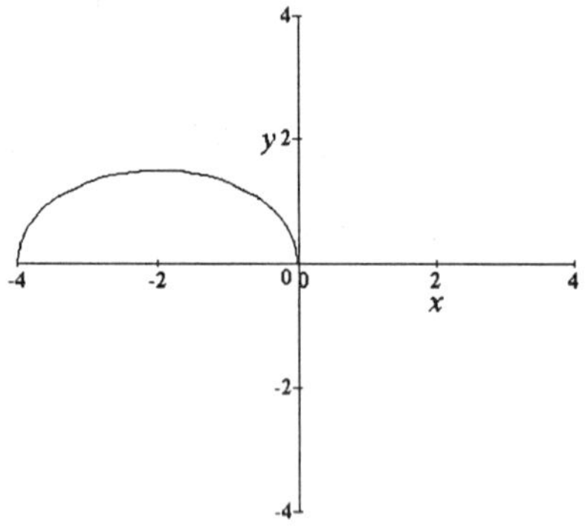

The graph in part (b) is obtained by reflecting the graph in part (a) in the y-axis. The graph in part (c) is obtained by reflecting the graph in part (a) in the y-axis and then reflecting in the x-axis. The graph in part (d) is obtained by shrinking horizontally the graph in part (a) by a factor of 2 and then reflecting in the y-axis. The graph in part (e) is obtained by enlarging horizontally the graph in part (a) by a factor of $\frac{4}{3}$ and then reflecting in the y-axis.

18. Determine whether $f(x) = x^4 - 2x^2$ is even, odd, or neither. If f is even or odd, use symmetry to sketch its graph.

Answer: medium

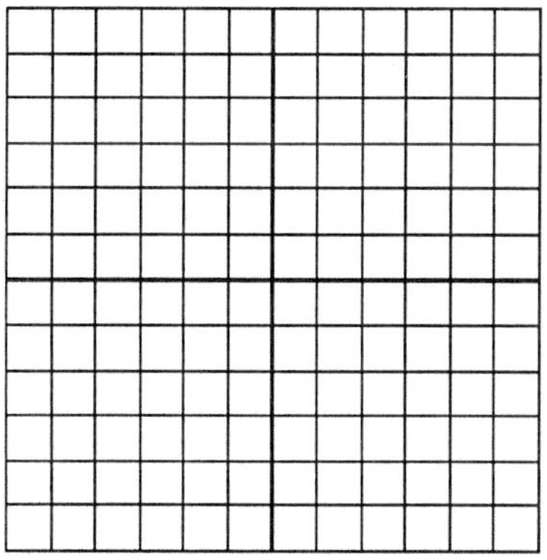

$f(x) = x^4 - 2x^2$. $f(-x) = (-x)^4 - 2(-x)^2 = x^4 - 2x^2$.
$-f(x) = -x^4 + 2x^2$. Since $f(x) = f(-x)$, f is even,
and since $f(-x) \neq -f(x)$, f is not odd.

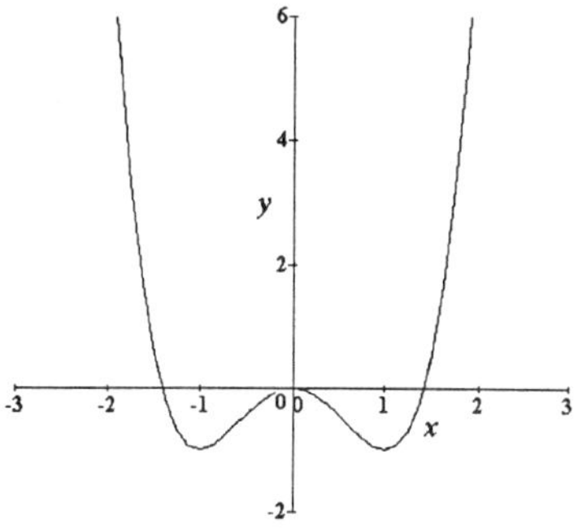

19. Determine whether $f(x) = 2x^3 + x^2 - 1$ is even, odd, or neither. If f is even or odd, use symmetry to sketch its graph.

Answer: medium

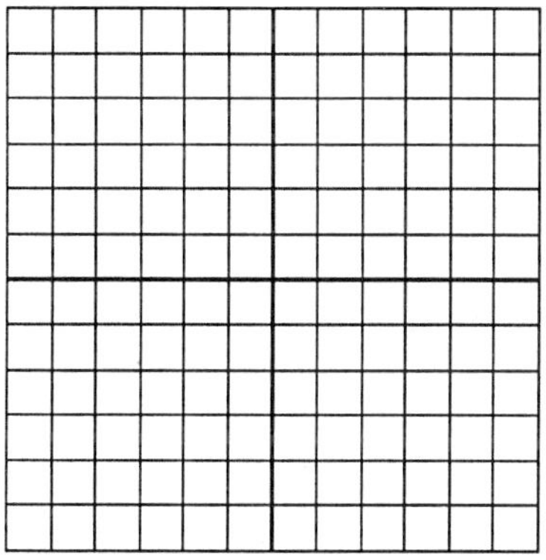

$f(x) = 2x^3 + x^2 - 1$.
$f(-x) = 2(-x)^3 + (-x)^2 - 1 = -2x^3 + x^2 - 1$.
$-f(x) = -2x^3 - x^2 + 1$. Since $f(x) \neq f(-x)$,
f is not even, and since $f(-x) \neq -f(x)$, f is not odd.

20. Determine whether f is even, odd, or neither. If f is even or odd, use symmetry to sketch its graph. $f(x) = x - 4/x$

Answer: medium

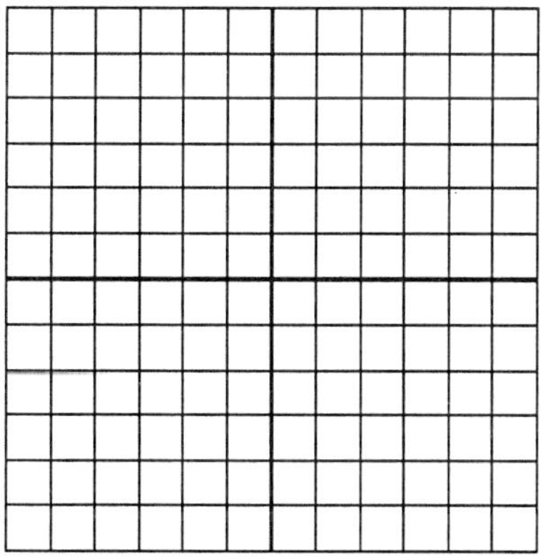

$f(x) = x - 4/x$. $f(-x) = -x + 4/x = -(x - 4/x) = -f(x)$.
Since $f(x) \neq f(-x)$, f is not even, and since
$f(-x) = -f(x)$, f is odd.

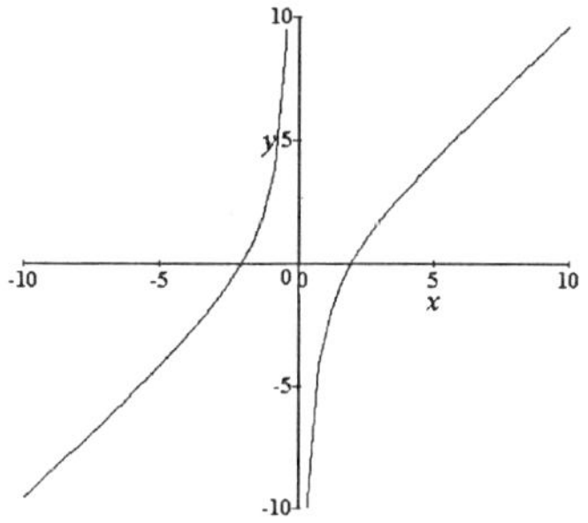

21. Sketch the graph of the function $f(x) = 2 - |x - 1|$, not by plotting points, but by starting with the graph of a standard function and applying transformations.

Answer: medium

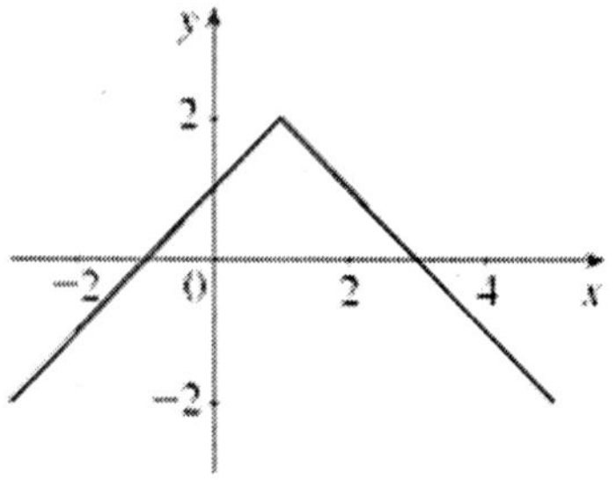

To graph the function $f(x) = 2 - |x - 1|$, shift the standard function $f(x) = |x|$ one unit to the right, reflect about the x-axis, and shift up two units.

GC 22. Graph the given functions on the same screen using the viewing rectangle $[-10, 10]$ by $[-20, 20]$.

(a) $y = x^2$

(b) $y = \frac{1}{2}x^2$

(c) $y = \frac{1}{2}(x - 3)^2$

(d) $y = \frac{1}{2}(x - 3)^2 - 3$

(e) $y = -\frac{1}{2}(x - 3)^2 - 3$

Answer: easy

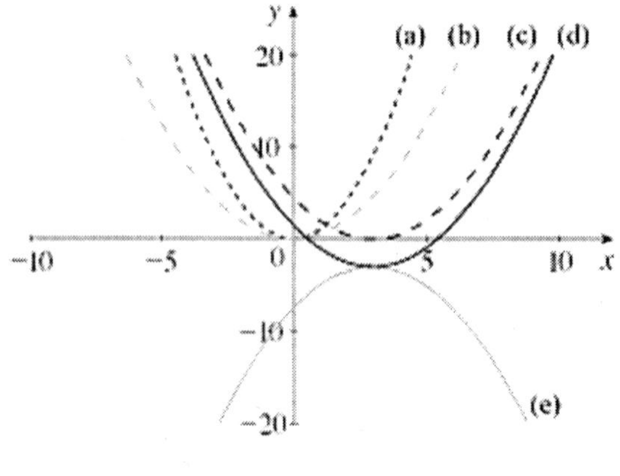

23. The graph of $f(x) = x^3 - 2x^2$ is shown. Use this graph to sketch the graph of $g(x) = |x^3 - 2x^2|$.

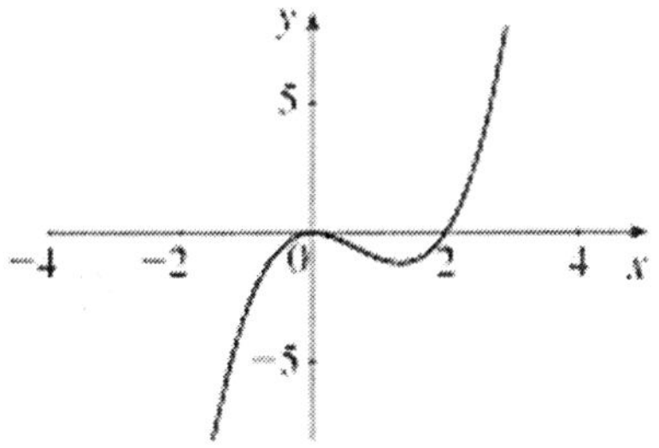

Answer: easy
The values of $g(x)$ must be positive for $x \in (-\infty, 2)$.

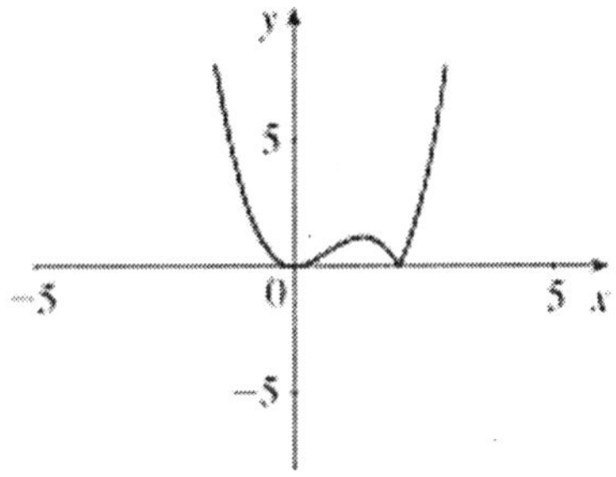

24. The graph of $f(x) = -x^4 + 8x^3 - 20x^2 + 16x$ is shown. Use this graph to sketch the graph of $g(x) = |-x^4 + 8x^3 - 20x^2 + 16x|$.

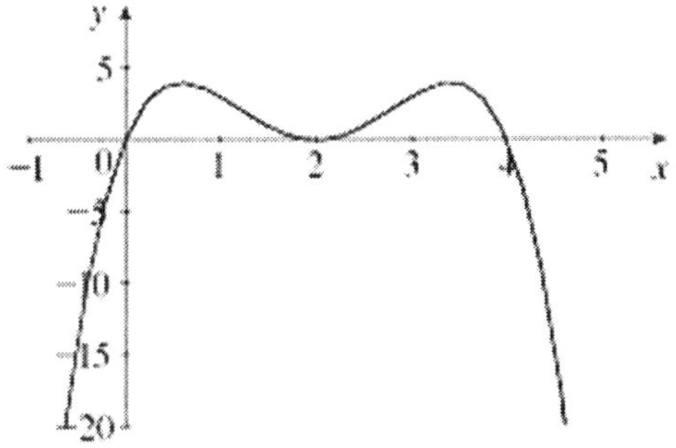

Answer: easy
The values of $g(x)$ must be positive for $x \in (-\infty, 0)$ and $x \in (4, \infty)$.

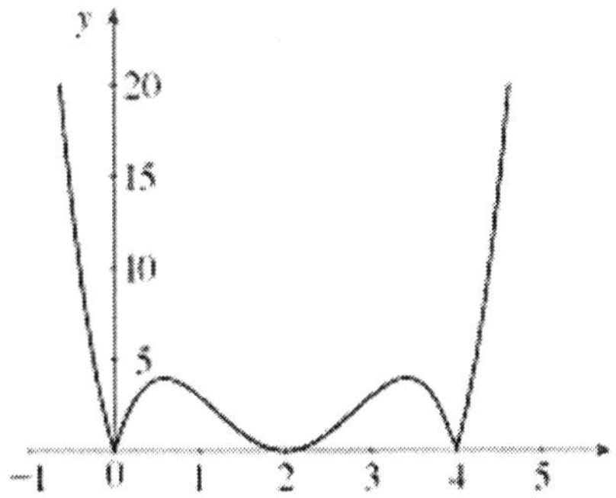

25. The graph of $f(x) = |x - 1| - 3$ is shown. Use this graph to sketch the graph of $g(x) = ||x - 1| - 3|$.

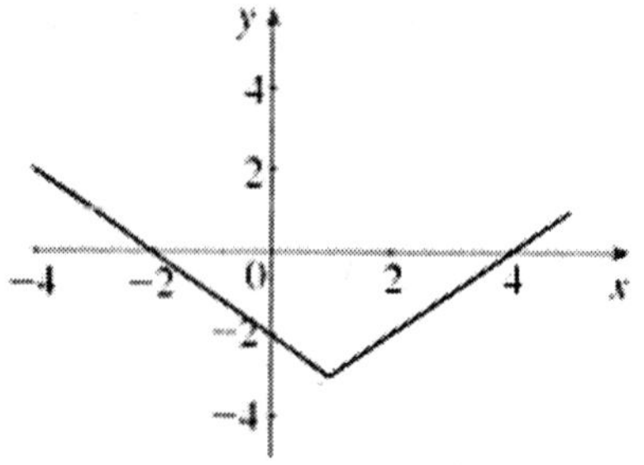

Answer: easy

The values of $g(x)$ must be positive for $x \in (-2, 4)$.

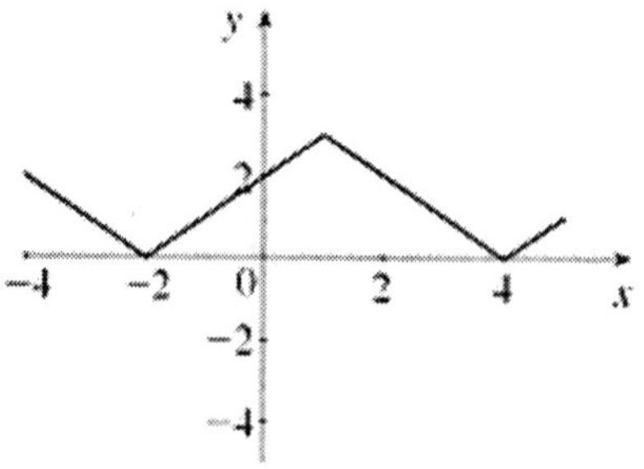

SECTION 4.6

1. Sketch the graph of the parabola $y = x^2 + 2$ and state the coordinates of its vertex and its intercepts.

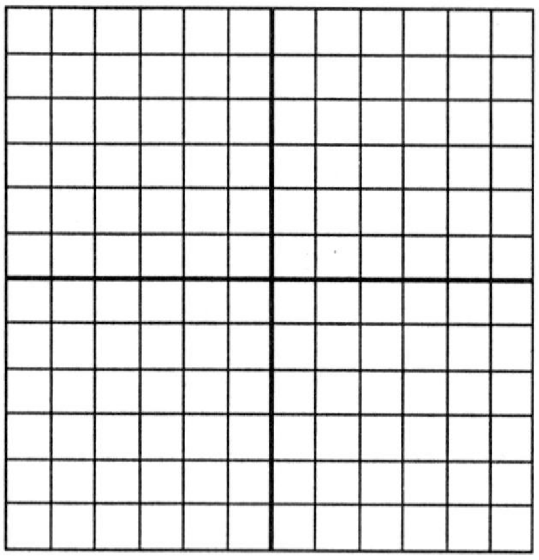

Answer: medium

The vertex is at $(0, 2)$, there is no x-intercept, and the y-intercept is 2.

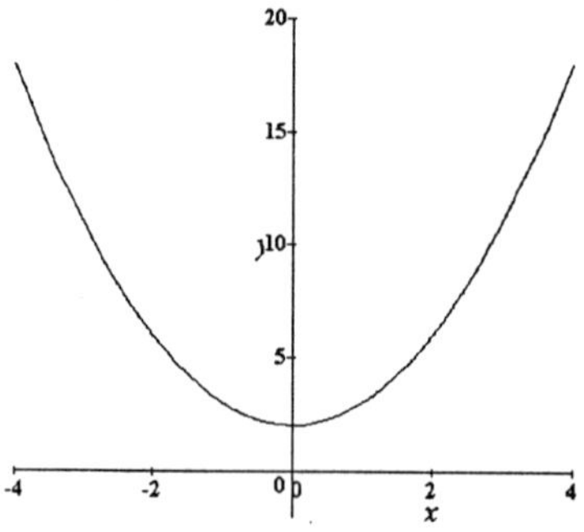

2. Sketch the graph of the parabola $y = \frac{1}{4}x^2 + \frac{3}{2}x$ and state the coordinates of its vertex and its intercepts.

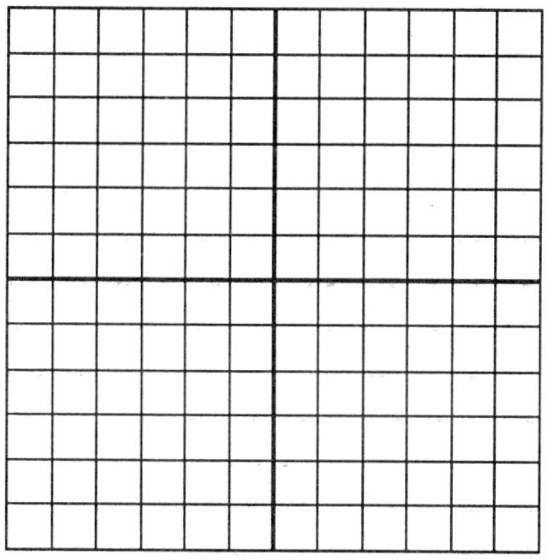

Answer: hard

$y = \frac{1}{4}x^2 + \frac{3}{2}x \Leftrightarrow y + \frac{9}{4} = \frac{1}{4}(x+3)^2$, the vertex is at $\left(-3, -\frac{9}{4}\right)$, the x-intercepts are 0 and -6, the y-intercept is 0.

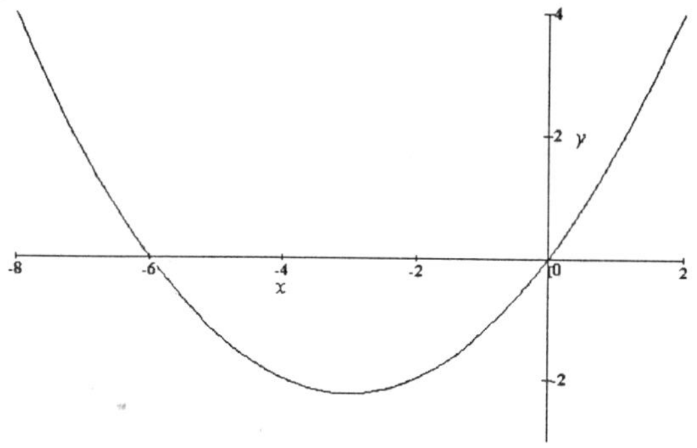

3. Sketch the graph of the parabola $y = -\frac{2}{3}x^2$ and state the coordinates of its vertex and its intercepts.

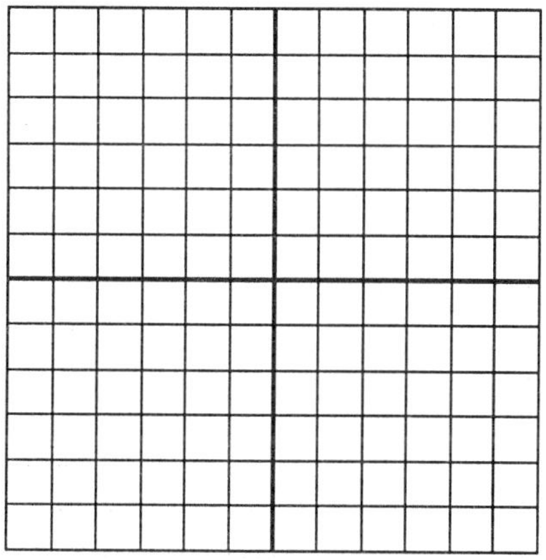

Answer: medium

$y = -\frac{2}{3}x^2$. The vertex is at $(0, 0)$, the x- and y-intercepts are 0.

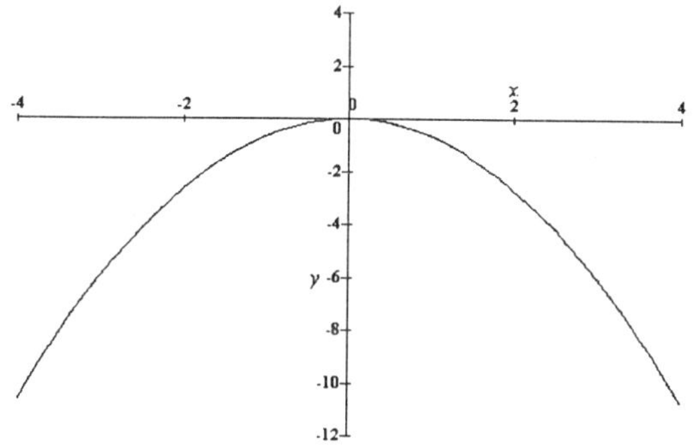

4. Sketch the graph of the parabola $y = x^2 + x + \frac{1}{2}$ and state the coordinates of its vertex and its intercepts.

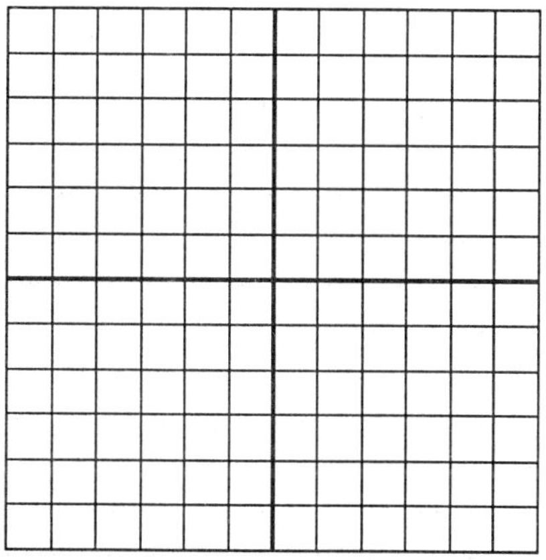

Answer: medium

$y = x^2 + x + \frac{1}{2} \Leftrightarrow y - \frac{1}{4} = \left(x + \frac{1}{2}\right)^2$, the vertex is at $\left(-\frac{1}{2}, \frac{1}{4}\right)$, there is no x-intercept, and the y-intercept is $\frac{1}{2}$.

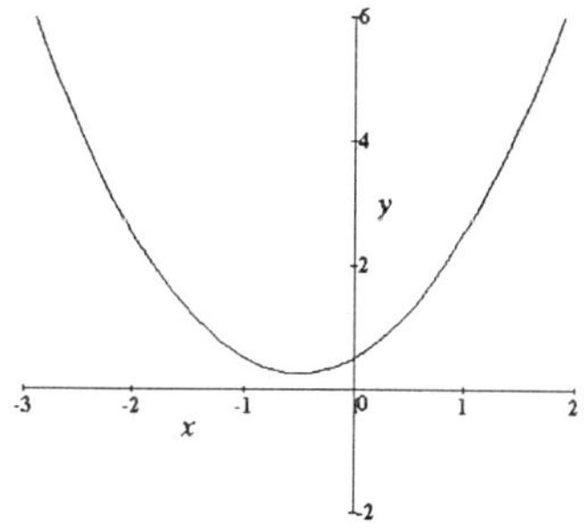

5. Sketch the graph of the parabola $y = -3x^2 + 6x - 2$ and state the coordinates of its vertex and its intercepts.

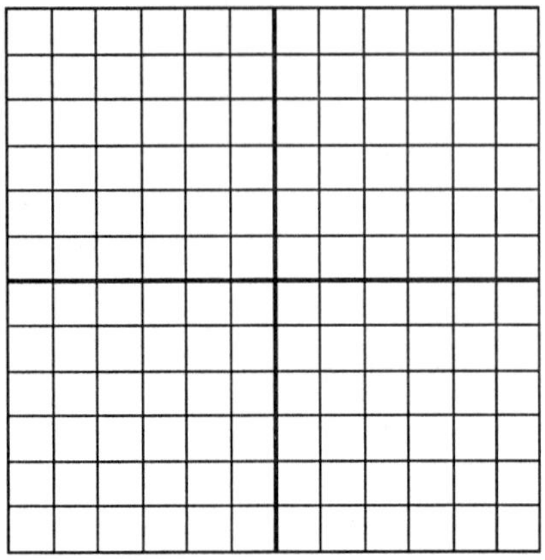

Answer: hard

$$y = -3x^2 + 6x - 2 = -3(x^2 - 2x + 1) - 2 + 3$$
$$= -3(x - 1)^2 + 1.$$

Vertex: $(1, 1)$; x-intercept: $y = 0 \quad \Leftrightarrow \quad -3(x-1)^2 + 1 = 0$

$\Leftrightarrow \quad x - 1 = \pm\sqrt{1/3} \quad \Leftrightarrow \quad x = 1 + \sqrt{1/3}$,

$x = 1 - \sqrt{1/3}$; y-intercept: $x = 0 \quad \Leftrightarrow \quad y = -2$

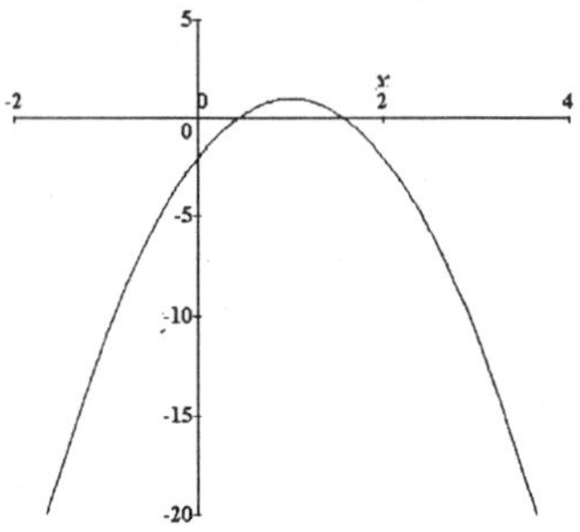

6. (a) Express the quadratic function $f(x) = 2x + x^2$ in standard form.
   (b) Sketch the graph of $f(x)$.
   (c) Find the maximum or minimum value of $f(x)$.

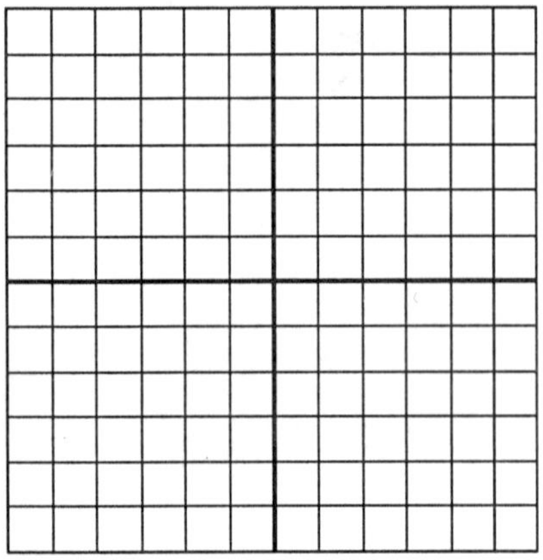

Answer: medium

(a) $f(x) = 2x + x^2 = (x^2 + 2x + 1) - 1 = (x + 1)^2 - 1$.

(b)

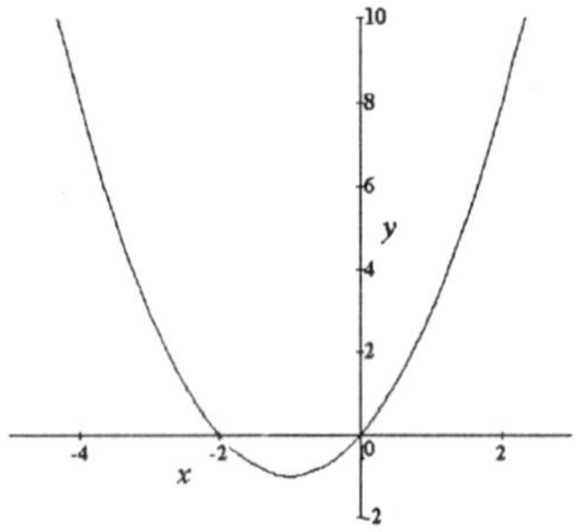

(c) The minimum value is $f(-1) = -1$.

7. (a) Express the quadratic function $f(x) = x^2 - 4x + 2$ in standard form.
   (b) Sketch the graph of $f(x)$.
   (c) Find the maximum or minimum value of $f(x)$.

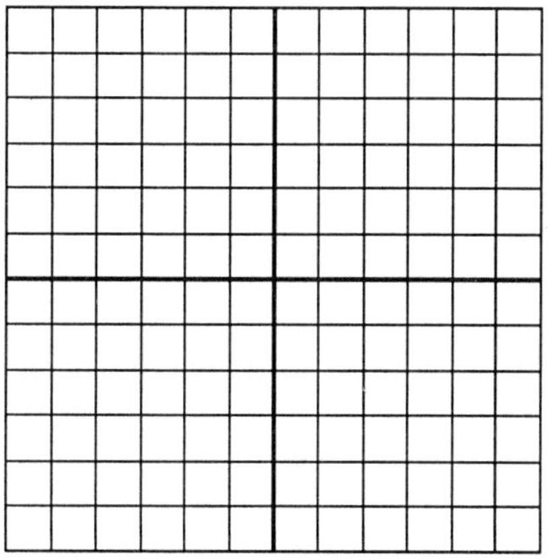

Answer: medium

(a) $f(x) = x^2 - 4x + 2 = (x^2 - 4x + 4) - 2 = (x - 2)^2 - 2$.

(b)

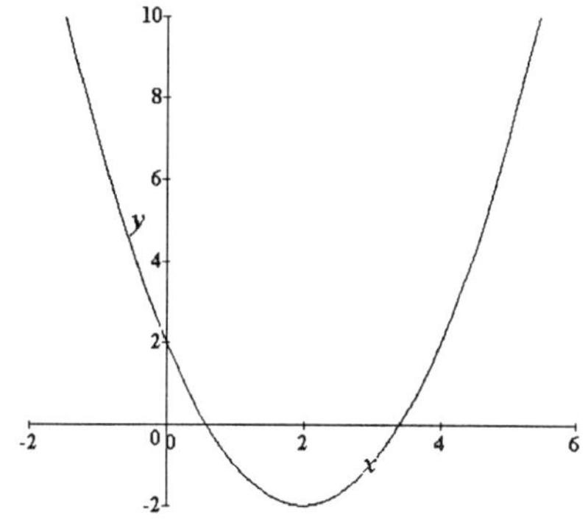

(c) The minimum value is $f(2) = -2$.

8. (a) Express the quadratic function $f(x) = 2 - 4x + 2x^2$ in standard form.
   (b) Sketch the graph of $f(x)$.
   (c) Find the maximum or minimum value of $f(x)$.

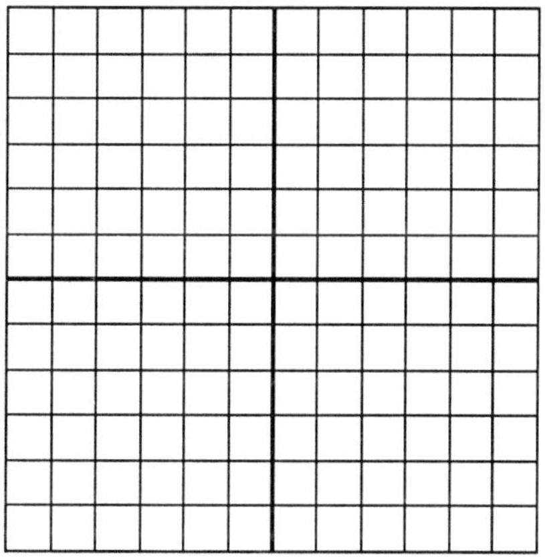

Answer: medium

(a) $f(x) = 2 - 4x + 2x^2 = 2(x^2 - 2x + 1) = 2(x - 1)^2$.

(b)

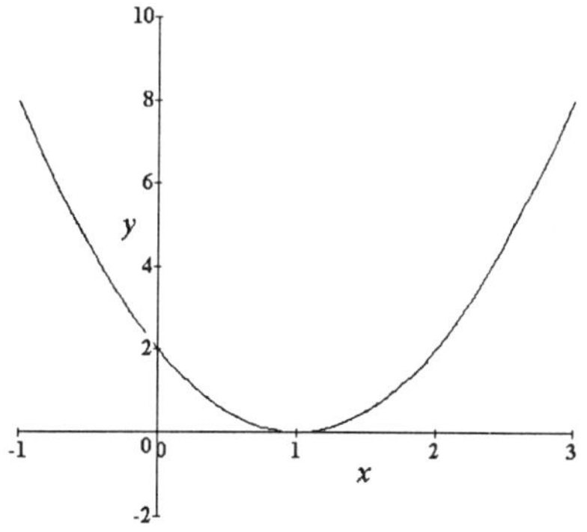

(c) The minimum value is $f(1) = 0$.

9. (a) Express the quadratic function $f(x) = 5 - 8x - 2x^2$ in standard form.

(b) Sketch the graph of $f(x)$.

(c) Find the maximum or minimum value of $f(x)$.

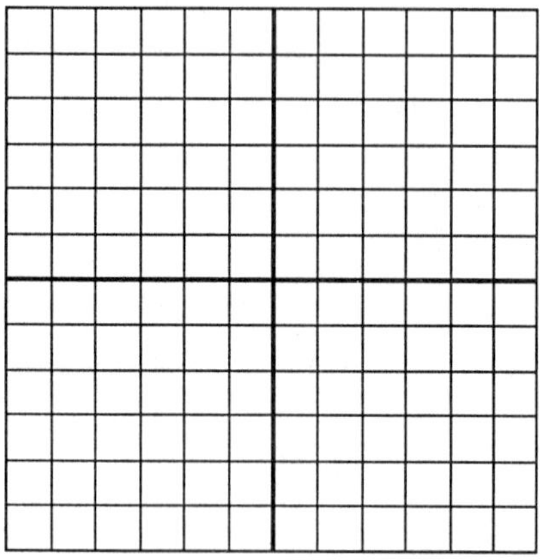

Answer: medium

(a) $f(x) = 5 - 8x - 2x^2 = -2(x^2 + 4x + 4) + 13$

$= -2(x + 2)^2 + 13$

(b)

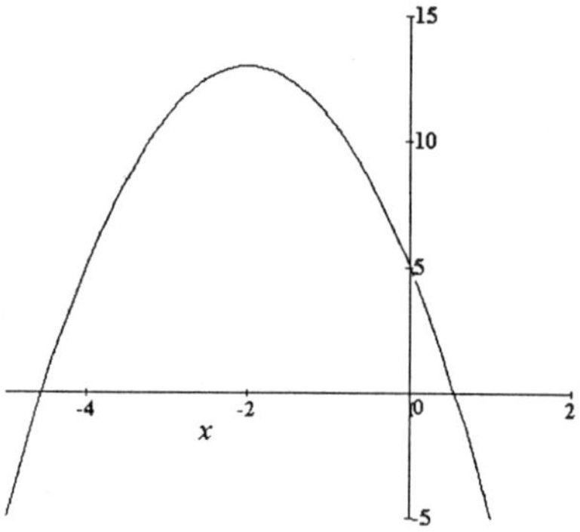

(c) The maximum value is $f(-2) = 13$.

10. Find the maximum or minimum value of the function $f(x) = x^2 - 8x + 8$.

Answer: medium

$f(x) = x^2 - 8x + 8 \quad \Rightarrow \quad a = 1$ and $b = -8$, so the minimum value is $f\left(-\frac{-8}{2}\right) = f(4) = -8$.

11. Find the maximum or minimum value of the function $f(x) = 1 - 6x - x^2$.

Answer: medium

$f(x) = 1 - 6x - x^2 \quad \Rightarrow \quad a = -1$ and $b = -6$, so the maximum value is $f\left(-\frac{-6}{-2}\right) = f(-3) = 10$.

12. Find the maximum or minimum value of the function $f(x) = 2x^2 + 8x + 11$.

Answer: medium

$f(x) = 2x^2 + 8x + 11 \quad \Rightarrow \quad a = 2$ and $b = 8$, so the minimum value is $f\left(-\frac{8}{4}\right) = f(-2) = 3$.

13. Find the maximum or minimum value of the function $f(x) = 3 - 4x - 4x^2$.

Answer: medium

$f(x) = 3 - 4x - 4x^2 \quad \Rightarrow \quad a = -4$ and $b = -4$, so the maximum value is $f\left(-\frac{-4}{-8}\right) = f\left(-\frac{1}{2}\right) = 4$.

14. Find a function whose graph is a parabola with vertex $(-3, 4)$ and that passes through the point $(1, -2)$.

Answer: hard

Since the vertex is $(-3, 4)$, the function is of the form $y = a(x + 3)^2 + 4$. Also, since the parabola passes through the point $(1, -2)$ it must satisfy $-2 = a(1 + 3)^2 + 4 \quad \Leftrightarrow \quad -6 = 16a \quad \Leftrightarrow \quad a = -\frac{3}{8}$. Thus, the function is $y = -\frac{3}{8}(x + 3)^2 + 4 = -\frac{3}{8}x^2 - \frac{9}{4}x - \frac{39}{8}$.

15. Find the domain and range of the function $f(x) = x^2 - 4x - 2$.

Answer: medium

$f(x) = x^2 - 4x - 2 = (x^2 - 4x + 4) - 4 - 2 = (x - 2)^2 - 6$. Then the domain of the function is $\mathbb{R}$ and since the minimum value of the function is $f(2) = -6$, the range of the function is the interval $[-6, \infty)$.

16. The effectiveness of a billboard depends on how many times it is seen. After some experiments, an advertising agency found that if the effectiveness E is measured on a scale of 0 to 10, then $E(n) = n - \frac{1}{40}n^2$ where n is the number of times a person sees the billboard. For a billboard to have maximum effectiveness, how many times should it be seen?

Answer: medium

$E(n) = n - \frac{1}{40}n^2 = -\frac{1}{40}(n^2 - 40n + 400) + 10 = -\frac{1}{40}(n - 20)^2 + 10$. Since the maximum of the function occurs when $n = 20$, the person should see the billboard 20 times.

17. Find two positive numbers whose sum is 50 and the sum of whose squares is a minimum.

Answer: medium

Let the two positive numbers be x and y. Since their sum is 50, $x + y = 50 \quad \Leftrightarrow \quad y = 50 - x$. Then the sum of squares is $S = x^2 + y^2 = x^2 + (50 - x)^2 = 2x^2 - 100x + 2500 = 2(x^2 - 50x + 625) + 1250 = 2(x - 25)^2 + 1250$. So the minimum of 1250 occurs when $x = 25$ and $y = 50 - 25 = 25$. Therefore both numbers are 25.

18. Among all rectangles that have a perimeter of 40 ft, find the dimensions of the one with the largest area.

Answer: medium

Let w and l be the width and the length in feet of a rectangle. Then $2w + 2l = 40 \quad \Leftrightarrow \quad l = 20 - w$. The area of a rectangle is given by $A = l \cdot w = (20 - w)w = 20w - w^2 = -(w^2 - 20w + 100) + 100 = -(w - 10)^2 + 100$. So the area is maximized when $w = 10$ and hence the largest rectangle is a square with dimension 10 ft.

19. Find the area of the largest rectangle that can be inscribed in a right triangle with legs of lengths 6 cm and 8 cm if two sides of the rectangle lie along the legs.

Answer: hard

Let w be the width of the triangle and h be the height. If we consider the triangle to be oriented on the positive x- and y-axes with the right angle at the origin, then the point (w, h) lies on the line segment joining the points $(8, 0)$ and $(0, 6)$. Since this line segment has slope $-\frac{3}{4}$ and y-intercept 6, $h = -\frac{3}{4}w + 6$. Therefore the area of the triangle is $A = w \cdot h = w(-\frac{3}{4}w + 6) = -\frac{3}{4}(w^2 - 8w + 16) + 12 = -\frac{3}{4}(w - 4)^2 + 12$. So the area of the largest triangle is 12 cm^2. (Another way to set up this area function is to use similar triangles.)

20. A student makes and sells hacky-sacks at the beach during the summer months. The material for each hacky-sack costs him \$4 and he has been selling about 15 per day at \$10 each. He has been wondering whether or not to lower his prices, so he takes a survey and finds that for every dollar decrease he would gain three sales per day. What should his selling price be in order to maximize profits?

Answer: hard

Let x be the selling price in dollars. Thus, $10 - x$ is the price decrease, and the number of hacky-sacks sold is $15 + 3(10 - x) = 45 - 3x$. The revenue from the sales is $x(45 - 3x)$ and the expenses are $4(45 - 3x)$. So the profit is $P = x(45 - 3x) - 4(45 - 3x) = (x - 4)(45 - 3x) = -3x^2 + 57x - 180 = -3(x^2 - 19x + 60) = -3\left(x - \frac{19}{2}\right)^2 + \frac{363}{4}$. This is maximized when $x = 9.5$, so he should set the selling price at \$9.50.

GC 21. (a) Use a graphing calculator or computer to find the maximum value of the quadratic function $f(x) = 1 + x - \sqrt{3}x^2$ correct to two decimal places.

(b) Find the exact maximum value of f and compare with your answer to part (a).

Answer: medium

(a) $f(x) = 1 + x - \sqrt{3}x^2$

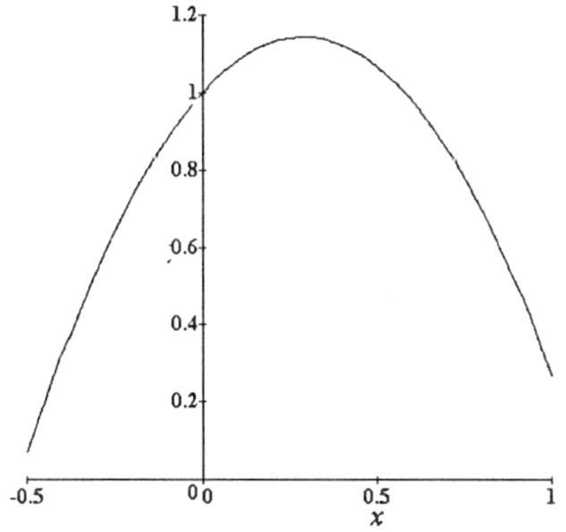

From the graph we see that the maximum value is approximately 1.14.

(b) $f(x) = -\sqrt{3}\left(x^2 - \frac{1}{\sqrt{3}}x\right) + 1 = -\sqrt{3}\left(x - \frac{1}{2\sqrt{3}}\right)^2 + \frac{\sqrt{3}}{12} + 1$

$\Rightarrow$ the maximum value is $f\left(\frac{1}{2\sqrt{3}}\right) = \frac{\sqrt{3}}{12} + 1 \approx 1.14$.

GC 22. Find the local maximum and minimum values of the function $f(x) = 3 - x - x^2 + x^3$ and the values of x at which they occur. State your answers to two decimal places.

Answer: medium

$f(x) = 3 - x - x^2 + x^3$

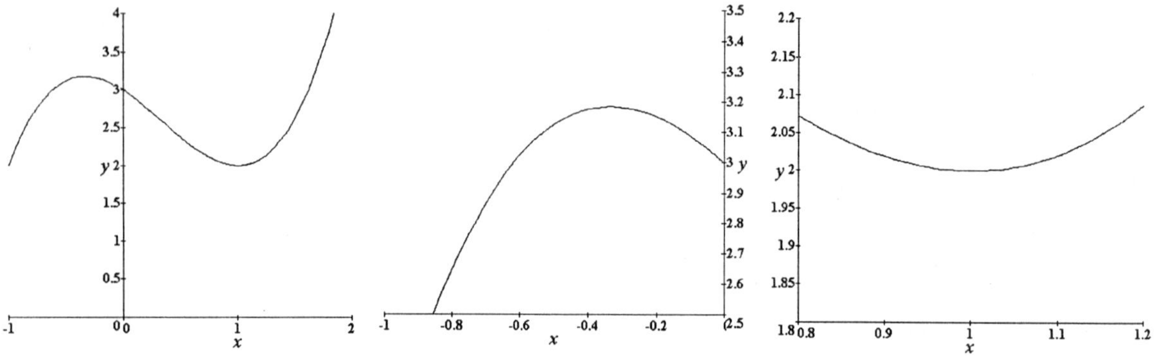

From the graphs, f has a local maximum of about 3.19 when $x \approx -0.33$ and a local minimum of about 2.00 when $x \approx 1.00$.

GC 23. Find the local maximum and minimum values of the function $g(x) = x^5 - 9x^3 + 22x$ and the values of x at which they occur. State your answers to two decimal places.

Answer: medium

$g(x) = x^5 - 9x^3 + 22x$

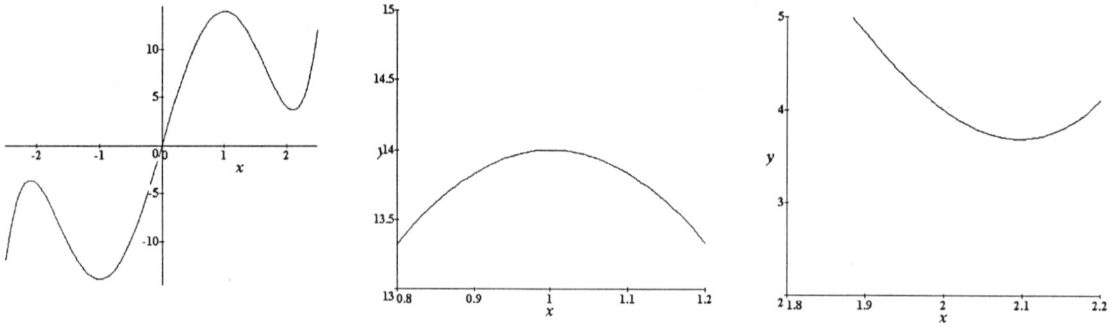

From the graphs, f has a local maximum ≈ 14.00 when $x \approx 1.00$ and a local minimum ≈ 3.69 when $x \approx 2.10$. Then, by symmetry (f is odd), f has a local minimum ≈ -14.00 when $x \approx -1.00$ and a local maximum ≈ -3.69 when $x \approx -2.10$.

GC 24. Find the local maximum and minimum values of the function $U(x) = 2x\sqrt{x - x^2}$ and the values of x at which they occur. State your answers to two decimal places.

Answer: medium

$U(x) = 2x\sqrt{x - x^2}$

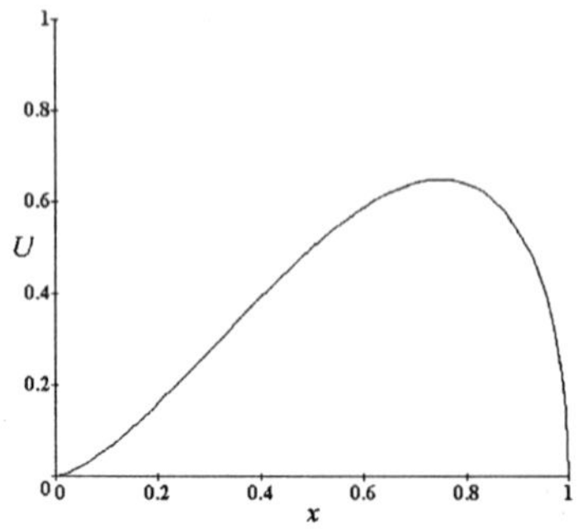

From the graph, the local maximum value of $U \approx 0.65$ when $x \approx 0.75$.

GC 25. Find the local maximum and minimum values of the function $V(x) = \dfrac{1}{x^2 - 3x + 3}$ and the values of x at which they occur. State your answers to two decimal places.

Answer: medium

$$V(x) = \frac{1}{x^2 - 3x + 3}$$

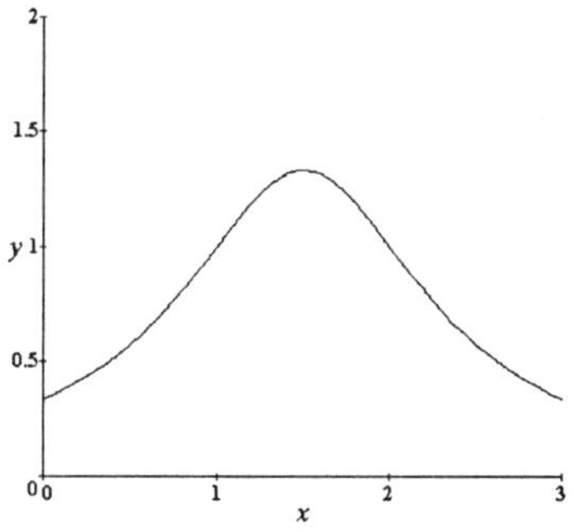

From the graph, the local maximum of $V \approx 1.33$ when $x \approx 1.50$.

GC 26. A highway engineer wants to estimate the maximum number of cars that can safely use a particular highway at a given speed. He assumes that each car is 16 ft long, travels at a speed s, and follows the car in front of it at the "safe following distance" for that speed. The safe following distance at 20 mi/h is one car length. Using the fact that the distance required to stop is proportional to the square of the speed and the fact that 1 mi/h = 88 ft/s, he finds that the number N of cars per minute that pass a given point is a function of the speed given by $N(s) = \dfrac{88s}{16 + 16\left(\frac{1}{20}s\right)^2}$

At what speed can the largest number of cars use the highway safely?

Answer: medium

$$N(s) = \frac{88s}{16 + 16\left(\frac{1}{20}s\right)^2}$$

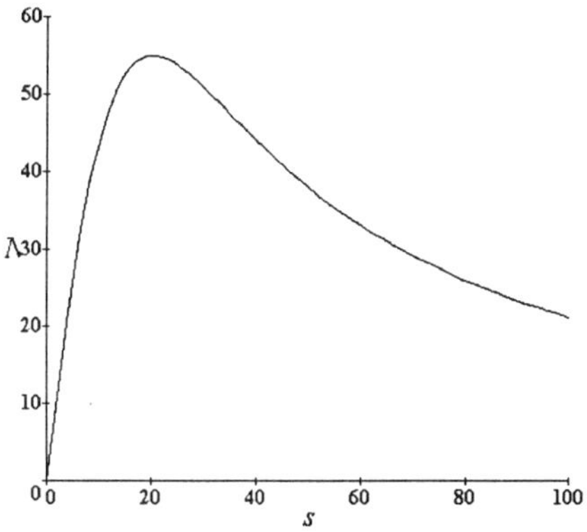

From the graph we see that at the speed of 20 mi/h the largest number of cars that can use the highway safely is 55.

GC 27. A box with an open top is to be constructed from a rectangular piece of cardboard with dimensions 16 inches by 24 inches by cutting out equal squares of side x at each corner and then folding up the sides. Find the largest volume that such a box can have.

Answer: hard

Since the length of the corner squares is x, then the height of the box is x, the width of the box is $16 - 2x$, and the length of the box is $24 - 2x$. Therefore, the volume of the box is $V = x(16 - 2x)(24 - 2x) = 4x^3 - 80x^2 + 384x$, $0 < x < 8$.

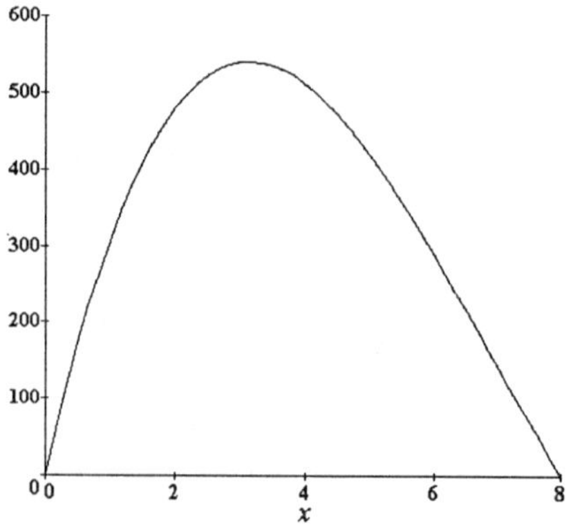

From the graph we see that the largest volume that such a box can have is about 541 in^3.

GC 28. Find the area of the largest rectangle that can be inscribed in a semicircle of radius 2 ft.

Answer: hard

Let w be the width of the rectangle. By the Pythagorean Theorem, $\left(\frac{1}{2}w\right)^2 + h^2 = r^2 \quad \Leftrightarrow \quad \frac{1}{4}w^2 + h^2 = r^2 \quad \Leftrightarrow$ $w^2 = 4(r^2 - h^2) \quad \Leftrightarrow \quad w = 2\sqrt{r^2 - h^2}$ since $w > 0$. Therefore, because the radius is 2 ft, the area of the rectangle is $A = wh = 2h\sqrt{r^2 - h^2} = 2h\sqrt{4 - h^2}$.

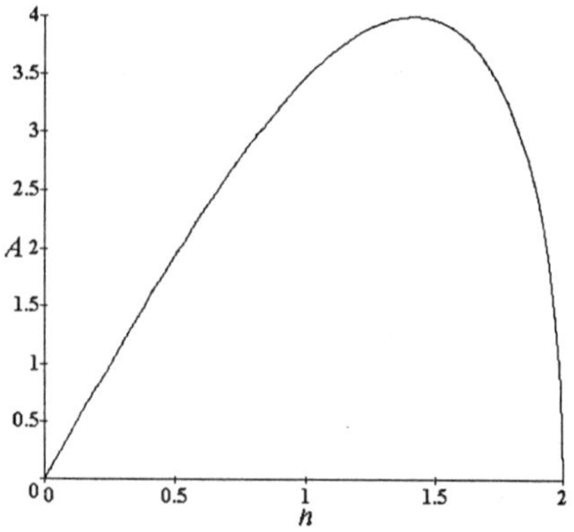

From the graph we see that the area of the largest rectangle that can be inscribed in such a semicircle is 4 ft^2.

GC 29. A man is at a point A on a bank of a straight river, 3 mi wide, and wants to reach point B, 7 mi downstream on the opposite bank, by first rowing his boat to a point P on the opposite bank and then walking the remaining distance x to B. He can row at a speed of 5 mi/h and walk at a speed of 4 mi/h. Where should he land in order to reach B as soon as possible?

Answer: hard

Let t_1 represent the time, in hours, spent walking, and let t_2 represent the time spent rowing.

Since the distance walked is x and the walking speed is 4 mi/h, then the time spent walking is $t_1 = \frac{1}{4}x$. By the Pythagorean Theorem, the distance rowed is $d = \sqrt{3^2 + (7-x)^2} = \sqrt{x^2 - 14x + 58}$ and so the time spent rowing is $t_2 = \frac{1}{5}\sqrt{x^2 - 14x + 58}$. So the total time is $T = \frac{1}{5}\sqrt{x^2 - 14x + 58} + \frac{1}{4}x$.

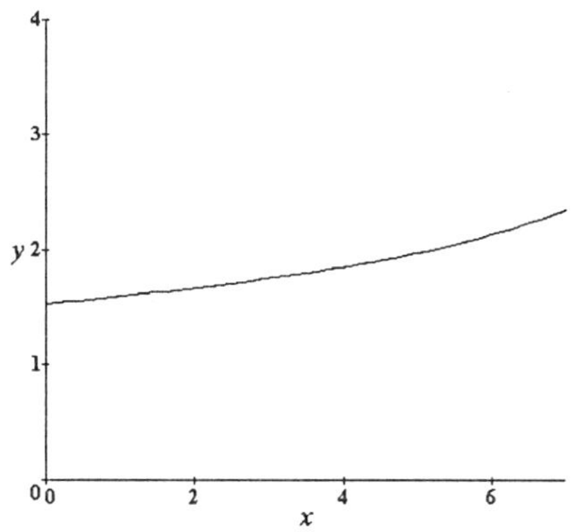

From the graph we see that T is minimized when $x = 0$. He should row directly to point B in order to reach point B as soon as possible.

30. Find the maximum value of the function $f(x) = 2 + 6x^2 - x^4$. (Hint: Let $t = x^2$.)

Answer: hard

$f(x) = 2 + 6x^2 - x^4$. Substitute $t = x^2$ and then $f(\sqrt{t}) = 2 + 6t - t^2 = -(t^2 - 6t + 9) + 9 + 2 = -(t-3)^2 + 11$. Therefore the function $f(\sqrt{t}) = f(x)$ has a maximum value of 11.

31. Find the maximum or the minimum value of the function $f(x) = -\dfrac{x^2}{7} + 3x + 5$.

Answer: medium

$f(x) = -\dfrac{x^2}{7} + 3x + 5 \;\Rightarrow\; a = -\frac{1}{7}$ and $b = 3$, so the maximum value is $f\left(\dfrac{-3}{2\left(-\frac{1}{7}\right)}\right) = \dfrac{83}{4}$.

32. Find the maximum or the minimum value of the function $f(x) = \dfrac{x}{4}\left(1 - \dfrac{x}{4}\right) + \dfrac{5}{12}$.

Answer: medium

$f(x) = \dfrac{x}{4}\left(1 - \dfrac{x}{4}\right) + \dfrac{5}{12} = \frac{5}{12} + \frac{1}{4}x - \frac{1}{16}x^2 \;\Rightarrow\; a = -\frac{1}{16}$ and $b = \frac{1}{4}$, so the maximum value is $f\left(\dfrac{-\frac{1}{4}}{2\left(-\frac{1}{16}\right)}\right) = \dfrac{2}{3}$.

33. Find a function whose graph is a parabola with vertex $(-1, -4)$ passing through the point $\left(\frac{1}{3}, \frac{124}{27}\right)$.

Answer: medium

The standard form of a parabola with vertex (h, k) is $y = a(x-h)^2 + k$. Vertex $(-1, -4) \;\Rightarrow\; y = a(x+1)^2 - 4$. Passing through the point $\left(\frac{1}{3}, \frac{124}{27}\right) \;\Rightarrow\; \frac{124}{27} = a\left(\frac{1}{3} + 1\right)^2 - 4 \;\Rightarrow\; a = \frac{29}{6}$. So $y = \frac{29}{6}(x+1)^2 - 4$.

34. Find a quadratic function whose graph is a parabola with x-intercepts -2 and 4 and highest point $(h, 4)$.

Answer: medium

In order to find h, find the midpoint of $(-2, 0)$ and $(4, 0)$: $h = \dfrac{-2+4}{2} = 1 \;\Rightarrow\; y = a(x-1)^2 + 4$. Since the function passes through the point $(4, 0)$, we can find a by substituting these values in the equation $y = a(x-1)^2 + 4 \;\Rightarrow$ $0 = a(4-1)^2 + 4 \;\Rightarrow\; a = -\frac{4}{9}$. Thus the quadratic function is $y = f(x) = -\frac{4}{9}(x-1)^2 + 4 = -\frac{4}{9}x^2 + \frac{8}{9}x + \frac{32}{9}$.

35. Find a quadratic function whose graph is a parabola with x-intercepts 1 and $\frac{4}{3}$ and lowest point $(h, -4)$.
Answer: medium
In order to find h, find the midpoint of $(1, 0)$ and $\left(\frac{4}{3}, 0\right)$: $h = \dfrac{1 + \frac{4}{3}}{2} = \dfrac{7}{6}$ $\Rightarrow$ $y = a\left(x - \frac{7}{6}\right)^2 - 4$. Since the function passes through the point $(1, 0)$, we can find a by substituting these values in the equation $y = a(x - 1)^2 + 4$ $\Rightarrow$ $0 = a\left(1 - \frac{7}{6}\right)^2 - 4$ $\Rightarrow$ $a = 144$. Thus the quadratic function is $y = f(x) = 144\left(x - \frac{7}{6}\right)^2 - 4 = 144x^2 - 336x + 192$.

36. Find the minimum value of a parabola with x-intercepts 3 and 12 and y-intercept 18.
Answer: medium
A quadratic function that has x-intercepts 3 and 12 can be written as $y = (x - 3)(x - 12) = x^2 - 15x + 36$. However, this particular equation does not meet the condition of the given y-intercept $(0, 18)$. One way to meet that condition is to multiply the quadratic function by $\frac{1}{2}$, so that $y = f(x) = \frac{1}{2}\left(x^2 - 15x + 36\right) = \frac{1}{2}x^2 - \frac{15}{2}x + 18$. Now find the minimum value. $a = \frac{1}{2}$ and $b = -\frac{15}{2}$, so the minimum value is $f\left(\dfrac{-\left(-\frac{15}{2}\right)}{2\left(\frac{1}{2}\right)}\right) = -\dfrac{81}{8}$.

37. The height (in feet) of a circus performer t seconds after being fired out of a cannon is given by the equation $h(t) = -16t^2 + 80.0t + 8.0$. If all goes well he should land in a pool of water at the other end of the performing area. What does h represent when $t = 0$? Find the time it takes for the performer to reach a maximum height. What is his maximum height?
Answer: hard
When $t = 0$, $h = 8$ ft, the vertical distance from the ground to the tip of the cannon. The time to reach a maximum height is $\dfrac{-80.0}{2(-16)} = 2.5$ s. Maximum height is $h(2.5) = 108.0$ ft.

SECTION 4.7

1. Use $f(x) = 2x - 4$ and $g(x) = 3 - x^2$ to evaluate the expression $g(f(1))$.
(a) -3 (b) -2 (c) -1 (d) 0 (e) 1
Answer: easy (c)
$f(1) = 2 \cdot 1 - 4 = -2$. $g(f(1)) = g(-2) = 3 - (-2)^2 = 3 - 4 = -1$

2. Use $f(x) = 2x - 4$ and $g(x) = 3 - x^2$ to evaluate the expression $g(g(3))$.
(a) 11 (b) 3 (c) -12 (d) -33 (e) -40
Answer: medium (d)
$g(3) = 3 - 3^2 = -6$. $g(g(3)) = g(-6) = 3 - (-6)^2 = 3 - 36 = -33$

3. Use $f(x) = 2x - 4$ and $g(x) = 3 - x^2$ to evaluate the expression $(g \circ f)(-2)$.
(a) -141 (b) -61 (c) 0 (d) 1 (e) 1729
Answer: medium (b)
$f(-2) = -8$. $(g \circ f)(-2) = g(-8) = 3 - (-8)^2 = -61$

4. Use $f(x) = 2x - 4$ and $g(x) = 3 - x^2$ to evaluate the expression $(g \circ g)(2)$.
(a) 2 (b) 4 (c) 9 (d) 11 (e) 114
Answer: medium (a)
$g(2) = 3 - 2^2 = -1$. $(g \circ g)(2) = g(-1) = 3 - (-1)^2 = 2$

5. Use $f(x) = 2x - 4$ and $g(x) = 3 - x^2$ to evaluate the expression $(g \circ f)(x)$.
(a) $-x^2 - x + 13$ (b) $-4x^2 + 16x - 13$ (c) $2x^2 - 3x + 5$ (d) $-5x^2 + 8x - 7$ (e) $3x^2 + 2x + 2$
Answer: medium (b)
$f(x) = 2x - 4$. $(g \circ f)(x) = g(2x - 4) = 3 - (2x - 4)^2 = 3 - (4x^2 - 16x + 16) = -4x^2 + 16x - 13$

6. Use $f(x) = 2x - 4$ and $g(x) = 3 - x^2$ to evaluate the expression $(g \circ g)(x)$.
(a) $x^4 - 12x^2 - 2$ (b) $x^6 - 4x^4 + 16x^2$ (c) $2x^4 - x^3 + x$ (d) $3x^4 - x^2 + 1$ (e) $-x^4 + 6x^2 - 6$
Answer: medium (e)
$g(x) = 3 - x^2$. $(g \circ g)(x) = g(3 - x^2) = 3 - (3 - x^2)^2 = 3 - (9 - 6x^2 + x^4) = -x^4 + 6x^2 - 6$

7. Using the functions $f(x) = 4x - 6$ and $g(x) = \frac{1}{3}x$ find the functions $f \circ g$, $g \circ f$, $f \circ f$, and $g \circ g$, and their domains.
Answer: medium
$f(x) = 4x - 6,\ g(x) = \frac{1}{3}x$
$(f \circ g)(x) = f\left(\frac{1}{3}x\right) = 4 \cdot \frac{1}{3}x - 6 = \frac{4}{3}x - 6,\ D \text{ is } (-\infty, \infty)$
$(g \circ f)(x) = g(4x - 6) = \frac{1}{3}(4x - 6) = \frac{4}{3}x - 2,\ D \text{ is } (-\infty, \infty)$
$(f \circ f)(x) = f(4x - 6) = 4(4x - 6) - 6 = 16x - 30,\ D \text{ is } (-\infty, \infty)$
$(g \circ g)(x) = g\left(\frac{1}{3}x\right) = \frac{1}{3}\left(\frac{1}{3}x\right) = \frac{1}{9}x,\ D \text{ is } (-\infty, \infty)$

8. Using the functions $f(x) = \dfrac{1}{x}$, $g(x) = x^2 + 3x$, find the functions $f \circ g$, $g \circ f$, $f \circ f$, and $g \circ g$, and their domains.
Answer: medium
$f(x) = \dfrac{1}{x},\ D = \{x \mid x \neq 0\}.\ g(x) = x^2 + 3x$
$$(f \circ g)(x) = f(x^2 + 3x) = \frac{1}{x^2 + 3x},\ D = \{x \mid x \neq 0, -3\}$$
$$(g \circ f)(x) = g\left(\frac{1}{x}\right) = \left(\frac{1}{x}\right)^2 + 3 \cdot \frac{1}{x} = \frac{1}{x^2} + \frac{3}{x},\ D = \{x \mid x \neq 0\}$$
$$(f \circ f)(x) = f\left(\frac{1}{x}\right) = \frac{1}{1/x} = x,\ D = \{x \mid x \neq 0\}$$
$$(g \circ g)(x) = g(x^2 + 3x) = (x^2 + 3x)^2 + 3(x^2 + 3x) = x^4 + 6x^3 + 9x^2 + 3x^2 + 9x = x^4 + 6x^3 + 12x^2 + 9x,\ D = (-\infty, \infty)$$

9. Using the functions $f(x) = \dfrac{1}{x+1}$ and $g(x) = \dfrac{x+1}{x-1}$, find the functions $f \circ g$, $g \circ f$, $f \circ f$, and $g \circ g$, and their domains.
Answer: medium
$f(x) = \dfrac{1}{x+1},\ D = \{x \mid x \neq -1\}.\ g(x) = \dfrac{x+1}{x-1},\ D = \{x \mid x \neq 1\}$
$$(f \circ g)(x) = f\left(\frac{x+1}{x-1}\right) = \frac{1}{\dfrac{x+1}{x-1} + 1} = \frac{x-1}{2x},\ D = \{x \mid x \neq 0, -1\}$$
$$(g \circ f)(x) = g\left(\frac{1}{x+1}\right) = \frac{\dfrac{1}{x+1} + 1}{\dfrac{1}{x+1} - 1} = \frac{\dfrac{1 + (x+1)}{x+1}}{\dfrac{1 - (x+1)}{x+1}} = -\frac{x+2}{x},\ D = \{x \mid x \neq -1, 0\}$$
$$(f \circ f)(x) = f\left(\frac{1}{x+1}\right) = \frac{1}{\dfrac{1}{x+1} + 1} = \frac{1}{\dfrac{1 + (x+1)}{x+1}} = \frac{x+1}{x+2},\ D = \{x \mid x \neq -2, -1\}$$
$$(g \circ g)(x) = g\left(\frac{x+1}{x-1}\right) = \frac{\dfrac{x+1}{x-1} + 1}{\dfrac{x+1}{x-1} - 1} = \frac{x + 1 + (x-1)}{x + 1 - (x-1)} = x,\ D = \{x \mid x \neq 1\}$$

10. Using the functions $f(x)=\sqrt{x^2-4}$ and $g(x)=\sqrt{2-x}$, find the functions $f\circ g$, $g\circ f$, $f\circ f$, and $g\circ g$, and their domains.

Answer: hard

$f(x)=\sqrt{x^2-4}$, D is $(-\infty,-2]\cup[2,\infty)$. $g(x)=\sqrt{2-x}$, D is $(-\infty,2]$.

$(f\circ g)(x)=f\left(\sqrt{2-x}\right)=\sqrt{\left(\sqrt{2-x}\right)^2-4}=\sqrt{2-x-4}=\sqrt{-x-2}$, D is $(-\infty,-2]$.

$(g\circ f)(x)=g\left(\sqrt{x^2-4}\right)=\sqrt{2-\sqrt{x^2-4}}$. Now we must have $2-\sqrt{x^2-4}\geq 0 \quad\Rightarrow\quad \sqrt{x^2-4}\leq 2 \quad\Rightarrow$ $-2\sqrt{2}\leq x\leq 2\sqrt{2}$. Thus the domain is D is $\left[-2\sqrt{2},-2\right]\cup\left[2,2\sqrt{2}\right]$.

$(f\circ f)(x)=f\left(\sqrt{x^2-4}\right)=\sqrt{\left(\sqrt{x^2-4}\right)^2-4}=\sqrt{x^2-8}$, D is $\left(-\infty,-2\sqrt{2}\right]\cup\left[2\sqrt{2},\infty\right)$.

$(g\circ g)(x)=g\left(\sqrt{2-x}\right)=\sqrt{2-\sqrt{2-x}}$. Now we must have $2-\sqrt{2-x}\geq 0 \quad\Rightarrow\quad \sqrt{2-x}\leq 2 \quad\Rightarrow\quad x\geq -2$. Thus the domain is D is $[-2,2]$.

11. Using the functions $f(x)=\dfrac{1}{2\sqrt{x}}$ and $g(x)=x^2-2x$ find the functions $f\circ g$, $g\circ f$, $f\circ f$, and $g\circ g$, and their domains.

Answer: hard

$f(x)=\dfrac{1}{2\sqrt{x}}$, $D=\{x\mid x>0\}$. $g(x)=x^2-2x$, $D=(-\infty,\infty)$

$(f\circ g)(x)=f(x^2-2x)=\dfrac{1}{2\sqrt{x^2-2x}}$, since $x^2-2x>0 \quad\Leftrightarrow\quad x(x-2)>0$

Case (i): $x>0$ and $x>2 \quad\Leftrightarrow\quad x>2$. Case (ii): $x<0$ and $x<2 \quad\Leftrightarrow\quad x<0$. Therefore the domain is $D=(-\infty,0)\cup(2,\infty)$

$$(g\circ f)(x)=g\left(\frac{1}{2\sqrt{x}}\right)=\left(\frac{1}{2\sqrt{x}}\right)^2-2\cdot\frac{1}{2\sqrt{x}}=\frac{1}{4x}-\frac{1}{\sqrt{x}},\quad D \text{ is } (0,\infty)$$

$$(f\circ f)(x)=f\left(\frac{1}{2\sqrt{x}}\right)=\frac{1}{2\sqrt{\frac{1}{2\sqrt{x}}}}=\frac{\sqrt{2}\,x^{1/4}}{2},\quad D \text{ is } (0,\infty)$$

$(g\circ g)(x)=g(x^2-2x)=(x^2-2x)^2-2(x^2-2x)=x^4-4x^3+4x^2-2x^2+4x=x^4-4x^3+2x^2+4x$, D is $(-\infty,\infty)$

12. Using the functions $f(x)=\dfrac{3}{x}$, $g(x)=x^2$ and $h(x)=x^3+2x$, find $f\circ g\circ h$.

Answer: medium

$(g\circ h)(x)=g(x^3+2x)=(x^3+2x)^2=x^6+4x^4+4x^2$.

$(f\circ g\circ h)(x)=f(x^6+4x^4+4x^2)=\dfrac{3}{x^6+4x^4+4x^2}$.

13. Using the functions $f(x)=2\sqrt{x}$, $g(x)=\dfrac{x}{x+1}$ and $h(x)=\sqrt[4]{x}$ find $f\circ g\circ h$.

Answer: medium

$$(g\circ h)(x)=g\left(\sqrt[4]{x}\right)=\frac{\sqrt[4]{x}}{\sqrt[4]{x}+1}.\quad (f\circ g\circ h)(x)=f\left(\frac{\sqrt[4]{x}}{\sqrt[4]{x}+1}\right)=2\sqrt{\frac{\sqrt[4]{x}}{\sqrt[4]{x}+1}}$$

14. Express the function $F(x)=\sqrt{x}+4$ in the form $f\circ g$.

Answer: easy

$F(x)=\sqrt{x}+4$. If $f(x)=x+4$ and $g(x)=\sqrt{x}$ then $F(x)=(f\circ g)(x)$.

15. Express the function $G(x)=\dfrac{1}{x+2}$ in the form $f\circ g$.

Answer: medium

$G(x)=\dfrac{1}{x+2}$. If $f(x)=\dfrac{1}{x}$ and $g(x)=x+2$ then $G(x)=(f\circ g)(x)$.

16. Express the function $H(x)=\sqrt{2+\sqrt{x}}$ in the form $f\circ g$.

Answer: medium

$H(x)=\sqrt{2+\sqrt{x}}$. If $f(x)=\sqrt{2+x}$ and $g(x)=\sqrt{x}$ then $H(x)=(f\circ g)(x)$.

17. Express the function $F(x) = \sqrt[4]{\sqrt{x}+1}$ in the form $f \circ g \circ h$.

Answer: medium

$F(x) = \sqrt[4]{\sqrt{x}+1}$. If $g(x) = x+1$ and $h(x) = \sqrt{x}$ then $(g \circ h)(x) = \sqrt{x}+1$, and if $f(x) = \sqrt[4]{x}$, then $(f \circ g \circ h)(x) = F(x)$. Other solutions are possible.

18. Express the function $G(x) = \dfrac{1}{\left(1-\sqrt{x}\right)^2}$ in the form $f \circ g \circ h$.

Answer: hard

$G(x) = \dfrac{1}{\left(1-\sqrt{x}\right)^2}$. If $g(x) = 1-x$ and $h(x) = \sqrt{x}$ then $(g \circ h)(x) = 1-\sqrt{x}$, and if $f(x) = \dfrac{1}{x^2}$, then $(f \circ g \circ h)(x) = G(x)$. Other solutions are possible.

19. A spherical balloon is being inflated. If the radius of the balloon is increasing at a rate of 2 cm/s, express the volume of the balloon as a function of time t (in seconds).

Answer: hard

Let r be the radius of the spherical balloon in cm. Since the radius is increasing at a rate of 2 cm/s, the radius is $r = 2t$ after t seconds. Therefore the volume of the balloon can be written $V = \frac{4}{3}\pi r^3 = \frac{4}{3}\pi(2t)^3 = \frac{32}{3}\pi t^3$.

20. If $f(x) = x-2$ and $h(x) = 2x+1$, find a function g such that $g \circ f = h$.

Answer: hard

$f(x) = x-2$ and $h(x) = 2x+1$. If $(g \circ f)(x) = h(x)$ then $g(f(x)) = g(x-2) = 2x+1 \quad \Leftrightarrow$ $g(x) = 2(x+2)+1 = 2x+5$.

21. Suppose g is an odd function and let $h = f \circ g$. Is h always an odd function? What if f is odd? What if f is even?

Answer: hard

g is an odd function $\Rightarrow$ $g(-x) = -g(x)$. Consider $h(-x) = (f \circ g)(-x) = f(g(-x)) = f(-g(x)) \Rightarrow$ $h(-x) = f(g(x)) = h(x)$ if f is even and $h(-x) = -f(g(x)) = -h(x)$ if f is odd. Therefore $h(x)$ is not always an odd function. But h is odd if f is odd and h is even if f is even.

22. For the functions $f(x) = 2x^2 - 2x$ and $g(x) = x$, find $f+g$, $f-g$, fg, $\dfrac{f}{g}$, and their domains.

Answer: easy

Function	Domain
$f+g = 2x^2 - 2x + x = 2x^2 - x$	$(-\infty, \infty)$
$f-g = 2x^2 - 2x - x = 2x^2 - 3x$	$(-\infty, \infty)$
$fg = \left(2x^2 - 2x\right)x = 2x^3 - 2x^2$	$(-\infty, \infty)$
$\dfrac{f}{g} = \dfrac{2x^2 - 2x}{x} = 2x - 2$	$(-\infty, 0) \cup (0, \infty)$

23. For the functions $f(x) = 3x^2 - x^3$ and $g(x) = x^2 - 1$, find $f+g$, $f-g$, fg, $\dfrac{f}{g}$, and their domains.

Answer: easy

Function	Domain
$f+g = 4x^2 - x^3 - 1$	$(-\infty, \infty)$
$f-g = 2x^2 - x^3 + 1$	$(-\infty, \infty)$
$fg = \left(3x^2 - x^3\right)\left(x^2 - 1\right) = 3x^4 - 3x^2 - x^5 + x^3$	$(-\infty, \infty)$
$\dfrac{f}{g} = \dfrac{3x^2 - x^3}{x^2 - 1}$	$(-\infty, -1) \cup (-1, 1) \cup (1, \infty)$

24. For the functions $f(x) = \sqrt{25 - x^2}$ and $g(x) = \sqrt{x^2 + 1}$, find $f + g$, $f - g$, fg, $\frac{f}{g}$, and their domains.

Answer: easy

Function	Domain
$f + g = \sqrt{25 - x^2} + \sqrt{x^2 + 1}$	$[-5, 5]$
$f - g = \sqrt{25 - x^2} - \sqrt{x^2 + 1}$	$[-5, 5]$
$fg = \sqrt{25 - x^2}\sqrt{x^2 + 1}$	$[-5, 5]$
$\frac{f}{g} = \frac{\sqrt{25 - x^2}}{\sqrt{x^2 + 1}}$	$[-5, 5]$

25. For the functions $f(x) = \frac{1}{x}$ and $g(x) = \frac{x}{x - 1}$, find $f + g$, $f - g$, fg, $\frac{f}{g}$, and their domains.

Answer: easy

Function	Domain
$f + g = \frac{1}{x} + \frac{x}{x - 1} = \frac{x - 1 + x^2}{x(x - 1)}$	$(-\infty, 0) \cup (0, 1) \cup (1, \infty)$
$f - g = \frac{1}{x} - \frac{x}{x - 1} = -\frac{-x + 1 + x^2}{x(x - 1)}$	$(-\infty, 0) \cup (0, 1) \cup (1, \infty)$
$fg = \left(\frac{1}{x}\right)\left(\frac{x}{x - 1}\right) = \frac{1}{x - 1}$	$(-\infty, 1) \cup (1, \infty)$
$\frac{f}{g} = \frac{\frac{1}{x}}{\frac{x}{x - 1}} = \frac{x - 1}{x^2}$	$(-\infty, 0) \cup (0, \infty)$

26. For the functions $f(x) = \frac{1}{\sqrt{x - 1}}$ and $g(x) = x\sqrt{x + 3}$, find $f + g$, $f - g$, fg, $\frac{f}{g}$, and their domains.

Answer: easy

Function	Domain
$f(x) + g(x) = \frac{1}{\sqrt{x - 1}} + x\sqrt{x + 3} = \frac{1 + x\sqrt{x + 3}\sqrt{x - 1}}{\sqrt{x - 1}}$	$(1, \infty)$
$f(x) - g(x) = \frac{1}{\sqrt{x - 1}} - x\sqrt{x + 3} = -\frac{-1 + x\sqrt{x + 3}\sqrt{x - 1}}{\sqrt{x - 1}}$	$(1, \infty)$
$f(x)\,g(x) = \frac{x\sqrt{x + 3}}{\sqrt{x - 1}}$	$(1, \infty)$
$\frac{f(x)}{g(x)} = \frac{1}{x\sqrt{x - 1}\sqrt{x + 3}}$	$(1, \infty)$

SECTION 4.8

1. Determine whether or not the function $f(x) = x^2 - 4x + 5$ is one-to-one.

   Answer: easy

   $f(x) = x^2 - 4x + 5 = (x^2 - 4x + 4) - 4 + 5 = (x - 2)^2 + 1$. Thus, $f(0) = 5 = f(4)$, so f is not one-to-one. [Or use the Horizontal Line Test.]

2. Determine whether or not the function $g(x) = |x + 1|$ is one-to-one.

   Answer: medium

   $g(x) = |x + 1|$. Since every number and its negative have the same absolute value, e.g., $|-2| = 2 = |2|$, g is not a one-to-one function.

3. Determine whether or not the function $h(x) = x^4 + 2, 0 \leq x \leq 2$ is one-to-one.

   Answer: medium

   $h(x) = x^4 + 2, 0 \leq x \leq 2$. If $x_1 \neq x_2$ then $x_1^4 \neq x_2^4$ because two different positive numbers cannot have the same fourth power. Thus, $x_1^4 + 2 \neq x_2^4 + 2$. Therefore h is one-to-one.

GC 4. Determine whether or not the function $f(x) = x^3 + 2x$ is one-to-one.

   Answer: medium

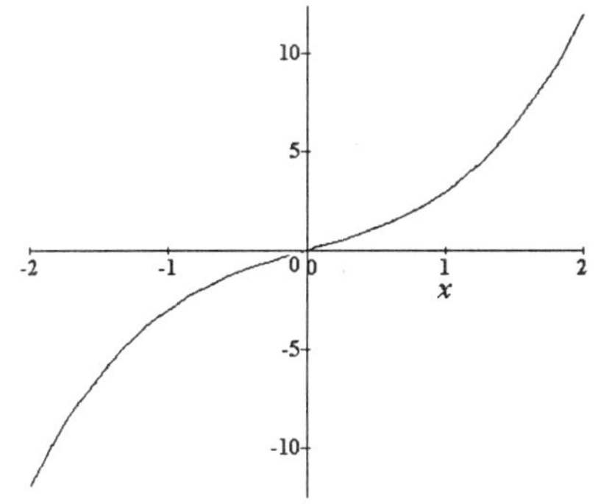

   Using a graphing calculator and the horizontal line test we see that $f(x) = x^3 + 2x$ is one-to-one.

GC 5. Determine whether or not the function $f(x) = \sqrt{x^3 - 4x + 1}$ is one-to-one.

   Answer: medium

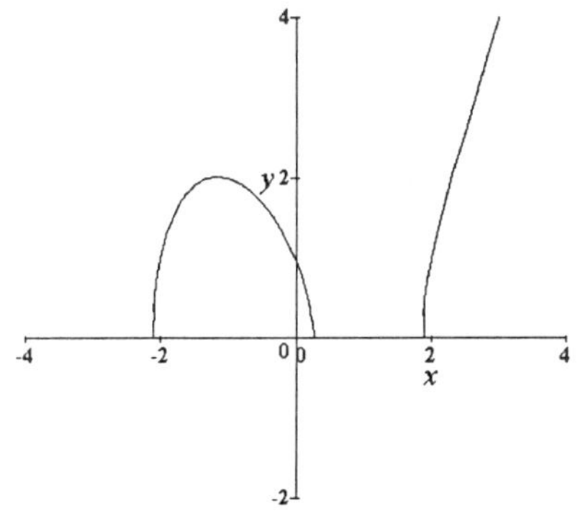

   Using a graphing calculator and the horizontal line test we see that $f(x) = \sqrt{x^3 - 4x + 1}$ is not one-to-one.

6. Find $f^{-1}(3)$ if f is a one-to-one function and $f(2) = 3$.

   (a) $f^{-1}(3) = 3$ (b) $f^{-1}(3) = 2$ (c) $f^{-1}(3) = 1$ (d) $f^{-1}(3) = 0$ (e) $f^{-1}(3) = -1$

   Answer: easy (b)

   $f(2) = 3$. Since f is one-to-one, $f^{-1}(3) = 2$.

7. If $g(x) = x^2 + 2x, x \geq 0$, find $g^{-1}(3)$.

   (a) $g^{-1}(3) = -3$ (b) $g^{-1}(3) = -1$ (c) $g^{-1}(3) = 1$ (d) $g^{-1}(3) = 3$ (e) $g^{-1}(3) = 5$

   Answer: medium (c)

   $g(x) = x^2 + 2x$. Since $g(1) = 3$, we have $g^{-1}(3) = 1$.

8. Show that the function $f(x) = \dfrac{x+2}{x-2}$ is one-to-one and find its inverse function.

   Answer: hard

   $f(x) = \dfrac{x+2}{x-2}$. $f(x_1) = f(x_2) \Leftrightarrow \dfrac{x_1+2}{x_1-2} = \dfrac{x_2+2}{x_2-2} \Leftrightarrow (x_1+2)(x_2-2) = (x_1-2)(x_2+2) \Leftrightarrow$ $x_1x_2 + 2x_2 - 2x_1 - 4 = x_1x_2 - 2x_2 + 2x_1 - 4 \Leftrightarrow 4x_1 = 4x_2 \Leftrightarrow x_1 = x_2$. Therefore f is one-to-one. $y = \dfrac{x+2}{x-2}$ $\Leftrightarrow y(x-2) = x+2 \Leftrightarrow xy - x = 2 + 2y \Leftrightarrow x(y-1) = 2(y+1) \Leftrightarrow x = \dfrac{2(y+1)}{y-1}$. So the inverse function is $f^{-1}(x) = \dfrac{2(x+1)}{x-1}$.

9. Show that the function $f(x) = 4 - 3x^3$ is one-to-one and find its inverse function.
Answer: medium
$f(x) = 4 - 3x^3$. If $x_1 \neq x_2$ then $x_1^3 \neq x_2^3$ (since no two numbers have the same cube) and $4 - 3x_1^3 \neq 4 - 3x_2^3$. Therefore f is one-to-one. $y = 4 - 3x^3 \quad \Leftrightarrow \quad 3x^3 = 4 - y \quad \Leftrightarrow \quad x^3 = \frac{1}{3}(4 - y) \quad \Leftrightarrow \quad x = \sqrt[3]{\frac{1}{3}(4 - y)}$. So $f^{-1}(x) = \sqrt[3]{\frac{1}{3}(4 - x)}$.

10. Show that the function $f(x) = x^2 + 4x$, $x \geq -2$ is one-to-one and find its inverse function.
Answer: medium
$f(x) = x^2 + 4x = (x^2 + 4x + 4) - 4 = (x + 2)^2 - 4$. For $x \geq -2 \quad \Leftrightarrow \quad x + 2 \geq 0$, no numbers $x_1 + 2$ and $x_2 + 2$ have the same square and so f is a one-to-one function. $y = (x + 2)^2 - 4 \quad \Leftrightarrow \quad (x + 2)^2 = y + 4 \quad \Leftrightarrow \quad x + 2 = \sqrt{y + 4} \quad \Leftrightarrow$ $x = \sqrt{y + 4} - 2$. So the inverse function is $f^{-1}(x) = \sqrt{x + 4} - 2$.

11. Find the inverse function of $f(x) = \dfrac{1}{x + 1}$, $x > -1$ and then verify that f^{-1} and f satisfy the equations:
$f^{-1}(f(x)) = x$ for every x in A and $f(f^{-1}(x)) = x$ for every x in B.
Answer: medium
$f(x) = \dfrac{1}{x + 1}$, $x > -1$. $y = \dfrac{1}{x + 1} \quad \Leftrightarrow \quad x = \dfrac{1}{y} - 1$. So $f^{-1}(x) = \dfrac{1}{x} - 1$.
$f(f^{-1}(x)) = f\left(\dfrac{1}{x} - 1\right) = \dfrac{1}{1/x - 1 + 1} = \dfrac{1}{1/x} = x$. $f^{-1}(f(x)) = f^{-1}\left(\dfrac{1}{x + 1}\right) = \dfrac{1}{1/(x + 1)} - 1 = x + 1 - 1 = x$.

12. Find the inverse function of $f(x) = \sqrt{3x - 1}$ and then verify that f^{-1} and f satisfy the equations:
$f^{-1}(f(x)) = x$ for every x in A and $f(f^{-1}(x)) = x$ for every x in B.
Answer: medium
$f(x) = \sqrt{3x - 1}$. $y = \sqrt{3x - 1} \quad \Leftrightarrow \quad 3x - 1 = y^2 \quad \Leftrightarrow \quad x = \frac{1}{3}(y^2 + 1)$. So the inverse function is $f^{-1}(x) = \frac{1}{3}(x^2 + 1)$.
$f(f^{-1}(x)) = f\left(\frac{1}{3}(x^2 + 1)\right) = \sqrt{3 \cdot \frac{1}{3}(x^2 + 1) - 1} = \sqrt{x^2 + 1 - 1} = x$.
$f^{-1}(f(x)) = f^{-1}\left(\sqrt{3x - 1}\right) = \frac{1}{3}\left[\left(\sqrt{3x - 1}\right)^2 + 1\right] = x - \frac{1}{3} + \frac{1}{3} = x$.

13. Find the inverse function of $f(x) = (1 - x^3)^7$ and then verify that f^{-1} and f satisfy the equations:
$f^{-1}(f(x)) = x$ for every x in A and $f(f^{-1}(x)) = x$ for every x in B.
Answer: hard
$f(x) = (1 - x^3)^7$. $y = (1 - x^3)^7 \quad \Leftrightarrow \quad 1 - x^3 = \sqrt[7]{y} \quad \Leftrightarrow \quad x^3 = 1 - \sqrt[7]{y} \quad \Leftrightarrow \quad x = \sqrt[3]{1 - \sqrt[7]{y}}$. So the inverse function is $f^{-1}(x) = \sqrt[3]{1 - \sqrt[7]{x}}$. $f(f^{-1}(x)) = f\left(\sqrt[3]{1 - \sqrt[7]{x}}\right) = \left[1 - \left(\sqrt[3]{1 - \sqrt[7]{x}}\right)^3\right]^7 = \left(1 - 1 + \sqrt[7]{x}\right)^7 = x$.
$f^{-1}(f(x)) = f^{-1}\left((1 - x^3)^7\right) = \sqrt[3]{1 - \sqrt[7]{(1 - x^3)^7}} = \sqrt[3]{1 - 1 + x^3} = x$.

14. For the function $f(x) = 6 - x$:
    (a) sketch the graph of f.
    (b) use the graph of f to sketch the graph of f^{-1}.
    (c) find f^{-1}.

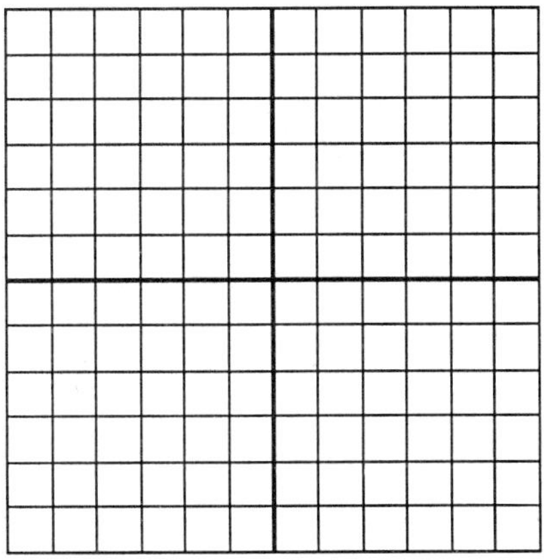

Answer: easy

(a),(b) $f(x) = 6 - x$

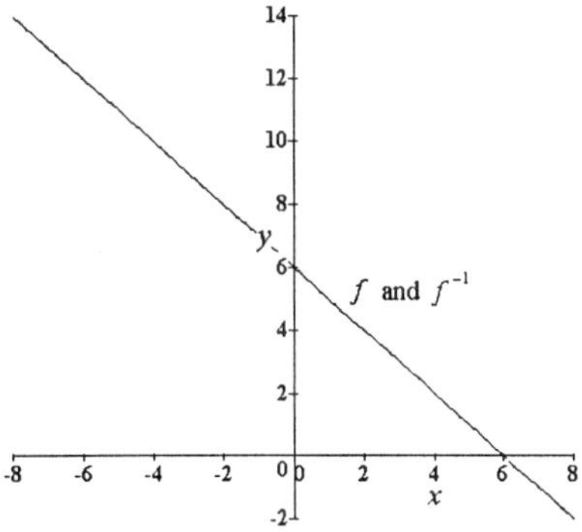

(c) $f(x) = 6 - x$. $y = 6 - x \quad \Leftrightarrow \quad x = 6 - y$. So $f^{-1}(x) = 6 - x$.

15. For the function $f(x) = 9 - x^2$, $0 \le x \le 3$:
    (a) sketch the graph of f.
    (b) use the graph of f to sketch the graph of f^{-1}.
    (c) find an equation for f^{-1}.

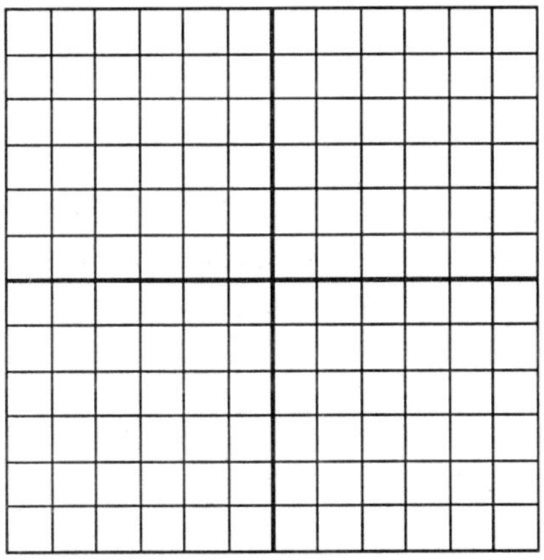

Answer: medium

(a), (b) $f(x) = 9 - x^2$, $0 \le x \le 3$

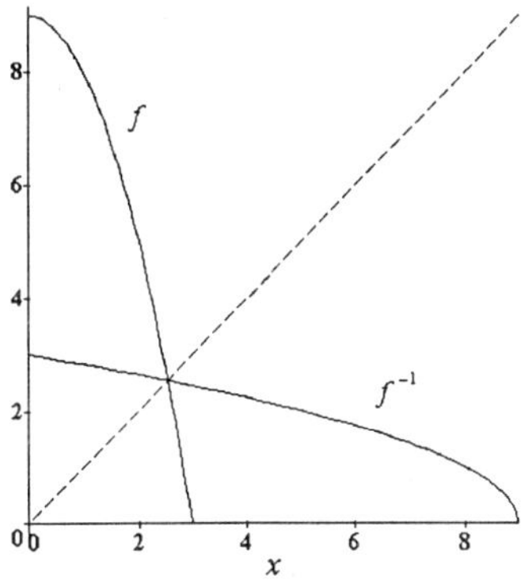

(c) $f(x) = 9 - x^2$, $0 \le x \le 3$. $y = 9 - x^2 \quad \Leftrightarrow \quad x^2 = 9 - y \quad \Leftrightarrow \quad x = \sqrt{9 - y}$. So the inverse function is $f^{-1}(x) = \sqrt{9 - x}$.

GC 16. Draw a graph of $f(x) = x^5 + x$ and use it to determine whether the function is one-to-one.

Answer: easy

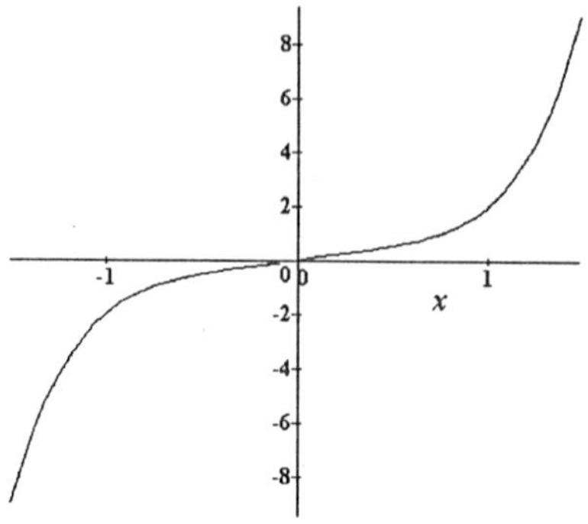

From the graph, we see that $f(x) = x^5 + x$ is one-to-one.

GC 17. Draw a graph of $f(x) = x^5 - x^3$ and use it to determine whether the function is one-to-one.

Answer: easy

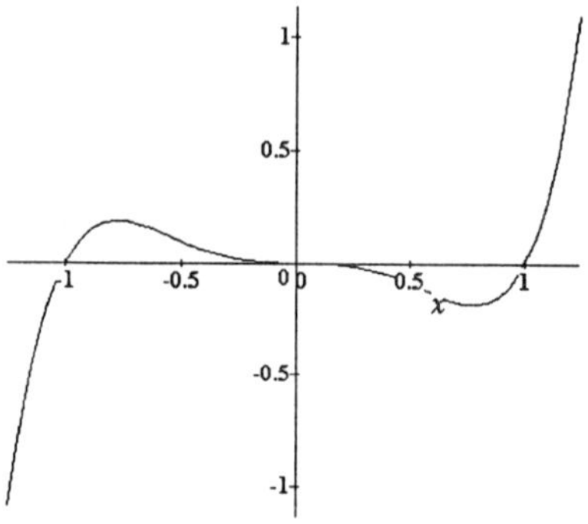

From the graph, we see that $f(x) = x^5 - x^3$ is not one-to-one.

18. Show that it is possible to restrict the domain of the function $f(x) = x^2 + 2x + 2$ in such a way that the resulting function is one-to-one.

Answer: hard

$f(x) = x^2 + 2x + 2 = (x+1)^2 + 1 \Rightarrow$ the vertex is at $(-1, 1)$. Hence it is possible to restrict the domain of f to $(-\infty, -1]$ or $[-1, \infty)$ to make f one-to-one.

19. Under what condition on the constants a, b, c, and d is the function $f(x) = \dfrac{ax+b}{cx+d}$ one-to-one?

Answer: hard

$f(x) = \dfrac{ax+b}{cx+d}$ is one-to-one if $f(x_1) = f(x_2) \Rightarrow x_1 = x_2$. Now $f(x_1) = f(x_2) \Leftrightarrow \dfrac{ax_1+b}{cx_1+d} = \dfrac{ax_2+b}{cx_2+d} \Leftrightarrow$
$(ax_1+b)(cx_2+d) = (ax_2+b)(cx_1+d) \Leftrightarrow acx_1x_2 + adx_1 + bcx_2 + bd = acx_1x_2 + adx_2 + bcx_1 + bd \Leftrightarrow$
$adx_1 + bcx_2 = adx_2 + bcx_1 \Leftrightarrow (ad-bc)x_1 = (ad-bc)x_2 \Rightarrow x_1 = x_2$ if $ad - bc \neq 0$.
So f is one-to-one $\Leftrightarrow ad \neq bc$.

20. Determine whether or not the function $h(x) = x^3 - x$ is one-to-one.

Answer: medium

$h(x) = x^3 - x = x(x-1)(x+1)$. Since $h(-1) = h(0) = h(1) = 0$, h is not one-to-one. (Or sketch the graph to see that the graph of the function fails the Horizontal Line Test.)

21. Determine whether or not the function $g(x) = 1 - x^2$, $x \geq 0$ is one-to-one.
Answer: medium
$g(x) = 1 - x^2$, $x \geq 0$. If $x_1 \neq x_2$, then $x_1^2 \neq x_2^2$. Two different positive numbers cannot have the same square. Thus $1 - (x_1)^2 \neq 1 - (x_2)^2$. Therefore y is one-to-one. (You can also use the Horizontal Line Test to show that no horizontal line intersects the graph of $g(x) = 1 - x^2$, $x \geq 0$ more than once.)

22. Find the inverse function of $f(x) = \frac{1}{2}x + 1$ and then verify that f^{-1} and f are inverses of each other by using the property of inverse functions to show that $f^{-1}(f(x)) = x$ for every x in the domain of f and $f(f^{-1}(x)) = x$ for every x in the domain of f^{-1}.
Answer: medium
$f(x) = \frac{1}{2}x + 1 \quad \Rightarrow \quad y = \frac{1}{2}x + 1 \quad \Leftrightarrow \quad x = 2y - 2$. So $f^{-1}(x) = 2x - 2$.
$f^{-1}(f(x)) = f^{-1}\left(\frac{1}{2}x + 1\right) = 2\left(\frac{1}{2}x + 1\right) - 2 = x$.
$f(f^{-1}(x)) = f(2x - 2) = \frac{1}{2}(2x - 2) + 1 = x$.

23. Find the inverse function of $f(x) = \dfrac{3x - 5}{5}$ and then verify that f^{-1} and f are inverses of each other by using the property of inverse functions to show that $f^{-1}(f(x)) = x$ for every x in the domain of f and $f(f^{-1}(x)) = x$ for every x in the domain of f^{-1}.
Answer: medium
$f(x) = \dfrac{3x - 5}{5} \quad \Rightarrow \quad y = \dfrac{3x - 5}{5} \quad \Leftrightarrow \quad x = \frac{5}{3}y + \frac{5}{3}$. So $f^{-1}(x) = \frac{5}{3}x + \frac{5}{3}$.
$$f^{-1}(f(x)) = f^{-1}\left(\frac{3x - 5}{5}\right) = \frac{5}{3}\left(\frac{3x - 5}{5}\right) + \frac{5}{3} = x.$$
$$f(f^{-1}(x)) = f\left(\tfrac{5}{3}x + \tfrac{5}{3}\right) = \frac{3\left(\frac{5}{3}x + \frac{5}{3}\right) - 5}{5} = x.$$

24. Find the inverse function of $f(x) = 2 - \dfrac{1}{x}$, $x \neq 0$ and then verify that f^{-1} and f are inverses of each other by using the property of inverse functions to show that $f^{-1}(f(x)) = x$ for every x in the domain of f and $f(f^{-1}(x)) = x$ for every x in the domain of f^{-1}.
Answer: medium
$f(x) = 2 - \dfrac{1}{x} \quad \Rightarrow \quad y = 2 - \dfrac{1}{x} \quad \Leftrightarrow \quad x = -\dfrac{1}{y - 2}$. So $f^{-1}(x) = -\dfrac{1}{x - 2}$.
$$f^{-1}(f(x)) = f^{-1}\left(2 - \frac{1}{x}\right) = -\frac{1}{\left(2 - \frac{1}{x}\right) - 2} = x.$$
$$f(f^{-1}(x)) = f\left(-\frac{1}{x - 2}\right) = 2 - \frac{1}{-\dfrac{1}{x - 2}} = x.$$

25. Find the inverse function of $f(x) = \sqrt{25 - x^2}$, $0 \leq x \leq 5$ and then verify that f^{-1} and f are inverses of each other by using the property of inverse functions to show that $f^{-1}(f(x)) = x$ for every x in the domain of f and $f(f^{-1}(x)) = x$ for every x in the domain of f^{-1}.
Answer: medium
$f(x) = \sqrt{25 - x^2} \quad \Rightarrow \quad y = \sqrt{25 - x^2} \quad \Leftrightarrow \quad x = \pm\sqrt{25 - y^2}$. But $0 \leq x \leq 5$, so $x = \sqrt{25 - y^2} \quad \Rightarrow$ $f^{-1}(x) = \sqrt{25 - x^2}$.
$$f^{-1}(f(x)) = f^{-1}\left(\sqrt{25 - x^2}\right) = \sqrt{25 - \left(\sqrt{25 - x^2}\right)^2} = x.$$
$$f(f^{-1}(x)) = f\left(\sqrt{25 - x^2}\right) = \sqrt{25 - \left(\sqrt{25 - x^2}\right)^2} = x.$$

26. Find the inverse function of $f(x) = x^2 - 4x + 9$, $x \geq 2$ and then verify that f^{-1} and f are inverses of each other by using the property of inverse functions to show that $f^{-1}(f(x)) = x$ for every x in the domain of f and $f(f^{-1}(x)) = x$ for every x in the domain of f^{-1}.
Answer: medium
$f(x) = x^2 - 4x + 9 \;\Rightarrow\; y = x^2 - 4x + 9$. Compete the square and write in standard form $\Rightarrow\; y = (x-2)^2 + 5$. Solve for x: $x - 2 = \pm\sqrt{y-5}$. But $x - 2 \geq 0$, so $x = 2 + \sqrt{y-5} \;\Rightarrow\; f^{-1}(x) = 2 + \sqrt{x-5}$.
$f^{-1}(f(x)) = f^{-1}((x-2)^2 + 5) = 2 + \sqrt{[(x-2)^2 + 5] - 5} = x$.
$f(f^{-1}(x)) = f(2 + \sqrt{x-5}) = [(2 + \sqrt{x-5}) - 2]^2 + 5 = x$.

GC 27. Determine whether or not the function $f(x) = ||x-2| + |x-6||$ is one-to-one.
Answer: medium
Using a graphing calculator and the Horizontal Line Test, we see that $f(x) = ||x-2| + |x-6||$ is not one-to-one.

GC 28. Determine whether or not the function $f(x) = |x^5|(x-1)$ is one-to-one.
Answer: medium
Using a graphing calculator with the appropriate viewing rectangle (for example, $[-1, 1]$ by $[-0.03, 0.03]$) and the Horizontal Line Test, we see that $f(x) = |x^5|(x-1)$ is not one-to-one.

CHAPTER 4 REVIEW

1. Find the domain of the function $f(x) = 2 + \sqrt{x+3}$.
(a) $D = (-\infty, 3]$ (b) $D = (-\infty, 3)$ (c) $D = [-3, \infty)$ (d) $D = (-3, \infty)$ (e) $D = (3, \infty)$
Answer: easy (c)
$f(x) = 2 + \sqrt{x+3}$. Then $\sqrt{x+3} \geq 0 \;\Leftrightarrow\; x + 3 \geq 0 \;\Leftrightarrow\; x \geq -3$. So the domain of f is $[-3, \infty)$

2. Find the domain of the function $f(x) = \dfrac{2x-1}{2x+1}$.
(a) $D = \mathbb{R}$ (b) $D = \{x \mid x \neq \frac{1}{2}\}$ (c) $D = (-\infty, \frac{1}{2})$ (d) $D = (\frac{1}{2}, \infty)$ (e) $D = \{x \mid x \neq -\frac{1}{2}\}$
Answer: easy (e)
$f(x) = \dfrac{2x-1}{2x+1}$. Then $2x + 1 \neq 0 \;\Leftrightarrow\; x \neq -\frac{1}{2}$. So the domain of f is $D = \{x \mid x \neq -\frac{1}{2}\}$.

3. Find the domain of the function $f(x) = \dfrac{\sqrt[3]{x+2}}{\sqrt[3]{x}+1}$.
(a) $\{x \mid x \neq -1\}$ (b) $\{x \mid x \neq -1 \text{ or } 1\}$ (c) $(-\infty, 1)$ (d) $(1, \infty)$ (e) $(-1, \infty)$
Answer: medium (a)
$f(x) = \dfrac{\sqrt[3]{x+2}}{\sqrt[3]{x}+1}$. Then $\sqrt[3]{x} + 1 \neq 0 \;\Leftrightarrow\; x \neq -1$. So the domain of f is $\{x \mid x \neq -1\}$.

4. Find the domain of the function $h(x) = \sqrt{6-x} + \sqrt{x^2-4}$.
(a) $(-\infty, 6]$ (b) $(-\infty, -6] \cup [0, 6]$ (c) $(-\infty, 0] \cup [2, 6]$
(d) $(-\infty, -2] \cup [2, 6]$ (e) $[2, 6]$
Answer: medium (d)
$h(x) = \sqrt{6-x} + \sqrt{x^2-4}$. Then $6 - x \geq 0$ and $x^2 - 4 \geq 0 \;\Leftrightarrow\; x \leq 6$ and $x^2 \geq 4 \;\Leftrightarrow\; x \leq 6$ and $x \geq 2$ or $x \leq -2$.
So the domain of h is $(-\infty, -2] \cup [2, 6]$.

5. Sketch the graph of the function $f(x) = \frac{1}{4}(x-3)$, $1 \le x \le 5$.

Answer: easy

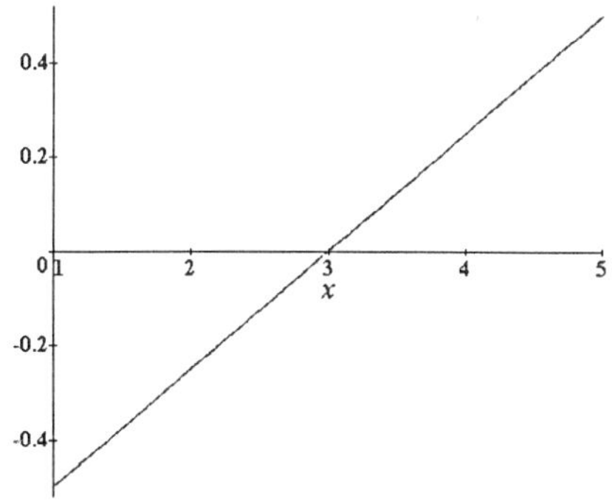

6. Sketch the graph of the function $G(x) = \dfrac{1}{(x-1)^2}$.

Answer: medium

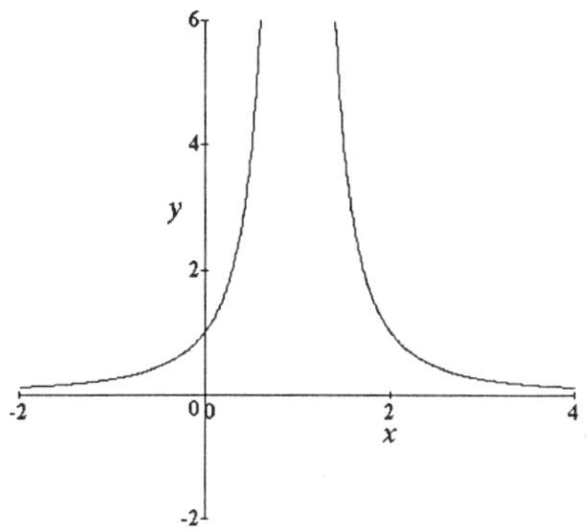

7. Sketch the graph of the function $H(x) = x^3 - 2x + 3$.

Answer: medium

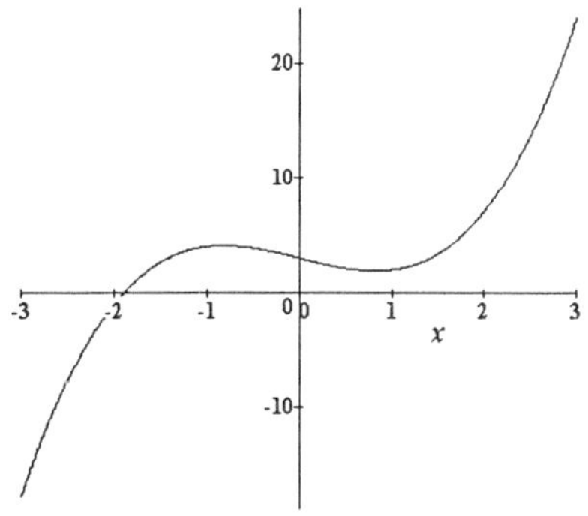

8. Sketch the graph of the function $g(t) = 2t^2 - t$.

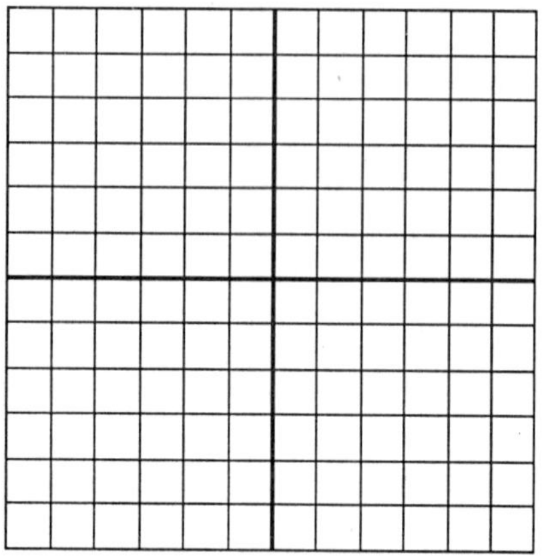

Answer: medium

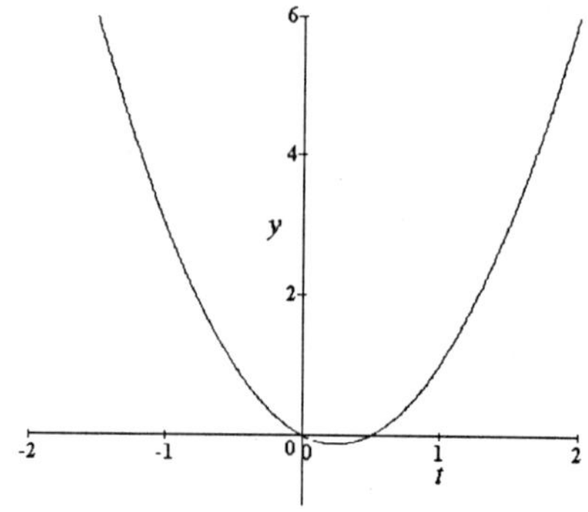

9. Sketch the graph of the function $f(x) = 2 - 6x - x^2$.

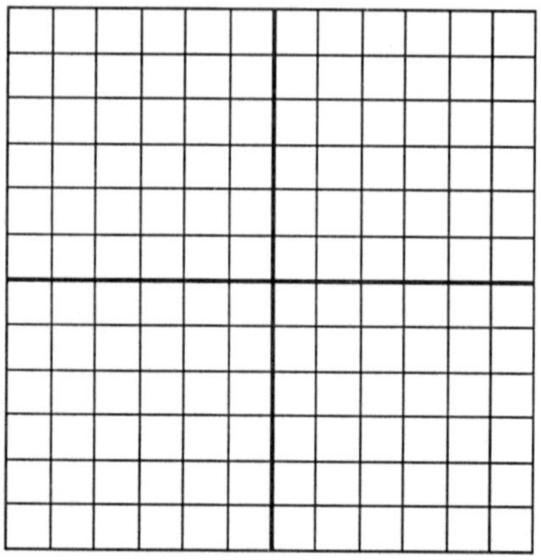

Answer: medium

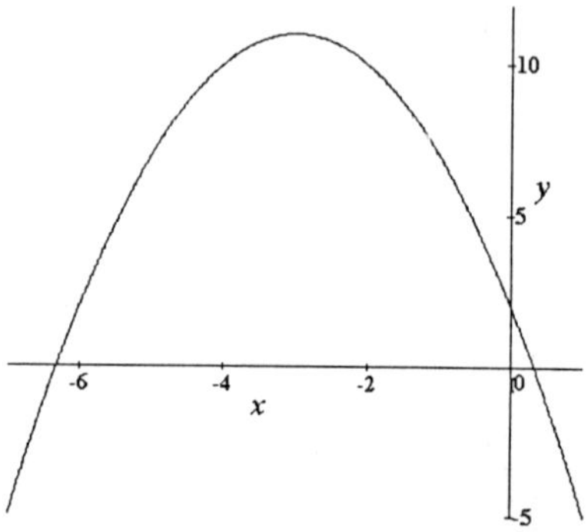

10. Sketch the graph of the function $y = -|x + 2|$.

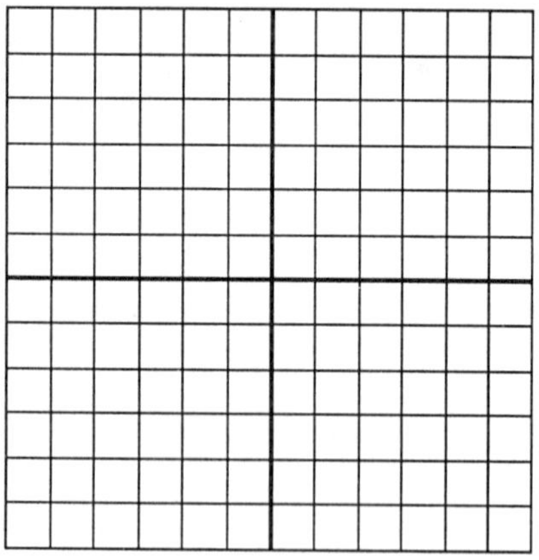

Answer: medium

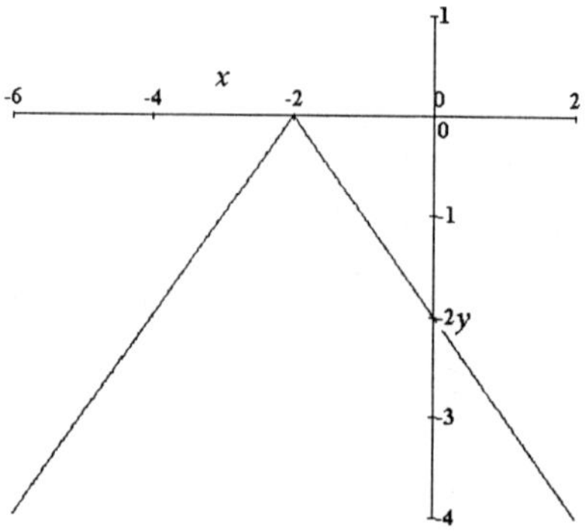

11. Sketch the graph of the function $y = 1 + \sqrt{x+2}$.

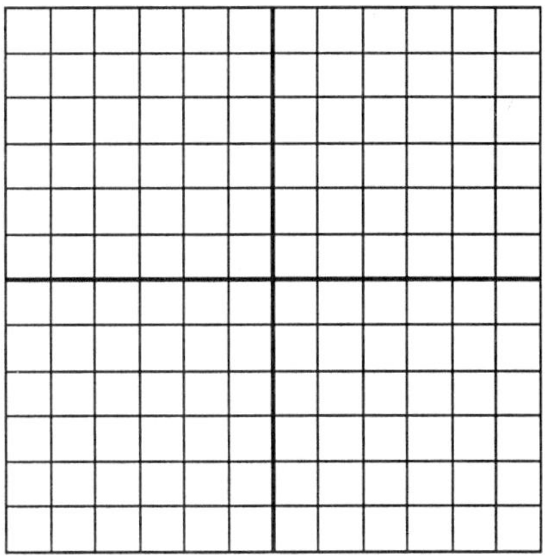

Answer: medium

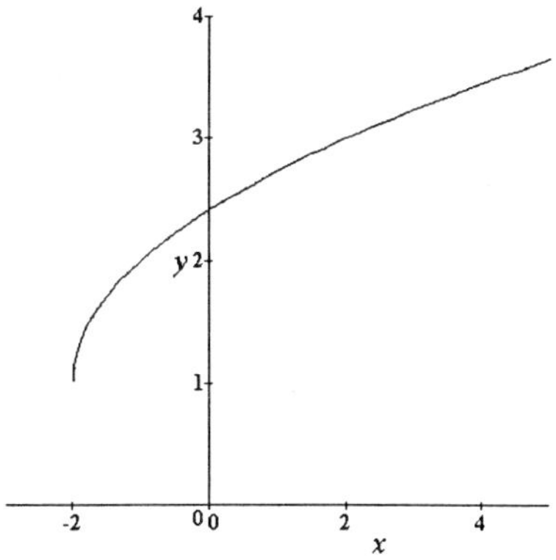

12. Sketch the graph of the function $f(x) = \begin{cases} 2 - x & \text{if } x \leq 0 \\ x - 2 & \text{if } x > 0 \end{cases}$

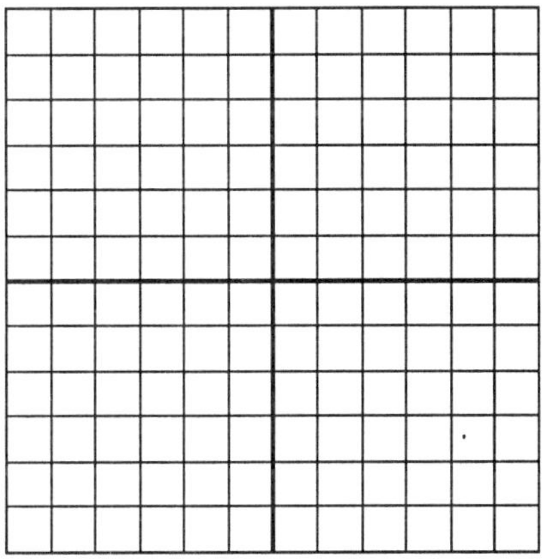

Answer: medium

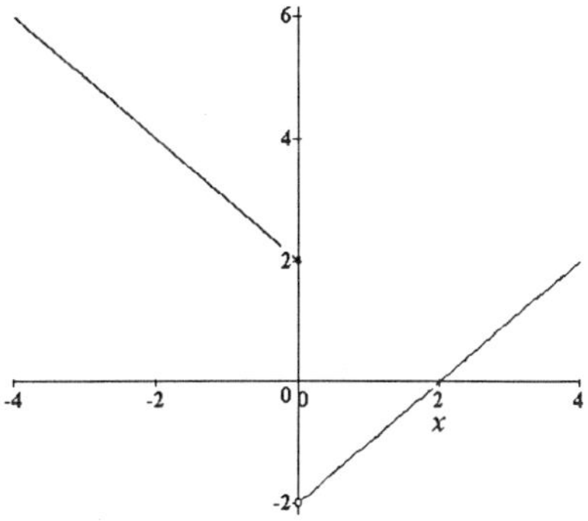

13. Sketch the graph of the function $f(x) = \begin{cases} x & \text{if } x < 0 \\ 3x - x^2 & \text{if } 0 \leq x < 2 \\ 2 & \text{if } x \geq 2 \end{cases}$

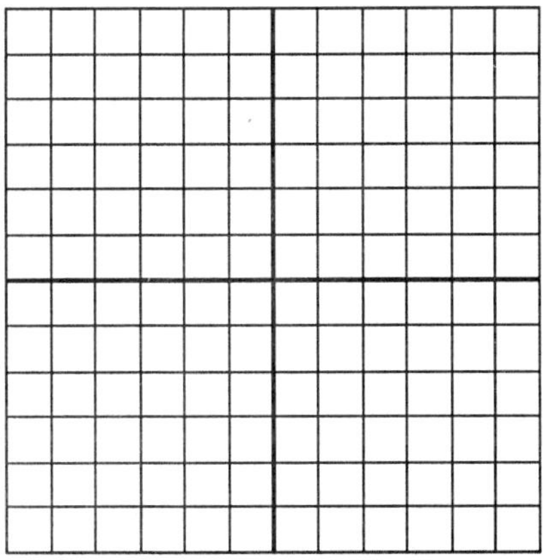

Answer: medium

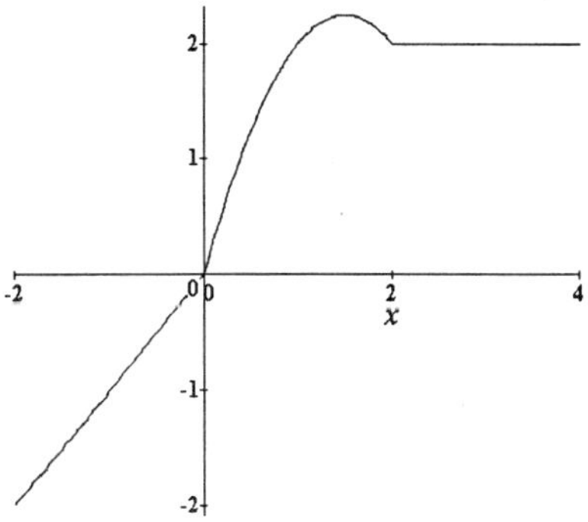

GC 14. Decide which of the following viewing rectangles produces the most appropriate graph of the function $f(x) = \sqrt{120 - x^3}$.

(a) $[-2, 4]$ by $[-2, 4]$ (b) $[-10, 12]$ by $[-12, 20]$ (c) $[-8, 6]$ by $[0, 25]$
(d) $[-120, 100]$ by $[-20, 50]$ (e) $[-60, 300]$ by $[-4, 4]$

Answer: easy (c)

$[-8, 6]$ by $[0, 25]$

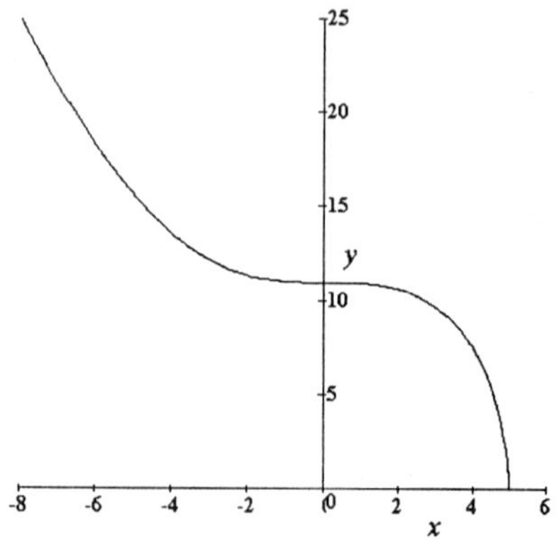

GC 15. Draw the graph of the function $f(x) = 1.2x^3 - 9.4x^2 - 2x + 3.2$ in an appropriate viewing rectangle.

Answer: medium

$[-3, 9]$ by $[-100, 100]$

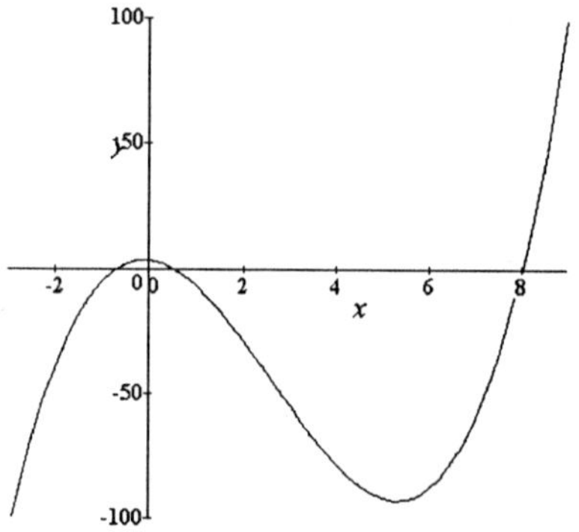

GC 16. Draw the graph of the function $y = |x(x + 2)(x + 4)|$ in an appropriate viewing rectangle.

Answer: medium

$[-5, 1]$ by $[0, 10]$

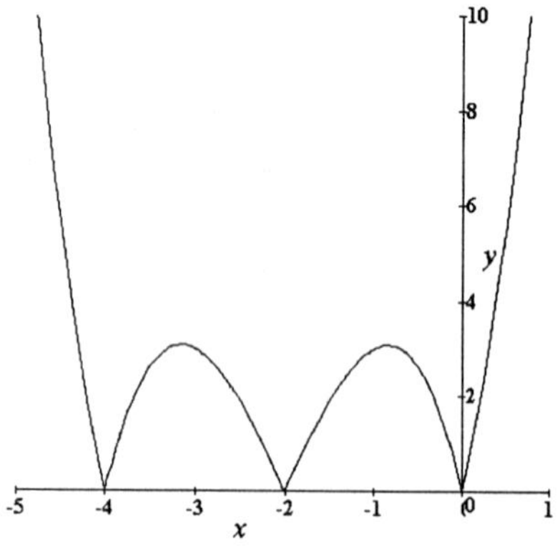

GC 17. Use a graphing device to solve the equation $x^3 - x + 4 = 0$ correct to two decimals.

Answer: easy

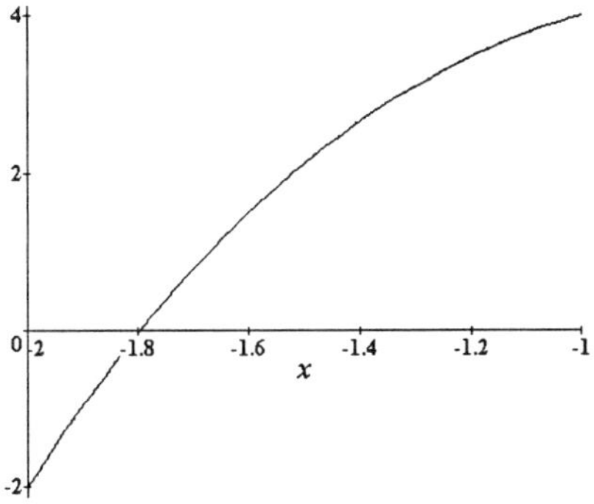

From the graph we see that the only solution of $x^3 - x + 4 = 0$ is $x \approx -1.80$.

GC 18. Use a graphing device to solve the inequality $3 - x^2 \geq 2x - \sqrt{x^2 + 1}$ correct to two decimals.

Answer: medium

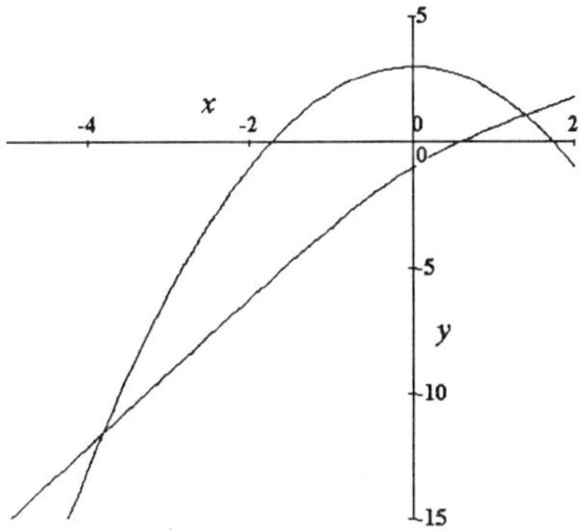

From the graphs we see that $3 - x^2 \geq 2x - \sqrt{x^2 + 1}$ holds for approximately $x \in (-3.82, 1.39)$.

GC 19. Find the range of the function $f(x) = x^4 - 2x^2 + 13x - 2$ correct to one decimal.

Answer: easy

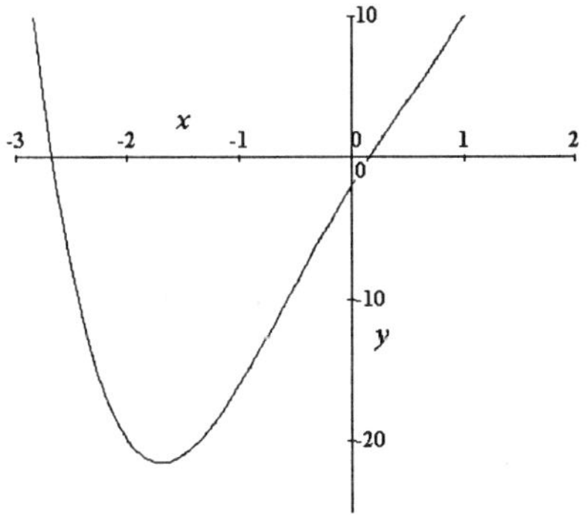

From the graph we see that the range of $f(x) = x^4 - 2x^2 + 13x - 2$ is approximately $[-21.5, \infty)$.

20. Determine whether $f(x) = x^{-3}$ is even, odd, or neither even nor odd.

Answer: medium

$f(x) = x^{-3}$. $f(-x) = (-x)^{-3} = -x^{-3}$. $-f(x) = -x^{-3}$. Since $f(x) \neq f(-x)$, f is not even, and since $f(-x) = -f(x)$, f is odd.

21. Express the quadratic function $f(x) = -x^2 + 4x - 12$ in standard form.

(a) $-(x-2)^2 - 8$ (b) $-(x+2)^2 - 14$ (c) $(x+2)^2 + 12$ (d) $-(x-4)^2 - 2$ (e) $(x-4)^2 - 24$

Answer: medium (a)

$f(x) = -x^2 + 4x - 12 = -(x^2 - 4x + 4) - 8 = -(x-2)^2 - 8$

22. Find the minimum value of the function $g(x) = 2x^2 + 3x - 5$.

Answer: medium

$g(x) = 2x^2 + 3x - 5 = 2\left(x^2 + \frac{3}{2}x + \frac{9}{16}\right) - 5 - \frac{9}{8} = 2\left(x + \frac{3}{4}\right)^2 - \frac{49}{8}$. Therefore the minimum value of the function is $g\left(-\frac{3}{4}\right) = -\frac{49}{8}$.

23. Find the local maximum and minimum values of the function $f(x) = x^{1/3}(6-x)^{2/3}$ and the values of x at which they occur. State your answers to two decimal places.

Answer: medium

$f(x) = x^{1/3}(6-x)^{2/3}$

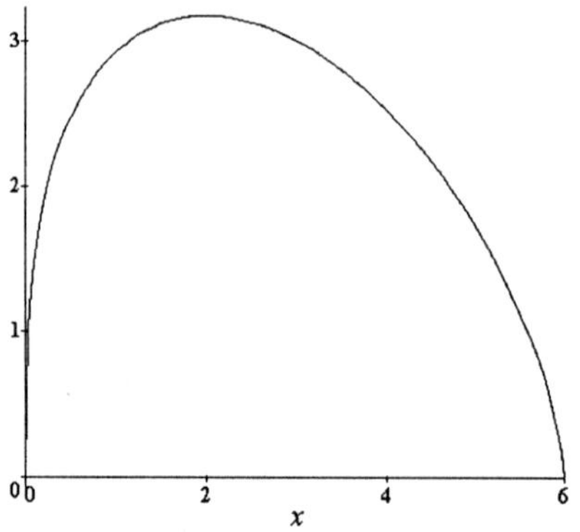

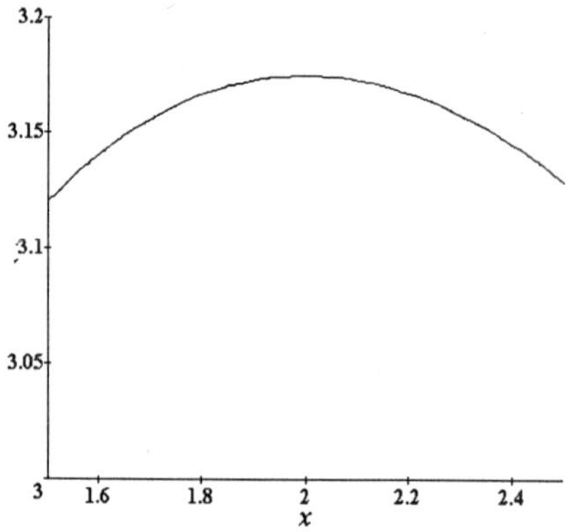

From the graphs, f has a local maximum of about 3.18 when $x \approx 2.00$ and local minima of 0 at $x = 0$ and at $x = 6$.

24. If $f(x) = x^2 - 2x + 1$ and $g(x) = 3 - 2x$, find:

(a) $f + g$ (b) $f - g$ (c) fg (d) f/g (e) $f \circ g$ (f) $g \circ f$

Answer: medium

$f(x) = x^2 - 2x + 1$ and $g(x) = 3 - 2x$.

(a) $(f+g)(x) = x^2 - 2x + 1 + 3 - 2x = x^2 - 4x + 4$

(b) $(f-g)(x) = x^2 - 2x + 1 - (3 - 2x) = x^2 - 2$

(c) $(fg)(x) = (x^2 - 2x + 1)(3 - 2x) = 3x^2 - 6x + 3 - 2x^3 + 4x^2 - 2x = -2x^3 + 7x^2 - 8x + 3$

(d) $(f/g)(x) = \dfrac{x^2 - 2x + 1}{3 - 2x}$, $x \neq \frac{3}{2}$

(e) $(f \circ g)(x) = f(3 - 2x) = (3 - 2x)^2 - 2(3 - 2x) + 1 = 9 - 12x + 4x^2 - 6 + 4x + 1 = 4x^2 - 8x + 4$

(f) $(g \circ f)(x) = g(x^2 - 2x + 1) = 3 - 2(x^2 - 2x + 1) = -2x^2 + 4x + 1$

25. For the functions $f(x) = \sqrt{x}$ and $g(x) = \dfrac{1}{x-3}$, find the functions $f \circ g$, $g \circ f$, $f \circ f$, $g \circ g$, and their domains.

Answer: medium

$f(x) = \sqrt{x}$, $D = \{x \mid x \geq 0\}$. $g(x) = \dfrac{1}{x-3}$, $D = \{x \mid x \neq 3\}$

$(f \circ g)(x) = f\left(\dfrac{1}{x-3}\right) = \sqrt{\dfrac{1}{x-3}}$, so that $\dfrac{1}{x-3} \geq 0 \quad \Leftrightarrow \quad x > 3$, D is $(3, \infty)$

$(g \circ f)(x) = g(\sqrt{x}) = \dfrac{1}{\sqrt{x} - 3}$, so that $\sqrt{x} - 3 \neq 0 \quad \Leftrightarrow \quad x \neq 9$, D is $[0, 9) \cup (9, \infty)$

$(f \circ f)(x) = f(\sqrt{x}) = \sqrt{\sqrt{x}} = x^{1/4}$, D is $[0, \infty)$

$(g \circ g)(x) = g\left(\dfrac{1}{x-3}\right) = \dfrac{1}{\frac{1}{x-3} - 3} = \dfrac{x-3}{1 - 3(x-3)} = \dfrac{x-3}{10 - 3x}$, so that $10 - 3x \neq 0$, $D = \left\{x \mid x \neq \frac{10}{3}, 3\right\}$

26. If $T(x) = \dfrac{1}{\sqrt{3 + \sqrt{2x}}}$, find functions f, g, and h such that $f \circ g \circ h = T$.

Answer: hard

If $h(x) = \sqrt{2x}$ and $g(x) = 3 + x$ then $(g \circ h)(x) = g\left(\sqrt{2x}\right) = 3 + \sqrt{2x}$

If $f(x) = \dfrac{1}{\sqrt{x}}$ then $(f \circ g \circ h)(x) = f(3 + \sqrt{2x}) = \dfrac{1}{\sqrt{3 + \sqrt{2x}}} = T(x)$. (Other solutions are possible.)

27. Determine whether or not the function $g(x) = 8 - 4x + x^2$ is one-to-one.

Answer: medium

$g(x) = 8 - 4x + x^2 = (x^2 - 4x + 4) + 4 = (x - 2)^2 + 4$. Since $g(0) = g(4)$, as is true for all pairs of numbers equidistant from 2, g is not a one-to-one function.

28. Determine whether or not the function $r(x) = 1 + \sqrt{x + 2}$ is one-to-one.

Answer: medium

$r(x) = 1 + \sqrt{x + 2}$. If $x_1 \neq x_2$ then $x_1 + 2 \neq x_2 + 2$, so $\sqrt{x_1 + 2} \neq \sqrt{x_2 + 2}$ and $1 + \sqrt{x_1 + 2} \neq 1 + \sqrt{x_2 + 2}$. Thus r is one-to-one.

GC 29. Determine whether or not the function $q(x) = 2.4 + 1.7x + 1.5x^3$ is one-to-one.

Answer: easy

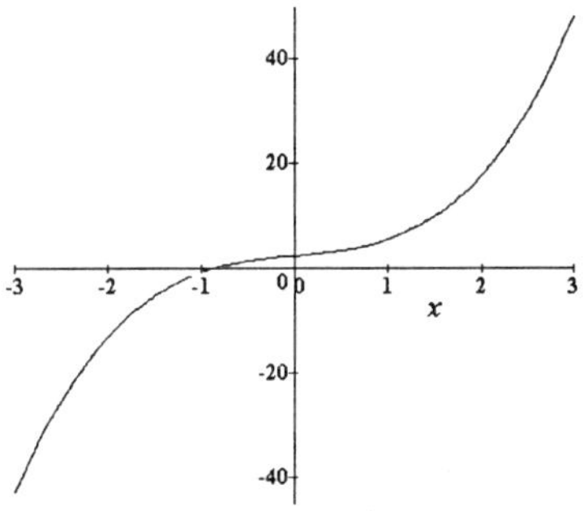

By using a graphing calculator and the horizontal line test, we see that $q(x) = 2.4 + 1.7x + 1.5x^3$ is one-to-one.

30. If $f(x) = 1 + \sqrt[3]{x - 1}$, find f^{-1}.

Answer: medium

$f(x) = 1 + \sqrt[3]{x - 1}$. $y = 1 + \sqrt[3]{x - 1} \quad \Leftrightarrow \quad x - 1 = (y - 1)^3 \quad \Leftrightarrow \quad x = 1 + (y - 1)^3$. So $f^{-1}(x) = 1 + (x - 1)^3$.

31. (a) Show that the function $f(x) = 1 + \sqrt[3]{x}$ is one-to-one.

(b) Sketch the graph of f.

(c) Use part (b) to sketch the graph of f^{-1}.

(d) Find an equation for f^{-1}.

Answer: hard

$f(x) = 1 + \sqrt[3]{x}$

(a) If $x_1 \neq x_2$ then $\sqrt[3]{x_1} \neq \sqrt[3]{x_2}$ ($\sqrt[3]{x}$ is increasing) and so $1 + \sqrt[3]{x_1} \neq 1 + \sqrt[3]{x_2}$. Therefore f is a one-to-one function.

(b), (c)

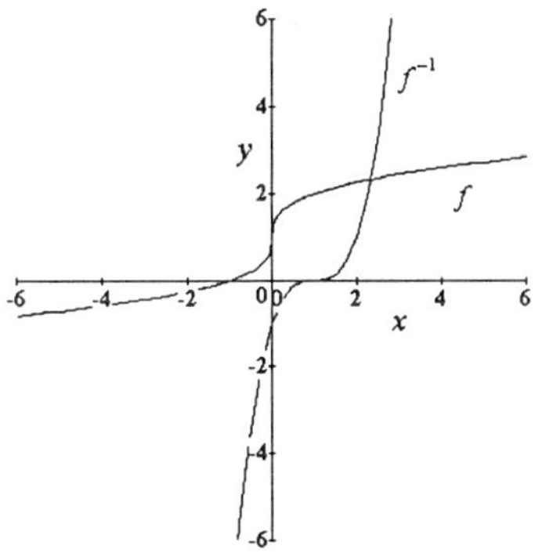

(d) $y = 1 + \sqrt[3]{x} \quad \Leftrightarrow \quad \sqrt[3]{x} = y - 1 \quad \Leftrightarrow \quad x = (y - 1)^3$.

So $f^{-1}(x) = (x - 1)^3$.

32. Suppose that z is inversely proportional to y. If $y = 12$, then $z = 4$. Use this information to write an expression for z.
Answer: medium
$z = k/y \quad \Rightarrow \quad 4 = k/12 \quad \Rightarrow \quad k = 48 \quad \Rightarrow \quad z = 48/y$

33. The number of Mother's Day cards sold by a greeting card store depends on the time of year. Sketch a rough graph of the number of Mother's Day cards sold as a function of the time of year.

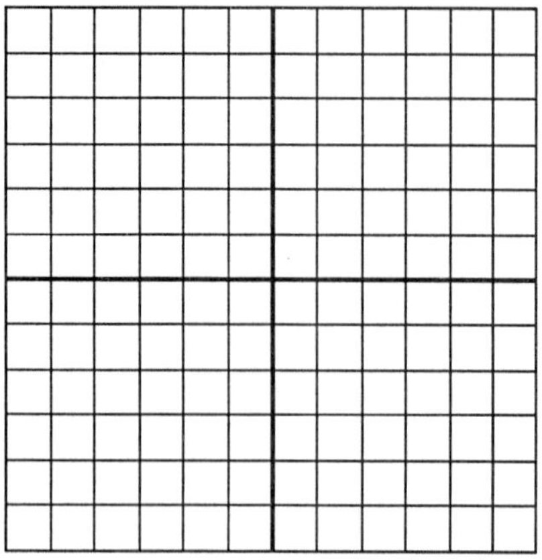

Answer: easy

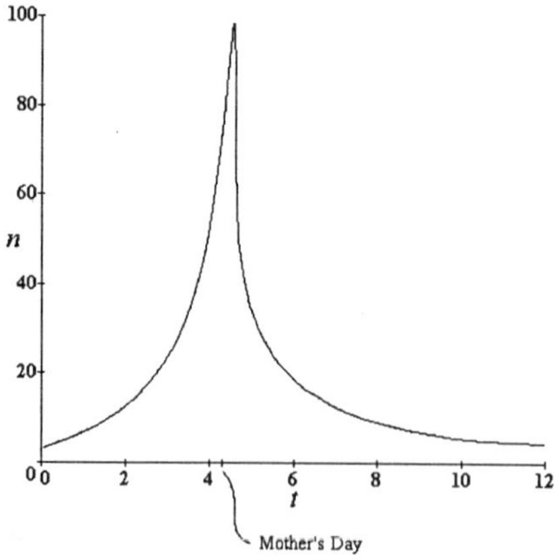

The sales of Mother's Day cards will increase approximately from January to May and then decrease until December.

34. A rectangle is inscribed in an equilateral triangle with a perimeter of 60 cm. One side of the rectangle, of length ℓ, lies along a portion of a side of the triangle.
(a) Express the area A of the rectangle as a function of the width of the rectangle.
(b) Find the dimensions of the rectangle with the largest area.
Answer: hard
Let $\ell = 2x$ be the width of the rectangle in cm and h be the height of the rectangle. If we orient the triangle on the xy-plane with the center of the base at the origin, the coordinates of the vertices are $(-10, 0)$, $(10, 0)$, and $(0, 10\sqrt{3})$ by applying the Pythagorean Theorem. Then the point (x, h) must lie on the line segment joining the points $(10, 0)$ and $(0, 10\sqrt{3})$. This line segment has slope $-\sqrt{3}$ and intercept $10\sqrt{3}$ and hence $h = -\sqrt{3}x + 10\sqrt{3} = \sqrt{3}(10 - x)$.
(a) The area of the rectangle is $A = \ell h = 2xh = 2x \cdot \sqrt{3}(10 - x) = 2\sqrt{3}x(10 - x)$.
(b) $A = -\sqrt{3}x^2 + 10\sqrt{3}x = -\sqrt{3}(x^2 - 10x + 25) + 25\sqrt{3} = -\sqrt{3}(x - 5)^2 + 25\sqrt{3}$. So the area is maximized when $x = 5$ and by substitution $h = \sqrt{3}(10 - 5) = 5\sqrt{3}$. Hence the dimensions of the rectangle with the largest area are 10 cm by $5\sqrt{3}$ cm.

35. A baseball team plays in a stadium that holds 65,000 spectators. With ticket prices at \$10, the average attendance has been 34,000. A market survey indicates that for every dollar that ticket prices are lowered, attendance will increase by 4,000. How should ticket prices be set to maximize revenue?
Answer: hard
Let x represent the lowered price in dollars. Then the change in price is $10 - x$. For the changed price the number of tickets sold is $n = 34{,}000 + 4{,}000(10 - x) = 64{,}000 - 4000x$. Thus the revenue is
$R = x \cdot n = x(64{,}000 - 4000x) = -4000(x^2 - 16x) = -4000[x^2 - 16x + 64] + 4000(64) = -4000(x - 8)^2 + 4000(64)$.
The revenue is maximized when $x = 8$ and so the price should be set at \$8.00.

GC 36. A bird is released from point A on an island 4 mi from the nearest point B on a straight shoreline. It flies to a point C on the shore, and then flies along the shoreline to its nesting area D, 10 miles along the shore from point B. Suppose the bird requires 8 kcal/mi of energy to fly over land and 12 kcal/mi to fly over water. If the bird instinctively chooses a path that minimizes its energy expenditure, to what point does it fly?

Answer: hard

Let x be the distance from point B to C. Hence $AC = \sqrt{x^2 + 16}$ and the energy used in flying from A to C, and then from C to D is $f(x) = 12\sqrt{x^2 + 16} + 8(10 - x)$.

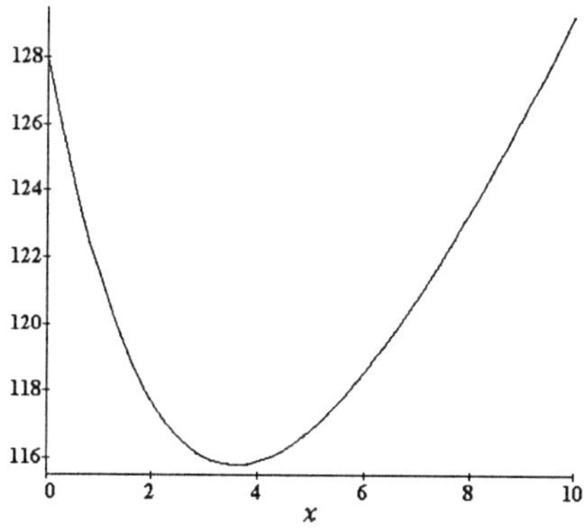

By using a graphing calculator, we see that $BC \approx 3.6$ mi to minimize its energy expenditure.

CHAPTER FIVE

SECTION 5.1

1. Sketch the graph of the function $y = (x+3)^3$ by transforming the graph of an appropriate function of the form $y = x^n$. Indicate all x- and y-intercepts on your graph.

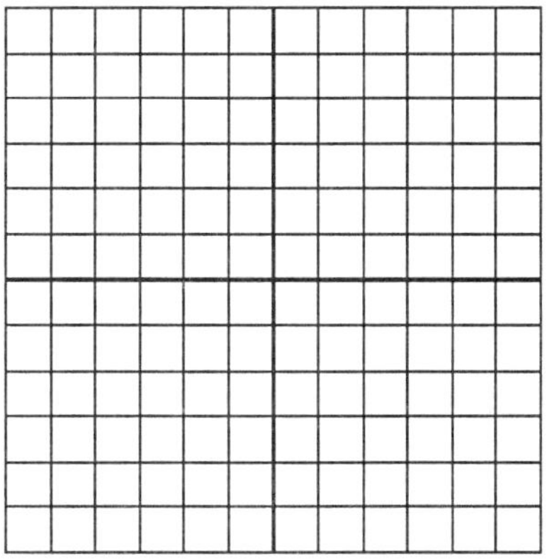

Answer: easy

$y = (x+3)^3$

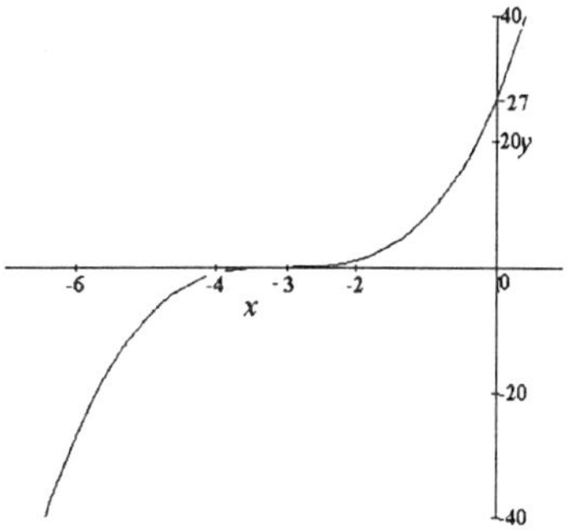

2. Sketch the graph of the function $y = -x^6 + 8$ by transforming the graph of an appropriate function of the form $y = x^n$. Indicate all x- and y-intercepts on your graph.

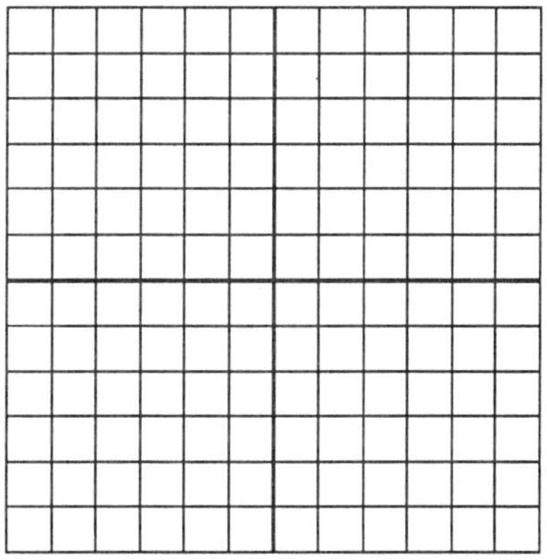

Answer: easy

$y = -x^6 + 8$

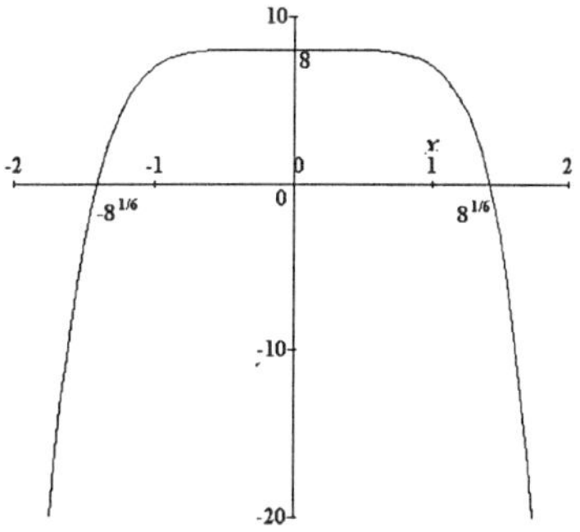

3. Sketch the graph of the function $y = 4x^5 - 7$ by transforming the graph of an appropriate function of the form $y = x^n$. Indicate all x- and y-intercepts on your graph.

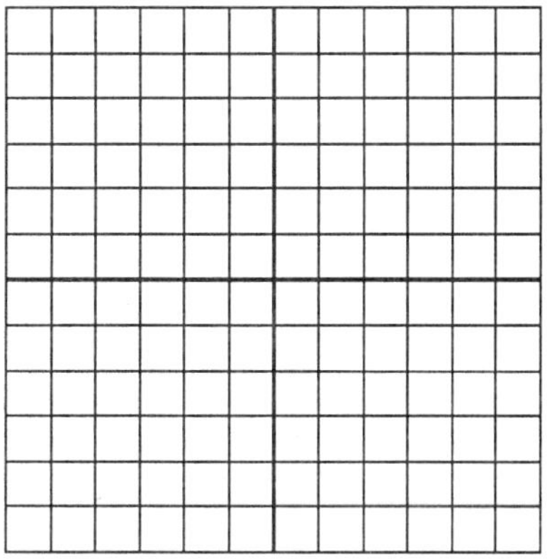

Answer: medium

$y = 4x^5 - 7$

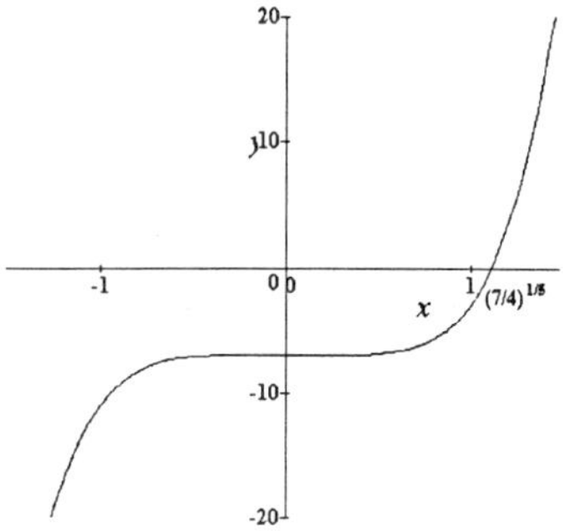

4. Sketch the graph of the function $y = 4x^4 - 64$ by transforming the graph of an appropriate function of the form $y = x^n$. Indicate all x- and y-intercepts on your graph.

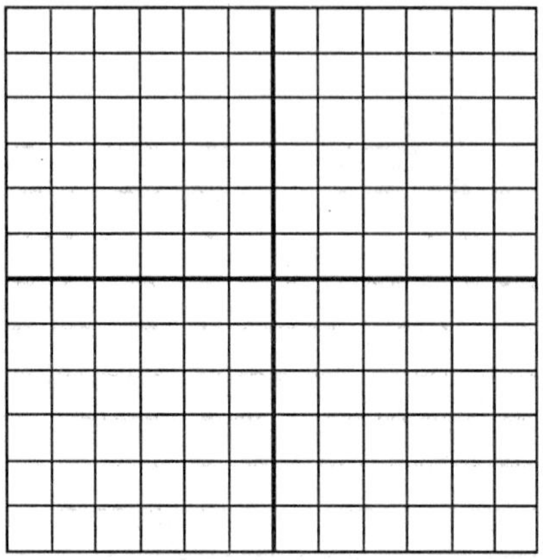

Answer: medium

$y = 4x^4 - 64$

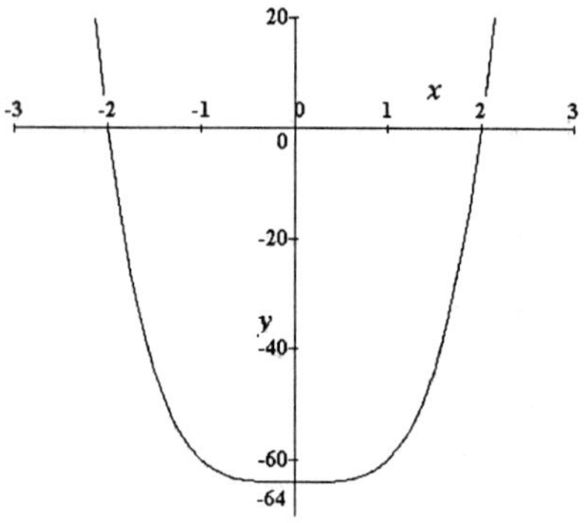

5. Sketch the graph of the function $y = (2x - 3)(x - 5)(x + 1)$ by first plotting all x-intercepts, the y-intercept, and sufficiently many other points to detect the shape of the curve, and then filling in the rest of the graph.

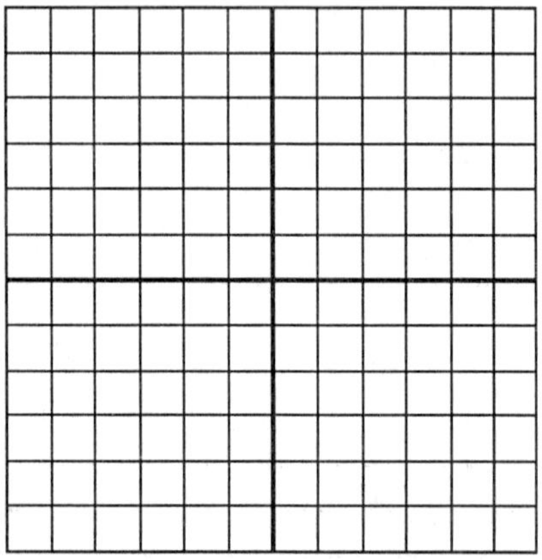

Answer: easy

$y = (2x - 3)(x - 5)(x + 1)$

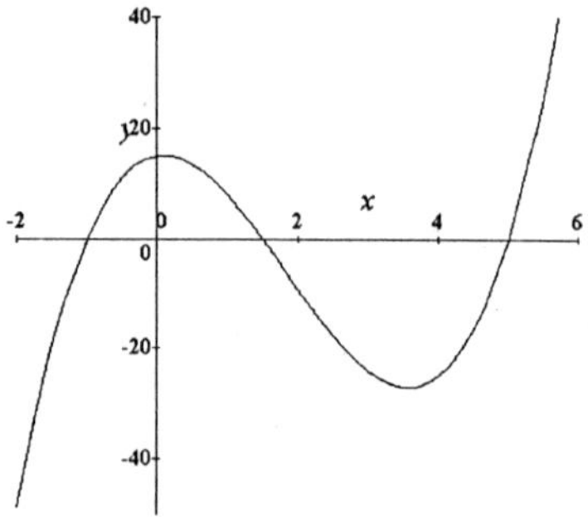

6. Sketch the graph of the function $y = \frac{1}{4}(x + 1)^2(x - 3)^2$ by first plotting all x-intercepts, the y-intercept, and sufficiently many other points to detect the shape of the curve, and then filling in the rest of the graph.

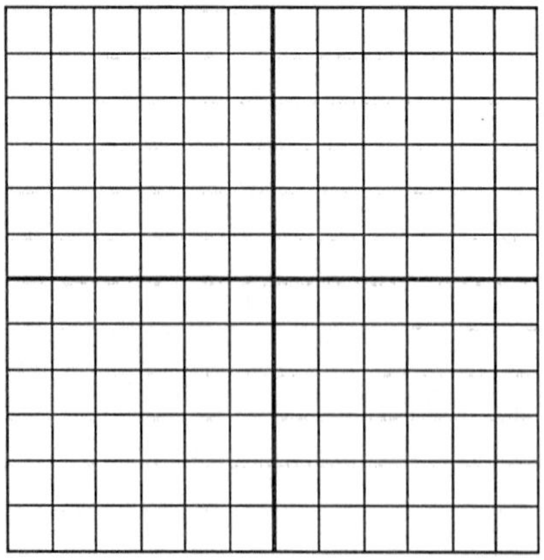

Answer: medium

$y = \frac{1}{4}(x + 1)^2(x - 3)^2$

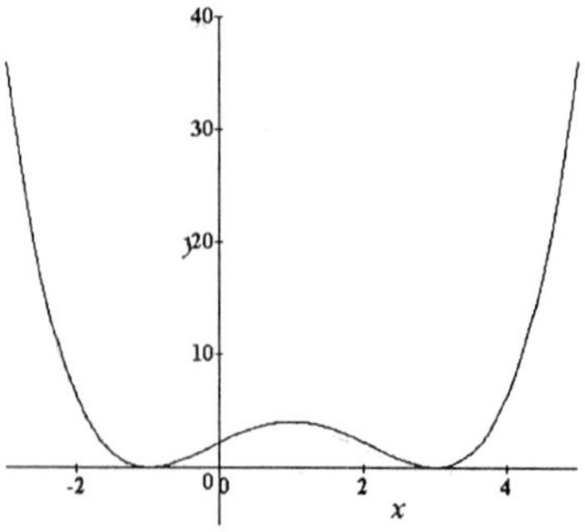

7. Sketch the graph of the function $y = x^3 - 4x^2 - 7x + 10$ by first plotting all x-intercepts, the y-intercept, and sufficiently many other points to detect the shape of the curve, and then filling in the rest of the graph.

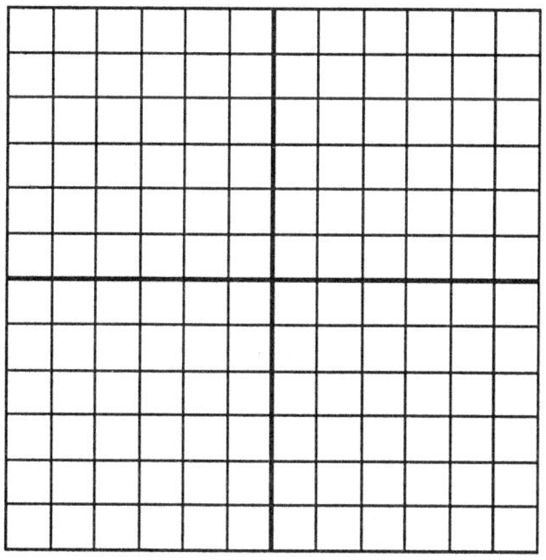

Answer: medium

$y = x^3 - 4x^2 - 7x + 10$

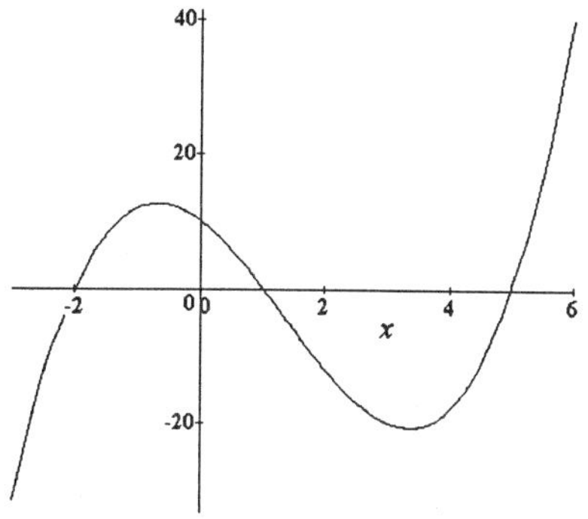

8. Sketch the graph of the function $y = 3x^3 - 14x^2 + 17x - 6$ by first plotting all x-intercepts, the y-intercept, and sufficiently many other points to detect the shape of the curve, and then filling in the rest of the graph.

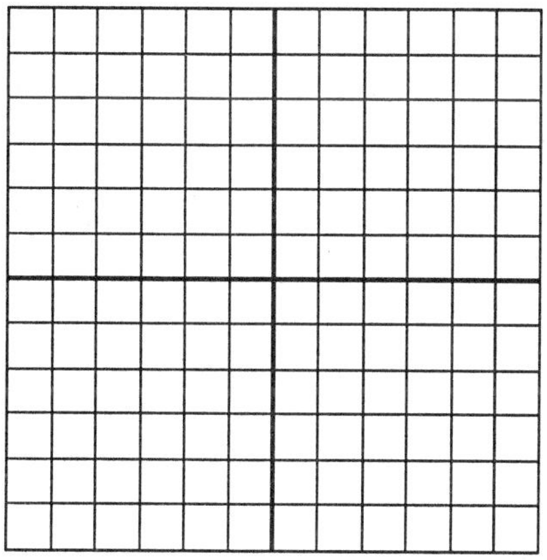

Answer: medium

$$\begin{aligned} y &= 3x^3 - 14x^2 + 17x - 6 \\ &= (3x - 2)(x - 1)(x - 3) \end{aligned}$$

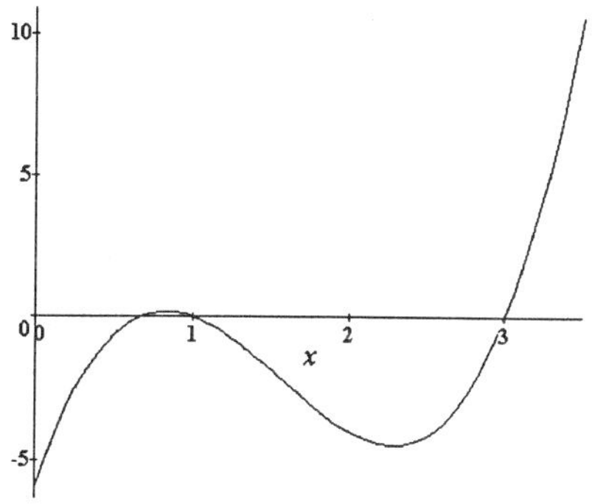

9. Sketch the graph of the function $y = 3x^4 + 4x^3 + x^2$ by first plotting all x-intercepts, the y-intercept, and sufficiently many other points to detect the shape of the curve, and then filling in the rest of the graph.

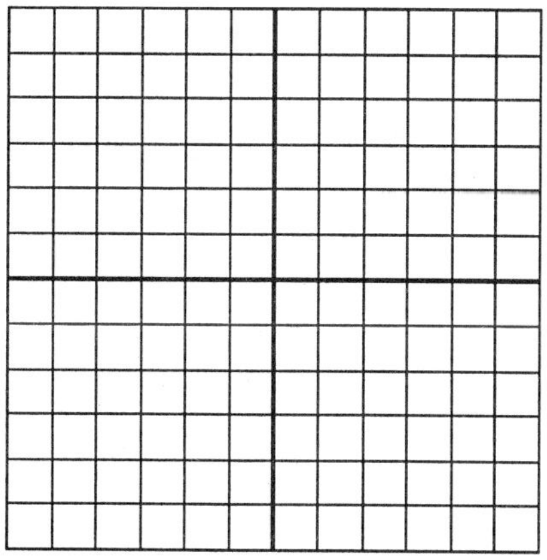

Answer: medium

$y = 3x^4 + 4x^3 + x^2$

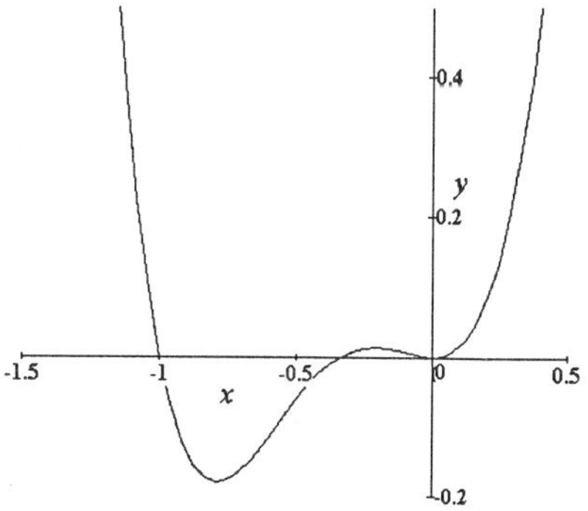

10. Sketch the graph of the function $y = x^4 - 2x^3 - 6x^2 + 2x + 5$ by first plotting all x-intercepts, the y-intercept, and sufficiently many other points to detect the shape of the curve, and then filling in the rest of the graph.

Answer: medium

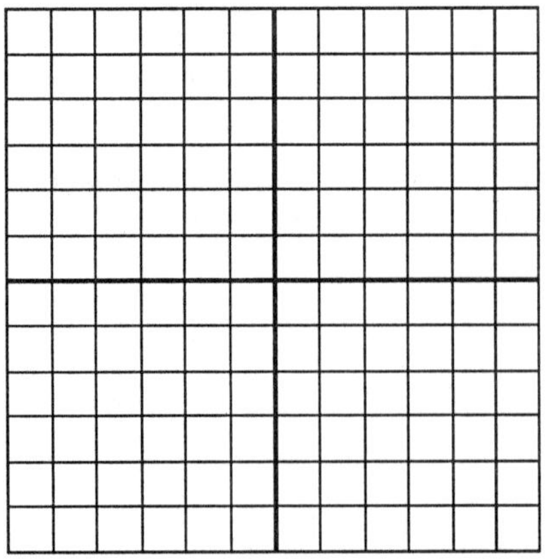

$$y = x^4 - 2x^3 - 6x^2 + 2x + 5$$
$$= \left(x - 1 - \sqrt{6}\right)\left(x - 1 + \sqrt{6}\right)(x - 1)(x + 1)$$

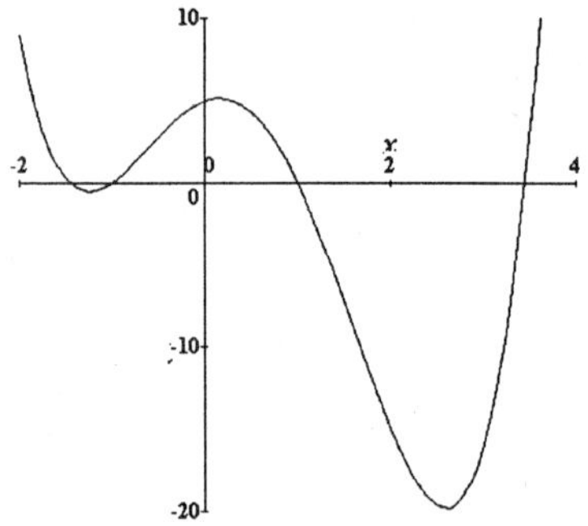

11. Sketch the graph of the function $y = x^4 + 2x^3 - x - 2$ by first plotting all x-intercepts, the y-intercept, and sufficiently many other points to detect the shape of the curve, and then filling in the rest of the graph.

Answer: medium

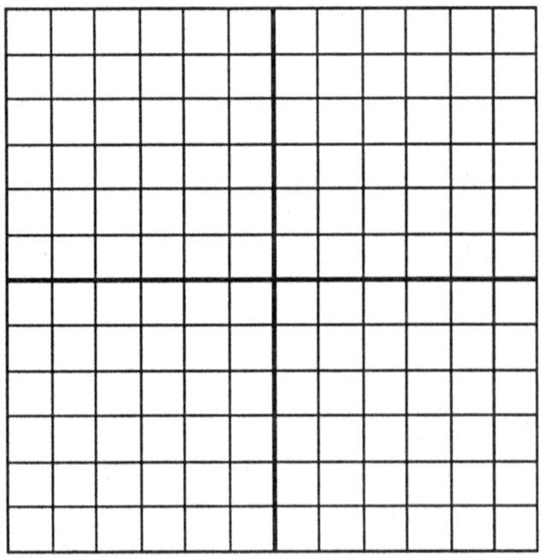

$$y = x^4 + 2x^3 - x - 2 = (x + 2)(x^3 - 1)$$
$$= (x + 2)(x - 1)(x^2 + x + 1)$$

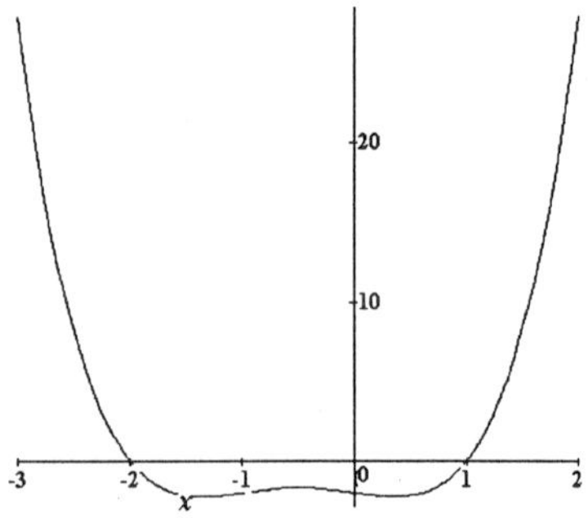

12. Sketch the graph of the function $y = x^4 + x^2 - 2$ by first plotting all x-intercepts, the y-intercept, and sufficiently many other points to detect the shape of the curve, and then filling in the rest of the graph.

Answer: medium

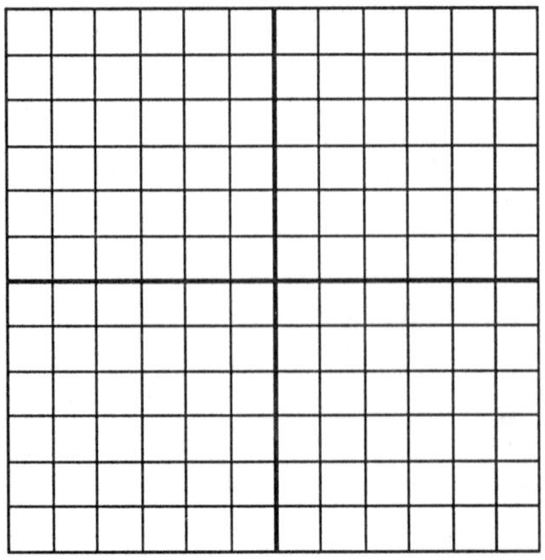

$$y = x^4 + x^2 - 2 = (x - 1)(x + 1)(x^2 + 2)$$

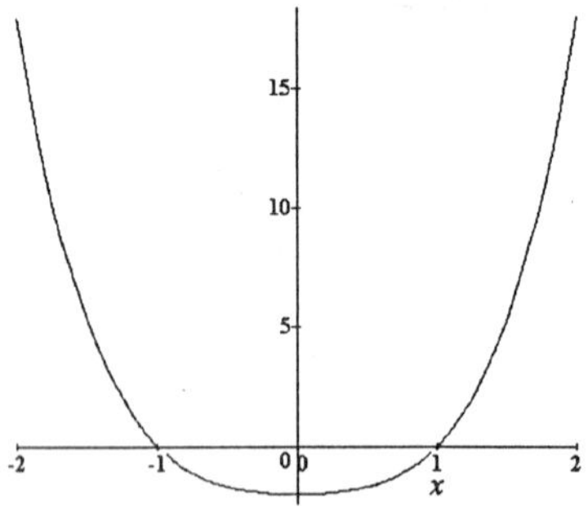

13. Recall that a function f is **odd** if $f(-x) = -f(x)$ and is **even** if $f(-x) = f(x)$, for all real x.
   (a) Show that if f and g are both odd, then so is the function $f + g$.
   (b) Show that if f and g are both even, then so is the function $f + g$.
   (c) Show that if f is odd and g is even, and neither has constant value zero, then the function $f + g$ is neither even nor odd.
   (d) Show that a polynomial $P(x)$ that contains only odd powers of x is an odd function.
   (e) Show that a polynomial $P(x)$ that contains only even powers of x is an even function.
   (f) Show that if polynomial $P(x)$ contains both odd and even powers of x, then it is neither an odd nor an even function.
   (g) Express the function $P(x) = x^5 + 6x^3 - x^2 - 2x + 5$ as the sum of an odd function and an even function.

   Answer: hard

   (a) Since f and g are odd, $f(-x) = -f(x)$ and $g(-x) = -g(x)$. Therefore, $(f+g)(-x) = f(-x) + g(-x) = -f(x) - g(x) = -[f(x) + g(x)] = -(f+g)(x)$, and so the function $f + g$ is also odd.

   (b) Since f and g are even $f(-x) = f(x)$ and $g(-x) = g(x)$. Therefore, $(f+g)(-x) = f(-x) + g(-x) = f(x) + g(x) = (f+g)(x)$, and so the function $f + g$ is also even.

   (c) Since f is odd and g is even, $f(-x) = -f(x)$ and $g(-x) = g(x)$. Therefore, $(f+g)(-x) = f(-x) + g(-x) = -f(x) + g(x) \neq -[f(x) + g(x)]$ (since neither f nor g is constantly 0) and so the function $f + g$ is not odd. Also, $-f(x) + g(x) \neq f(x) + g(x)$ (for the same reason) and so the function $f + g$ is not even. Thus, $f + g$ is neither even nor odd.

   (d) Let $P(x)$ be a polynomial containing only odd powers of x. Then, each term of $P(x)$ can be written as Cx^{2n+1} for some constant C and non-negative integer n. Since $C(-x)^{2n+1} = -Cx^{2n+1}$, each term of $P(x)$ is an odd function. Thus, by part (a), $P(x)$ is an odd function.

   (e) Let $P(x)$ be a polynomial containing only even powers of x. Then each term of $P(x)$ can be written as Cx^{2n} for some constant C and non-negative integer n. Since $C(-x)^{2n} = Cx^{2n}$, each term of $P(x)$ is an even function. Thus, by part (b), $P(x)$ is an even function.

   (f) Since $P(x)$ contains both even and odd powers of x, we can write it in the form $P(x) = R(x) + Q(x)$, where $R(x)$ contains all the even-powered terms in P and $Q(x)$ contains all the odd. By part (d), Q is an odd function, and by part (e), R is an even function. Thus, since neither Q nor R are constantly 0 (by assumption), $P = Q + R$ is neither even nor odd, by part (c).

   (g) $P(x) = x^5 + 6x^3 - x^2 - 2x + 5 = (x^5 + 6x^3 - 2x) + (-x^2 + 5) = Q(x) + R(x)$ where $Q(x) = x^5 + 6x^3 - 2x$ and $R(x) = -x^2 + 5$. Since $Q(x)$ contains only odd powers of x, it is an odd function, and since $R(x)$ contains only even powers of x, it is an even function.

GC 14. Graph the polynomial $y = x^3 - 4x^2 - 7x - 12$ in the viewing rectangle $[-3, 6]$ by $[-50, 20]$. Find the x-intercept(s), if any, the y-intercept, and the coordinates of all local extrema correct to two decimals.

   Answer: easy

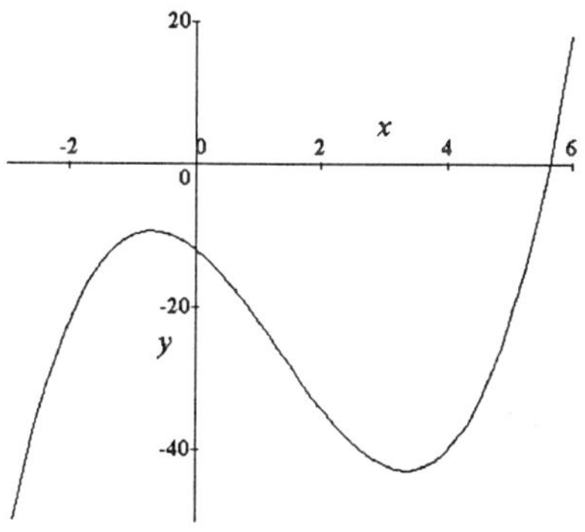

$y = x^3 - 4x^2 - 7x - 12$, $[-3, 6]$ by $[-50, 20]$
y-intercept $= -12$, x-intercept ≈ 5.62,
local maximum is about $(-0.69, -9.40)$,
local minimum is about $(3.36, -42.75)$.

GC 15. Graph the polynomial $y = x^4 - 32x + 12$ in the viewing rectangle $[-2, 4]$ by $[-40, 140]$. Find the x-intercept(s), if any, the y-intercept, and the coordinates of all local extrema correct to two decimals.
Answer: easy

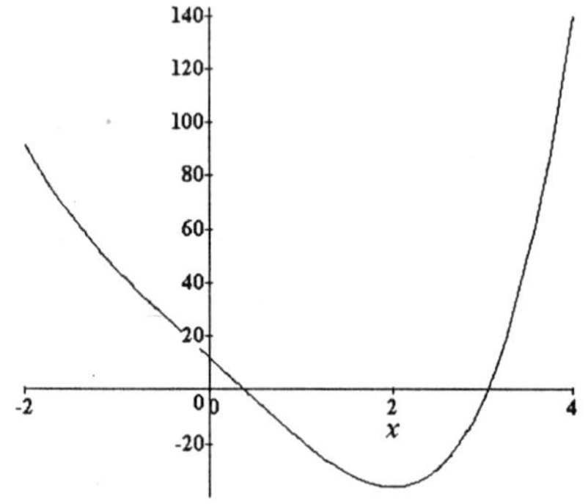

$y = x^4 - 32x + 12$, $[-2, 4]$ by $[-40, 140]$
y-intercept $= 12$, x-intercepts ≈ 0.38 and 3.04, local minimum is $(2, -36)$.

GC 16. Graph the polynomial $y = x^5 - 3x^2 + 4$ in the viewing rectangle $[-2, 2]$ by $[-20, 20]$. Find the x-intercept(s), if any, the y-intercept, and the coordinates of all local extrema correct to two decimals.
Answer: easy

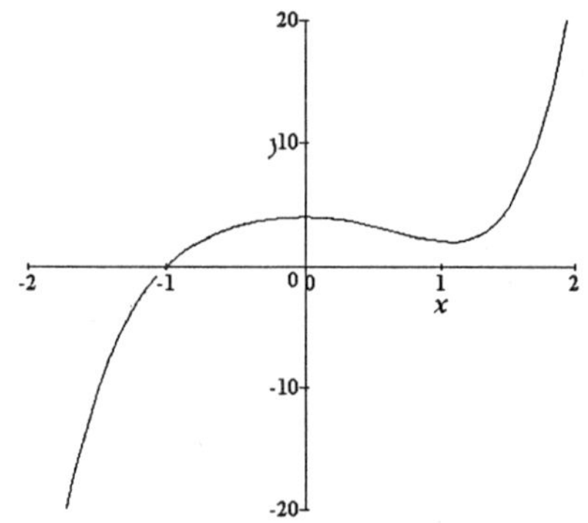

$y = x^5 - 3x^2 + 4$, $[-2, 2]$ by $[-20, 20]$
y-intercept $= 4$, x-intercept $= -1$, local maximum is $(0, 4)$,
local minimum is approximately $(1.06, 1.97)$

GC 17. Find all local extrema of the polynomial $y = 4x^3 - 2x + 3$ correct to two decimals.
Answer: medium

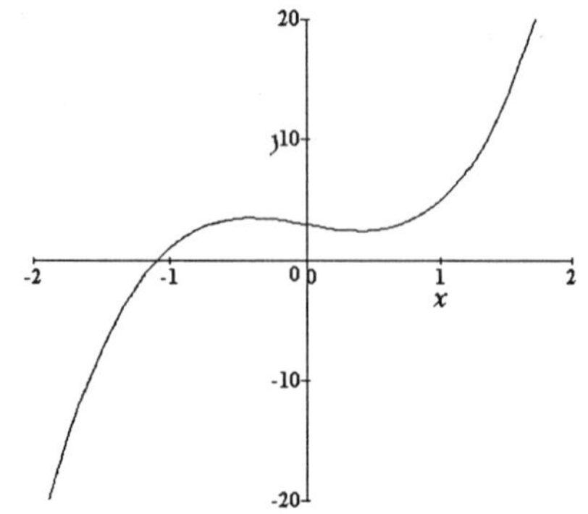

$y = 4x^3 - 2x + 3$ has local maximum $(-0.41, 3.54)$ and local minimum $(0.41, 2.46)$.

GC 18. Find all local extrema of the polynomial $y = 1.4x^5 + 3.25x^4 - 4x^3 - 12x^2 + 9x$ correct to two decimals.
Answer: medium

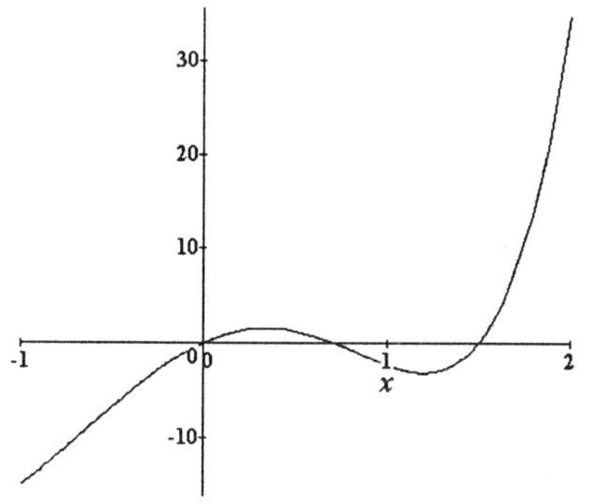

$y = 1.4x^5 + 3.25x^4 - 4x^3 - 12x^2 + 9x$ has local maximum $(0.34, 1.57)$ and local minimum $(1.20, -3.16)$.

GC 19. Find all local extrema of the polynomial $y = (x^3 - 1)^2$ correct to two decimals.
Answer: medium

$y = (x^3 - 1)^2$ has local minimum $(1, 0)$.

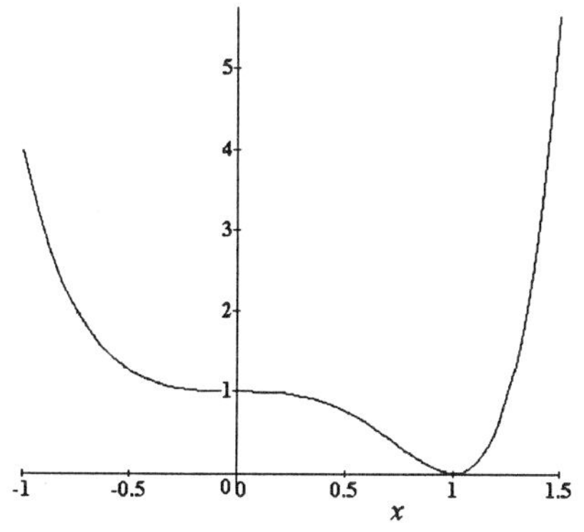

GC 20. Find all local extrema of the polynomial $y = (x+1)^2(x-2)^2$ correct to two decimals.
Answer: medium

$y = (x+1)^2(x-2)^2$ has local minima $(-1, 0)$ and $(2, 0)$, and local maximum $(0.5, 5.06)$.

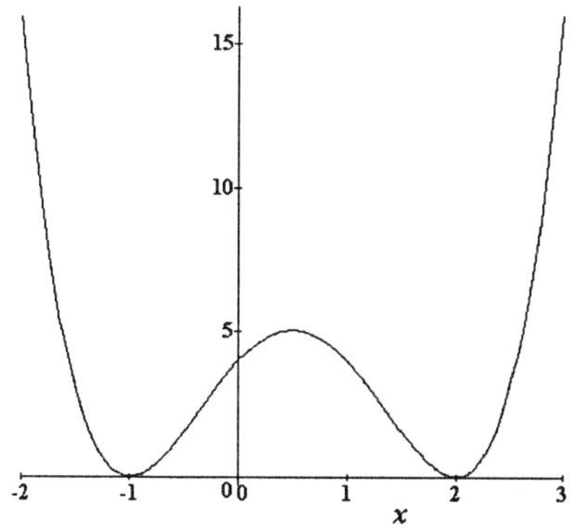

GC 21. Graph the polynomial $y = -\frac{1}{4}x^3 + \frac{1}{3}x^2 + 8x$ and describe its end behavior.

Answer: medium

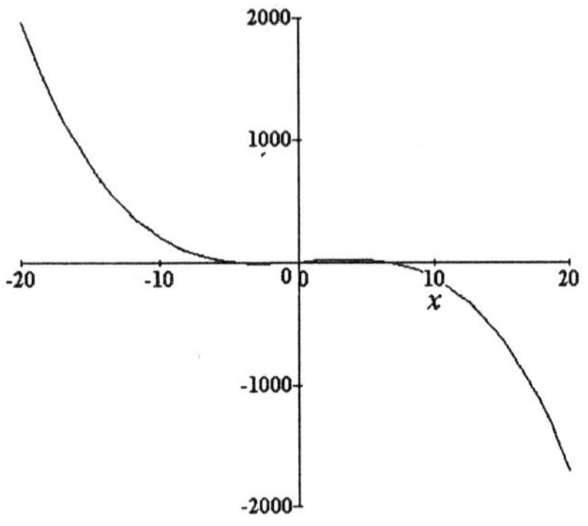

$y = -\frac{1}{4}x^3 + \frac{1}{3}x^2 + 8x$, $[-20, 20]$ by $[-2000, 2000]$

$y \to \infty$ as $x \to -\infty$

$y \to -\infty$ as $x \to \infty$

GC 22. Graph the polynomial $y = (2 - x)^5$ and describe its end behavior.

Answer: medium

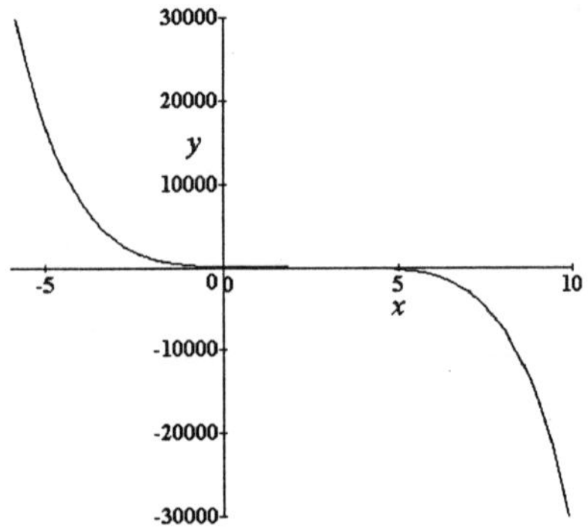

$y = (2 - x)^5$, $[-6, 10]$ by $[-30000, 30000]$

$y \to -\infty$ as $x \to \infty$

$y \to \infty$ as $x \to -\infty$

GC 23. Graph the polynomial $y = 8x^2 - x^{10}$ and describe its end behavior.

Answer: medium

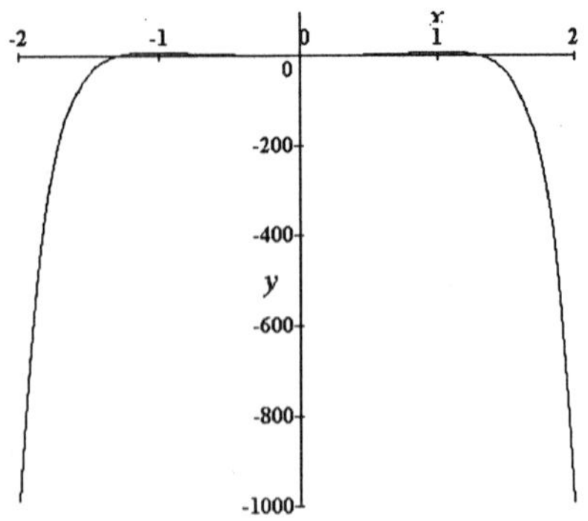

$y = 8x^2 - x^{10}$, $[-2, 2]$ by $[-1000, 100]$

$y \to -\infty$ as $x \to \pm\infty$

GC 24. Solve the inequality $3x^3 + 2x^2 - 12x - 6 \le 0$ and express the solution in interval notation.

Answer: medium

$3x^3 + 2x^2 - 12x - 6 \le 0$ has the solution $(-\infty, -2.11] \cup [-0.49, 1.93]$.

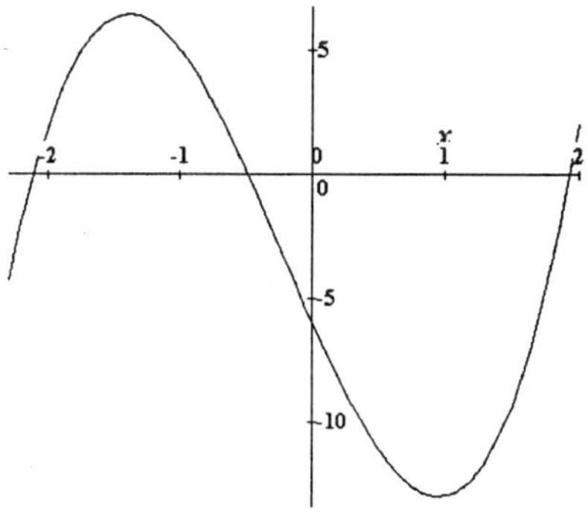

GC 25. Solve the inequality $x^4 - 4x^3 + 5x > 0$ and express your answer using interval notation.

Answer: medium

$x^4 - 4x^3 + 5x > 0$ has the solution $(-\infty, -1) \cup (0, 1.38) \cup (3.61, \infty)$.

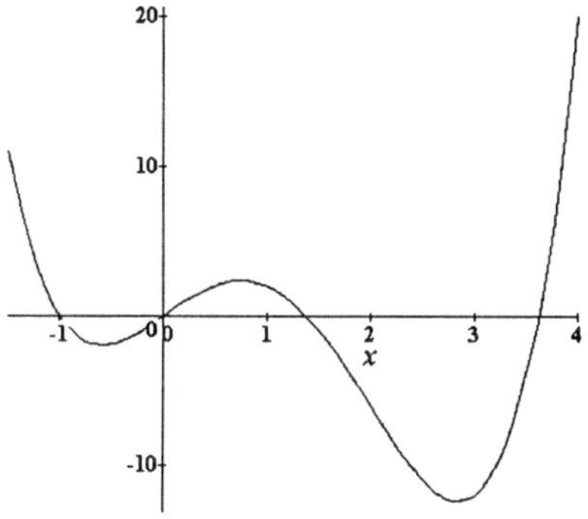

GC 26. Solve the inequality $x^4 + x^2 \ge x^3 + 8x$ and express your answer using interval notation.

Answer: medium

$x^4 + x^2 \ge x^3 + 8x$ has the solution $(-\infty, 0] \cup [2.20, \infty)$.

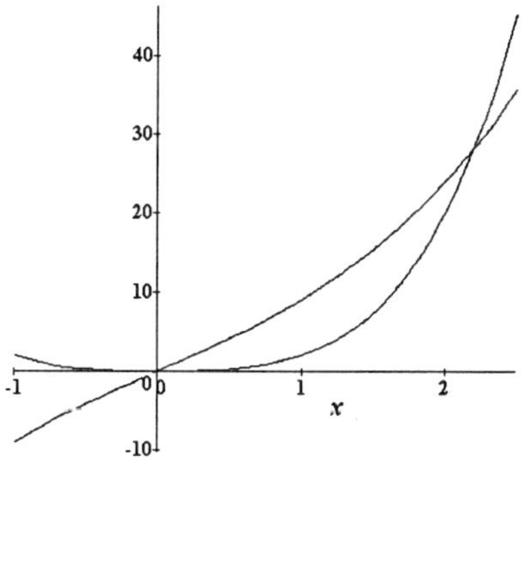

GC 27. Give an example of a polynomial that has 8 local extrema.

Answer: medium

$P(x) = (x-1)(x-2)(x-3)(x-4)(x-5)(x-6)(x-7)(x-8)(x-9)$

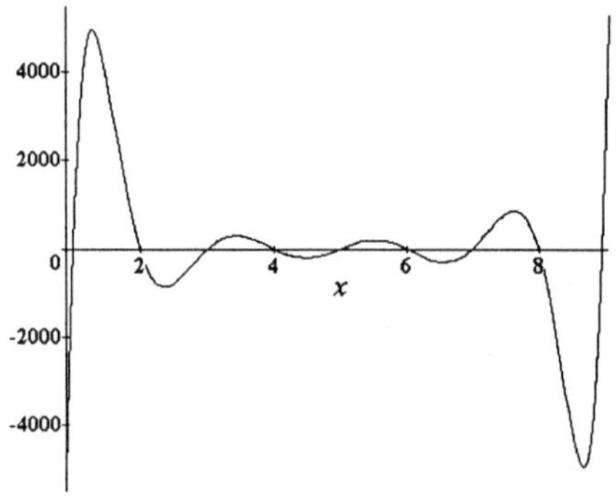

GC 28. Is it possible for a third degree polynomial to have exactly one local extremum? Explain your answer. (Hint: Think about the end behavior of such a polynomial.)

Answer: hard

No, it is impossible. The end behavior of a third degree polynomial is the same as that of $y = kx^3$, and for this function the values of y go off in opposite directions as $x \to \infty$ and as $x \to -\infty$. But for a function with just one extremum, the values of y would head off in the same direction (either both up or both down) on either side of the extremum.

GC 29. Graph the family of polynomials $P(x) = (x+c)^3$ in the same viewing rectangle, using the values $c = -1, 0, 1, 2$. Explain how changing the value of c affects the graph.

Answer: medium

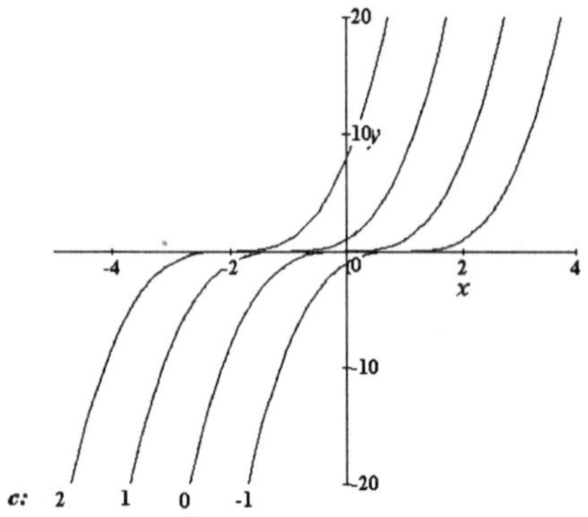

As c increases, the graph moves to the left.

GC 30. Graph the family of polynomials $P(x) = x^3 + cx^2$ in the same viewing rectangle, using the values $c = -8, -4, 0, 4, 8$. Explain how changing the value of c affects the graph.

Answer: medium

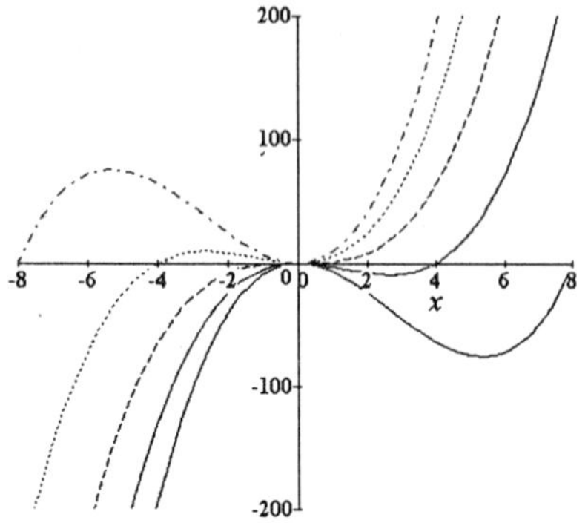

As c increases from -8, the minimum to the right of the origin moves towards the origin. As c increases through 0, this minimum disappears and instead a maximum appears to the left of the origin, and moves further from the origin as c increases.

31. Sketch the graph of the function $p(x) = -x^4 + 8x^3 - 16x^2$ by plotting the y-intercept and all x-intercepts. Make sure you plot enough points so that your graph exhibits the proper end behavior.
Answer: easy

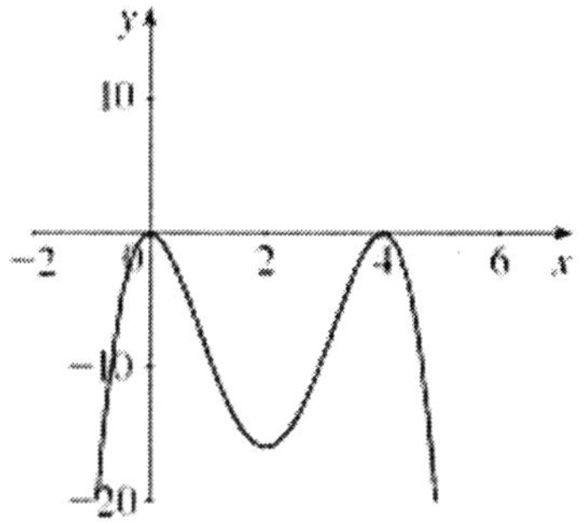

x-intercepts: $-x^4 + 8x^3 - 16x^2 = 0 \Rightarrow -x^2(x-4)^2 = 0 \Rightarrow x = 0$ or $x = 4 \Rightarrow (0,0), (4,0)$. y-intercept: $(0,0)$

32. Sketch the graph of the function $p(x) = -\left(x^6 - 5x^3 + 6\right)$ by plotting the y-intercept and all x-intercepts. Make sure you plot enough points so that your graph exhibits the proper end behavior.
Answer: easy

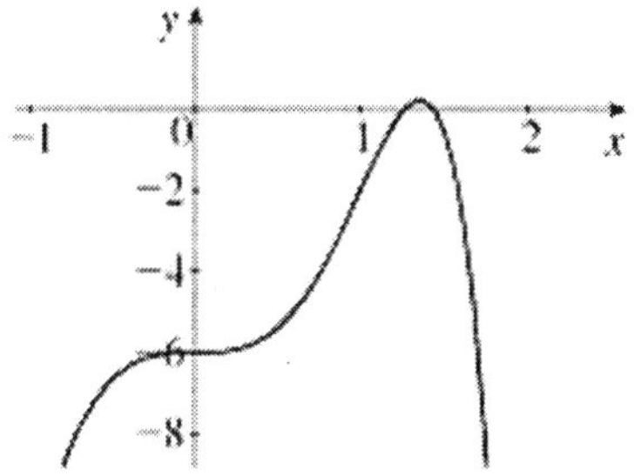

x-intercepts: $-\left(x^6 - 5x^3 + 6\right) = 0 \Rightarrow -\left(x^3-2\right)\left(x^3-3\right) = 0 \Rightarrow x = \sqrt[3]{2}$ or $x = \sqrt[3]{3} \Rightarrow \left(\sqrt[3]{2}, 0\right), \left(\sqrt[3]{3}, 0\right)$.
y-intercept: $(0,-6)$

GC 33. On the same coordinate axes, sketch the graphs (as accurately as possible) of the functions $f(x) = x^3 - 3x^2 - x + 3$ and $g(x) = -4x^2 + 3x + 3$. Based on your sketch, at how many points do the graphs appear to intersect? Find the coordinates of all intersection points.
Answer: medium

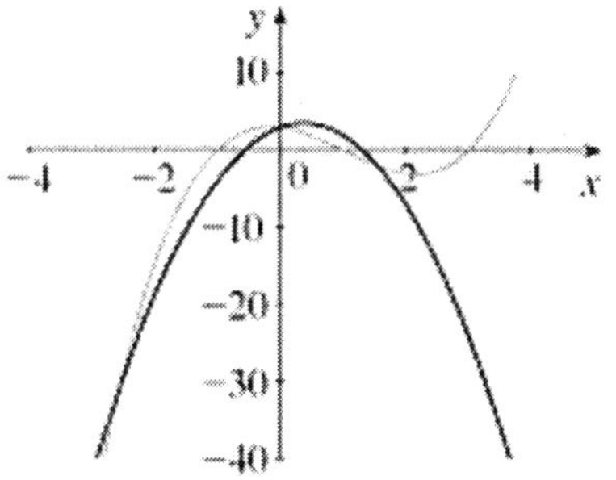

Points of intersection: $f(x) = g(x) \Rightarrow x^3 - 3x^2 - x + 3 = -4x^2 + 3x + 3 \Rightarrow x^3 + x^2 - 4x = 0 \Rightarrow x\left(x^2 + x - 4\right) = 0 \Rightarrow x = 0$ or $x = -\frac{1}{2} \pm \frac{1}{2}\sqrt{17}$. So the points are $(0,3)$, $\left(-\frac{1}{2} - \frac{1}{2}\sqrt{17}, -\frac{33}{2} - \frac{7}{2}\sqrt{17}\right)$, and $\left(-\frac{1}{2} + \frac{1}{2}\sqrt{17}, -\frac{33}{2} + \frac{7}{2}\sqrt{17}\right)$.

34. An open box is to be constructed from a piece of sheet metal 15 in. by 25 in. by cutting squares of length x from each corner and folding up the sides.
(a) Express the volume V of the box as a function of x.
(b) What is the domain of V?
(c) Draw a graph of the function V and use it to estimate the maximum volume for the box.
Answer: medium
(a) $V(x) = (15 - 2x)(25 - 2x)x = 375x - 80x^2 + 4x^3$
(b) Domain: $\left(0, \frac{15}{2}\right)$
(c)

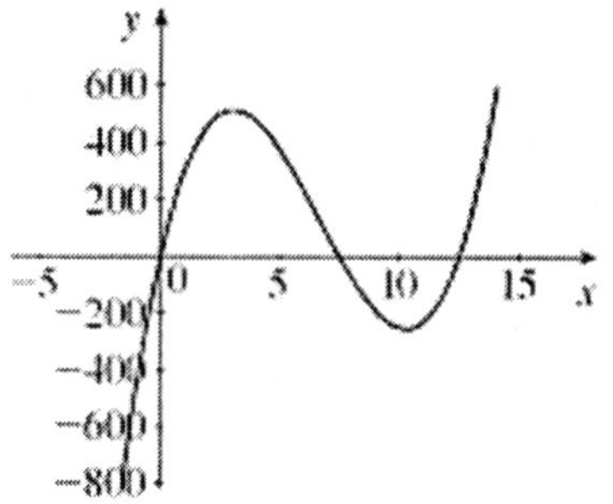

The maximum volume is about 513 in^2.

GC 35. A business consultant analyzing a small computer company finds that if the company produces and sells x (in hundreds) microprocessors it has a total profit function (in thousands of dollars) given by $p(x) = 0.4x^2 - 0.0043x^3 + 7x - 244$, with x (in hundreds) being the number of processors produced and sold annually.
(a) Graph the profit function and determine how many processors must be produced and sold in order to break even.
(b) What is domain of the profit function where the company actually loses money?
(c) Estimate the number of processors that must be produced in order to give the company the largest possible profit.
(d) Estimate the largest profit.
Answer: medium
(a)

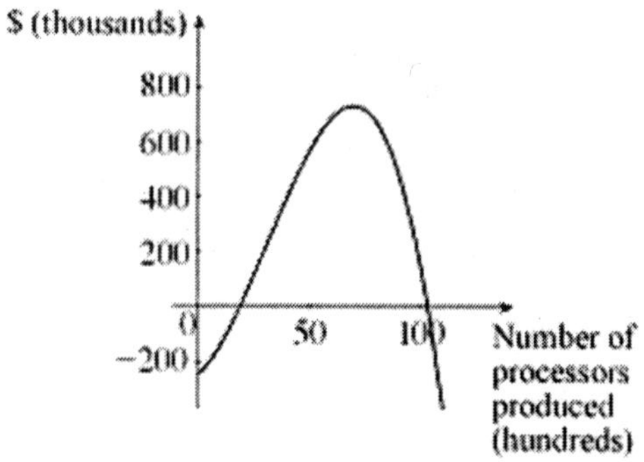

To break even the company would have to produce and sell about 1878 processors.
(b) $p(x) < 0 \quad \Leftrightarrow \quad x \in [0, 1878) \cup (10346, \infty)$, so the company loses money if they produce and sell fewer than 1878 processors or more than 10,346 processors.
(c) About 7000 processors
(d) About \$730,000

SECTION 5.2

1. For $\dfrac{x^4 + 30x + 12}{2x + 6}$ find the quotient and remainder.

(a) $Q(x) = \frac{1}{4}(x^3 - 6x^2 + 4x + 4), R(x) = 2$
(b) $Q(x) = \frac{1}{2}(x^3 - 4x^2 + 2x + 1), R(x) = 3$
(c) $Q(x) = \frac{1}{2}(x^3 - 3x^2 + 9x + 3), R(x) = 3$
(d) $Q(x) = \frac{1}{4}(x^3 + 5x^2 - 2x - 7), R(x) = 0$
(e) $Q(x) = 2(x^3 + 2x^2 + 3x + 1), R(x) = 2x + 1$

Answer: easy (c)

$$\frac{x^4 + 30x + 12}{2x + 6} = \frac{1}{2}\left(\frac{x^4 + 30x + 12}{x + 3}\right)$$

$$\begin{array}{r|rrrrr} & 1 & 0 & 0 & 30 & 12 \\ \hline -3 & 1 & -3 & 9 & 3 & 3 \end{array}$$

Therefore, $Q(x) = \frac{1}{2}(x^3 - 3x^2 + 9x + 3)$, and $R(x) = 3$.

2. For $\dfrac{x^5 - 3x^4 + 3x^3 + 20x - 6}{x^2 + 2x - 6}$ find the quotient and remainder.

(a) $Q(x) = x^3 + 2x^2 - 11x - 64, R(x) = 110x - 325$
(b) $Q(x) = x^3 - 2x^2 + 5x + 12, R(x) = 104x - 216$
(c) $Q(x) = 2x^3 - 3x^2 + 10x - 366, R(x) = 20x^2 - 12x$
(d) $Q(x) = x^3 - 15x^2 + 4x - 12, R(x) = 244x + 140$
(e) $Q(x) = x^3 - 5x^2 + 19x - 68, R(x) = 270x - 414$

Answer: medium (e)

$$\begin{array}{rrrrrrr}
 & x^3 & -5x^2 & +19x & -68 & & \\ \cline{2-7}
x^2 + 2x - 6 & x^5 & -3x^4 & +3x^3 & +0x^2 & +20x & -6 \\
 & x^5 & +2x^4 & -6x^3 & & & \\ \cline{2-4}
 & & -5x^4 & +9x^3 & & & \\
 & & -5x^4 & -10x^3 & +30x^2 & & \\ \cline{3-5}
 & & & 19x^3 & -30x^2 & & \\
 & & & 19x^3 & +38x^2 & -114x & \\ \cline{4-6}
 & & & & -68x^2 & +134x & \\
 & & & & -68x^2 & -136x & +408 \\ \cline{5-7}
 & & & & & 270x & -414
\end{array}$$

Therefore, $Q(x) = x^3 - 5x^2 + 19x - 68$, and $R(x) = 270x - 414$.

3. For $\dfrac{2x^3 - x^2 - 5}{x - 3/2}$ find the quotient and remainder.

(a) $Q(x) = 2x^2 - 2x + 1$, and $R(x) = -1$
(b) $Q(x) = 3x^2 + x + 4$, and $R(x) = 0$
(c) $Q(x) = 2x^2 + x + 6$, and $R(x) = 2$
(d) $Q(x) = 2x^2 + 2x + 3$, and $R(x) = -\frac{1}{2}$
(e) $Q(x) = 2x^2 + 3x - 4$, and $R(x) = -1$

Answer: easy (d)

$$\begin{array}{r|rrrr} & 2 & -1 & 0 & -5 \\ \hline \frac{3}{2} & 2 & 2 & 3 & -\frac{1}{2} \end{array}$$

Therefore, $Q(x) = 2x^2 + 2x + 3$, and $R(x) = -\frac{1}{2}$.

4. For $\dfrac{15x-7}{5x+12}$ find the quotient and remainder.
(a) $Q(x)=3$, and $R(x)=-43$
(b) $Q(x)=4$, and $R(x)=-115$
(c) $Q(x)=2x$, and $R(x)=11$
(d) $Q(x)=3$, and $R(x)=-22$
(e) $Q(x)=2$, and $R(x)=15$
Answer: easy (a)

$$\begin{array}{r|rr} & 3 & \\ \hline 5x+12 & 15x & -7 \\ & 15x & +36 \\ \hline & & -43 \end{array}$$

Therefore, $Q(x)=3$, and $R(x)=-43$.

5. Find the value $P(-3)$ of the polynomial $P(x)=x^4+4x^3+7x^2+10x+15$ using the Remainder Theorem.
(a) 6 (b) 13 (c) 15 (d) 21 (e) 33
Answer: easy (d)
$P(x)=x^4+4x^3+7x^2+10x+15$; $P(-3)$

$$\begin{array}{r|rrrrr} & 1 & 4 & 7 & 10 & 15 \\ \hline -3 & 1 & 1 & 4 & -2 & 21 \end{array}$$

Therefore, $P(-3)=21$.

6. Use the Factor Theorem to show that $x+4$ is a factor of the polynomial $P(x)=x^5+4x^4-7x^3-23x^2+23x+12$.
Answer: medium
$x+4$ is a factor of $P(x)=x^5+4x^4-7x^3-23x^2+23x+12$ if and only if $P(-4)=0$.

$$\begin{array}{r|rrrrrr} & 1 & 4 & -7 & -23 & 23 & 12 \\ \hline -4 & 1 & 0 & -7 & 5 & 3 & 0 \end{array}$$

Since $P(-4)=0$, $x+4$ is a factor of the polynomial.

7. Find a polynomial of degree 3 with constant coefficient 12 that has zeros $-\frac{1}{2}$, 2, and 3.
Answer: medium
Since the zeros are $-\frac{1}{2}$, 2, and 3, a factorization is
$P(x)=C\left(x+\frac{1}{2}\right)(x-2)(x-3)=\frac{1}{2}C(2x+1)(x^2-5x+6)=\frac{1}{2}C(2x^3-10x^2+12x+x^2-5x+6)$
$=\frac{1}{2}C(2x^3-9x^2+7x+6)$.
Since the constant coefficient is 12, $C=4$, and so the polynomial is $P(x)=4x^3-18x^2+14x+12$.

8. Does there exist a polynomial of degree 4 with integers coefficients that has zeros i, $2i$, $3i$, and $4i$? If so, find it. If not, explain why not.
Answer: hard
No, there is no polynomial of degree 4 with integer coefficients that has zeros i, $2i$,$3i$ and $4i$, since the imaginary roots of polynomial equations with real coefficients come in complex conjugate pairs.

9. If we divide the polynomial $P(x)=x^4+kx^2-kx+2$ by $x+2$, the remainder is 72. What must the value of k be?
Answer: hard
Since division of $P(x)=x^4+kx^2-kx+2$ by $x+2$ leaves a remainder of 72, it follows that $P(-2)=72$. Now,
$P(-2)=(-2)^4+k(-2)^2-k(-2)+2=16+4k+2k+2=18+6k=72 \quad\Leftrightarrow\quad 6k=54 \quad\Leftrightarrow\quad k=9$.

10. The quadratic equation $x^2-2kx+16=0$ has two positive roots, one of which is four times the other. What is the value of k, and what are the two roots of the equation?
Answer: hard
Let m and $4m$ be the two roots. Then the polynomial is $(x-m)(x-4m)=x^2-5mx+4m^2=x^2-2kx+16$. Comparing coefficients, we see that $4m^2=16$, so $m=2$ (since $m>0$), and $2k=5m \Rightarrow k=\frac{5}{2}m=5$. So $k=5$, the roots are 2, 8.

11. For $\dfrac{x^4 - 2x^3 - x + 2}{x + 2}$, find the quotient and remainder using synthetic division.
Answer: easy

$$\begin{array}{r|rrrrr} -2 & 1 & -2 & 0 & -1 & 2 \\ & & -2 & 8 & -16 & 34 \\ \hline & 1 & -4 & 8 & -17 & 36 \end{array}$$

Therefore the quotient is $x^3 - 4x^2 + 8x - 17$, and the remainder is 36. Thus
$\dfrac{x^4 - 2x^3 - x + 2}{x + 2} = x^3 - 4x^2 + 8x - 17 + \dfrac{36}{x + 2}$.

12. For $\dfrac{x^4 - 256}{x + 4}$, find the quotient and remainder using synthetic division.
Answer: easy

$$\begin{array}{r|rrrrr} -4 & 1 & 0 & 0 & 0 & -256 \\ & & -4 & 16 & -64 & 256 \\ \hline & 1 & -4 & 16 & -64 & 0 \end{array}$$

Therefore the quotient is $x^3 - 4x^2 + 16x - 64$, and the remainder is 0. Thus $\dfrac{x^4 - 256}{x + 4} = x^3 - 4x^2 + 16x - 64$.

13. For $\dfrac{x^3 + 8x + 4}{3x - 2}$, find the quotient and remainder using synthetic division.
Answer: easy

$$\begin{array}{r|rrrr} \frac{2}{3} & 1 & 0 & 8 & 4 \\ & & \frac{2}{3} & \frac{4}{9} & \frac{152}{27} \\ \hline & 1 & \frac{2}{3} & \frac{76}{9} & \frac{260}{27} \end{array}$$

Therefore the quotient is $x^2 + \frac{2}{3}x + \frac{76}{9}$, and the remainder is $\frac{260}{27}$. Thus $\dfrac{x^3 + 8x + 4}{3x - 2} = x^2 + \frac{2}{3}x + \frac{76}{9} + \dfrac{260}{27(3x - 2)}$.

14. Use synthetic division and the Remainder Theorem to evaluate $p\left(\frac{1}{2}\right)$ for $p(x) = 2x^2 - 2x + 2$.
Answer: easy

$$\begin{array}{r|rrr} \frac{1}{2} & 2 & -2 & 2 \\ & & 1 & -\frac{1}{2} \\ \hline & 2 & -1 & \frac{3}{2} \end{array}$$

Therefore, $p\left(\frac{1}{2}\right) = \frac{3}{2}$.

15. Use synthetic division and the Remainder Theorem to evaluate $p(-4)$ for $p(x) = 2x^3 - 2x^2 + 11x - 100$.
Answer: easy

$$\begin{array}{r|rrrr} -4 & 2 & -2 & 11 & -100 \\ & & -8 & 40 & -204 \\ \hline & 2 & -10 & 51 & -304 \end{array}$$

Therefore, $p(-4) = -304$.

16. Use synthetic division and the remainder theorem to evaluate $p(-1)$ for $p(x) = -2x^7 + 3x^6 - 12x^5 - 8x^4 + 3x^2 + x - 1$.
Answer: medium

$$\begin{array}{r|rrrrrrrr} -1 & -2 & 3 & -12 & -8 & 0 & 3 & 1 & -1 \\ \hline & & 2 & -5 & 17 & -9 & 9 & -12 & 11 \\ \hline & -2 & 5 & -17 & 9 & -9 & 12 & -11 & 10 \end{array}$$

Therefore, $p(-1) = 10$.

17. Use the Factor Theorem to show that $x + \frac{2}{3}$ is a factor of the polynomial $P(x) = x^3 + \frac{7}{6}x^2 - \frac{1}{6}x - \frac{1}{3}$.
Answer: easy
$x + \frac{2}{3}$ is a factor of $P(x) = x^3 + \frac{7}{6}x^2 - \frac{1}{6}x - \frac{1}{3}$ if and only if $P\left(-\frac{2}{3}\right) = 0$.
$P\left(-\frac{2}{3}\right) = \left(-\frac{2}{3}\right)^3 + \frac{7}{6}\left(-\frac{2}{3}\right)^2 - \frac{1}{6}\left(-\frac{2}{3}\right) - \frac{1}{3} = 0$. Since $P\left(-\frac{2}{3}\right) = 0$, $x + \frac{2}{3}$ is a factor of the polynomial.

18. Use the Factor Theorem to show that $x + 5$ is a factor of the polynomial $P(x) = x^5 + x^4 - 36x^3 - 16x^2 + 320x$.
Answer: easy
$x + 5$ is a factor of $P(x) = x^5 + x^4 - 36x^3 - 16x^2 + 320x$ if and only if $P(-5) = 0$.
$P(-5) = (-5)^5 + (-5)^4 - 36(-5)^3 - 16(-5)^2 + 320(-5) = 0$. Since $P(-5) = 0$, $x + 5$ is a factor of the polynomial.

19. Use the Factor Theorem to show that $2 - x$ is a factor of the polynomial $P(x) = x^5 - 13x^4 + 57x^3 - 83x^2 - 34x + 120$.
Answer: easy
$2 - x$ is a factor of $P(x) = x^5 - 13x^4 + 57x^3 - 83x^2 - 34x + 120$ if and only if $P(2) = 0$.
$P(2) = (2)^5 - 13(2)^4 + 57(2)^3 - 83(2)^2 - 34(2) + 120 = 0$. Since $P(2) = 0$, $2 - x$ is a factor of the polynomial.

20. Let $p(x) = -x^3 - 3x^2 + 33x + 35$. Show that $p(-7) = 0$, and use this fact to find all the zeros of $p(x)$.
Answer: medium
Substituting, we see that $p(-7) = -(-7)^3 - 3(-7)^2 + 33(-7) + 35 = 0$. By the Factor Theorem, this means that $x + 7$ is a factor of $p(x)$. Using synthetic division:

$$\begin{array}{r|rrrr} -7 & -1 & -3 & 33 & 35 \\ \hline & & 7 & -28 & -35 \\ \hline & -1 & 4 & 5 & 0 \end{array}$$

We see that $p(x) = (x + 7)(-x^2 + 4x + 5) = -(x + 1)(x - 5)(x + 7)$. Therefore the zeros are -1, 5, and -7.

21. Let $p(x) = x^4 - 10x^3 + 37x^2 - 60x + 36$. Show that $p(2) = 0$ and $p(3) = 0$, and use this fact to find all the zeros of $p(x)$.
Answer: medium
Substituting, we see that $p(2) = (2)^4 - 10(2)^3 + 37(2)^2 - 60(2) + 36 = 0$ and
$p(3) = (3)^4 - 10(3)^3 + 37(3)^2 - 60(3) + 36 = 0$. By the Factor Theorem, this means that $x - 2$ and $x - 3$ are factors of $p(x)$. Using synthetic division:

$$\begin{array}{r|rrrrr} 2 & 1 & -10 & 37 & -60 & 36 \\ \hline & & 2 & -16 & 42 & 36 \\ \hline 3 & 1 & -8 & 21 & -18 & 0 \\ \hline & & 3 & -15 & 18 & \\ \hline & 1 & -5 & 6 & 0 & \end{array}$$

We see that $p(x) = (x - 2)(x - 3)(x^2 - 5x + 6) = (x - 2)(x - 3)(x - 2)(x - 3)$. Therefore the zeros are 2 and 3.

22. Find the polynomial of degree 3 that has zeros -2, -1, and 3.
Answer: easy
$P(x) = (x+2)(x+1)(x-3) = x^3 - 7x - 6$

23. Find the polynomial of degree 4 that has zeros -1, 0, 1, and 6.
Answer: easy
$P(x) = (x+1)(x)(x-1)(x-6) = x^4 - 6x^3 - x^2 + 6x$

24. Find the polynomial of degree 5 that has zeros 1 (twice), 2 (twice), and 3.
Answer: medium
$P(x) = (x-1)^2(x-2)^2(x-3) = x^5 - 9x^4 + 31x^3 - 51x^2 + 40x - 12$

25. Find the remainder of $p(x) = 8x^{242} + 18x^{63} - 12x^2 - 20$ divided by $x - 1$.
Answer: easy
If $p(x)$ is divided by $x-1$ then its remainder is equal to $p(1) = 8(1)^{242} + 18(1)^{63} - 12(1)^2 - 20 = -6$.

26. Find the remainder of $p(x) = 22x^{2004} - 5x^{569} + x^{104} - 27x + 22x^{2003} - 5x^{568} + x^{103} - 27$ divided by $x+1$. What is the significance of the remainder?
Answer: medium
If $p(x)$ is divided by $x+1$ then its remainder is equal to
$p(-1) = 22(-1)^{2004} - 5(-1)^{569} + (-1)^{104} - 27(-1) + 22(-1)^{2003} - 5(-1)^{568} + (-1)^{103} - 27 = 0$. Because the remainder is 0, $x = -1$ is a zero, and $x+1$ is a factor of $p(x)$.

SECTION 5.3

1. For $P(x) = x^{30} - 4x^{25} + 6x^2 + 60$ list all possible rational zeros given by the Rational Roots Theorem (but do not check to see which values are actually roots).
(a) ± 2, ± 3, ± 4, ± 5, ± 6, ± 8, ± 10, ± 12, ± 15, ± 20, ± 30, and ± 60
(b) ± 1, ± 3, ± 4, ± 5, ± 6, ± 12, ± 15, ± 20, ± 30, and ± 60
(c) ± 1, ± 2, ± 3, ± 4, ± 5, ± 6, ± 10, ± 12, ± 15, ± 18, ± 20, ± 24, ± 30, and ± 60
(d) ± 1, ± 2, ± 3, ± 4, ± 5, ± 6, ± 10, ± 12, ± 15, ± 20, ± 30, and ± 60
(e) ± 1, ± 2, ± 4, ± 6, ± 10, ± 12, ± 15, ± 18, ± 20, ± 24, ± 30, and ± 60
Answer: easy (d)
$P(x) = x^{30} - 4x^{25} + 6x^2 + 60$ has possible rational zeros ± 1, ± 2, ± 3, ± 4, ± 5, ± 6, ± 10, ± 12, ± 15, ± 20, ± 30, and ± 60.

2. For $P(x) = 6x^4 - x^3 + x^2 - 24$ list all possible rational zeros given by the Rational Roots Theorem (but do not check to see which values are actually roots)..
(a) ± 1, ± 2, ± 4, ± 6, ± 8, ± 12, ± 18, $\pm\frac{3}{2}$, $\pm\frac{1}{3}$, $\pm\frac{2}{3}$, $\pm\frac{4}{3}$, $\pm\frac{8}{3}$, $\pm\frac{1}{6}$, and $\pm\frac{1}{8}$
(b) ± 1, ± 2, ± 3, ± 4, ± 6, ± 8, ± 12, ± 24, $\pm\frac{1}{2}$, $\pm\frac{3}{2}$, $\pm\frac{1}{3}$, $\pm\frac{2}{3}$, $\pm\frac{4}{3}$, $\pm\frac{8}{3}$, and $\pm\frac{1}{6}$
(c) ± 1, ± 2, ± 3, ± 6, ± 9, ± 12, ± 15, ± 24, and $\pm\frac{1}{6}$
(d) ± 2, ± 3, ± 4, ± 6, ± 8, ± 12, ± 24, $\pm\frac{1}{2}$, $\pm\frac{3}{2}$, $\pm\frac{2}{3}$, $\pm\frac{8}{3}$, and $\pm\frac{1}{6}$
(e) ± 1, ± 2, ± 3, ± 4, ± 5, ± 6, ± 8, ± 9, ± 12, ± 18, ± 24, $\pm\frac{1}{2}$, $\pm\frac{3}{2}$, $\pm\frac{1}{3}$, $\pm\frac{2}{3}$, $\pm\frac{4}{3}$, $\pm\frac{8}{3}$, $\pm\frac{16}{3}$, and $\pm\frac{1}{6}$
Answer: easy (b)
$P(x) = 6x^4 - x^3 + x^2 - 24$ has possible rational zeros ± 1, ± 2, ± 3, ± 4, ± 6, ± 8, ± 12, ± 24, $\pm\frac{1}{2}$, $\pm\frac{3}{2}$, $\pm\frac{1}{3}$, $\pm\frac{2}{3}$, $\pm\frac{4}{3}$, $\pm\frac{8}{3}$, and $\pm\frac{1}{6}$.

3. Find all rational roots of the polynomial $P(x) = x^{11} - 1$, and then find the irrational roots, if any.

Answer: medium

$x^{11} - 1 = 0$. The possible rational roots are ± 1. Since $P(x)$ has 1 variation in sign, there is 1 positive real root. By inspection, the root is 1. Since $P(-x) = -x^{11} - 1$ has no variation in sign, there are no negative real roots. The only real root is 1.

4. Find all rational roots of the polynomial $P(x) = x^3 - x^2 - 8x + 12$, and then find the irrational roots, if any.

(a) $-1, -2$ and 3 (b) -3 and $\sqrt{3}$ (c) -3 and 2 (d) -3 and $\sqrt{2}$ (e) 2 and $\sqrt{3}$

Answer: medium (c)

$x^3 - x^2 - 8x + 12 = 0$. The possible rational roots are $\pm 1, \pm 2, \pm 3, \pm 4, \pm 6, \pm 12$. $P(x)$ has 2 variations in sign and hence 0 or 2 positive real roots. $P(-x) = -x^3 - x^2 + 8x + 12$ has 1 variation in sign and hence 1 negative root.

	1	-1	-8	12	
1	1	0	-8	4	
2	1	1	-6	0	$\Rightarrow$ $x = 2$ is a root.

Thus, $(x-2)(x^2 + x - 6) = 0 \quad \Leftrightarrow \quad (x-2)(x+3)(x-2) = 0$, and so the roots are -3 and 2.

5. Find all rational roots of the polynomial $P(x) = x^4 - x^3 - 23x^2 - 3x + 90$, and then find the irrational roots, if any.

(a) -2, $\sqrt{3}$ and $\sqrt{5}$ (b) -1, $\sqrt{2}$ and 5 (c) -3, 2 and 5 (d) -2 and 5 (e) -3 and $\sqrt{2}$

Answer: medium (c)

$x^4 - x^3 - 23x^2 - 3x + 90 = 0$. The possible rational roots are $\pm 1, \pm 2, \pm 3, \pm 5, \pm 6, \pm 9, \pm 10, \pm 15, \pm 18, \pm 30, \pm 45, \pm 90$. Since $P(x)$ has 2 variations in sign, $P(x)$ has 0 or 2 positive real roots. Since $P(-x) = x^4 + x^3 - 23x^2 + 3x + 90$ has 2 variations in sign, $P(x)$ has 0 or 2 negative real roots.

	1	-1	-23	-3	90	
1	1	0	-23	-26	64	
2	1	1	-21	-45	0	$\Rightarrow$ $x = 2$ is a root, and so $(x-2)(x^3 + x^2 - 21x - 45) = 0$.

	1	1	-21	-45	
3	1	4	-9	-72	
5	1	6	9	0	$\Rightarrow$ $x = 5$ is a root, and so $(x-2)(x-5)(x^2 + 6x + 9) = 0 \quad \Leftrightarrow$

$(x-2)(x-5)(x+3)^2 = 0$. Therefore, the roots are -3, 2, and 5.

6. Find all rational roots of the polynomial $P(x) = 4x^3 - 7x + 3$, and then find the irrational roots, if any.

(a) $-\frac{1}{2}$ and 1 (b) 2 and $\frac{\sqrt{2}}{2}$ (c) $-\frac{3}{2}$ and 1 (d) $-\frac{1}{2}$ and 2 (e) $-\frac{3}{2}, \frac{1}{2}$ and 1

Answer: medium (e)

$4x^3 - 7x + 3 = 0$. The possible rational roots are $\pm 1, \pm 3, \pm \frac{1}{2}, \pm \frac{3}{2}, \pm \frac{1}{4}, \pm \frac{3}{4}$. Since $P(x)$ has 2 variations in sign, there are 0 or 2 positive roots. Since $P(-x) = -4x^3 + 7x + 3$ has 1 variation in sign, there is 1 negative root.

	4	0	-7	3	
$\frac{1}{2}$	4	2	-6	0	$\Rightarrow$ $x = \frac{1}{2}$ is a root, and so $\left(x - \frac{1}{2}\right)(4x^2 + 2x - 6) = 0 \quad \Leftrightarrow$

$(2x-1)(2x^2 + x - 3) = 0$

$\Leftrightarrow \quad (2x-1)(x-1)(2x+3) = 0$. Thus, the roots are $-\frac{3}{2}, \frac{1}{2}$ and 1.

7. Use Descartes' Rule of Signs to determine how many positive and how many negative real zeros the polynomial polynomial $P(x) = 3x^5 - 4x^4 + 8x^3 - 5$ can have, and then determine the possible total number of real zeros.

Answer: easy

$P(x) = 3x^5 - 4x^4 + 8x^3 - 5$. Since $P(x)$ has 3 variations in sign, $P(x)$ can have 3 or 1 positive real zeros. Since $P(-x) = -3x^5 - 4x^4 - 8x^3 - 5$ has 0 variations in sign, $P(x)$ has 0 negative real zeros. Thus, $P(x)$ has 1 or 3 real zeros.

8. Use Descartes' Rule of Signs to determine how many positive and how many negative real zeros the polynomial polynomial $P(x) = x^4 + x^3 + x^2 + x + 22$ can have, and then determine the possible total number of real zeros.

Answer: easy

$P(x) = x^4 + x^3 + x^2 + x + 22$. Since $P(x)$ has 0 variations in sign, $P(x)$ has 0 positive real zeros. Since $P(-x) = x^4 - x^3 + x^2 - x + 22$ has 4 variations in sign, $P(x)$ has 4, 2, or 0 negative real zeros. Therefore, $P(x)$ has 0, 2, or 4 real zeros.

9. Use Descartes' Rule of Signs to determine how many positive and how many negative real zeros the polynomial polynomial $P(x) = 125x^{32} + 76x^{25} - 12x^{12} - 1$ can have, and then determine the possible total number of real zeros.

Answer: medium

$P(x) = 125x^{32} + 76x^{25} - 12x^{12} - 1$. Since $P(x)$ has 1 variation in sign, $P(x)$ has 1 positive real zero. Since $P(-x) = 125x^{32} - 76x^{25} - 12x^{12} - 1$ has 1 variation in sign, $P(x)$ has 1 negative real zero. Therefore, $P(x)$ has 2 real zeros.

10. Use Descartes' Rule of Signs to determine how many positive and how many negative real zeros the polynomial polynomial $P(x) = x^8 - x^5 + x^4 - \sqrt{3}x^3 + x^2 - x + 100$ can have, and then determine the possible total number of real zeros.

Answer: medium

$P(x) = x^8 - x^5 + x^4 - \sqrt{3}x^3 + x^2 - x + 100$. Since $P(x)$ has 6 variations in sign, $P(x)$ has 6, 4, 2, or 0 positive real zeros. Since $P(-x) = x^8 + x^5 + x^4 + \sqrt{3}x^3 + x^2 + x + 100$ has 0 variation in sign, $P(x)$ has 0 negative real zeros. Therefore, $P(x)$ has 0, 2, 4, or 6 real zeros.

11. Show that the given values for a and b are lower and upper bounds, respectively, for the real roots of the polynomial.
$P(x) = 3x^4 - 17x^3 + 24x^2 - 9x + 1$; $a = 0, b = 6$

Answer: medium

$3x^4 - 17x^3 + 24x^2 - 9x + 1 = 0$; $a = 0, b = 6$. Since $P(-x) = 3x^4 + 17x^3 + 24x^2 + 9x + 1$ has 0 variations in sign, $P(x)$ has 0 negative real zeros, and so by Descartes' Rule of Signs, $a = 0$ is a lower bound.

	3	−17	24	−9	1	
6	3	1	30	171	1027	⇒ all positive. Therefore, $b = 6$ is an upper bound.

12. Find integers that are upper and lower bounds for the real roots of the polynomial $P(x) = 4x^4 + 20x^3 + 35x^2 + 32x + 15$.

Answer: medium

$4x^4 + 20x^3 + 35x^2 + 32x + 15 = 0$

	4	20	35	32	15	⇒ no positive roots, so 0 is an upper bound
−1	4	16	19	13	2	
−2	4	12	11	10	−5	
−3	4	8	11	−1	18	
−4	4	4	19	−44	191	
−5	4	0	35	−143	730	⇒ alternately non-positive and non-negative.

Therefore, a lower bound is −5 and an upper bound is 0.

13. Find integers that are upper and lower bounds for the real roots of the polynomial $P(x) = x^5 - x^4 + 1$.

Answer: medium

$x^5 - x^4 + 1 = 0$

	1	−1	0	0	0	1	
1	1	0	0	0	0	1	⇒ all non-negative, so 1 is an upper bound
−1	1	−2	2	−2	2	−1	⇒ alternately positive and negative

Therefore, a lower bound is −1 and an upper bound is 1.

14. Find all rational roots of the polynomial $P(x) = 4x^4 - 25x^2 + 36$, and then find the irrational roots, if any.

(a) ± 1 and ± 2 (b) ± 2 and $\pm\frac{3}{2}$ (c) ± 1 and $\sqrt{2}$ (d) ± 2 and $\sqrt{3}$ (e) ± 1 and $1 \pm \sqrt{2}$

Answer: medium (b)

$4x^4 - 25x^2 + 36 = 0$ has possible rational roots $\pm 1, \pm 2, \pm 3, \pm 4, \pm 6, \pm 9, \pm 12, \pm 18, \pm 36, \pm\frac{1}{2}, \pm\frac{1}{4}, \pm\frac{3}{2}, \pm\frac{3}{4}, \pm\frac{9}{2}, \pm\frac{9}{4}$. Since $P(x)$ has 2 variations in sign, there are 0 or 2 positive real roots. Since $P(-x) = 4x^4 - 25x^2 + 36$ has 2 variations in sign, there are 0 or 2 negative real roots.

	4	0	−25	0	36	
1	4	4	−21	−21	15	
2	4	8	−9	−18	0	⇒ $x = 2$ is a root, and so $(x-2)(4x^3 + 8x^2 - 9x - 18) = 0$.

	4	8	−9	−18	
2	4	16	23	28	⇒ all positive
$\frac{1}{2}$	4	10	−4	−20	
$\frac{1}{4}$	4	9	$-\frac{27}{4}$	$-\frac{315}{16}$	
$\frac{3}{2}$	4	14	12	0	⇒ $x = \frac{3}{2}$ is a root, and so $(x-2)(2x-3)(2x^2 + 7x + 6) = 0$ ⇔

$(x-2)(2x-3)(2x+3)(x+2) = 0$ Therefore, the roots are $\pm 2, \pm\frac{3}{2}$.

15. Find all rational roots of the polynomial $P(x) = 2x^4 + 15x^3 + 31x^2 + 20x + 4$, and then find the irrational roots, if any.

(a) -2 and $\frac{-3 \pm \sqrt{13}}{2}$ (b) -1, 1 and $\pm\sqrt{2}$ (c) -2, -1 and $\frac{2 \pm \sqrt{7}}{2}$

(d) $-\frac{1}{2}$, -2, and $\frac{-5 \pm \sqrt{17}}{2}$ (e) -2, $\frac{1}{2}$ and $1 \pm \sqrt{3}$

Answer: medium (d)

$2x^4 + 15x^3 + 31x^2 + 20x + 4 = 0$. The possible rational roots are $\pm 1, \pm 2, \pm 4, \pm\frac{1}{2}$. Since all of the coefficients are positive, there are no positive roots. Since $P(-x) = 2x^4 - 15x^3 + 31x^2 - 20x + 4$ has 4 variations in sign, there are 0, 2, or 4 negative real roots.

	2	15	31	20	4
-1	2	13	18	2	2
-2	2	11	9	2	0

$\Rightarrow$ $x = -2$ is a root, and so $(x+2)(2x^3 + 11x^2 + 9x + 2) = 0$.

	2	11	9	2
-4	2	3	-3	14
$-\frac{1}{2}$	2	10	4	0

$\Rightarrow$ $x = -\frac{1}{2}$ is a root, and so $(x+2)(2x+1)(x^2 + 5x + 2) = 0$.

Now, $x^2 + 5x + 2 = 0$ when $x = \frac{-5 \pm \sqrt{25 - 4 \cdot 1 \cdot 2}}{2} = \frac{-5 \pm \sqrt{17}}{2}$. Thus, the roots are $-\frac{1}{2}$, -2, and $\frac{-5 \pm \sqrt{17}}{2}$.

16. Find all rational roots of the polynomial $P(x) = 5x^4 - 27x^3 + 40x^2 - 22x + 4$, and then find the irrational roots, if any.

(a) $\frac{2}{5}$, 1, and $2 \pm \sqrt{2}$ (b) $\frac{2}{3}$ and $1 \pm \sqrt{2}$ (c) $-\frac{1}{2}$, $\frac{1}{4}$ and $\frac{3 \pm \sqrt{5}}{2}$ (d) $\frac{1}{3}$ and 1 (e) $-\frac{2}{3}$, $\frac{5}{2}$ and $3 \pm \sqrt{3}$

Answer: medium (a)

$5x^4 - 27x^3 + 40x^2 - 22x + 4 = 0$. The possible rational roots are $\pm 1, \pm 2, \pm 4, \pm\frac{1}{5}, \pm\frac{2}{5}, \pm\frac{4}{5}$. Since $P(x)$ has 4 variations in sign, there are 0, 2, or 4 positive real roots. Since $P(-x) = 5x^4 + 27x^3 + 40x^2 + 22x + 4$ has 0 variations in sign, there are no negative real roots.

	5	-27	40	-22	4
1	5	-22	18	-4	0

$\Rightarrow$ $x = 1$ is a zero, and so $(x-1)(5x^3 - 22x^2 + 18x - 4) = 0$.

	5	-22	18	-4
1	5	-17	1	-3
2	5	-12	-6	-16
4	5	-2	10	36
$\frac{2}{5}$	5	-20	10	0

$\Rightarrow$ $x = \frac{2}{5}$ is a root, and so $(x-1)(5x-2)(x^2 - 4x + 2) = 0$.

Now, $x^2 - 4x + 2 = 0$ when $x = \frac{4 \pm \sqrt{16 - 4 \cdot 2 \cdot 1}}{2} = 2 \pm \sqrt{2}$. Thus, the roots are $\frac{2}{5}$, 1, $2 \pm \sqrt{2}$.

17. Find all rational roots of the polynomial $P(x) = x^4 - \frac{11}{4}x^3 - \frac{11}{2}x^2 + \frac{5}{4}x + 3$, and then find the irrational roots, if any.

(a) -2 and 3 (b) -1 and 3 (c) -2, $\frac{1}{4}$ and 3 (d) ± 2 and $\pm\frac{1}{4}$ (e) -1, $\frac{3}{4}$ and 4

Answer: medium (e)

$x^4 - \frac{11}{4}x^3 - \frac{11}{2}x^2 + \frac{5}{4}x + 3 = 0 \Leftrightarrow 4x^4 - 11x^3 - 22x^2 + 5x + 12 = 0$. The possible rational roots are $\pm 1, \pm 2, \pm 3, \pm 4, \pm 6, \pm 12, \pm\frac{1}{2}, \pm\frac{3}{2}, \pm\frac{1}{4}, \pm\frac{3}{4}$. Since $P(x)$ has 2 variations in sign, there are 0 or 2 positive roots. Since $P(-x) = 4x^4 + 11x^3 - 22x^2 - 5x + 12$ has 2 variations in sign, there are 0 or 2 negative real roots.

	4	-11	-22	5	12
1	4	-7	-29	-24	-12
2	4	-3	-28	-51	-90
3	4	1	-19	-52	-144
4	4	5	-2	-3	0

$\Rightarrow$ $x = 4$ is a root, and so $(x-4)(4x^3 + 5x^2 - 2x - 3) = 0$.

	4	5	-2	-3	
4	4	21	82	325	$\Rightarrow$ all positive
$\frac{3}{4}$	4	8	4	0	

$\Rightarrow$ $x = \frac{3}{4}$ is a root, and so $(x-4)(4x-3)(x^2 + 2x + 1) = 0 \Leftrightarrow$ $(x-4)(4x-3)(x+1)^2 = 0$. Therefore, the roots are -1, $\frac{3}{4}$, and 4.

18. Find all rational roots of the polynomial $P(x) = 8x^5 - 14x^4 - 22x^3 + 57x^2 - 35x + 6$, and then find the irrational roots, if any.

(a) $2, \frac{1}{4}, -2$, and $\frac{3\pm\sqrt{2}}{2}$ (b) $1, \frac{3}{4}, -2$, and $\frac{2\pm\sqrt{2}}{2}$ (c) $-\frac{3}{4}$, 2, and $1\pm\sqrt{2}$
(d) $\pm 1, \pm 2$, and $\sqrt{2}$ (e) $-1, \frac{1}{4}, \frac{3}{4}$, and 2

Answer: medium (b)

$8x^5 - 14x^4 - 22x^3 + 57x^2 - 35x + 6 = 0$. The possible rational roots are $\pm 1, \pm 2, \pm 3, \pm 6, \pm\frac{1}{2}, \pm\frac{3}{2}, \pm\frac{1}{4}, \pm\frac{3}{4}, \pm\frac{1}{8}, \pm\frac{3}{8}$. Since $P(x)$ has 4 variations in sign, there are 0, 2, or 4 positive real roots. Since $P(-x) = -8x^5 - 14x^4 + 22x^3 + 57x^2 + 35x + 6$ has 1 variation in sign, there is 1 negative real root.

	8	−14	−22	57	−35	6
−1	8	−22	0	57	−92	98
−2	8	−30	38	−19	3	0

$\Rightarrow$ $x = -2$ is a root, and so

$(x + 2)(8x^4 - 30x^3 + 38x^2 - 19x + 3) = 0$. All the other real roots are positive.

	8	−30	38	−19	3
1	8	−22	16	−3	0

$\Rightarrow$ $x = 1$ is a root, and so $(x + 2)(x - 1)(8x^3 - 22x^2 + 16x - 3) = 0$.

	8	−22	16	−3
1	8	−14	2	−1
$\frac{1}{2}$	8	−18	7	$\frac{1}{2}$
$\frac{3}{2}$	8	−10	1	$-\frac{3}{2}$
$\frac{1}{4}$	8	−20	11	$-\frac{1}{4}$
$\frac{3}{4}$	8	−16	4	0

$\Rightarrow$ $x = \frac{3}{4}$ is a root, and so $(x + 2)(x - 1)(4x - 3)(2x^2 - 4x + 1) = 0$.

Now, $2x^2 - 4x + 1 = 0$ when $x = \frac{4\pm\sqrt{16-4\cdot 2\cdot 1}}{4} = \frac{2\pm\sqrt{2}}{2}$. Thus the roots are $1, \frac{3}{4}, -2$, and $\frac{2\pm\sqrt{2}}{2}$.

19. Find all rational roots of the polynomial $P(x) = \frac{2}{3}x^6 - x^5 - \frac{13}{3}x^4 + \frac{29}{3}x^3 - 9x^2 + \frac{32}{3}x - 4$, and then find the irrational roots, if any.

(a) $-3, \frac{1}{2}$, and 2 (b) $-3, -\frac{1}{2}$, and 1 (c) 1, 2, and $\frac{5}{2}$ (d) $\frac{2}{3}, \frac{4}{3}$, and 2 (e) $-\frac{1}{2}$, 1, and 3

Answer: medium (a)

$\frac{2}{3}x^6 - x^5 - \frac{13}{3}x^4 + \frac{29}{3}x^3 - 9x^2 + \frac{32}{3}x - 4 = 0 \quad\Leftrightarrow\quad 2x^6 - 3x^5 - 13x^4 + 29x^3 - 27x^2 + 32x - 12 = 0$. The possible rational roots are $\pm 1, \pm 2, \pm 3, \pm 4, \pm 6, \pm 12, \pm\frac{1}{2}, \pm\frac{3}{2}$. Since $P(x)$ has 5 variations in sign, there are 1, 3, or 5 positive real roots. Since $P(-x) = 2x^6 + 3x^5 - 13x^4 - 29x^3 - 27x^2 - 32x - 12$ has 1 variation in sign, there is 1 negative real root.

	2	−3	−13	29	−27	32	−12
−1	2	−5	−8	37	−64	96	−108
−2	2	−7	1	27	−81	194	−400
−3	2	−9	14	−13	12	−4	0

$\Rightarrow$ $x = -3$ is a root, and so

$(x + 3)(2x^5 - 9x^4 + 14x^3 - 13x^2 + 12x - 4) = 0$.

	2	−9	14	−13	12	−4
1	2	−7	7	−6	6	2
2	2	−5	4	−5	2	0

$\Rightarrow$ $x = 2$ is a root, and so

$(x + 3)(x - 2)(2x^4 - 5x^3 + 4x^2 - 5x + 2) = 0$

	2	−5	4	−5	2
2	2	−1	2	−1	0

$\Rightarrow$ $x = 2$ is a root, and so $(x + 3)(x - 2)^2(2x^3 - x^2 + 2x - 1) = 0 \Leftrightarrow$ $(x + 3)(x - 2)^2(2x - 1)(x^2 + 1) = 0$. Therefore, the real roots are $-3, \frac{1}{2}$, and 2.

20. Show that the polynomial $P(x) = 4x^4 - x^3 + x + 4$ does not have any rational zeros.

Answer: medium

$P(x) = 4x^4 - x^3 + x + 4$ has possible rational zeros $\pm 1, \pm 2, \pm 4, \pm\frac{1}{2}, \pm\frac{1}{4}$.

	4	−1	0	1	4	
$\frac{1}{2}$	4	1	$\frac{1}{2}$	$\frac{5}{4}$	$\frac{37}{8}$	$\Rightarrow$ all positive, so $\frac{1}{2}$ is an upper bound
$\frac{1}{4}$	4	0	0	1	$\frac{17}{4}$	
$-\frac{1}{4}$	4	−2	$\frac{1}{2}$	$\frac{7}{8}$	$\frac{121}{32}$	
$-\frac{1}{2}$	4	−3	$\frac{3}{2}$	$\frac{1}{4}$	$\frac{31}{8}$	
−1	4	−5	5	−4	8	$\Rightarrow$ alternately positive and negative, so −1 is a lower bound.

Therefore, there are no rational zeros.

21. Show that the polynomial $P(x) = x^{50} - 5x^{25} + x^2 - 1$ does not have any rational zeros.

Answer: medium

$P(x) = x^{50} - 5x^{25} + x^2 - 1$ has possible rational roots ± 1. Since $P(1) = 1 - 5 + 1 - 1 = -4$ and $P(-1) = 1 + 5 + 1 - 1 = 6$, there are no rational zeros.

22. Consider the polynomial $P(x) = x^3 + kx + 4$, where k is a real number.

(a) Find all values of k for which this polynomial has at least one rational root.

(b) For which values of k does this polynomial have one rational root and two irrational roots?

(c) What is the largest number of rational roots this polynomial can have?

Answer: hard

$x^3 + kx + 4 = 0$ has possible rational roots $\pm 1,\ \pm 2,\ \pm 4$.

	1	0	k	4
1	1	1	$k+1$	$k+5$
2	1	2	$k+4$	$2k+12$
4	1	4	$k+16$	$4k+68$
-1	1	-1	$k+1$	$-k+3$
-2	1	-2	$k+4$	$-2k-4$
-4	1	-4	$k+16$	$-4k-60$

(a) The equation will have a rational root if at least one of the remainders in the right-hand column is zero. This gives the following possible values for k: $-5, -6, -17, 3, -2$, and -15.

(b) If k is positive, then by Descartes' Rule, the equation has no positive and 1 negative real root. If k is negative, then the equation has 2 or 0 positive real roots and 1 negative real root. Thus from the list in part (a), $k = -5, -6$, or -17 will satisfy the conditions of part (b), while $k = 3$ will not. It remains to check $k = -2$ and $k = -15$.

$k = -2$: $x^3 - 2x + 4 = (x+2)(x^2 - 2x + 2)$. The zeros of the second factor are $x = \frac{2 \pm \sqrt{4-8}}{2}$ which are not real, so there are no irrational roots.

$k = -15$: $x^3 - 15x + 4 = (x+4)(x^2 - 4x + 1)$. The zeros of the second factor are $x = \frac{4 \pm \sqrt{16-4}}{2} = 2 \pm \sqrt{3}$, both of which are irrational. Thus, $k = -5, -6, -15$, or -17.

(c) One

23. Find all real zeros of the polynomial $p(x) = x^3 - 6x^2 + 11x - 6$.

Answer: easy

The possible rational zeros are ± 1, ± 2, ± 3, and ± 6. Use Descartes□Rule of Signs or a graphing calculator to help determine the zeros.

1	1	-6	11	-6
		1	-5	6
	1	-5	6	0

Factoring the quotient $x^2 - 5x + 6$, gives us $(x-2)(x-3)$. Thus $p(x) = x^3 - 6x^2 + 11x - 6 = (x-1)(x-2)(x-3)$. All real zeros of $p(x)$ are 1, 2, and 3.

24. Find all real zeros of the polynomial $p(x) = x^4 - 9x^3 + 15x^2 - 7x$.

Answer: medium

First factor x from $p(x)$ to get $x(x^3 - 9x^2 + 15x - 7)$. The possible rational zeros of $x^3 - 9x^2 + 15x - 7$ are ± 1 and ± 7. Use Descartes□Rule of Signs or a graphing calculator to help determine the zeros.

7	1	-9	15	-7
		7	-14	7
	1	-2	1	0

Factoring the quotient $x^2 - 2x + 1$, gives us $(x-1)(x-1)$. Thus $p(x) = x^4 - 9x^3 + 15x^2 - 7x = x(x-7)(x-1)^2$. All real zeros of $p(x)$ are 0, 1, and 7.

25. Find all real zeros of the polynomial $p(x) = 18x^4 - 33x^3 - 7x^2 + 32x - 12$.
Answer: medium
The possible rational zeros of $18x^4 - 33x^3 - 7x^2 + 32x - 12$ are $\pm1, \pm2, \pm3, \pm4, \pm6, \pm12, \pm\frac{1}{2}, \pm\frac{1}{3}, \pm\frac{1}{4}, \pm\frac{1}{6}, \pm\frac{1}{9}, \pm\frac{1}{18}, \pm\frac{2}{3}, \pm\frac{2}{9}, \pm\frac{3}{2}, \pm\frac{3}{4}, \pm\frac{4}{9}$, and $\pm\frac{4}{3}$. Use Descartes’ Rule of Signs and/or a graphing calculator to help determine the zeros.

-1	18	-33	-7	32	-12
		-18	51	-44	12
$\frac{2}{3}$	18	-51	44	-12	0
		12	-26	12	
	18	-39	18	0	

Factoring the quotient $18x^2 - 39x + 18$, gives us $3(3x-2)(2x-3)$. Thus
$p(x) = 18x^4 - 33x^3 - 7x^2 + 32x - 12 = (x+1)(2x-3)(3x-2)^2$. All real zeros of $p(x)$ are $x = -1, \frac{3}{2}$, and $\frac{2}{3}$.

26. Find all real zeros of the polynomial $p(x) = x^6 + 2x^5 - 7x^4 - 20x^3 - 12x^2$.
Answer: medium
First factor x^2 from $p(x)$ to get $x^2(x^4 + 2x^3 - 7x^2 - 20x - 12)$. The possible rational zeros of $x^4 + 2x^3 - 7x^2 - 20x - 12$ are $\pm1, \pm2, \pm3, \pm4, \pm6$, and ±12. Use Descartes’ Rule of Signs or a graphing calculator to help determine the zeros.

3	1	2	-7	-20	-12
		3	15	24	12
-1	1	5	8	4	0
		-1	-4	-4	
	1	4	4	0	

Factoring the quotient $x^2 + 4x + 4$, gives us $(x+2)(x+2)$. Thus
$p(x) = x^6 + 2x^5 - 7x^4 - 20x^3 - 12x^2 = x^2(x+1)(x-3)(x+2)^2$. All real zeros of $p(x)$ are $x = -2, -1, 0$, and 3.

27. Find all real zeros of the polynomial $p(x) = x^4 - 3x^3 - 2x^2 + 6x + 4$.
Answer: medium
The possible rational zeros of $x^4 - 3x^3 - 2x^2 + 6x + 4$ are $\pm1, \pm2$, and ±4. Use Descartes’ Rule of Signs or a graphing calculator to help determine the zeros.

-1	1	-3	-2	6	4
		-1	4	-2	-4
2	1	-4	2	4	0
		2	-4	-4	
	1	-2	-2	0	

Use the quadratic formula to find the zeros of the quotient $x^2 - 2x - 2 \Rightarrow x = 1 + \sqrt{3}, x = 1 - \sqrt{3}$ .Thus all real zeros of $p(x)$ are $-1, 2, 1 + \sqrt{3}$, and $1 - \sqrt{3}$.

28. Find all real zeros of the polynomial $p(x) = -x^3 + 9x^2 - 24x + 16$ and sketch the graph of p.
Answer: medium
The possible rational zeros of $-x^3 + 9x^2 - 24x + 16$ are ±1, ±2, ±4, ±8, and ±16. According to Descartes' Rule of Signs, this polynomial only has positive real zeros.

-1	-1	9	-24	16
		-1	8	-16
	-1	8	-16	

Factoring the quotient $-x^2 + 8x - 16$ gives us $-(x-4)^2$. Thus $p(x) = -x^3 + 9x^2 - 24x + 16 = -(x-1)(x-4)^2$. All real zeros of $p(x)$ are 1 and 4.

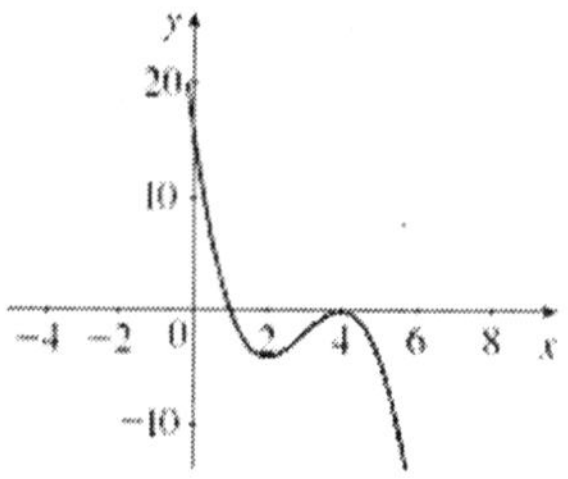

29. Find all real zeros of the polynomial $p(x) = x^5 - 4x^4 - 13x^3 + 70x^2 - 90x + 36$ and sketch the graph of p.
Answer: medium
The possible rational zeros of $x^5 - 4x^4 - 13x^3 + 70x^2 - 90x + 36$ are ±1, ±2, ±3, ±4, ±6, ±9, ±12, ±18, and ±36. A graphing device is helpful in determining the zeros.

1	1	-4	-13	70	-90	36
		1	-3	-16	54	-36
1	1	-3	-16	54	-36	0
		1	-2	-18	36	
2	1	-2	-18	36	0	
		2	0	-36		
	1	0	-18	0		

Solving the quotient $x^2 - 18 = 0 \;\Rightarrow\; x = \pm3\sqrt{2}$. Thus all real zeros of $p(x)$ are 1, 2, and $\pm3\sqrt{2}$.

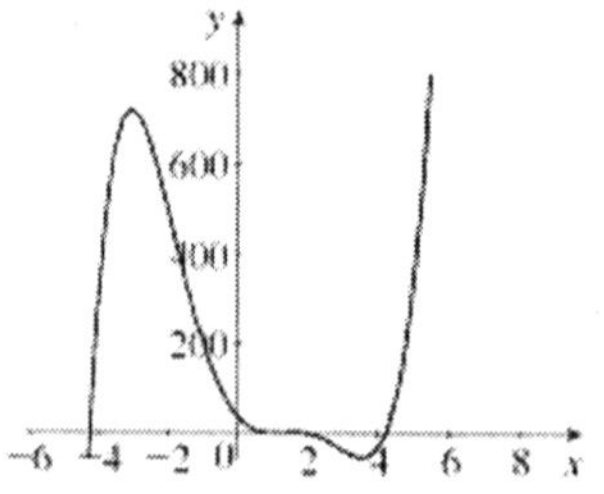

GC 30. Use a graphing device to find all real solutions of the equation $x^3 - 5x^2 - 4 = 0$.
Answer: easy
$x \approx 5.2$

GC 31. Use a graphing device to find all real solutions of the equation $3x^4 + 8x^3 + 2x^2 - 5x - 3 = 0$.
Answer: easy
$x \approx -2.07$, $x \approx 0.82$

GC 32. Use a graphing device to find all real solutions of the equation $x^5 - 1.9x^4 - 0.85x^3 - 4.5x^2 + 9.00x + 5.00 = 0$.
Answer: easy
$x \approx -0.45$, $x \approx 1.90$, $x \approx 2.09$

SECTION 5.4

1. Find all solutions of the polynomial $P(x) = 2x^2 - 3x + 3$.
(a) $\frac{2}{3} \pm \frac{i\sqrt{13}}{2}$ (b) $\frac{3}{4} \pm \frac{i\sqrt{7}}{16}$ (c) $\frac{3}{4} \pm \frac{i\sqrt{15}}{4}$ (d) $\frac{2}{3} \pm \frac{i\sqrt{13}}{4}$ (e) $\pm\frac{3}{4}$
Answer: easy (c)
$$2x^2 - 3x + 3 = 0 \Leftrightarrow x = \frac{-(-3) \pm \sqrt{(-3)^2 - 4 \cdot 2 \cdot 3}}{2 \cdot 2} = \frac{3 \pm \sqrt{-15}}{4} = \frac{3}{4} \pm \frac{i\sqrt{15}}{4}$$

2. Find all solutions of the polynomial $P(x) = x^3 + 8 = 0$.
(a) ± 2 and $1 \pm i\sqrt{2}$ (b) 2 and $-1 \pm i\sqrt{2}$ (c) -2 and $-1 \pm i\sqrt{2}$ (d) ± 2 and $-2 \pm i\sqrt{3}$ (e) -2 and $1 \pm i\sqrt{3}$
Answer: medium (e)
$$x^3 + 8 = 0 \quad \Leftrightarrow \quad (x+2)(x^2 - 2x + 4) = 0 \quad \Leftrightarrow \quad x = -2 \text{ or } \frac{2 \pm \sqrt{4 - 16}}{2} = 1 \pm i\sqrt{3}$$

3. Find all solutions of the polynomial $P(\theta) = \theta^4 + \theta^3 + \theta^2 = 0$.
(a) 0, 1 and $\pm i$ (b) 0 and $-\frac{1}{2} \pm \frac{i\sqrt{3}}{2}$ (c) $0, \pm\frac{1}{2}$ and $\sqrt{2}i$ (d) 0 and $\pm\sqrt{2}i$ (e) $\pm\sqrt{2}$ and $\pm\sqrt{2}i$
Answer: easy (b)
$$\theta^4 + \theta^3 + \theta^2 = 0 \quad \Leftrightarrow \quad \theta^2(\theta^2 + \theta + 1) = 0 \quad \Leftrightarrow \quad \theta = 0 \text{ or } \theta = \frac{-1 \pm \sqrt{1 - 4 \cdot 1 \cdot 1}}{2 \cdot 1} = \frac{-1 \pm \sqrt{-3}}{2} = -\frac{1}{2} \pm \frac{i\sqrt{3}}{2}$$

4. Find all solutions of the polynomial $P(x) = x^4 + 4x^2 + 4$.
(a) $\pm i\sqrt{2}$ (b) $\pm 2i$ (c) $\pm\sqrt{2}$ and $\pm i\sqrt{2}$ (d) ± 2 and $\pm i\sqrt{2}$ (e) $\pm 2i$ and $\pm i\sqrt{2}$
Answer: medium (a)
$$x^4 + 4x^2 + 4 = 0 \quad \Leftrightarrow \quad (x^2 + 2)^2 = 0 \quad \Leftrightarrow \quad x^2 + 2 = 0 \quad \Leftrightarrow \quad x = \pm\sqrt{-2} = \pm i\sqrt{2}$$

5. Find all solutions of the polynomial $P(z) = z^3 - iz^2 + z - i = 0$.
(a) $\pm\sqrt{2}i$ (b) -1 and $\pm i$ (c) ± 1 and $-i$ (d) 1 and $\pm i$ (e) $\pm i$
Answer: medium (e)
$$z^3 - iz^2 + z - i = 0 \quad \Leftrightarrow \quad (z - i)(z^2 + 1) = 0 \quad \Leftrightarrow \quad (z - i)(z - i)(z + i) = 0 \quad \Leftrightarrow \quad z = \pm i$$

6. Find a polynomial with integer coefficients that satisfy the conditions: $Q(x)$ has degree 5 and zeros $1 + i\sqrt{3}$ and 4, -1 is a zero of multiplicity 2, and $Q(0) = 64$.
Answer: medium
Since $1 + i\sqrt{3}$ is a zero, $1 - i\sqrt{3}$ is also a zero. Since the other zeros are 4 and -1 (which has multiplicity 2), a factorization is $Q(x) = C\left(x - \left[1 + i\sqrt{3}\right]\right)\left(x - \left[1 - i\sqrt{3}\right]\right)(x - 4)(x + 1)^2 = C(x^2 - 2x + 4)(x - 4)(x^2 + 2x + 1)$
$= C(x^2 - 2x + 4)(x^3 - 2x^2 - 7x - 4) = C(x^5 - 4x^4 + x^3 + 2x^2 - 20x - 16)$.
Since $Q(0) = -16C = 64$, $C = -4$. Therefore, $Q(x) = -4x^5 + 16x^4 - 4x^3 - 8x^2 + 80x + 64$.

7. Find the polynomial $S(x)$ with real coefficients that has degree 4 and zeros $1 + i$ and $3 - 4i$, and in which the coefficient of x^2 is 7.
Answer: medium
Since the zeros are $1 + i$ and $3 - 4i$, the remaining zeros are $1 - i$ and $3 + 4i$. Thus, a factorization is
$$\begin{aligned} S(x) &= C(x - [1 + i])(x - [1 - i])(x - [3 - 4i])(x - [3 + 4i]) \\ &= C(x^2 - 2x + [1 - i][1 + i])(x^2 - 6x + [3 - 4i][3 + 4i]) = C(x^2 - 2x + 2)(x^2 - 6x + 25) \\ &= C(x^4 - 6x^3 + 25x^2 - 2x^3 + 12x^2 - 50x + 2x^2 - 12x + 50) = C(x^4 - 8x^3 + 39x^2 - 62x + 50). \end{aligned}$$
Since the coefficient of x^2 is 7, it follows that $C = \frac{7}{39}$ and so $S(x) = \frac{7}{39}x^4 - \frac{56}{39}x^3 + 7x^2 - \frac{434}{39}x + \frac{350}{39}$.

8. Find the polynomial $U(x)$ with real coefficients that has degree 3, with $U(3i) = 0$, $U(2) = 0$, and $U(1) = -5$.

Answer: medium

Since $U(x)$ has zeros $3i$, $-3i$, and 2, a factorization is
$U(x) = C(x-3i)(x+3i)(x-2) = C(x^2-9i^2)(x-2) = C(x^2+9)(x-2) = C(x^3-2x^2+9x-18)$. Since
$U(1) = -5 = C(1^3 - 2\cdot 1^2 + 9\cdot 1 - 18) = -10C \quad \Leftrightarrow \quad C = \frac{1}{2}$. Therefore, $U(x) = \frac{1}{2}x^3 - x^2 + \frac{9}{2}x - 9$.

9. Show that $x = \frac{5}{2} - \frac{\sqrt{3}}{2}i$ is a root of the polynomial $P(x) = x^4 - 13x^2 + 10x + 35$, and then find all roots.

Answer: medium

$x^4 - 13x^2 + 10x + 35 = 0$. If $x = \frac{5}{2} - \frac{\sqrt{3}}{2}i$ is a solution then so is $\overline{x} = \frac{5}{2} + \frac{\sqrt{3}}{2}i$. This will be the case only if
$\left[x - \left(\frac{5}{2} - \frac{\sqrt{3}}{2}i\right)\right]\left[x - \left(\frac{5}{2} + \frac{\sqrt{3}}{2}i\right)\right] = x^2 - 5x + 7$ divides the polynomial evenly.

$$\begin{array}{r|lllll}
 & x^2 & +5x & +5 & & \\
\hline
x^2-5x+7 & x^4 & +0x^3 & -13x^2 & +10x & +35 \\
 & x^4 & -5x^3 & +7x^2 & & \\
\hline
 & & 5x^3 & -20x^2 & & \\
 & & 5x^3 & -25x^2 & +35x & \\
\hline
 & & & 5x^2 & -25x & \\
 & & & 5x^2 & -25x & +35 \\
\hline
 & & & & & 0
\end{array}$$

Since the remainder is 0, both x and $\overline{x}$ must be roots. The other two roots are $\dfrac{-5 \pm \sqrt{25 - 4\cdot 5}}{2} = \dfrac{-5 \pm \sqrt{5}}{2}$.

10. Show that $x = 3i$ is a solution of the equation $2x^6 - x^5 + 30x^4 + 25x^3 + 88x^2 + 306x - 180 = 0$, and then find all solutions.

Answer: hard

$2x^6 - x^5 + 30x^4 + 25x^3 + 88x^2 + 306x - 180 = 0$.

$P(3i) = 2(3i)^6 - (3i)^5 + 30(3i)^4 + 25(3i)^3 + 88(3i)^2 + 306(3i) - 180$
$= -1458 - 243i + 2430 - 675i - 792 + 918i - 180 = 0$.

Thus, solutions are $\pm 3i$, and $(x-3i)(x+3i) = x^2 + 9$ divides the polynomial evenly.

$$\begin{array}{r|lllllll}
 & 2x^4 & -x^3 & +12x^2 & +34x & -20 & & \\
\hline
x^2+9 & 2x^6 & -x^5 & +30x^4 & +25x^3 & +88x^2 & +306x & -180 \\
 & 2x^6 & & +18x^4 & & & & \\
\hline
 & & -x^5 & +12x^4 & & & & \\
 & & -x^5 & & -9x^3 & & & \\
\hline
 & & & 12x^4 & +34x^3 & & & \\
 & & & 12x^4 & & +108x^2 & & \\
\hline
 & & & & 34x^3 & -20x^2 & & \\
 & & & & 34x^3 & & +306x & \\
\hline
 & & & & & -20x^2 & & \\
 & & & & & -20x^2 & & -180 \\
\hline
 & & & & & & & 0
\end{array}$$

Hence, a factorization is $(x^2+9)(2x^4 - x^3 + 12x^2 + 34x - 20)$. The possible rational roots of
$2x^4 - x^3 + 12x^2 + 34x - 20 = 0$ are $\pm 1, \pm 2, \pm 4, \pm 5, \pm 10, \pm 20, \pm\frac{1}{2}, \pm\frac{5}{2}$.

	2	−1	12	34	−20		
1	2	1	13	47	27	$\Rightarrow$	upper bound
$\frac{1}{2}$	2	0	12	40	0	$\Rightarrow$	$x = \frac{1}{2}$ is a zero, and so $(x^2+9)(2x-1)(x^3+6x+20) = 0$.

	1	0	6	20		
$\frac{1}{2}$	1	$\frac{1}{2}$	$\frac{25}{4}$	$\frac{185}{8}$	$\Rightarrow$	upper bound
−1	1	−1	7	13		
−2	1	−2	10	0	$\Rightarrow$	$x = -2$ is a zero, and so $(x^2+9)(2x-1)(x+2)(x^2-2x+10) = 0$.

Now, $x^2 - 2x + 10 = 0$ when $x = \dfrac{2 \pm \sqrt{4 - 4\cdot 1\cdot 10}}{2} = \dfrac{2 \pm \sqrt{-36}}{2} = 1 \pm 3i$. Therefore, the solutions are $\pm 3i$, $\frac{1}{2}$, -2, and $1 \pm 3i$.

11. Find all roots of the polynomial $P(x) = x^3 - 27$.

(a) $-3, \dfrac{-3 \pm 2i\sqrt{3}}{2}$ (b) $3, \dfrac{-3 \pm 3i\sqrt{3}}{2}$ (c) $\sqrt{3}, \dfrac{-3 \pm 3i\sqrt{2}}{2}$ (d) $3, \dfrac{-3 \pm 2i\sqrt{3}}{2}$ (e) $3\sqrt{3}, \dfrac{-3 \pm 2i\sqrt{3}}{2}$

Answer: easy (b)

$x^3 - 27 = 0 \quad \Leftrightarrow \quad (x-3)(x^2+3x+9) = 0 \quad \Leftrightarrow \quad x - 3 = 0$ or $x^2 + 3x + 9 = 0 \quad \Leftrightarrow \quad x = 3$ or $x = \dfrac{-3 \pm \sqrt{9 - 4 \cdot 9}}{2} = \dfrac{-3 \pm \sqrt{-27}}{2} = \dfrac{-3 \pm 3i\sqrt{3}}{2}$. Therefore, the solutions are 3, $\dfrac{-3 \pm 3i\sqrt{3}}{2}$.

12. Find all roots of the polynomial $P(x) = x^4 + 9$.

(a) $\dfrac{\pm\sqrt{3} \pm i\sqrt{3}}{2}$ (b) $\dfrac{-\sqrt{6} \pm i\sqrt{6}}{2}$ (c) $\pm\sqrt{3}i$ and $\pm\sqrt{3}$ (d) $\dfrac{\pm\sqrt{6} \pm i\sqrt{6}}{2}$ (e) $\dfrac{\pm\sqrt{3} \pm i\sqrt{6}}{2}$

Answer: medium (d)

$x^4 + 9 = 0 \quad \Leftrightarrow \quad x^4 + 6x^2 + 9 - 6x^2 = 0 \quad \Leftrightarrow \quad (x^2+3)^2 - \left(\sqrt{6}x\right)^2 = 0 \quad \Leftrightarrow$

$\left(x^2 + 3 + \sqrt{6}x\right)\left(x^2 + 3 - \sqrt{6}x\right) = 0 \quad \Leftrightarrow \quad x^2 + \sqrt{6}x + 3 = 0$ or $x^2 - \sqrt{6}x + 3 = 0 \quad \Leftrightarrow$

$x = \frac{-\sqrt{6} \pm \sqrt{6 - 4 \cdot 3 \cdot 1}}{2} = \frac{-\sqrt{6} \pm i\sqrt{6}}{2}$ or $x = \dfrac{\sqrt{6} \pm \sqrt{6 - 4 \cdot 3 \cdot 1}}{2} = \dfrac{\sqrt{6} \pm i\sqrt{6}}{2}$. Thus, the solutions are $x = \dfrac{\pm\sqrt{6} \pm i\sqrt{6}}{2}$.

13. Find all roots of the polynomial $P(x) = x^6 - 7x^3 - 8$.

(a) $2, -1, -1 \pm i\sqrt{3}$, and $\dfrac{1 \pm i\sqrt{3}}{2}$ (b) $-2, 1, -2 \pm i\sqrt{2}$, and $\dfrac{1 \pm i\sqrt{2}}{2}$ (c) $2, -1$, and $\frac{\pm 1 \pm i\sqrt{3}}{2}$

(d) $\pm 2, -2 \pm i\sqrt{2}$, and $\dfrac{1 \pm i\sqrt{2}}{2}$ (e) $2, -1, -2 \pm i\sqrt{3}$, and $\dfrac{2 \pm i\sqrt{3}}{2}$

Answer: medium (a)

$x^6 - 7x^3 - 8 = \left(x^3 - 8\right)\left(x^3 + 1\right) = (x-2)\left(x^2 + 2x + 4\right)(x+1)\left(x^2 - x + 1\right) = 0$. Therefore, $x = 2, -1$, $\frac{-2 \pm \sqrt{4 - 4 \cdot 4}}{2}$, or $\dfrac{1 \pm \sqrt{1 - 4 \cdot 1}}{2}$, and so the solutions are $2, -1, -1 \pm i\sqrt{3}$, and $\dfrac{1 \pm i\sqrt{3}}{2}$.

14. Find all roots of the polynomial $P(x) = x^3 + 6x^2 + 14x + 12$.

(a) -1 and $-1 \pm i\sqrt{2}$ (b) ± 2 and $1 - \sqrt{2}$ (c) -2 and $-2 \pm i\sqrt{2}$ (d) -3 and $-1 \pm i\sqrt{3}$ (e) -1 and $-1 \pm i\sqrt{3}$

Answer: medium (c)

$x^3 + 6x^2 + 14x + 12 = 0$ has possible rational roots $\pm 1, \pm 2, \pm 3, \pm 4, \pm 6, \pm 12$. Since all of the coefficients are positive, there are no positive real roots.

	1	6	14	12	
-1	1	5	9	3	
-2	1	4	6	0	$\Rightarrow$ $x = -2$ is a zero, and so $(x+2)(x^2 + 4x + 6) = 0$.

Now, $x^2 + 4x + 6 = 0$ when $x = \dfrac{-4 \pm \sqrt{16 - 4 \cdot 1 \cdot 6}}{2} = \dfrac{-4 \pm i\sqrt{8}}{2} = -2 \pm i\sqrt{2}$. Therefore, the solutions are -2 and $-2 \pm i\sqrt{2}$.

15. Find all roots of the polynomial $P(x) = 8x^4 + 36x^3 + 42x^2 + 9x + 10$.

(a) $-\frac{5}{2}, -2$, and $\pm\frac{1}{2}i$ (b) ± 2 and $\pm\frac{3}{2}i$ (c) $\pm\frac{5}{2}$ and $\pm 2i$ (d) $-1, 3$ and $\pm i$ (e) $-2, 3$ and $-1 \pm i\sqrt{2}$

Answer: hard (a)

$8x^4 + 36x^3 + 42x^2 + 9x + 10 = 0$ has possible rational roots $\pm 1, \pm 2, \pm 5, \pm 10, \pm\frac{1}{2}, \pm\frac{5}{2}, \pm\frac{1}{4}, \pm\frac{5}{4}, \pm\frac{1}{8}, \pm\frac{5}{8}$. Since all of the coefficients are positive, there are no positive real roots.

	8	36	42	9	10	
-1	8	28	14	-5	15	
-2	8	20	2	5	0	$\Rightarrow$ $x = -2$ is a zero and an upper bound, so $(x+2)(8x^3 + 20x^2 + 2x + 5) = 0$.
-5	8	-20	102	-505		$\Rightarrow$ lower bound
$-\frac{5}{2}$	8	0	2	0		$\Rightarrow$ $x = -\frac{5}{2}$ is a zero, and so $(x+2)(2x+5)(4x^2+1) = 0$.

Now, $4x^2 + 1 = 0$ when $x = \pm\frac{1}{2}i$. Therefore, the solutions are $-\frac{5}{2}$, -2, and $\pm\frac{1}{2}i$.

16. Find all roots of the polynomial $P(x) = 8x^6 - 24x^5 - 8x^4 + 23x^3 + 3x^2 + x - 3$.

(a) $\pm 1,\ 2,\ \frac{2}{3}$, and $\frac{-2 \pm i\sqrt{5}}{4}$ (b) $\pm 2,\ -1,\ \frac{1}{4}$, and $\frac{-1 \pm i\sqrt{5}}{4}$ (c) $\pm 1,\ \pm\frac{1}{2}$, and $\frac{-3 \pm i\sqrt{2}}{4}$

(d) $\pm 1,\ 3,\ \frac{1}{2}$, and $\frac{-1 \pm i\sqrt{3}}{4}$ (e) $\pm\frac{1}{2},\ \frac{-1 \pm i\sqrt{3}}{4}$, and $\frac{-3 \pm i\sqrt{5}}{4}$

Answer: hard (d)

$8x^6 - 24x^5 - 8x^4 + 23x^3 + 3x^2 + x - 3 = 0$ has possible rational roots $\pm 1, \pm 3, \pm\frac{1}{2}, \pm\frac{3}{2}, \pm\frac{1}{4}, \pm\frac{3}{4}, \pm\frac{1}{8}, \pm\frac{3}{8}$.

	8	−24	−8	23	3	1	−3
1	8	−16	−24	−1	2	3	0

$\Rightarrow$ $x = 1$ is a root, and so $(x - 1)(8x^5 - 16x^4 - 24x^3 - x^2 + 2x + 3) = 0$

	8	−16	−24	−1	2	3
1	8	−8	−32	−33	−31	−28
3	8	8	0	−1	−1	0

$\Rightarrow$ $x = 3$ is a root, and so $(x - 1)(x - 3)(8x^4 + 8x^3 - x - 1) = 0$.

By inspection $8x^4 + 8x^3 - x - 1 = (x + 1)(8x^3 - 1)$, and so $(x - 1)(x - 3)(x + 1)(8x^3 - 1) = 0$. Now,

$8x^3 - 1 = (2x - 1)(4x^2 + 2x + 1)$ and $4x^2 + 2x + 1 = 0$ when $x = \frac{-2 \pm \sqrt{4 - 4 \cdot 4 \cdot 1}}{2 \cdot 4} = \frac{-2 \pm \sqrt{-12}}{2 \cdot 4} = \frac{-1 \pm i\sqrt{3}}{4}$.

Thus, the solutions are $\pm 1,\ 3,\ \frac{1}{2}$, and $\frac{-1 \pm i\sqrt{3}}{4}$.

17. Find the complete factorization of the polynomial $x^4 + x^3 + 7x^2 + 9x - 18$.

Answer: medium

$P(x) = x^4 + x^3 + 7x^2 + 9x - 18$ has possible zeros $\pm 1, \pm 2, \pm 3, \pm 6, \pm 9, \pm 18$.

	1	1	7	9	−18
1	1	2	9	18	0

$\Rightarrow$ $x = 1$ is a zero, and so $(x - 1)(x^3 + 2x^2 + 9x + 18) = 0$.

By inspection $x^3 + 2x^2 + 9x + 18 = (x^2 + 9)(x + 2)$, and so $(x - 1)(x + 2)(x^2 + 9) = (x + 2)(x - 1)(x - 3i)(x + 3i)$.

18. Find the complete factorizations of the polynomial $4x^4 - 625$.

Answer: medium

$P(x) = 4x^4 - 625 = (2x^2 - 25)(2x^2 + 25) = 4\left(x - \frac{5}{\sqrt{2}}\right)\left(x + \frac{5}{\sqrt{2}}\right)\left(x - \frac{5}{\sqrt{2}}i\right)\left(x + \frac{5}{\sqrt{2}}i\right)$.

19. Find the complete factorization of the polynomial $x^5 - x^4 + 3x^3 - 3x^2 + 2x - 2$.

Answer: hard

$P(x) = x^5 - x^4 + 3x^3 - 3x^2 + 2x - 2 = 0$ has possible rational roots $\pm 1, \pm 2$.

	1	−1	3	−3	2	−2
1	1	0	3	0	2	0

$\Rightarrow$ $x = 1$ is a root, and so $P(x) = (x - 1)(x^4 + 3x^2 + 2)$.

Now, $x^4 + 3x^2 + 2 = (x^2 + 1)(x^2 + 2)$, and so

$P(x) = (x - 1)(x^2 + 1)(x^2 + 2) = (x - 1)(x - i)(x + i)\left(x - i\sqrt{2}\right)\left(x + i\sqrt{2}\right)$.

20. Find the value of the polynomial $2x^3 + 3ix^2 - 2ix - (15 + 3i)$ at the complex number 3.

Answer: easy

$P(x) = 2x^3 + 3ix^2 - 2ix - (15 + 3i)$. Then

$P(3) = 2 \cdot 3^3 + 3i \cdot 3^2 - 2i \cdot 3 - (15 + 3i) = 54 + 27i - 6i - 15 - 3i = 39 + 18i$.

21. (a) Show that $2i$ and $1 - i$ are both roots of the polynomial $P(x) = x^2 - (1 + i)x + (2 + 2i)$ but that $-2i$ and $1 + i$ are not.

(b) Explain why this does not violate the Conjugate Roots Theorem.

Answer: hard

(a) $P(x) = x^2 - (1 + i)x + (2 + 2i)$. Then, $P(2i) = (2i)^2 - 2i(1 + i) + 2 + 2i = -4 - 2i + 2 + 2 + 2i = 0$, and $P(1 - i) = (1 - i)^2 - (1 + i)(1 - i) + (2 + 2i) = 1 - 2i - 1 - 1 - 1 + 2 + 2i = 0$. Therefore, $2i$ and $1 - i$ are solutions of the equation $x^2 - (1 + i)x + (2 + 2i) = 0$. However,
$P(-2i) = (-2i)^2 - (1 + i)(-2i) + 2 + 2i = -4 + 2i - 2 + 2 + 2i = -4 + 4i$, and
$P(1 + i) = (1 + i)^2 - (1 + i)(1 + i) + 2 + 2i = 2 + 2i$. Since, $P(-2i) \neq 0$ and $P(1 + i) \neq 0$, $-2i$ and $1 + i$ are not solutions.

(b) This does not violate the Conjugate Roots Theorem because the coefficients of the polynomial $P(x)$ are not all real.

22. Factor the polynomial $p(x) = x^2 + 4$ completely and find all its zeros. State the multiplicity of each zero.
Answer: easy
Solving the polynomial $x^2 + 4 = 0$ directly gives us $x = \pm 2i$, each zero of multiplicity one. Therefore $p(x) = x^2 + 4 = (x - 2i)(x + 2i)$.

23. Factor the polynomial $p(x) = x^6 - 6x^4 + 9x^2$ completely and find all its zeros. State the multiplicity of each zero.
Answer: easy
$p(x) = x^2(x^2 - 6x + 9) = x^2(x - 3)(x - 3)$. The zeros are $x = 0$ and $x = 3$, each of multiplicity two.

24. Factor the polynomial $p(x) = x^3 - 8$ completely and find all its zeros. State the multiplicity of each zero.
Answer: easy
$p(x) = x^3 - 8 = (x - 2)(x^2 + 2x + 4)$. Using the quadratic formula, $x^2 + 2x + 4$ has zeros of $x = -1 + i\sqrt{3}$ and $x = -1 - i\sqrt{3}$. The complete factorization is $p(x) = (x - 2)(x + 1 - i\sqrt{3})(x + 1 + i\sqrt{3})$, and zeros are $x = 2$, $x = -1 + i\sqrt{3}$, and $x = -1 - i\sqrt{3}$, each of multiplicity one.

25. Factor the polynomial $p(x) = 256x^4 - 25$ completely and find all its zeros. State the multiplicity of each zero.
Answer: easy
$p(x) = 256x^4 - 25 = (16x^2 - 5)(16x^2 + 5) = (4x - \sqrt{5})(4x + \sqrt{5})(4x - i\sqrt{5})(4x + i\sqrt{5})$. Zeros are $x = \frac{1}{4}\sqrt{5}$, $x = -\frac{1}{4}\sqrt{5}$, $x = \frac{1}{4}i\sqrt{5}$, and $x = -\frac{1}{4}i\sqrt{5}$, each of multiplicity one.

26. Factor the polynomial $p(x) = x^8 + 16x^5 + 64x^2$ completely and find all its zeros. State the multiplicity of each zero.
Answer: medium
$p(x) = x^8 + 16x^5 + 64x^2 = x^2(x^6 + 16x^3 + 64) = x^2(x^3 + 8)(x^3 + 8) = x^2(x + 2)^2(x^2 - 2x + 4)^2$. At this point use the quadratic formula to find the zeros of $x^2 - 2x + 4$ to be $x = 1 - i\sqrt{3}$ and $x = 1 + i\sqrt{3}$. Thus, $p(x) = x^8 + 16x^5 + 64x^2 = x^2(x + 2)^2\left[x - (1 - i\sqrt{3})\right]^2\left[x - (1 + i\sqrt{3})\right]^2$ has zeros $x = 0$, $x = -2$, $x = 1 - i\sqrt{3}$, and $x = 1 + i\sqrt{3}$, each of multiplicity two.

27. Find a polynomial P of degree two with integer coefficients that has a zero of $1 - 2i$.
Answer: easy
If $1 - 2i$ is a zero of $P(x)$, then its complex conjugate $1 + 2i$ is also a zero. Thus $P(x) = [x - (1 - 2i)][x - (1 + 2i)] = x^2 - 2x + 5$.

28. Find a polynomial P of degree three with integer coefficients that has zeros of $1 - \sqrt{3}i$ and 0.
Answer: easy
If $1 - \sqrt{3}i$ is a zero of $P(x)$, then its complex conjugate $1 + \sqrt{3}i$ is also a zero. Thus $P(x) = x\left[x - (1 - \sqrt{3}i)\right]\left[x - (1 + \sqrt{3}i)\right] = x^3 - 2x^2 + 4x$.

29. Find a polynomial P of degree 4 with integer coefficients that has zeros of $-4i$ and $5i$, and that has 3 as the coefficient of its fourth-degree term.
Answer: medium
If $-4i$ and $5i$ are zeros of $P(x)$, then their complex conjugates $4i$ and $-5i$ are also zeros. Thus $P(x) = 3(x - 4i)(x + 4i)(x - 5i)(x + 5i) = 3x^4 + 123x^2 + 1200$.

30. Find a polynomial P of degree 5 with integer coefficients that has zeros of $-\frac{1}{2}$, 2, and i, with 2 being a zero of multiplicity two.
Answer: medium
$P(x) = \left(x + \frac{1}{2}\right)(x - 2)^2(x - i)(x + i) = x^5 + 3x^3 - \frac{7}{2}x^4 - \frac{3}{2}x^2 + 2x + 2$

31. Find a polynomial P of degree 5 with integer coefficients that has zeros of i and 2, with i being a zero of multiplicity two.
Answer: medium
$P(x) = 2(x-i)^2(x+i)^2(x-2) = 2x^5 - 4x^4 + 4x^3 - 8x^2 + 2x - 4$

SECTION 5.5

1. Find the x- and y-intercepts of the function $y = \dfrac{2}{x^3 - 2x + 8}$.
(a) x-intercept $\frac{1}{2}$; y-intercept $\frac{1}{3}$ (b) No x-intercept; y-intercept $\frac{1}{3}$ (c) No x-intercept; y-intercept $\frac{1}{4}$
(d) x-intercepts $\pm\frac{1}{4}$; y-intercept $\frac{1}{4}$ (e) x-intercept $\frac{1}{2}$; y-intercepts $\pm\frac{1}{4}$
Answer: easy (c)
$y = \dfrac{2}{x^3 - 2x + 8}$. When $x = 0$, $y = \dfrac{2}{0 - 0 + 8} = \dfrac{1}{4}$, and so the y-intercept is $\dfrac{1}{4}$. Since it is impossible for y to equal 0, there is no x-intercept.

2. Find the x- and y-intercepts of the function $y = \dfrac{x^4 + 16}{x^2 - 2x - 15}$.
(a) No x-intercept; y-intercept $-\frac{16}{15}$ (b) No x-intercept; y-intercepts $\pm\frac{14}{15}$ (c) No x-intercept; y-intercept $\frac{1}{15}$
(d) x-intercepts $\pm\frac{2}{15}$; y-intercept $\frac{16}{15}$ (e) x-intercept $\frac{2}{3}$; y-intercepts $\pm\frac{1}{3}$
Answer: medium (a)
$y = \dfrac{x^4 + 16}{x^2 - 2x - 15}$. When $x = 0$, $y = \dfrac{0 + 16}{0 - 0 - 15} = -\dfrac{16}{15}$, and so the y-intercept is $-\dfrac{16}{15}$. Since $x^4 + 16 > 0$ for all x, it is impossible for y to equal 0, and so there is no x-intercept.

3. Find the x- and y-intercepts of the function $y = \dfrac{x^2 + 12}{3x}$.
(a) No x-intercept; y-intercept $\frac{1}{4}$ (b) No x-intercept; y-intercepts $\pm\frac{1}{3}$ (c) x-intercept $\frac{2}{3}$; no y-intercept
(d) No intercept (e) x-intercept $\frac{2}{3}$; y-intercepts $\pm\frac{1}{3}$
Answer: easy (d)
$y = \dfrac{x^2 + 12}{3x}$. When $x = 0$, $y = \dfrac{0 + 12}{0}$ which is undefined, and so there is no y-intercept. Since $x^2 + 12 > 0$ for all x, it is impossible for y to equal 0, and so there is no x-intercept.

4. Find all asymptotes (including vertical, horizontal, and slant) of the function $y = \dfrac{3x + 1}{x - 3}$.
(a) No horizontal; vertical: $x = 3$ (b) Horizontal: $y = \frac{1}{3}$; vertical: $x = 3$ (c) Horizontal: $y = \frac{1}{3}$; vertical: $x = \frac{1}{3}$
(d) Horizontal: $y = 3$; vertical: $x = \pm 3$ (e) Horizontal: $y = 3$; vertical: $x = 3$
Answer: easy (e)
$y = \dfrac{3x + 1}{x - 3} = \dfrac{3 + 1/x}{1 - 3/x} \to 3$ as $x \to \infty$. The horizontal asymptote is $y = 3$. There is a vertical asymptote when $x - 3 = 0$ $\Leftrightarrow$ $x = 3$, and so the vertical asymptote is $x = 3$.

5. Find all asymptotes (including vertical, horizontal, and slant) of the function $y = \dfrac{2x - 5}{x^2 + x + 1}$.
(a) No asymptote (b) Horizontal: $y = 0$; vertical: $x = \frac{5}{2}$ (c) Horizontal: $y = 2$; vertical: $x = \frac{5}{2}$
(d) Horizontal: $y = 0$; no vertical (e) Horizontal: $y = \pm 1$; vertical: $x = -1$
Answer: easy (d)
$y = \dfrac{2x - 5}{x^2 + x + 1}$. The vertical asymptotes occur when $x^2 + x + 1 = 0$ $\Leftrightarrow$ $x = \dfrac{-1 \pm \sqrt{1 - 4 \cdot 1 \cdot 1}}{2} = \dfrac{-1 \pm \sqrt{-3}}{2}$ which is imaginary, and so there are no vertical asymptotes. Since $y = \dfrac{2x - 5}{x^2 + x + 1} = \dfrac{2/x - 5/x^2}{1 + 1/x + 1/x^2} \to 0$ as $x \to \infty$, $y = 0$ is the horizontal asymptote.

6. Find all asymptotes (including vertical, horizontal, and slant) of the function $y = \dfrac{(x-1)(x-2)(x-3)}{(x-4)(x-5)}$.

(a) No horizontal; vertical: $x = 4$ and $x = 5$
(b) No horizontal; vertical: $x = 4$ and $x = 5$; slant: $y = x + 3$
(c) Horizontal: $y = 1$; vertical: $x = 4$ and $x = 5$
(d) No horizontal; vertical: $x = 4$ and $x = 5$; slant: $y = x - 3$
(e) Horizontal: $y = -4$ and $y = -5$; vertical: $x = 1$, $x = 2$ and $x = 3$

Answer: medium (b)

$y = \dfrac{(x-1)(x-2)(x-3)}{(x-4)(x-5)}$. The vertical asymptotes occur when $(x-4)(x-5) = 0 \quad \Leftrightarrow \quad x = 4$ or $x = 5$, and so the vertical asymptotes are $x = 4$ and $x = 5$. Since $y = \dfrac{(x-1)(x-2)(x-3)}{(x-4)(x-5)}$ $= \frac{x^3-6x^2+11x-6}{x^2-9x+20} = x + 3 + \frac{18x-66}{x^2-9x+20} \to \infty$ as $x \to \infty$, there is no horizontal asymptote but $y = x + 3$ is the slant asymptote.

7. Find all asymptotes (including vertical, horizontal, and slant) of the function $y = \dfrac{x^3+3x^2}{x^2+x+16}$.

(a) No horizontal; no vertical; slant: $y = x + 2$
(b) Horizontal: $y = 2$; vertical: $x = 1$; slant: $y = x + 1$
(c) No horizontal; vertical: $x = 1$; slant: $y = x + 1$
(d) Horizontal: $y = 1$; no vertical; slant: $y = x$
(e) No asymptote

Answer: medium (a)

$y = \dfrac{x^3+3x^2}{x^2+x+16} = x + 2 - \dfrac{18x+32}{x^2+x+16} \to \infty$ as $x \to \infty$. Hence, there is no horizontal asymptote but $y = x + 2$ is the slant asymptote. The vertical asymptotes occur when $x^2 + x + 16 = 0 \quad \Leftrightarrow \quad x = \dfrac{-1 \pm \sqrt{1 - 4 \cdot 1 \cdot 16}}{2}$ which is imaginary and so there are no vertical asymptotes.

8. Find all asymptotes (including vertical, horizontal, and slant) of the function $y = \dfrac{6x^4+x^2-1}{x^2+64}$.

(a) Horizontal: $y = 4$; vertical: $x = -4$
(b) Horizontal: $y = -4$; vertical: $x = 2$
(c) No horizontal; vertical: $x = \sqrt{3}$
(d) No horizontal; no vertical; slant: $y = 2x$
(e) No asymptote

Answer: medium (e)

$y = \dfrac{6x^4+x^2-1}{x^2+64}$. Since $x^2 + 64 \geq 64$ for all x, there are no vertical asymptotes. There are also no horizontal or slant asymptotes.

9. Find intercepts and asymptotes, and then graph the rational function $y = \dfrac{x+6}{x-2}$.

Answer: easy

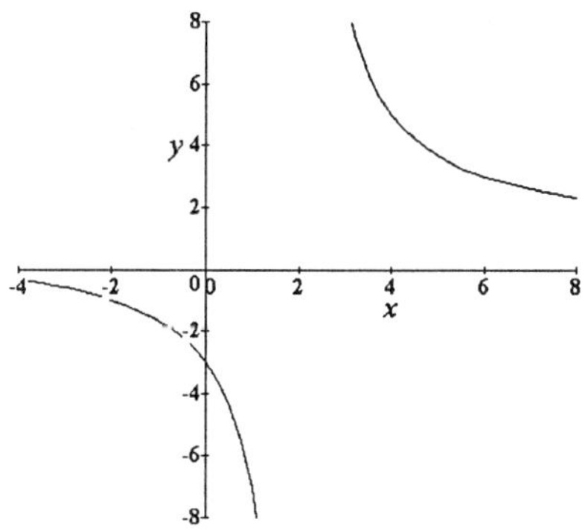

$y = \dfrac{x+6}{x-2}$. When $x = 0$, $y = \dfrac{0+6}{0-2} = -3$, and so the y-intercept is -3.

When $y = 0$, $x + 6 = 0 \quad \Leftrightarrow \quad x = -6$, and so the x-intercept is -6.

Since $y = \frac{x+6}{x-2} = \frac{1+6/x}{1-2/x} \to 1$ as $x \to \infty$, $y = 1$ is the horizontal asymptote.

$x - 2 = 0 \quad \Leftrightarrow \quad x = 2$ is the vertical asymptote.

10. Find intercepts and asymptotes, and then graph the rational function $y = \dfrac{3x-7}{x^2-6x-7}$.

Answer: easy

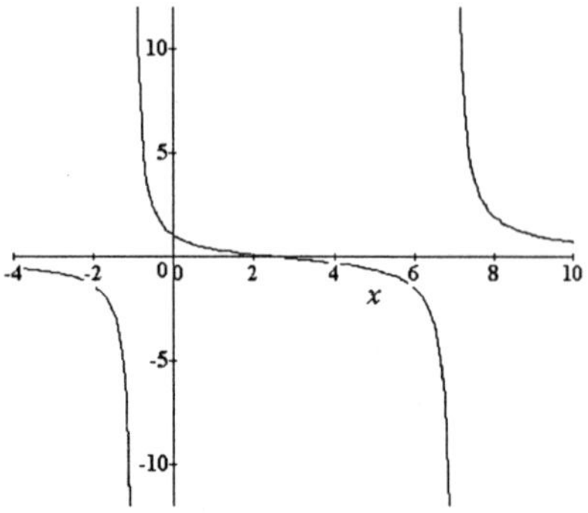

$y = \dfrac{3x-7}{x^2-6x-7} = \dfrac{3x-7}{(x-7)(x+1)}$. When $x = 0$, $y = \dfrac{0-7}{0-0-7} = 1$, and so the y-intercept is 1. When $y = 0$, $3x - 7 = 0 \quad \Leftrightarrow \quad x = \frac{7}{3}$, and so the x-intercept is $\frac{7}{3}$. The horizontal asymptote is $y = 0$ and the vertical asymptotes are $x = -1$ and $x = 7$.

11. Find intercepts and asymptotes, and then graph the rational function $y = \dfrac{x-2}{(x+2)^2}$.

Answer: easy

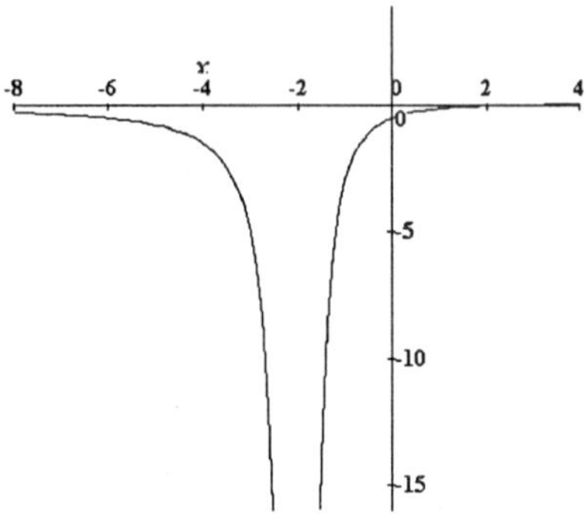

$y = \dfrac{x-2}{(x+2)^2}$. When $x = 0$, $y = \dfrac{0-2}{(0+2)^2} = -\dfrac{1}{2}$, and so the y-intercept is $-\frac{1}{2}$. When $y = 0$, $x - 2 = 0 \quad \Leftrightarrow \quad x = 2$, and so the x-intercept is 2. The horizontal asymptote is $y = 0$ and the vertical asymptote is $x + 2 = 0$ $\Leftrightarrow \quad x = -2$. (Note that the graph approaches $y = 0$ from above as $x \to \infty$.)

12. Find intercepts and asymptotes, and then graph the rational function $y = \dfrac{x(x+4)}{(x^2-1)(x-2)}$.

Answer: medium

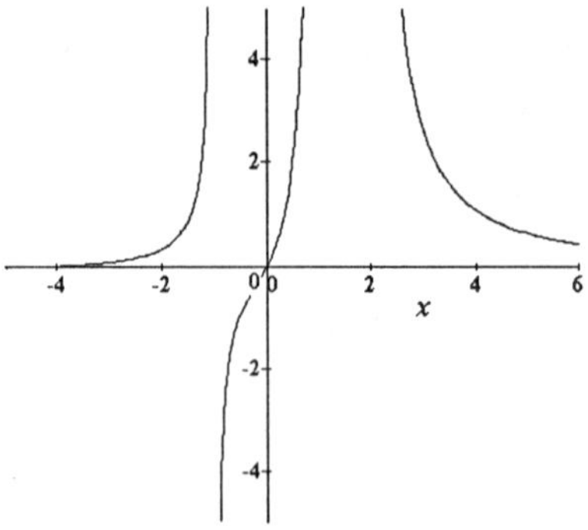

$y = \dfrac{x(x+4)}{(x^2-1)(x-2)}$. When $x = 0$, $y = \dfrac{0(0+4)}{(0-1)(0-2)} = 0$, and so the y-intercept is 0. When $y = 0$, $x = 0$ or $x + 4 = 0 \quad \Leftrightarrow \quad x = -4$, and so the x-intercepts are -4 and 0. The vertical asymptotes occur when $(x^2-1)(x-2) = 0 \quad \Leftrightarrow \quad x = \pm 1$ or $x = 2$, and so the vertical asymptotes are $x = -1$, $x = 1$, and $x = 2$. The horizontal asymptote is $y = 0$. (Note that the graph approaches $y = 0$ from below as $x \to -\infty$.)

13. Find intercepts and asymptotes, and then graph the rational function $y = \dfrac{x^3 - 8}{x^3 + 8}$.

Answer: medium

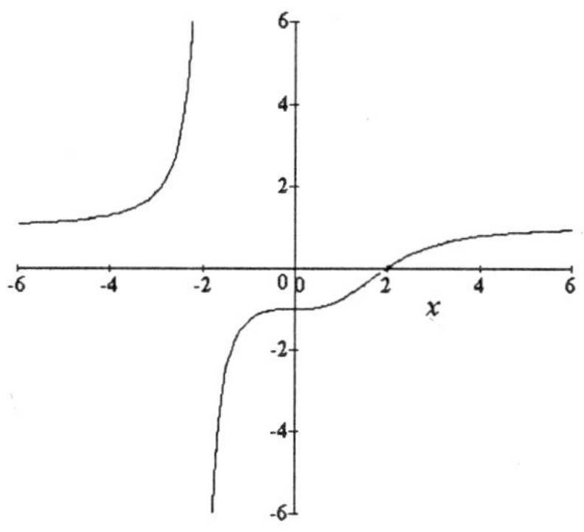

$y = \dfrac{x^3 - 8}{x^3 + 8} = \dfrac{1 - 8/x^3}{1 + 8/x^3} \to 1$ as $x \to \infty$. When $x = 0$, $y = \dfrac{0 - 8}{0 + 8} = -1$, and so the y-intercept is -1. When $y = 0$, $x^3 - 8 = 0 \quad \Leftrightarrow \quad x = 2$, and so the x-intercept is 2. The horizontal asymptote is $y = 1$ and the vertical asymptote is $x^3 + 8 = 0 \quad \Leftrightarrow \quad x = -2$.

14. Find intercepts and asymptotes, and then graph the rational function $y = \dfrac{6x^2 + 5x - 4}{3x^2 + 3x - 6}$.

Answer: medium

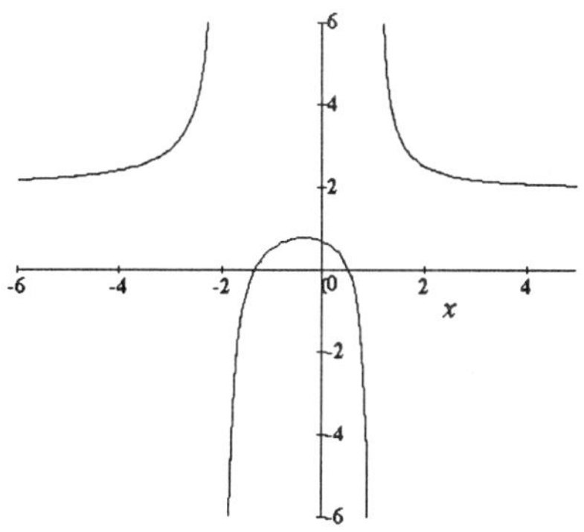

$y = \dfrac{6x^2 + 5x - 4}{3x^2 + 3x - 6} = \dfrac{(3x + 4)(2x - 1)}{3(x - 1)(x + 2)} \to 2$ as $x \to \infty$. When $x = 0$, $y = \dfrac{(0 + 4)(0 - 1)}{3(0 - 1)(0 + 2)} = \dfrac{2}{3}$, and so the y-intercept is $\frac{2}{3}$. When $y = 0$, $3x + 4 = 0 \quad \Leftrightarrow \quad x = -\frac{4}{3}$ or $2x - 1 = 0 \quad \Leftrightarrow \quad x = \frac{1}{2}$, and so the x-intercepts are $-\frac{4}{3}$ and $\frac{1}{2}$. The horizontal asymptote is $y = 2$, and the vertical asymptotes are $x - 1 = 0 \quad \Leftrightarrow \quad x = 1$, and $x + 2 = 0 \quad \Leftrightarrow \quad x = -2$.

15. Find intercepts and asymptotes, and then graph the rational function $y = \dfrac{x^2 + x + 2}{2x - 1}$.

Answer: medium

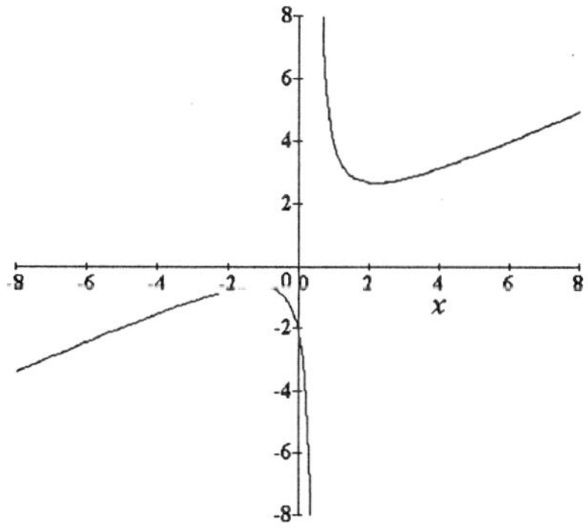

$$y = \frac{x^2 + x + 2}{2x - 1} = \frac{1}{2}\left(\frac{x^2 + x + 2}{x - \frac{1}{2}}\right) = \frac{1}{2}\left(x + \frac{3}{2} + \frac{11/4}{x - 1/2}\right).$$

When $x = 0$, $y = \dfrac{0 + 0 + 2}{0 - 1} = -2$, and so the y-intercept is -2.

Since $x^2 + x + 2 \neq 0$ for any x, there is no x-intercept.

The vertical asymptote is $2x - 1 = 0 \quad \Leftrightarrow \quad x = \frac{1}{2}$ and the slant asymptote is $y = \frac{1}{2}\left(x + \frac{3}{2}\right) = \frac{1}{2}x + \frac{3}{4}$.

16. Find intercepts and asymptotes, and then graph the rational function $y = \dfrac{x^3+2}{x+2}$.

Answer: medium

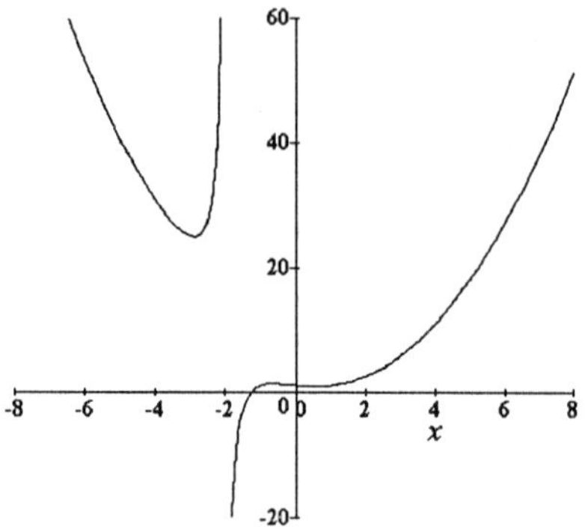

$y = \dfrac{x^3+2}{x+2}$. When $x = 0$, $y = \dfrac{0+2}{0+2} = 1$, and so the y-intercept is 1. When $y = 0$, $x^3 + 2 = 0 \quad \Leftrightarrow \quad x = \sqrt[3]{-2}$, and so the x-intercept is $\sqrt[3]{-2} \approx -1.26$. The vertical asymptote is $x + 2 = 0 \quad \Leftrightarrow \quad x = -2$.

17. Find intercepts and asymptotes, and then graph the rational function $y = \dfrac{x^3-8}{x^2-8}$.

Answer: medium

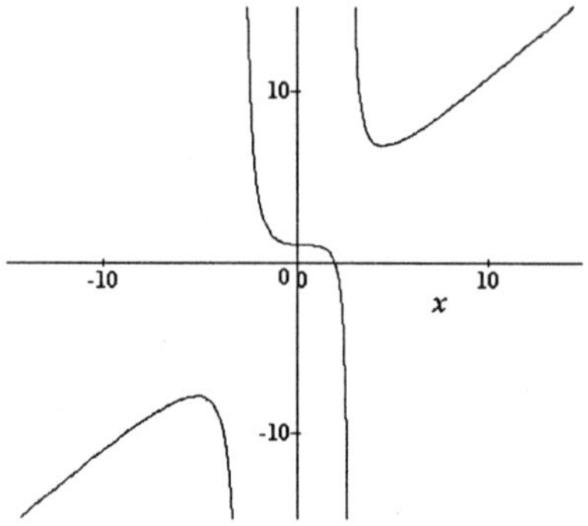

$y = \dfrac{x^3-8}{x^2-8} = x + \dfrac{8x-8}{x^2-8}$. When $x = 0$, $y = \dfrac{0-8}{0-8} = 1$, and so the y-intercept is 1. When $y = 0$, $x^3 - 8 = 0 \quad \Leftrightarrow \quad x = 2$, and so the x-intercept is 2. The vertical asymptotes are $x^2 - 8 = 0 \quad \Leftrightarrow \quad x = \pm 2\sqrt{2}$, and the slant asymptote is $y = x$.

18. Find intercepts and asymptotes, and then graph the rational function $y = \dfrac{16}{1-x^4}$.

Answer: medium

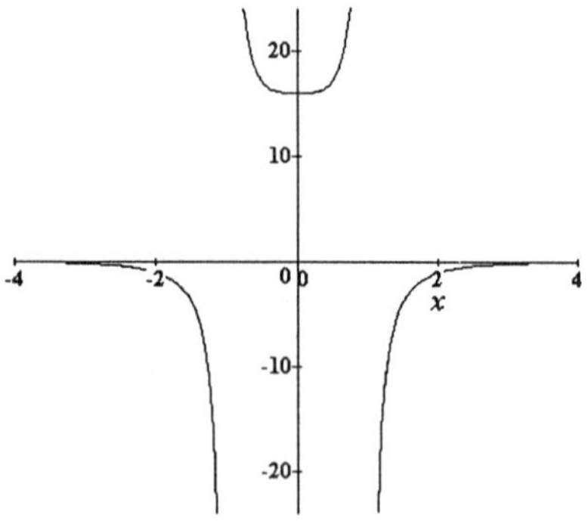

$y = \dfrac{16}{1-x^4}$. When $x = 0$, $y = \dfrac{16}{1-0} = 16$, and so the y-intercept is 16. Since $16 > 0$ for all x, there is no x-intercept. The horizontal asymptote is $y = 0$ and the vertical asymptotes are $1 - x^4 = 0 \quad \Leftrightarrow \quad x = \pm 1$.

19. Find intercepts and asymptotes, and then graph the rational function $y = \dfrac{x^3 - 3x^2 - x + 3}{3x^3 - 12x}$.

Answer: medium

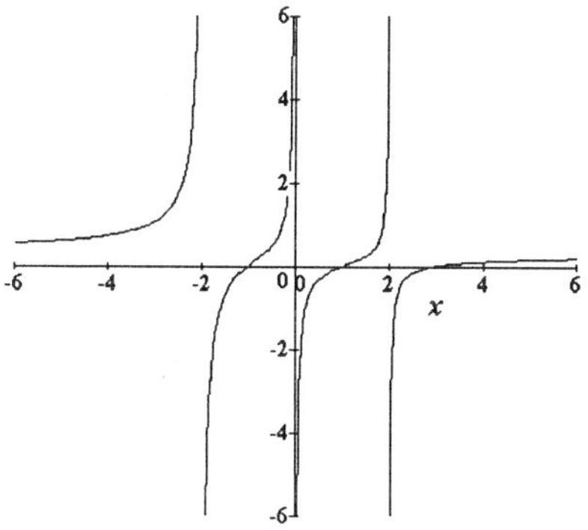

$y = \dfrac{x^3 - 3x^2 - x + 3}{3x^3 - 12x} = \dfrac{(x+1)(x-3)(x-1)}{3x(x^2-4)}$. When $x = 0$, y is undefined, and so there is no y-intercept. When $y = 0$, $x + 1 = 0 \Leftrightarrow x = -1$, or $x - 3 = 0 \Leftrightarrow x = 3$, or $x - 1 = 0 \Leftrightarrow x = 1$, and so the x-intercepts are -1, 1, and 3. The horizontal asymptote is $y = \frac{1}{3}$ and the vertical asymptotes are $3x = 0 \Leftrightarrow x = 0$, and $x^2 - 4 = 0 \Leftrightarrow x = \pm 2$.

20. Find intercepts and asymptotes, and then graph the rational function $y = \dfrac{x^3 - x^2 - x + 1}{x^2 + 4x + 4}$.

Answer: hard

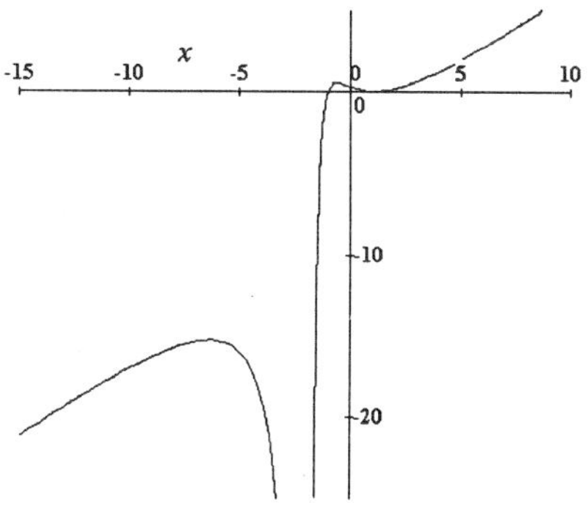

$y = \dfrac{x^3 - x^2 - x + 1}{x^2 + 4x + 4} = \dfrac{(x-1)(x^2-1)}{(x+2)^2} = x - 5 + \dfrac{15x + 21}{(x+2)^2}$. When $x = 0$, $y = \dfrac{0 - 0 - 0 + 1}{0 + 0 + 4} = \dfrac{1}{4}$, and so the y-intercept is $\frac{1}{4}$. When $y = 0$, $x - 1 = 0 \Leftrightarrow x = 1$, or $x^2 - 1 = 0 \Leftrightarrow x = \pm 1$, and so the x-intercepts are ± 1. The slant asymptote is $y = x - 5$ and the vertical asymptote is $x + 2 = 0 \Leftrightarrow x = -2$.

21. Find intercepts and asymptotes, and then graph the rational function $y = \dfrac{6x^3 - 24}{2x^3 + 11x^2 + 20x + 12}$.

Answer: hard

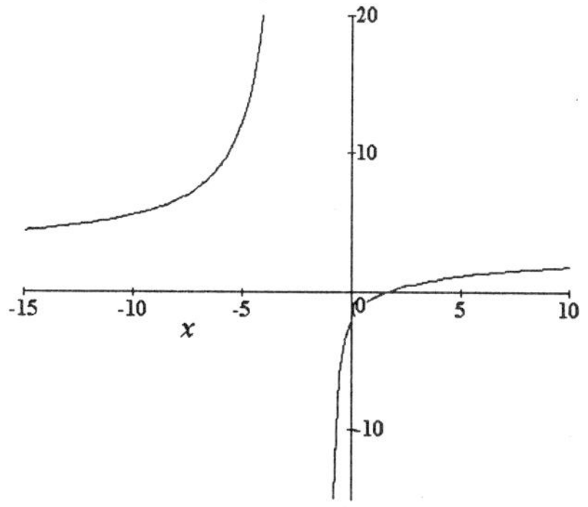

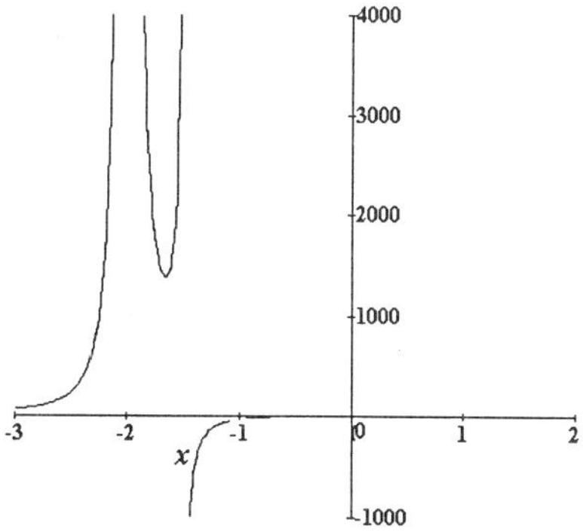

$y = \dfrac{6x^3 - 24}{2x^3 + 11x^2 + 20x + 12} = \dfrac{6(x^3 - 4)}{(x+2)(x+2)(2x+3)}$. When $x = 0$, $y = \dfrac{0 - 24}{0 + 0 + 0 + 12} = -2$, and so the y-intercept is -2. When $y = 0$, $x^3 - 4 = 0 \Leftrightarrow x = \sqrt[3]{4}$, and so the x-intercept is $\sqrt[3]{4}$. The horizontal asymptote is $y = 3$ and the vertical asymptotes are $x + 2 = 0 \Leftrightarrow x = -2$, and $2x + 3 = 0 \Leftrightarrow x = -\frac{3}{2}$.

22. Graph the rational function $y = \dfrac{x^2 + x - 20}{x + 5}$

Answer: medium

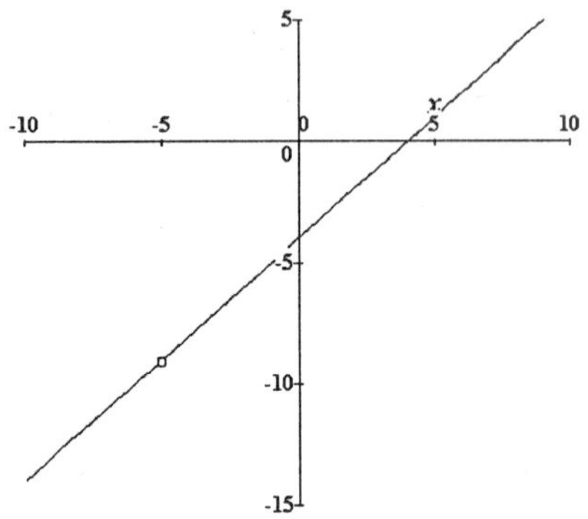

$y = \dfrac{x^2 + x - 20}{x + 5} = \dfrac{(x + 5)(x - 4)}{x + 5} = x - 4$, and $x + 5 \neq 0 \Leftrightarrow x \neq -5$. Therefore, the graph is the line $y = x - 4$ with the point $(-5, -9)$ removed.

23. Graph the rational function $y = \dfrac{x^2 - 3x + 2}{x^2 - 4x + 4}$

Answer: medium

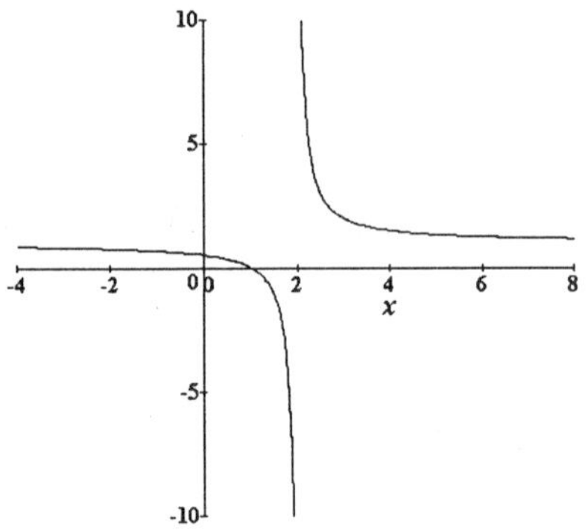

$y = \dfrac{x^2 - 3x + 2}{x^2 - 4x + 4} = \dfrac{(x - 2)(x - 1)}{(x - 2)(x - 2)} = \dfrac{x - 1}{x - 2}$, and $x - 2 \neq 0 \Leftrightarrow x \neq 2$. However, the function is not defined anyway for $x = 2$, so the graph is just that of $y = \dfrac{x - 1}{x - 2}$.

24. Construct a rational function $y = P(x)/Q(x)$ that has the indicated properties, and in which the degrees of P and Q are as small as possible. The function has vertical asymptote $x = 4$, horizontal asymptote $y = 0$, and y-intercept $-\frac{1}{4}$, and never crosses the x-axis.

Answer: medium

Since the vertical asymptote is $x = 4$, $Q(x) = x - 4$. The horizontal asymptote is $y = 0$ and hence the highest power term in $P(x)$ is at most $1 - 1 = 0$. Therefore, $P(x)$ is a constant, $P(x) = C$, and so $y = C/(x - 4)$. Since the y-intercept is $-\frac{1}{4}$, it follows that $-\dfrac{1}{4} = \dfrac{C}{0 - 4} \Leftrightarrow C = 1$. Thus, the function is $y = \dfrac{1}{x - 4}$.

25. Construct a rational function $y = P(x)/Q(x)$ that has the indicated properties, and in which the degrees of P and Q are as small as possible. The function has horizontal asymptote $y = 1$ but no vertical asymptotes. The origin is the only x-intercept, and i is a zero of $Q(x)$.

Answer: hard

Since i is a zero of $Q(x)$, $Q(x) = x^2 + 1$. Since the horizontal asymptote is $y = 1$, $P(x)$ must be a quadratic polynomial with x^2 as its leading term. Since 0 is the only y-intercept, we let $P(x) = x^2$. Therefore the function is $y = \dfrac{x^2}{x^2 + 1}$.

26. Show that the function $y = \dfrac{x^8 + 12}{x^6 + 4x^4 + 8x^2 + 15}$ has no horizontal, vertical, or slant asymptotes, and no x-intercepts.

Answer: hard

$y = \dfrac{x^8 + 12}{x^6 + 4x^4 + 8x^2 + 15}$. Since $x^8 + 12 \geq 12$ for all x, there are no x-intercepts. Since the highest power of x in the numerator is greater than the highest power of x in the denominator, there is no horizontal asymptote, and furthermore, because the difference in powers is greater than 1, there is no slant asymptote. Since $x^6 + 4x^4 + 8x^2 + 15 \geq 15$ for all x, there is no vertical asymptote.

CHAPTER 5 REVIEW

1. Graph the polynomial $y = x^3 - 4x^2 - 3x + 14$. Show clearly all x- and y-intercepts.

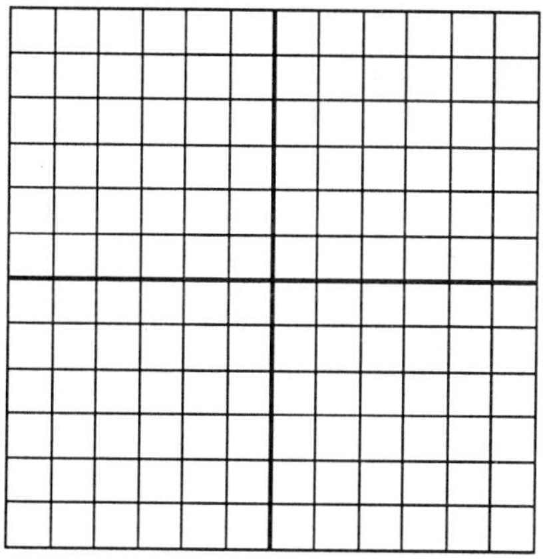

Answer: easy

$y = x^3 - 4x^2 - 3x + 14$

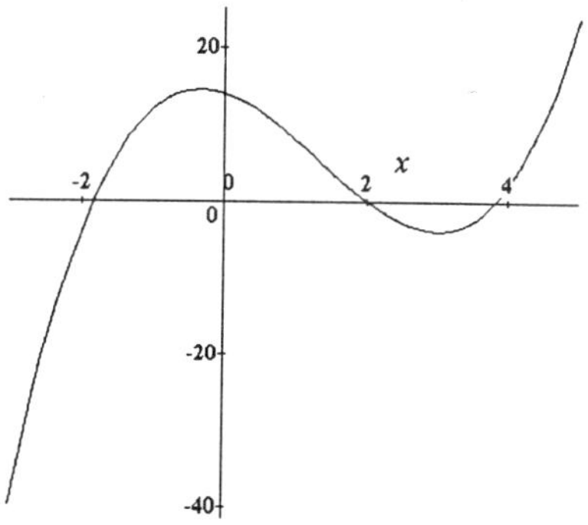

2. Graph the polynomial $y = 2x^3 + 7x^2 + 2x - 3$. Show clearly all x- and y-intercepts.

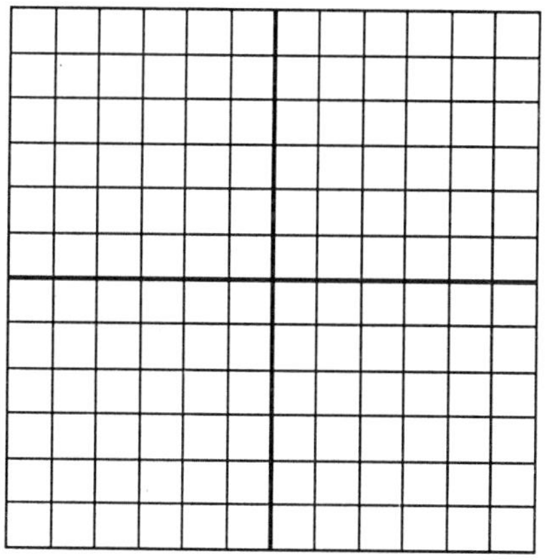

Answer: easy

$y = 2x^3 + 7x^2 + 2x - 3$

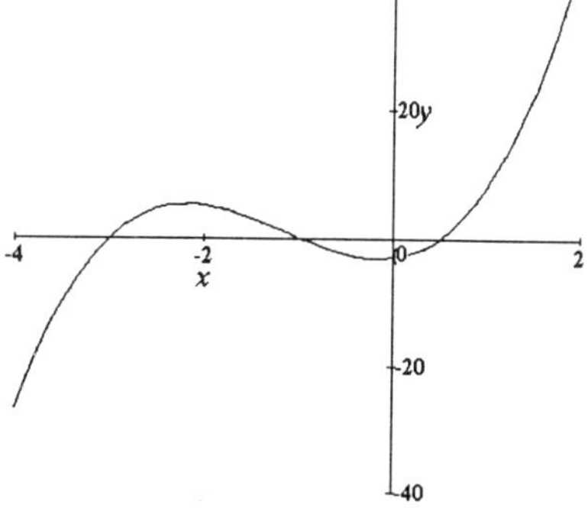

GC 3. Use a graphing device to graph the polynomial $y = x^4 - 12x^2 + 8$. Find the x- and y-intercepts and the coordinates of all local extrema, correct to the nearest decimal. Describe the end behavior of the function.

Answer: medium

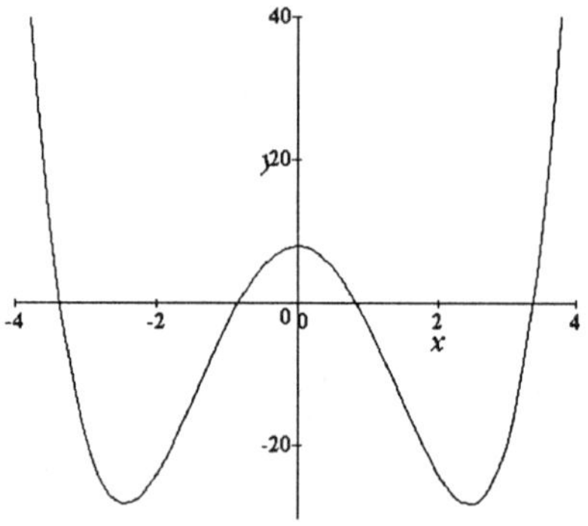

$y = x^4 - 12x^2 + 8$

The x-intercepts are $x \approx \pm 3.4$ and $x \approx \pm 0.8$;
the y-intercept is 8, the local maximum is $(0, 8)$;
the local minima are $(\pm 2.4, -28)$; and
$y \to \infty$ as $x \to \pm\infty$.

GC 4. Use a graphing device to graph the polynomial $y = 3x^5 - x^4 - 4x$. Find the x- and y-intercepts and the coordinates of all local extrema, correct to the nearest decimal. Describe the end behavior of the function.

Answer: medium

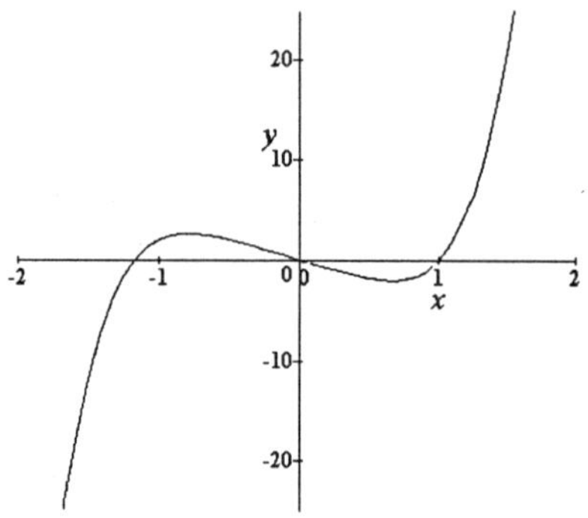

$y = 3x^5 + x^4 - 4x$

The x-intercepts are 0, 1, and approximately -1.2; the y-intercept is $y = 0$;
the local maximum is $(-0.8, 2.6)$; the local minimum is $(0.7, -2.1)$;
$y \to -\infty$ as $x \to -\infty$, and $y \to \infty$ as $x \to \infty$.

5. Show that the equation $x^3 - 5x + 2 = 0$ has a root between 0 and 1, and then find that root correct to the nearest tenth.

Answer: medium

$x^3 - 5x + 2 = 0$

x	$x^3 - 5x + 2$
0	2
1	-2
0.4	0.064
0.5	-0.375
0.45	-0.1589

Hence, the root is about 0.4.

6. Show that the equation $x^6 - 3x^4 - 12 = 0$ has a root between 1 and 2, and then find that root correct to two decimal places.

Answer: medium

$x^6 - 3x^4 - 12 = 0$

x	$x^6 - 3x^4 - 12$
1	-14
2	4
1.9	-4.05
1.95	-0.397
1.96	0.420
1.955	0.008

Hence, the root is about 1.95.

7. Find a polynomial of degree 3 with constant coefficient 6 and with zeros $-\frac{3}{2}$, 1, and 2.
(a) $P(x) = 2x^3 + 5x^2 - 5x + 2$ (b) $P(x) = 2x^3 - 3x^2 - 5x + 6$ (c) $P(x) = 3x^3 + x^2 - 12x + 3$
(d) $P(x) = x^3 - 5x^2 - 3x + 12$ (e) $P(x) = 4x^3 - 2x^2 - 3x + 1$
Answer: medium (b)
Since the zeros are $-\frac{3}{2}$, 1, and 2, a factorization is
$P(x) = C\left(x + \frac{3}{2}\right)(x - 1)(x - 2) = \frac{1}{2}C(2x + 3)(x^2 - 3x + 2) = \frac{1}{2}C(2x^3 - 3x^2 - 5x + 6)$. Since the constant coefficient is 6, C = 2, and so the polynomial is $P(x) = 2x^3 - 3x^2 - 5x + 6$.

8. Does there exist a polynomial of degree 4 with integer coefficients that has zeros i, $2i$, $3i$, and $4i$? If so, find it. If not, explain why.
Answer: hard
No, there is no polynomial of degree 4 with integer coefficients that has zeros i, $2i$,$3i$ and $4i$, since the imaginary roots of polynomial equations with real coefficients come in complex conjugate pairs.

9. For the polynomial $x^5 - 6x^4 + 11x^3 - 2x^2 - 12x + 8 = 0$ find all rational, irrational, and imaginary roots (and state their multiplicities). Use Descartes' Rule of Signs, the Upper and Lower Bounds Theorem, the quadratic formula, or other factoring techniques to help you if possible.
Answer: medium
$x^5 - 6x^4 + 11x^3 - 2x^2 - 12x + 8 = 0$ has possible rational roots ±1, ±2, ±4, and ±8.

	1	-6	11	-2	-12	8	
1	1	-5	6	4	-8	0	$\Rightarrow$ $x = 1$ is a root, and so $(x-1)(x^4 - 5x^3 + 6x^2 + 4x - 8) = 0$
2	1	-3	0	4	0		$\Rightarrow$ $x = 2$ is a root, and so $(x-1)(x-2)(x^3 - 3x^2 + 4) = 0$
2	1	-1	-2	0			$\Rightarrow$ $x = 2$ is a root, and so $(x-1)(x-2)^2(x^2 - x - 2) = 0$.

Now $(x-1)(x-2)^2(x^2 - x - 2) = (x-1)(x-2)^3(x+1) = 0$. Therefore, the roots are 2 (multiplicity 3), -1, and 1.

10. For the polynomial $x^6 = 729$ find all rational, irrational, and imaginary roots (and state their multiplicities). Use Descartes' Rule of Signs, the Upper and Lower Bounds Theorem, the quadratic formula, or other factoring techniques to help you if possible.
Answer: medium
$x^6 = 729 \quad\Leftrightarrow\quad x^6 - 729 = 0 \quad\Leftrightarrow\quad (x^3 - 27)(x^3 + 27) = 0 \quad\Leftrightarrow\quad (x-3)(x^2 + 3x + 9)(x+3)(x^2 - 3x + 9)$. Now $x^2 + 3x + 9 = 0$ when $x = \dfrac{-3 \pm \sqrt{9 - 36}}{2} = \dfrac{-3 \pm 3i\sqrt{3}}{2}$, and $x^2 - 3x + 9 = 0$ when $x = \dfrac{3 \pm \sqrt{9 - 36}}{2} = \dfrac{3 \pm 3i\sqrt{3}}{2}$. So the roots are ±3 and $\pm\frac{3}{2} \pm \frac{3}{2}i\sqrt{3}$.

11. Show that $3 - i\sqrt{2}$ is a root of $x^4 - 7x^3 + 15x^2 + x - 22 = 0$, and find all other roots.
Answer: medium
$3 - i\sqrt{2}$ will be a root if and only if $3 + i\sqrt{2}$ is a root. Thus $\left(x - 3 + i\sqrt{2}\right)\left(x - 3 - i\sqrt{2}\right) = x^2 - 6x + 11$ will divide the polynomial evenly if these numbers are roots. By long division, $x^4 - 7x^3 + 15x^2 + x - 22 = (x^2 - 6x + 11)(x^2 - x - 2)$ and so $x^2 - 6x + 11$ does divide the polynomial evenly. Since $x^2 - x - 2 = (x-2)(x+1)$, the other roots are 2, -1, and $3 + i\sqrt{2}$.

12. The polynomial $x^4 - 2x^2 + 2x - 2 = 0$ has one positive real root. Show that it is irrational, and find a decimal approximation for it, correct to two decimal places.

Answer: medium

$x^4 - 2x^2 + 2x - 2 = 0$ has possible rational roots ± 1, ± 2 and one positive real root.

	1	0	−2	2	−2
1	1	1	−1	1	−1
2	1	2	2	6	10

Therefore, the positive real root is irrational and lies between 1 and 2.

x	$x^4 - 2x^2 + 2x - 2$
1.2	−0.4064
1.3	0.0761
1.28	−0.0324
1.29	0.0210
1.285	−0.0059

Therefore, the positive irrational root is 1.29 (correct to two decimals).

GC 13. Use a graphing device to find all real solutions of the equation $x^3 - x^2 - 14x + 24 = 0$.

Answer: medium

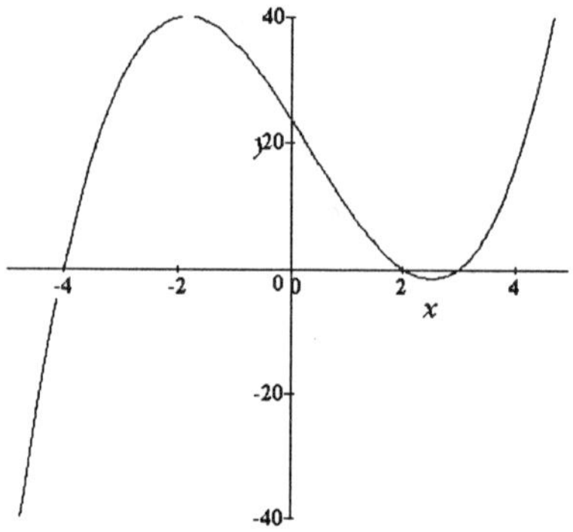

$x^3 - x^2 - 14x + 24 = 0$ By using a graphing device, the real solutions are -4, 2 and 3.

GC 14. Use a graphing device to find all real solutions of the equation $x^5 = x + 4$ correct to two decimals.

Answer: medium

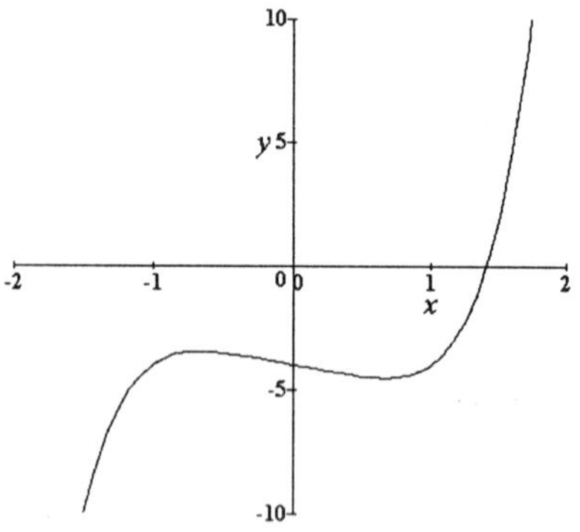

$x^5 = x + 4 \quad \Leftrightarrow \quad x^5 - x - 4 = 0$ By using a graphing device, we find that the real solution is about 1.40.

15. Graph the rational function $y = \dfrac{2x-8}{x+3}$. Show clearly all x- and y-intercepts and asymptotes.

Answer: easy

$$y = \frac{2x-8}{x+3}$$

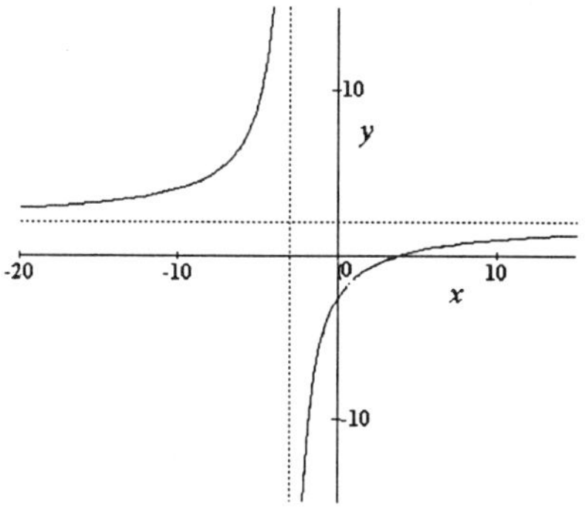

16. Graph the rational function $y = \dfrac{2x^2-5x-6}{x-3}$. Show clearly all x- and y-intercepts and asymptotes.

Answer: medium

$$y = \frac{2x^2-5x-6}{x-3}$$

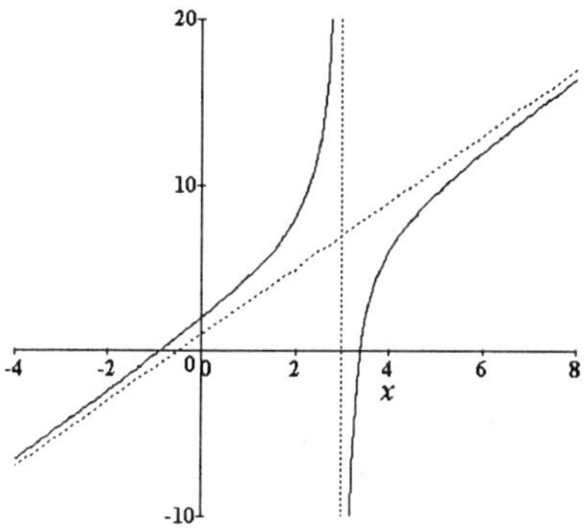

17. Graph the rational function $y = \dfrac{x^3+8}{x+1}$. Show clearly all x- and y-intercepts and asymptotes.

Answer: hard

$$y = \frac{x^3+8}{x+1}$$

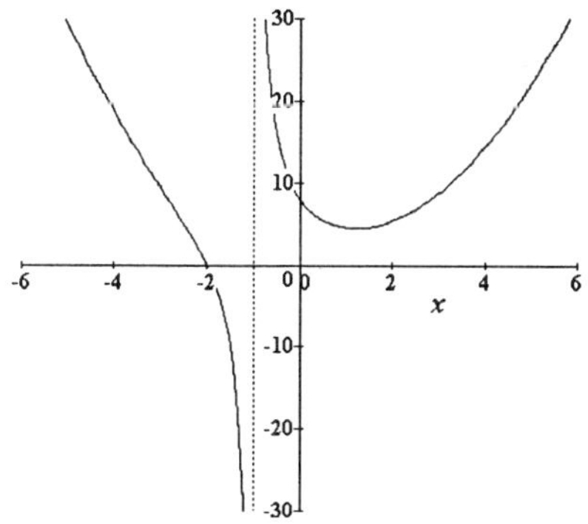

GC 18. Use a graphing device to analyze the graph of the rational function $y = \frac{3x - 7}{x^2 + 6}$. Find all x- and y-intercepts, vertical, horizontal, and slant asymptotes, and the coordinates of local extrema, correct to two decimals. If the function has no horizontal or slant asymptote, find a polynomial that has the same end behavior.

Answer: medium

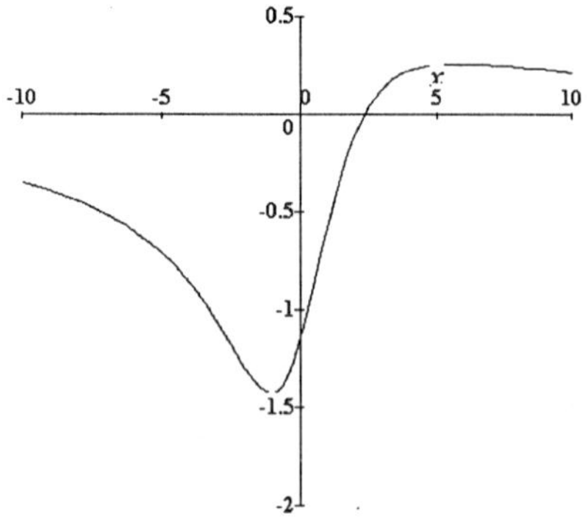

$y = \frac{3x - 7}{x^2 + 6}$. By using a graphing device, we see that the y-intercept is -1.17, the x-intercept is 2.33, the horizontal asymptote is $y = 0$, the local minimum is $(-1.05, -1.43)$, and the local maximum is $(5.72, 0.26)$.

GC 19. Use a graphing device to analyze the graph of the rational function $y = \frac{3x^3 - x^2}{x + 2}$. Find all x- and y-intercepts, vertical, horizontal, and slant asymptotes, and the coordinates of local extrema, correct to two decimals. If the function has no horizontal or slant asymptote, find a polynomial that has the same end behavior.

Answer: medium

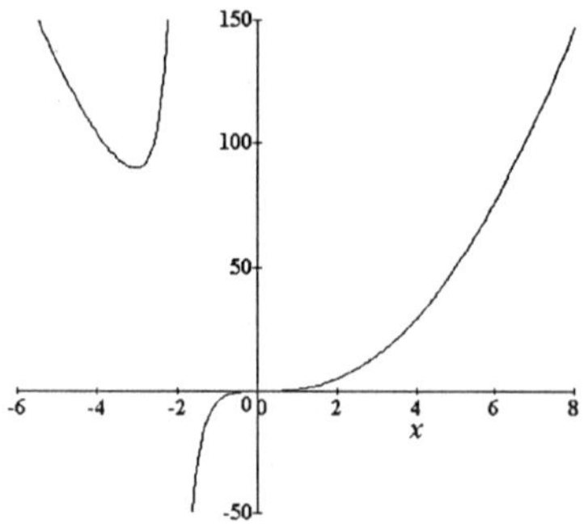

$y = \frac{3x^3 - x^2}{x + 2}$. By using a graphing device, we see that the x-intercepts are 0 and $\frac{1}{3}$, the y-intercept is 0, the vertical asymptote is $x = -2$, and the local minimum is $(-3.05, 89.92)$. The polynomial that has the same end behavior as y is $3x^2 - 7x - 14$.

20. Find the coordinates of all points of intersection of the graphs of $y = x^4 + 11x^2 + 32x$ and $y = 8x^3 + 60$.

Answer: medium

The graphs $y = x^4 + 11x^2 + 32x$ and $y = 8x^3 + 60$ intersect when $x^4 + 11x^2 + 32x = 8x^3 + 60 \quad \Leftrightarrow$ $x^4 - 8x^3 + 11x^2 + 32x - 60 = 0$. The possible rational roots are $\pm1, \pm2, \pm3, \pm4, \pm5, \pm6, \pm10, \pm12, \pm15, \pm20, \pm30$, and ±60.

	1	-8	11	32	-60	
1	1	-7	4	36	-24	
2	1	-6	-1	30	0	$\Rightarrow$ $x = 2$ is a root, and so $(x - 2)(x^3 - 6x^2 - x + 30) = 0$
3	1	-3	-10	0		$\Rightarrow$ $x = 3$ is a root, and so $(x - 3)(x - 2)(x^2 - 3x - 10) = 0$.

However, $x^2 - 3x - 10 = (x - 5)(x + 2)$, and so the graphs intersect when $(x - 2)(x - 3)(x - 5)(x + 2) = 0$. Hence, the points of intersection are $(2, 124)$, $(3, 276)$, $(5, 1060)$, and $(-2, -4)$.

CHAPTER SIX

SECTION 6.1

1. Sketch the graph of the function $f(x) = \left(\frac{2}{3}\right)^x$ by making a table of values. Use a calculator if necessary.

Answer: easy

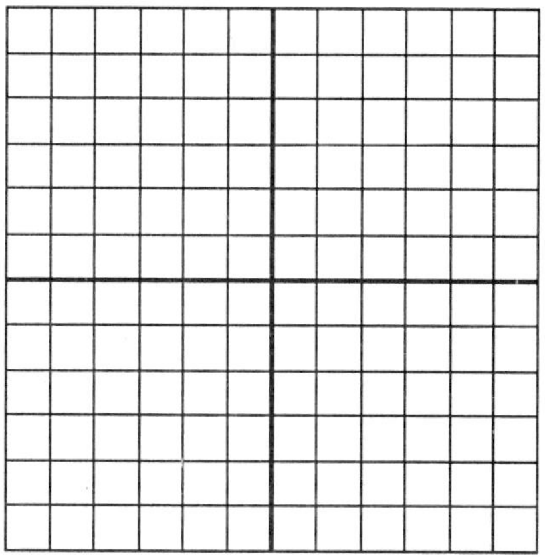

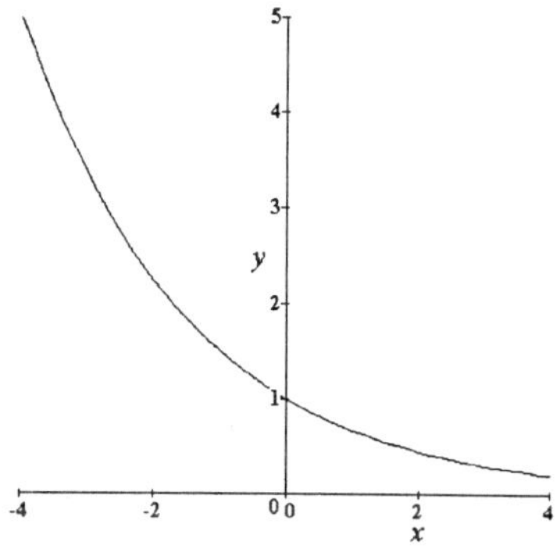

2. Sketch the graph of the function $h(x) = (1.2)^x$ by making a table of values. Use a calculator if necessary.

Answer: easy

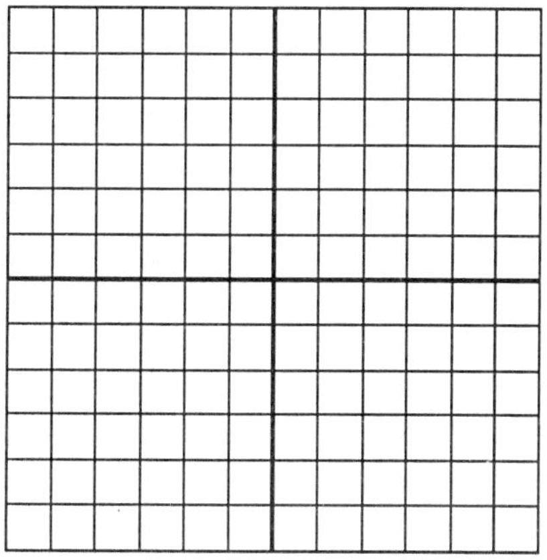

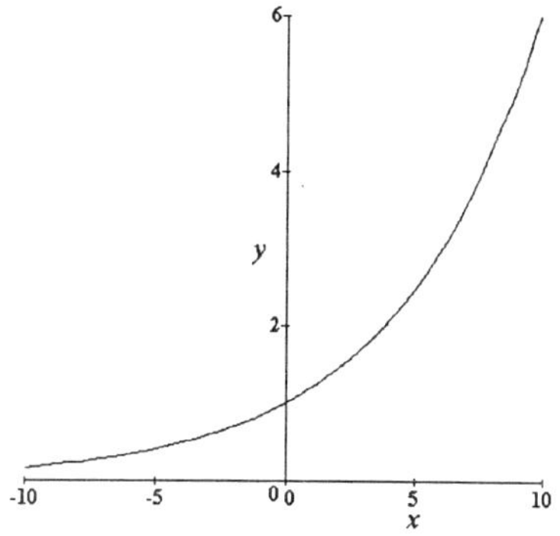

3. Graph both functions on one set of axes: $y = \left(\frac{3}{4}\right)^x$ and $y = \left(\frac{5}{4}\right)^x$.

Answer: easy

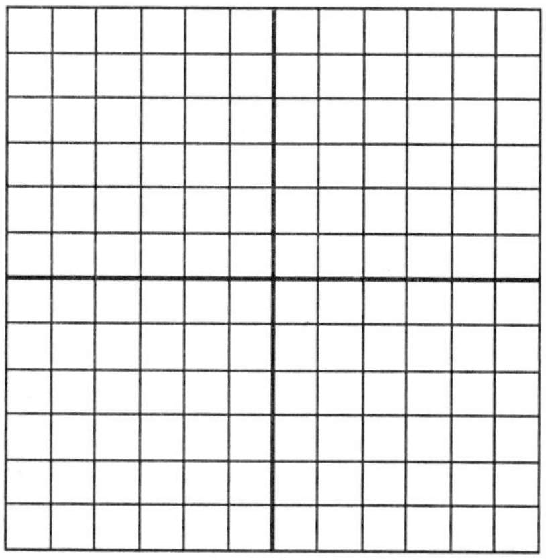

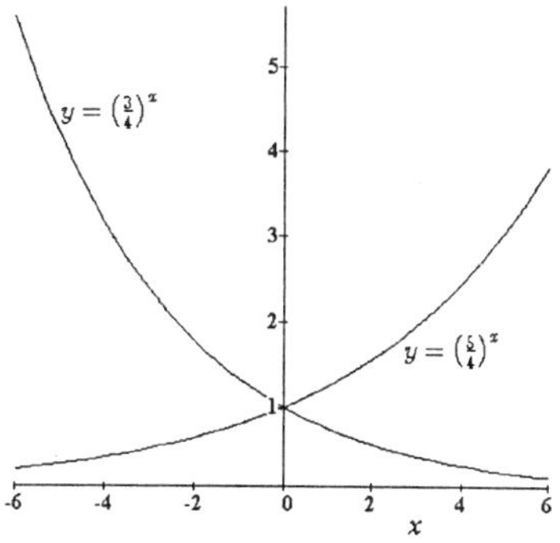

4. Graph the function $f(x) = 3^{-x}$, not by plotting points but by applying your knowledge of the general shape of graphs of the form a^x. State the domain, range, and asymptote of the function.

Answer: medium

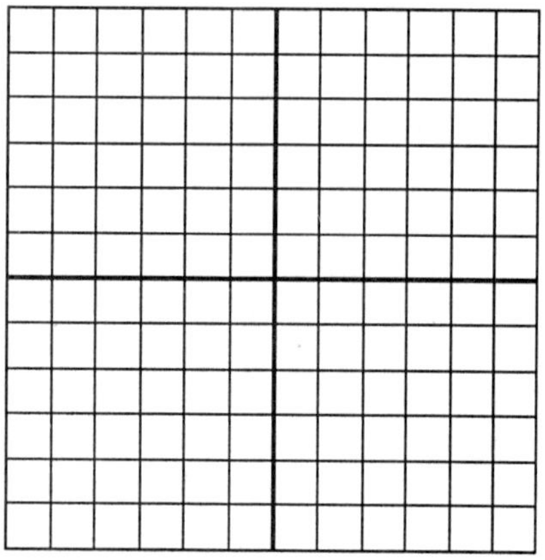

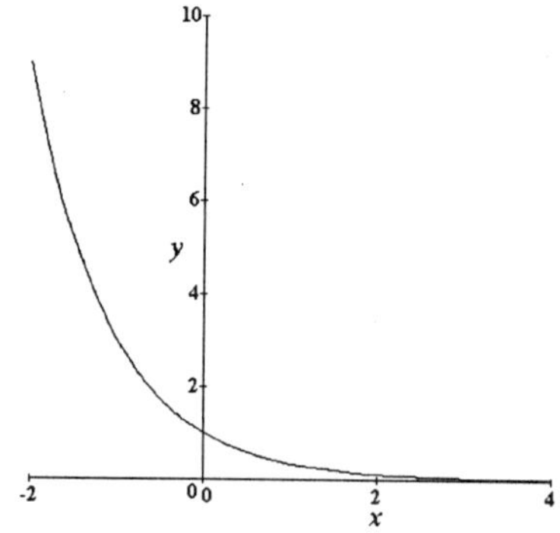

$f(x) = 3^{-x}$. D: $(-\infty, \infty)$, R: $(0, \infty)$, A: $y = 0$

5. Graph the function $g(x) = 3^{x-2}$, not by plotting points but by applying your knowledge of the general shape of graphs of the form a^x. State the domain, range, and asymptote of the function.

Answer: medium

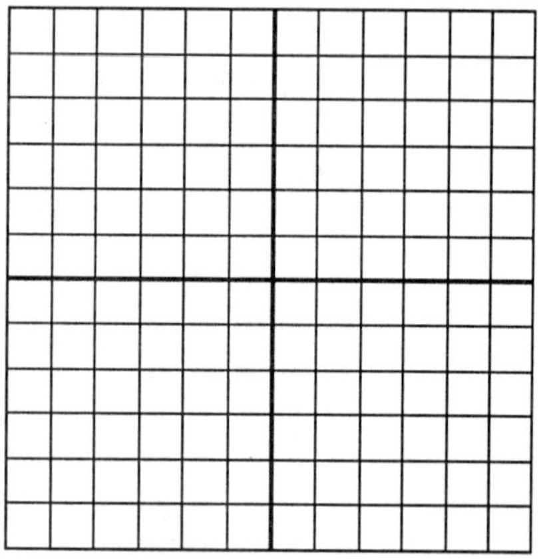

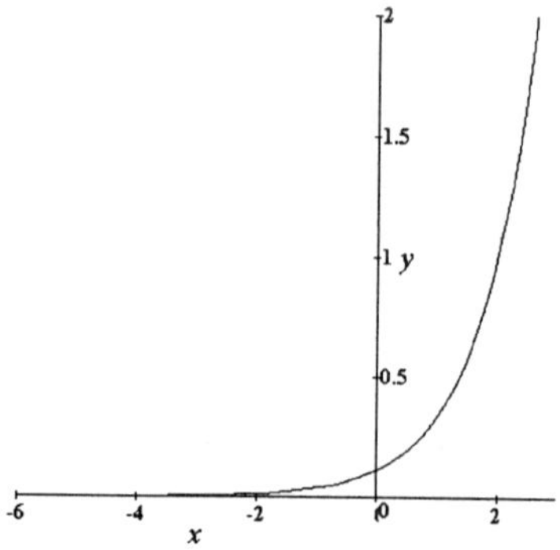

$g(x) = 3^{x-2}$. D: $(-\infty, \infty)$, R: $(0, \infty)$, A: $y = 0$

6. Graph the function $y = 4 - 2^x$, not by plotting points but by applying your knowledge of the general shape of graphs of the form a^x. State the domain, range, and asymptote of the function.

Answer: medium

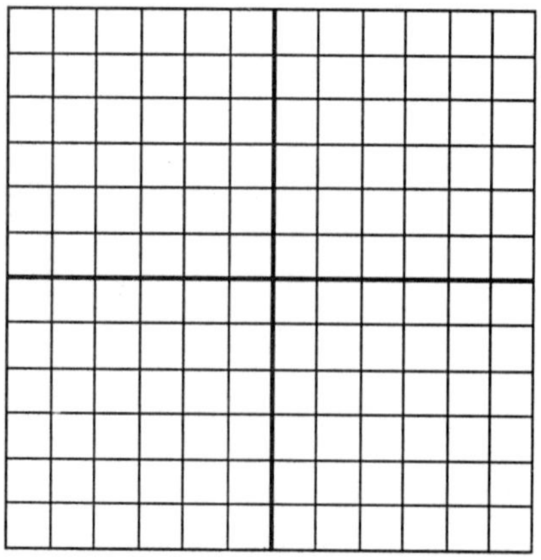

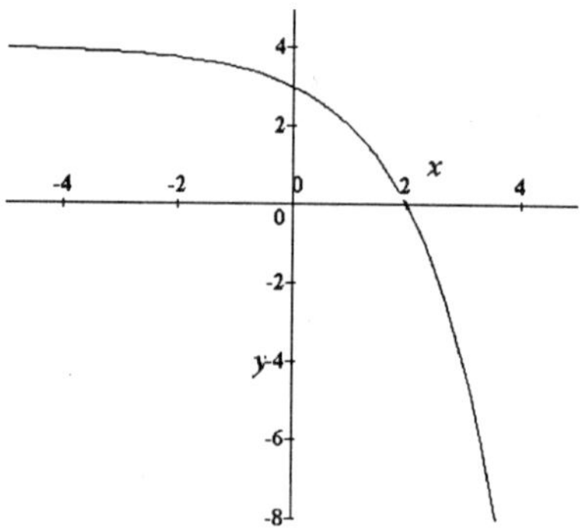

$y = 4 - 2^x$. D: $(-\infty, \infty)$, R: $(-\infty, 4)$, A: $y = 4$

7. Graph the function $y = -\left(\frac{1}{4}\right)^x$, not by plotting points but by applying your knowledge of the general shape of graphs of the form a^x. State the domain, range, and asymptote of the function.

Answer: medium

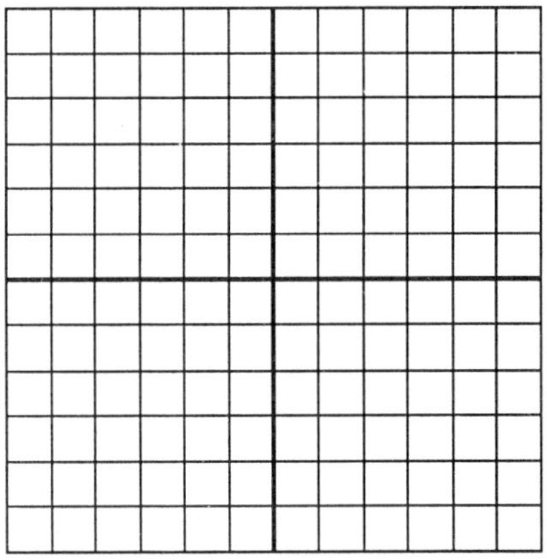

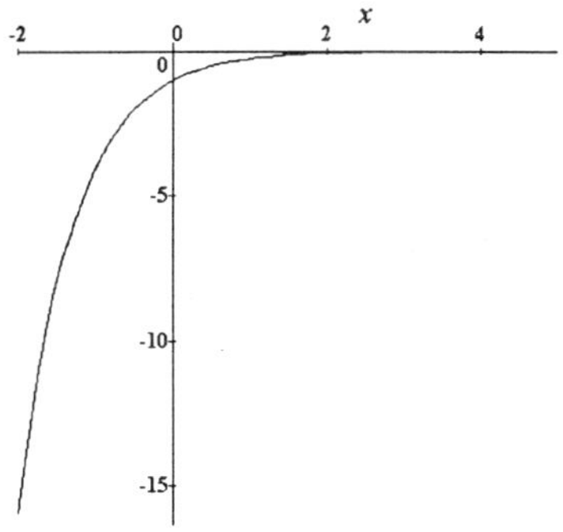

$y = -\left(\frac{1}{4}\right)^x$. D: $(-\infty, \infty)$, R: $(-\infty, 0)$, A: $y = 0$

8. Graph the function $y = 1 + 2^{x-1}$, not by plotting points but by applying your knowledge of the general shape of graphs of the form a^x. State the domain, range, and asymptote of the function.

Answer: medium

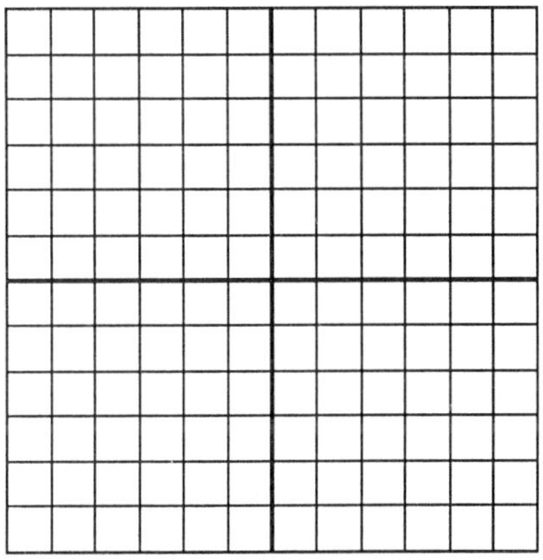

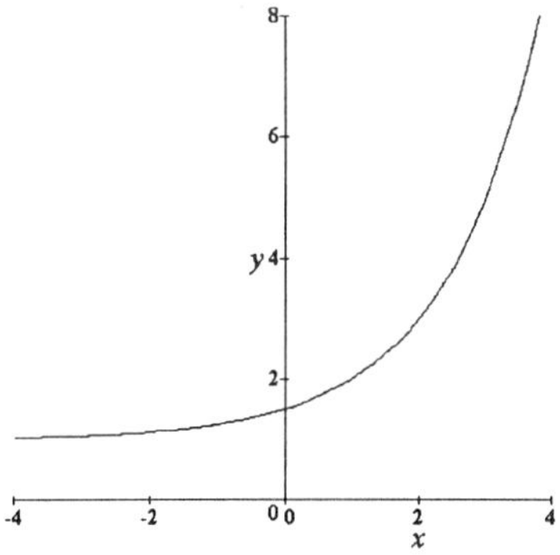

$y = 1 + 2^{x-1}$. D: $(-\infty, \infty)$, R: $(1, \infty)$, A: $y = 1$

9. Graph the function $y = 2 + 5(1 - 10^{-x})$, not by plotting points but by applying your knowledge of the general shape of graphs of the form a^x. State the domain, range, and asymptote of the function.

Answer: hard

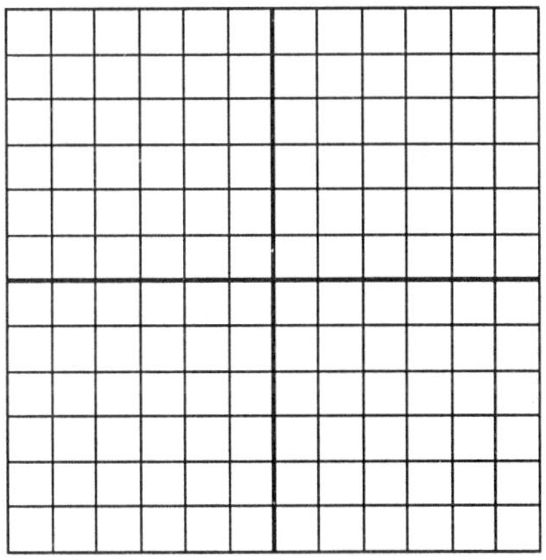

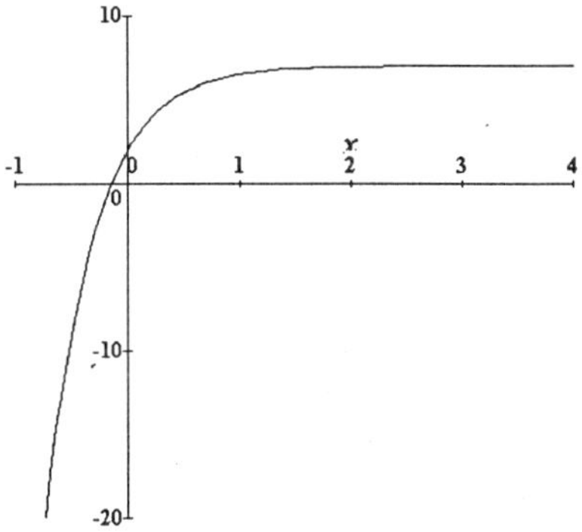

$y = 2 + 5(1 - 10^{-x})$. D: $(-\infty, \infty)$, R: $(-\infty, 7)$, A: $y = 7$

10. Graph the function $y = 3^{-|x|}$, not by plotting points but by applying your knowledge of the general shape of graphs of the form a^x. State the domain, range, and asymptote of the function.

Answer: hard

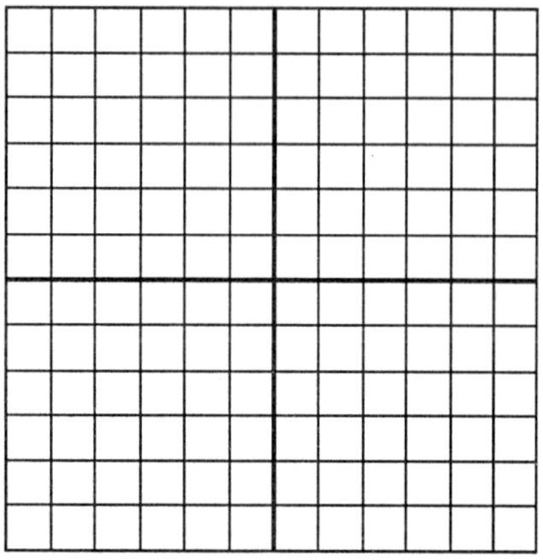

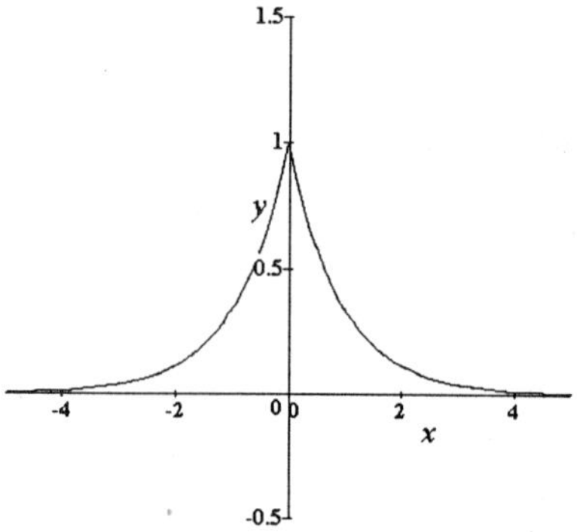

$y = 3^{-|x|}$. D: $(-\infty, \infty)$, R: $(0, 1]$, A: $y = 0$

11. Compare the functions $f(x) = x^2$ and $g(x) = 2^x$ by evaluating each of them for $x = 0, 1, 2, 3, 4, 5, 6, 7, 8, 9, 10, 15$, and 20. Then draw the graphs of f and g on the same set of axes for $-4 \le x \le 6$.

Answer:

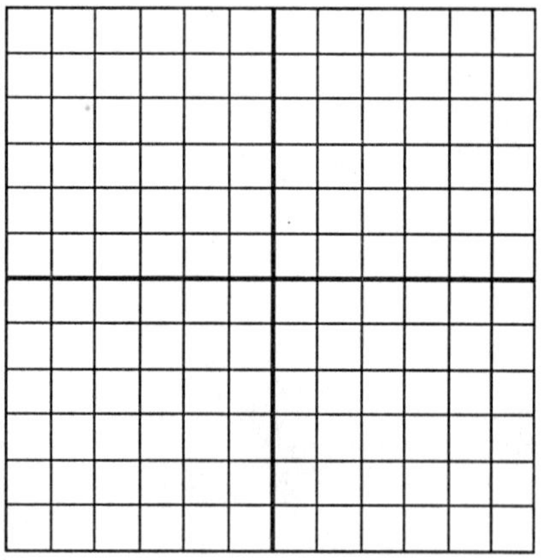

medium

x	$f(x)$	$g(x)$
0	0	1
1	1	2
2	4	4
3	9	8
4	16	16
5	25	32
6	36	64
7	49	128
8	64	256
9	81	512
10	100	1024
15	225	32,768
20	400	1,048,576

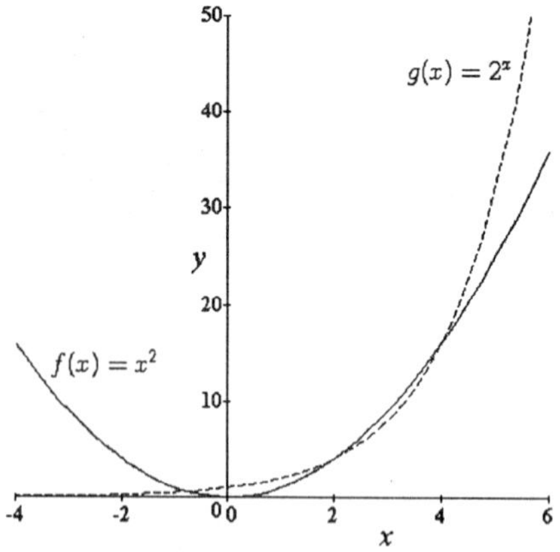

12. If $f(x) = 10^x$, show that $\dfrac{f(x+h) - f(x)}{h} = 10^x\left(\dfrac{10^h - 1}{h}\right)$.

Answer: medium

$f(x) = 10^x$. Then $\dfrac{f(x+h) - f(x)}{h} = \dfrac{10^{x+h} - 10^x}{h} = \dfrac{10^x \cdot 10^h - 10^x}{h} = 10^x\left(\dfrac{10^h - 1}{h}\right)$

GC 13. (a) Compare the rates of growth of the functions $f(x) = 2^x$ and $g(x) = x^4$ by drawing the graphs of both functions in the following viewing rectangles:

(i) $[-2, 2]$ by $[0, 10]$ (ii) $[0, 15]$ by $[0, 50000]$ (iii) $[0, 20]$ by $[0, 10^6]$

(b) Find the solutions of the equation $2^x = x^4$ correct to two decimal places.

Answer: medium

(a) $f(x) = 2^x$ and $g(x) = x^4$

(i) $[-2, 2]$ by $[0, 10]$ (ii) $[0, 15]$ by $[0, 50000]$ (iii) $[0, 20]$ by $[0, 10^6]$

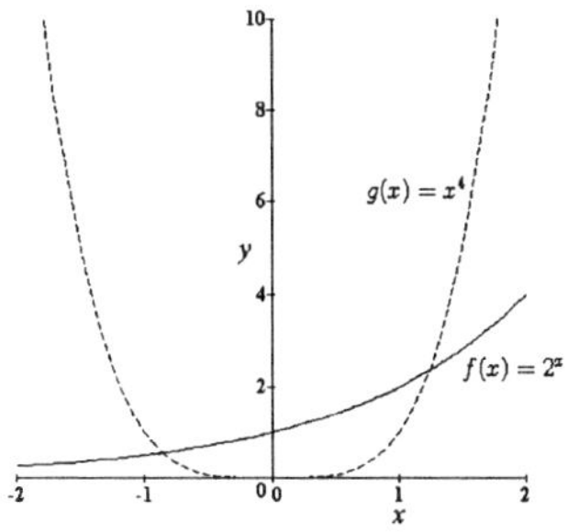

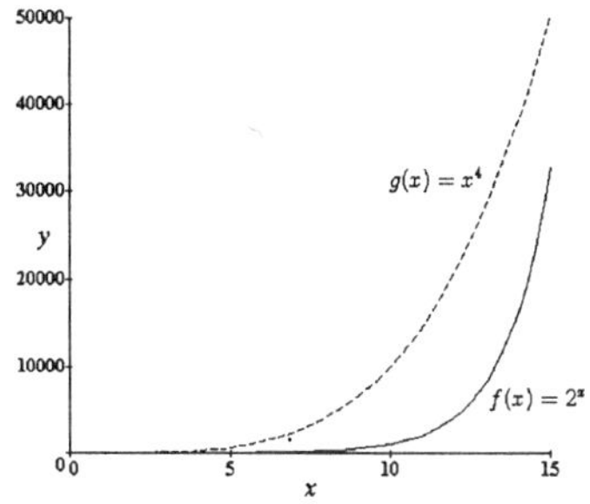

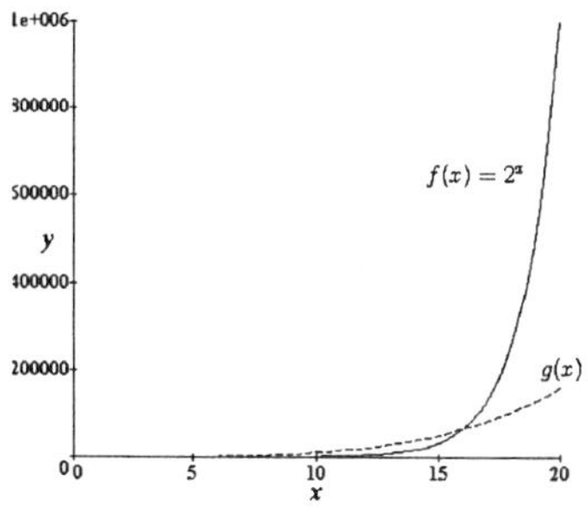

Note that, eventually, $f(x) = 2^x$ grows much more rapidly than $g(x) = x^4$.

(b) From the graphs in parts (i) and (iii) we see that the solutions of $2^x = x^4$ are $x \approx -0.86$, $x \approx 1.24$ and $x = 16$.

GC 14. Graph the function $y = \dfrac{3^x}{x}$ and comment on vertical and horizontal asymptotes.

Answer: medium

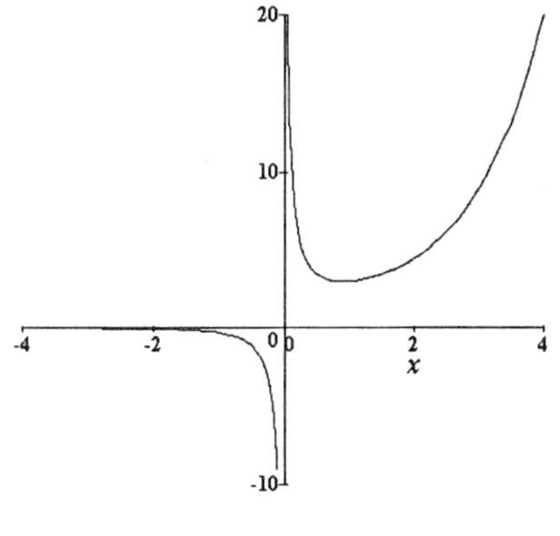

$y = \dfrac{3^x}{x}$. VA: $x = 0$. HA: $y = 0$.

GC 15. Find the local maximum and minimum values of the function $f(x) = \sqrt[x]{2x}$ and the values of x at which they occur. State your answers correct to two decimal places.

Answer: medium

$f(x) = \sqrt[x]{2x}$. From the graph, the local maximum is $f(1.36) \approx 2.09$.

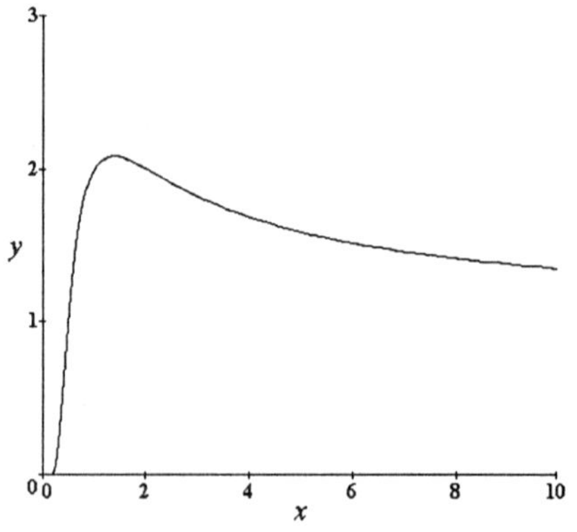

GC 16. For the function $y = x3^x$ find, approximately, (a) the intervals on which the function is increasing or decreasing and (b) the range of the function,

Answer: medium

$y = x3^x$, $[-2, 1]$ by $[-1, 3]$

(a) Decreasing on $(-\infty, -0.91)$, increasing on $(-0.91, \infty)$.

(b) $[-0.33, \infty)$

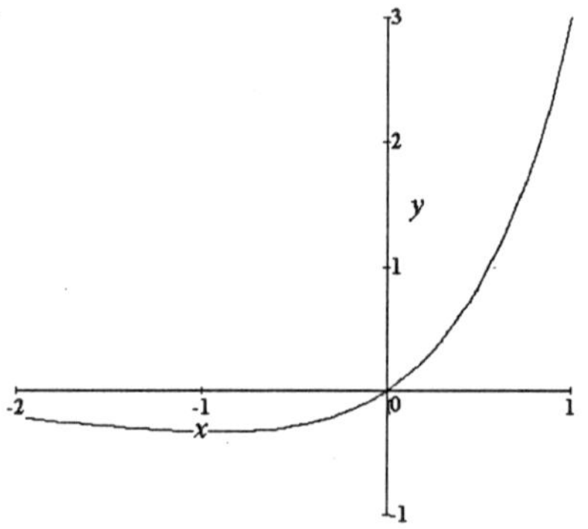

17. Sketch the functions $g(x) = 4^{-x}$ and $h(x) = -4^{-x}$ on one set of axes.

Answer: easy

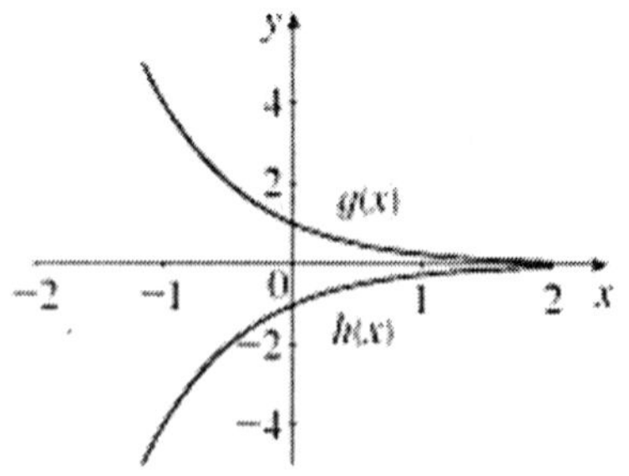

18. If $f(x) = 5^{x+h}$, show that $\dfrac{f(x+h) - f(x)}{5^h} = 5^x (5^h - 1)$.

Answer: easy

$$\frac{f(x+h) - f(x)}{5^h} = \frac{5^{x+h+h} - 5^{x+h}}{5^h} = \frac{5^x 5^{2h}}{5^h} - \frac{5^x 5^h}{5^h} = 5^{x+h} - 5^x = 5^x (5^h - 1)$$

SECTION 6.2

1. Complete the table and use it to graph the function $f(x) = 2e^x$.

x	$f(x) = 2e^x$
-2	
-1.5	
-1	
-0.5	
0	
0.5	
1	
1.5	
2	

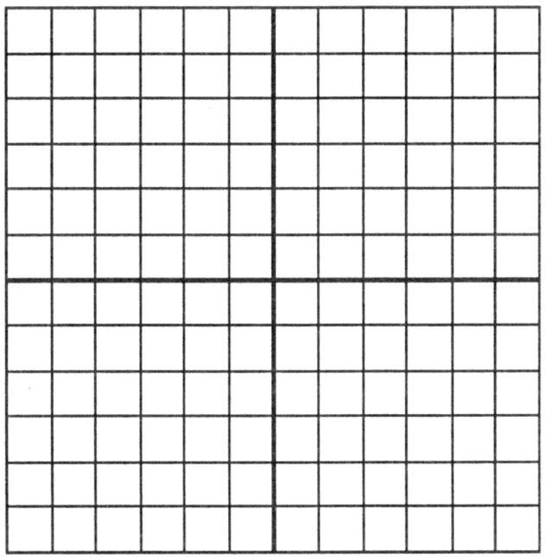

Answer: medium

x	$f(x) = 2e^x$
-2	0.2707
-1.5	0.4463
-1	0.7358
-0.5	1.2131
0	2
0.5	3.2974
1	5.4366
1.5	8.9634
2	14.7781

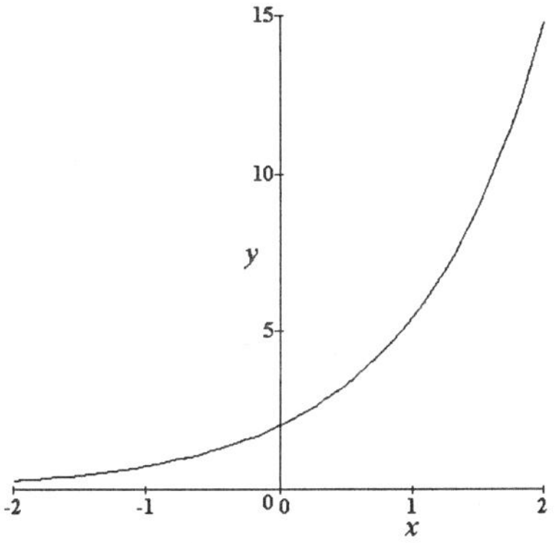

2. Graph the function $y = 2 - e^x$, not by plotting points, but by starting with the graph of $y = e^x$. State the domain, range, and asymptote.

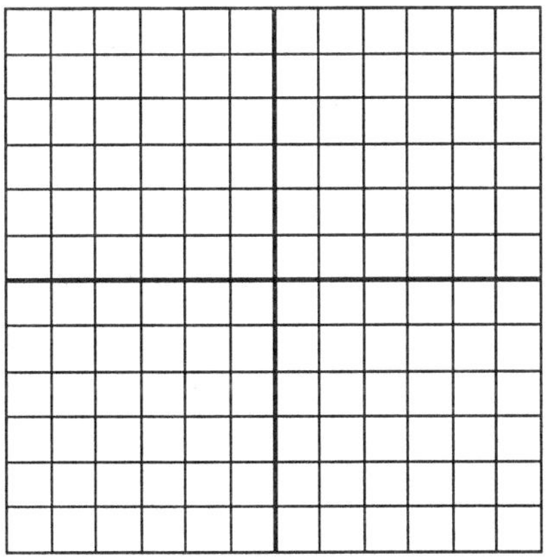

Answer: medium

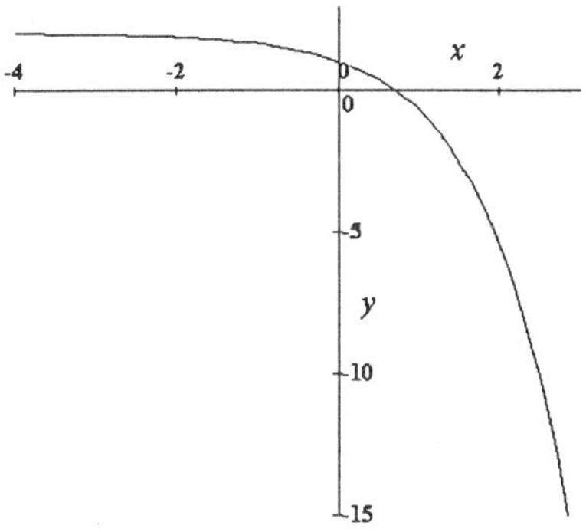

$f(x) = 2 - e^x$. D: $(-\infty, \infty)$, R: $(-\infty, 2)$, A: $y = 2$

3. Graph the function $y = -e^{-2x}$, not by plotting points, but by starting with the graph of $y = e^x$. State the domain, range, and asymptote.

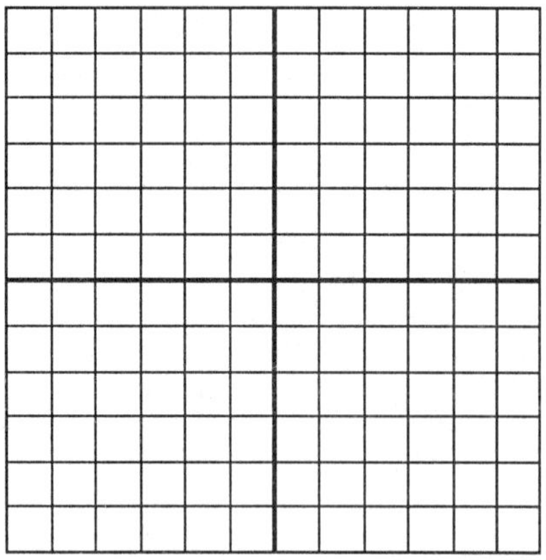

Answer: medium

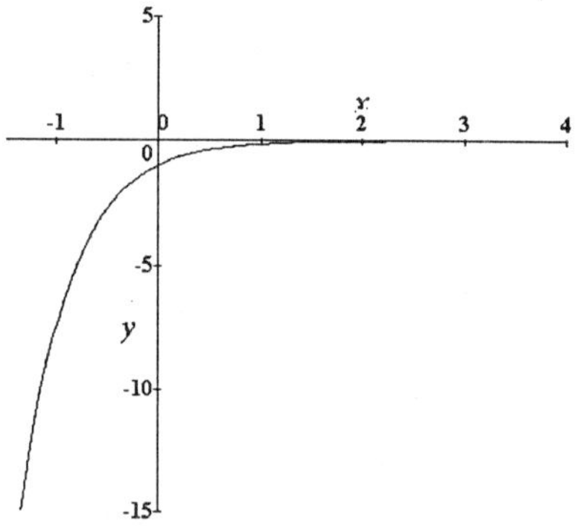

$f(x) = -e^{-2x}$. D: $(-\infty, \infty)$, R: $(-\infty, 0)$, A: $y = 0$

4. Graph the function $y = e^{x+2} - 1$, not by plotting points, but by starting with the graph of $y = e^x$. State the domain, range, and asymptote.

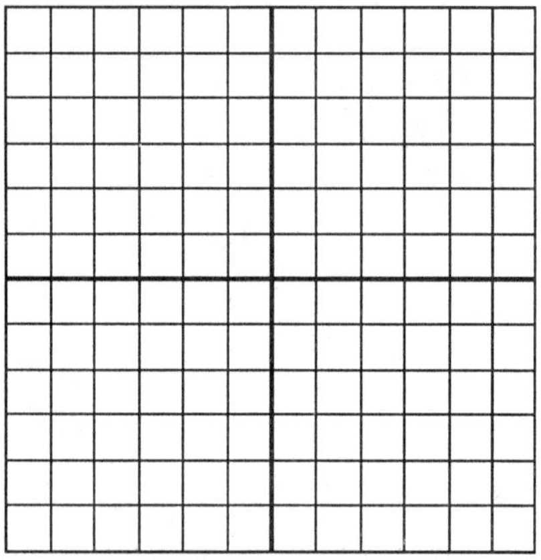

Answer: medium

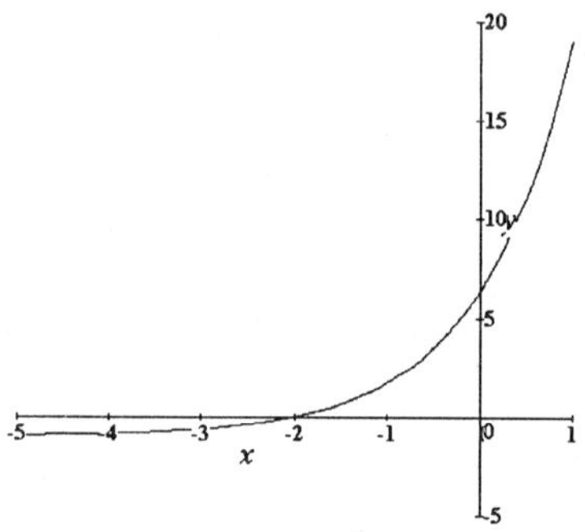

$f(x) = e^{x+2} - 1$. D: $(-\infty, \infty)$, R: $(-1, \infty)$, A: $y = -1$

5. If \$4000 is borrowed at a rate of 14% interest per year, compounded quarterly, find the amount due at the end of (a) 4 years, (b) 6 years, and (c) 8 years.

Answer: medium

$P = 4000$, $r = 0.14$, and $n = 4$. Then $A(t) = 4000\left(1 + \frac{0.14}{4}\right)^{4t} = 4000 \cdot 1.035^{4t}$

(a) $A(4) = 4000 \cdot 1.035^{16} \approx 6935.94$ and so the amount due is \$6935.94.

(b) $A(6) = 4000 \cdot 1.035^{24} \approx 9133.31$ and so the amount due is \$9133.31.

(c) $A(8) = 4000 \cdot 1.035^{32} \approx 12{,}026.83$ and so the amount due is \$12,026.83.

6. Which of the following would be the better investment:

(a) An account paying $8\frac{1}{4}\%$ per year compounded semiannually, or

(b) An account paying 8% per year compounded continuously?

Answer: hard

(a) If $r = 0.0825$ and $n = 2$ then the return is $A_1 = P\left(1 + \frac{0.0825}{2}\right)^{2t} = P \cdot (1.04125)^{2t}$

(b) If $r = 0.08$ and interest is compounded continuously, $A_2 = Pe^{0.08t}$. If $A_1 > A_2$ then $(1.04125)^{2t} > e^{0.08t} \Leftrightarrow 1.04125^2 > e^{0.08} \Leftrightarrow 1.08420 > 1.08329$, which is true and so $A_1 > A_2$. Thus the account paying $8\frac{1}{4}\%$ per year compounded semiannually is the better investment.

7. The doubling period of a bacteria population is 15 min. At time $t = 1.25$ h, the population was estimated to be 80,000.
   (a) What was the initial population?
   (b) What is the population after t hours?
   (c) What is the population after 3 h?
   (d) What is the population after 40 min?
   (e) Sketch the graph of the population function.

   Answer: medium

   $d = \frac{1}{4}$, $n\left(\frac{5}{4}\right) = 80{,}000$

   (a) $n(t) = n_0 \cdot 2^{t/d} = n_0 \cdot 2^{4t}$. Then $n\left(\frac{5}{4}\right) = n_0 \cdot 2^5 = 80{,}000 \quad \Leftrightarrow \quad n_0 = \frac{80{,}000}{32} = 2500.$

   (b) $n(t) = 2500 \cdot 2^{4t}$.

   (c) After 3 hours, $t = 3$ and so $n(3) = 2500 \cdot 2^{12} \approx 1.02 \times 10^7$

   (d) After 40 minutes, $t = \frac{2}{3}$ and so $n\left(\frac{2}{3}\right) = 2500 \cdot 2^{8/3} \approx 15874$.

   (e)

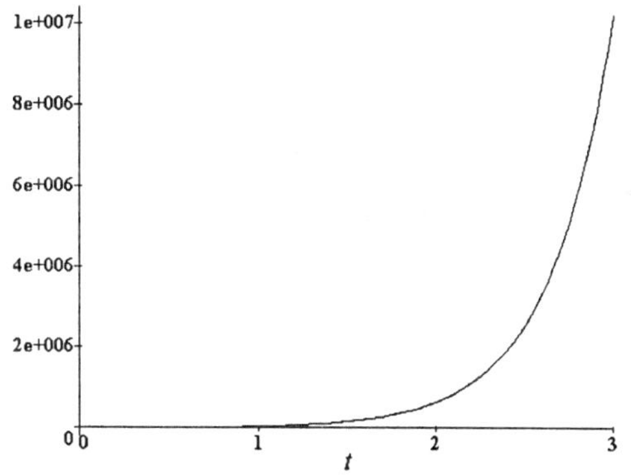

8. A bacteria culture starts with 5000 bacteria. After 3 h, the population is estimated to be 80,000.
   (a) Find the doubling period.
   (b) What is the population after 3.5 h?

   Answer: medium

   $n_0 = 5000$, $n(3) = 80{,}000$

   (a) $n(t) = n_0 \cdot 2^{t/d} = 5000 \cdot 2^{t/d} \quad \Leftrightarrow \quad n(3) = 5000 \cdot 2^{3/d} = 80{,}000 \quad \Leftrightarrow \quad 2^{3/d} = 16 \quad \Leftrightarrow \quad 2^{3/d} = 2^4 \quad \Leftrightarrow \quad 4 = \frac{3}{d}$
   $\Leftrightarrow \quad d = \frac{3}{4}$. Therefore the doubling period is 45 minutes.

   (b) $n\left(\frac{7}{2}\right) = 5000 \cdot 2^{4t/3} = 5000 \cdot 2^{14/3} \approx 126{,}992$

9. The population of a certain city grows at a rate of 5% per year. The population in 1995 was 632,000. If the growth rate remains constant, find the projected population of the city for the years 2000 and 2030.

   Answer: medium

   $P_0 = 632{,}000$ and $r = 0.05$ $P(t) = P_0(1 + r)^t = 632{,}000(1 + 0.05)^t = 632{,}000 \cdot 1.05^t$. In the year 2000, $t = 2000 - 1995 = 5$, so $P(2000 - 1995) = 632{,}000 \cdot 1.05^5 \approx 806{,}610$. In the year 2030, $t = 2030 - 1995 = 35$, so $P(2030 - 1995) = 632{,}000 \cdot 1.05^{35} \approx 3{,}486{,}122$.

10. An isotope of sodium, ^{24}Na, has a half-life of 15 h. Find the amount remaining from a 2 g sample after
    (a) 5 h. (b) 20 h. (c) 4 days.

    Answer: medium

    $h = 15$ hours and $m_0 = 2$ g Then $m(t) = m_0 \cdot 2^{-t/h} = 2 \cdot 2^{-t/15} = 2^{1-t/15}$ where t is measured in hours.

    (a) $m(5) = 2^{1-5/15} = 2^{2/3} \approx 1.59$ g.

    (b) $m(20) = 2^{1-20/15} = 2^{-1/3} \approx 0.79$ g.

    (c) $m(4\,\text{days}) = m(96) = 2^{1-96/15} = 2^{-81/15} = 2^{-27/5} \approx 0.024$ g.

11. A certain amount of Vanadium-48, $\left({}^{48}\mathrm{V}\right)$, decays to $\frac{1}{8}$ of its original mass in 48 days. What is the half-life of ${}^{48}\mathrm{V}$?

Answer: medium

$m(48) = \frac{1}{8}m_0$. Since $m(48) = m_0 \cdot 2^{-t/h} = m_0 \cdot 2^{-48/h}$, then $\frac{1}{8}m_0 = m_0 \cdot 2^{-48/h} \quad \Leftrightarrow \quad \frac{1}{8} = 2^{-3} = 2^{-48/h} \quad \Leftrightarrow$ $-3 = -48/h \quad \Leftrightarrow \quad h = 16$, and so the half-life is 16 days.

GC 12. A bacteria culture starts with 10,000 bacteria and the doubling period is 40 min. If the number of bacteria after t minutes is $n(t)$, draw the graph of $n(t)$ and use it to determine when the population of bacteria will reach 50,000.

Answer: medium

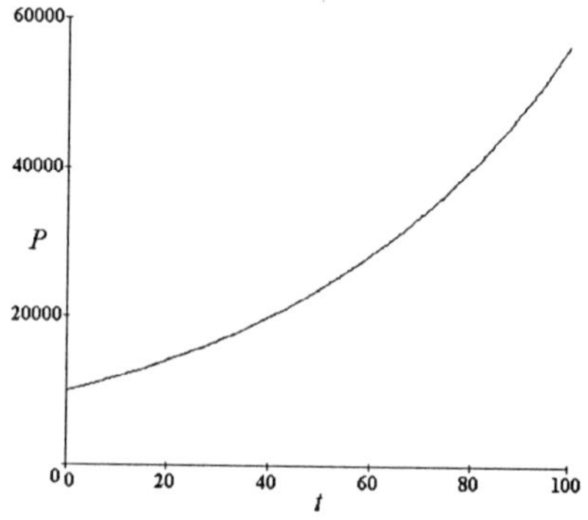

$n_0 = 10{,}000$, $d = 40$; $n(t) = n_0 \cdot 2^{t/d} = 10{,}000 \cdot 2^{t/40}$.

From the graph we see that the population reaches 50,000 when $t \approx 93$ min.

GC 13. A sum of \$10,000 is invested at an interest rate of 8% per year compounded semiannually.

(a) Find the value $A(t)$ of the investment after t years.

(b) Draw the graph of $A(t)$.

(c) Use the graph of $A(t)$ to determine when this investment will amount to \$25,000.

Answer: medium

$P = 10{,}000$, $r = 0.08$, and $n = 2$

(a) $A(t) = 10{,}000\left(1 + \frac{0.08}{2}\right)^{2t} = 10{,}000 \cdot 1.04^{2t}$

(b)

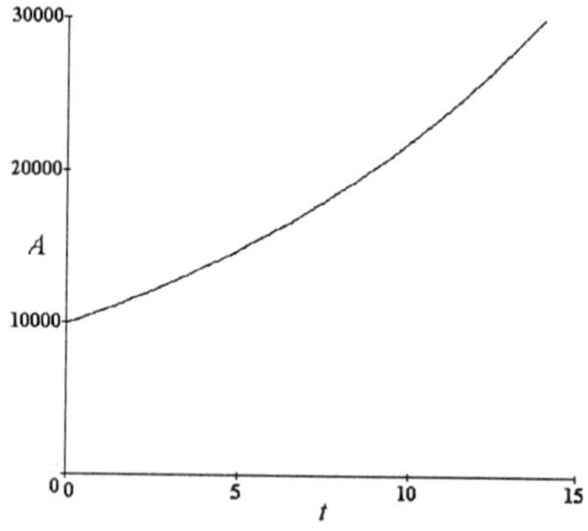

(c) $t \approx 11.68$ yr

14. The hyperbolic sine function is defined by $\sinh(x) = \dfrac{e^x - e^{-x}}{2}$. Graph this function using graphical addition.

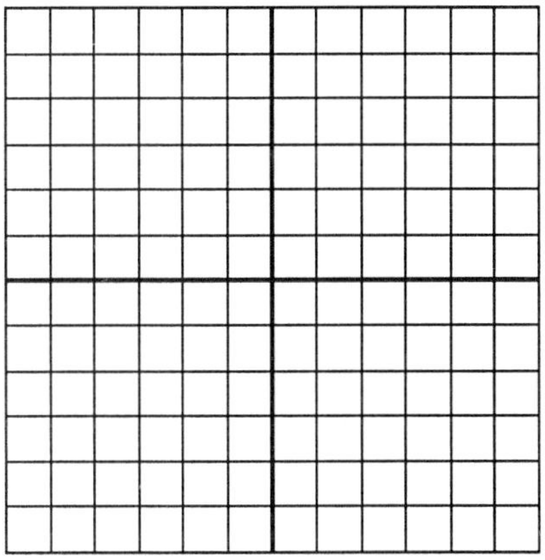

Answer: medium

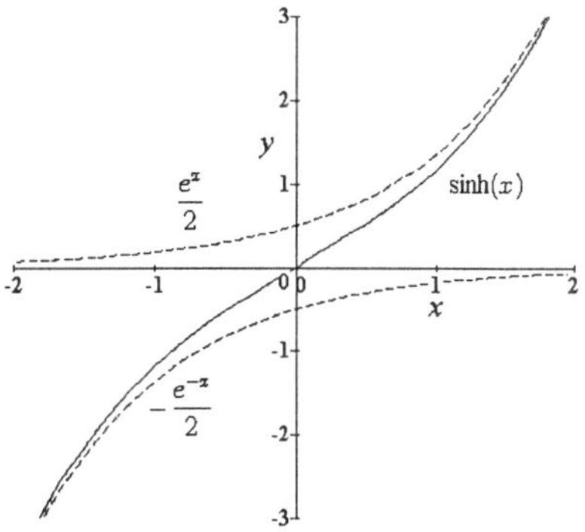

15. Use the definition $\sinh(x) = \dfrac{e^x - e^{-x}}{2}$ to prove the identity $\sinh(-x) = -\sinh(x)$.

Answer: medium

$$\sinh(-x) = \frac{e^{-x} - e^{-(-x)}}{2} = \frac{-e^x + e^{-x}}{2} = -\frac{e^x - e^{-x}}{2} = -\sinh(x)$$

16. Use the definitions $\cosh(x) = \dfrac{e^x + e^{-x}}{2}$ and $\sinh(x) = \dfrac{e^x - e^{-x}}{2}$ to prove the identity $\sinh(x + y) = \sinh(x)\cosh(y) + \cosh(x)\sinh(y)$.

Answer: hard

$$\begin{aligned}\sinh(x)\cosh(y) + \cosh(x)\sinh(y) &= \frac{e^x - e^{-x}}{2} \cdot \frac{e^y + e^{-y}}{2} + \frac{e^x + e^{-x}}{2} \cdot \frac{e^y - e^{-y}}{2} \\ &= \frac{e^{x+y} + e^{x-y} - e^{y-x} - e^{-(x+y)}}{4} + \frac{e^{x+y} - e^{x-y} + e^{y-x} - e^{-(x+y)}}{4} = \frac{e^{x+y} - e^{-(x+y)}}{2} \\ &= \sinh(x+y)\end{aligned}$$

GC 17. Investigate the behavior of the function $f(x) = \left(1 - \dfrac{1}{x}\right)^{-x}$ as $x \to \infty$ by drawing its graph and the line $y = e$ on the same screen using the viewing rectangle $[0, 100]$ by $[0, 5]$.

Answer: medium

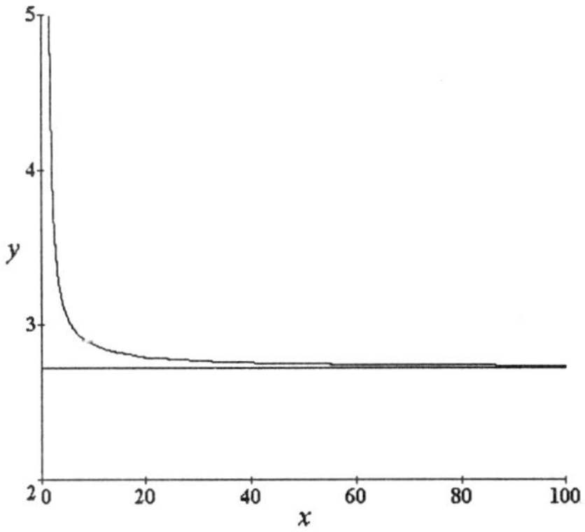

From the graph, we see that $f(x) \to e$ as $x \to \infty$.

GC 18. Find the local maximum and minimum values of the function $f(x) = e^x + e^{-2x}$ and the values of x at which they occur. State your answers correct to two decimal places.

Answer: medium

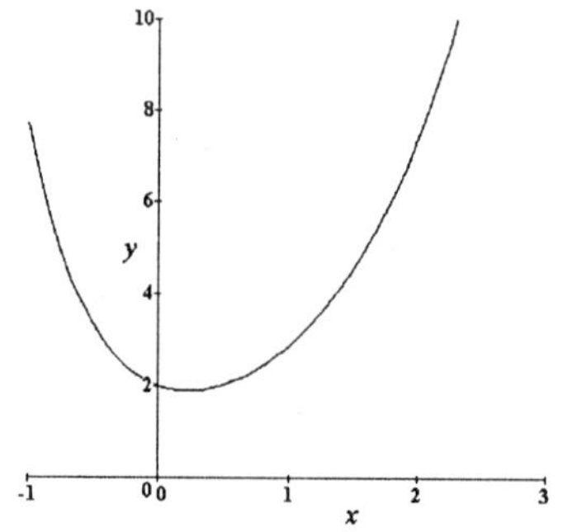

$f(x) = e^x + e^{-2x}$. From the graph, the local minimum is $f(0.23) = 1.89$.

19. The number of bacteria in a culture is given by the formula $n(t) = 1200e^{0.35t}$, where t is measured in hours.

(a) What is the relative rate of growth of this bacterial population?

(b) What is the initial population of the culture?

(c) How many bacteria will the culture contain after 3 hours?

Answer: medium

(a) At any time t, the growth rate is 35% of the population.

(b) When $t = 0$, $n(0) = 1200e^{0.35(0)} = 1200$.

(c) When $t = 3$, $n(3) = 1200e^{0.35(3)} \approx 3429$.

20. The population of a certain town has a relative growth rate of 2.3% per year. The population in 1992 was 6,723. Find the projected population of the town for the years 2000 and 2010.

Answer: medium

$n_0 = 6,723$ and $r = 0.023$, so $n(t) = 6723e^{0.023t}$. For the year 2000: $n(8) = 6723e^{0.023(8)} \approx 8081$. For the year 2010: $n(18) = 6723e^{0.023(18)} \approx 10{,}171$.

21. The relative growth rate for a certain bacteria population is 112% per hour. Two and half hours after a culture is formed, a count shows approximately 12,600 bacteria.

(a) Find the initial number of bacteria in the culture.

(b) Estimate the number of bacteria 4 hours from the time the culture was started.

Answer: medium

(a) $12{,}600 = n_0e^{1.12(2.5)} \quad \Leftrightarrow \quad n_0 \approx 766$

(b) $n(t) = 766e^{1.12(t)} \quad \Rightarrow \quad n(4) = 766e^{1.12(4)} \approx 67{,}588$

22. A particular radioactive substance decays in such a way that the amount of mass remaining after t days is given by the function $m(t) = 8e^{-0.018t}$, where $m(t)$ is measured in kilograms.

(a) Find the initial mass.

(b) What percent of mass remains after 37 days?

(c) Use a graphing device to help approximate the time it takes for the substance to decay to a mass of 2 kilograms.

Answer: hard

(a) When $t = 0$, $m(0) = 8e^{-0.018(0)} = 8$ kilograms.

(b) After 37 days ,$m(37) = 8e^{-0.018(37)} \approx 4.11$ kilos or 51.4%.

(c)

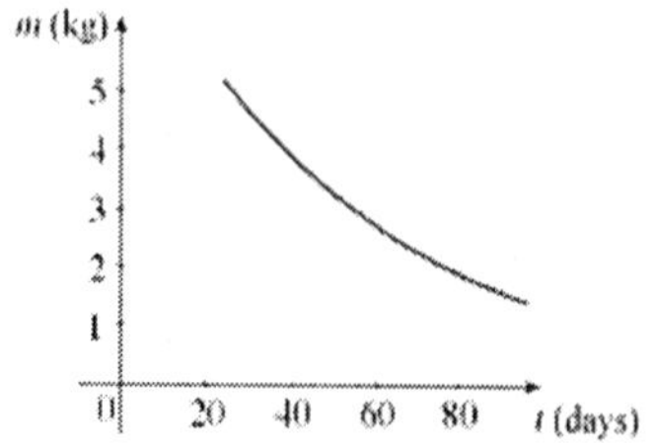

From the graph, the substance decays to a mass of 2 kilos after about 77 days.

GC 23. Compare the functions $f(x) = e^x$ and $g(x) = x^4 + 1$ by drawing their graphs in each of the following viewing rectangles: $[-0.5, 1.5] \times [0, 4]$ and $[7, 10] \times [4000, 10{,}000]$. Find the solutions of the equation $e^x = x^4 + 1$ in each viewing rectangle, correct to two decimal places.
Answer: medium

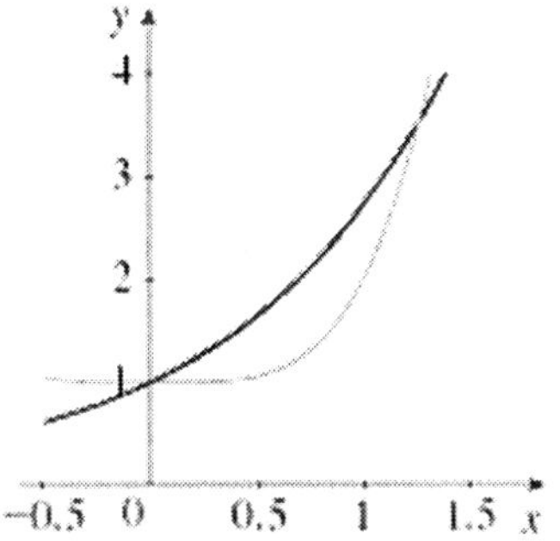

For $[-0.5, 1.5] \times [0, 4]$ the solutions are $x = 0$, $x \approx 1.26$.

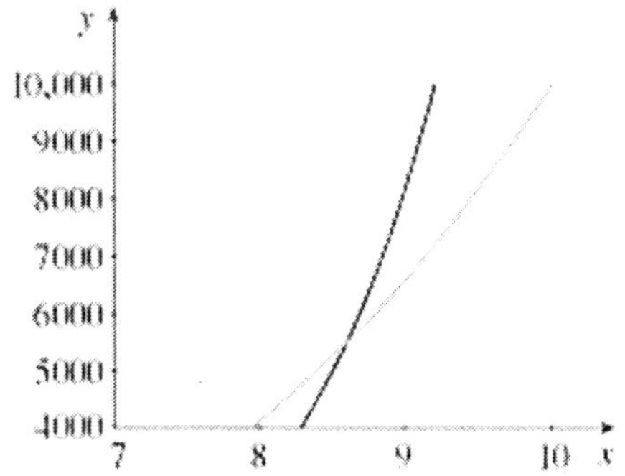

For $[7, 10] \times [4000, 10{,}000]$ the solution is $x \approx 8.62$.

SECTION 6.3

1. Express the equation $\log_6 1 = 0$ in exponential form.
(a) $6^1 = 0$ (b) $0^6 = 1$ (c) $6^0 = 1$ (d) $1^6 = 1$ (e) $6^1 = 6$
Answer: easy (c)
$\log_6 1 = 0 \quad \Leftrightarrow \quad 6^0 = 1$

2. Express the equation $\log_{27} 9 = \frac{2}{3}$ in exponential form.
(a) $3^3 = 27$ (b) $\dfrac{9^2}{3} = 27$ (c) $9 = 3^2$ (d) $27^{2/3} = 9$ (e) $9^{3/2} = 27$
Answer: easy (d)
$\log_{27} 9 = \frac{2}{3} \quad \Leftrightarrow \quad 27^{2/3} = 9$

3. Express the equation $\log_2\left(\frac{1}{8}\right) = -3$ in exponential form.
(a) $2^3 = 8$ (b) $2^{-3} = \frac{1}{8}$ (c) $8^{-1/3} = 2$ (d) $8^{1/3} = 2$ (e) $-3^8 = 2$
Answer: easy (b)
$\log_2\left(\frac{1}{8}\right) = -3 \quad \Leftrightarrow \quad 2^{-3} = \frac{1}{8}$

4. Express the equation $\log_r v = w$ in exponential form.
(a) $v^w = r$ (b) $r^w = v$ (c) $v^r = w$ (d) $w^r = v$ (e) $w^v = r$
Answer: easy (b)
$\log_r v = w \quad \Leftrightarrow \quad r^w = v$

5. Express the equation $10^5 = 100{,}000$ in logarithmic form.
(a) $\log_{10} 10{,}000 = 4$ (b) $\log_5 100{,}000 = 10$ (c) $\log_{10} 10 = 5$ (d) $\log_{100{,}000} 5 = 10$ (e) $\log_{10} 100{,}000 = 5$
Answer: easy (e)
$10^5 = 100{,}000 \quad \Leftrightarrow \quad \log_{10} 100{,}000 = 5$

6. Express the equation $16^{1/2} = 4$ in logarithmic form.
(a) $\log_4 \frac{1}{2} = \frac{1}{16}$ (b) $\log_4 16 = 2$ (c) $\log_4 2 = \frac{1}{16}$ (d) $\log_4 2 = 2$ (e) $\log_{16} 4 = \frac{1}{2}$
Answer: easy (e)
$16^{1/2} = 4 \quad \Leftrightarrow \quad \log_{16} 4 = \frac{1}{2}$

7. Express the equation $5^{-1} = \frac{1}{5}$ in logarithmic form.
(a) $\log_5\left(-\frac{1}{5}\right) = -1$ (b) $\log_5 5 = 1$ (c) $\log_1 \frac{1}{5} = -\frac{1}{5}$ (d) $\log_5 \frac{1}{5} = -1$ (e) $\log_5 1 = -5$
Answer: easy (d)
$5^{-1} = \frac{1}{5} \quad \Leftrightarrow \quad \log_5 \frac{1}{5} = -1$

8. Express the equation $10^m = n$ in logarithmic form.
(a) $\log_{10} n = m$ (b) $\log_{10} m = n$ (c) $\log_m 10 = n$ (d) $\log_n m = 10$ (e) $\log_n 10 = m$
Answer: easy (a)
$10^m = n \quad \Leftrightarrow \quad \log_{10} n = m$

9. Evaluate the expression $\log_2 16$.
(a) 3 (b) 4 (c) 5 (d) 6 (e) 7
Answer: easy (b)
$\log_2 16 = \log_2 2^4 = 4$

10. Evaluate the expression $\log_7 7^{13}$.
(a) 7 (b) 11 (c) 13 (d) 17 (e) 20
Answer: easy (c)
$\log_7 7^{13} = 13$

11. Evaluate the expression $\log_5 1$.
(a) -1 (b) 1 (c) 3 (d) 5 (e) 0
Answer: easy (e)
$\log_5 1 = \log_5 5^0 = 0$

12. Evaluate the expression $\log_5 625$.
(a) 3 (b) 4 (c) 5 (d) 125 (e) 625
Answer: easy (b)
$\log_5 625 = \log_5 5^4 = 4$

13. Evaluate the expression $3^{\log_3 7}$.
(a) 7 (b) 10 (c) 11 (d) 24 (e) 49
Answer: easy (a)
$3^{\log_3 7} = 7$

14. Evaluate the expression $\log_8 16$.
(a) $\frac{1}{2}$ (b) $\frac{2}{3}$ (c) 1 (d) $\frac{4}{3}$ (e) 2
Answer: medium (d)
$\log_8 16 = \log_8 8^{4/3} = \frac{4}{3}$

15. Solve the equation $\log_3 x = 4$ for x.
Answer: easy
$\log_3 x = 4 \quad \Leftrightarrow \quad x = 3^4 = 81$

16. Solve the equation $\log_3(2 - x) = 3$ for x.
Answer: medium
$\log_3(2 - x) = 3 \quad \Leftrightarrow \quad 2 - x = 3^3 = 27 \quad \Leftrightarrow \quad x = -25$

17. Solve the equation $\log_x 5 = \frac{1}{2}$ for x.
Answer: medium
$\log_x 5 = \frac{1}{2} \quad \Leftrightarrow \quad 5 = x^{1/2} \quad \Leftrightarrow \quad 25 = x$

18. Use a calculator to evaluate the expression $\ln\sqrt{3}$.
(a) 0.2465 (b) 0.5493 (c) 0.6489 (d) 0.9954 (e) 1.066
Answer: easy (b)
$\ln\sqrt{3} \approx 0.5493$

19. Use a calculator to evaluate the expression $\ln 0.5$.

(a) -2.2241 (b) -1.1042 (c) -1.0314 (d) -0.9421 (e) -0.6931

Answer: easy (e)

$\ln 0.5 \approx -0.6931$

20. Use a calculator to evaluate the expression $\ln \pi$.

(a) 0.2465 (b) 1.1211 (c) 1.1447 (d) 1.3043 (e) 2.0104

Answer: easy (c)

$\ln \pi \approx 1.1447$

21. Use a calculator to evaluate the expression $\ln 107.9$.

(a) 2.0302 (b) 3.1660 (c) 3.9025 (d) 4.6812 (e) 5.0029

Answer: easy (d)

$\ln 107.9 \approx 4.6812$

22. Graph the function $f(x) = -\log_3 x$, not by plotting points but by applying your knowledge of the general shape of the logarithmic function. State the domain, range, and asymptote.

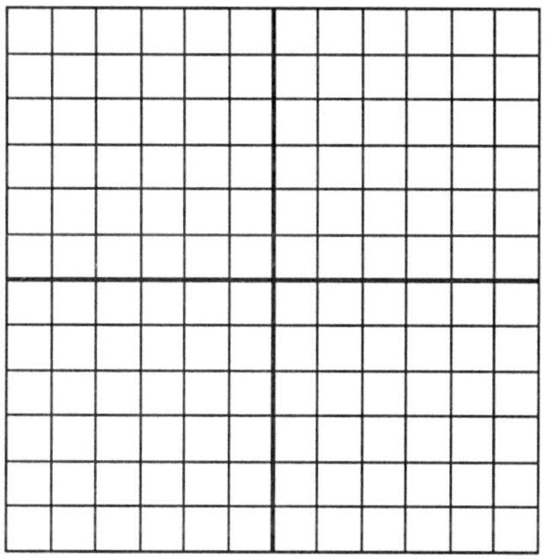

Answer: medium

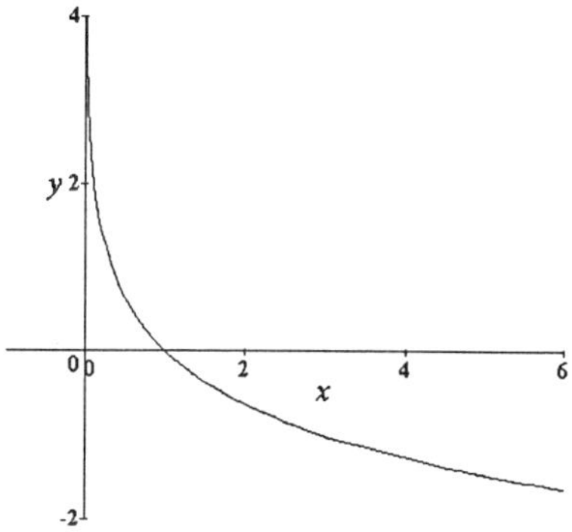

$f(x) = -\log_3 x$. D: $(0, \infty)$, R: $(-\infty, \infty)$, A: $x = 0$

23. Graph the function $g(x) = \ln(x + 3)$, not by plotting points but by applying your knowledge of the general shape of the logarithmic function. State the domain, range, and asymptote.

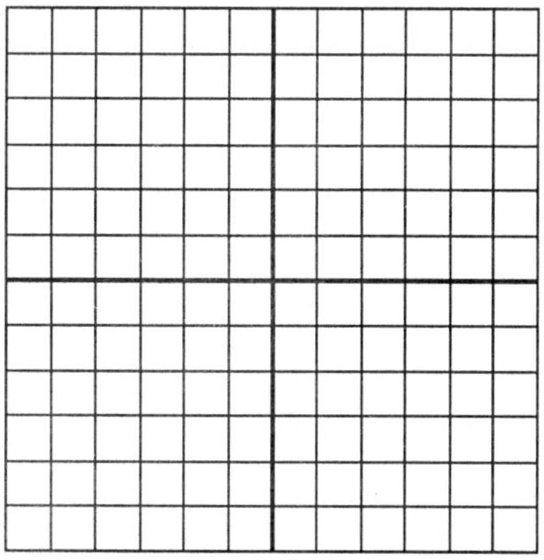

Answer: medium

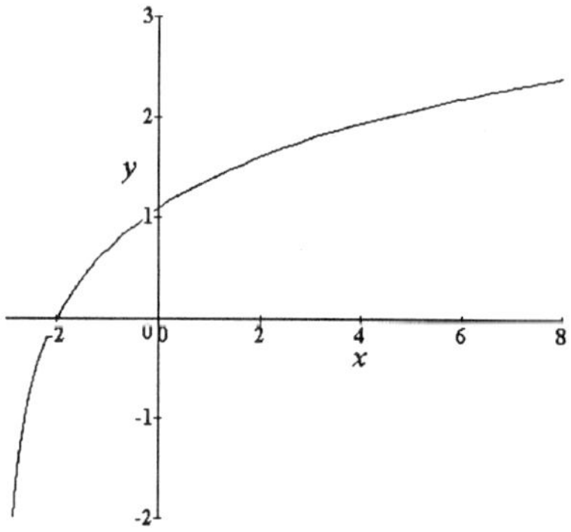

$g(x) = \ln(x + 3)$. D: $(-3, \infty)$, R: $(-\infty, \infty)$, A: $x = -3$

24. Graph the function $y = \log_4(x - 1) - 3$, not by plotting points but by applying your knowledge of the general shape of the logarithmic function. State the domain, range, and asymptote.

Answer: medium

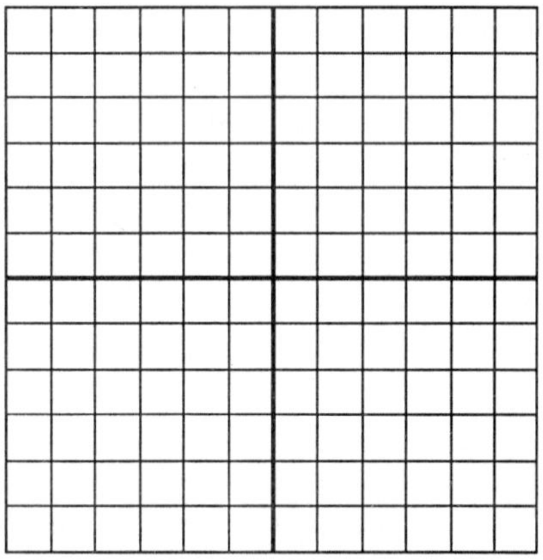

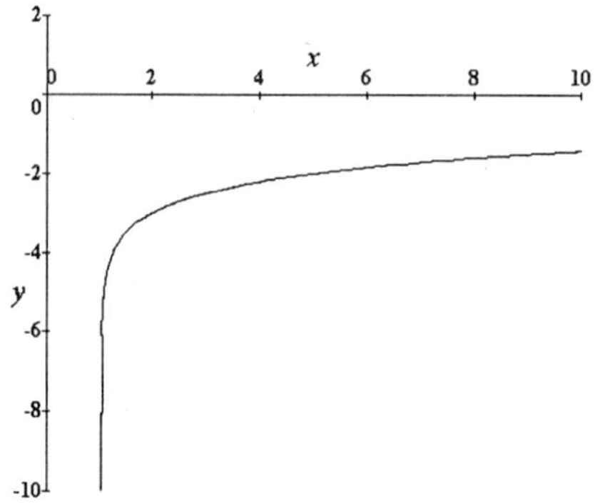

$f(x) = \log_4(x - 1) - 3$. D: $(1, \infty)$, R: $(-\infty, \infty)$, A: $x = 1$

25. Graph the function $y = 2 - \ln(-x)$, not by plotting points but by applying your knowledge of the general shape of the logarithmic function. State the domain, range, and asymptote.

Answer: medium

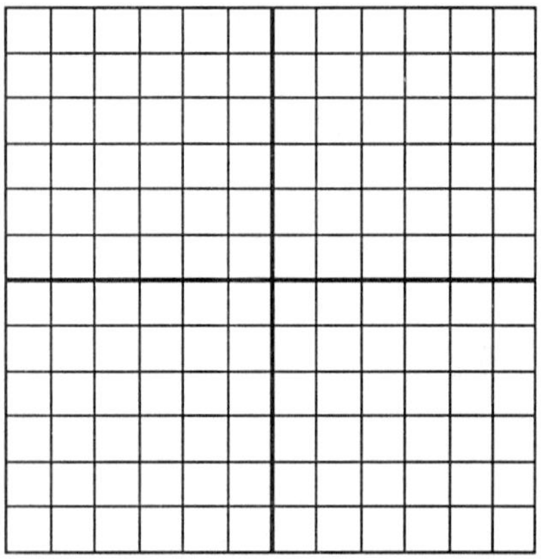

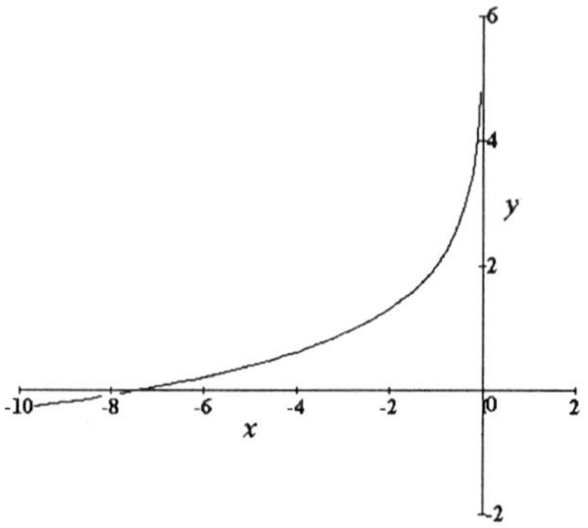

$f(x) = 2 - \ln(-x)$. D: $(-\infty, 0)$, R: $(-\infty, \infty)$, A: $x = 0$

26. Graph the function $y = \ln|2x|$, not by plotting points but by applying your knowledge of the general shape of the logarithmic function. State the domain, range, and asymptote.

Answer: medium

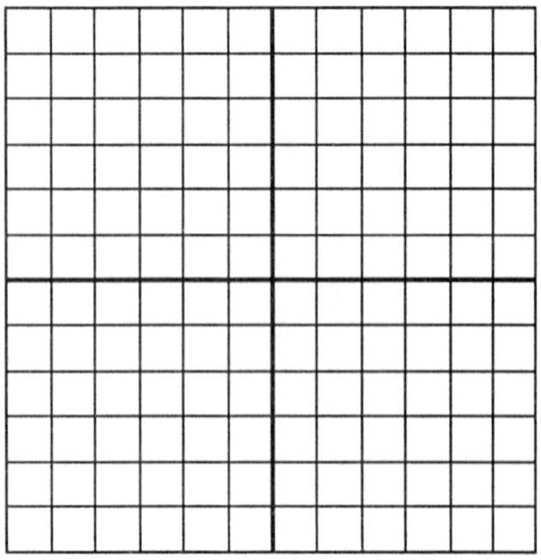

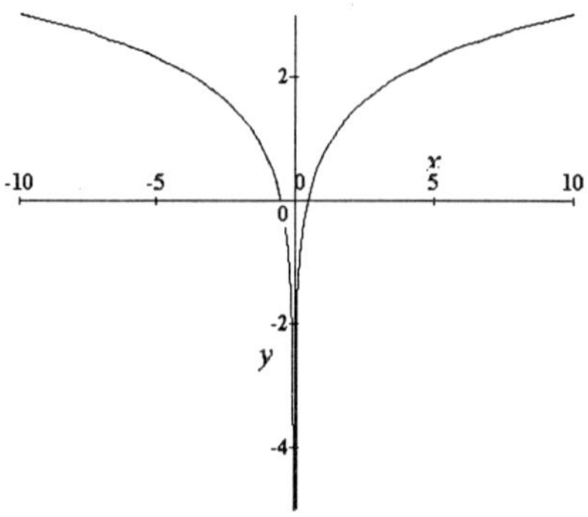

$f(x) = \ln|2x|$. D: $(-\infty, 0) \cup (0, \infty)$, R: $(-\infty, \infty)$, A: $x = 0$

27. Find the domain of the function $f(x) = \log_2(10 - 2x)$.

Answer: medium

$f(x) = \log_2(10 - 2x)$. Then $10 - 2x > 0 \Leftrightarrow x < 5$ and so D is $(-\infty, 5)$

28. Find the domain of the function $g(x) = \ln(x^2 - x)$.

Answer: medium

$g(x) = \ln(x^2 - x)$. Then $x^2 - x > 0 \quad \Leftrightarrow \quad x(x-1) > 0$.

Case (i): $x > 0$ and $x - 1 > 0 \quad \Leftrightarrow \quad x > 1$.

Case (ii): $x < 0$ and $x - 1 < 0 \quad \Leftrightarrow \quad x < 0$. Combining the two cases gives a domain of $(-\infty, 0) \cup (1, \infty)$.

29. Find the domain of the function $h(x) = \sqrt{x-3} - \log_4(10 - 3x)$.

Answer: medium

$h(x) = \sqrt{x-3} - \log_4(10 - 3x)$. Then $x - 3 \geq 0$ and $10 - 3x > 0 \quad \Leftrightarrow \quad x \geq 3$ and $x < \frac{10}{3}$, so the domain is $\left[3, \frac{10}{3}\right)$.

GC 30. Draw the graph of the function $f(x) = \ln(x - x^2)$ in a suitable viewing rectangle and use it to find the domain, the asymptotes, and the local maximum and minimum values.

Answer: medium

$f(x) = \ln(x - x^2)$. D: $(0, 1)$, A: $x = 0, 1$; local maximum $f(0.5) \approx -1.386$

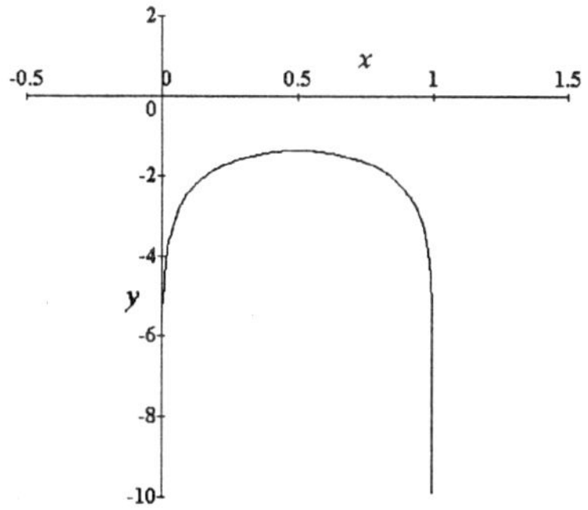

GC 31. Draw the graph of the function $f(x) = x^2 \ln x$ in a suitable viewing rectangle and use it to find the domain, the asymptotes, and the local maximum and minimum values.

Answer: medium

$f(x) = x^2 \ln x$. D: $(0, \infty)$, no asymptotes, local minimum $f(0.61) \approx -0.18$

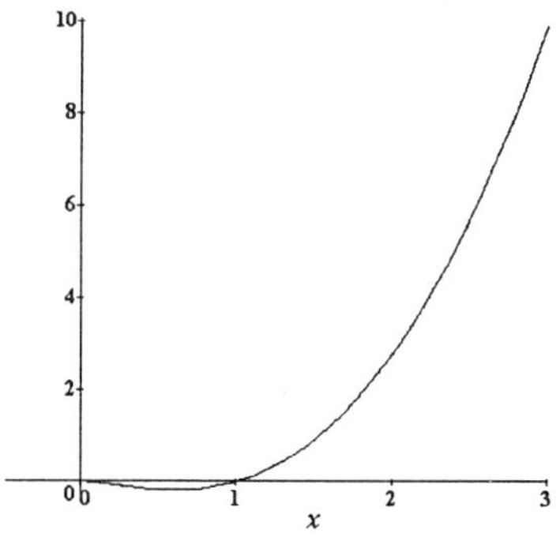

GC 32. Draw the graph of the function $f(x) = x \log_4(x + 4)$ in a suitable viewing rectangle and use it to find the domain, the asymptotes, and the local maximum and minimum values.

Answer: medium

$f(x) = x \log_4(x + 4)$.

D: $(-4, \infty)$, A: $x = -4$, local minimum $f(-1.78) \approx -1.02$

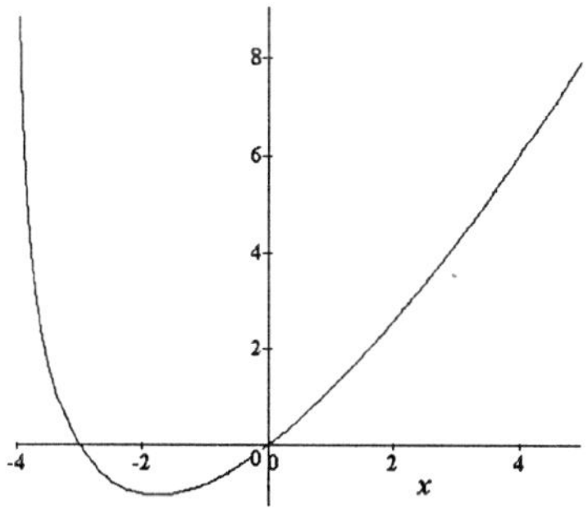

GC 33. (a) By drawing the graphs of the functions $f(x) = 2 + \ln(1 + x)$ and $g(x) = \sqrt{x}$ in a suitable viewing rectangle show that, even when a logarithmic function starts out higher than a root function, it is ultimately overtaken by the root function.

(b) Find the solutions of the equation $\sqrt{x} = 2 + \ln(1 + x)$ correct to two decimal places.

Answer: medium

(a)

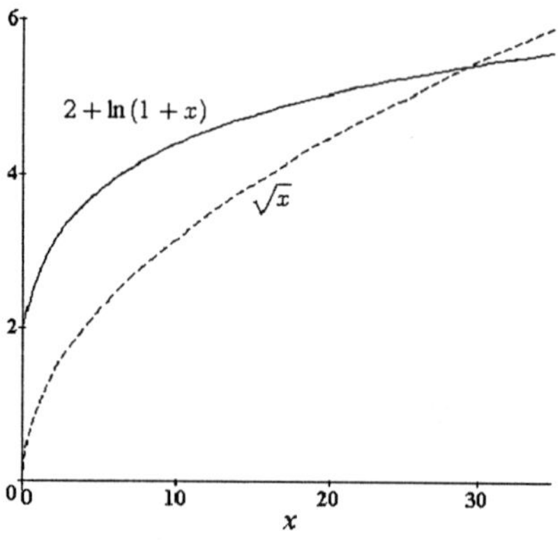

(b) $\sqrt{x} = 2 + \ln(1 + x)$ has solution $x \approx 29.27$.

34. Which is larger, $\log_5 26$ or $\log_6 35$?

Answer: hard

Since $\log_5 x$ is increasing, $\log_5 26 > \log_5 25 = 2$. Also, because $\log_6$ is increasing, $\log_6 35 < \log_6 36 = 2$. Therefore $\log_6 35 < 2 < \log_5 26$ and so $\log_5 26$ is larger.

35. (a) Find the domain of the function $f(x) = \ln(\ln(\ln(\ln x)))$.

(b) Find the inverse function of f.

Answer: hard

(a) $f(x) = \ln(\ln(\ln(\ln x)))$. Then $(\ln(\ln(\ln x))) > 0 \quad \Leftrightarrow \quad e^{(\ln(\ln(\ln x)))} > e^0 \quad \Leftrightarrow \quad \ln(\ln x) > 1 \quad \Leftrightarrow \quad e^{\ln(\ln x)} > e^1 \quad \Leftrightarrow$ $\ln x > e \quad \Leftrightarrow \quad e^{\ln x} = x > e^e$ and so the domain is (e^e, ∞).

(b) $y = \ln(\ln(\ln(\ln x))) \quad \Leftrightarrow \quad e^y = (\ln(\ln(\ln x))) \quad \Leftrightarrow \quad e^{e^y} = \ln(\ln x) \quad \Leftrightarrow \quad e^{e^{e^y}} = \ln x \quad \Leftrightarrow \quad e^{e^{e^{e^y}}} = x$. Therefore the inverse function is $f^{-1}(x) = e^{e^{e^{e^x}}}$.

36. (a) Find the inverse of the function $f(x) = \dfrac{3^x}{1 + 3^x}$.

(b) What is the domain of the inverse function?

Answer: hard

(a) $y = \dfrac{3^x}{1 + 3^x} \quad \Leftrightarrow \quad y(1 + 3^x) = 3^x \quad \Leftrightarrow \quad 3^x(y - 1) = -y \quad \Leftrightarrow \quad 3^x = \dfrac{-y}{y - 1} \quad \Leftrightarrow \quad x = \log_3\left(\dfrac{y}{1 - y}\right)$. Therefore the inverse function is $f^{-1}(x) = \log_3\left(\dfrac{x}{1 - x}\right)$.

(b) $f^{-1}(x) = \log_3\left(\dfrac{x}{1 - x}\right)$. Then $\dfrac{x}{1 - x} > 0$.

Case (i): $x > 0$ and $1 - x > 0 \quad \Leftrightarrow \quad x > 0$ and $x < 1 \quad \Leftrightarrow \quad 0 < x < 1$.

Case (ii): $x < 0$ and $1 - x < 0 \quad \Leftrightarrow \quad x < 0$ and $x > 1$, which is impossible.

Combining the two cases gives a domain of $(0, 1)$.

37. Use the definition of the logarithmic function to solve the equation $\log_x\left(\frac{1}{10{,}000}\right) = 4$ for x.

Answer: easy

$\log_x\left(\frac{1}{10{,}000}\right) = 4 \quad \Leftrightarrow \quad x^4 = \frac{1}{10{,}000} \quad \Leftrightarrow \quad x^4 = \left(\frac{1}{10}\right)^4 \quad \Leftrightarrow \quad x = \left(\frac{1}{10}\right)$

38. Use the definition of the logarithmic function to solve the equation $\log_{25}\sqrt{5} = x$ for x.
Answer: easy
$\log_{25}\sqrt{5} = x \quad\Leftrightarrow\quad 25^x = \sqrt{5} \quad\Leftrightarrow\quad 5^{2x} = 5^{1/2} \quad\Leftrightarrow\quad x = \frac{1}{4}$

39. Use the definition of the logarithmic function to solve the equation $\log_7 49 = \frac{1}{2}x$ for x.
Answer: easy
$\log_7 49 = \frac{1}{2}x \quad\Leftrightarrow\quad 7^{x/2} = 49 \quad\Leftrightarrow\quad 7^{x/2} = 7^2 \quad\Leftrightarrow\quad \frac{1}{2}x = 2 \quad\Leftrightarrow\quad x = 4$

40. Use the definition of the logarithmic function to solve the equation $\log_x\left(\frac{1}{512}\right) = -3$ for x.
Answer: medium
$\log_x\left(\frac{1}{512}\right) = -3 \quad\Leftrightarrow\quad x^{-3} = \left(\frac{1}{512}\right) \quad\Leftrightarrow\quad x^{-3} = 2^{-9} \quad\Leftrightarrow\quad x = 2^3 = 8$

41. Graph the function $f(x) = 3 + \ln(2 - x)$, not by plotting points but applying your knowledge of the shape of the log function. State domain, range, and asymptote.
Answer: easy

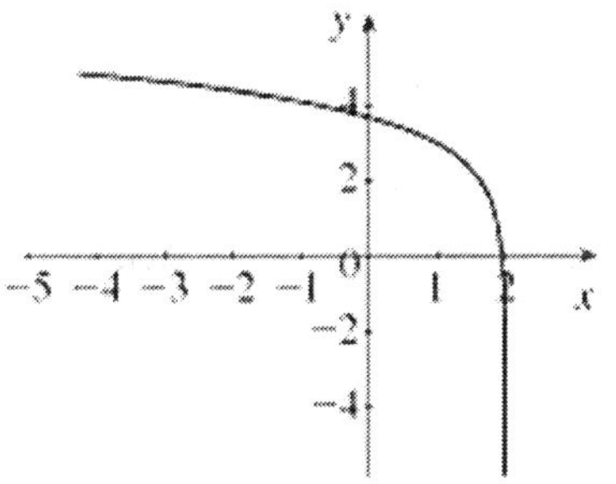

$f(x) = 3 + \ln(2 - x)$
Domain $(-\infty, 2)$, range $(-\infty, \infty)$, asymptote $x = 2$

42. Graph the function $h(x) = -\log_{10}|x + 1|$, not by plotting points but applying your knowledge of the shape of the log function. State domain, range, and asymptote.
Answer: medium

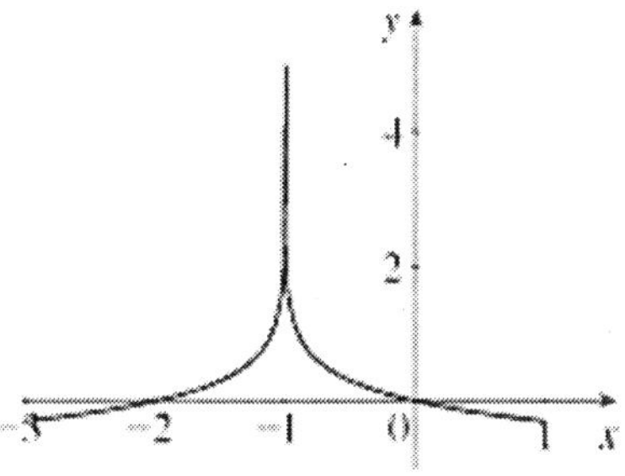

$h(x) = -\log_{10}|x + 1|$
Domain $(-\infty, -1) \cup (-1, \infty)$, range $(-\infty, \infty)$, asymptote $x = -1$

43. Find the domain of the function $h(x) = \dfrac{1}{x+2} + \log\left(x^2 + x\right)$.
Answer: medium
The domain for $\dfrac{1}{x+2}$ is $(-\infty, -2) \cup (-2, \infty)$. The domain for $\log\left(x^2 + x\right)$ is found as follows: $x^2 + x > 0 \quad\Leftrightarrow$ $x(x+1) > 0$, so the domain is $(-\infty, -1) \cup (-1, 0) \cup (0, \infty)$. Thus, the domain for $h(x) = \dfrac{1}{x+2} + \log\left(x^2 + x\right)$ is $(-\infty, -2) \cup (-2, -1) \cup (-1, 0) \cup (0, \infty)$ or $\{x \mid x \neq -2, -1, 0\}$.

SECTION 6.4

1. Use the Laws of Logarithms to rewrite the expression $\log_3\left(\frac{x}{4}\right)$ in a form with no logarithms of products, quotients, or powers.

(a) $\log_3 x + \log_3 4$ (b) $\log_3 x - \log_3 4$ (c) $\log_3(x-4)$ (d) $\log_4 x - \log_4 3$ (e) $\frac{\log_3 x}{\log_3 4}$

Answer: easy (b)

$\log_3\left(\frac{x}{4}\right) = \log_3 x - \log_3 4$

2. Use the Laws of Logarithms to rewrite the expression $\ln(ex)$ in a form with no logarithms of products, quotients, or powers.

(a) $2 + 2\ln x$ (b) $1 - \ln x$ (c) $e\ln x$ (d) $-\ln x$ (e) $1 + \ln x$

Answer: easy (e)

$\ln(ex) = \ln e + \ln x = 1 + \ln x$

3. Use the Laws of Logarithms to rewrite the expression $\log_6 \sqrt[5]{13}$ in a form with no logarithms of products, quotients, or powers.

(a) $\frac{1}{6}\log_5 13$ (b) $\log_6 13 - \log_6 5$ (c) $\sqrt[5]{\log_6 13}$ (d) $\frac{1}{5}\log_6 13$ (e) $\frac{1}{3}\log_6 5$

Answer: easy (d)

$\log_6 \sqrt[5]{13} = \frac{1}{5}\log_6 13$

4. Use the Laws of Logarithms to rewrite the expression $\log_3(xy)^7$ in a form with no logarithms of products, quotients, or powers.

(a) $3(\log_7 x + \log_7 y)$ (b) $7(\log_3 x - \log_3 y)$ (c) $3(\log_3 x - \log_3 y)$ (d) $7(\log_3 x + \log_3 y)$ (e) $7\log_3 x + 3\log_7 y$

Answer: medium (d)

$\log_3(xy)^7 = 7 \cdot \log_3(xy) = 7(\log_3 x + \log_3 y)$

5. Use the Laws of Logarithms to rewrite the expression $\log_a \frac{x^3}{y^2z^2}$ in a form with no logarithms of products, quotients, or powers.

(a) $3\log_a x - 2(\log_a y + \log_a z)$ (b) $3(\log_a x - \log_a y - \log_a z)$ (c) $\log_a x - \log_a y - \log_a z$
(d) $\frac{3}{2}\log_a x - \frac{2}{3}(\log_a y + \log_a z)$ (e) $3x\log_a y + 3y\log_a z$

Answer: medium (a)

$\log_a \frac{x^3}{y^2z^2} = \log_a x^3 - \log_a y^2z^2 = 3\log_a x - 2(\log_a y + \log_a z)$

6. Use the Laws of Logarithms to rewrite the expression $\ln \sqrt[3]{4rs^4}$ in a form with no logarithms of products, quotients, or powers.

(a) $\frac{1}{4}(\ln 3 + \ln r + 3\ln s)$ (b) $\frac{1}{3}(4\ln r + \ln s)$ (c) $\ln 4 - \ln 3(\ln r + 4\ln s)$
(d) $\frac{1}{3}(\ln 4 + \ln r + 4\ln s)$ (e) $\frac{1}{3}\left(\ln 4 - \ln r - \frac{4}{3}\ln s\right)$

Answer: medium (d)

$\ln \sqrt[3]{4rs^4} = \frac{1}{3}\ln(4rs^4) = \frac{1}{3}(\ln 4r + 4\ln s) = \frac{1}{3}(\ln 4 + \ln r + 4\ln s)$

7. Use the Laws of Logarithms to rewrite the expression $\log \frac{a^3}{b\sqrt[3]{c}}$ in a form with no logarithms of products, quotients, or powers.

(a) $\frac{1}{3}(\log a - \log b - 3\log c)$ (b) $3\log a - (3\log b + \log c)$ (c) $3\log a - \left(\log b + \frac{1}{3}\log c\right)$
(d) $2\log a - \log b - \frac{1}{2}\log c$ (e) $\log a - (\log b)(\log c)$

Answer: medium (c)

$\log \frac{a^3}{b\sqrt[3]{c}} = \log a^3 - \log\left(b\sqrt[3]{c}\right) = 3\log a - \left(\log b + \frac{1}{3}\log c\right)$

8. Use the Laws of Logarithms to rewrite the expression $\log_4 \sqrt{\frac{x+1}{x-1}}$ in a form with no logarithms of products, quotients, or powers.

(a) $\frac{1}{2}[\log_4(x-1) + \log_4(x+1)]$ (b) $\frac{1}{4}[\log_3(x+2) - \log_3(x-2)]$ (c) $\frac{1}{2}[\log_4(x+1) + \log_4(x-1)]$
(d) $\log_2(x+1) - \log_2(x-1)$ (e) $\frac{1}{2}[\log_4(x+1) - \log_4(x-1)]$

Answer: medium (e)

$\log_4 \sqrt{\frac{x+1}{x-1}} = \frac{1}{2}\log_4\left(\frac{x+1}{x-1}\right) = \frac{1}{2}[\log_4(x+1) - \log_4(x-1)]$

9. Use the Laws of Logarithms to rewrite the expression $\ln\dfrac{4x^3}{(x-1)^7}$ in a form with no logarithms of products, quotients, or powers.

(a) $\ln 4 + 3\ln x - 7\ln(x-1)$ (b) $\ln 12 + \ln x - \ln 7 + \ln(x-1)$ (c) $3\ln 4 + \ln x - \ln 7 - \ln x$
(d) $\ln 4 - \ln 3 + \ln x - \ln(x-1)$ (e) $2 + x\ln 3 - \ln 7$

Answer: medium (a)

$\ln\dfrac{4x^3}{(x-1)^7} = \ln(4x^3) - \ln[(x-1)^7] = \ln 4 + 3\ln x - 7\ln(x-1)$

10. Use the Laws of Logarithms to rewrite the expression $\log\dfrac{1}{\sqrt[3]{1+x}}$ in a form with no logarithms of products, quotients, or powers.

(a) $\frac{1}{3}\log(-1-x)$ (b) $1 - \frac{1}{3}\log x$ (c) $-3\log x$ (d) $-\frac{1}{3}\log(1+x)$ (e) $1 + \frac{1}{3}\log(x-1)$

Answer: medium (d)

$\log\dfrac{1}{\sqrt[3]{1+x}} = \log 1 - \log\sqrt[3]{1+x} = 0 - \frac{1}{3}\log(1+x) = -\frac{1}{3}\log(1+x)$

11. Use the Laws of Logarithms to rewrite the expression $\log\sqrt{x\sqrt{y\sqrt{z\sqrt{w}}}}$ in a form with no logarithms of products, quotients, or powers.

(a) $2\log x + 4\log y + 6\log z + 8\log w$ (b) $2\log x + 4\log y + 8\log z + 16\log w$
(c) $\frac{1}{2}\log x + \frac{1}{4}\log y + \frac{1}{8}\log z + \frac{1}{16}\log w$ (d) $\frac{1}{2}\log\left(x + \frac{1}{2}y + \frac{1}{4}z + \frac{1}{8}w\right)$
(e) $\frac{1}{2}\log\left(\frac{1}{2}x + \frac{1}{4}y + \frac{1}{8}z + \frac{1}{16}w\right)$

Answer: medium (c)

$$\log\sqrt{x\sqrt{y\sqrt{z\sqrt{w}}}} = \tfrac{1}{2}\log\left(x\sqrt{y\sqrt{z\sqrt{w}}}\right) = \tfrac{1}{2}\left[\log x + \tfrac{1}{2}\log\left(y\sqrt{z\sqrt{w}}\right)\right] = \tfrac{1}{2}\left(\log x + \tfrac{1}{2}\left[\log y + \tfrac{1}{2}\log\left(z\sqrt{w}\right)\right]\right)$$
$$= \tfrac{1}{2}\left(\log x + \tfrac{1}{2}\left[\log y + \tfrac{1}{2}\left(\log z + \tfrac{1}{2}\log w\right)\right]\right) = \tfrac{1}{2}\log x + \tfrac{1}{4}\log y + \tfrac{1}{8}\log z + \tfrac{1}{16}\log w$$

12. Use the Laws of Logarithms to rewrite the expression $\log\dfrac{10^{2x}}{x(x^2-1)(x^3-2)}$ in a form with no logarithms of products, quotients, or powers.

(a) $2x - [\log x + \log(x^2-1) + \log(x^3-2)]$ (b) $20 - (6\log x - 3)$
(c) $20x - [\log x + 2\log(x-1) + 3\log(x-2)]$ (d) $2x - [\log x + \log(x+1) + 2\log(x-1)]$
(e) $2x - [3\log x + \log(x^3) - 2]$

Answer: medium (a)

$\log\dfrac{10^{2x}}{x(x^2-1)(x^3-2)} = \log 10^{2x} - \log\left(x(x^2-1)(x^3-2)\right) = 2x - \left[\log x + \log(x^2-1) + \log(x^3-2)\right]$

13. Evaluate the expression $\log_2 144 - \log_2 9$.

(a) 3 (b) 4 (c) 5 (d) 6 (e) 7

Answer: medium (b)

$\log_2 144 - \log_2 9 = \log_2\frac{144}{9} = \log_2 16 = \log_2 2^4 = 4$

14. Evaluate the expression $\log\sqrt[3]{0.001}$.

(a) -1 (b) 2 (c) -3 (d) $-\frac{1}{2}$ (e) $\frac{1}{3}$

Answer: medium (a)

$\log\sqrt[3]{0.001} = \frac{1}{3}\log(10^{-3}) = -1$

15. Evaluate the expression $\log_6 12 + \log_6 18$.

(a) 1 (b) 2 (c) 3 (d) 4 (e) 5

Answer: medium (c)

$\log_6 12 + \log_6 18 = \log_6(12\cdot 18) = \log_6 6^3 = 3$

16. Evaluate the expression $\log_2 16^{85}$.

Answer: medium

$\log_2 16^{85} = \log_2 (2^4)^{85} = \log_2 2^{340} = 340$

17. Rewrite the expression $\log 8 + \frac{1}{3}\log 9 - \log 2$ as a single logarithm.

Answer: medium

$\log 8 + \frac{1}{3}\log 9 - \log 2 = \log 8\sqrt[3]{9} - \log 2 = \log \frac{8\sqrt[3]{9}}{2} = \log\left(4\sqrt[3]{9}\right)$

18. Rewrite the expression $\log_4(x^2 - 1) - \log_4(x+1)$ as a single logarithm.

Answer: medium

$$\log_4(x^2 - 1) - \log_4(x+1) = \log_4 \frac{x^2 - 1}{x+1} = \log_4(x-1)$$

19. Rewrite the expression $\ln(a-b) - \ln(a+b) + 2\ln c$ as a single logarithm.

Answer: medium

$$\ln(a-b) - \ln(a+b) + 2\ln c = \ln \frac{a-b}{a+b} + \ln c^2 = \ln \frac{c^2(a-b)}{a+b}$$

20. Rewrite the expression $\frac{1}{3}[\log_4 x + 3\log_4 y - 2\log_4 z]$ as a single logarithm.

Answer: medium

$$\tfrac{1}{3}[\log_4 x + 3\log_4 y - 2\log_4 z] = \tfrac{1}{3}\log_4 \frac{xy^3}{z^2} = \log_4 \sqrt[3]{\frac{xy^3}{z^2}}$$

21. Rewrite the expression $c\log_a b + \log_a d - r\log_a s$ as a single logarithm.

Answer: medium

$$c\log_a b + \log_a d - r\log_a s = \log_a b^c d - \log_a s^r = \log_a \frac{b^c d}{s^r}$$

22. State whether or not the equation $\log_5 \frac{a}{b^2} = \log_5 a - 2\log_5 b$ is an identity.

Answer: medium

$\log_5 \frac{a}{b^2} = \log_5 a - \log_5 b^2 = \log_5 a - 2\log_5 b$ and so the equation is an identity.

23. State whether or not the equation $(\log P)(\log Q) = \log P + \log Q$ is an identity.

Answer: medium

$\log P + \log Q = \log(PQ) \neq (\log P)(\log Q)$ and so the equation is not an identity.

24. State whether or not the equation $(\log_2 7)^x = x\log_2 7$ is an identity.

Answer: medium

$x\log_2 7 = \log_2 7^x \neq (\log_2 7)^x$ and so the equation is not an identity.

25. State whether or not the equation $\log(x-y) = \dfrac{\log x}{\log y}$ is an identity.

Answer: medium

$\log(x-y) \neq \dfrac{\log x}{\log y}$ and so the equation is not an identity.

26. State whether or not the equation $r\ln s = \ln(s^r)$ is an identity.

Answer: medium

$r\ln s = \ln s^r$ and so the equation is an identity.

27. Use the change of base formula and a calculator to evaluate the logarithm $\log_5 3$ correct to six decimal places.

Answer: medium

$$\log_5 3 = \frac{\log 3}{\log 5} \approx 0.682606$$

28. Use the change of base formula and a calculator to evaluate the logarithm $\log_5 85$ correct to six decimal places.

Answer: medium

$$\log_5 85 = \frac{\log 85}{\log 5} \approx 2.760374$$

29. Use the change of base formula to show that $\log_n x = \dfrac{\ln x}{\ln n}$. Then use this fact to draw the graphs of the functions $y = \log_2 x$, $y = \ln x$, $y = \log_5 x$, and $y = \log_{10} x$ on the same screen using the viewing rectangle $[0, 5]$ by $[-3, 3]$.
Answer: medium

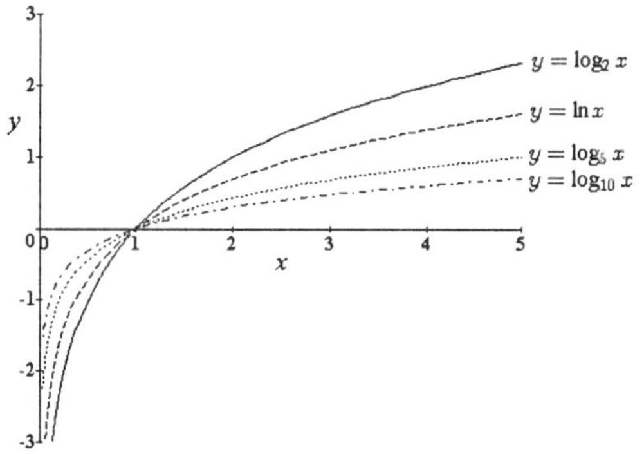

30. Simplify $(\log_3 6)(\log_6 9)$.
Answer: medium
$(\log_3 6)(\log_6 9) = \dfrac{\log 6}{\log 3} \cdot \dfrac{\log 9}{\log 6} = \dfrac{\log 9}{\log 3} = \log_3 9 = 2$

31. Use the Laws of Logarithms to rewrite the expression $\ln \dfrac{x^3}{y^3 \sqrt[3]{z}}$ in a form with no logarithms of products, quotient, or powers.
Answer: easy
$\ln\left(\dfrac{x^3}{y^3 \sqrt[3]{z}}\right) = \ln x^3 - \left(\ln y^3 + \ln z^{1/3}\right) = 3 \ln x - 3 \ln y - \frac{1}{3} \ln z$

32. Use the Laws of Logarithms to rewrite the expression $\ln \dfrac{e^{3x}\sqrt{x^2+1}}{x^3+y^3}$ in a form with no logarithms of products, quotient, or powers.
Answer: easy
$\ln \dfrac{e^{3x}\sqrt{x^2+1}}{x^3+y^3} = \ln e^{3x} + \ln \sqrt{(x^2+1)} - \ln\left(x^3+y^3\right) = 3x + \frac{1}{2}\ln\left(x^2+1\right) - \ln(x+y) - \ln\left(x^2 - xy + y^2\right)$

33. Use the Laws of Logarithms to rewrite the expression $\log \dfrac{1^{3x+1}}{x(x-2)\left(\sqrt[3]{x-2}\right)}$ in a form with no logarithms of products, quotient, or powers.
Answer: medium
$\log \dfrac{1^{3x+1}}{x(x-2)\left(\sqrt[3]{x-2}\right)} = \log 1^{3x+1} - \left[\log x + \log_3 (x-2) + \log \sqrt[3]{x-2}\right]$
$= 0 - \log x - \log(x-2) - \frac{1}{3}\log(x-2) = -\log x - \frac{4}{3}\log(x-2)$

34. Evaluate the expression $\log_5 625 - \log_5 25$.
Answer: easy
$\log_5 625 - \log_5 25 = \log_5\left(\frac{625}{25}\right) = \log_5 25 = 2$

35. Rewrite the expression $\log\left(x^2-4\right) + \log(x+2) - \log\left(x^4 - 8x^2 + 16\right)$ as a single logarithm.
Answer: easy
$\log\left(x^2-4\right) + \log(x+2) - \log\left(x^4 - 8x^2 + 16\right) = \log \dfrac{\left(x^2-4\right)(x+2)}{x^4 - 8x^2 + 16} = \log \dfrac{(x+2)(x-2)(x+2)}{(x-2)^2(x+2)^2}$
$= \log \dfrac{1}{x-2} = -\log(x-2)$

36. Rewrite the expression $\log\left(x^3-16x\right)+\log\left(x^2+10x+25\right)-\log\left(x^5-41x^3+400x\right)$ as a single logarithm.

Answer: easy

$$\log\left(x^3-16x\right)+\log\left(x^2+10x+25\right)-\log\left(x^5-41x^3+400x\right)=\log\frac{x\left(x^2-4\right)(x+5)(x+5)}{x^5-41x^3+400x}$$

$$=\log\frac{x\left(x^2-4\right)(x+5)(x+5)}{x(x+4)(x-4)(x+5)(x-5)}=\log\left(\frac{x+5}{x-5}\right)$$

SECTION 6.5

1. Find the solution of the equation $8^{1-x}=5$ correct to four decimal places.

(a) 0.2260 (b) 0.2756 (c) 0.3045 (d) 0.9384 (e) 1.7563

Answer: medium (a)

$8^{1-x}=5 \quad\Leftrightarrow\quad \log 8^{1-x}=\log 5 \quad\Leftrightarrow\quad (1-x)\log 8=\log 5 \quad\Leftrightarrow\quad 1-x=\dfrac{\log 5}{\log 8} \quad\Leftrightarrow\quad x=1-\dfrac{\log 5}{\log 8}\approx 0.2260$

2. Find the solution of the equation $3^{x/12}=0.1$ correct to four decimal places.

(a) -32.2566 (b) -29.0743 (c) -25.1508 (d) -19.0828 (e) -7.0523

Answer: medium (c)

$3^{x/12}=0.1 \quad\Leftrightarrow\quad \log 3^{x/12}=\log 0.1 \quad\Leftrightarrow\quad \frac{1}{12}x\log 3=-1 \quad\Leftrightarrow\quad x=-12/(\log 3)\approx -25.1508$

3. Find the solution of the equation $\left(\frac{1}{4}\right)^x=81$ correct to four decimal places.

(a) -3.1699 (b) -3.0501 (c) -2.7593 (d) -2.1175 (e) -1.7659

Answer: medium (a)

$\left(\frac{1}{4}\right)^x=81 \quad\Leftrightarrow\quad x\log\frac{1}{4}=\log 81 \quad\Leftrightarrow\quad x=-\frac{\log 81}{\log 4}\approx -3.1699$

4. Find the solution of the equation $10^{1-x}=4^x$ correct to four decimal places.

(a) -0.0365 (b) 0.0265 (c) 0.3785 (d) 0.3932 (e) 0.6242

Answer: medium (e)

$10^{1-x}=4^x \quad\Leftrightarrow\quad 1-x=x\log 4 \quad\Leftrightarrow\quad x=\dfrac{1}{1+\log 4}\approx 0.6242$

5. Find the solution of the equation $e^{2-5x}=8$ correct to four decimal places.

(a) -0.1875 (b) -0.0159 (c) 0.2056 (d) 0.3869 (e) 1.6532

Answer: medium (b)

$e^{2-5x}=8 \quad\Leftrightarrow\quad \ln e^{2-5x}=\ln 8 \quad\Leftrightarrow\quad 2-5x=\ln 8 \quad\Leftrightarrow\quad x=\frac{1}{5}(2-\ln 8)\approx -0.0159$

6. Find the solution of the equation $7^{x/3}=5^{2-x}$ correct to four decimal places.

(a) 1.0372 (b) 1.4255 (c) 1.6729 (d) 2.0410 (e) 2.0654

Answer: hard (b)

$7^{x/3}=5^{2-x} \quad\Leftrightarrow\quad \log 7^{x/3}=\log 5^{2-x} \quad\Leftrightarrow\quad \frac{1}{3}x\log 7=(2-x)\log 5 \quad\Leftrightarrow\quad x=\dfrac{2\log 5}{\frac{1}{3}\log 7+\log 5}\approx 1.4255$

7. Solve the equation $e^x=10$ for x.

(a) $x=-\frac{1}{10}$ (b) $x=2\ln 2$ (c) $x=\ln 5$ (d) $x=\ln 10$ (e) $x=2\ln 5$

Answer: medium (d)

$e^x=10 \quad\Leftrightarrow\quad x=\ln 10$

8. Solve the equation $e^{1-4x}=2$ for x.

(a) $x=\frac{1}{2}(\ln 2-1)$ (b) $x=2\ln 2-1$ (c) $x=\frac{3}{4}\ln\frac{1}{2}$ (d) $x=\frac{2}{3}(2-\ln 3)$ (e) $x=\frac{1}{4}(1-\ln 2)$

Answer: medium (e)

$e^{1-4x}=2 \quad\Leftrightarrow\quad \ln e^{1-4x}=\ln 2 \quad\Leftrightarrow\quad 1-4x=\ln 2 \quad\Leftrightarrow\quad x=\frac{1}{4}(1-\ln 2)$

9. Solve the equation $10^{5^x} = 3$ for x.
(a) $x = \log_3(\log_5 10)$ (b) $x = \log_3(\log_{10} 5)$ (c) $x = \log(\log 3)$ (d) $x = \log_5(\log_5 3)$ (e) $x = \log_5(\log_{10} 3)$
Answer: hard (e)
$10^{5^x} = 3 \quad \Leftrightarrow \quad \log_{10} 10^{5^x} = \log_{10} 3 \quad \Leftrightarrow \quad 5^x = \log_{10} 3 \quad \Leftrightarrow \quad x = \log_5(\log_{10} 3)$

10. Solve the equation $\log_2 x + \log_4 x + \log_8 x = 11$.
(a) $x = 27$ (b) $x = 64$ (c) $x = 128$ (d) $x = 2^{11}$ (e) $x = 8^{11}$
Answer: hard (b)
$\log_2 x + \log_4 x + \log_8 x = 11 \quad \Leftrightarrow \quad \dfrac{\log x}{\log 2} + \dfrac{\log x}{\log 4} + \dfrac{\log x}{\log 8} = 11 \quad \Leftrightarrow \quad (\log x)\left(\dfrac{1}{\log 2} + \dfrac{1}{\log 4} + \dfrac{1}{\log 8}\right) = 11 \quad \Leftrightarrow$
$11 \log x = 11(6 \log 2) \Leftrightarrow \log x = 6 \log 2 = \log 2^6 \quad \Leftrightarrow \quad x = 2^6 = 64.$

11. Solve the equation $2 \log x = \log 2 + \log(x + 4)$ for x.
Answer: medium
$2 \log x = \log 2 + \log(x + 4) \quad \Leftrightarrow \quad \log x^2 = \log(2(x + 4)) \quad \Leftrightarrow \quad x^2 = 2x + 8 \quad \Leftrightarrow \quad x^2 - 2x - 8 = 0 \quad \Leftrightarrow$
$(x - 4)(x + 2) = 0 \quad \Leftrightarrow \quad x = 4, -2$. But -2 is not a solution because negative numbers do not have logarithms. So $x = 4$ is the only solution.

12. Solve the equation $\log_4 x + \log_4(x + 1) = \log_4 30$ for x.
Answer: medium
$\log_4 x + \log_4(x + 1) = \log_4 30 = \log_4 x(x + 1) = \log_4 30 \quad \Leftrightarrow \quad x(x + 1) = 30 \quad \Leftrightarrow \quad x^2 + x - 30 = 0 \quad \Leftrightarrow$
$(x + 6)(x - 5) = 0 \quad \Leftrightarrow \quad x = -6$ or 5. But $x = -6$ is inadmissible, so $x = 5$ is the only solution.

13. Solve the equation $\log_{15} x + \log_{15}(x - 2) = 1$ for x.
Answer: medium
$\log_{15} x + \log_{15}(x - 2) = 1 \quad \Leftrightarrow \quad \log_{15} x(x - 2) = 1 \quad \Leftrightarrow \quad x(x - 2) = 15^1 \quad \Leftrightarrow \quad x^2 - 2x - 15 = 0 \quad \Leftrightarrow$
$(x - 5)(x + 3) = 0 \quad \Leftrightarrow \quad x = 5$ or -3. But $x = -3$ is inadmissible, so the only solution is $x = 5$.

14. Solve the equation $\ln(x + 1) + \ln(x - 2) = 1$ for x.
Answer: hard
$\ln(x + 1) + \ln(x - 2) = 1 \quad \Leftrightarrow \quad \ln((x + 1)(x - 2)) = 1 \quad \Leftrightarrow \quad (x + 1)(x - 2) = e \quad \Leftrightarrow \quad x^2 - x - (2 + e) = 0 \quad \Leftrightarrow$
$x = \dfrac{1 \pm \sqrt{1 + 4 \cdot 1(2 + e)}}{2 \cdot 1} = \dfrac{1 \pm \sqrt{9 + 4e}}{2}$, but $\dfrac{1 - \sqrt{9 + 4e}}{2}$ is inadmissible, so the only solution is $\dfrac{1 + \sqrt{9 + 4e}}{2}$.

15. For what value of x is it true that $(\log x)^4 = 4 \log x$?
Answer: hard
$(\log x)^4 = 4 \log x \quad \Leftrightarrow \quad (\log x)^3 = 4$ or $\log x = 0 \quad \Leftrightarrow \quad \log x = \sqrt[3]{4}$ or $x = 1 \quad \Leftrightarrow \quad x = 10^{\sqrt[3]{4}}$ or $x = 1$.

16. A **learning curve** is a graph of a function $P(t)$ that measures the performance of someone learning a skill as a function of the training time t. At first, the rate of learning is rapid. Then, as performance increases and approaches a maximal value M, the rate of learning decreases. It has been found that the function $P(t) = M - Ce^{-kt}$, where k and C are positive constants and $C < M$, is a reasonable model for learning.
(a) Sketch the graph of P.
(b) Express the learning time t as a function of the performance level P.
Answer: hard

(a)

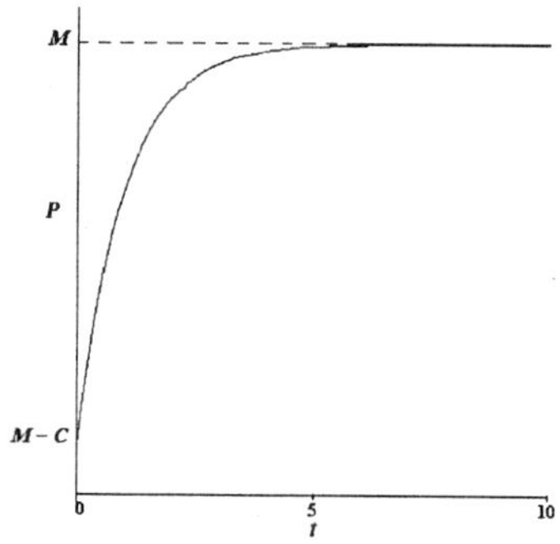

(b) $P(t) = M - Ce^{-kt} \quad \Leftrightarrow \quad Ce^{-kt} = M - P \quad \Leftrightarrow$
$e^{-kt} = \dfrac{1}{C}(M - P) \quad \Leftrightarrow \quad -kt = \ln\left(\dfrac{1}{C}(M - P)\right) \quad \Leftrightarrow$
$t = -\dfrac{1}{k}(-\ln C + \ln(M - P))$. Therefore the learning time as a function of the performance level is $t(P) = -\frac{1}{k}(-\ln C + \ln(M - P))$.

GC 17. Find the solutions of the equation $\log_{10} x = x^2 - 1$ correct to two decimal places.

Answer: medium

From the graph, $\log_{10} x = x^2 - 1$ has solutions $x \approx 0.10$ and $x = 1$.

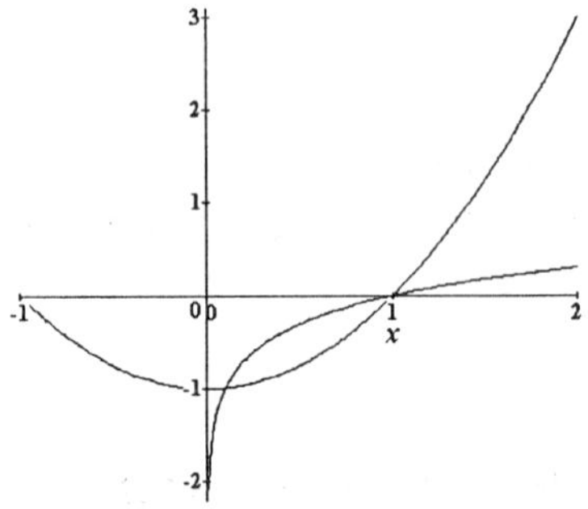

18. Find the solutions of the equation $x = \ln(9 - x^2)$ correct to two decimal places.

Answer: medium

From the graph, $x = \ln(9 - x^2)$ has solutions $x \approx -2.99$ and $x \approx 1.77$.

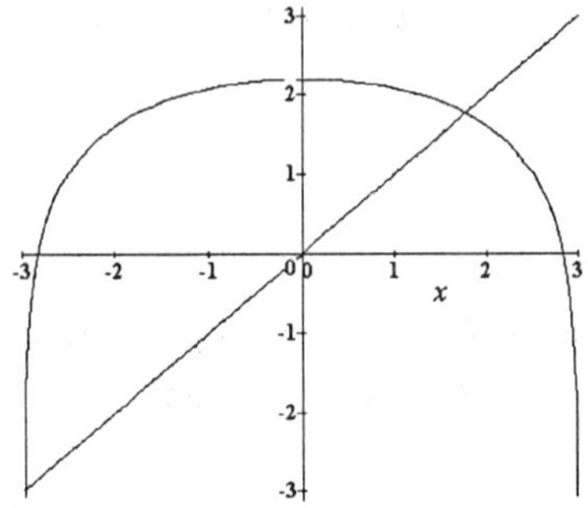

19. Solve the equation $2^{-x} = x - 2$. Give the solution correct to two decimal places.

Answer: medium

From the graph, $2^{-x} = x - 2$ has solution $x \approx 2.22$.

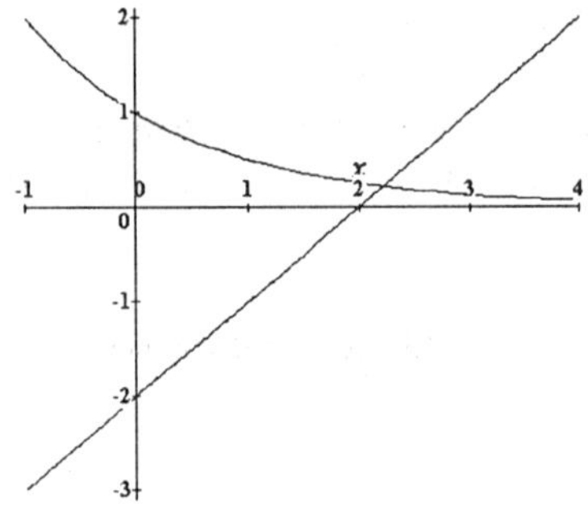

20. Solve the equation $e^{x^2} - 3 = x^3 - 2x$. Give the solution correct to two decimal places.

Answer: medium

From the graph, $e^{x^2} - 3 = x^3 - 2x$ has solutions $x \approx -1.15$ and

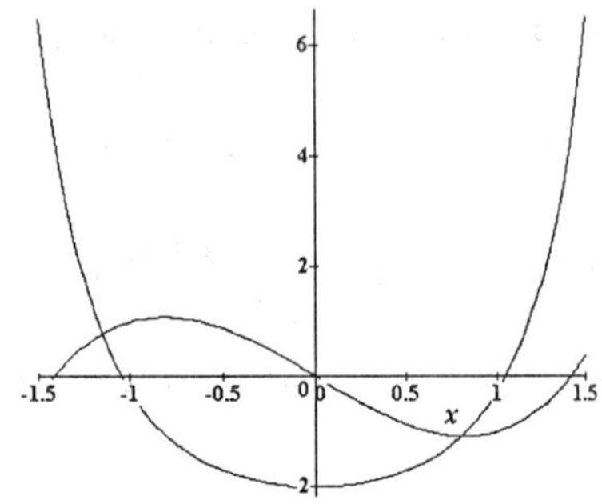

$x \approx 0.80$.

21. Solve the inequality: $3 \le \log_3 x \le 4$.

Answer: medium

$3 \le \log_3 x \le 4 \quad \Leftrightarrow \quad 3^3 \le 3^{\log_3 x} \le 3^4 \quad \Leftrightarrow \quad 3^3 \le x \le 3^4$ ($3^x > 0$ for all x) $\quad \Leftrightarrow \quad 27 \le x \le 81$.

22. Solve the equation $(x+1)^{\log(x+1)} = 100(x+1)$.

Answer: hard

$(x+1)^{\log(x+1)} = 100(x+1)$. Substitute $y = x+1$, then $y^{\log y} = 100y$. Take logarithms of both sides: $(\log y)(\log y) = \log(100y) = \log 100 + \log y \quad \Leftrightarrow \quad (\log y)^2 - \log y - 2 = 0 \quad \Leftrightarrow \quad (\log y - 2)(\log y + 1) = 0 \quad \Leftrightarrow$ $\log y = 2$ or $\log y = -1 \quad \Leftrightarrow \quad y = 10^2 = 100$ or $y = 10^{-1} = \frac{1}{10} \quad \Leftrightarrow \quad x = 99$ or $x = -\frac{9}{10}$.

23. Find the solution of the equation $e^{(x/4)-1} = 15$, correct to three decimal places.

Answer: easy

$e^{(x/4)-1} = 15 \quad \Leftrightarrow \quad \ln e^{(x/4)-1} = \ln 15 \quad \Leftrightarrow \quad \left(\frac{1}{4}x - 1\right) \ln e = \ln 15 \quad \Leftrightarrow \quad \dfrac{x}{4} = \ln 15 + 1 \quad \Leftrightarrow \quad x \approx 14.832$

24. Find the solution of the equation $32^{2x+1} = 8^{2-x}$ correct to three decimal places.

Answer: medium

$32^{2x+1} = 8^{2-x} \quad \Leftrightarrow \quad \ln 32^{2x+1} = \ln 8^{2-x} \quad \Leftrightarrow \quad (2x+1)\ln 32 = (2-x)\ln 8 \quad \Leftrightarrow \quad 2(\ln 32)x + \ln 32 = 2\ln 8 - (\ln 8)x$ $\Leftrightarrow \quad 2(\ln 32)x + (\ln 8)x = 2\ln 8 - \ln 32 \quad \Leftrightarrow \quad x = -\dfrac{\ln 32 - 2\ln 8}{2\ln 32 + \ln 8} \approx 0.076$.

25. Solve the equation $2x^2(3^x) - 5x(3^x) + 3^{x+1} = 0$ for x.

Answer: medium

$2x^2(3^x) - 5x(3^x) + 3^{x+1} \quad \Leftrightarrow \quad 3^x(2x^2 - 5x + 3) = 0 \quad \Leftrightarrow \quad 3^x(x-1)(2x-3) = 0$. $3^x = 0$ has no solution because $3^x > 0$ for all x. $x - 1 = 0 \quad \Leftrightarrow \quad x = 1$, and $2x - 3 = 0 \quad \Leftrightarrow \quad x = \frac{3}{2}$. Thus the two solutions are $x = 1$ and $x = \frac{3}{2}$.

26. Solve the equation $e^{4x} - 13e^{2x} + 12 = 0$ for x.

Answer: hard

$e^{4x} - 13e^{2x} + 12 = 0$. Let $u = e^{2x}$, so $u^2 = e^{4x}$. Then $u^2 - 13u + 12 = 0 \quad \Leftrightarrow \quad (u-1)(u-12) = 0 \quad \Leftrightarrow \quad u = 1$ or $u = 12$. Now since $u = e^{2x}$, we have $1 = e^{2x} \quad \Leftrightarrow \quad \ln 1 = \ln e^{2x} \quad \Leftrightarrow \quad x = 0$, or in the other case, $12 = e^{2x} \quad \Leftrightarrow$ $\ln 12 = \ln e^{2x} \quad \Leftrightarrow \quad x = \frac{1}{2}\ln 12 \approx 1.24$. The two solutions are $x = 0$ and $x = \frac{1}{2}\ln 12 \approx 1.24$.

27. The population of a particular city in the northeast has been declining for several years. It is known that in 1990 the population was $p(t) = 326{,}800e^{-0.0069t}$ with t in years. What was the population in 1995? Starting in 1990, how long (in years) would it take the population to become half of its original size?

Answer: medium

$p(5) = 326800e^{-0.0069(5)} \approx 315{,}668$. $\frac{1}{2} = e^{-0.0069t} \quad \Leftrightarrow \quad \ln\frac{1}{2} = \ln e^{-0.0069t} \quad \Leftrightarrow \quad t \approx 100.46$, about 100 years.

SECTION 6.6

1. Vince wants to invest \$3000 in savings certificates that bear an interest rate of 8.25% compounded semiannually. How long a time period should she choose in order to save an amount of \$5000?

(a) 2.54 yrs (b) 3.08 yrs (c) 6.32 yrs (d) 9.25 yrs (e) 10.36 yrs

Answer: medium (c)

$P = 3000$, $A(t) = 5000$, $r = 0.0825$, and $n = 2$. Since $A(t) = P\left(1 + \dfrac{r}{n}\right)^{nt}$, therefore $5000 = 3000\left(1 + \dfrac{0.0825}{2}\right)^{2t} \quad \Leftrightarrow$ $1.04125^{2t} = \frac{5}{3} \quad \Leftrightarrow \quad 2t = \dfrac{\log\frac{5}{3}}{\log 1.04125} \quad \Leftrightarrow \quad t = \dfrac{\log\frac{5}{3}}{2\log 1.04125} \approx 6.32$ and so he should choose a period of approximately 6.32 yrs.

2. A sum of \$1000 was invested for three years and the interest was compounded semiannually. If this sum amounted to \$1283.68 after the given time, what was the interest rate?

(a) 6.25% (b) 6.5% (c) 7% (d) 7.5% (e) 8.5%

Answer: medium (e)

$P = 1000$, $n = 2$, $t = 3$, and $A(3) = 1283.68$. Since $A(t) = P\left(1 + \frac{r}{n}\right)^{nt}$, therefore $1283.68 = 1000\left(1 + \frac{r}{2}\right)^6 \Leftrightarrow \left(1 + \frac{r}{2}\right)^6 = 1.28368 \Leftrightarrow 1 + \frac{r}{2} = \sqrt[6]{1.28368} \Leftrightarrow r = 2\left(\sqrt[6]{1.28368} - 1\right) \approx 0.085$ and so the interest rate was 8.5%.

3. If a bacteria culture starts with 5000 bacteria and doubles every 25 min, when will the population reach a level of 20,000?

(a) 42.68 min (b) 50 min (c) 62.83 min (d) 65 min (e) 66.82 min

Answer: medium (b)

$n_0 = 5000$ and $d = 25$. Then $n(t) = 5000 \cdot 2^{t/25} = 20{,}000 \Leftrightarrow 2^{t/25} = \frac{20{,}000}{5000} = 4 \Leftrightarrow \frac{t}{25} \cdot \ln 2 = \ln 4 \Leftrightarrow t = 25\frac{\ln 4}{\ln 2} = 50$. Thus the population will reach 20,000 in about 50 min.

4. The count in a bacteria culture was 600 after 2 hours and 38,400 after 6 hours.

(a) What was the initial size of the culture?

(b) Find the doubling period.

(c) Find the size after 4.5 h.

(d) When will the population be 1,500,000?

Answer: medium

$n(2) = 600$ and $n(6) = 38{,}400$. Then $n_0 \cdot 2^{2/d} = 600$ and $n_0 \cdot 2^{6/d} = 38{,}400$. Dividing the second equation by the first gives $2^{4/d} = 64 = 2^6 \Leftrightarrow 4/d = 6 \Leftrightarrow d = \frac{2}{3}$.

(a) $n_0 = 600/2^{2/d} = 600/2^3 = 75$ and so the initial population of the culture was 75.

(b) The doubling period is $\frac{2}{3}$ hours or 40 minutes.

(c) $n(4.5) = 75 \cdot 2^{9/2 \cdot 3/2} = 75 \cdot 2^{27/4} = 75 \cdot 2^6 \cdot 2^{3/4} \approx 8073$ and so the population after 4.5 hours is approximately 8073.

(d) $n(t) = 1{,}500{,}000 = 75 \cdot 2^{3t/2} \Leftrightarrow 2^{3t/2} = 20{,}000 \Leftrightarrow \frac{3}{2}t = \log_2 20{,}000 \Leftrightarrow t = \frac{2}{3}\log_2 20{,}000 \approx 9.53$. Hence the population will reach 150,000 after roughly 9 hours and 32 minutes.

5. The population of California was 10,586,223 in 1950 and 26,981,000 in 1986.

(a) Assuming exponential growth during that period, find the doubling period.

(b) Use these data to predict the population of California in the year 2010.

Answer: medium

Calculating dates relative to 1950 gives $n_0 = 10{,}586{,}223$ and $n(36) = 26{,}981{,}000$. Then $10{,}586{,}223 \cdot 2^{36/d} = 26{,}981{,}000 \Leftrightarrow 2^{36/d} = \frac{26{,}981{,}000}{10{,}586{,}223} \approx 2.54869 \Leftrightarrow \frac{36}{d}\log 2 = \log 2.54869 \Leftrightarrow d \approx 26.6715$

(a) The doubling period is about 27 years.

(b) $n(60) \approx 10{,}586{,}223 \cdot 2^{60/26.6715} \approx 50{,}342{,}617$ and so the population in the year 2010 will be approximately 50,000,000.

6. Radium-221 has a half-life of 30 s. How long will it take for 90% of a sample to decompose?

Answer: medium

$h = 30$ s and we want to find t such that $m(t) = 0.1 \cdot m_0$. Since $m(t) = m_0 \cdot 2^{-t/30}$ then $0.1 \cdot m_0 = m_0 \cdot 2^{-t/30} \Leftrightarrow 2^{-t/30} = 0.1 \Leftrightarrow t = -30\frac{\log 0.1}{\log 2} \approx 99.66$ s and so it will take about 100 s.

7. After three days a sample of Radon-222 decayed to 58% of its original amount.

(a) What is the half-life of Radon-222?

(b) How long will it take the sample to decay to 15% of its original amount?

Answer: medium

$m(3) = 0.58m_0$. Since $m(t) = m_0 \cdot 2^{-t/h}$ then $0.58m_0 = m_0 \cdot 2^{-3/h} \Leftrightarrow 2^{-3/h} = 0.58$.

(a) $h = -3\frac{\log 2}{\log 0.58} \approx 3.82$ days. Thus the half-life of Radon-222 is 3.82 days.

(b) $m(t) = 0.15m_0 = m_0 \cdot 2^{-t/h} \Leftrightarrow 2^{-t/h} = 0.15 \Leftrightarrow -\frac{t}{h} = \frac{\log 0.15}{\log 2} \Leftrightarrow t = -3.82\frac{\log 0.15}{\log 2} \approx 10.46$.

Therefore it will take roughly 10.5 days to decay to 15% of its original amount.

8. A kettle full of water is brought to a boil in a room with temperature 20°C. After 15 minutes the temperature of the water has decreased from 100°C to 70°C. Find the temperature after another 10 min. Illustrate by sketching the graph of the temperature function.

Answer: hard

$T_S = 20$ and $D_0 = 100 - 20 = 80$. Then $T(t) = 20 + 80e^{-kt}$. Since $T(15) = 70$, therefore $20 + 80e^{-15k} = 70 \quad \Leftrightarrow \quad 80e^{-15k} = 50 \quad \Leftrightarrow$ $-15k = \ln \frac{50}{80} = \ln \frac{5}{8} \quad \Leftrightarrow \quad k = -\frac{1}{15} \ln \frac{5}{8}$. Thus $T(25) = 20 + 80e^{[\ln(5/8)/15] \cdot 25} \approx 56.6$ and so the temperature after another 10 min is 57°C.

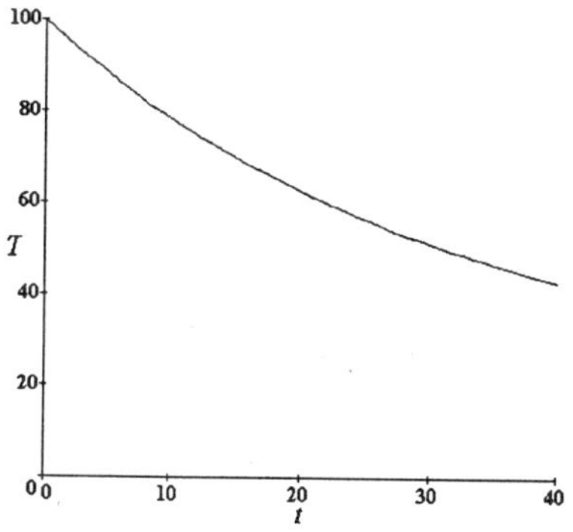

9. An unknown substance has a hydrogen ion concentration of $[H^+] = 2.4 \times 10^{-8}$ M. Find the pH and classify the substance as acidic or basic.

(a) 2.1; acidic (b) 4.6; acidic (c) 6.2; acidic (d) 7.6; basic (e) 9.4; basic

Answer: easy (d)

$\text{pH} = -\log [H^+] = -\log(2.4 \times 10^{-8}) \approx 7.6$ and the substance is basic.

10. The pH readings of glasses of beer and water are given. Find the hydrogen ion concentration in each liquid.

(a) beer: pH = 4.5 (b) water: pH = 7.2.

Answer: easy

(a) $\text{pH} = -\log [H^+] = 4.5 \quad \Leftrightarrow \quad [H^+] = 10^{-4.5}\text{ M} \approx 3.2 \times 10^{-5}\text{ M}$

(b) $\text{pH} = -\log [H^+] = 7.2 \quad \Leftrightarrow \quad [H^+] = 10^{-7.2}\text{ M} \approx 6.3 \times 10^{-8}\text{ M}$

11. The pH readings for wine vary from 2.7 to 3.9. Find the corresponding range of hydrogen ion concentrations.

(a) $[1.259 \times 10^{-4}, 1.995 \times 10^{-3}]$ (b) $[1.300 \times 10^{-4}, 2.035 \times 10^{-3}]$

(c) $[1.029 \times 10^{-4}, 1.702 \times 10^{-3}]$ (d) $[1.277 \times 10^{-4}, 2.106 \times 10^{-3}]$

(e) $[2.204 \times 10^{-4}, 3.025 \times 10^{-3}]$

Answer: easy (a)

$2.7 \le \text{pH} \le 3.9 \quad \Leftrightarrow \quad -2.7 \ge -\text{pH} \ge -3.9 \quad \Leftrightarrow \quad 10^{-2.7} \ge 10^{-\text{pH}} \ge 10^{-3.9} \quad \Leftrightarrow$ $1.995 \times 10^{-3} \ge [H^+] \ge 1.259 \times 10^{-4}$. The range of $[H^+]$ is 1.259×10^{-4} to 1.995×10^{-3}.

12. The 1906 earthquake in San Francisco had a magnitude of 8.3 on the Richter scale. At the same time in Japan there was an earthquake with magnitude 4.9 that caused only minor damage. How many times more intense was the San Francisco earthquake than the Japanese earthquake?

(a) 285 (b) 802 (c) 2512 (d) 7641 (e) 2.85×10^8

Answer: medium (c)

Let the subscript SF represent the San Francisco earthquake and J the Japan earthquake. Then $M_{SF} = \log \frac{I_{SF}}{S} = 8.3 \quad \Leftrightarrow$ $I_{SF} = S \cdot 10^{8.3}$ and $M_J = \log \frac{I_J}{S} = 4.9 \quad \Leftrightarrow \quad I_J = S \cdot 10^{4.9}$. So $\frac{I_{SF}}{I_J} = \frac{10^{8.3}}{10^{4.9}} = 10^{3.4} \approx 2511.9$ and so the San Francisco earthquake was 2512 times more intense than the Japan earthquake.

13. The intensity of the sound of traffic at a busy intersection was measured at 2.25×10^{-5} watt/m². Find the intensity level in decibels.

(a) 61 dB (b) 67 dB (c) 74 dB (d) 82 dB (e) 97 dB

Answer: medium (c)

$\beta = 10 \log \frac{I}{I_0} = 10 \log \frac{2.25 \times 10^{-5}}{1.0 \times 10^{-12}} = 10 \log(2.25 \times 10^7) = 10(\log 2.25 + \log 10^7) = 10(\log 2.25 + 7) \approx 74$. Therefore the intensity level was 74 dB.

14. The noise from a power mower was measured at 104 dB. The noise level at a rock concert was measured at 124 dB. Find the ratio of the intensity of the rock music to that of the power mower.

Answer: hard

Let the subscript PM represent the power mower and RC the rock concert. Then $104 = 10\log\frac{I_{\text{PM}}}{10^{-12}} \Leftrightarrow \log(10^{12}I_{\text{PM}}) = 10.4 \Leftrightarrow 10^{12}I_{\text{PM}} = 10^{10.4}$ and $124 = 10\log(10^{12}I_{\text{RC}}) \Leftrightarrow \log(10^{12}I_{\text{RC}}) = 12.4 \Leftrightarrow 10^{12}I_{\text{RC}} = 10^{12.4}$. So $\frac{I_{\text{RC}}}{I_{\text{PM}}} = \frac{10^{12.4}}{10^{10.4}} = 10^2 = 100$ and so the ratio of intensity is 100.

15. The table shows the mean distances d of the planets from the sun (taking the unit of measurement to be the distance from the earth to the sun) and their periods T (time of revolution in years). Try to discover a relationship between T and d. [Hint: Consider their logarithms.]

Planet	d	T
Mercury	0.387	0.241
Venus	0.723	0.615
Earth	1.000	1.000
Mars	1.523	1.881
Jupiter	5.203	11.861
Saturn	9.541	29.457
Uranus	19.190	84.008
Neptune	30.086	164.784
Pluto	39.507	248.35

Answer: hard

Planet	d	T	$\log d$	$\log T$	$\frac{\log d}{\log T}$
Mercury	0.387	0.241	−0.412	−0.618	0.667
Venus	0.723	0.615	−0.141	−0.211	0.668
Earth	1.000	1.000	0	0	undefined
Mars	1.523	1.881	0.183	0.274	0.668
Jupiter	5.203	11.861	0.716	1.074	0.667
Saturn	9.541	29.457	0.980	1.469	0.667
Uranus	19.190	84.008	1.283	1.924	0.667
Neptune	30.086	164.784	1.478	2.217	0.667
Pluto	39.507	248.35	1.597	2.395	0.667

Therefore the apparent relationship is $\frac{\log d}{\log T} = \frac{2}{3} \Leftrightarrow d = T^{2/3}$.

16. A sum of \$1250 was invested for 8 years, and the interest compounded quarterly. If this sum amounted to \$6583.32 in the given time, what was the interest rate?

Answer: medium

Use $A = P\left(1+\frac{r}{n}\right)^{nt} \Rightarrow 6583.32 = 1250\left(1+\frac{r}{4}\right)^{4\cdot 8}$, and solve for r: $6583.32 = 1250\left(1+\frac{r}{4}\right)^{4\cdot 8} \Leftrightarrow 5.27 = \left(1+\frac{r}{4}\right)^{4\cdot 8} \Leftrightarrow \ln 5.27 = \ln\left(1+\frac{r}{4}\right)^{4(8)} \Leftrightarrow \ln 5.27 = 32\ln\left(1+\frac{r}{4}\right) \Leftrightarrow 0.0519 = \ln\left(1+\frac{r}{4}\right) \Leftrightarrow e^{0.0519} = 1+\frac{r}{4} \Leftrightarrow r \approx 0.213$ or 21.3%.

17. A culture starts with 680 bacteria. After just 30 minutes the count is 3250. Find a formula for the number of bacteria after t hours. Find the number of bacteria after 3.5 hours. After how many minutes will the number of bacteria triple?

Answer: medium

$3250 = 680e^{r(0.5)} \Leftrightarrow \ln\frac{3250}{680} = r(0.5) \Leftrightarrow r \approx 3.128$. $N(t) = 680e^{3.128t}$, so after 3.5 hours, we have $N(3.5) = 680e^{3.128(3.5)} \approx 38{,}651{,}370$. Time to triple: $2040 = 680e^{3.128t} \Leftrightarrow 3 = e^{3.128t} \Leftrightarrow \ln 3 = \ln e^{3.128t} \Leftrightarrow$ $t = 0.35$ hours, or about 21 minutes.

18. A wooden plank salvaged from a shipwreck contains 96.6% of the carbon-14 that is present in living trees. How long ago was the ship built? (The half life of carbon-14 is 5730 years.)
Answer : medium
$\frac{1}{2} = e^{-r(5730)} \quad \Leftrightarrow \quad r = \frac{\ln(1/2)}{5730} \approx -0.000120.$ $0.966 = e^{-0.000120t} \quad \Leftrightarrow \quad t = \frac{\ln(0.966)}{-0.000120} \approx 288$ years ago.

19. The intensity level of sound at a crowded football game was measured at 96 dB. Find the intensity in watts/m^2.
Answer: medium
$$\beta = 10\log\frac{I}{I_0} \quad \Rightarrow \quad 96 = 10\log\frac{I}{10^{-12}} \quad \Leftrightarrow \quad 10^{9.6} = \frac{I}{10^{-12}} \quad \Leftrightarrow \quad I = 10^{9.6}\left(10^{-12}\right) = 10^{-2.4} \approx 3.98 \times 10^{-3} \text{ watts/m}^2$$

CHAPTER 6 REVIEW

1. Sketch the graph of the function $g(x) = 3^{x-1}$. State the domain, range, and asymptote.

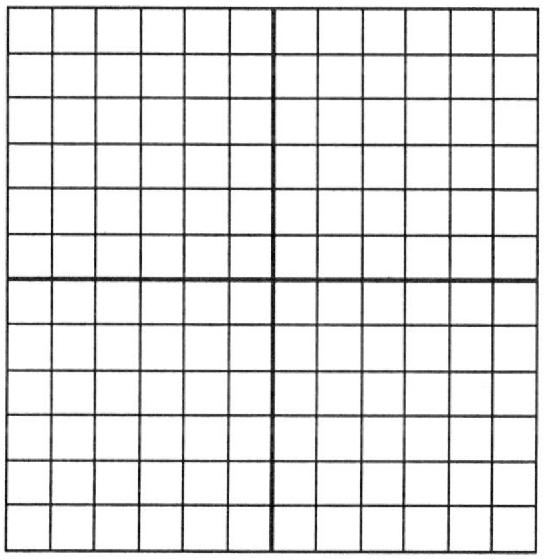

Answer: medium

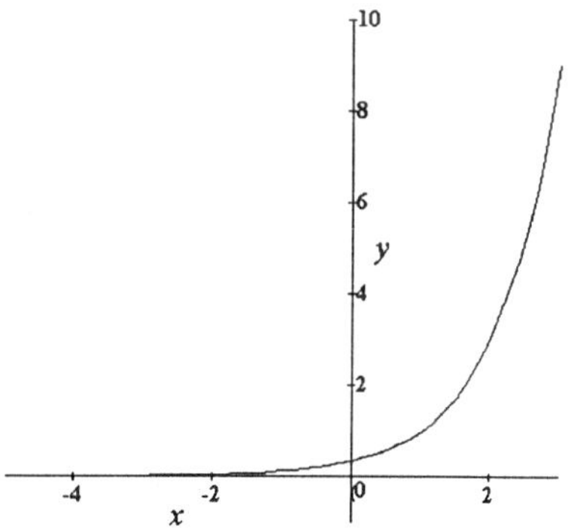

$g(x) = 3^{x-1}$. D: $(-\infty, \infty)$, R: $(0, \infty)$, A: $y = 0$

2. Sketch the graph of the function $y = 2 + 4^{-x}$. State the domain, range, and asymptote.

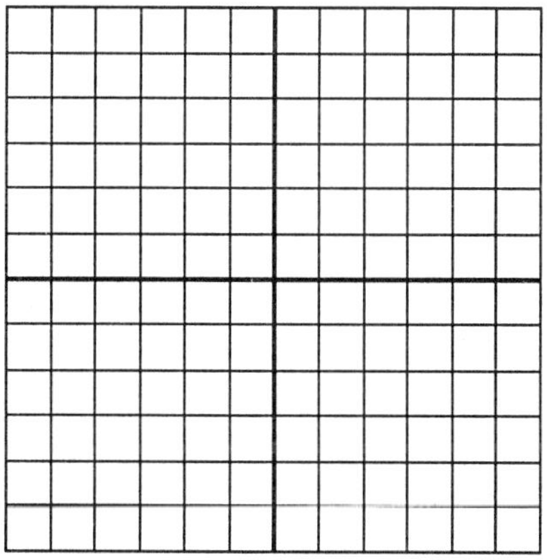

Answer: medium

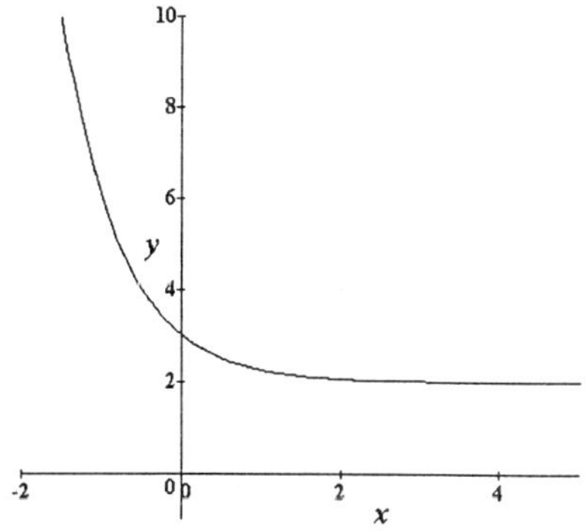

$y = 2 + 4^{-x}$. D: $(-\infty, \infty)$, R: $(2, \infty)$, A: $y = 2$

3. Sketch the graph of the function $g(x) = -\log(-x)$. State the domain, range, and asymptote.

Answer: medium

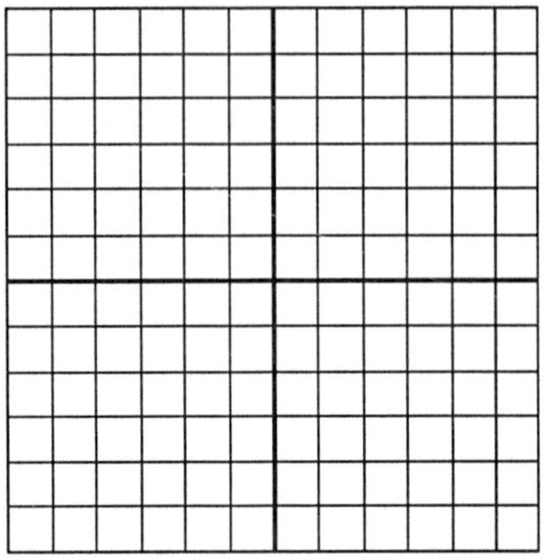

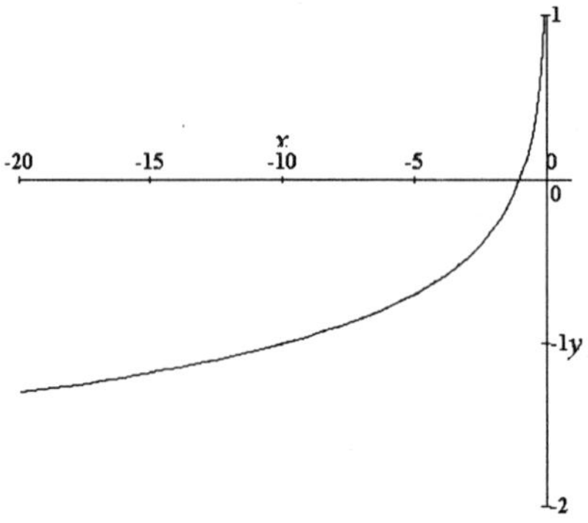

$g(x) = -\log(-x)$. D: $(-\infty, 0)$, R: $(-\infty, \infty)$, A: $x = 0$

4. Sketch the graph of the function $y = 2 + \log_4(x + 3)$. State the domain, range, and asymptote.

Answer: medium

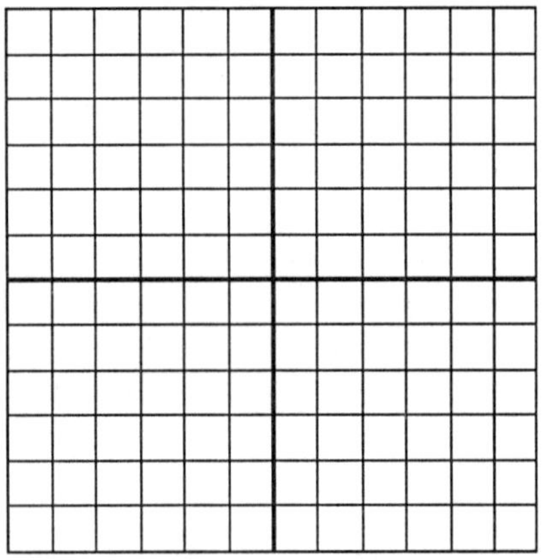

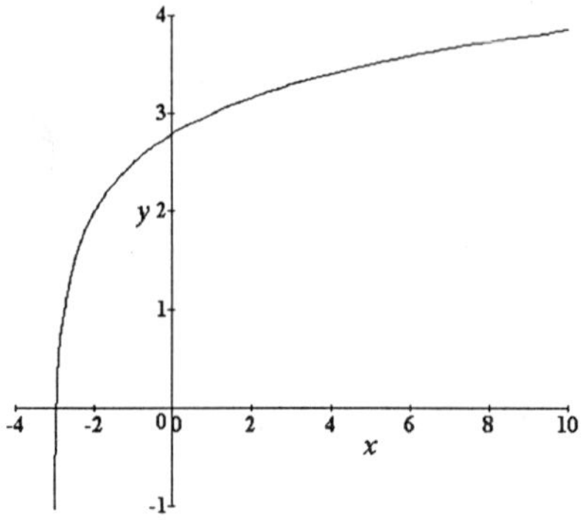

$y = 2 + \log_4(x + 3)$. D: $(-3, \infty)$, R: $(-\infty, \infty)$, A: $x = -3$

5. Sketch the graph of the function $G(x) = \frac{1}{3}e^{x-2}$. State the domain, range, and asymptote.

Answer: medium

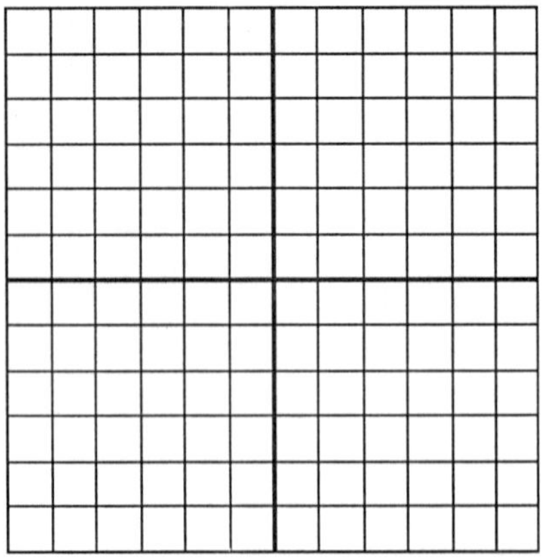

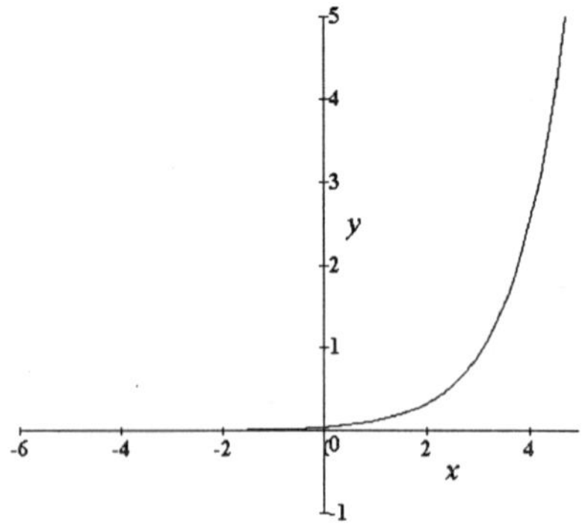

$G(x) = \frac{1}{3}e^{x-2}$. D: $(-\infty, \infty)$, R: $(0, \infty)$, A: $y = 0$

6. Sketch the graph of the function $y = \ln(x^3)$. State the domain, range, and asymptote.

Answer: medium

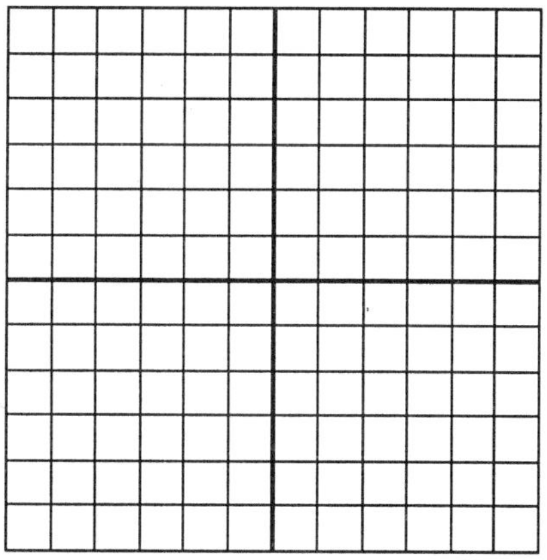

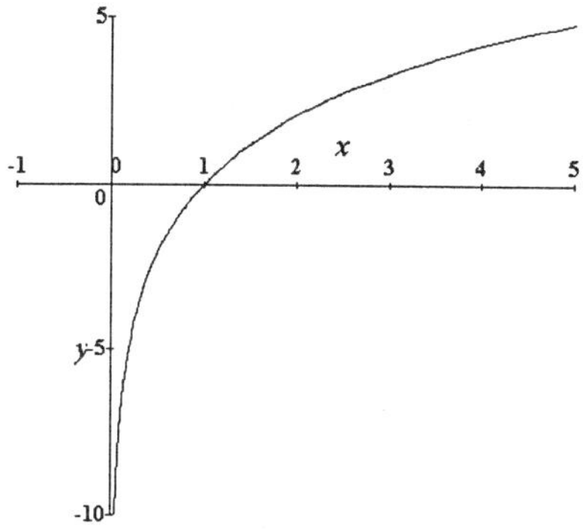

$y = \ln(x^3)$. D: $(0, \infty)$, R: $(-\infty, \infty)$, A: $x = 0$

7. Find the domain of the function $g(x) = \ln(2 - x - x^2)$.

(a) $[-2, 1]$ (b) $(-2, 1)$ (c) $(-\infty, \infty)$ (d) $(-1, 2)$ (e) $(-\infty, -2) \cup (1, \infty)$

Answer: medium (b)

$g(x) = \ln(2 - x - x^2)$. Then $2 - x - x^2 > 0$ (since $\ln y$ is defined only for $y > 0$) $\Leftrightarrow$ $x^2 + x - 2 < 0$ $\Leftrightarrow$ $(x + 2)(x - 1) < 0$.

Case (i): $x + 2 < 0$ $\Leftrightarrow$ $x < -2$ and $x - 1 > 0$ $\Leftrightarrow$ $x > 1$ which is impossible.

Case (ii): $x + 2 > 0$ $\Leftrightarrow$ $x > -2$ and $x - 1 < 0$ $\Leftrightarrow$ $x < 1$, so $x \in (-2, 1)$.

Combining the two cases gives a domain of $(-2, 1)$.

8. Write the equation $\log_5 34 = x$ in exponential form.

(a) $x^{34} = 5$ (b) $x^5 = 34$ (c) $5^x = 34$ (d) $5^{34} = x$ (e) $34^x = 5$

Answer: easy (c)

$\log_5 34 = x \quad \Leftrightarrow \quad 5^x = 34$

9. Write the equation $\ln c = 15$ in exponential form.

(a) $e^{15} = c$ (b) $e^c = 15$ (c) $15^c = e$ (d) $15^e = c$ (e) $c^{15} = e$

Answer: medium (a)

$\ln c = 15 \quad \Leftrightarrow \quad e^{15} = c$

10. Write the equation $81^{-1/2} = \frac{1}{9}$ in logarithmic form.

(a) $\log_9 81 = -2$ (b) $\log_{1/2} \frac{1}{9} = -\frac{1}{81}$ (c) $\log_9 \frac{1}{2} = -\frac{1}{9}$ (d) $\log_{81} \frac{1}{2} = -9$ (e) $\log_{81} \frac{1}{9} = -\frac{1}{2}$

Answer: easy (e)

$81^{-1/2} = \frac{1}{9} \quad \Leftrightarrow \quad \log_{81} \frac{1}{9} = -\frac{1}{2}$

11. Write the equation $e^{2k} = m$ in logarithmic form.

(a) $\ln k = 2m$ (b) $\ln m = 2k$ (c) $\ln(2m) = k$ (d) $\ln(2k) = m$ (e) $2 \ln k = m$

Answer: easy (b)

$e^{2k} = m \quad \Leftrightarrow \quad \ln m = 2k$

12. Evaluate the expression $\log_5 1$ without using a calculator.

(a) 2 (b) -2 (c) 1 (d) -1 (e) 0

Answer: easy (e)

$\log_5 1 = \log_5(5^0) = 0$

13. Evaluate the expression $\log 0.0000001$ without using a calculator.

(a) -10 (b) -9 (c) -8 (d) -7 (e) -6

Answer: easy (d)

$\log 0.0000001 = \log 10^{-7} = -7$

14. Evaluate the expression $\log_8 16$ without using a calculator.

(a) $\frac{3}{4}$ (b) $\frac{4}{3}$ (c) $\frac{5}{4}$ (d) $\frac{5}{3}$ (e) $\frac{7}{4}$

Answer: easy (b)

$\log_8 16 = \log_8 (8^{4/3}) = \frac{4}{3}$

15. Evaluate the expression $3^{\log_3 17}$ without using a calculator.

(a) 12 (b) 15 (c) 17 (d) 22 (e) 3^{17}

Answer: easy (c)

$3^{\log_3 17} = 17$

16. Evaluate the expression $e^{3\ln 4}$ without using a calculator.

(a) 64 (b) -16 (c) 2^8 (d) 8 (e) 2

Answer: easy (a)

$e^{3\ln 4} = (e^{\ln 4})^3 = 4^3 = 64$

17. Evaluate the expression $\log_3 \sqrt{27}$ without using a calculator.

(a) $\frac{1}{2}$ (b) $\frac{3}{2}$ (c) 1 (d) 2 (e) $\frac{5}{2}$

Answer: medium (b)

$\log_3 \sqrt{27} = \log_3 (3^{3/2}) = \frac{3}{2}$

18. Evaluate the expression $\log_5 2500 - \log_5 4$ without using a calculator.

(a) 1 (b) 2 (c) 3 (d) 4 (e) 5

Answer: medium (d)

$\log_5 2500 - \log_5 4 = \log_5 \frac{2500}{4} = \log_5 625 = \log_5 5^4 = 4$

19. Evaluate the expression $\log \log \log 10^{10^{10}}$ without using a calculator.

(a) -1 (b) 10 (c) 10^{10} (d) -10 (e) 1

Answer: medium (e)

$\log_{10}(\log_{10}(\log_{10} 10^{10^{10}})) = \log_{10}(\log_{10} 10^{10}) = \log_{10} 10 = 1$

20. Rewrite the expression $\log_2 (x^2 \sqrt{x+1})$ in a form with no logarithms of products, quotients, or powers.

(a) $2\log_2 x + \frac{1}{2}\log_2(x+1)$ (b) $4\log_2 x + 2\log_2(x+1)$

(c) $2\log_2 x - 2\log_2(x+2)$ (d) $\log_2 x + \log_2(x+1)$

(e) $2\log_2(x-1) + \frac{1}{2}\log_2 x$

Answer: medium (a)

$\log_2 (x^2 \sqrt{x+1}) = \log_2 x^2 + \log_2 \sqrt{x+1} = 2\log_2 x + \frac{1}{2}\log_2(x+1)$

21. Rewrite the expression $\log \dfrac{3x^4}{y^3(x-2)^4}$ in a form with no logarithms of products, quotients, or powers.

(a) $\log 3 + 4\log x - [3\log y + 4\log(x-2)]$ (b) $\log 12 + \log x - [4\log y + 3\log(x-2)]$

(c) $\log(3+4x) - [\log(3+y) + 4\log(x-2)]$ (d) $12\log x - 3\log y - \log(4-x)$

(e) $3\log x + 4 - 3\log y - 4\log x + 2$

Answer: medium (a)

$$\log \frac{3x^4}{y^3(x-2)^4} = \log(3x^4) - \log(y^3(x-2)^4) = \log 3 + 4\log x - [3\log y + 4\log(x-2)]$$

22. Rewrite the expression $\ln \dfrac{\sqrt[4]{x^3+12}}{(x+12)\sqrt{x-2}}$ in a form with no logarithms of products, quotients, or powers.

(a) $\frac{1}{3}\ln(x^3+8) - [\ln(x+8) + \frac{1}{3}\ln(x-1)]$ (b) $\frac{1}{4}\ln 3 + 3\ln x + 12 - \ln(x+12) + \ln(x-1)$

(c) $\frac{1}{4}\ln(x^3+12) - [\ln(x+12) + \frac{1}{2}\ln(x-2)]$ (d) $\ln(\frac{3}{4}x+9) - [\ln(x+12) + \frac{1}{2}\ln x - 1]$

(e) $\frac{1}{3}\ln(x^4+16) - [\frac{3}{4}\ln(x+16) + \frac{1}{3}\ln(x-\frac{2}{3})]$

Answer: medium (c)

$$\ln \frac{\sqrt[4]{x^3+12}}{(x+12)\sqrt{x-2}} = \frac{1}{4}\ln(x^3+12) - [\ln(x+12) + \frac{1}{2}\ln(x-2)]$$

23. Rewrite the expression $\log x + \log xy^2 + 2\log y$ as a single logarithm.

(a) $\log(xy^8)$ (b) $\log(x^2 + y^4)$ (c) $2\log(xy)$ (d) $\log(8xy)$ (e) $\log(x^2y^4)$

Answer: medium (e)

$\log x + \log xy^2 + 2\log y = \log(x \cdot xy^2 \cdot y^2) = \log(x^2y^4)$

24. Rewrite the expression $\log_5 3 + \log_5(x+2) - \frac{1}{4}\log_5(2x+3)$ as a single logarithm.

(a) $\log_5 \frac{3x+2}{\sqrt{2x+3}}$ (b) $\log_5 \frac{x+5}{\frac{1}{2}x+4}$ (c) $\log_5 \frac{x+3}{8x+12}$ (d) $\log_5 \frac{3(x+2)}{\sqrt[4]{2x+3}}$ (e) $\log_5 \frac{x+6}{\sqrt[4]{x+6}}$

Answer: medium (d)

$\log_5 3 + \log_5(x+2) - \frac{1}{4}\log_5(2x+3) = \log_5(3(x+2)) - \log_5(2x+3)^{1/4} = \log_5 \frac{3(x+2)}{\sqrt[4]{2x+3}}$

25. Rewrite the expression $\frac{1}{3}[\ln(x-2) + 5\ln(x^2+2x)]$ as a single logarithm.

Answer: medium

$\frac{1}{3}[\ln(x-2) + 5\ln(x^2+2x)] = \frac{1}{3}\ln\left((x-2)(x^2+2x)^5\right) = \ln\sqrt[3]{(x-2)(x^2+2x)^5}$

26. Use a calculator to find the solution of the equation $2^{4x-7} = 3$ correct to two decimal places.

Answer: medium

$2^{4x-7} = 3 \quad \Leftrightarrow \quad (4x-7)\log 2 = \log 3 \quad \Leftrightarrow \quad x = \frac{1}{4}\left(7 + \frac{\log 3}{\log 2}\right) \approx 2.15$

27. Use a calculator to find the solution of the equation $e^{-13k} = 20{,}000$ correct to two decimal places.

Answer: medium

$e^{-13k} = 20{,}000 \quad \Leftrightarrow \quad -13k = \ln 20{,}000 \quad \Leftrightarrow \quad k = -\frac{1}{13}\ln 20{,}000 \approx -0.76$

GC 28. Draw the graph of the function $f(x) = x^2 - \ln x$ and use it to determine the asymptotes and local maximum and minimum values.

Answer: medium

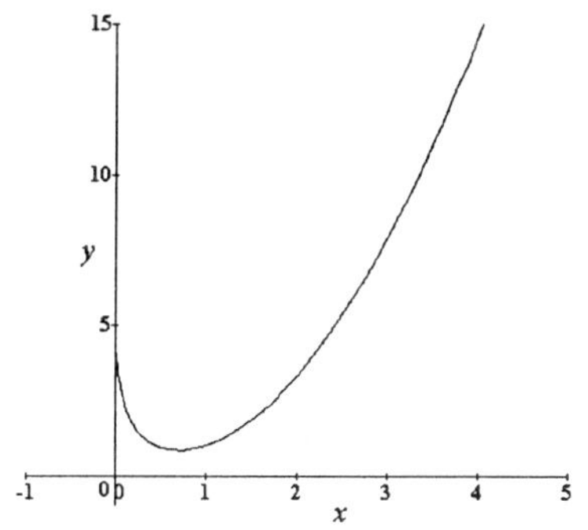

$f(x) = x^2 - \ln x$. A: $x = 0$, local minimum $f(0.707) \approx 0.847$

29. Draw the graph of the function $f(x) = 8^x - 4^x$ and use it to determine the asymptotes and local maximum and minimum values.

Answer: medium

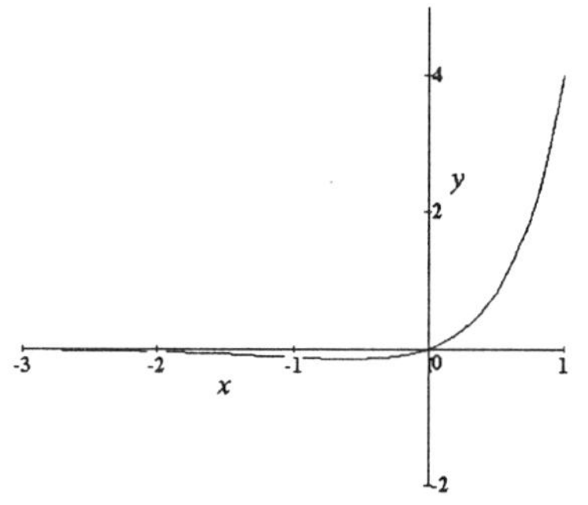

$f(x) = 8^x - 4^x$. A: $y = 0$, local minimum $f(-0.585) \approx -0.148$

30. Find the solutions of the equation $3 - x^2 = e^{-3x}$ correct to two decimal places.

Answer: medium

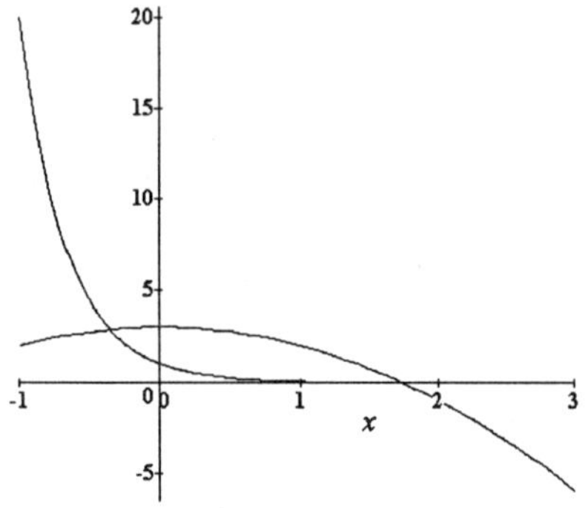

$3 - x^2 = e^{-3x}$ has solutions $x \approx -0.35$ and 1.73

31. Solve the inequality $e^x < 3x^2$ graphically.

Answer: medium

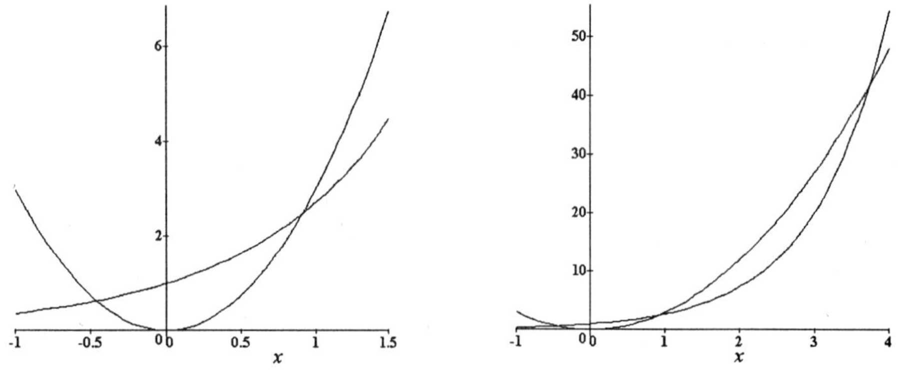

$e^x < 3x^2$ has solution set $(-\infty, -0.46) \cup (0.91, 3.73)$.

32. Solve the inequality $0.4 \le \log x < 4$.

Answer: medium

$0.4 \le \log x < 4 \quad \Leftrightarrow \quad 10^{0.4} \le x < 10^4 \quad \Leftrightarrow \quad 10^{2/5} \le x < 10{,}000$

33. Find the inverse function of the function $f(x) = 3^{2^x}$ and state its domain and range.

Answer: medium

$f(x) = 3^{2^x}$. Then $y = 3^{2^x} \quad \Leftrightarrow \quad \log_3 y = 2^x \quad \Leftrightarrow \quad \log_2(\log_3 y) = x$ and so the inverse function is $f^{-1}(x) = \log_2(\log_3 x)$. Since $\log_2 y$ is defined only when $y > 0$ then $\log_3 x > 0 \quad \Leftrightarrow \quad x > 1$. Therefore the domain is $(1, \infty)$ and the range is $(-\infty, \infty)$.

34. A bacteria culture contains 5,000 bacteria initially. After an hour the bacteria count is 20,000.

(a) Find the doubling period.

(b) Find the population after 3 hours.

Answer: medium

$n_0 = 5{,}000$ and $n(1) = 20{,}000$. Then $n(t) = n_0 \cdot 2^{t/d} = 5{,}000 \cdot 2^{t/d}$.

(a) Since $n(1) = 20{,}000$ then $20{,}000 = 5{,}000 \cdot 2^{1/d} \quad \Leftrightarrow \quad 2^{1/d} = \dfrac{20{,}000}{5{,}000} = 4 \quad \Leftrightarrow \quad 2^{1/d} = 4 = 2^2 \quad \Leftrightarrow \quad 1/d = 2 \quad \Leftrightarrow$ $d = \frac{1}{2}$, and so the doubling period is 30 min.

(b) $n(3) = 5{,}000 \cdot 2^{3/0.5} = 320{,}000$ and so the population after 3 hours is about 320,000.

35. A sample of Bismuth-210 decayed to 33% of its original mass after 8 days.

(a) Find the half-life of this element.

(b) Find the mass remaining after 10 days.

Answer: medium

$m(8) = 0.33m_0$ and $m(t) = m_0 \cdot 2^{-t/h}$

(a) $m(8) = 0.33m_0 = m_0 \cdot 2^{-8/h} \quad \Leftrightarrow \quad 2^{-8/h} = 0.33 \quad \Leftrightarrow \quad -\frac{8}{h} = \frac{\log(0.33)}{\log 2} \quad \Leftrightarrow \quad h = \frac{-8\log 2}{\log(0.33)} \approx 5.002$, and so the half-life of this element is roughly 5 days.

(b) $m(10) \approx m_0 \cdot 2^{-10/5.002} \approx 0.25m_0$ and so 25% of the original mass remains.

36. The pH of lime juice is 2.0. Find the hydrogen ion concentration.

Answer: easy

$\text{pH} = 2.0 = -\log[\text{H}^+]$. Then $[\text{H}^+] = 10^{-2.0} = 1.0 \times 10^{-2}$ M

37. The noise from a jackhammer was measured at 128 dB. The sound of whispering was measured at 25 dB. Find the ratio of the intensity of the jackhammer to that of the whispering.

Answer: hard

Let the subscript JH represent the jackhammer and W the whispering. $\beta_{\text{JH}} = 128 = 10\log\frac{I_{\text{JH}}}{I_0} \quad \Leftrightarrow \quad \log\frac{I_{\text{JH}}}{I_0} = 12.8 \quad \Leftrightarrow$ $\frac{I_{\text{JH}}}{I_0} = 10^{12.8}$. Similarly $\beta_{\text{W}} = 25 = 10\log\frac{I_{\text{W}}}{I_0} \quad \Leftrightarrow \quad \log\frac{I_{\text{W}}}{I_0} = 2.5 \quad \Leftrightarrow \quad \frac{I_{\text{W}}}{I_0} = 10^{2.5}$. So $\frac{I_{\text{JH}}}{I_{\text{W}}} = \frac{10^{12.8}}{10^{2.5}} = 10^{10.3} \approx 2.00 \times 10^{10}$ and so the ratio of intensities is 2.00×10^{10}.

CHAPTER SEVEN

SECTION 7.1

1. Find the radian measure of the angle with the degree measure $15°$.
   Answer: easy
   $15° = 15\left(\frac{\pi}{180}\right)$ rad $= \frac{\pi}{12}$ rad.

2. Find the radian measure of the angle with the degree measure $-75°$.
   Answer: easy
   $-75° = -75\left(\frac{\pi}{180}\right)$ rad $= -\frac{5\pi}{12}$ rad.

3. Find the radian measure of the angle with the degree measure $105°$.
   Answer: easy
   $105° = 105\left(\frac{\pi}{180}\right)$ rad $= \frac{7\pi}{12}$ rad.

4. Find the radian measure of the angle with the degree measure $750°$.
   Answer: easy
   $750° = 750\left(\frac{\pi}{180}\right)$ rad $= \frac{25\pi}{6}$ rad.

5. Find the radian measure of the angle with the degree measure $150°$.
   Answer: easy
   $150° = 150\left(\frac{\pi}{180}\right)$ rad $= \frac{5\pi}{6}$ rad.

6. Find the degree measure of the angle with the radian measure $\frac{7\pi}{15}$ rad.
   Answer: easy
   $\frac{7\pi}{15}$ rad $= \frac{7\pi}{15}\left(\frac{180}{\pi}\right)^{\circ} = 7 \cdot 12° = 84°$.

7. Find the degree measure of the angle with the radian measure $\frac{11\pi}{12}$ rad.
   Answer: easy
   $\frac{11\pi}{12}$ rad $= \frac{11\pi}{12}\left(\frac{180}{\pi}\right)^{\circ} = 11 \cdot 15° = 165°$.

8. Find the degree measure of the angle with the radian measure $-\frac{13\pi}{9}$ rad.
   Answer: easy
   $-\frac{13\pi}{9}$ rad $= -\frac{13\pi}{9}\left(\frac{180}{\pi}\right)^{\circ} = -13 \cdot 20° = -260°$.

9. Find the degree measure of the angle with the radian measure 3.5 rad.
   Answer: easy
   3.5 rad $= 3.5\left(\frac{180}{\pi}\right)^{\circ} = \frac{3.5 \cdot 180}{\pi}^{\circ} = \frac{630}{\pi}^{\circ}$.

10. Find the degree measure of the angle with the radian measure $\frac{\pi}{15}$ rad.
    Answer: easy
    $\frac{\pi}{15}$ rad $= \frac{\pi}{15}\left(\frac{180}{\pi}\right)^{\circ} = \frac{180}{15}^{\circ} = 12°$.

11. Find two positive angles that are coterminal with the angle given by $49°$.
    Answer: easy
    $49° + 360° = 409°$ and $49° + 720° = 769°$ are both coterminal.

12. Find two negative angles that are coterminal with the angle given by $\frac{7\pi}{3}$.
    Answer: easy
    $\frac{7\pi}{3} - 4\pi = -\frac{5\pi}{3}$ and $\frac{7\pi}{3} - 6\pi = -\frac{11\pi}{3}$ are both coterminal.

13. Find two negative angles that are coterminal with the angle given by $-370°$.

Answer: easy

$-370° + 360° = -10°$ and $-370° - 360° = -730°$ are both coterminal.

14. Determine whether the angles given by $30°$ and $390°$ are coterminal.

Answer: easy

$390° - 30° = 360°$, so they are coterminal.

15. Determine whether the angles given by $\frac{18\pi}{7}$ and $\frac{34\pi}{7}$ are coterminal.

Answer: easy

$\frac{34\pi}{7} - \frac{18\pi}{7} = \frac{16\pi}{7}$ which is not a multiple of 2π, so they are not coterminal.

16. Determine whether the angles given by $-152°$ and $208°$ are coterminal.

Answer: easy

$208° - (-152°) = 360°$, so they are coterminal.

17. Find an angle between $0°$ and $360°$ that is coterminal with the angle given by $-5043°$.

Answer: medium

$-5043° + 15 \cdot 360° = -5043° + 5400° = 357°$.

18. Find an angle between $0°$ and $360°$ that is coterminal with the angle given by $839°$.

Answer: medium

$839° - 720° = 119°$.

19. Find an angle between 0 and 2π that is coterminal with the angle given by $\frac{27\pi}{2}$.

Answer: medium

$\frac{27\pi}{2} - 12\pi = \frac{3\pi}{2}$.

20. Find an angle between 0 and 2π that is coterminal with the angle given by $-\frac{77\pi}{5}$.

Answer: medium

$-\frac{77\pi}{5} + 16\pi = \frac{3\pi}{5}$.

21. Find the length of an arc that subtends a central angle of $75°$ in a circle of radius 3 m.

Answer: medium

$75\left(\frac{\pi}{180}\right) \cdot 3 = \frac{5\pi}{12} \cdot 3 = \frac{5\pi}{4}$ m.

22. Find the length of an arc that subtends a central angle of $135°$ in a circle of radius 12 ft.

Answer: medium

$135\left(\frac{\pi}{180}\right) \cdot 12 = \frac{3\pi}{4} \cdot 12 = 9$ ft.

23. Find the length of an arc that subtends a central angle of $\frac{5\pi}{24}$ in a circle of radius 8 m.

Answer: medium

$\frac{5\pi}{24} \cdot 8 = \frac{5\pi}{3}$ m.

24. Find the radian measure of an angle in a circle of radius 12 m which is subtended by an arc of length 3 m.

Answer: medium

$3 = 12\theta \quad \Rightarrow \quad \theta = \frac{3}{12} = \frac{1}{4}$.

25. Find the radius of a circle in which an angle of $\frac{\pi}{6}$ radians is subtended by an arc of length 4 m.

Answer: medium

$4 = r \cdot \frac{\pi}{6} \quad \Rightarrow \quad r = \frac{2}{3\pi}$.

26. Two cities, A and B lie on the same meridian, one is found at $20°$ S and the other at $27°$ S. Find the distance between these cities given that the circumference of the earth is 3960 mi.

Answer: hard

The angle between the cities is $7°$, so the length of the arc joining them is $7\left(\frac{\pi}{180}\right) \cdot 3960 \approx 483.8$ mi.

27. Find the area of a sector with central angle $\frac{3\pi}{2}$ in a circle with radius 5 m.

Answer: medium

$A = \frac{1}{2}5^2\frac{3\pi}{2} = \frac{75\pi}{4}$

28. Find the area of a sector with central angle 120° in a circle with radius 12 m.

Answer: medium

$A = \frac{1}{2}12^2 \cdot 120\left(\frac{\pi}{180}\right) = \frac{144\pi}{3} = 48\pi$

29. Find the radius of a circle in which a sector with central angle 3 rad has an area of 24 m^2.

Answer: hard

$A = \frac{1}{2}r^2\theta \quad \Rightarrow \quad 24 = \frac{1}{2}r^2 3 \quad \Rightarrow \quad r^2 = 16 \quad \Rightarrow \quad r = 4$, since r is positive.

SECTION 7.2

1. Find the values of the six trigonometric ratios of the angle θ in the triangle.

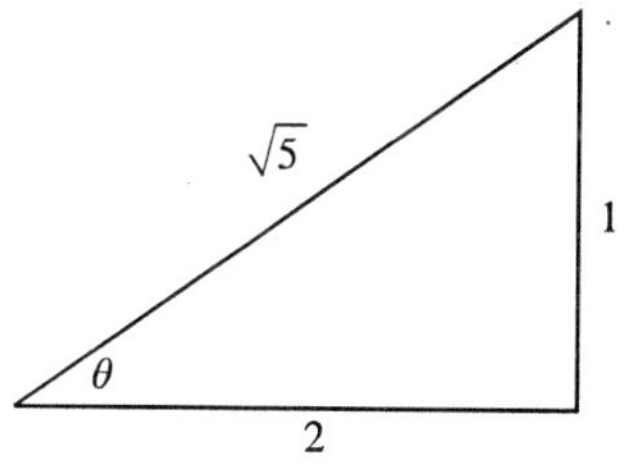

Answer: easy

$\sin\theta = \frac{1}{\sqrt{5}}$; $\cos\theta = \frac{2}{\sqrt{5}}$; $\tan\theta = \frac{1}{2}$;

$\csc\theta = \sqrt{5}$; $\sec\theta = \frac{\sqrt{5}}{2}$; $\cot\theta = 2$

2. Find the values of the six trigonometric ratios of the angle θ in the triangle.

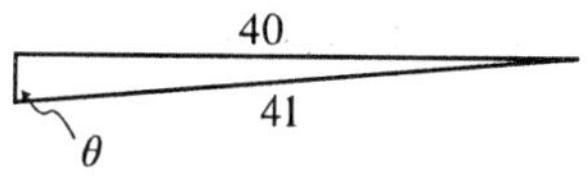

Answer: easy

$\sin\theta = \frac{40}{41}$; $\cos\theta = \frac{8}{41}$; $\tan\theta = \frac{40}{8} = 5$;

$\csc\theta = \frac{41}{40}$; $\sec\theta = \frac{41}{8}$; $\cot\theta = \frac{1}{5}$

3. Find $\sin\alpha$, $\cos\beta$, $\tan\alpha$, $\cot\beta$, $\csc\alpha$, and $\sec\beta$.

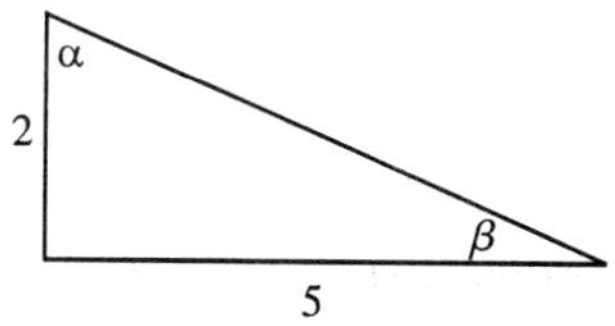

Answer: easy

$\sin\alpha = \frac{5}{\sqrt{29}}$; $\cos\beta = \frac{5}{\sqrt{29}}$; $\tan\alpha = \frac{5}{2}$;

$\cot\beta = \frac{5}{2}$; $\csc\alpha = \frac{\sqrt{29}}{5}$; $\sec\beta = \frac{\sqrt{29}}{5}$

4. Find the side labeled x.

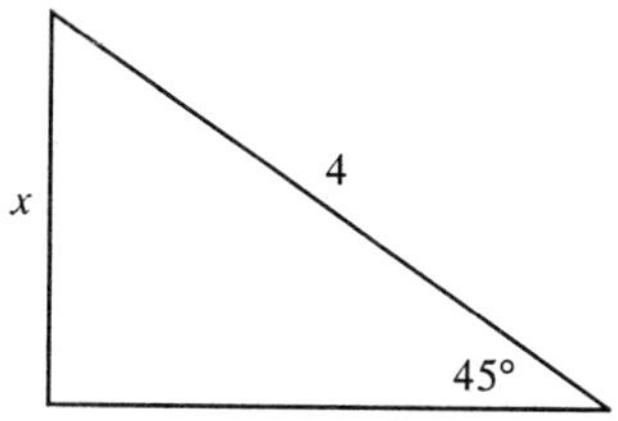

Answer: easy

$\sin 45° = \frac{x}{4} \quad \Leftrightarrow \quad x = 4 \cdot \frac{\sqrt{2}}{2} = 2\sqrt{2}$

5. Find the side labeled x.

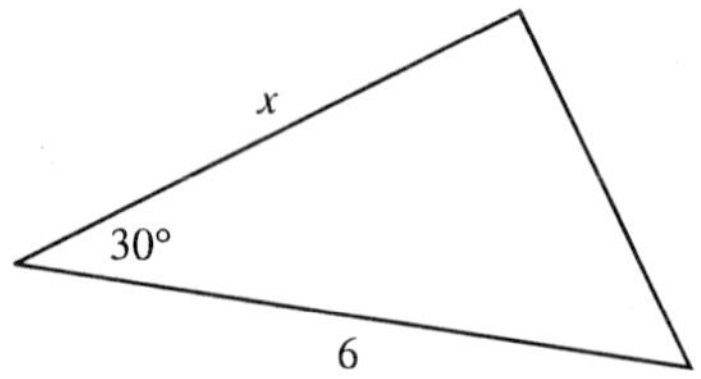

Answer: easy

$\sec 30° = \frac{x}{6} \quad \Leftrightarrow \quad x = 6 \cdot \frac{2\sqrt{3}}{3} = 4\sqrt{3}$

6. Find the side labeled x, correct to 5 decimal places.

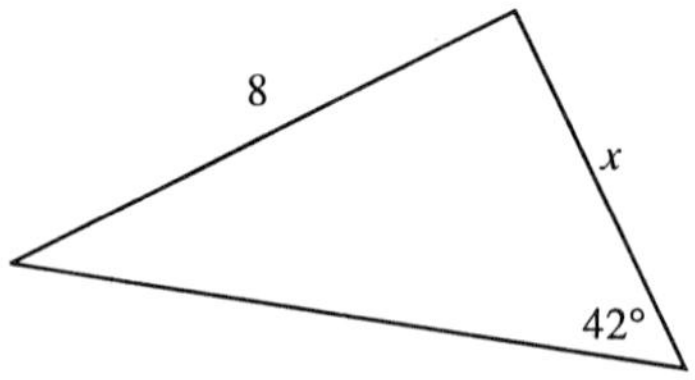

Answer: easy

$\cot 42° = \frac{x}{8} \quad \Leftrightarrow \quad x = 8 \cdot \cot 42° \approx 8.88490$

7. Express x and y in terms of trigonometric ratios of θ.

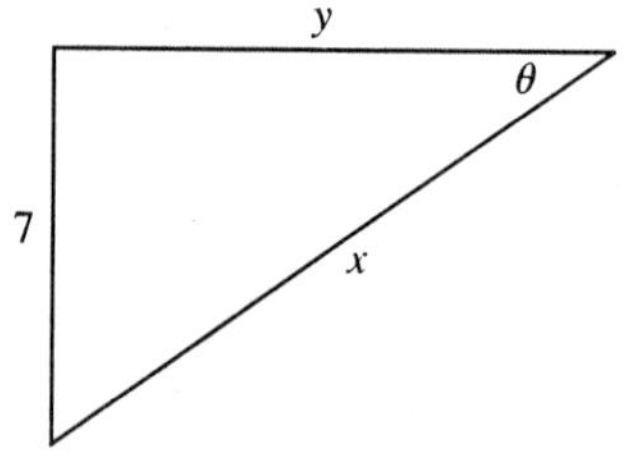

Answer: medium

$\csc\theta = \frac{x}{7} \quad \Rightarrow \quad x = 7\csc\theta$

$\cot\theta = \frac{y}{7} \quad \Rightarrow \quad y = 7\cot\theta$

8. Sketch a triangle that has angle θ, where $\cos\theta = \frac{3}{7}$, and use it to find the other five trigonometric ratios of θ.

Answer: medium

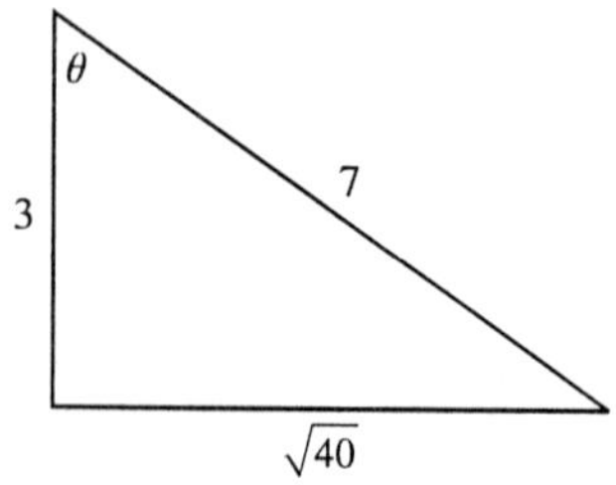

$\sin\theta = \frac{\sqrt{40}}{7}$; $\tan\theta = \frac{\sqrt{40}}{3}$; $\csc\theta = \frac{7}{\sqrt{40}}$;

$\sec\theta = \frac{7}{3}$; $\cot\theta = \frac{3}{\sqrt{40}}$

9. Sketch a triangle that has angle θ, where $\tan\theta = \sqrt{15}$, and use it to find the other five trigonometric ratios of θ.

Answer: medium

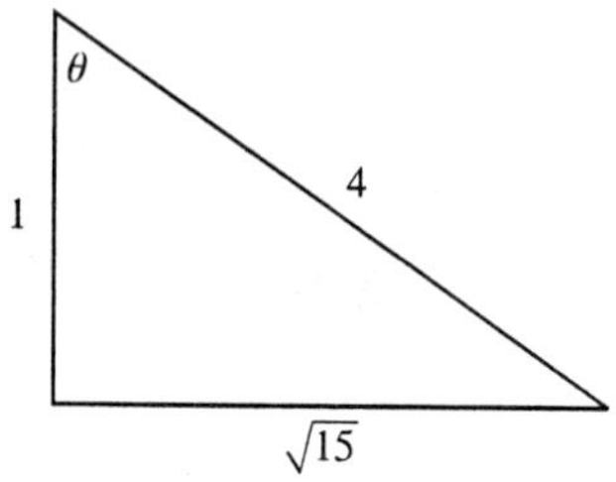

$\sin\theta = \frac{\sqrt{15}}{4}$; $\cos\theta = \frac{1}{4}$; $\csc\theta = \frac{4}{\sqrt{15}}$;

$\sec\theta = 4$; $\cot\theta = \frac{1}{\sqrt{15}}$

10. Evaluate $\sin 30° (\cos 30°)^2$.

Answer: easy

$\sin 30° (\cos 30°)^2 = \frac{1}{2} \cdot \left(\frac{\sqrt{3}}{2}\right)^2 = \frac{1}{2} \cdot \frac{3}{4} = \frac{3}{8}$

11. Evaluate $\sin 60° + (\cos 30°)^2$.

Answer: easy

$$\sin 60° + (\cos 30°)^2 = \frac{\sqrt{3}}{2} + \left(\frac{\sqrt{3}}{2}\right)^2 = \frac{\sqrt{3}}{2} + \frac{3}{4} = \frac{2\sqrt{3}+3}{4}$$

12. Evaluate $\sin \frac{\pi}{3} \sin \frac{\pi}{4} + \cos \frac{\pi}{3} \cos \frac{\pi}{4}$.

Answer: medium

$$\sin \frac{\pi}{3} \sin \frac{\pi}{4} + \cos \frac{\pi}{3} \cos \frac{\pi}{4} = \frac{\sqrt{3}}{2} \cdot \frac{\sqrt{2}}{2} + \frac{1}{2} \cdot \frac{\sqrt{2}}{2} = \frac{\sqrt{6}+\sqrt{2}}{4}$$

13. Evaluate $\sin \frac{\pi}{3} + \cos \frac{\pi}{4} + \cos \frac{\pi}{6}$.

Answer: medium

$$\sin \frac{\pi}{3} + \cos \frac{\pi}{4} + \cos \frac{\pi}{6} = \frac{\sqrt{3}}{2} + \frac{\sqrt{2}}{2} + \frac{\sqrt{3}}{2} = \frac{2\sqrt{3}+\sqrt{2}}{2}$$

14. Solve the given right triangle.

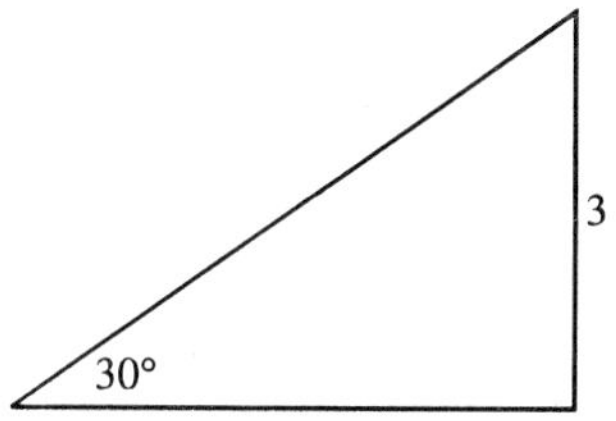

Answer: medium

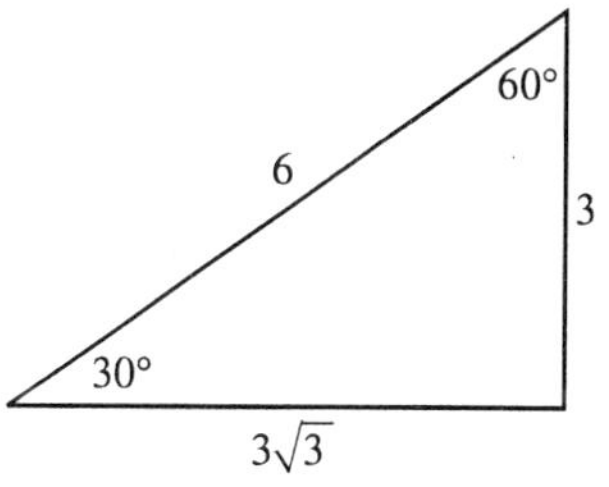

15. Solve the given right triangle.

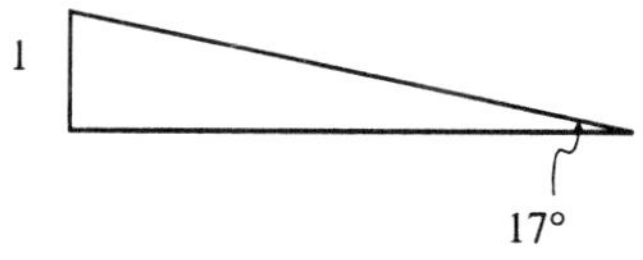

Answer: medium

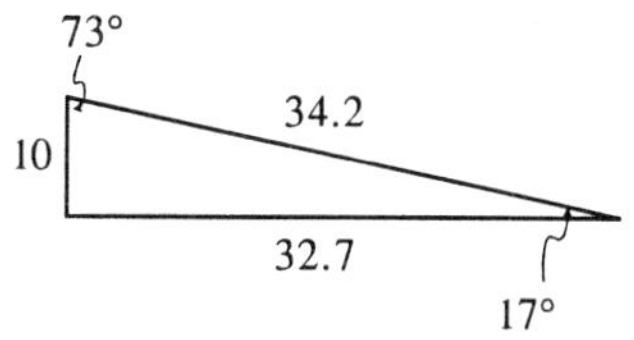

16. From the top of a 250 ft lighthouse, the angle of depression to a ship in the ocean is 18°. How far is the ship from the base of the lighthouse?

Answer: medium

$\tan 18° = \frac{250}{d}$, where d is the distance from the lighthouse, so $d = 250 \cot 18° \approx 769$ ft.

17. A 50 ft tree casts a shadow that is 100 ft long. What is the angle of elevation of the sun?

Answer: medium

$\tan \theta = \frac{50}{100} = \frac{1}{2}$, where θ is the angle of elevation of the sun, so $\theta = \tan^{-1} \frac{1}{2} \approx 26.6°$.

18. A 75 ft tree casts a shadow that is 45 ft long. What is the angle of elevation of the sun?

Answer: medium

$\tan \theta = \frac{75}{45} = \frac{5}{3}$, where θ is the angle of elevation of the sun, so $\theta = \tan^{-1} \frac{5}{3} \approx 59.0°$.

19. A 200 ft guy wire is attached to the top of a tower. If the wire makes a 55° angle with the ground, how tall is the tower?
Answer: medium
$\sin 55° = \frac{h}{200}$, where h is the height of the tower, so $h = 200 \sin 55° \approx 163.8$ ft.

20. A helicopter is dropping water on a forest fire from a height of 300 ft. If one side of the fire makes an angle of depression of 40°, and the other side makes an angle of depression of 85° on the other side of the helicopter, how wide is the fire.
Answer: hard
The width of fire on either side of the helicopter is given by $\cot\theta = \frac{w}{300}$, where θ is the angle of depression, and w the width of the fire. Thus the total width is $w_1 + w_2 = 300\cot\theta_1 + 300\cot\theta_2 = 300(\cot 40° + \cot 85°) \approx 278$ ft.

SECTION 7.3

1. Find the reference angle for $\theta = 39°$.
Answer: easy
$\overline{\theta} = \theta = 39°$

2. Find the reference angle for $\theta = 139°$.
Answer: easy
$\overline{\theta} = 180° - \theta = 180° - 139° = 41°$

3. Find the reference angle for $\theta = -202°$.
Answer: easy
$\overline{\theta} = -180° - \theta = -180° - (-202°) = 22°$.

4. Find the reference angle for $\theta = \frac{7\pi}{6}$.
Answer: easy
$\overline{\theta} = \theta - \pi = \frac{7\pi}{6} - \pi = \frac{\pi}{6}$.

5. Find the reference angle for $\theta = -\frac{3\pi}{5}$.
Answer: easy
$\overline{\theta} = \theta + \pi = -\frac{3\pi}{5} + \pi = \frac{2\pi}{5}$

6. Find the reference angle for $\theta = \frac{17\pi}{5}$.
Answer: easy
$\overline{\theta} = \theta - 3\pi = \frac{17\pi}{5} - 3\pi = \frac{2\pi}{5}$.

7. Find the exact value of $\sin 420°$.
Answer: easy
$\sin 420° = \frac{\sqrt{3}}{2}$

8. Find the exact value of $\sin 210°$.
Answer: easy
$\sin 210° = \frac{1}{2}$

9. Find the exact value of $\cos -135°$.
Answer: easy
$\cos -135° = -\frac{\sqrt{2}}{2}$

10. Find the exact value of $\tan 150°$.
Answer: easy
$\tan 150° = -\frac{\sqrt{3}}{3}$

11. Find the exact value of $\sec 330°$.

    Answer: easy

    $\sec 330° = \frac{2\sqrt{3}}{3}$

12. Find the exact value of $\cot -405°$.

    Answer: easy

    $\cot -405° = -1$

13. Find the exact value of $\cot 210°$.

    Answer: easy

    $\cot 210° = \sqrt{3}$

14. Find the exact value of $\csc \frac{7\pi}{4}$.

    Answer: easy

    $\csc \frac{7\pi}{4} = -\sqrt{2}$

15. Find the exact value of $\sin -\frac{8\pi}{3}$.

    Answer: easy

    $\sin -\frac{8\pi}{3} = -\frac{\sqrt{3}}{2}$

16. Find the exact value of $\csc \frac{11\pi}{3}$.

    Answer: easy

    $\csc \frac{11\pi}{3} = -\frac{2\sqrt{3}}{3}$

17. Find the exact value of $\cot \frac{18\pi}{4}$.

    Answer: easy

    $\cot \frac{18\pi}{4} = 0$

18. Find the exact value of $\sec \frac{14\pi}{6}$.

    Answer: easy

    $\sec \frac{14\pi}{6} = 2$

19. Find the exact value of $\tan -\frac{8\pi}{3}$.

    Answer: easy

    $\tan -\frac{8\pi}{3} = \sqrt{3}$

20. Find the exact value of $\cos \frac{6\pi}{4}$.

    Answer: easy

    $\cos \frac{6\pi}{4} = 0$

21. Find the exact value of $\csc -\frac{\pi}{3}$.

    Answer: easy

    $\csc -\frac{\pi}{3} = -\frac{2\sqrt{3}}{3}$

22. Find the quadrant in which θ lies given that $\sin\theta < 0$ and $\sec\theta > 0$.

    Answer: medium

    θ lies in quadrant IV.

23. Find the quadrant in which θ lies given that $\tan\theta > 0$ and $\csc\theta > 0$.

    Answer: medium

    θ lies in quadrant I.

24. Write $\tan\theta$ in terms of $\sin\theta$ in quadrant II.

    Answer: medium

    $\tan\theta = \frac{\sin\theta}{\cos\theta} = \frac{\sin\theta}{\pm\sqrt{1-\sin^2\theta}} = -\frac{\sin\theta}{\sqrt{1-\sin^2\theta}}$, since tan is negative in quadrant II.

25. Write $\csc\theta$ in terms of $\cot\theta$ in quadrant III.

Answer: medium

$\csc\theta = \pm\sqrt{1+\cot^2\theta} = \sqrt{1+\cot^2\theta}$ since sin is positive in quadrant II.

26. Find the values of the five other trigonometric functions of θ given that $\cos\theta = -\frac{3}{5}$ and θ lies in quadrant II.

Answer: medium

$\sin\theta = \frac{4}{5}$; $\tan\theta = -\frac{4}{3}$; $\csc\theta = \frac{5}{4}$; $\sec\theta = -\frac{5}{3}$; $\cot\theta = -\frac{3}{4}$

27. Find the values of the five other trigonometric functions of θ given that $\csc\theta = 2$ and $\tan\theta < 0$.

Answer: medium

$\sin\theta = \frac{1}{2}$; $\cos\theta = -\frac{\sqrt{3}}{2}$; $\tan\theta = -\frac{\sqrt{3}}{3}$; $\sec\theta = -\frac{2\sqrt{3}}{3}$; $\cot\theta = -\sqrt{3}$

28. Find the area of a triangle with sides of length 4 and 11 and included angle 35°.

Answer: medium

$A = \frac{1}{2}\cdot 4\cdot 11\cdot \sin 35° \approx 12.62$.

29. Find the area of a triangle with sides of length 8 and 13 and included angle 82°.

Answer: medium

$A = \frac{1}{2}\cdot 8\cdot 13\cdot \sin 82° \approx 51.49$.

SECTION 7.4

1. Use the Law of Sines to find x.

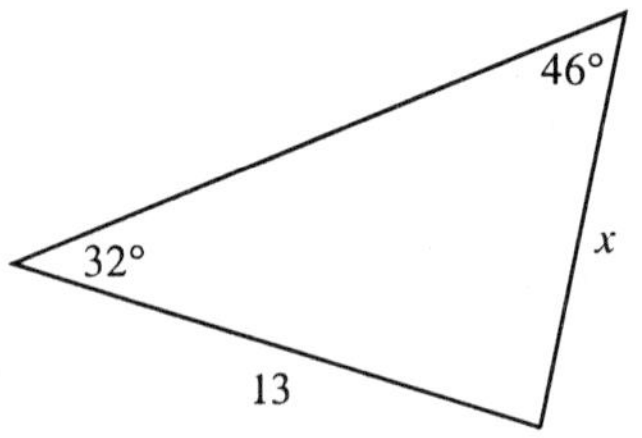

Answer: easy

$$\frac{\sin 32}{x} = \frac{\sin 46}{13} \quad\Leftrightarrow\quad x = 13\frac{\sin 32}{\sin 46} \quad\Rightarrow\quad x \approx 9.58$$

2. Use the Law of Sines to find θ.

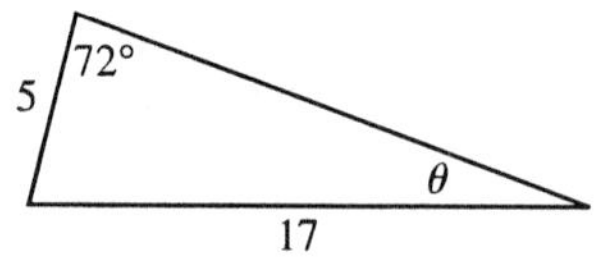

Answer: easy

$$\frac{\sin\theta}{5} = \frac{\sin 72}{17} \quad\Leftrightarrow\quad \sin\theta = 5\frac{\sin 72}{17} \quad\Rightarrow\quad \theta \approx 16.2$$

3. Solve the triangle using the Law of Sines.

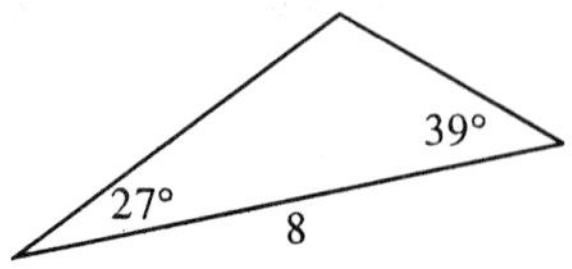

Answer: easy

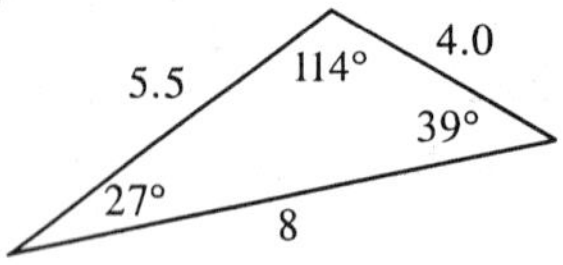

4. Sketch the triangle given by $\angle A = 23°$, $\angle B = 78°$, $c = 88$, and solve it using the Law of Sines.

Answer: easy

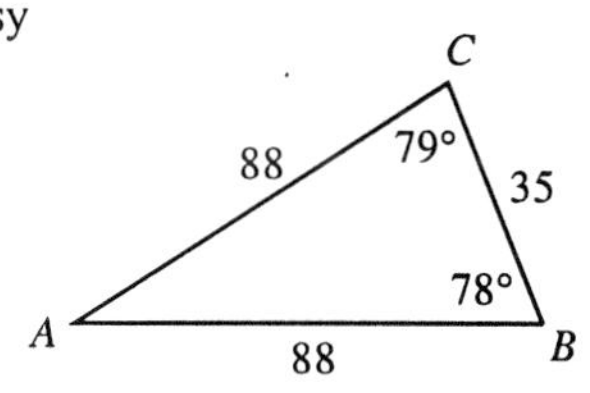

5. Sketch the triangle given by $\angle A = 53°$, $\angle B = 43°$, $c = 50$, and solve it using the Law of Sines.

Answer: easy

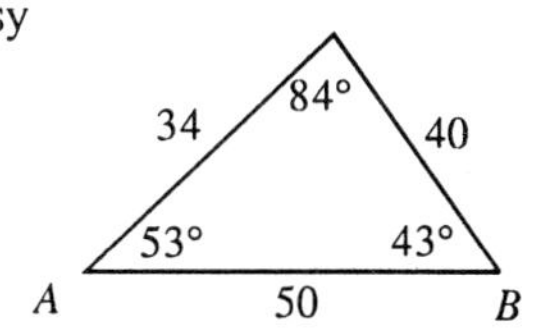

6. Sketch the triangle given by $\angle A = 113°$, $\angle B = 37°$, $c = 25$, and solve it using the Law of Sines.

Answer: easy

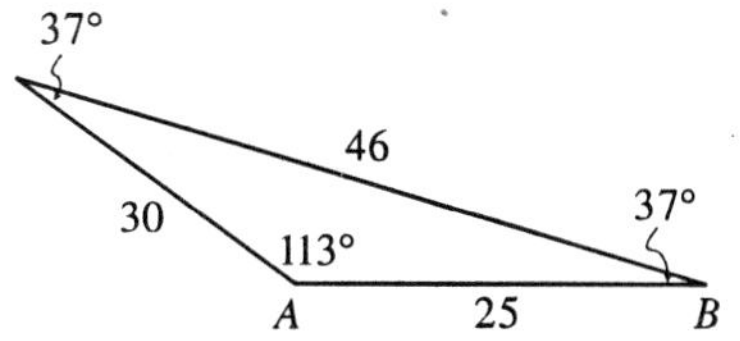

7. Use the Law of Sines to solve for all possible triangles that satisfy $a = 18$, $b = 32$, $\angle A = 83°$.

Answer: medium

$$\frac{\sin B}{32} = \frac{\sin 83}{18} \quad \Leftrightarrow \quad \sin B = 32 \cdot \frac{\sin 83}{18} \approx 1.76 > 1, \text{ so there is no such triangle.}$$

8. Use the Law of Sines to solve for all possible triangles that satisfy $a = 12$, $b = 19$, $\angle A = 65°$.

Answer: medium

$$\frac{\sin B}{19} = \frac{\sin 65}{12} \quad \Leftrightarrow \quad \sin B = 19 \cdot \frac{\sin 65}{12} \approx 1.43 > 1, \text{ so there is no such triangle.}$$

9. Use the Law of Sines to solve for all possible triangles that satisfy $a = 12$, $b = 21$, $\angle B = 57°$.

Answer: hard

$\dfrac{\sin A}{12} = \dfrac{\sin 57}{21} \quad \Leftrightarrow \quad \sin A = 12 \cdot \dfrac{\sin 57}{21} \approx 0.479$. Possible angles with $\sin A = 0.479$ are $\angle A = 28.6°$ and $\angle A = 151.4°$.

Since $\angle B = 57°$, it is not possible that $\angle A = 151.4°$, so $\angle A = 28.6°$. Thus the remaining angle is $\angle C = 180° - 57° - 28.6° = 94.4°$, and the remaining side is $c = \sin 94.4 \dfrac{21}{\sin 57} \approx 25.0$.

10. Use the Law of Sines to solve for all possible triangles that satisfy $a = 9$, $b = 13$, $\angle B = 67°$.

Answer: hard

$\dfrac{\sin A}{9} = \dfrac{\sin 67}{13} \quad \Leftrightarrow \quad \sin A = 9 \cdot \dfrac{\sin 67}{13} \approx 0.637$. Possible angles with $\sin A = 0.637$ are $\angle A = 39.6°$ and $\angle A = 140.4°$.

Since $\angle B = 67°$, it is not possible that $\angle A = 140.4°$, so $\angle A = 39.6°$. Thus the remaining angle is $\angle C = 180° - 67° - 39.6° = 73.4°$, and the remaining side is $c = \sin 73.4 \dfrac{13}{\sin 67} \approx 13.5$.

11. Use the Law of Sines to solve for all possible triangles that satisfy $b = 13$, $c = 17$, $\angle B = 43°$.

Answer: hard

$\dfrac{\sin C}{17} = \dfrac{\sin 43}{13} \quad \Leftrightarrow \quad \sin C = 17 \cdot \dfrac{\sin 43}{13} \approx 0.892$. Possible angles with $\sin C = 0.892$ are $\angle C = 116.9°$ and $\angle C = 63.1°$.

Both of these angles give possible triangles.

(i) $\angle C = 116.9° \quad \Rightarrow \quad \angle A = 180 - 116.9 - 43 = 20.1°$, and $a = \sin 20.1 \dfrac{17}{\sin 116.9} \approx 6.6$.

(ii) $\angle C = 63.1° \quad \Rightarrow \quad \angle A = 180 - 63.1 - 43 = 73.9°$, and $a = \sin 73.9 \dfrac{17}{\sin 116.9} \approx 18.3$.

12. To find the distance across a river a surveyor chooses points A and B, 100 m apart and a reference point C on the other side of the river. Approximate the distance, b, between A and C given that $\angle A = 93°$ and $\angle B = 62°$.

Answer: medium

$\dfrac{b}{\sin 62} = \dfrac{100}{\sin(180 - 93 - 62)} = \dfrac{100}{\sin 25} \quad \Rightarrow \quad b = \sin 62 \dfrac{100}{\sin 25} \approx 208.9$ m.

13. To find the distance across a river a surveyor chooses points A and B, 50 m apart and a reference point C on the other side of the river. Approximate the distance, b, between A and C given that $\angle A = 54°$ and $\angle B = 77°$.

Answer: medium

$\dfrac{b}{\sin 77} = \dfrac{50}{\sin(180 - 54 - 77)} = \dfrac{50}{\sin 49} \quad \Rightarrow \quad b = \sin 77 \dfrac{50}{\sin 49} \approx 64.6$ m.

14. A satellite passes over two tracking stations, A and B, 300 km apart. When the satellite is between the two stations the angles of elevation at the stations are measured as 84.5° and 88.2° respectively. What is the distance, a, between the satellite and station B.

Answer: medium

$\dfrac{a}{\sin 88.2} = \dfrac{300}{\sin(180 - 84.5 - 88.2)} = \dfrac{300}{\sin 7.3} \quad \Rightarrow \quad a = \sin 88.2 \dfrac{300}{\sin 7.3} \approx 2359$ km.

15. A satellite passes over two tracking stations, A and B, 100 km apart. When the satellite is between the two stations the angles of elevation at the stations are measured as 89° and 87° respectively. What is the distance, a, between the satellite and station B.

Answer: medium

$\dfrac{a}{\sin 89} = \dfrac{100}{\sin(180 - 89 - 87)} = \dfrac{100}{\sin 4} \quad \Rightarrow \quad a = \sin 89 \dfrac{100}{\sin 4} \approx 1433$ km.

SECTION 7.5

1. Use the Law of Cosines to find x.

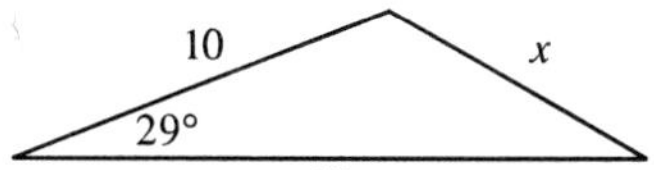

Answer: easy
$x^2 = 10^2 + 17^2 - 2 \cdot 10 \cdot 17 \cdot \cos 29 \quad \Rightarrow$
$x \approx \sqrt{389 - 297} = \sqrt{92} \approx 9.59$

2. Use the Law of Cosines to find x.

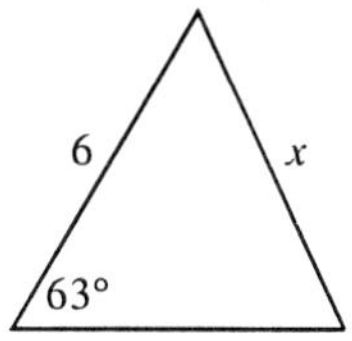

Answer: easy
$x^2 = 5^2 + 6^2 - 2 \cdot 5 \cdot 6 \cdot \cos 63 \quad \Rightarrow$
$x \approx \sqrt{61 - 27} = \sqrt{34} \approx 5.83$

3. Use the Law of Cosines to find x.

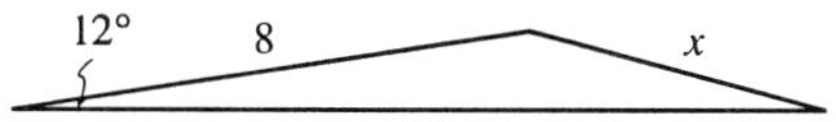

Answer: easy
$x^2 = 8^2 + 12^2 - 2 \cdot 8 \cdot 12 \cdot \cos 12 \quad \Rightarrow$
$x \approx \sqrt{208 - 188} = \sqrt{20} \approx 4.47$

4. Use the Law of Cosines to find θ.

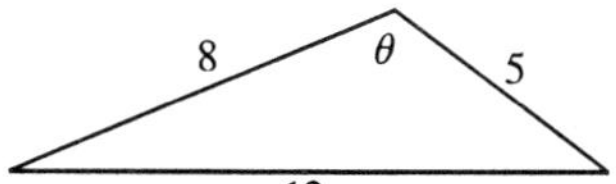

Answer: easy
$$\cos\theta = \frac{5^2 + 8^2 - 12^2}{2 \cdot 5 \cdot 8} = -0.6875 \quad \Rightarrow \quad \theta \approx 133°.$$

5. Use the Law of Cosines to find θ.

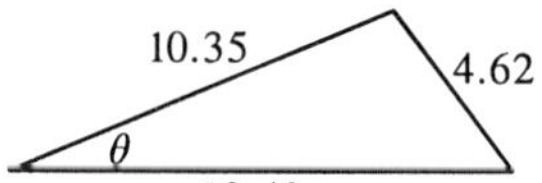

Answer: easy
$$\cos\theta = \frac{10.35^2 + 13.42^2 - 4.62^2}{2 \cdot 10.35 \cdot 13.42} \approx 0.9571 \quad \Rightarrow \quad \theta \approx 16.8°.$$

6. Use the Law of Cosines to find θ.

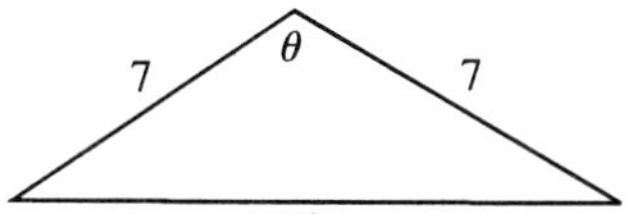

Answer: easy
$$\cos\theta = \frac{7^2 + 7^2 - 12^2}{2 \cdot 7 \cdot 7} = -0.4694 \quad \Rightarrow \quad \theta \approx 118°.$$

7. Solve the triangle $\angle A = 23°$, $b = 78$, $c = 88$.

Answer: easy

$a^2 = 78^2 + 88^2 - 2 \cdot 78 \cdot 88 \cos 23 = 13828 - 13728 \cos 23 \quad \Rightarrow \quad a \approx 34.5.$

$\cos B = \dfrac{34.5^2 + 88^2 - 78^2}{2 \cdot 34.5 \cdot 88} \approx 0.469 \quad \Rightarrow \quad B \approx 62°.$

$\cos C = \dfrac{34.5^2 + 78^2 - 88^2}{2 \cdot 34.5 \cdot 78} \approx -0.087 \quad \Rightarrow \quad C \approx 95°.$

8. Solve the triangle $\angle A = 45°$, $b = 18$, $c = 24$.

Answer: easy

$a^2 = 18^2 + 24^2 - 2 \cdot 18 \cdot 24 \cos 45 = 900 - 864 \cos 45 \quad \Rightarrow \quad a \approx 17.0.$

$\cos B = \dfrac{17.0^2 + 24^2 - 18^2}{2 \cdot 17.0 \cdot 24} \approx 0.663 \quad \Rightarrow \quad B \approx 48°.\ C = 180 - 45 - 48 = 87°.$

9. Solve the triangle $\angle A = 53°$, $b = 11$, $c = 23$.

Answer: easy

$a^2 = 11^2 + 23^2 - 2 \cdot 11 \cdot 23 \cos 53 \approx 345.5 \quad \Rightarrow \quad a \approx 18.6.$

$\cos B = \dfrac{18.6^2 + 23^2 - 11^2}{2 \cdot 18.6 \cdot 23} \approx 0.881 \quad \Rightarrow \quad B \approx 28°.\ C = 180 - 53 - 28 = 99°.$

10. Solve the triangle $\angle B = 67°$, $a = 3.42$, $c = 2.56$.

Answer: easy

$b^2 = 3.42^2 + 2.56^2 - 2 \cdot 3.42 \cdot 2.56 \cos 67 \approx 11.41 \quad \Rightarrow \quad b \approx 3.38.$

$\cos A = \dfrac{3.38^2 + 2.56^2 - 3.42^2}{2 \cdot 3.38 \cdot 2.56} \approx 0.363 \quad \Rightarrow \quad A \approx 69°.\ C = 180 - 67 - 69 = 44°.$

11. Solve the triangle $\angle C = 34°$, $a = 3$, $b = 2$.

Answer: easy

$c^2 = 3^2 + 2^2 - 2 \cdot 3 \cdot 2 \cos 34 \approx 3.05 \quad \Rightarrow \quad c \approx 1.75.$

$\cos A = \dfrac{1.75^2 + 2^2 - 3^2}{2 \cdot 1.75 \cdot 2} \approx -0.277 \quad \Rightarrow \quad A \approx 106°.\ B = 180 - 34 - 106 = 40°.$

12. Solve the triangle $\angle B = 88°$, $a = 15$, $c = 7$.

Answer: easy

$b^2 = 15^2 + 7^2 - 2 \cdot 15 \cdot 7 \cos 88 \approx 266.67 \quad \Rightarrow \quad b \approx 16.33.$

$\cos A = \dfrac{16.33^2 + 7^2 - 15^2}{2 \cdot 16.33 \cdot 7} \approx 0.397 \quad \Rightarrow \quad A \approx 67°.\ C = 180 - 88 - 67 = 25°.$

13. Solve the triangle $\angle C = 52°$, $a = 1.54$, $b = 1.87$.

Answer: easy

$c^2 = 1.54^2 + 1.87^2 - 2 \cdot 1.54 \cdot 1.87 \cos 52 \approx 2.32 \quad \Rightarrow \quad c \approx 1.52.$

$\cos A = \dfrac{1.52^2 + 1.87^2 - 1.54^2}{2 \cdot 1.52 \cdot 1.87} \approx 0.604 \quad \Rightarrow \quad A \approx 53°.\ B = 180 - 52 - 53 = 75°.$

14. Use either the Law of Sines or the Law of Cosines to find x.

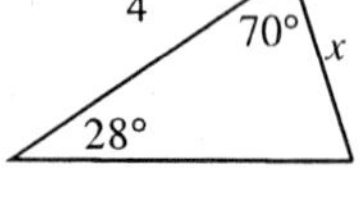

Answer: medium

$\dfrac{\sin 28}{x} = \dfrac{\sin(180 - 70 - 28)}{4} \quad \Leftrightarrow$

$x = 4\dfrac{\sin 28}{\sin 82} \quad \Rightarrow \quad x \approx 1.90$

15. Use either the Law of Sines or the Law of Cosines to find x.

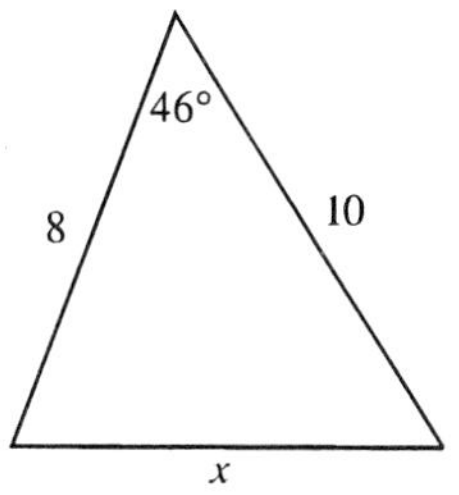

Answer: medium
$x^2 = 8^2 + 10^2 - 2 \cdot 8 \cdot 10 \cdot \cos 46 \quad \Rightarrow$
$x = \sqrt{164 - 169 \cos 46} \approx 6.83$

16. Use either the Law of Sines or the Law of Cosines to find θ.

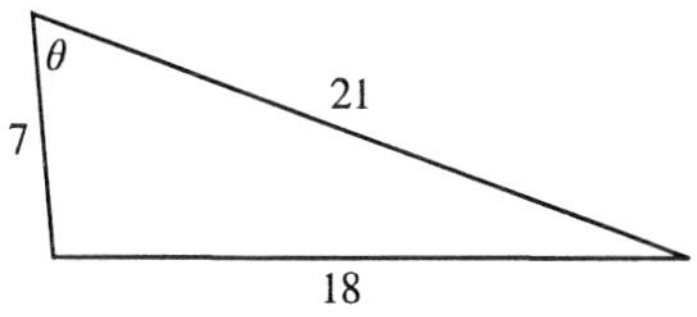

Answer: medium
$\cos \theta = \dfrac{21^2 + 7^2 - 18^2}{2 \cdot 21 \cdot 7} = 0.565 \quad \Rightarrow \quad \theta \approx 56°.$

17. Use either the Law of Sines or the Law of Cosines to find θ.

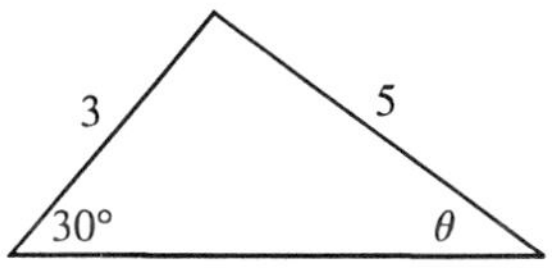

Answer: medium
$\dfrac{\sin \theta}{3} = \dfrac{\sin 30}{5} \quad \Leftrightarrow \quad \sin \theta = 3 \dfrac{\sin 30}{5} \quad \Rightarrow \quad \theta \approx 17.5$

18. A car travels along a straight road, heading west for three hours, and then travels northeast on another road for two hours. If the car has maintained a constant speed of 55 mi/h, how far is it from its starting position?
Answer: hard
The angle between the roads is 45°, and using the Law of Cosines gives, where d is the distance from starting position, $d^2 = (2 \cdot 55)^2 + (3 \cdot 55)^2 - 2(2 \cdot 55)(3 \cdot 55) \cos 45 = 39325 - 36300 \frac{\sqrt{2}}{2} \quad \Rightarrow \quad d \approx 117$. So the car is approximately 117 mi from its starting location.

19. Two tugboats are pulling a barge with cables 50 m and 62 m long. If the angle between the cables is 33° approximately how far apart are the barges?
Answer: medium
Using the Law of Cosines gives $d^2 = 50^2 + 62^2 - 2 \cdot 50 \cdot 62 \cdot \cos 33 = 6344 - 6200 \cos 33 \quad \Rightarrow \quad d \approx 33.8$ m, where d is the distance between the tugboats.

20. A girl is flying two kites at the same time. If the strings are 200 ft and 230 ft long and the kites are 100 ft apart, what angle do the strings make in her hands?
Answer: medium
Let A be the angle, then using the Law of Cosines gives $\cos A = \dfrac{200^2 + 230^2 - 100^2}{2 \cdot 200 \cdot 230} = \dfrac{829}{920} \quad \Rightarrow \quad A \approx 25.7°.$

CHAPTER 7 REVIEW

1. Find the radian measure of the angle with the degree measure 135°

   Answer: easy

   $135° = 135\left(\frac{\pi}{180}\right)$ rad $= \frac{9\pi}{12}$ rad $= \frac{3\pi}{4}$ rad.

2. Find the degree measure of the angle with the radian measure $-\frac{3\pi}{10}$ rad

   Answer: easy

   $-\frac{3\pi}{10}$ rad $= -\frac{3\pi}{10}\left(\frac{180}{\pi}\right)° = -3 \cdot 18° = -54°$.

3. Find a negative angle and a different positive angle that are coterminal with the angle given by $\frac{3\pi}{4}$.

   Answer: easy

   $\frac{3\pi}{4} - 2\pi = -\frac{\pi}{4}$ and $\frac{3\pi}{4} + 2\pi = \frac{7\pi}{4}$ are both coterminal.

4. Determine whether the angles given by $-77°$ and $283°$ are coterminal.

   Answer: easy

   $283° - (-77°) = 360°$, so they are coterminal.

5. Find an angle between 0 and 2π that is coterminal with the angle given by $-\frac{35\pi}{6}$.

   Answer: easy

   $-\frac{35\pi}{6} + 6\pi = \frac{\pi}{6}$.

6. Find the length of an arc that subtends a central angle of $\frac{2\pi}{7}$ in a circle of radius 3 m.

   Answer: easy

   $\frac{2\pi}{7} \cdot 3 = \frac{6\pi}{7}$ m.

7. Find the values of the six trigonometric ratios of the angle θ in the triangle.

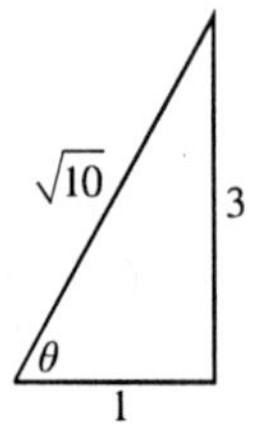

Answer: medium

$\sin\theta = \frac{3}{\sqrt{10}}$; $\cos\theta = \frac{1}{\sqrt{10}}$; $\tan\theta = 3$;

$\csc\theta = \frac{\sqrt{10}}{3}$; $\sec\theta = \sqrt{10}$; $\cot\theta = \frac{1}{3}$

8. Find the side labeled x.

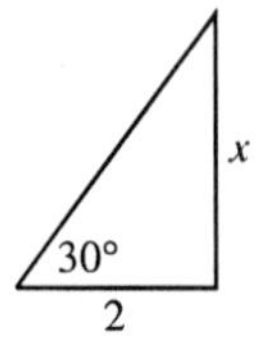

Answer: easy

$\tan 30° = \frac{x}{2} \quad \Leftrightarrow \quad x = 2 \cdot \frac{\sqrt{3}}{3} = \frac{2\sqrt{3}}{3}$

9. Express x and y in terms of trigonometric ratios of θ.

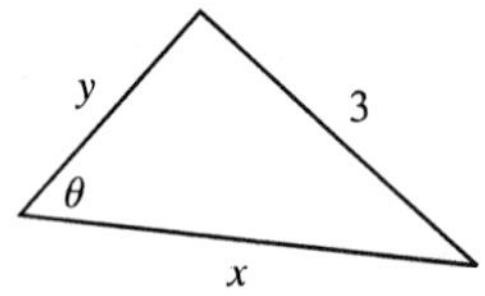

Answer: easy

$\csc\theta = \frac{x}{3} \quad \Rightarrow \quad x = 3\csc\theta$

$\cot\theta = \frac{y}{3} \quad \Rightarrow \quad y = 3\cot\theta$

10. Evaluate $\sin \frac{\pi}{3} \cos \frac{\pi}{4} + \cos \frac{\pi}{3} \sin \frac{\pi}{4}$.

Answer: medium

$$\sin \frac{\pi}{3} \cos \frac{\pi}{4} + \cos \frac{\pi}{3} \sin \frac{\pi}{4} = \frac{\sqrt{3}}{2} \cdot \frac{\sqrt{2}}{2} + \frac{1}{2} \cdot \frac{\sqrt{2}}{2} = \frac{\sqrt{6} + \sqrt{2}}{4}$$

11. Solve the given right triangle.

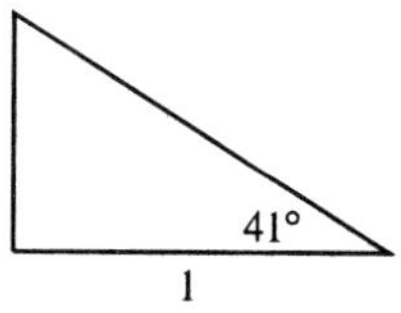

Answer: medium

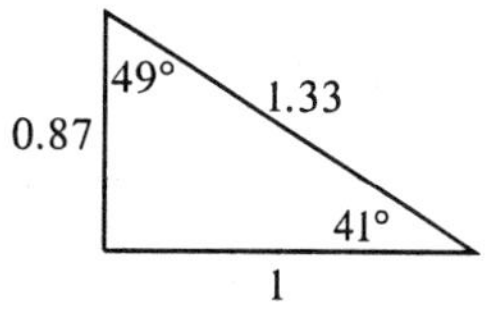

12. A 12 ft tree casts a shadow that is 7 ft long, what is the angle of elevation of the sun?

Answer: easy

$\tan \theta = \frac{12}{7}$, where θ is the angle of elevation of the sun, so $\theta = \tan^{-1} \frac{12}{7} \approx 59.7°$.

13. Find the reference angle for $\theta = 233°$.

Answer: easy

$\bar{\theta} = \theta - 180° = 233° - 180° = 53°$

14. Find the reference angle for $\theta = -\frac{13\pi}{4}$.

Answer: easy

$\bar{\theta} = -\theta - 3\pi = \frac{13\pi}{4} - 3\pi = \frac{\pi}{4}$.

15. Find the exact value of $\sin -135°$.

Answer: easy

$\sin -135° = -\frac{\sqrt{2}}{2}$

16. Find the exact value of $\tan 315°$.

Answer: easy

$\tan 315° = -1$

17. Find the quadrant in which θ lies given that $\tan \theta < 0$ and $\sec \theta > 0$.

Answer: medium

θ lies in quadrant II.

18. Find the area of a triangle with sides of length 2 and 5 and included angle 44°.

Answer: medium

$A = \frac{1}{2} \cdot 2 \cdot 5 \cdot \sin 44° \approx 3.47$.

19. Use the Law of Sines to find x.

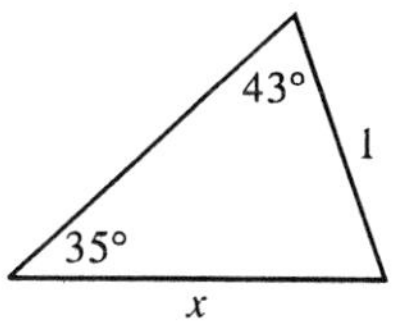

Answer: easy

$$\frac{\sin 43}{x} = \frac{\sin 35}{1} \quad \Leftrightarrow \quad x = \frac{\sin 43}{\sin 35} \quad \Rightarrow \quad x \approx 1.19$$

20. Solve the triangle using the Law of Sines.

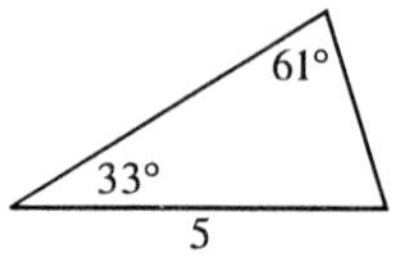

Answer: easy

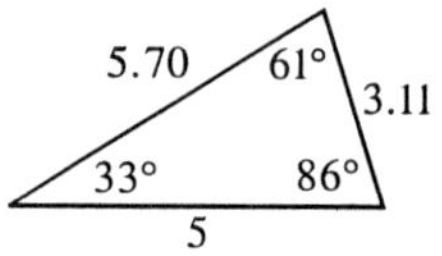

21. Sketch the triangle given by $\angle A = 78°$, $\angle B = 47°$, $c = 4$, and solve it using the Law of Sines.

Answer: easy

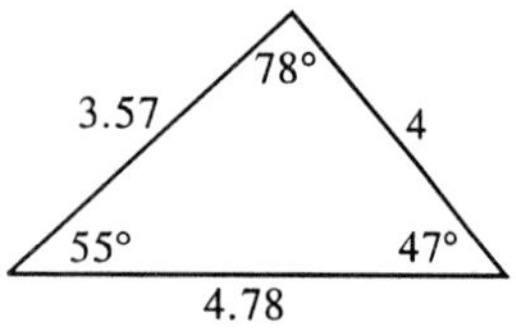

22. Use the Law of Sines to solve for all possible triangles that satisfy $a = 3$, $b = 7$, $\angle A = 79°$.

Answer: medium

$\dfrac{\sin B}{7} = \dfrac{\sin 79}{3} \quad \Leftrightarrow \quad \sin B = 7 \cdot \dfrac{\sin 79}{3} \approx 2.29 > 1$, so there is no such triangle.

23. Use the Law of Sines to solve for all possible triangles that satisfy $a = 8$, $b = 12$, $\angle B = 65°$.

Answer: medium

$\dfrac{\sin A}{8} = \dfrac{\sin 65}{12} \quad \Leftrightarrow \quad \sin A = 8 \cdot \dfrac{\sin 65}{12} \approx 0.604$. Possible angles with $\sin A = 0.637$ are $\angle A = 37.2°$ and $\angle A = 142.8°$. Since $\angle B = 65°$, it is not possible that $\angle A = 142.8°$, so $\angle A = 37.2°$. Thus the remaining angle is $\angle C = 180° - 65° - 37.2° = 77.8°$, and the remaining side is $c = \sin 77.8 \dfrac{12}{\sin 65} \approx 12.9$.

24. A satellite passes over two tracking stations, A and B, 150 km apart. When the satellite is between the two stations the angles of elevation at the stations are measured as 85° and 83° respectively. What is the distance, a, between the satellite and station B.

Answer: medium

$\dfrac{a}{\sin 85} = \dfrac{150}{\sin(180 - 85 - 83)} = \dfrac{150}{\sin 12} \quad \Rightarrow \quad a = \sin 85 \dfrac{150}{\sin 12} \approx 719\,\text{km}.$

25. Use the Law of Cosines to find x.

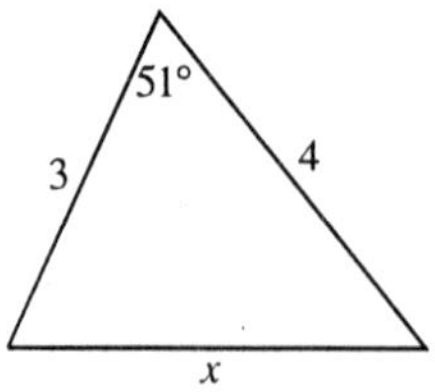

Answer: easy

$x^2 = 4^2 + 3^2 - 2 \cdot 3 \cdot 4 \cdot \cos 51 \quad \Rightarrow$

$x = \sqrt{25 - 24 \cos 51} \approx 3.15$

26. Use the Law of Cosines to find θ.

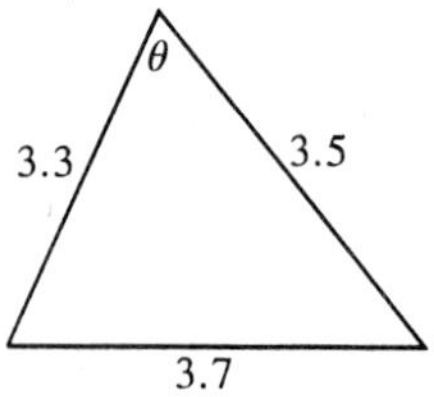

Answer: easy

$\cos\theta = \dfrac{3.5^2 + 3.3^2 - 3.7^2}{2 \cdot 3.5 \cdot 3.3} \approx 0.4091 \quad \Rightarrow \quad \theta \approx 65.9°.$

27. Solve the triangle $\angle A = 47°$, $b = 12$, $c = 11$.

Answer: medium

$a^2 = 12^2 + 11^2 - 2 \cdot 12 \cdot 11 \cos 47 = 265 - 132 \cos 47 \quad \Rightarrow \quad a \approx 13.2.$

$\cos B = \dfrac{13.2^2 + 11^2 - 12^2}{2 \cdot 13.2 \cdot 11} \approx 0.521 \quad \Rightarrow \quad B \approx 59°$, and $C = 180 - 47 - 59 = 74°$.

28. Solve the triangle $\angle B = 34°$, $a = 1.23$, $c = 2.11$.

Answer: medium

$b^2 = 1.23^2 + 2.11^2 - 2 \cdot 1.23 \cdot 2.11 \cos 34 \approx 1.66 \quad \Rightarrow \quad b \approx 1.29.$

$\cos A = \dfrac{1.29^2 + 2.11^2 - 1.23^2}{2 \cdot 1.29 \cdot 2.11} \approx 0.846 \quad \Rightarrow \quad A \approx 32°.\ C = 180 - 32 - 34 = 114°.$

29. Use either the Law of Sines or the Law of Cosines to find x.

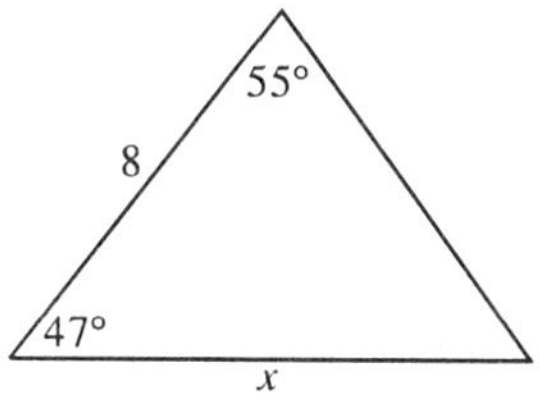

Answer: medium

$\dfrac{\sin 55}{x} = \dfrac{\sin(180 - 55 - 47)}{8} \quad \Leftrightarrow$

$x = 8\dfrac{\sin 55}{\sin 98} \quad \Rightarrow \quad x \approx 6.62$

30. Use either the Law of Sines or the Law of Cosines to find θ.

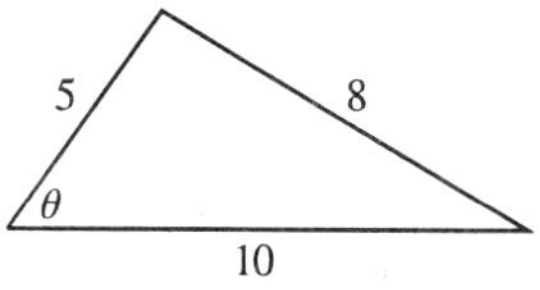

Answer: medium

$\cos\theta = \dfrac{10^2 + 5^2 - 8^2}{2 \cdot 10 \cdot 5} = 0.61 \quad \Rightarrow \quad \theta \approx 52°.$

CHAPTER EIGHT

SECTION 8.1

1. Show that $\left(-\frac{1}{2}, \frac{\sqrt{3}}{2}\right)$ is on the unit circle.

   Answer: easy

   $(-\frac{1}{2})^2 + \left(\frac{\sqrt{3}}{2}\right)^2 = \frac{1}{4} + \frac{3}{4} = 1.$

2. Show that $\left(\frac{1}{\sqrt{5}}, -\frac{2}{\sqrt{5}}\right)$ is on the unit circle.

   Answer: easy

   $\left(\frac{1}{\sqrt{5}}\right)^2 + \left(-\frac{2}{\sqrt{5}}\right)^2 = \frac{1}{5} + \frac{4}{5} = 1.$

3. Show that $\left(-\frac{1}{8}, \frac{3\sqrt{7}}{8}\right)$ is on the unit circle.

   Answer: easy

   $(-\frac{1}{8})^2 + \left(\frac{3\sqrt{7}}{8}\right)^2 = \frac{1}{64} + \frac{63}{64} = 1.$

4. Find $P(x, y)$ on the unit circle, given that the x-coordinate of P is $-\frac{1}{2}$ and P is in quadrant III.

   Answer: medium

   P is in quadrant III, so y is negative. $y = -\sqrt{1 - (-\frac{1}{2})^2} = -\sqrt{\frac{3}{4}} = -\frac{\sqrt{3}}{2}$. So $P = \left(-\frac{1}{2}, -\frac{\sqrt{3}}{2}\right)$.

5. Find $P(x, y)$ on the unit circle, given that the x-coordinate of P is $\frac{1}{3}$ and P is in quadrant IV.

   Answer: medium

   P is in quadrant IV, so y is negative. $y = -\sqrt{1 - (\frac{1}{3})^2} = -\sqrt{\frac{8}{9}} = -\frac{2\sqrt{2}}{3}$. So $P = \left(\frac{1}{3}, -\frac{2\sqrt{2}}{3}\right)$.

6. Find $P(x, y)$ on the unit circle, given that the x-coordinate of P is $\frac{1}{8}$ and P is in quadrant I.

   Answer: medium

   P is in quadrant I, so y is positive. $y = \sqrt{1 - (\frac{1}{8})^2} = \sqrt{\frac{63}{64}} = \frac{3\sqrt{7}}{8}$. So $P = \left(\frac{1}{8}, \frac{3\sqrt{7}}{8}\right)$.

7. On the unit circle, determine which quadrant contains the terminal point of the arc whose length is $\frac{\pi}{3}$.

   Answer: medium

   First quadrant

8. On the unit circle, determine which quadrant contains the terminal point of the arc whose length is -1.56.

   Answer: medium

   Fourth quadrant

9. Find the terminal point $P(x, y)$ on the unit circle determined by the value $t = \frac{3\pi}{2}$.

   Answer: easy

   $(0, -1)$

10. Find the terminal point $P(x, y)$ on the unit circle determined by the value $t = -\frac{3\pi}{4}$.

    Answer: $\left(-\frac{\sqrt{2}}{2}, -\frac{\sqrt{2}}{2}\right)$

11. Find the terminal point $P(x, y)$ on the unit circle determined by the value $t = -\frac{7\pi}{6}$

    Answer: easy

    $\left(-\frac{\sqrt{3}}{2}, \frac{1}{2}\right)$

12. Find the terminal point $P(x, y)$ on the unit circle determined by the value $t = \frac{17\pi}{4}$

Answer: easy

$\left(\frac{\sqrt{2}}{2}, \frac{\sqrt{2}}{2}\right)$

13. Find the terminal point $P(x, y)$ on the unit circle determined by the value $t = -\frac{2\pi}{3}$

Answer: easy

$\left(-\frac{1}{2}, -\frac{\sqrt{3}}{2}\right)$

14. Suppose that the terminal point determined by t is the point $\left(\frac{5}{13}, \frac{12}{13}\right)$ on the unit circle. Find the terminal point determined by (a) $\pi + t$, (b) $-t$, (c) $2\pi - t$, (d) $t - \pi$.

Answer: medium

(a) $\left(-\frac{5}{13}, -\frac{12}{13}\right)$ (b) $\left(\frac{5}{13}, -\frac{12}{13}\right)$ (c) $\left(\frac{5}{13}, -\frac{12}{13}\right)$ (d) $\left(-\frac{5}{13}, -\frac{12}{13}\right)$

15. Suppose that the terminal point determined by t is the point $\left(-\frac{1}{\sqrt{5}}, \frac{2}{\sqrt{5}}\right)$ on the unit circle. Find the terminal point determined by (a) $-\pi - t$, (b) $\pi + t$, (c) $2\pi + t$, (d) $t - 7\pi$.

Answer: medium

(a) $\left(\frac{1}{\sqrt{5}}, \frac{2}{\sqrt{5}}\right)$ (b) $\left(\frac{1}{\sqrt{5}}, -\frac{2}{\sqrt{5}}\right)$ (c) $\left(-\frac{1}{\sqrt{5}}, \frac{2}{\sqrt{5}}\right)$ (d) $\left(\frac{1}{\sqrt{5}}, -\frac{2}{\sqrt{5}}\right)$

16. Find the reference number for $t = \frac{\pi}{2}$

Answer: easy

$\bar{t} = t = \frac{\pi}{2}$

17. Find the reference number for $t = \frac{7\pi}{8}$

Answer: easy

$\bar{t} = \pi - t = \frac{\pi}{8}$

18. Find the reference number for $t = -\frac{6\pi}{5}$

Answer: easy

$\bar{t} = -\pi - t = \frac{6\pi}{5} - \pi = \frac{\pi}{5}$

19. Find the reference number for $t = 3.53$

Answer: easy

$\bar{t} = t - \pi = 3.53 - \pi \approx 0.39$

20. Find the reference number and the terminal point $P(x, y)$ determined by $t = \frac{17\pi}{3}$.

Answer: medium

$\bar{t} = 6\pi - \frac{17\pi}{3} = \frac{\pi}{3}$ and $P(x, y) = \left(\frac{1}{2}, -\frac{\sqrt{3}}{2}\right)$.

21. Find the reference number and the terminal point $P(x, y)$ determined by $t = \frac{11\pi}{6}$.

Answer: medium

$\bar{t} = 2\pi - \frac{11\pi}{6} = \frac{\pi}{6}$ and $P(x, y) = \left(\frac{\sqrt{3}}{2}, -\frac{1}{2}\right)$.

22. Find the reference number and the terminal point $P(x, y)$ determined by $t = -\frac{7\pi}{4}$.

Answer: medium

$\bar{t} = 2\pi + t = 2\pi - \frac{7\pi}{4} = \frac{\pi}{4}$ and $P(x, y) = \left(\frac{\sqrt{2}}{2}, \frac{\sqrt{2}}{2}\right)$.

SECTION 8.2

1. Find the exact value of $\sin\left(\frac{2\pi}{3}\right)$ and $\cos\left(\frac{2\pi}{3}\right)$.

   Answer: easy

   $\sin\left(\frac{2\pi}{3}\right) = \frac{\sqrt{3}}{2}$ and $\cos\left(\frac{2\pi}{3}\right) = -\frac{1}{2}$

2. Find the exact value of $\sin\left(\frac{7\pi}{4}\right)$ and $\cos\left(\frac{7\pi}{4}\right)$.

   Answer: easy

   $\sin\left(\frac{7\pi}{4}\right) = -\frac{\sqrt{2}}{2}$ and $\cos\left(\frac{7\pi}{4}\right) = \frac{\sqrt{2}}{2}$

3. Find the exact value of $\sin\left(\frac{9\pi}{2}\right)$ and $\cos\left(-\frac{9\pi}{2}\right)$.

   Answer: easy

   $\sin\left(\frac{9\pi}{2}\right) = 1$ and $\cos\left(-\frac{9\pi}{2}\right) = 0$

4. Find the exact value of $\sin\left(\frac{19\pi}{6}\right)$ and $\cos\left(-\frac{5\pi}{6}\right)$.

   Answer: easy

   $\sin\left(\frac{19\pi}{6}\right) = -\frac{1}{2}$ and $\cos\left(-\frac{5\pi}{6}\right) = -\frac{\sqrt{3}}{2}$

5. Find the exact value of $\sin\left(-\frac{7\pi}{3}\right)$ and $\cos\left(\frac{5\pi}{3}\right)$.

   Answer: easy

   $\sin\left(-\frac{7\pi}{3}\right) = -\frac{\sqrt{3}}{2}$ and $\cos\left(\frac{5\pi}{3}\right) = \frac{1}{2}$

6. Find the exact value of $\tan\left(\frac{5\pi}{6}\right)$ and $\tan\left(\frac{3\pi}{4}\right)$.

   Answer: easy

   $\tan\left(\frac{5\pi}{6}\right) = -\frac{\sqrt{3}}{3}$ and $\tan\left(\frac{3\pi}{4}\right) = -1$.

7. Find the exact value of $\tan\left(-\frac{8\pi}{3}\right)$ and $\cot\left(-\frac{\pi}{3}\right)$.

   Answer: easy

   $\tan\left(-\frac{8\pi}{3}\right) = \sqrt{3}$ and $\cot\left(-\frac{\pi}{3}\right) = -\frac{\sqrt{3}}{3}$

8. Find the exact value of $\sec\left(\frac{5\pi}{6}\right)$ and $\csc\left(\frac{7\pi}{4}\right)$.

   Answer: easy

   $\sec\left(\frac{5\pi}{6}\right) = -\frac{2\sqrt{3}}{3}$ and $\csc\left(\frac{7\pi}{4}\right) = -\sqrt{2}$

9. If the terminal point determined by t is $\left(-\frac{4}{5}, \frac{3}{5}\right)$, find $\sin t$, $\cos t$ and $\tan t$.

   Answer: easy

   $\frac{3}{5}, -\frac{4}{5}, -\frac{3}{4}$

10. If the terminal point determined by t is $\left(\frac{1}{2}, -\frac{\sqrt{3}}{2}\right)$, find $\sin t$, $\cos t$ and $\tan t$.

    Answer: easy

    $-\frac{\sqrt{3}}{2}, \frac{1}{2}, -\sqrt{3}$

11. If the terminal point determined by t is $\left(\frac{1}{\sqrt{5}}, -\frac{2}{\sqrt{5}}\right)$, find $\sin t$, $\cos t$ and $\tan t$.

    Answer: easy

    $-\frac{2}{\sqrt{5}}, \frac{1}{\sqrt{5}}, -2$

12. If the terminal point determined by t is $\left(-\frac{1}{8}, \frac{3\sqrt{7}}{8}\right)$, find $\sin t$, $\cos t$ and $\tan t$.

    Answer: easy

    $\frac{3\sqrt{7}}{8}, -\frac{1}{8}, -3\sqrt{7}$

13. Find the approximate value of $\sin 2.5$ using a calculator.

    Answer: medium

    0.59

14. Find the approximate value of $\cos(-0.9)$ using a calculator.
Answer: medium
0.62

15. Find the sign of $\cos t \sin t$ if the terminal point determined by t is in quadrant II.
Answer: medium
negative

16. Find the sign of $\dfrac{\tan t}{\sec t}$ if the terminal point determined by t is in quadrant IV.
Answer: medium
negative

17. Given that $\sin t > 0$ and $\cos t < 0$, find the quadrant in which the terminal point determined by t lies.
Answer: easy
Quadrant II.

18. Given that $\tan t < 0$ and $\sec t > 0$, find the quadrant in which the terminal point determined by t lies.
Answer: easy
Quadrant IV.

19. Write $\sin t$ in terms of $\cos t$ given that t lies in quadrant IV
Answer: medium
$\sin t = -\sqrt{1 - \cos^2 t}$, since sin is negative in quadrant IV.

20. Write $\tan t$ in terms of $\cos t$ given that t lies in quadrant II
Answer: medium
$\tan t = \dfrac{\sin t}{\cos t} = \dfrac{\sqrt{1 - \cos^2 t}}{\cos t}$, since sin is positive in quadrant II.

21. Write $\tan t$ in terms of $\sec t$ given that t lies in quadrant I
Answer: medium
$\tan t = \sqrt{\sec^2 t - 1}$, since tan is positive in quadrant I.

22. Write $\tan^2 t$ in terms of $\cos t$.
Answer: medium
$\tan^2 t = \dfrac{\sin^2 t}{\cos^2 t} = \dfrac{1 - \cos^2 t}{\cos^2 t}$.

23. Indicate if the function $f(x) = \sin^2 x - \cos x$ is even, odd, or neither.
Answer: easy
even

24. Indicate if the function $f(x) = x + \cos^2 x$ is even, odd, or neither.
Answer: easy
neither

25. Indicate if the function $f(x) = x - x^2 \sin x$ is even, odd, or neither.
Answer: easy
odd

SECTION 8.3

1. Sketch the graph of $y = 2 - \sin x$

Answer: medium

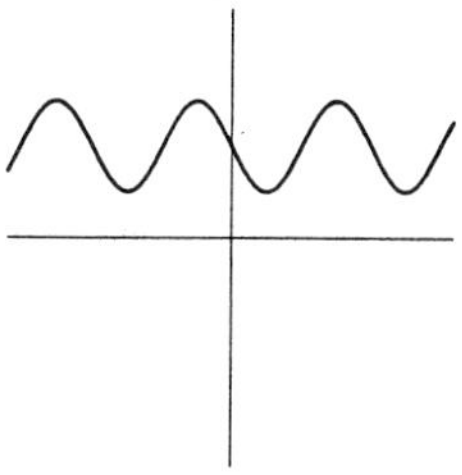

2. Sketch the graph of $y = 3 + \cos x$

Answer: medium

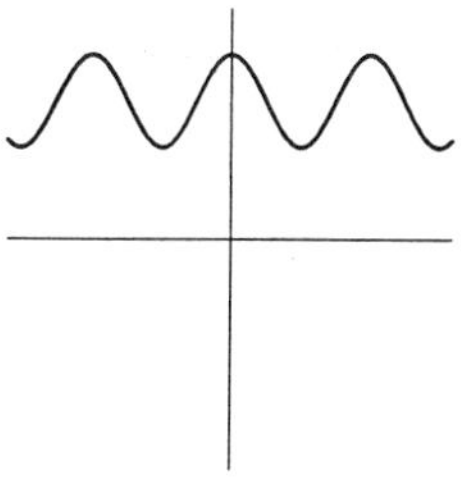

3. Sketch the graph of $y = 2 - \cos x$

Answer: medium

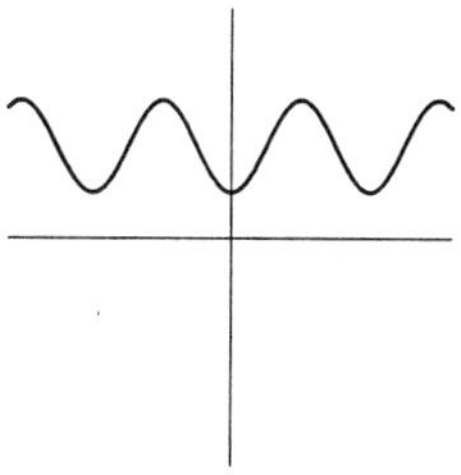

4. Sketch the graph of $y = -2\cos x$

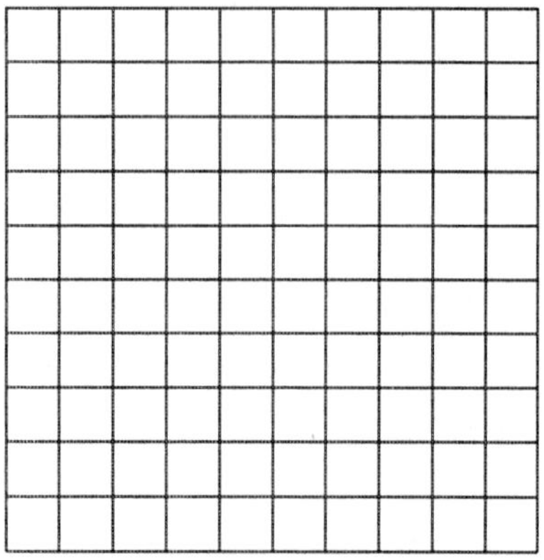

Answer: medium

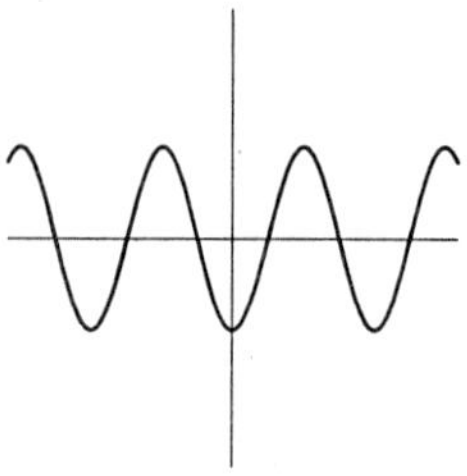

5. Sketch the graph of $y = 3 - 2\sin x$

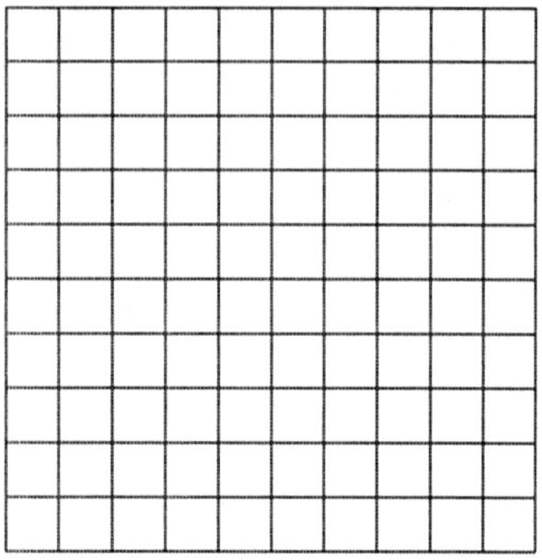

Answer: medium

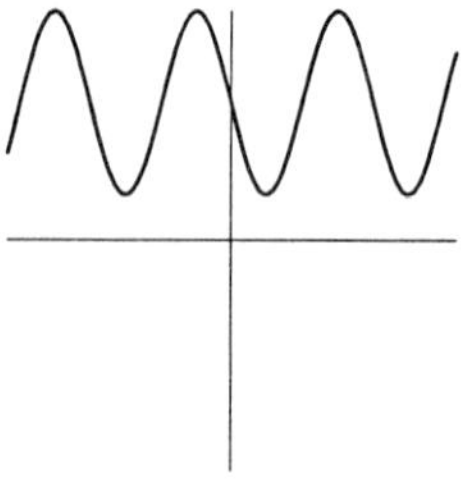

6. Sketch the graph of $y = |2\cos x|$

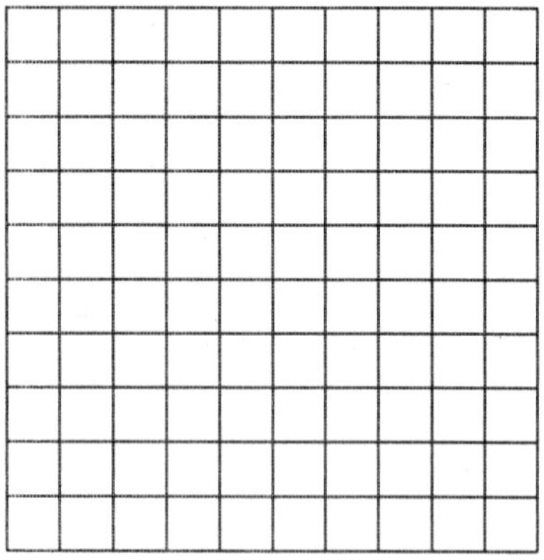

Answer: medium

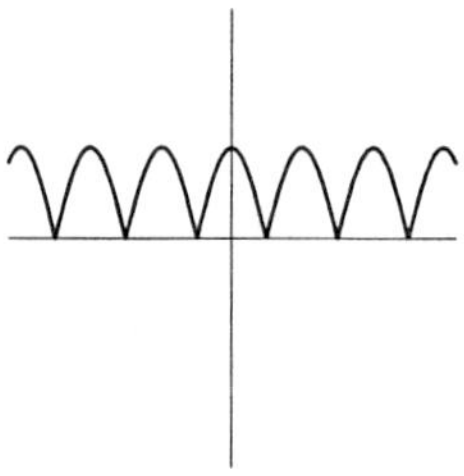

7. Find the amplitude and the period of $y = \sin 2x$ and sketch its graph.

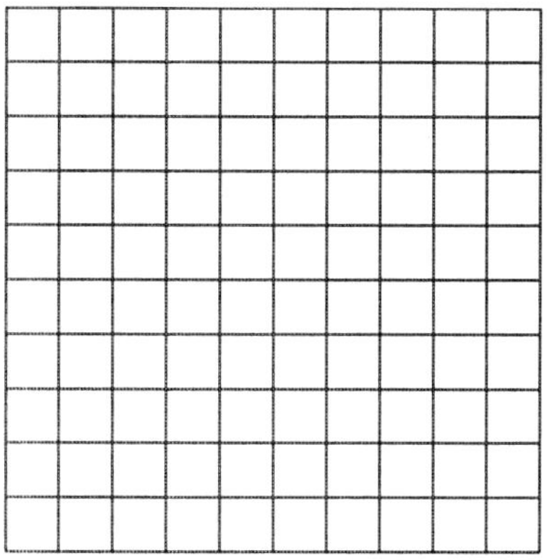

Answer: medium

Amplitude: 1; Period: π

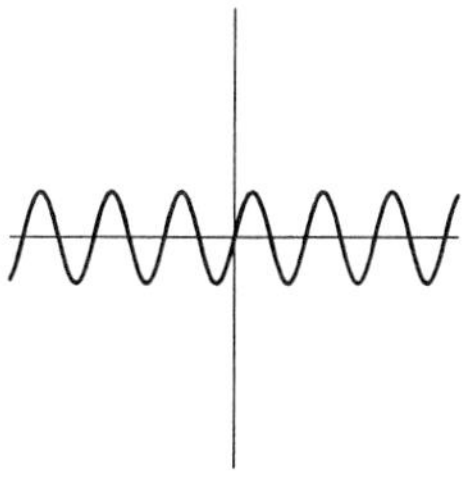

8. Find the amplitude and the period of $y = -3\cos 4x$ and sketch its graph.

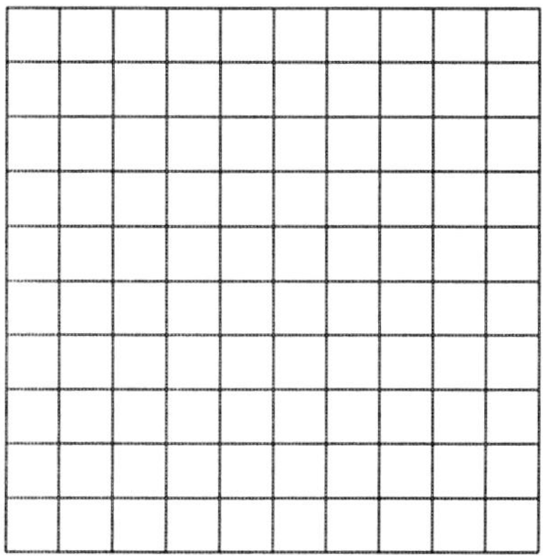

Answer: medium

Amplitude: 3; Period: $\frac{\pi}{2}$

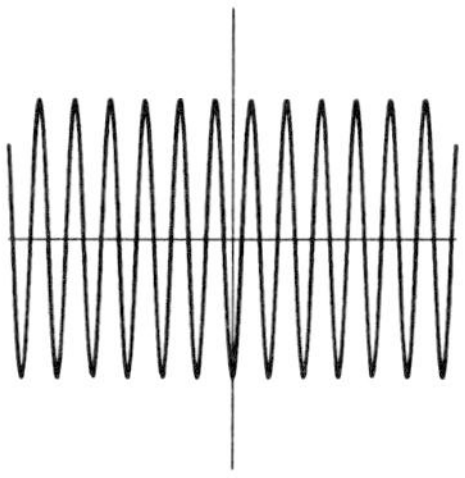

9. Find the amplitude and the period of $y = 2\sin(-x)$ and sketch its graph.

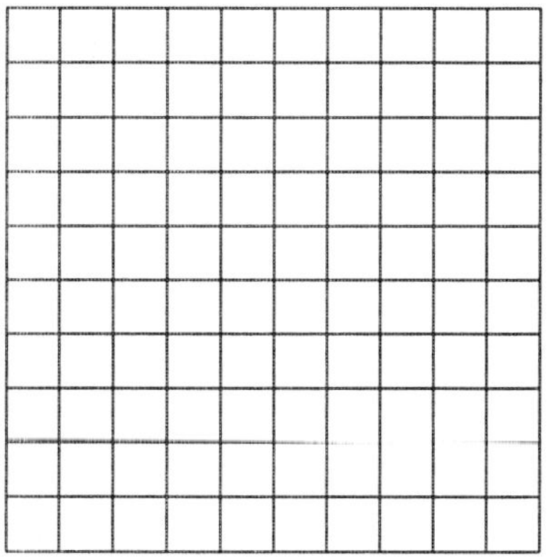

Answer: medium

Amplitude: 2; Period: 2π

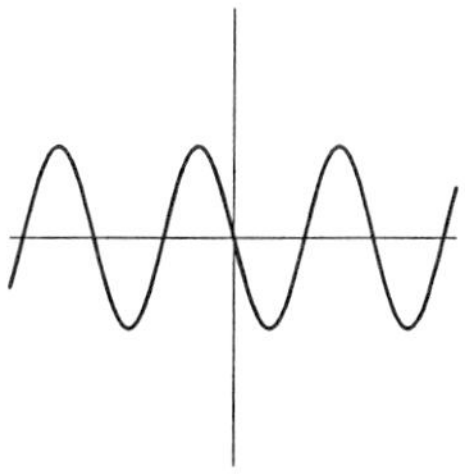

10. Find the amplitude and the period of $y = \cos \frac{\pi}{2}x$ and sketch its graph.

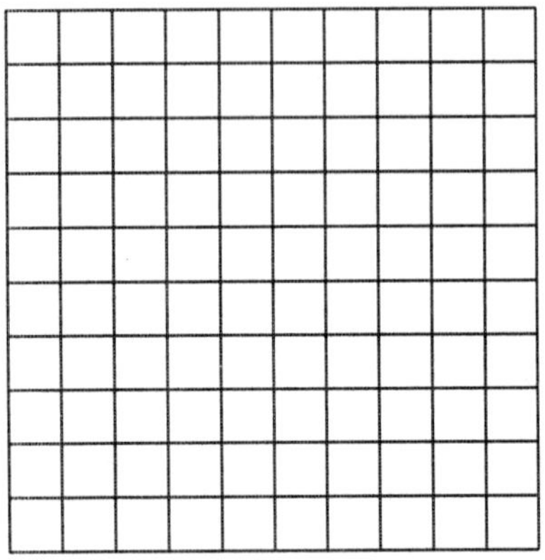

Answer: medium

Amplitude: 1; Period: 4

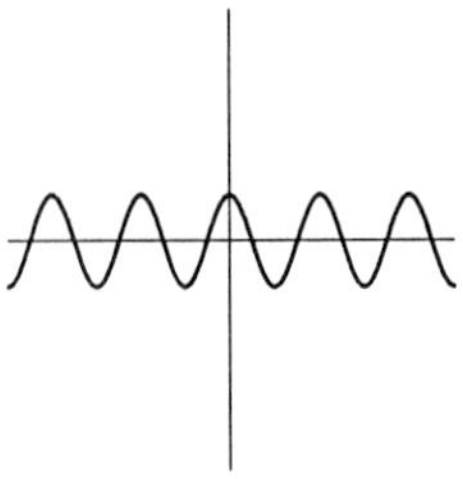

11. Find the amplitude, period and phase shift of $y = \cos(x + 2)$ and sketch its graph.

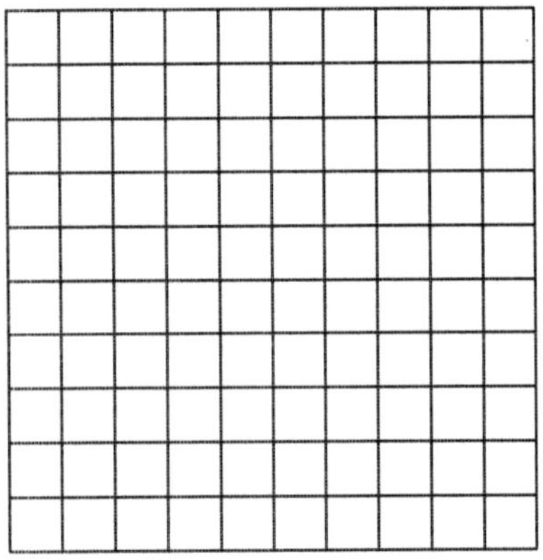

Answer: medium

Amplitude: 1; Period: 2π; Phase shift: -2

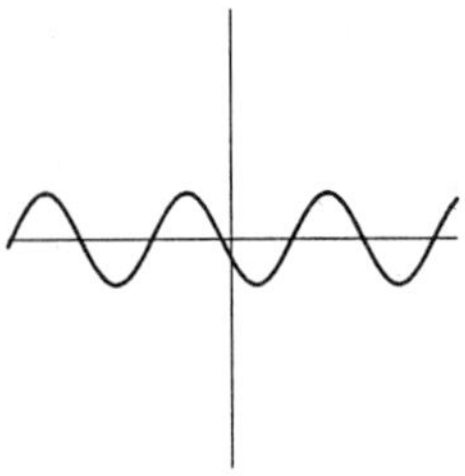

12. Find the amplitude, period and phase shift of $y = 2\sin\left(x - \frac{2\pi}{3}\right)$ and sketch its graph.

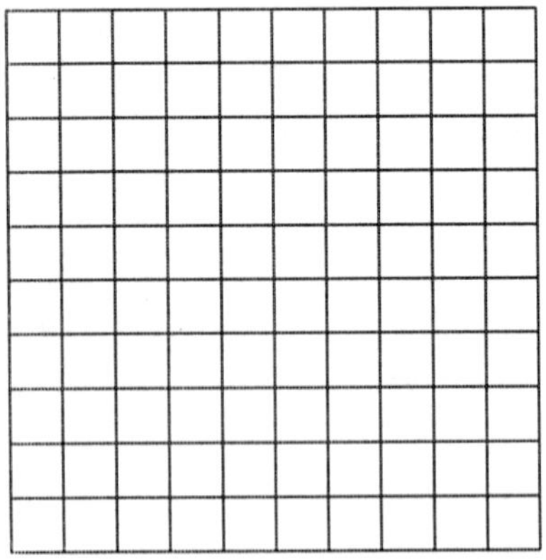

Answer: medium

Amplitude: 2; Period: 2π; Phase shift: $\frac{2\pi}{3}$

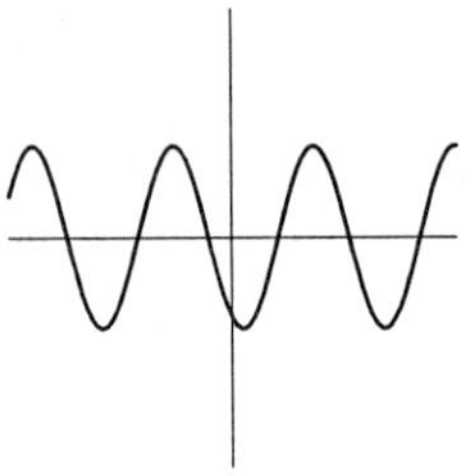

13. Find the amplitude, period and phase shift of $y = 3\cos(2x)$ and sketch its graph.

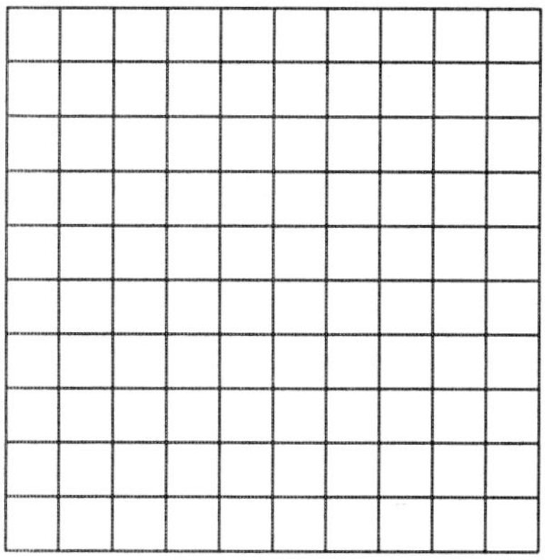

Answer: medium

Amplitude: 3; Period: π; Phase shift: 0

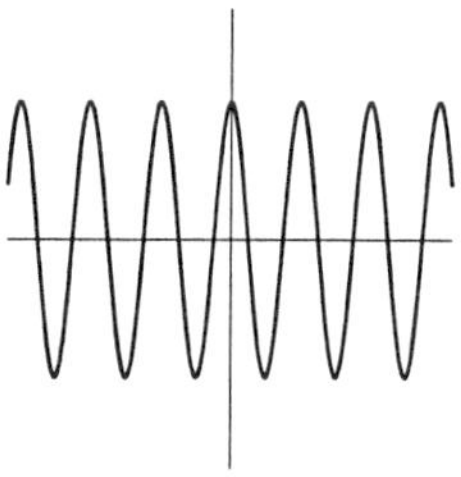

14. Find the amplitude, period and phase shift of $y = \pi\sin(\pi x + \pi)$ and sketch its graph.

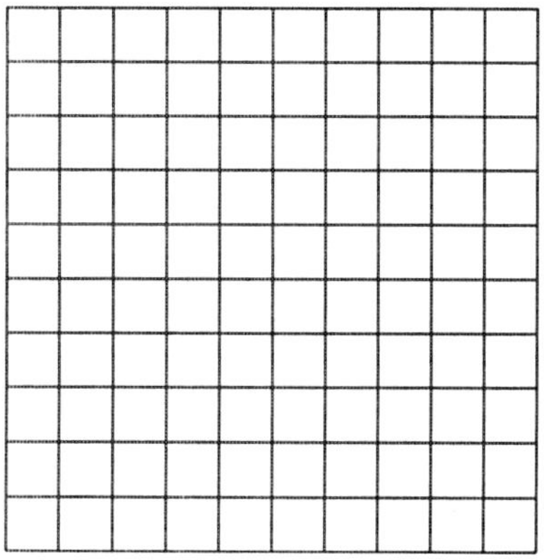

Answer: medium

Amplitude: π; Period: 2; Phase shift: -1

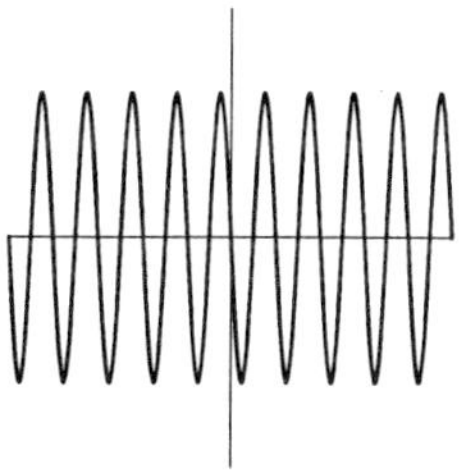

15. Find the amplitude, period and phase shift of $y = 2\cos\left(4x - \frac{\pi}{2}\right)$ and sketch its graph.

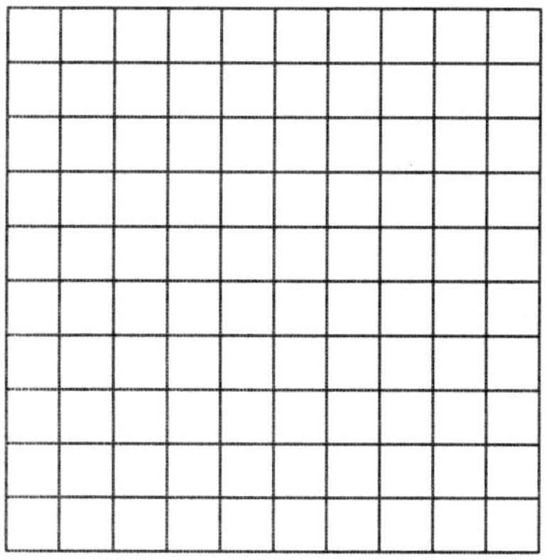

Answer: medium

Amplitude: 2; Period: $\frac{\pi}{2}$; Phase shift: $\frac{\pi}{8}$

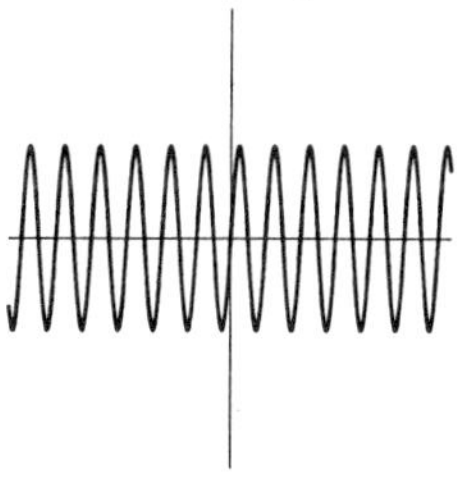

16. Find the amplitude, period and phase shift of $y = -2\sin\left(2x - \frac{3}{2}\right)$ and sketch its graph.

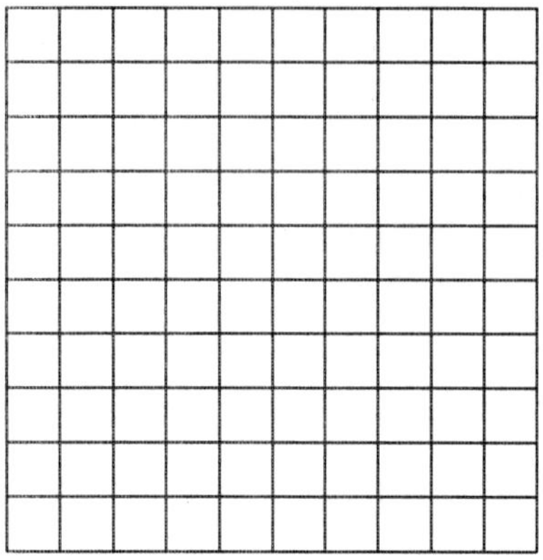

Answer: medium

Amplitude: 2; Period: π; Phase shift: $\frac{3}{4}$

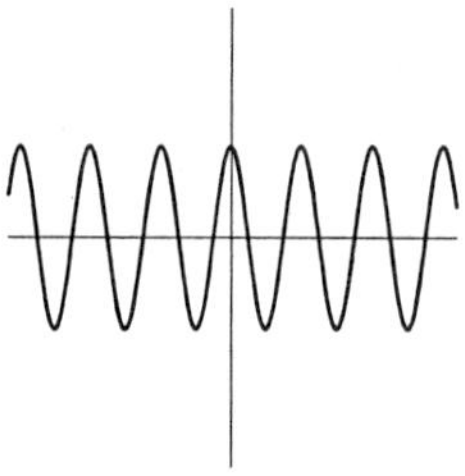

17. Find the amplitude, period and phase shift of the graph given below, and write an equation that represents the curve.

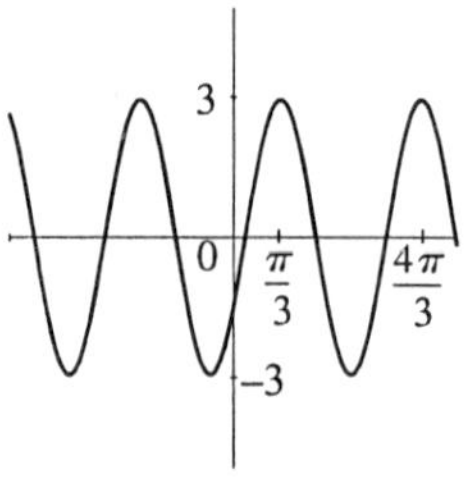

Answer: medium

Amplitude: 3; Period: π; Phase shift: $\frac{\pi}{3}$

$y = 3\cos\left(2x - \frac{2\pi}{3}\right)$

18. Find the amplitude, period and phase shift of the graph given below, and write an equation that represents the curve.

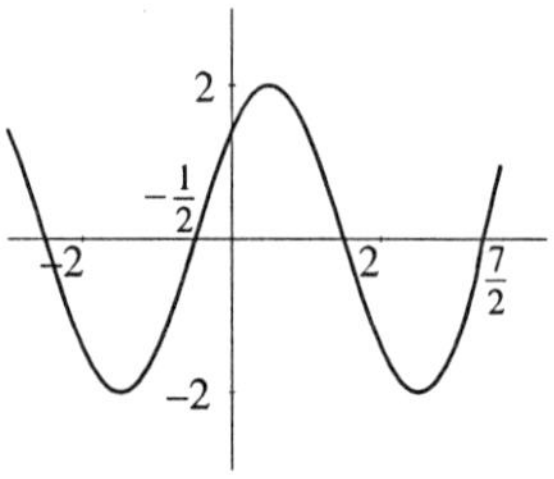

Answer: medium

Amplitude: 2; Period: 4; Phase shift: $-\frac{1}{2}$

$y = 2\sin\left(\frac{\pi}{2}\left(x + \frac{1}{2}\right)\right)$

19. Determine an appropriate viewing rectangle and use it to draw the graph of $f(x) = \sin 35x$

Answer: medium

Amplitude: 1

Period $\frac{2\pi}{35} \approx 0.18$

So use $[-0.2, 0.2]$ by $[-1, 1]$

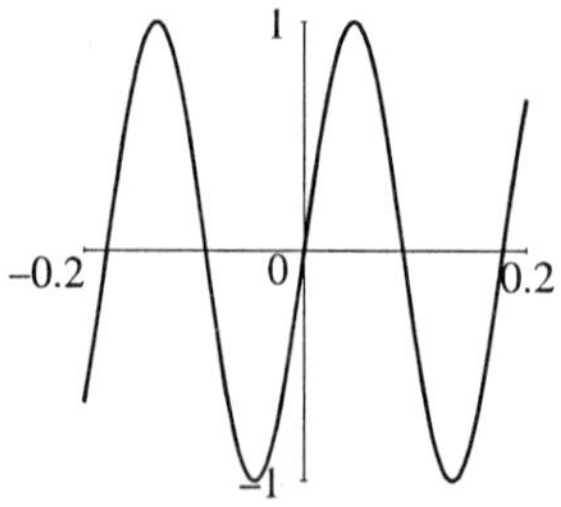

20. Determine an appropriate viewing rectangle and use it to draw the graph of $y = 6\cos 150x$

Answer: medium

Amplitude: 6

Period $\frac{2\pi}{150} \approx 0.04$

So use $[-0.05, 0.05]$ by $[-6, 6]$

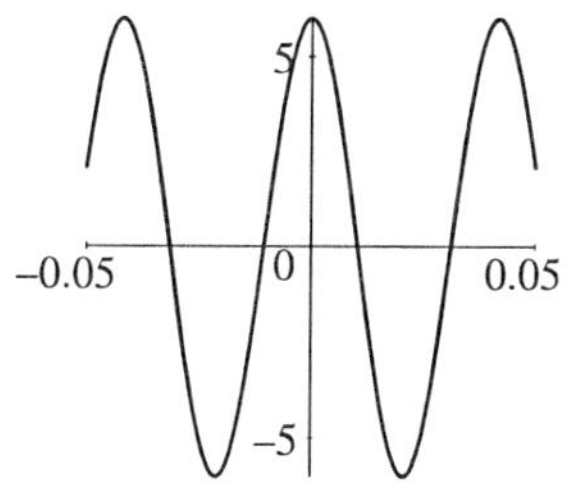

21. Determine an appropriate viewing rectangle and use it to draw the graph of $y = \frac{1}{2}\csc\frac{x}{12}$

Answer: hard

Period $12 \cdot 2\pi \approx 75$

$|\csc x|$, use large y-axis values

So use $[-75, 75]$ by $[-20, 20]$

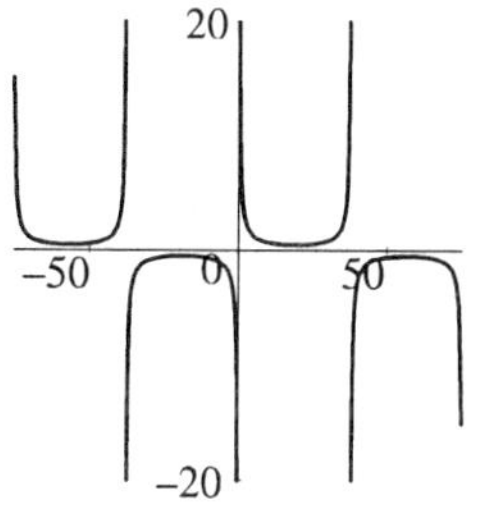

22. Graph $3e^x$, $-3e^x$ and $-3e^x\cos 5x$ on a common screen and explain their relation.

Answer: medium

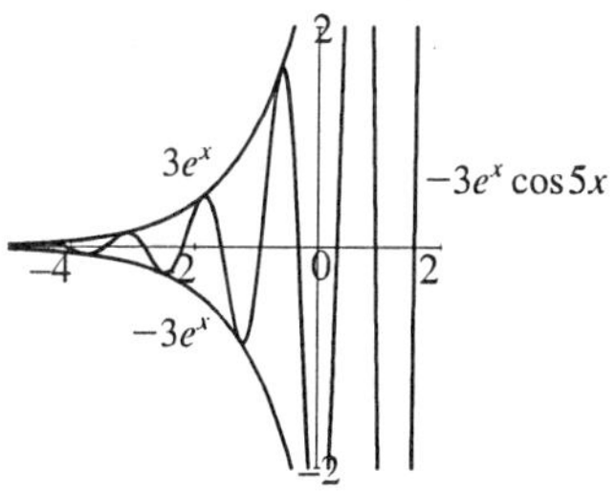

The graph of $-3e^x\cos 5x$ lies between the graphs of $3e^x$ and $-3e^x$.

23. Graph $x^4 - 3x^2$, $-x^4 + 3x^2$ and $(x^4 - 3x^2)\sin 4x$ on a common screen and explain their relation.

Answer: medium

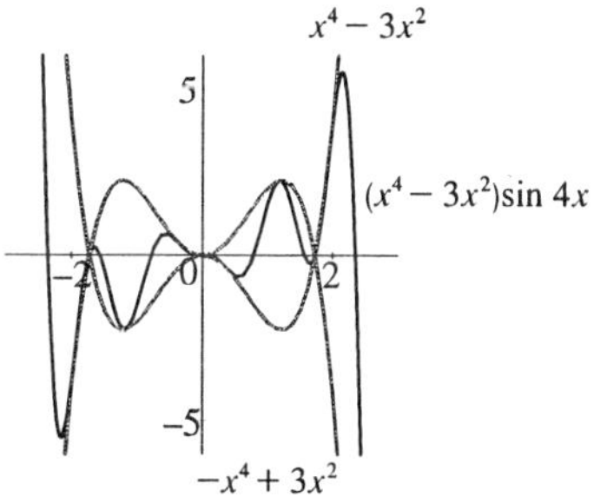

The graph of $(x^4 - 3x^2)\sin 4x$ lies between the graphs of $x^4 - 3x^2$ and $-x^4 + 3x^2$.

24. Graph $y = -|\cos|x||$ and determine whether the function is odd, even or periodic. If it is periodic determine its period.
Answer: medium

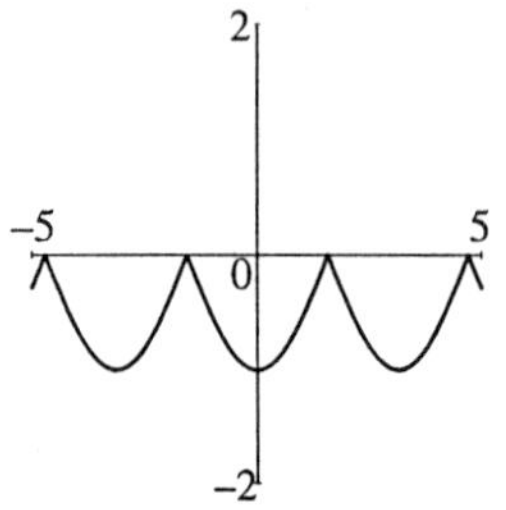

It is an even function, and periodic with period π.

25. Graph $y = \sin e^x$ and determine whether the function is odd, even or periodic. If it is periodic determine its period.
Answer: medium

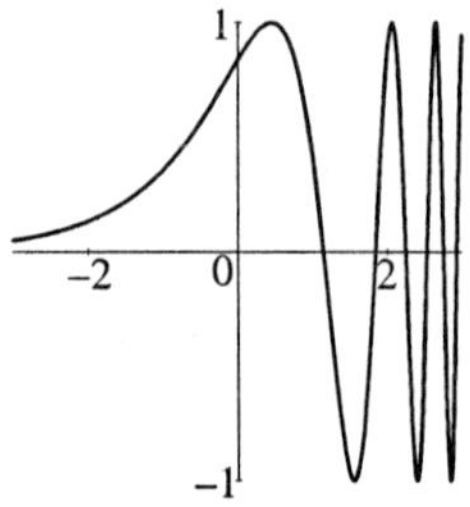

It is neither odd, even, nor periodic.

26. Use a graphing calculator or computer to find the maximum and minimum values of $f(x) = \sin x - \cos 2x$.
Answer: medium

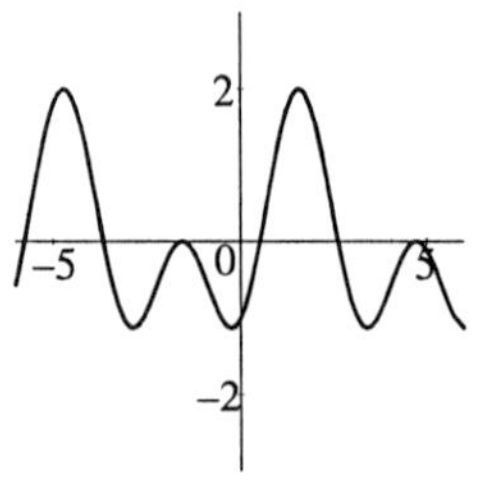

From the graph it is seen that the function is periodic and that the maximum value is 2 and that the minimum value is 1.

27. Use a graphing calculator or computer to find all solutions to $\cos 2x = 0.3$ that lie in the interval $\left[0, \frac{\pi}{2}\right]$, correct to two decimal places.
Answer: medium
$x = 0.63$

SECTION 8.4

1. Find the period of the function $y = 2\tan x$ and sketch its graph.
   Answer: easy
   period $= \pi$

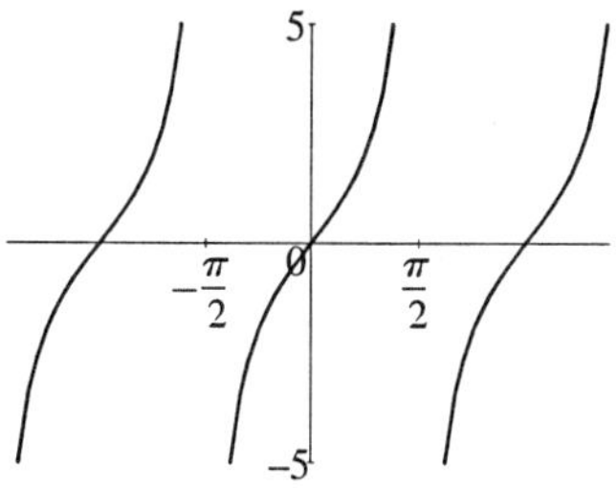

2. Find the period of the function $y = \frac{1}{5}\cot x$ and sketch its graph.
   Answer: easy

   period $= \pi$

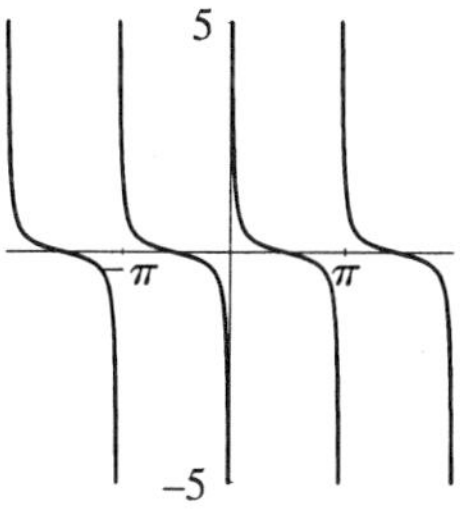

3. Find the period of the function $y = 3\sec x$ and sketch its graph.
   Answer: easy

   period $= 2\pi$

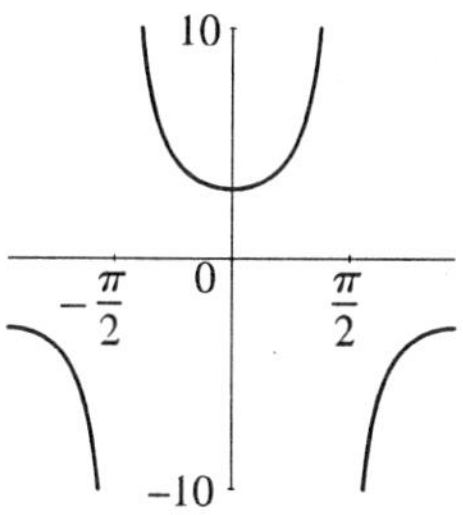

4. Find the period of the function $y = \frac{1}{3}\csc x$ and sketch its graph.
   Answer: easy

   period $= 2\pi$

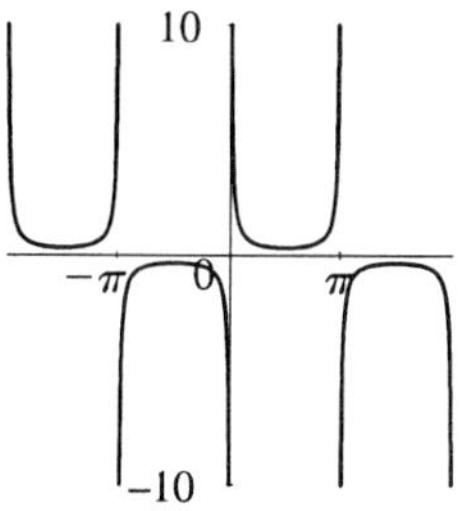

5. Find the period of the function $y = \frac{1}{3}\sec\left(x - \frac{\pi}{3}\right)$ and sketch its graph.
Answer: easy

period $= 2\pi$

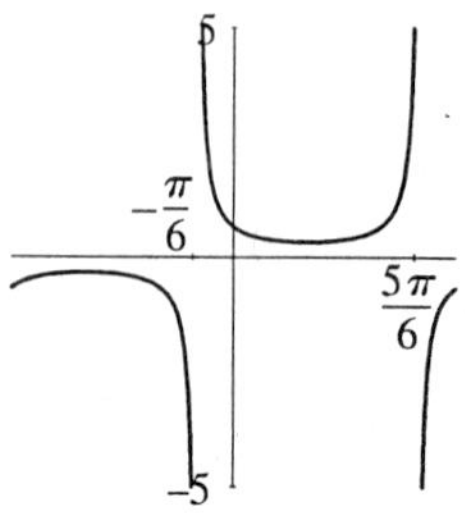

6. Find the period of the function $y = 4\csc\left(x + \frac{\pi}{2}\right)$ and sketch its graph.
Answer: easy

period $= 2\pi$

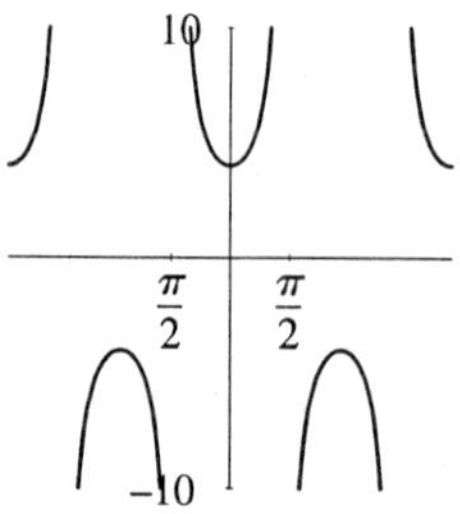

7. Find the period of the function $y = \sec 7x$ and sketch its graph.
Answer: easy

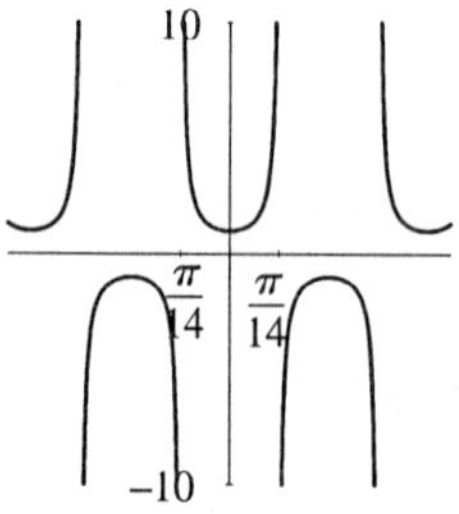

period $= \frac{2\pi}{7}$

8. Find the period of the function $y = \csc \frac{2}{\pi}x$ and sketch its graph.
Answer: easy

period $= \pi^2$

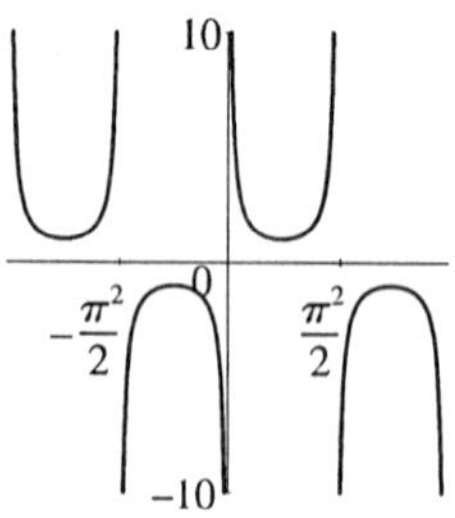

9. Find the period of the function $y = \sec\left(\frac{1}{2}x - \frac{\pi}{4}\right)$ and sketch its graph.

Answer: easy

period $= 4\pi$

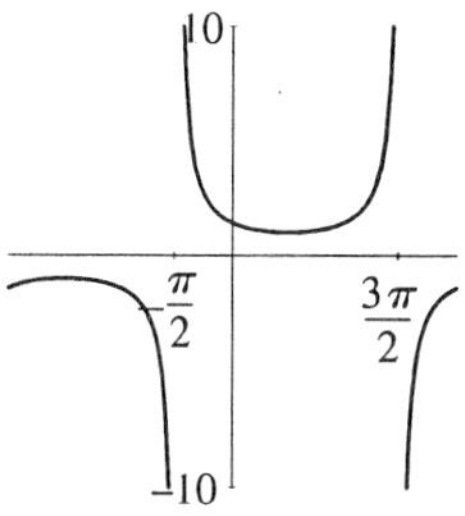

10. Find the period of the function $y = 3\sec\left(\frac{\pi}{2}\left(x - \frac{1}{2}\right)\right)$ and sketch its graph.

Answer: easy

period $= 4$

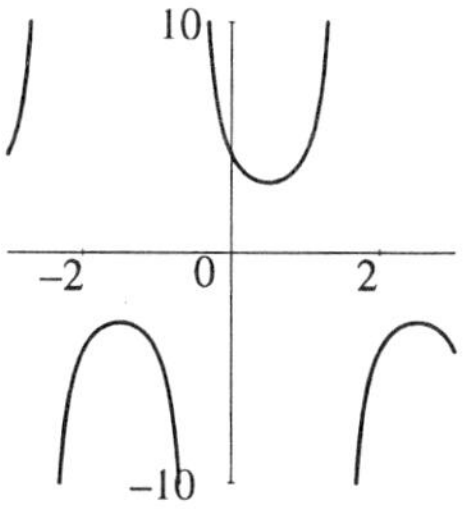

11. Find the period of the function $y = \frac{1}{2}\csc\left(2\left(x + \frac{\pi}{3}\right)\right)$ and sketch its graph.

Answer: easy

period $= \pi$

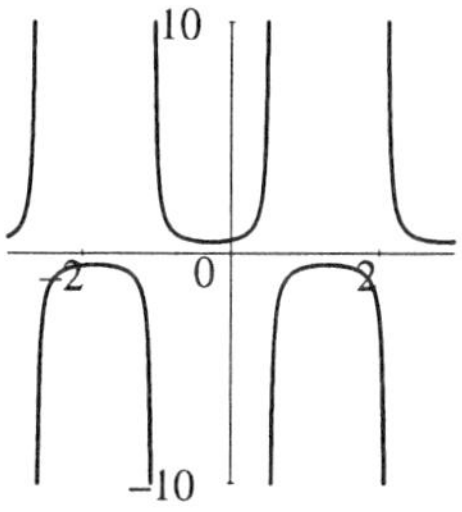

CHAPTER 8 REVIEW

1. Show that $\left(-\frac{1}{2}, \frac{\sqrt{3}}{2}\right)$ is on the unit circle.

   Answer: easy

   $\left(-\frac{1}{2}\right)^2 + \left(\frac{\sqrt{3}}{2}\right)^2 = \frac{1}{4} + \frac{3}{4} = 1.$

2. Find $P(x, y)$ given that the x-coordinate of P is $-\frac{3}{5}$ and P is in quadrant II.

   Answer: easy

   P is in quadrant II, so y is positive. $y = \sqrt{1 - \left(-\frac{3}{5}\right)^2} = \sqrt{\frac{16}{25}} = \frac{4}{5}$. So $P = \left(-\frac{3}{5}, \frac{4}{5}\right)$.

3. Find the terminal point $P(x, y)$ on the unit circle determined by the value $t = -\frac{5\pi}{6}$.

   Answer: easy

   $\left(-\frac{\sqrt{3}}{2}, -\frac{1}{2}\right)$

4. Find the terminal point $P(x, y)$ on the unit circle determined by the value $t = \frac{3\pi}{4}$.

   Answer: easy

   $\left(-\frac{\sqrt{2}}{2}, \frac{\sqrt{2}}{2}\right)$

5. Find the reference number for $t = \frac{2\pi}{3}$.

   Answer: easy

   $\bar{t} = \pi - t = \frac{\pi}{3}$

6. Find the reference number and the terminal point $P(x, y)$ determined by $t = \frac{8\pi}{3}$.

   Answer: easy

   $\bar{t} = 3\pi - \frac{8\pi}{3} = \frac{\pi}{3}$ and $P(x, y) = \left(-\frac{1}{2}, \frac{\sqrt{3}}{2}\right)$.

7. Find the exact value of $\sin\left(-\frac{4\pi}{3}\right)$ and $\cos\left(\frac{4\pi}{3}\right)$.

   Answer: easy

   $\sin\left(-\frac{4\pi}{3}\right) = \frac{\sqrt{3}}{2}$ and $\cos\left(\frac{4\pi}{3}\right) = -\frac{1}{2}$

8. Find the exact value of $\tan\left(\frac{\pi}{6}\right)$ and $\tan\left(\frac{5\pi}{4}\right)$.

   Answer: easy

   $\tan\left(\frac{\pi}{6}\right) = \frac{\sqrt{3}}{3}$ and $\tan\left(\frac{5\pi}{4}\right) = 1$.

9. If the terminal point determined by t is $\left(-\frac{\sqrt{2}}{2}, \frac{\sqrt{2}}{2}\right)$, find $\sin t$, $\cos t$ and $\tan t$.

   Answer: easy

   $\frac{\sqrt{2}}{2}, -\frac{\sqrt{2}}{2}, -1$

10. Find the sign of $\tan t \sin t$ if the terminal point determined by t is in quadrant III.

    Answer: medium

    negative

11. Write $\cos t$ in terms of $\sin t$ given that t lies in quadrant III.

    Answer: medium

    $\cos t = -\sqrt{1 - \sin^2 t}$, since cos is negative in quadrant III.

12. Indicate if the function $f(x) = x^2 - x^2 \cos x$ is even, odd, or neither.

    Answer: hard

    even

13. Sketch the graph of $y = 1 + \sin x$.

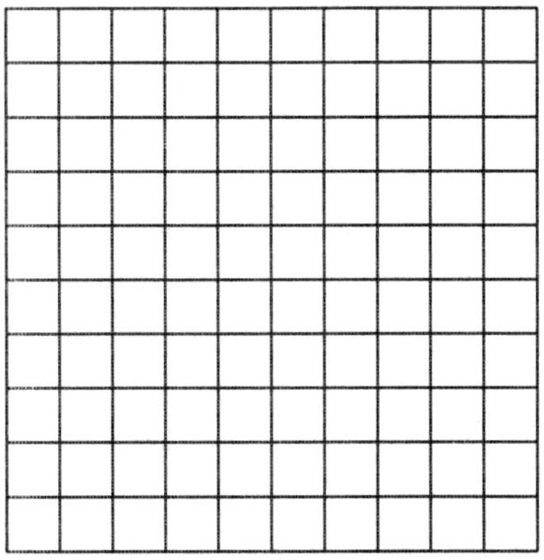

Answer: medium

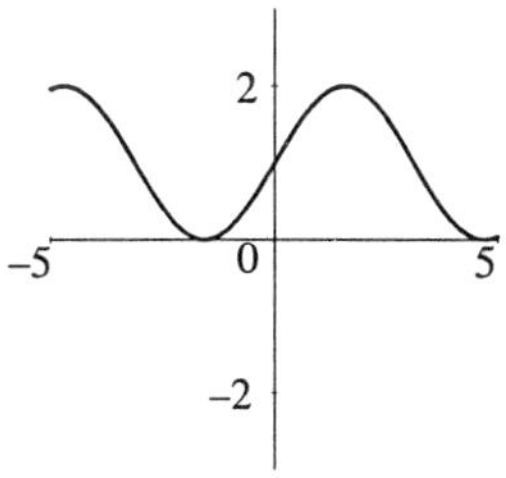

14. Sketch the graph of $y = 2 - \cos x$

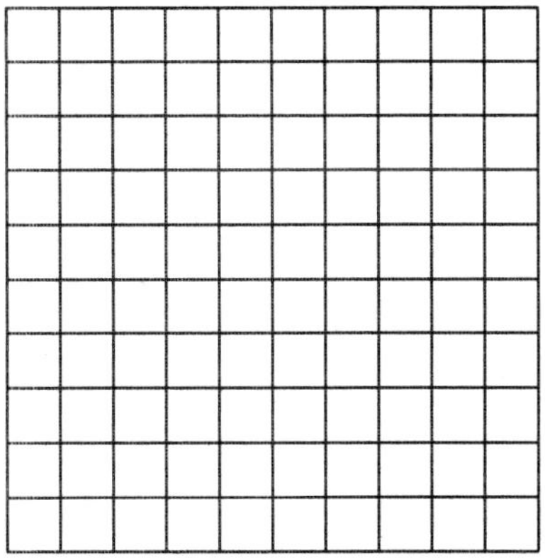

Answer: medium

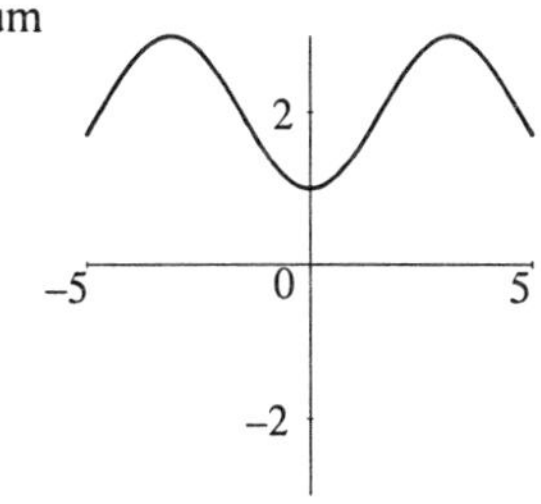

15. Find the amplitude and the period of $y = 3 \sin x$ and sketch its graph.

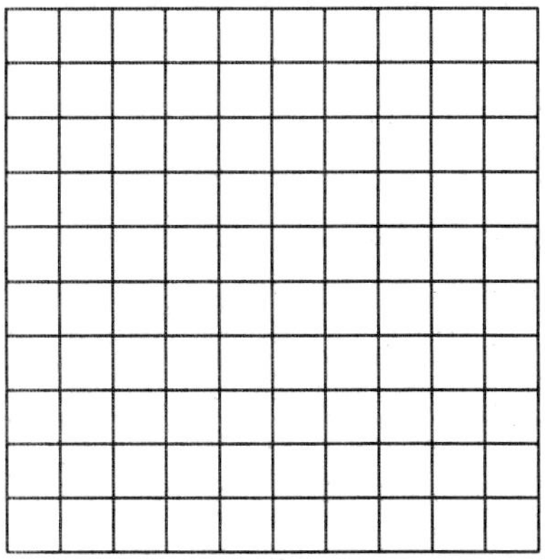

Answer: medium

Amplitude: 3; Period: 2π

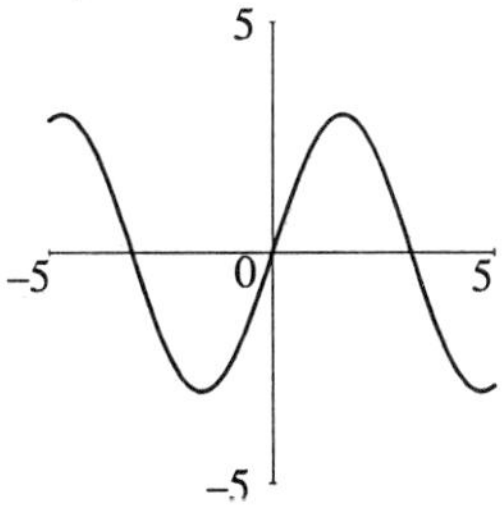

16. Find the amplitude and the period of $y = 2\cos\frac{\pi}{3}x$ and sketch its graph.

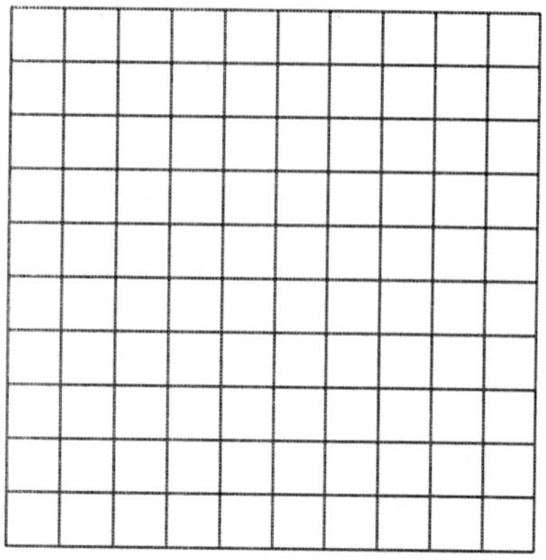

Answer: medium

Amplitude: 2; Period: 6

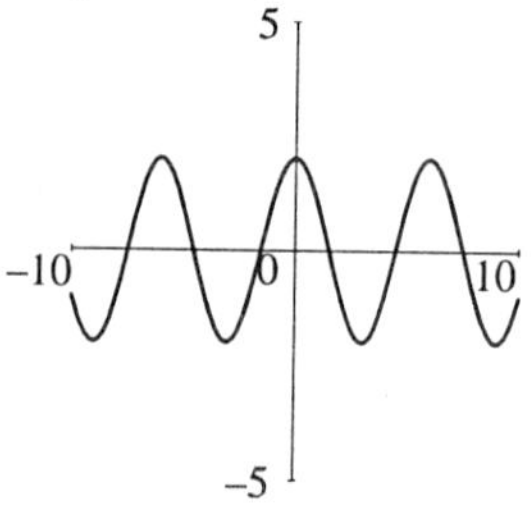

17. Find the amplitude, period and phase shift of $y = 2\cos 3(x-1)$ and sketch its graph.

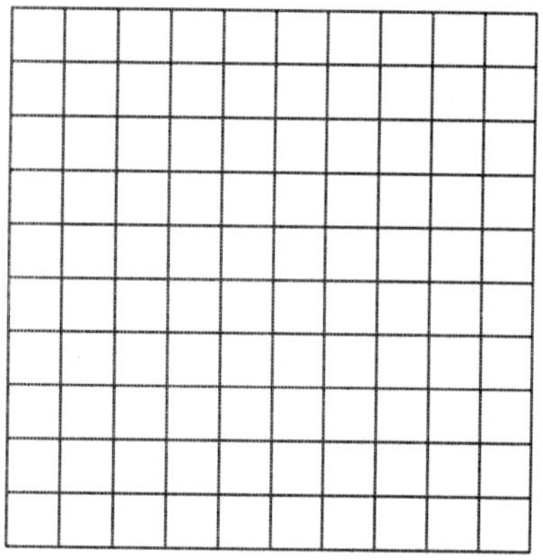

Answer: medium

Amplitude: 2; Period: $\frac{2\pi}{3}$; Phase shift: 1

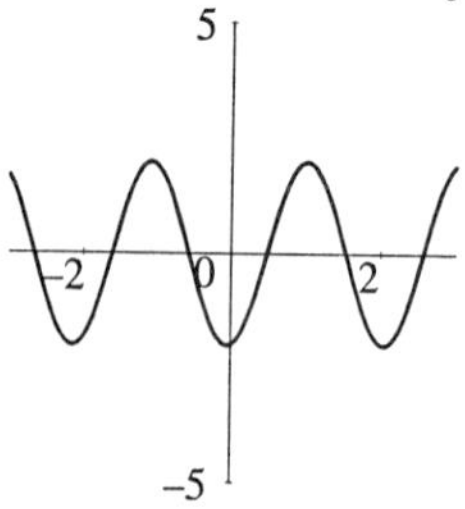

18. Find the amplitude, period and phase shift of the graph given below, and write an equation that represents the curve.

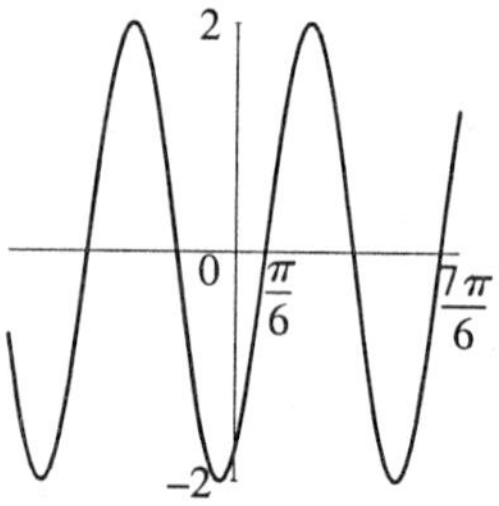

Answer: hard

Amplitude: 2; Period: π; Phase shift: $\frac{\pi}{6}$

$y = 2\sin\left(2x - \frac{\pi}{3}\right)$

19. Determine an appropriate viewing rectangle and use it to draw the graph of $y = 2\sin 75x$

Answer: medium

Amplitude: 2

Period $\frac{2\pi}{75} \approx 0.08$

So use $[-0.09, 0.09]$ by $[-2, 2]$

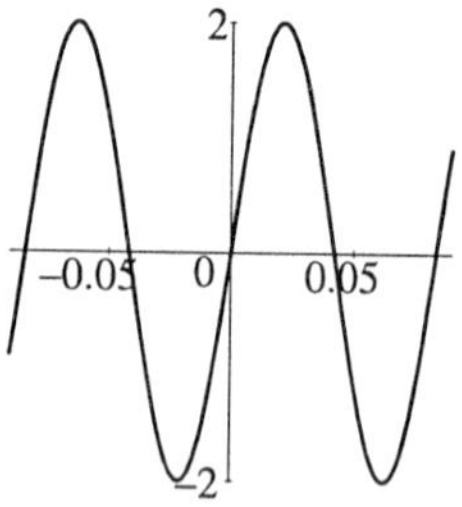

20. Find the period and sketch the graph of the function $y = \sec 3x$.
Answer: medium

period $= \frac{2\pi}{3}$

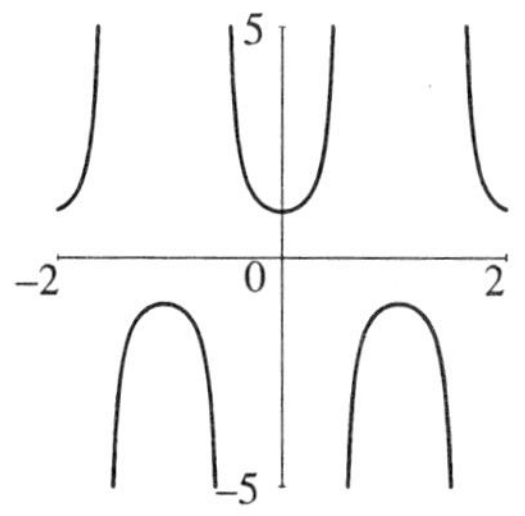

21. Find the period and sketch the graph of the function $y = 3\csc 2\left(x - \frac{\pi}{6}\right)$.
Answer: medium

period $= \pi$

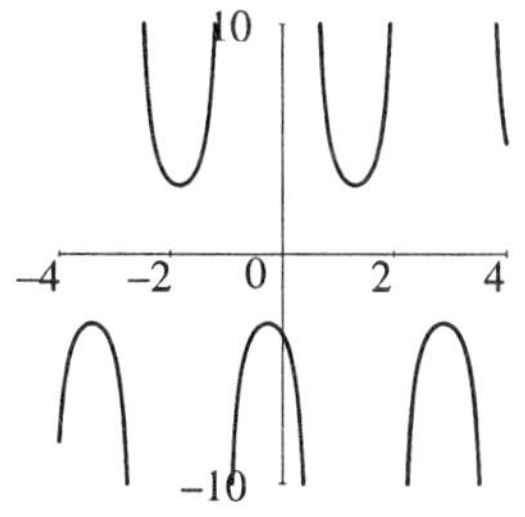

CHAPTER NINE

SECTION 9.1

1. Write $\cos x \sin x \sec x$ in terms of $\sin x$ and $\cos x$ and simplify.

   Answer: easy

   $\cos x \sin x \sec x = \dfrac{\cos x \sin x}{\cos x} = \sin x$

2. Write $\sin^2 x(1 + \cot^2 x)$ in terms of $\sin x$ and $\cos x$ and simplify.

   Answer: easy

   $\sin^2 x(1 + \cot^2 x) = \sin^2 x \csc^2 x = 1$

3. Write $\dfrac{\sec x - \cos x}{\tan x}$ in terms of $\sin x$ and $\cos x$ and simplify.

   Answer: easy

   $\dfrac{\sec x - \cos x}{\tan x} = \dfrac{\cos x(\sec x - \cos x)}{\sin x} = \dfrac{1 - \cos^2 x}{\sin x} = \dfrac{\sin^2 x}{\sin x} = \sin x$

4. Simplify $\dfrac{\sin x \csc x}{\tan x}$.

   Answer: easy

   $\dfrac{\sin x \csc x}{\tan x} = \dfrac{1}{\tan x} = \cot x$

5. Simplify $\dfrac{1 + \cos x}{1 + \sec x}$.

   Answer: easy

   $\dfrac{1 + \cos x}{1 + \sec x} = \cos x \dfrac{1 + \cos x}{\cos x + 1} = \cos x$

6. Simplify $\dfrac{\cot x}{\csc(-x)}$.

   Answer: easy

   $\dfrac{\cot x}{\csc(-x)} = -\sin x \dfrac{\cos x}{\sin x} = -\cos x$

7. Simplify $\dfrac{1 + \sec x}{\sin x + \tan x}$.

   Answer: easy

   $\dfrac{1 + \sec x}{\sin x + \tan x} = \dfrac{\cos x + 1}{\cos x \sin x + \sin x} = \dfrac{1}{\sin x} - \csc x$

8. Simplify $\cot x \sin x \sec x$.

   Answer: easy

   $\cot x \sin x \sec x = \dfrac{\cos x}{\sin x} \sin x \dfrac{1}{\cos x} = 1$

9. Simplify $\dfrac{2 + \cot^2 x}{\csc^2 x}$.

   Answer: easy

   $\dfrac{2 + \cot^2 x}{\csc^2 x} = \dfrac{1 + \csc^2 x}{\csc^2 x} = \sin^2 x + 1$

10. Simplify $\dfrac{\sin x}{\csc x + \cot x}$.

    Answer: easy

    $\dfrac{\sin x}{\csc x + \cot x} = \dfrac{\sin^2 x}{1 + \cos x} = \dfrac{1 - \cos^2 x}{1 + \cos x} = 1 - \cos x$

11. Simplify $\dfrac{\csc x}{\sin x} + \dfrac{\sec x}{\cos x}$.

Answer: medium

$$\frac{\csc x}{\sin x} + \frac{\sec x}{\cos x} = \frac{1}{\sin^2 x} + \frac{1}{\cos^2 x} = \frac{\sin^2 x + \cos^2 x}{\sin^2 x \cos^2 x} = \csc^2 x \sec^2 x$$

12. Simplify $\dfrac{1+\cos x}{\sin x} + \dfrac{\sin x}{1+\cos x}$.

Answer: medium

$$\frac{1+\cos x}{\sin x} + \frac{\sin x}{1+\cos x} = \frac{1 + 2\cos x + \cos^2 x + \sin^2 x}{\sin x(1+\cos x)} = \frac{2(1+\cos x)}{\sin x(1+\cos x)} = 2\csc x$$

13. Simplify $\dfrac{1-\cos^2 x}{\cos^2 x}$.

Answer: easy

$$\frac{1-\cos^2 x}{\cos^2 x} = \frac{\sin^2 x}{\cos^2 x} = \tan^2 x$$

14. Simplify $\dfrac{1}{\csc^2 x} + \dfrac{1}{\sec^2 x}$.

Answer: easy

$$\frac{1}{\csc^2 x} + \frac{1}{\sec^2 x} = \sin^2 x + \cos^2 x = 1$$

15. Simplify $\sec(-x)\tan(-x)\cos x$.

Answer: easy

$$\sec(-x)\tan(-x)\cos x = \sec x\,(-\tan x)\cos x = -\tan x$$

16. Simplify $\dfrac{1+\tan x}{1+\cot x}$.

Answer: medium

$$\frac{1+\tan x}{1+\cot x} = \tan x\frac{1+\tan x}{\tan x + 1} = \tan x$$

17. Simplify $\sec^2 x(1-\sin^2 x)$.

Answer: easy

$$\sec^2 x(1-\sin^2 x) = \sec^2 x\cos^2 x = 1$$

18. Simplify $\dfrac{\sec^2 x - 2}{\tan x + 1}$.

Answer: medium

$$\frac{\sec^2 x - 2}{\tan x + 1} = \frac{\tan^2 x - 1}{\tan x + 1} = \tan x - 1$$

19. Verify the identity $\sec(-x) - \tan(-x) = \sec x + \tan x$.

Answer: easy

$$\sec(-x) - \tan(-x) = \sec x - (-\tan x) = \sec x + \tan x$$

20. Verify the identity $\tan x\cos x = \sin x$.

Answer: easy

$$\tan x\cos x = \frac{\sin x}{\cos x}\cos x = \sin x$$

21. Verify the identity $\cos x + \sin x\tan x = \sec x$.

Answer: easy

$$\cos x + \sin x\tan x = \frac{\cos^2 x}{\cos x} + \frac{\sin^2 x}{\cos x} = \frac{1}{\cos x} = \sec x$$

22. Verify the identity $\sin(-x)\tan(-x) = \sin^2 x\sec x$.

Answer: easy

$$\sin(-x)\tan(-x) = -\sin x\,(-\tan x) = \sin^2 x\sec x$$

23. Verify the identity $(\tan x + \cot x)^2 = \sec^2 x + \csc^2 x$.

Answer: easy

$(\tan x + \cot x)^2 = \tan^2 x + 2\tan x \cot x + \cot^2 x = \tan^2 x + 2 + \cot^2 x = \sec^2 x + \csc^2 x$

24. Verify the identity $\dfrac{1-\cos x}{1+\cos x} = (\sec x - \cot x)^2$.

Answer: easy

$$\frac{1-\cos x}{1+\cos x} = \frac{(1-\cos x)^2}{1-\cos^2 x} = \frac{(1-\cos x)^2}{\sin^2 x} = (\sec x - \cot x)^2$$

25. Verify the identity $\dfrac{1}{1-\sin x} = \sec x(\sec x + \tan x)$.

Answer: easy

$$\frac{1}{1-\sin x} = \frac{1+\sin x}{1-\sin^2 x} = \frac{1+\sin x}{\cos^2 x} = \sec x(\sec x + \tan x)$$

26. Verify the identity $\csc x - \cot x = \dfrac{1}{\csc x + \cot x}$.

Answer: easy

$$\csc x - \cot x = \frac{(\csc x - \cot x)(\csc x + \cot x)}{\csc x + \cot x} = \frac{\csc^2 x - \cot^2 x}{\csc x + \cot x} = \frac{1}{\csc x + \cot x}$$

27. Verify the identity $\dfrac{1+\sin x}{1-\sin x} = (\sec x + \tan x)^2$.

Answer: easy

$$\frac{1+\sin x}{1-\sin x} = \frac{(1+\sin x)^2}{1-\sin^2 x} = \frac{(1+\sin x)^2}{\cos^2 x} = (\sec x + \tan x)^2$$

28. Verify the identity $\cot^2 x - \cos^2 x = \cot^2 x \cos^2 x$.

Answer: easy

$\cot^2 x - \cos^2 x = \cos^2 x \left(\csc^2 x - 1\right) = \cot^2 x \cos^2 x$

29. Substitute $\tan\theta$ for x in $\dfrac{x}{\sqrt{1+x^2}}$ and simplify. Assume that $0 \le \theta < \dfrac{\pi}{2}$.

Answer: medium

$$\frac{x}{\sqrt{1+x^2}} = \frac{\tan\theta}{\sqrt{1+\tan^2\theta}} = \frac{\tan\theta}{\sec\theta} = \sin\theta$$

30. Substitute $2\sin\theta$ for x in $\dfrac{1}{\sqrt{4-x^2}}$ and simplify. Assume that $0 \le \theta < \dfrac{\pi}{2}$.

Answer: medium

$$\frac{1}{\sqrt{4-x^2}} = \frac{1}{\sqrt{4-4\sin^2\theta}} = \frac{1}{2\sqrt{\cos^2\theta}} = \frac{1}{2}\sec\theta$$

31. Show that $\tan x^2 = 2\tan x$ is not an identity.

Answer: medium

$\tan 1^2 \neq 0$, so $\tan 1^2 = \tan 1 \neq 2\tan 1$, and thus $\tan x^2 = 2\tan x$ is not an identity.

SECTION 9.2

1. Use addition or subtraction formulas to find the exact value of $\cos 105°$.
   Answer: easy

   $\cos 105° = \cos(60+45) = \cos 60 \cos 45 - \sin 60 \sin 45 = \frac{1}{2}\frac{\sqrt{2}}{2} - \frac{\sqrt{3}}{2}\frac{\sqrt{2}}{2} = \frac{\sqrt{2}\left(1-\sqrt{3}\right)}{4}.$

2. Use addition or subtraction formulas to find the exact value of $\tan -15°$.
   Answer: easy

   $\tan -15° = \tan(30-45) = \frac{\tan 30 - \tan 45}{1+\tan 30 \tan 45} = \frac{\frac{\sqrt{3}}{3}-1}{1+\frac{\sqrt{3}}{3}\cdot 1} = \frac{1-\sqrt{3}}{1+\sqrt{3}} = \frac{1-2\sqrt{3}+3}{1-\sqrt{3}^2} = \sqrt{3}-2.$

3. Use addition or subtraction formulas to find the exact value of $\sin \frac{\pi}{12}$.
   Answer: easy

   $\sin\frac{\pi}{12} = \sin\left(\frac{\pi}{3}-\frac{\pi}{4}\right) = \sin\frac{\pi}{3}\cos\frac{\pi}{4} - \cos\frac{\pi}{3}\sin\frac{\pi}{4} = \frac{\sqrt{3}}{2}\frac{\sqrt{2}}{2} - \frac{1}{2}\frac{\sqrt{2}}{2} = \frac{\sqrt{6}-\sqrt{2}}{4}.$

4. Use addition or subtraction formulas to find the exact value of $\tan \frac{7\pi}{12}$.
   Answer: easy

   $\tan\frac{7\pi}{12} = \tan\left(\frac{\pi}{3}+\frac{\pi}{4}\right) = \frac{\tan\frac{\pi}{3}-\tan\frac{\pi}{4}}{1+\tan\frac{\pi}{3}\tan\frac{\pi}{4}} = \frac{\sqrt{3}-1}{1+\sqrt{3}\cdot 1} = -\frac{1-\sqrt{3}}{1+\sqrt{3}} = -\frac{1-2\sqrt{3}+3}{1-\sqrt{3}^2} = 2-\sqrt{3}.$

5. Use an addition or subtraction formula to write $\cos 75 \cos 15 + \sin 75 \sin 15$ and find its exact value.
   Answer: easy

   $\cos 75 \cos 15 + \sin 75 \sin 15 = \cos(75-15) = \cos 60 = \frac{1}{2}.$

6. Use an addition or subtraction formula to write $\cos 12 \cos 18 - \sin 12 \sin 18$ and find its exact value.
   Answer: easy

   $\cos 12 \cos 18 - \sin 12 \sin 18 = \cos(12+18) = \cos 30 = \frac{\sqrt{3}}{2}.$

7. Use an addition or subtraction formula to write $\sin 74 \cos 14 - \cos 74 \sin 14$ and find its exact value.
   Answer: easy

   $\sin 74 \cos 14 - \cos 74 \sin 14 = \sin(74-14) = \sin 60 = \frac{\sqrt{3}}{2}.$

8. Use an addition or subtraction formula to write $\sin 23 \cos 22 + \cos 23 \sin 22$ and find its exact value.
   Answer: easy

   $\sin 23 \cos 22 + \cos 23 \sin 22 = \sin(23+22) = \sin 45 = \frac{\sqrt{2}}{2}.$

9. Use an addition or subtraction formula to write $\frac{\tan 10 + \tan 20}{1 - \tan 10 \tan 20}$ and find its exact value.
   Answer: easy

   $\frac{\tan 10 + \tan 20}{1 - \tan 10 \tan 20} = \tan(10+20) = \tan 30 = \frac{\sqrt{3}}{3}.$

10. Prove $\cos\left(x-\frac{\pi}{2}\right) + \cos\left(x+\frac{\pi}{2}\right) = 0$.
    Answer: medium

    $\cos\left(x-\frac{\pi}{2}\right) + \cos\left(x+\frac{\pi}{2}\right) = \cos x \cos\frac{\pi}{2} + \sin x \sin\frac{\pi}{2} + \cos x \cos\frac{\pi}{2} - \sin x \sin\frac{\pi}{2} = 2\cos x \cdot 0 = 0.$

11. Prove $\cos\left(x+\frac{\pi}{4}\right) = -\sin\left(x-\frac{\pi}{4}\right)$.
    Answer: medium

    $\cos\left(x+\frac{\pi}{4}\right) = \cos x \cos\frac{\pi}{4} - \sin x \sin\frac{\pi}{4} = \frac{\sqrt{2}}{2}\cos x - \frac{\sqrt{2}}{2}\sin x = \sin\frac{\pi}{4}\cos x - \cos\frac{\pi}{4}\sin x = \sin\left(\frac{\pi}{4}-x\right) = -\sin\left(x-\frac{\pi}{4}\right).$

12. Prove $\tan\left(x+\frac{\pi}{4}\right)=\dfrac{1+\tan x}{1-\tan x}$.

Answer: medium

$\tan\left(x+\frac{\pi}{4}\right)=\dfrac{\tan x+\tan\frac{\pi}{4}}{1-\tan x\tan\frac{\pi}{4}}=\dfrac{1+\tan x}{1-\tan x}$.

13. Prove $\sin(x+y)+\sin(x-y)=2\sin x\cos y$.

Answer: medium

$\sin(x+y)+\sin(x-y)=\sin x\cos y+\cos x\sin y+\sin x\cos y-\cos x\sin y=2\sin x\cos y$.

14. Prove $\cos(x+y)-\cos(x-y)=-2\sin x\sin y$.

Answer: medium

$\cos(x+y)-\cos(x-y)=\cos x\cos y-\sin x\sin y-\cos x\cos y-\sin x\sin y=-2\sin x\sin y$.

15. Prove $\dfrac{\sin(x-y)}{\cos(x+y)}=\dfrac{\tan x-\tan y}{1-\tan x\tan y}$.

Answer: hard

$\dfrac{\sin(x-y)}{\cos(x+y)}=\dfrac{\sin x\cos y-\cos x\sin y}{\cos x\cos y-\sin x\sin y}=\dfrac{\frac{1}{\cos x\cos y}}{\frac{1}{\cos x\cos y}}\,\dfrac{\sin x\cos y-\cos x\sin y}{\cos x\cos y-\sin x\sin y}=\dfrac{\tan x-\tan y}{1-\tan x\tan y}$.

16. Prove $\dfrac{\sin(\pi-1)x}{\sin x\cos x}=\dfrac{\sin\pi x}{\sin x}-\dfrac{\cos\pi x}{\cos x}$.

Answer: hard

$\dfrac{\sin(\pi-1)x}{\sin x\cos x}=\dfrac{\sin(\pi x-x)}{\sin x\cos x}=\dfrac{\sin\pi x\cos x-\cos\pi x\sin x}{\sin x\cos x}=\dfrac{\sin\pi x}{\sin x}-\dfrac{\cos\pi x}{\cos x}$.

17. Prove $\cos(x+y+z)=\cos x\cos y\cos z-\sin x\sin y\cos z-\sin x\cos y\sin z-\cos x\sin y\sin z$.

Answer: hard

$$\begin{aligned}\cos(x+y+z)&=\cos(x+y)\cos z-\sin(x+y)\sin z\\&=(\cos x\cos y-\sin x\sin y)\cos z-(\sin x\cos y+\cos x\sin y)\sin z\\&=\cos x\cos y\cos z-\sin x\sin y\cos z-\sin x\cos y\sin z-\cos x\sin y\sin z\end{aligned}$$

18. Prove $\cos\left(x-\frac{\pi}{3}\right)=\frac{1}{2}\left(\cos x+\sqrt{3}\sin x\right)$.

Answer: medium

$\cos\left(x-\frac{\pi}{3}\right)=\cos x\cos\frac{\pi}{3}+\sin x\sin\frac{\pi}{3}=\frac{1}{2}\cos x+\frac{\sqrt{3}}{2}\sin x=\frac{1}{2}\left(\cos x+\sqrt{3}\sin x\right)$.

19. Prove $\tan\left(x-\frac{\pi}{2}\right)=-\cot x$.

Answer: medium

$\tan\left(x-\frac{\pi}{2}\right)=\dfrac{\sin\left(x-\frac{\pi}{2}\right)}{\cos\left(x-\frac{\pi}{2}\right)}=\dfrac{\sin x\cos\frac{\pi}{2}-\cos x\sin\frac{\pi}{2}}{\cos x\cos\frac{\pi}{2}+\sin x\sin\frac{\pi}{2}}=\dfrac{-\cos x}{\sin x}=-\cot x$.

20. Write $\sin x\,\frac{\sqrt{3}}{2}-\cos x\,\frac{1}{2}$ in terms of sine only.

Answer: medium

$\sin x\,\frac{\sqrt{3}}{2}-\cos x\,\frac{1}{2}=\sin x\cos\frac{\pi}{6}-\cos x\sin\frac{\pi}{6}=\sin\left(x-\frac{\pi}{6}\right)$.

21. Write $3\sin x+7\cos x$ in terms of sine only.

Answer: medium

$3\sin x+7\cos x=\sqrt{3^2+7^2}\sin(x+\phi)\approx\sqrt{58}\sin(x+66.8°)$.

22. Write $8\sin 3x+2\cos 3x$ in terms of sine only.

Answer: medium

$8\sin 3x+2\cos 3x=\sqrt{8^2+2^2}\sin(3x+\phi)\approx 2\sqrt{17}\sin(x+14.0°)$.

SECTION 9.3

1. Find $\sin 2x$, $\cos 2x$ and $\tan 2x$, given that $\sin x = \frac{3}{5}$ and x is in quadrant II.

   Answer: easy

   $\cos x = -\frac{4}{5}$, so $\sin 2x = 2\sin x\cos x = 2\cdot\frac{3}{5}\cdot\left(-\frac{4}{5}\right) = -\frac{24}{25}$, $\cos 2x = \cos^2 x - \sin^2 x = \left(-\frac{4}{5}\right)^2 - \left(\frac{3}{5}\right)^2 = \frac{7}{25}$, and $\tan 2x = \dfrac{\sin 2x}{\cos 2x} = -\dfrac{24}{7}$.

2. Find $\sin 2x$, $\cos 2x$ and $\tan 2x$, given that $\cos x = \frac{5}{13}$ and x is in quadrant IV.

   Answer: easy

   $\sin x = -\frac{12}{13}$, so $\sin 2x = 2\sin x\cos x = 2\cdot\left(-\frac{12}{13}\right)\cdot\frac{5}{13} = -\frac{120}{169}$, $\cos 2x = \cos^2 x - \sin^2 x = \left(\frac{5}{13}\right)^2 - \left(-\frac{12}{13}\right)^2 = -\frac{119}{169}$, and $\tan 2x = \dfrac{\sin 2x}{\cos 2x} = \dfrac{120}{119}$.

3. Find $\sin 2x$, $\cos 2x$ and $\tan 2x$, given that $\sec x = -4$ and $\sin x < 0$.

   Answer: easy

   $\cos x = -\frac{1}{4}$, $\sin x = -\frac{\sqrt{15}}{4}$, so $\sin 2x = 2\sin x\cos x = 2\cdot\left(-\frac{\sqrt{15}}{4}\right)\cdot\left(-\frac{1}{4}\right) = \frac{2\sqrt{15}}{16}$,

   $\cos 2x = \cos^2 x - \sin^2 x = \left(-\frac{1}{4}\right)^2 - \left(-\frac{\sqrt{15}}{4}\right)^2 = -\frac{15}{16}$, and $\tan 2x = \dfrac{\sin 2x}{\cos 2x} = -\dfrac{2\sqrt{15}}{16}$.

4. Find $\sin 2x$, $\cos 2x$ and $\tan 2x$, given that $\cot x = -\frac{3}{4}$ and $\sin x < 0$.

   Answer: easy

   $\cos x = \frac{3}{5}$, $\sin x = -\frac{4}{5}$, so $\sin 2x = 2\sin x\cos x = 2\cdot\left(-\frac{4}{5}\right)\cdot\left(\frac{3}{5}\right) = -\frac{24}{25}$, $\cos 2x = \cos^2 x - \sin^2 x = \left(\frac{3}{5}\right)^2 - \left(-\frac{4}{5}\right)^2 = -\frac{7}{25}$, and $\tan 2x = \dfrac{\sin 2x}{\cos 2x} = \dfrac{24}{7}$.

5. Use the formulas for lowering powers to rewrite $\sin^6 x$ in terms of the first power of cosine.

   Answer: medium

$$
\begin{aligned}
\sin^6 x &= \left(\frac{1-\cos 2x}{2}\right)^3 \\
&= \frac{1}{8}\left(1 - 3\cos 2x + 3\cos^2 2x - \cos^3 2x\right) \\
&= \frac{1}{8}\left(1 - 3\cos 2x + 3\left(\frac{1-\cos 4x}{2}\right) - \cos 2x\left(\frac{1-\cos 4x}{2}\right)\right) \\
&= \frac{1}{16}(2 - 6\cos 2x + 3 - 3\cos 4x - \cos 2x + \cos 2x\cos 4x) \\
&= \frac{1}{16}(5 - 7\cos 2x - 3\cos 4x + \cos 2x\cos 4x)
\end{aligned}
$$

6. Use the formulas for lowering powers to rewrite $\cos^6 x\sin^2 x$ in terms of the first power of cosine.

   Answer: medium

$$
\begin{aligned}
\cos^6 x\sin^2 x &= \left(\frac{1+\cos 2x}{2}\right)^3\left(\frac{1-\cos 2x}{2}\right) \\
&= \frac{1}{16}\left(1 + 2\cos 2x + \cos^2 2x\right)\left(1 - \cos^2 2x\right) \\
&= \frac{1}{16}\left(1 + 2\cos 2x + \frac{1+\cos 4x}{2}\right)\left(1 - \frac{1+\cos 4x}{2}\right) \\
&= \frac{1}{64}(2 + 4\cos 2x + 1 + \cos 4x)(2 - 1 - \cos 4x) \\
&= \frac{1}{64}(3 + 4\cos 2x + \cos 4x)(1 - \cos 4x) \\
&= \frac{1}{64}\left(3 + 4\cos 2x + \cos 4x - 3\cos 4x - 4\cos 2x\cos 4x + \cos^2 4x\right) \\
&= \frac{1}{64}\left(3 + 4\cos 2x - 2\cos 4x - 4\cos 2x\cos 4x + \frac{1+\cos 8x}{2}\right)
\end{aligned}
$$

7. Use a half angle formula to find the exact value of $\sin -15°$.
Answer: easy

$$\sin(-15°) = -\sqrt{\frac{1-\cos(-30)}{2}} = -\sqrt{\frac{1-\frac{\sqrt{3}}{2}}{2}} = -\sqrt{\frac{2-\sqrt{3}}{4}} = \frac{\sqrt{\sqrt{3}-2}}{2}.$$

8. Use a half angle formula to find the exact value of $\cos -15°$.
Answer: easy

$$\cos(-15°) = \sqrt{\frac{1+\cos(-30)}{2}} = \sqrt{\frac{1+\frac{\sqrt{3}}{2}}{2}} = \sqrt{\frac{2+\sqrt{3}}{4}} = \frac{\sqrt{2+\sqrt{3}}}{2}.$$

9. Use a half angle formula to find the exact value of $\sin \frac{\pi}{8}$.
Answer: easy

$$\sin\frac{\pi}{8} = \sqrt{\frac{1-\cos\frac{\pi}{4}}{2}} = \sqrt{\frac{1-\frac{\sqrt{2}}{2}}{2}} = \frac{\sqrt{2-\sqrt{2}}}{2}.$$

10. Use a half angle formula to find the exact value of $\cos \frac{\pi}{8}$.
Answer: easy

$$\cos\frac{\pi}{8} = \sqrt{\frac{1+\cos\frac{\pi}{4}}{2}} = \sqrt{\frac{1+\frac{\sqrt{2}}{2}}{2}} = \frac{\sqrt{2+\sqrt{2}}}{2}.$$

11. Use a double angle formula or a half angle formula to simplify $2\sin 12\cos 12$.
Answer: easy
$2\sin 12\cos 12 = \sin 24$.

12. Use a double angle formula or a half angle formula to simplify $\cos^2 2\theta - \sin^2 2\theta$.
Answer: easy
$\cos^2 2\theta - \sin^2 2\theta = \cos 4\theta$.

13. Use a double angle formula or a half angle formula to simplify $\frac{2\tan 32}{1-\tan^2 32}$.
Answer: easy
$$\frac{2\tan 32}{1-\tan^2 32} = \tan 64.$$

14. Use a double angle formula or a half angle formula to simplify $\sqrt{\frac{1-\cos 6\theta}{2}}$.
Answer: easy
$$\sqrt{\frac{1-\cos 6\theta}{2}} = \sin 3\theta.$$

15. Use a double angle formula or a half angle formula to simplify $\sqrt{\frac{1+\cos 4\pi}{2}}$.
Answer: easy
$$\sqrt{\frac{1+\cos 4\pi}{2}} = \cos 2\pi = 1.$$

16. Use a double angle formula or a half angle formula to simplify $\frac{1-\cos 52}{\sin 52}$.
Answer: easy
$$\frac{1-\cos 52}{\sin 52} = \tan 26.$$

17. Use a double angle formula or a half angle formula to simplify $\frac{\sin 14\theta}{1+\cos 14\theta}$.
Answer: easy
$$\frac{\sin 14\theta}{1+\cos 14\theta} = \tan 7\theta.$$

18. Use a double angle formula or a half angle formula to simplify $1 - 2\sin^2 47$.
Answer: easy
$1 - 2\sin^2 47 = \cos 94$.

19. Use a double angle formula or a half angle formula to simplify $2\cos^2 13 - 1$.
Answer: easy
$2\cos^2 13 - 1 = \cos 26$.

20. Find $\sin\frac{x}{2}$, $\cos\frac{x}{2}$ and $\tan\frac{x}{2}$, given that $\sin x = \frac{4}{5}$ and $0° < x < 90°$.
Answer: medium
$\cos x = \dfrac{3}{5}$, so $\sin\dfrac{x}{2} = \sqrt{\dfrac{1 - \frac{3}{5}}{2}} = \dfrac{\sqrt{5}}{5}$, $\cos\dfrac{x}{2} = \sqrt{\dfrac{1 + \frac{3}{5}}{2}} = \dfrac{2\sqrt{5}}{5}$ and $\tan\dfrac{x}{2} = \dfrac{1 - \frac{3}{5}}{\frac{4}{5}} = \dfrac{1}{2}$.

21. Find $\sin\frac{x}{2}$, $\cos\frac{x}{2}$ and $\tan\frac{x}{2}$, given that $\cos x = -\frac{3}{5}$ and $90° < x < 180°$.
Answer: medium
$\sin x = \dfrac{4}{5}$, so $\sin\dfrac{x}{2} = \sqrt{\dfrac{1 + \frac{3}{5}}{2}} = \dfrac{2}{\sqrt{5}}$, $\cos\dfrac{x}{2} = \sqrt{\dfrac{1 - \frac{3}{5}}{2}} = \dfrac{1}{\sqrt{5}}$ and $\tan\dfrac{x}{2} = \dfrac{1 + \frac{3}{5}}{\frac{4}{5}} = \dfrac{1}{2}$.

22. Find $\sin\frac{x}{2}$, $\cos\frac{x}{2}$ and $\tan\frac{x}{2}$, given that $\tan x = 1$ and $180° < x < 270°$.
Answer: medium
$\cos x = -\dfrac{\sqrt{2}}{2}$, and $\sin x = -\dfrac{\sqrt{2}}{2}$, so $\sin\dfrac{x}{2} = \sqrt{\dfrac{1+\frac{\sqrt{2}}{2}}{2}} = \dfrac{\sqrt{2+\sqrt{2}}}{2}$, $\cos\dfrac{x}{2} = -\sqrt{\dfrac{1 - \frac{\sqrt{2}}{2}}{2}} = \dfrac{\sqrt{2-\sqrt{2}}}{2}$ and
$\tan\dfrac{x}{2} = \dfrac{1 + \frac{\sqrt{2}}{2}}{-\frac{\sqrt{2}}{2}} = -\left(\dfrac{2}{\sqrt{2}} + 1\right) = -\sqrt{2} - 1$.

23. Write $\sin 4x \cos x$ as a sum.
Answer: medium
$\sin 4x \cos x = \frac{1}{2}\sin 5x + \frac{1}{2}\sin 3x$.

24. Write $\sin 2x \sin 3x$ as a sum.
Answer: medium
$\sin 2x \sin 3x = \frac{1}{2}\cos(-x) - \frac{1}{2}\cos 5x = \frac{1}{2}\cos x - \frac{1}{2}\cos 5x$.

25. Write $\sin 4x + \sin 2x$ as a product.
Answer: medium
$\sin 4x + \sin 2x = 2\sin 3x \cos x$.

26. Write $\sin 3x - \sin x$ as a product.
Answer: medium
$\sin 3x - \sin x = 2\cos 2x \sin x$.

27. Write $\cos 4x + \cos 2x$ as a product.
Answer: medium
$\cos 4x + \cos 2x = 2\cos 3x \cos x$.

28. Find the value of $2\sin 75 \sin 15$.
Answer: medium
$2\sin 75 \sin 15 = \cos(75 - 15) - \cos(75 + 15) = \cos 60 - \cos 90 = \frac{1}{2}$.

29. Find the value of $\sin\frac{\pi}{12} - \sin\frac{5\pi}{12}$.
Answer: medium
$\sin\frac{\pi}{12} - \sin\frac{5\pi}{12} = 2\cos\left(\frac{3\pi}{12}\right)\sin\left(-\frac{2\pi}{12}\right) = 2\cos\frac{\pi}{4}\left(-\sin\frac{\pi}{6}\right) = -2\cdot\frac{\sqrt{2}}{2}\cdot\frac{1}{2} = -\frac{\sqrt{2}}{2}$.

30. Prove that $2\sin\frac{x}{2}\cos\frac{x}{2} = \sin x$.

Answer: hard

$2\sin\frac{x}{2}\cos\frac{x}{2} = \sin\left(\frac{x}{2}+\frac{x}{2}\right) + \sin\left(\frac{x}{2}-\frac{x}{2}\right) = \sin x$.

31. Prove that $\dfrac{\cos x - \cos y}{\sin x + \sin y} = -\tan\left(\dfrac{x-y}{2}\right)$.

Answer: hard

$$\frac{\cos x - \cos y}{\sin x + \sin y} = \frac{-2\sin\left(\frac{x+y}{2}\right)\sin\left(\frac{x-y}{2}\right)}{2\sin\left(\frac{x+y}{2}\right)\cos\left(\frac{x-y}{2}\right)} = -\tan\left(\frac{x-y}{2}\right).$$

32. Prove that $(1-\cos 4x)(2+\tan^2 x+\cot^2 x) = 8$.

Answer: hard

$(1-\cos 4x)(2+\tan^2 x+\cot^2 x) = (1-(1-2\sin^2 2x))(2+\tan^2 x+\cot^2 x) = 2\sin^2 2x(\sec^2 x+\csc^2 x)$
$= 8\sin^2 x\cos^2 x\,(\sec^2 x+\csc^2 x) = 8(\sin^2 x+\cos^2 x) = 8$.

33. Prove that $\dfrac{\sin(x+y)-\sin(x-y)}{\cos(x-y)-\cos(x+y)} = \cot x$.

Answer: hard

$$\frac{\sin(x+y)-\sin(x-y)}{\cos(x-y)-\cos(x+y)} = \frac{2\cos x\sin y}{2\sin x\sin y} = \cot x$$

SECTION 9.4

1. Find the value of $\sin^{-1}\left(-\frac{1}{2}\right)$ if it is defined.

Answer: easy

$\sin^{-1}\left(-\frac{1}{2}\right) = -\frac{\pi}{6}$.

2. Find the value of $\sin^{-1}\left(-\frac{5}{4}\right)$ if it is defined.

Answer: easy

$-\frac{5}{4} < -1$, so $\sin^{-1}\left(-\frac{5}{4}\right)$ is not defined.

3. Find the value of $\sin^{-1}\left(-\frac{\sqrt{3}}{2}\right)$ if it is defined.

Answer: easy

$\sin^{-1}\left(-\frac{\sqrt{3}}{2}\right) = -\frac{\pi}{3}$.

4. Find the value of $\cos^{-1}\left(-\frac{1}{2}\right)$ if it is defined.

Answer: easy

$\cos^{-1}\left(-\frac{1}{2}\right) = -\frac{2\pi}{3}$.

5. Find the value of $\cos^{-1}\sqrt{3}$ if it is defined.

Answer: easy

$\sqrt{3} > 1$, so $\cos^{-1}\sqrt{3}$ is not defined.

6. Find the value of $\tan^{-1}\sqrt{3}$ if it is defined.

Answer: easy

$\tan^{-1}\sqrt{3} = \frac{\pi}{3}$.

7. Find the value of $\tan^{-1}\frac{1}{\sqrt{3}}$ if it is defined.

Answer: easy

$\tan^{-1}\frac{1}{\sqrt{3}} = \frac{\pi}{6}$.

8. Find an approximate value of $\sin^{-1}(0.6)$ correct to five decimal places.
Answer: easy
$\sin^{-1}(0.6) \approx 0.64350$.

9. Find an approximate value of $\cos^{-1}(-0.8732)$ correct to five decimal places.
Answer: easy
$\cos^{-1}(-0.8732) \approx 2.63253$.

10. Find an approximate value of $\tan^{-1}12$ correct to five decimal places.
Answer: easy
$\tan^{-1}12 \approx 1.48766$.

11. Find the value of $\sin\left(\sin^{-1}\left(\frac{5}{7}\right)\right)$ if it is defined.
Answer: easy
$\sin\left(\sin^{-1}\left(\frac{5}{7}\right)\right) = \frac{5}{7}$.

12. Find the value of $\cos\left(\cos^{-1}\left(-\frac{3}{4}\right)\right)$ if it is defined.
Answer: easy
$\cos\left(\cos^{-1}\left(-\frac{3}{4}\right)\right) = -\frac{3}{4}$.

13. Find the value of $\tan(\tan^{-1}169)$ if it is defined.
Answer: easy
$\tan(\tan^{-1}169) = 169$.

14. Find the value of $\cos(\cos^{-1}(-13))$ if it is defined.
Answer: easy
$-13 < 1$, so $\cos^{-1}(-13)$ is not defined.

15. Find the value of $\sin^{-1}\left(\sin\frac{\pi}{15}\right)$ if it is defined.
Answer: easy
$\sin^{-1}\left(\sin\frac{\pi}{15}\right) = \frac{\pi}{15}$.

16. Find the value of $\cos^{-1}\left(\cos\frac{12\pi}{13}\right)$ if it is defined.
Answer: easy
$\cos^{-1}\left(\cos\frac{12\pi}{13}\right) = \frac{12\pi}{13}$.

17. Find the value of $\tan^{-1}\left(\tan\left(-\frac{1}{4}\right)\right)$ if it is defined.
Answer: easy
$\tan^{-1}\left(\tan\left(-\frac{1}{4}\right)\right) = -\frac{1}{4}$.

18. Find the value of $\tan(\cos^{-1}(-3))$ if it is defined.
Answer: medium
$-3 < -1$, so $\cos^{-1}(-3)$ is not defined.

19. Find the value of $\tan^{-1}\left(2\cos\frac{\pi}{3}\right)$ if it is defined.
Answer: easy
$\tan^{-1}\left(2\cos\frac{\pi}{3}\right) = \tan^{-1}1 = \frac{\pi}{4}$.

20. Find the value of $\sin\left(\cos^{-1}\left(\frac{\sqrt{3}}{2}\right)\right)$ if it is defined.
Answer: easy
$\sin\left(\cos^{-1}\left(\frac{\sqrt{3}}{2}\right)\right) = \sin\left(\frac{\pi}{6}\right) = \frac{1}{2}$.

21. Find the value of $\cos\left(\tan^{-1}\left(\sqrt{3}\right)\right)$ if it is defined.
Answer: easy
$\cos\left(\tan^{-1}\left(\sqrt{3}\right)\right) = \cos\left(\frac{\pi}{3}\right) = \frac{1}{2}$.

22. Find the value of $\tan\left(\cos^{-1}\left(\sqrt{3}\right)\right)$ if it is defined.

Answer: medium

$\sqrt{3} > 1$, so $\cos^{-1}\left(\sqrt{3}\right)$ is not defined.

23. Evaluate $\cos\left(\sin^{-1}\frac{3}{5}\right)$ by drawing a triangle.

Answer: easy

$\theta = \sin^{-1}\frac{3}{5}$, so $\cos\left(\sin^{-1}\frac{3}{5}\right) = \frac{4}{5}$.

24. Evaluate $\tan\left(\cos^{-1}\frac{5}{13}\right)$ by drawing a triangle.

Answer: easy

$\theta = \cos^{-1}\frac{5}{13}$, so $\tan\left(\cos^{-1}\frac{5}{13}\right) = \frac{12}{5}$.

25. Evaluate $\sec\left(\tan^{-1}3\right)$ by drawing a triangle.

Answer: easy

$\theta = \tan^{-1}3$, so $\sec\left(\tan^{-1}3\right) = \sqrt{10}$.

26. Rewrite $\sin\left(2\cos^{-1}x\right)$ as an algebraic expression in x.

Answer: medium

$\sin\left(2\cos^{-1}x\right) = \sin^2(\cos^{-1}x) + \cos^2(\cos^{-1}x) = \sqrt{1-\cos^2 x(\cos^{-1}x)} + x^2 = \sqrt{1-x^2} + x^2$.

27. Rewrite $\tan\left(2\sin^{-1}x\right)$ as an algebraic expression in x.

Answer: medium

$$\begin{aligned}\tan\left(2\sin^{-1}x\right) &= \tan\left(\sin^{-1}x + \sin^{-1}x\right)\\ &= \frac{2\tan\left(\sin^{-1}x\right)}{1-\tan^2(\sin^{-1}x)}\\ &= 2\frac{\sin\left(\sin^{-1}x\right)}{\cos\left(\sin^{-1}x\right)}\cdot\frac{\cos^2(\sin^{-1}x)}{\cos^2(\sin^{-1}x)-\sin^2(\sin^{-1}x)}\\ &= 2\frac{x}{\sqrt{1-x^2}}\cdot\frac{1-x^2}{1-2x^2}\\ &= 2\frac{x\sqrt{1-x^2}}{1-2x^2}\end{aligned}$$

28. Rewrite $\sin\left(\cos^{-1}x - \sin^{-1}x\right)$ as an algebraic expression in x.

Answer: medium

$$\begin{aligned}\sin\left(\cos^{-1}x - \sin^{-1}x\right) &= \sin\left(\cos^{-1}x\right)\cos\left(\sin^{-1}x\right) - \cos\left(\cos^{-1}x\right)\sin\left(\sin^{-1}x\right).\\ &= \sqrt{1-x^2}\sqrt{1-x^2} - x^2\\ &= 1-2x^2\end{aligned}$$

29. Rewrite $\tan\left(\frac{1}{2}\sin^{-1}x\right)$ as an algebraic expression in x.

Answer: medium

$$\tan\left(\frac{1}{2}\sin^{-1}x\right) = \frac{1-\cos\left(\sin^{-1}x\right)}{\sin\left(\sin^{-1}x\right)} = \frac{1-\sqrt{1-x^2}}{x}.$$

SECTION 9.5

1. Find all solutions of $2\cos x - \sqrt{3} = 0$.

Answer: medium

$2\cos x - \sqrt{3} = 0 \quad\Leftrightarrow\quad \cos x = \frac{\sqrt{3}}{2}$, solutions in $[0, 2\pi)$ are $x = \frac{\pi}{6}$ and $x = 2\pi - \frac{\pi}{6} = \frac{11\pi}{6}$, so the general solutions are $x = \frac{\pi}{6} + 2k\pi$ and $x = \frac{11\pi}{6} + 2k\pi$.

2. Find all solutions of $2\sin x - \sqrt{2} = 0$.
Answer: medium
$2\sin x - \sqrt{2} = 0 \quad\Leftrightarrow\quad \sin x = \frac{\sqrt{2}}{2}$, solutions in $[0, 2\pi)$ are $x = \frac{\pi}{4}$ and $x = \pi - \frac{\pi}{4} = \frac{3\pi}{4}$, so the general solutions are $x = \frac{\pi}{4} + 2k\pi$ and $x = \frac{3\pi}{4} + 2k\pi$.

3. Find all solutions of $4\sin^2 x - 1 = 0$.
Answer: medium
$4\sin^2 x - 1 = 0 \quad\Leftrightarrow\quad \sin^2 x = \frac{1}{4} \quad\Leftrightarrow\quad \sin x = \pm\frac{1}{2}$, solutions to $\sin x = \frac{1}{2}$ in $[0, 2\pi)$ are $x = \frac{\pi}{6}$ and $x = \pi - \frac{\pi}{6} = \frac{5\pi}{6}$, and solutions to $\sin x = -\frac{1}{2}$ in $[0, 2\pi)$ are $x = \frac{7\pi}{6}$ and $x = \frac{11\pi}{6}$ so the general solutions are $x = \frac{\pi}{6} + 2k\pi$, $x = \frac{5\pi}{6} + 2k\pi$, $x = \frac{7\pi}{6} + 2k\pi$ and $x = \frac{11\pi}{6} + 2k\pi$.

4. Find all solutions of $\sec^2 x - 4 = 0$.
Answer: medium
$\sec^2 x - 4 = 0 \quad\Leftrightarrow\quad \sec x = \pm 2 \quad\Leftrightarrow\quad \sin x = \pm\frac{1}{2}$, solutions to $\sin x = \frac{1}{2}$ in $[0, 2\pi)$ are $x = \frac{\pi}{6}$ and $x = \pi - \frac{\pi}{6} = \frac{5\pi}{6}$, and solutions to $\sin x = -\frac{1}{2}$ in $[0, 2\pi)$ are $x = \frac{7\pi}{6}$ and $x = \frac{11\pi}{6}$ so the general solutions are $x = \frac{\pi}{6} + 2k\pi$, $x = \frac{5\pi}{6} + 2k\pi$, $x = \frac{7\pi}{6} + 2k\pi$ and $x = \frac{11\pi}{6} + 2k\pi$.

5. Find all solutions of $\sin x\,(2\cos x + 1) = 0$.
Answer: medium
$\sin x\,(2\cos x + 1) = 0 \quad\Leftrightarrow\quad \sin x = 0$ or $2\cos x + 1 = 0$. $\sin x = 0 \quad\Leftrightarrow\quad x = k\pi$, and $2\cos x + 1 = 0 \quad\Leftrightarrow\quad \cos x = -\frac{1}{2}$, which has solutions $x = \left(\pi - \frac{\pi}{6}\right) + 2k\pi$ and $x = \left(\pi + \frac{\pi}{6}\right) + 2k\pi$, so the general solutions are $x = k\pi$, $x = \frac{5\pi}{6} + 2k\pi$ and $x = \frac{7\pi}{6} + 2k\pi$.

6. Find all solutions of $\left(\tan x - \sqrt{3}\right)\left(\sin x - \sqrt{3}\right) = 0$.
Answer: medium
$\left(\tan x - \sqrt{3}\right)\left(\sin x - \sqrt{3}\right) = 0 \quad\Leftrightarrow\quad \left(\tan x - \sqrt{3}\right) = 0$ or $\left(\sin x - \sqrt{3}\right) = 0$. $\sqrt{3} > 1$, so $\sin x - \sqrt{3} = 0$ has no solution. $\tan x - \sqrt{3} = 0 \quad\Leftrightarrow\quad \tan x = \sqrt{3}$ has solution the solution $x = \frac{\pi}{3}$ in $(-\frac{\pi}{2}, \frac{\pi}{2})$, so the general solution is $x = \frac{\pi}{3} + k\pi$.

7. Find all solutions of $\tan x \cos x - \cos x = 0$.
Answer: medium
$\tan x \cos x - \cos x = 0 \quad\Leftrightarrow\quad (\tan x - 1)\cos x = 0 \quad\Leftrightarrow\quad \tan x = 1$ or $\cos x = 0$. $\cos x = 0$ has solution $x = \frac{\pi}{2} + k\pi$, and $\tan x = 1$ has solution $x = \frac{\pi}{4} + k\pi$, so the general solutions are $x = \frac{\pi}{2} + k\pi$ and $x = \frac{\pi}{4} + k\pi$.

8. Find all solutions of $4\sin^2 x - 2\left(\sqrt{3} + \sqrt{2}\right)\sin x + \sqrt{6} = 0$.
Answer: medium
$4\sin^2 x - 2\left(\sqrt{3} + \sqrt{2}\right)\sin x + \sqrt{6} = 0 \quad\Leftrightarrow\quad \left(2\sin x - \sqrt{3}\right)\left(2\sin x - \sqrt{2}\right) = 0 \quad\Leftrightarrow\quad 2\sin x - \sqrt{3} = 0$ or $2\sin x - \sqrt{2} = 0 \quad\Leftrightarrow\quad \sin x = \frac{\sqrt{3}}{2}$ or $\sin x = \frac{\sqrt{2}}{2}$. $\sin x = \frac{\sqrt{3}}{2}$ has solutions $x = \frac{\pi}{3}$ and $x = \frac{2\pi}{3}$ in $[0, 2\pi)$ and $\sin x = \frac{\sqrt{2}}{2}$ has solutions $x = \frac{\pi}{4}$ and $x = \frac{3\pi}{4}$ in $[0, 2\pi)$. So the general solutions are $x = \frac{\pi}{3} + 2k\pi$, $x = \frac{2\pi}{3} + 2k\pi$, $x = \frac{\pi}{4} + 2k\pi$ and $x = \frac{2\pi}{4} + 2k\pi$.

9. Find all solutions of $\cos^2 x = \cos x + 6$.
Answer: medium
$\cos^2 x = \cos x + 6 \quad\Leftrightarrow\quad \cos^2 x - \cos x - 6 = 0 \quad\Leftrightarrow\quad (\cos x - 3)(\cos x + 2) = 0$, but this has no solutions since $3 > 1$ and $-2 < -1$.

10. Find all solutions of $\tan^2 x + \left(\sqrt{3} - 1\right)\tan x - \sqrt{3} = 0$.
Answer: medium
$\tan^2 x + \left(\sqrt{3} - 1\right)\tan x - \sqrt{3} = 0 \quad\Leftrightarrow\quad (\tan x - 1)\left(\tan x + \sqrt{3}\right) = 0 \quad\Leftrightarrow\quad \tan x - 1 = 0$ or $\tan x + \sqrt{3} = 0$. These have solutions $x = \frac{\pi}{4}$ and $x = -\frac{\pi}{3}$ in $(-\frac{\pi}{2}, \frac{\pi}{2})$, so the general solutions are $x = \frac{\pi}{4} + k\pi$ and $x = k\pi - \frac{\pi}{3}$.

11. Find all solutions of $\cos^2 x = 1 - 2\sin^2 x$.
Answer: medium
$\cos^2 x = 1 - 2\sin^2 x \Leftrightarrow \cos^2 x + \sin^2 x = 1 - \sin^2 x \Leftrightarrow 1 = 1 - \sin^2 x \Leftrightarrow \sin^2 x = 0 \Leftrightarrow \sin x = 0$, so the general solution is $x = \frac{k\pi}{2}$.

12. Find all solutions of $2\cos 3x - 1 = 0$.
Answer: medium
$2\cos 3x - 1 = 0 \Leftrightarrow \cos 3x = \frac{1}{2}$, which has solutions $3x = \frac{\pi}{3}$ and $3x = \frac{5\pi}{3}$ in $[0, 2\pi)$, and general solutions $3x = \frac{\pi}{3} + 2k\pi \Leftrightarrow x = \frac{\pi}{9} + \frac{2k\pi}{3}$ and $3x = \frac{5\pi}{3} + 2k\pi \Leftrightarrow x = \frac{5\pi}{9} + \frac{2k\pi}{3}$.

13. Find all solutions of $\tan\frac{x}{3} - 1 = 0$.
Answer: medium
$\tan\frac{x}{3} - 1 = 0 \Leftrightarrow \tan\frac{x}{3} = 1$, which has general solution $\frac{x}{3} = \frac{\pi}{4} + k\pi \Leftrightarrow x = \frac{3\pi}{4} + 3k\pi$.

14. Find all solutions of $\tan^4 x \sin x - 9\sin x = 0$.
Answer: hard
$\tan^4 x \sin x - 9\sin x = 0 \Leftrightarrow \sin x(\tan^4 x - 9) = 0 \Leftrightarrow \sin x = 0$ or $\tan^4 x - 9 = 0$. The general solution for $\sin x = 0$ is $x = k\pi$. $\tan^4 x - 9 = 0 \Leftrightarrow (\tan^2 x - 3)(\tan^2 x + 3) = 0 \Leftrightarrow \left(\tan x - \sqrt{3}\right)\left(\tan x + \sqrt{3}\right)(\tan^2 x + 3) = 0$. $\tan^2 x + 3 = 0$ has no solution, $\tan x - \sqrt{3} = 0$ has the general solution $x = \frac{\pi}{3} + k\pi$, and $\tan x + \sqrt{3} = 0$ has the general solution $x = -\frac{\pi}{3} + k\pi$, so the general solutions are $x = k\pi$, $x = \frac{\pi}{3} + k\pi$ and $x = -\frac{\pi}{3} + k\pi$.

15. Find all solutions of $4\sin x\cos x - 2(\sin x + \cos x) + 1 = 0$.
Answer: hard
$4\sin x\cos x - 2(\sin x + \cos x) + 1 = 0 \Leftrightarrow (2\sin x - 1)(2\cos x - 1) = 0 \Leftrightarrow \sin x = \frac{1}{2}$ or $\cos x = \frac{1}{2} \Leftrightarrow x = \frac{\pi}{6} + 2k\pi$, $x = \frac{5\pi}{6} + 2k\pi$ or $x = \frac{\pi}{3} + 2k\pi$, $x = \frac{5\pi}{3} + 2k\pi$.

16. Find all solutions of $\cos^2 \pi x - \sin^2 \pi x = 0$.
Answer: hard
$\cos^2 \pi x - \sin^2 \pi x = 0 \Leftrightarrow 2\cos^2 \pi x - 1 = 0 \Leftrightarrow \cos^2 \pi x = \frac{1}{2} \Leftrightarrow \cos \pi x = \pm\frac{\sqrt{2}}{2}$, which has general solutions $\pi x = \frac{\pi}{4} + 2k\pi$, $\pi x = \frac{7\pi}{4} + 2k\pi$, $\pi x = \frac{3\pi}{4} + 2k\pi$, and $\pi x = \frac{5\pi}{4} + 2k\pi$, these can be written $x = \frac{1}{4} + 2k$, $x = \frac{7}{4} + 2k$, $x = \frac{3}{4} + 2k$, $x = \frac{5}{4} + 2k$, or briefly as $x = \frac{1}{4} + \frac{k}{2}$, where k is an integer.

17. Find all solutions of $\sec 2x = \cos 2x$.
Answer: hard
$\sec 2x = \cos 2x \Leftrightarrow 1 = \cos^2 2x \Leftrightarrow \cos 2x = \pm 1$. Solutions to $\cos 2x = 1$ are $2x = 2k\pi$ and solutions to $\cos 2x = -1$ are $2x = \pi + 2k\pi$, so the general solution is $2x = k\pi$, or $x = \frac{k\pi}{2}$.

18. Find all solutions of $\csc x - \cot x = \sin x$.
Answer: medium
$\csc x - \cot x = \sin x \Leftrightarrow 1 - \cos x = \sin^2 x \Leftrightarrow 1 - \sin^2 x = \cos x \Leftrightarrow \cos^2 x = \cos x \Leftrightarrow \cos x = 0$ or $\cos x = 1$, which have general solutions $x = \frac{\pi}{2} + k\pi$ and $x = 2k\pi$.

19. Find all solutions of $2\sin 4x = 1$ in the interval $[0, 2\pi]$.
Answer: hard
$2\sin 4x = 1 \Leftrightarrow \sin 4x = \frac{1}{2}$. The general solutions to this are $4x = \frac{\pi}{6} + 2k\pi$ and $4x = \frac{5\pi}{6} + 2k\pi$, or $x = \frac{\pi}{24} + \frac{k\pi}{2}$ and $x = \frac{5\pi}{24} + \frac{k\pi}{2}$. Solutions which lie in the interval $[0, 2\pi]$ are $\frac{\pi}{24}$, $\frac{\pi}{24} + \frac{\pi}{2}$, $\frac{\pi}{24} + \pi$, $\frac{\pi}{24} + \frac{3\pi}{2}$, $\frac{5\pi}{24}$, $\frac{5\pi}{24} + \frac{\pi}{2}$, $\frac{5\pi}{24} + \pi$ and $\frac{5\pi}{24} + \frac{3\pi}{2}$.

20. Find all solutions of $3\sec^2 x = 4$ in the interval $[0, 2\pi]$.
Answer: medium
$3\sec^2 x = 4 \Leftrightarrow \sec^2 x = \frac{4}{3} \Leftrightarrow \sin x = \pm\frac{\sqrt{3}}{2}$, whose solutions in the interval $[0, 2\pi]$ are $\frac{\pi}{3}$, $\frac{2\pi}{3}$, $\frac{4\pi}{3}$ and $\frac{5\pi}{3}$.

21. Find all solutions of $\tan x - \frac{1}{3}\cot x = 0$ in the interval $\left[-\frac{\pi}{2}, \frac{\pi}{2}\right]$.

Answer: medium

$\tan x - \frac{1}{3}\cot x = 0 \quad \Leftrightarrow \quad \tan x = \frac{1}{3}\cot x \quad \Leftrightarrow \quad \tan^2 x = \frac{1}{3} \quad \Leftrightarrow \quad \tan x = \pm\frac{\sqrt{3}}{3}$, whose solutions in the interval $\left[-\frac{\pi}{2}, \frac{\pi}{2}\right]$ are $\frac{\pi}{6}$ and $-\frac{\pi}{6}$.

22. Use a calculator to solve $\sin x = 0.3$ to five decimal places on the interval $[0, 2\pi]$, and find all solutions.

Answer: medium

$\sin x = 0.3 \quad \Leftrightarrow \quad x = \sin^{-1} 0.3 \quad \Rightarrow \quad x \approx 0.30469$ or $x \approx \pi - 0.30469 \approx 2.83690$. Thus the general solutions are $x \approx 0.30469 + 2k\pi$ and $x \approx 2.83690 + 2k\pi$.

23. Use a calculator to solve $\sec x - 7 = 0$ to five decimal places on the interval $[0, 2\pi]$, and find all solutions.

Answer: medium

$\sec x - 7 = 0 \quad \Leftrightarrow \quad x = \cos^{-1}\frac{1}{7} \quad \Rightarrow \quad x \approx 1.42745$ or $x \approx 2\pi - 1.42745 \approx 4.85574$. Thus the general solutions are $x \approx 1.42745 + 2k\pi$ and $x \approx 4.85574 + 2k\pi$.

24. Use a calculator to solve $4\cos x = 9\sin x$ to five decimal places on the interval $\left[-\frac{\pi}{2}, \frac{\pi}{2}\right]$, and find all solutions.

Answer: medium

$4\cos x = 9\sin x \quad \Leftrightarrow \quad \tan x = \frac{4}{9} \quad \Leftrightarrow \quad x = \tan^{-1}\frac{4}{9} \quad \Rightarrow \quad x \approx 0.41822$. Thus the general solution is $x \approx 0.41822 + k\pi$.

25. Use a calculator to solve $9\sin^2 x - 2 = 0$ to five decimal places on the interval $[0, 2\pi]$, and find all solutions.

Answer: medium

$9\sin^2 x - 2 = 0 \quad \Leftrightarrow \quad \sin^2 x = \frac{2}{9} \quad \Leftrightarrow \quad \sin x = \pm\frac{\sqrt{2}}{3} \quad \Rightarrow \quad x = \sin^{-1}\frac{\sqrt{2}}{3}$ or $x = \sin^{-1}\left(-\frac{\sqrt{2}}{3}\right) \quad \Rightarrow \quad x \approx 0.49088$, $x \approx \pi - 0.49088 \approx 2.65071$, $x \approx 2\pi - 0.49088 \approx 5.79230$ or $x \approx \pi + 0.49088 \approx 3.63248$. Thus the general solutions are $x \approx 0.49088 + 2k\pi$, $x \approx 2.65071 + 2k\pi$, $x \approx 5.79230 + 2k\pi$ and $x \approx 3.63248 + 2k\pi$.

26. Sketch the graphs of $f(x) = 2\sin x$ and $g(x) = \sin 2x$ on the same axes and find their points of intersection.

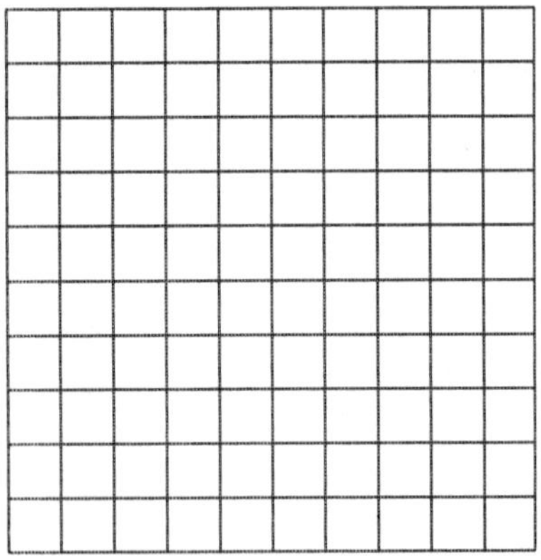

Answer: medium

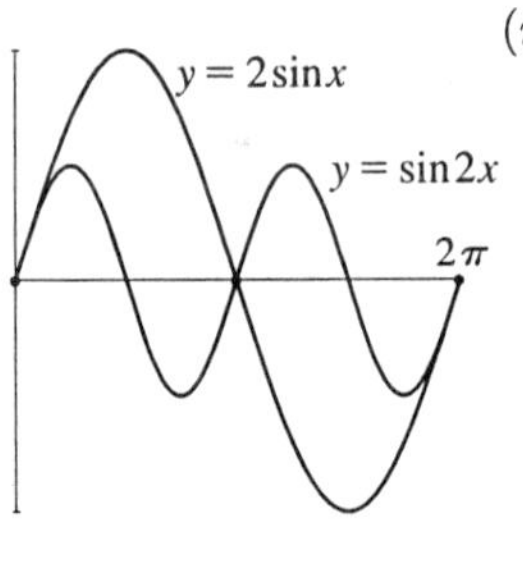

$(n\pi, 0)$, n an integer

27. Sketch the graphs of $f(x) = \sin^2 x$ and $g(x) = \cos^2 x$ on the same axes and find their points of intersection.

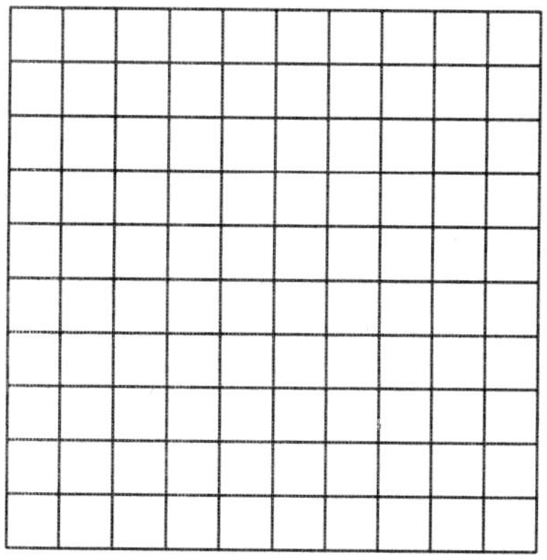

Answer: medium

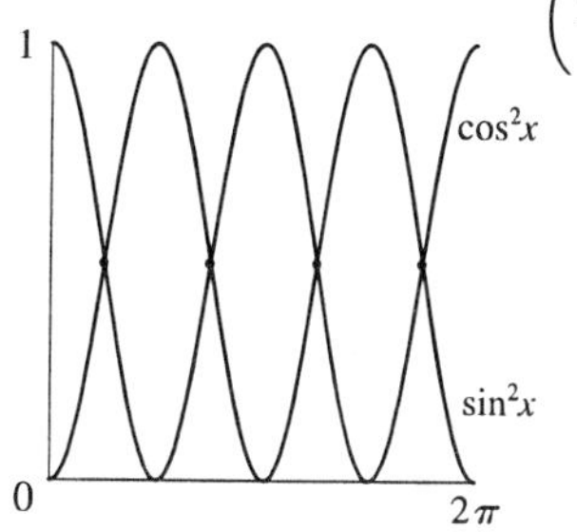

$\left(\dfrac{2n+1}{4}\pi, \dfrac{1}{2}\right)$, n an integer

28. Use an addition or subtraction formula to simplify $\sin x \cos 2x + \cos x \sin 2x = 0$ and then find all solutions in the interval $[0, 2\pi]$.
Answer: medium
$\sin x \cos 2x + \cos x \sin 2x = 0 \quad \Leftrightarrow \quad \sin(x + 2x) = 0 \quad \Leftrightarrow \quad \sin 3x = 0 \quad \Leftrightarrow \quad 3x = k\pi \quad \Leftrightarrow \quad x = \frac{k\pi}{3}$, so the solutions in the interval $[0, 2\pi]$ are $0, \frac{\pi}{3}, \frac{2\pi}{3}, \pi, \frac{4\pi}{3}, \frac{5\pi}{3}$ and 2π.

29. Use a double or half angle formula to simplify $2 \sin \pi x \cos \pi x = 0$ and then find all solutions in the interval $[0, 2\pi]$.
Answer: medium
$2 \sin \pi x \cos \pi x = 0 \quad \Leftrightarrow \quad \sin 2\pi x = 0 \quad \Leftrightarrow \quad 2\pi x = k\pi \quad \Leftrightarrow \quad x = \frac{k}{2}$, so all the solutions in the interval $[0, 2\pi]$ are $0, \frac{1}{2}, 1, \frac{3}{2}, 2, \frac{5}{2}$ and 3.

30. Solve $\cos 3x - \cos 5x = 0$ by first using a sum to product identity.
Answer: medium
$\cos 3x - \cos 5x = 0 \quad \Leftrightarrow \quad -2 \sin 4x \sin(-x) = 0 \quad \Leftrightarrow \quad \sin 4x \sin x = 0 \quad \Leftrightarrow \quad \sin x = 0$ or $\sin 4x = 0 \quad \Leftrightarrow \quad x = k\pi$ or $x = \frac{k\pi}{4}$, so the general solution is $x = \frac{k\pi}{4}$.

SECTION 9.6

1. Write $-1 - i$ in trigonometric form, with θ between 0 and 2π.
Answer: easy
$-1 - i = \sqrt{1^2 + 1^2}\left(\cos \frac{3\pi}{4} + i \sin \frac{3\pi}{4}\right) = \sqrt{2}\left(\cos \frac{3\pi}{4} + i \sin \frac{3\pi}{4}\right)$.

2. Write $1 + \sqrt{3}i$ in trigonometric form, with θ between 0 and 2π.
Answer: easy
$1 + \sqrt{3}i = \sqrt{1^2 + \sqrt{3}^2}\left(\cos \frac{\pi}{3} + i \sin \frac{\pi}{3}\right) = 2\left(\cos \frac{\pi}{3} + i \sin \frac{\pi}{3}\right)$.

3. Write $-2\sqrt{3} - 2i$ in trigonometric form, with θ between 0 and 2π.
Answer: easy
$-2\sqrt{3} - 2i = \sqrt{\left(2\sqrt{3}\right)^2 + 2^2}\left(\cos \frac{7\pi}{6} + i \sin \frac{7\pi}{6}\right) = 4\left(\cos \frac{7\pi}{6} + i \sin \frac{7\pi}{6}\right)$.

4. Write $3 - 3\sqrt{3}i$ in trigonometric form, with θ between 0 and 2π.
Answer: easy
$3 - 3\sqrt{3}i = \sqrt{3^2 + \left(3\sqrt{3}\right)^2}\left(\cos \frac{5\pi}{3} + i \sin \frac{5\pi}{3}\right) = 6\left(\cos \frac{5\pi}{3} + i \sin \frac{5\pi}{3}\right)$.

5. Write $\sqrt{2}i$ in trigonometric form, with θ between 0 and 2π.
   Answer: easy
   $\sqrt{2}i = \sqrt{\sqrt{2}^2}\left(\cos\frac{\pi}{2} + i\sin\frac{\pi}{2}\right) = \sqrt{2}\left(\cos\frac{\pi}{2} + i\sin\frac{\pi}{2}\right).$

6. Write 7 in trigonometric form, with θ between 0 and 2π.
   Answer: easy
   $7 = \sqrt{7^2}(\cos 0 + i\sin 0) = 7(\cos 0 + i\sin 0).$

7. Write $2 + 2i$ in trigonometric form, with θ between 0 and 2π.
   Answer: easy
   $2 + 2i = \sqrt{2^2 + 2^2}\left(\cos\frac{\pi}{4} + i\sin\frac{\pi}{4}\right) = 2\sqrt{2}\left(\cos\frac{\pi}{4} + i\sin\frac{\pi}{4}\right).$

8. Write $i\left(1 - \sqrt{3}i\right)$ in trigonometric form, with θ between 0 and 2π.
   Answer: easy
   $i\left(1 - \sqrt{3}i\right) = \sqrt{3} + i = \sqrt{\sqrt{3}^2 + 1^2}\left(\cos\frac{\pi}{6} + i\sin\frac{\pi}{6}\right) = 2\left(\cos\frac{\pi}{6} + i\sin\frac{\pi}{6}\right).$

9. Write $-3 + 4i$ in trigonometric form, with θ between 0 and 2π.
   Answer: easy
   $-3 + 4i = \sqrt{(-3)^2 + 4^2}\left(\cos\left(\pi + \tan^{-1}\left(-\frac{4}{3}\right)\right) + i\sin\left(\pi + \tan^{-1}\left(-\frac{4}{3}\right)\right)\right) = 5\left(-\frac{3}{5} + \frac{4}{5}\right).$

10. Write $-2(1 + i)$ in trigonometric form, with θ between 0 and 2π.
    Answer: easy
    $-2(1 + i) = \sqrt{(-2)^2 + (-2)^2}\left(\cos\frac{5\pi}{4} + i\sin\frac{5\pi}{4}\right) = 2\sqrt{2}\left(\cos\frac{5\pi}{4} + i\sin\frac{5\pi}{4}\right).$

11. Write $4\left(\sqrt{3} - i\right)$ in trigonometric form, with θ between 0 and 2π.
    Answer: easy
    $4\left(\sqrt{3} - i\right) = \sqrt{\left(4\sqrt{3}\right)^2 + 4^2}\left(\cos\frac{11\pi}{6} + i\sin\frac{11\pi}{6}\right) = 8\left(\cos\frac{11\pi}{6} + i\sin\frac{11\pi}{6}\right).$

12. Write $i\left(3 - \sqrt{3}i\right)$ in trigonometric form, with θ between 0 and 2π.
    Answer: easy
    $i\left(3 - \sqrt{3}i\right) = \sqrt{3} + 3i = \sqrt{\sqrt{3}^2 + 3^2}\left(\cos\frac{\pi}{3} + i\sin\frac{\pi}{3}\right) = 2\sqrt{3}\left(\cos\frac{\pi}{3} + i\sin\frac{\pi}{3}\right).$

13. Write $-\sqrt{2} + \sqrt{2}i$ in trigonometric form, with θ between 0 and 2π.
    Answer: easy
    $-\sqrt{2} + \sqrt{2}i = \sqrt{\sqrt{2}^2 + \sqrt{2}^2}\left(\cos\frac{3\pi}{4} + i\sin\frac{3\pi}{4}\right) = 2\sqrt{2}\left(\cos\frac{3\pi}{4} + i\sin\frac{3\pi}{4}\right).$

14. Write πi in trigonometric form, with θ between 0 and 2π.
    Answer: easy
    $\pi i = \sqrt{\pi^2}\left(\cos\frac{\pi}{2} + i\sin\frac{\pi}{2}\right) = \pi\left(\cos\frac{\pi}{2} + i\sin\frac{\pi}{2}\right).$

15. Write the product $z_1 z_2$, and quotient z_1/z_2, of $z_1 = \cos\frac{\pi}{4} + i\sin\frac{\pi}{4}$ and $z_2 = \cos\frac{\pi}{2} + i\sin\frac{\pi}{2}$ in trigonometric form.
    Answer: medium
    $z_1 z_2 = 1 \cdot 1\left(\cos\left(\frac{\pi}{4} + \frac{\pi}{2}\right) + i\sin\left(\frac{\pi}{4} + \frac{\pi}{2}\right)\right) = \cos\frac{3\pi}{4} + i\sin\frac{3\pi}{4};$
    $z_1/z_2 = 1/1\left(\cos\left(\frac{\pi}{4} - \frac{\pi}{2}\right) + i\sin\left(\frac{\pi}{4} - \frac{\pi}{2}\right)\right) = \cos -\frac{\pi}{4} + i\sin -\frac{\pi}{4} = \cos\frac{7\pi}{4} + i\sin\frac{7\pi}{4}.$

16. Write the product $z_1 z_2$, and quotient z_1/z_2, of $z_1 = 2\left(\cos\frac{5\pi}{6} + i\sin\frac{5\pi}{6}\right)$ and $z_2 = 5\left(\cos\frac{\pi}{3} + i\sin\frac{\pi}{3}\right)$ in trigonometric form.
    Answer: medium
    $z_1 z_2 = 2 \cdot 5\left(\cos\left(\frac{5\pi}{6} + \frac{\pi}{3}\right) + i\sin\left(\frac{5\pi}{6} + \frac{\pi}{3}\right)\right) = 10\left(\cos\frac{7\pi}{6} + i\sin\frac{7\pi}{6}\right);$
    $z_1/z_2 = 2/5\left(\cos\left(\frac{5\pi}{6} - \frac{\pi}{3}\right) + i\sin\left(\frac{5\pi}{6} - \frac{\pi}{3}\right)\right) = \frac{2}{5}\left(\cos\frac{\pi}{2} + i\sin\frac{\pi}{2}\right).$

17. Write the product z_1z_2, and quotient z_1/z_2, of $z_1 = \frac{3}{4}(\cos 35° + i \sin 35°)$ and $z_2 = \frac{4}{7}(\cos 115° + i \sin 115°)$ in trigonometric form.

Answer: medium

$z_1z_2 = \frac{3}{4} \cdot \frac{4}{7}(\cos(35° + 115°) + i \sin(35° + 115°)) = \frac{3}{7}(\cos 150° + i \sin 150°)$;

$z_1/z_2 = \frac{3}{4}/\frac{4}{7}(\cos(35° - 115°) + i \sin(35° - 115°)) = \frac{21}{16}(\cos -80° + i \sin -80°) = \frac{21}{16}(\cos 280° + i \sin 280°)$.

18. Write the product z_1z_2, and quotient z_1/z_2, of $z_1 = \frac{2}{3}(\cos 150° + i \sin 150°)$ and $z_2 = \frac{1}{3}(\cos 250° + i \sin 250°)$ in trigonometric form.

Answer: medium

$z_1z_2 = \frac{2}{3} \cdot \frac{1}{3}(\cos(150° + 250°) + i \sin(150° + 250°)) = \frac{2}{9}(\cos 400° + i \sin 400°) = \frac{2}{9}(\cos 40° + i \sin 40°)$;

$z_1/z_2 = \frac{2}{3}/\frac{1}{3}(\cos(150° - 250°) + i \sin(150° - 250°)) = 2(\cos(-100°) + i \sin(-100°)) = 2(\cos(260°) + i \sin(260°))$.

19. Write $z_1 = 1 - i$ and $z_2 = 1 - \sqrt{3}i$ in trigonometric form and then find z_1z_2, z_1/z_2, and $1/z_1$.

Answer: hard

$z_1 = 1 - i = \sqrt{1^2 + 1^2}(\cos \frac{7\pi}{4} + i \sin \frac{7\pi}{4}) = \sqrt{2}(\cos \frac{7\pi}{4} + i \sin \frac{7\pi}{4})$;

$z_2 = 1 - \sqrt{3}i = \sqrt{1^2 + \sqrt{3}^2}(\cos \frac{5\pi}{3} + i \sin \frac{5\pi}{3}) = 2(\cos \frac{5\pi}{3} + i \sin \frac{5\pi}{3})$;

$z_1z_2 = 2\sqrt{2}(\cos(\frac{7\pi}{4} + \frac{5\pi}{3}) + i \sin(\frac{7\pi}{4} + \frac{5\pi}{3})) = 2\sqrt{2}(\cos \frac{41\pi}{12} + i \sin \frac{41\pi}{12})$;

$z_1/z_2 = 2/\sqrt{2}(\cos(\frac{7\pi}{4} - \frac{5\pi}{3}) + i \sin(\frac{7\pi}{4} - \frac{5\pi}{3})) = \sqrt{2}(\cos \frac{\pi}{12} + i \sin \frac{\pi}{12})$;

$1/z_1 = 1/\sqrt{2}(\cos -\frac{7\pi}{4} + i \sin -\frac{7\pi}{4}) = \sqrt{2}/2(\cos \frac{\pi}{4} + i \sin \frac{\pi}{4})$.

20. Write $z_1 = \sqrt{2} + \sqrt{2}i$ and $z_2 = \sqrt{2} - \sqrt{2}i$ in trigonometric form and then find z_1z_2, z_1/z_2, and $1/z_1$.

Answer: hard

$z_1 = \sqrt{2} + \sqrt{2}i = \sqrt{\sqrt{2}^2 + \sqrt{2}^2}(\cos \frac{\pi}{4} + i \sin \frac{\pi}{4}) = 2\sqrt{2}(\cos \frac{\pi}{4} + i \sin \frac{\pi}{4})$;

$z_2 = \sqrt{2} - \sqrt{2}i = \sqrt{\sqrt{2}^2 + \sqrt{2}^2}(\cos \frac{7\pi}{4} + i \sin \frac{7\pi}{4}) = 2\sqrt{2}(\cos \frac{7\pi}{4} + i \sin \frac{7\pi}{4})$;

$z_1z_2 = (2\sqrt{2})(2\sqrt{2})(\cos(\frac{\pi}{4} + \frac{7\pi}{4}) + i \sin(\frac{\pi}{4} + \frac{7\pi}{4})) = 8(\cos 0 + i \sin 0)$;

$z_1/z_2 = (2\sqrt{2})/(2\sqrt{2})(\cos(\frac{\pi}{4} - \frac{7\pi}{4}) + i \sin(\frac{\pi}{4} - \frac{7\pi}{4})) = \cos \frac{\pi}{2} + i \sin \frac{\pi}{2}$;

$1/z_1 = 1/(2\sqrt{2})(\cos -\frac{\pi}{4} + i \sin -\frac{\pi}{4}) = \sqrt{2}/4(\cos \frac{7\pi}{4} + i \sin \frac{7\pi}{4})$.

21. Write $z_1 = 3$ and $z_2 = 6 + 6i$ in trigonometric form and then find z_1z_2, z_1/z_2, and $1/z_1$.

Answer: hard

$z_1 = 3 = 3(\cos 0 + i \sin 0)$; $z_2 = 6 + 6i = \sqrt{6^2 + 6^2}(\cos \frac{\pi}{4} + i \sin \frac{\pi}{4}) = 6\sqrt{2}(\cos \frac{\pi}{4} + i \sin \frac{\pi}{4})$;

$z_1z_2 = 3 \cdot 6\sqrt{2}(\cos(0 + \frac{\pi}{4}) + i \sin(0 + \frac{\pi}{4})) = 18\sqrt{2}(\cos \frac{\pi}{4} + i \sin \frac{\pi}{4})$;

$z_1/z_2 = 3/(6\sqrt{2})(\cos(0 - \frac{\pi}{4}) + i \sin(0 - \frac{\pi}{4})) = \sqrt{2}/4(\cos \frac{7\pi}{4} + i \sin \frac{7\pi}{4})$; $1/z_1 = 1/3(\cos 0 + i \sin 0)$.

22. Find $(3 - \sqrt{3}i)^3$ using DeMoivre's Theorem.

Answer: medium

$3 - \sqrt{3}i = \sqrt{3^2 + \sqrt{3}^2}(\cos \frac{5\pi}{3} + i \sin \frac{5\pi}{3}) = 2\sqrt{3}(\cos \frac{5\pi}{3} + i \sin \frac{5\pi}{3})$, so

$(3 - \sqrt{3}i)^3 = (2\sqrt{3})^3(\cos 3\frac{5\pi}{3} + i \sin 3\frac{5\pi}{3}) = (2\sqrt{3})^3(\cos 3\frac{5\pi}{3} + i \sin 3\frac{5\pi}{3}) = 8 \cdot 3^{3/2}(\cos 5\pi + i \sin 5\pi) =$
$8 \cdot 3^{3/2}(\cos \pi + i \sin \pi)$.

23. Find $(2\sqrt{3} - 2i)^{12}$ using DeMoivre's Theorem.

Answer: medium

$2\sqrt{3} - 2i = \sqrt{(2\sqrt{3})^2 + 2^2}(\cos \frac{11\pi}{6} + i \sin \frac{11\pi}{6}) = 4(\cos \frac{11\pi}{6} + i \sin \frac{11\pi}{6})$, so $(2\sqrt{3} - 2i)^{12} =$
$4^{12}(\cos 12\frac{11\pi}{6} + i \sin 12\frac{11\pi}{6}) = 4^{12}(\cos 22\pi + i \sin 22\pi) = 4^{12}(\cos 0 + i \sin 0)$.

24. Find $(-2+2i)^{16}$ using DeMoivre's Theorem.

Answer: medium

$-2+2i = \sqrt{2^2+2^2}\left(\cos\frac{3\pi}{4} + i\sin\frac{3\pi}{4}\right) = 2\sqrt{2}\left(\cos\frac{3\pi}{4} + i\sin\frac{3\pi}{4}\right)$, so

$(-2+2i)^{16} = \left(2\sqrt{2}\right)^{16}\left(\cos 16\frac{3\pi}{4} + i\sin 16\frac{3\pi}{4}\right) = 8^8(\cos 12\pi + i\sin 12\pi) = 8^8(\cos 0 + i\sin 0)$.

25. Find $(-1+i)^5$ using DeMoivre's Theorem.

Answer: medium

$-1+i = \sqrt{1^2+1^2}\left(\cos\frac{3\pi}{4} + i\sin\frac{3\pi}{4}\right) = \sqrt{2}\left(\cos\frac{3\pi}{4} + i\sin\frac{3\pi}{4}\right)$, so $(-1+i)^5 =$
$\sqrt{2}^5\left(\cos 5\frac{3\pi}{4} + i\sin 5\frac{3\pi}{4}\right) = \sqrt{2}^5\left(\cos\frac{15\pi}{4} + i\sin\frac{15\pi}{4}\right) = \sqrt{2}^5\left(\cos\frac{7\pi}{4} + i\sin\frac{7\pi}{4}\right)$.

26. Find $\left(3+\sqrt{3}i\right)^{-2}$ using DeMoivre's Theorem.

Answer: medium

$3+\sqrt{3}i = \sqrt{3^2+\sqrt{3}^2}\left(\cos\frac{\pi}{6} + i\sin\frac{\pi}{6}\right) = 2\sqrt{3}\left(\cos\frac{\pi}{6} + i\sin\frac{\pi}{6}\right)$, so $\left(3+\sqrt{3}i\right)^{-2} =$
$\left(2\sqrt{3}\right)^{-2}\left(\cos\left(-2\frac{\pi}{6}\right) + i\sin\left(-2\frac{\pi}{6}\right)\right) = \frac{1}{12}\left(\cos\left(-\frac{\pi}{3}\right) + i\sin\left(-\frac{\pi}{3}\right)\right) = \frac{1}{12}\left(\cos\frac{5\pi}{3} + i\sin\frac{5\pi}{3}\right)$.

27. Find and sketch the square roots of $3+\sqrt{3}i$.

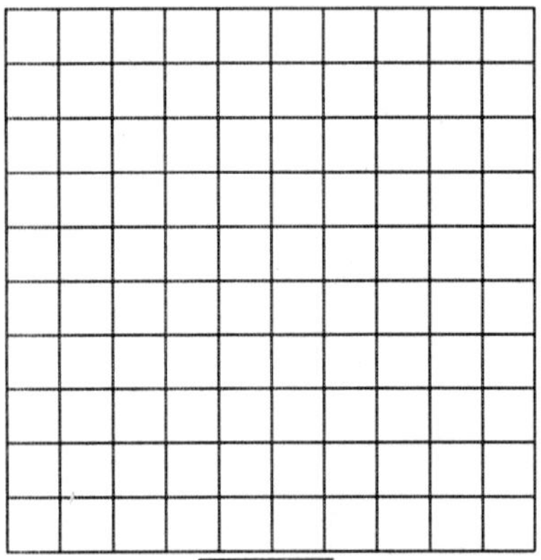

Answer: medium

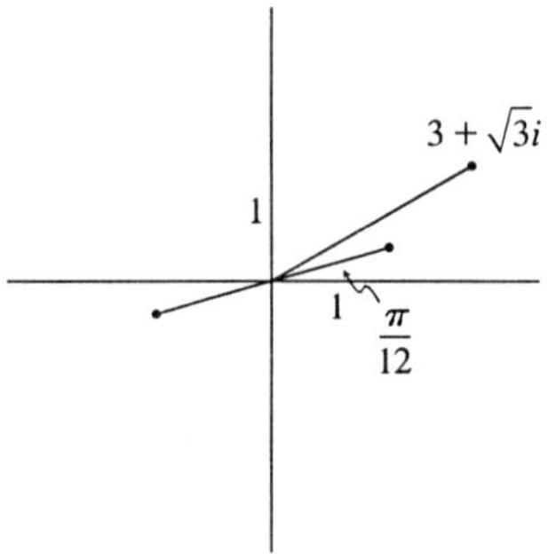

$3+\sqrt{3}i = \sqrt{3^2+\sqrt{3}^2}\left(\cos\frac{\pi}{6} + i\sin\frac{\pi}{6}\right) = 2\sqrt{3}\left(\cos\frac{\pi}{6} + i\sin\frac{\pi}{6}\right)$, so
$\left(3+\sqrt{3}i\right)^{1/2} = \left(2\sqrt{3}\right)^{1/2}\left(\cos\left(\frac{1}{2}\left(\frac{\pi}{6}+2k\pi\right)\right) + i\sin\left(\frac{1}{2}\left(\frac{\pi}{6}+2k\pi\right)\right)\right)$. There are 2 distinct square roots, and they are
$\sqrt{2}\sqrt[4]{3}\left(\cos\frac{\pi}{12} + i\sin\frac{\pi}{12}\right)$, $\sqrt{2}\sqrt[4]{3}\left(\cos\frac{13\pi}{12} + i\sin\frac{13\pi}{12}\right)$.

28. Find and sketch the fifth roots of 1.

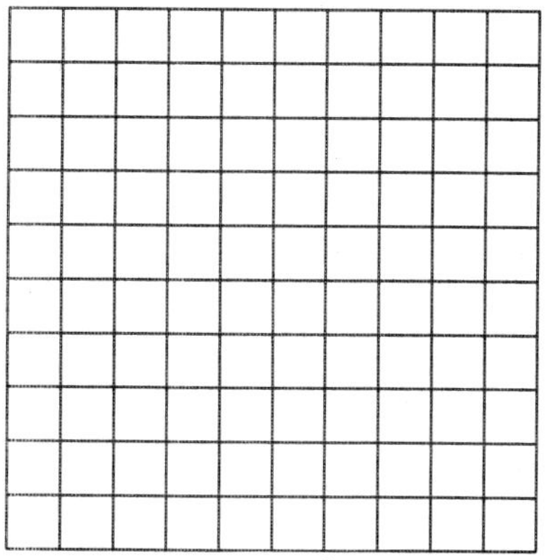

Answer: medium

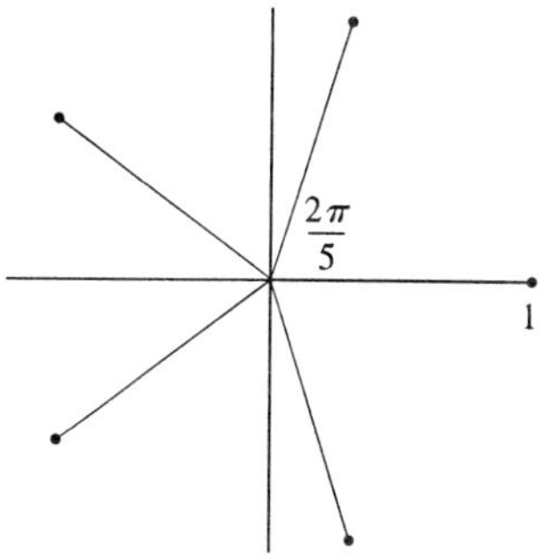

$1 = \cos 0 + i \sin 0$, so $1^{\frac{1}{5}} = \cos \frac{1}{5} 2k\pi + i \sin \frac{1}{5} 2k\pi$. There are 5 distinct fifth roots, and they are $\cos 0 + i \sin 0$, $\cos \frac{2}{5}\pi + i \sin \frac{2}{5}\pi$, $\cos \frac{4}{5}\pi + i \sin \frac{4}{5}\pi$, $\cos \frac{6}{5}\pi + i \sin \frac{6}{5}\pi$ and $\cos \frac{8}{5}\pi + i \sin \frac{8}{5}\pi$

29. Solve $z^6 + 1 = 0$.

Answer: medium

$z^6 + 1 = 0 \quad \Leftrightarrow \quad z^6 = -1 \quad \Leftrightarrow \quad z^6 = \cos \pi + i \sin \pi \quad \Leftrightarrow \quad z = \cos \frac{1}{6}(\pi + 2k\pi) + i \sin \frac{1}{6}(\pi + 2k\pi) \quad \Rightarrow$ $z = \cos \frac{\pi}{6} + i \sin \frac{\pi}{6}$, $\cos \frac{3\pi}{6} + i \sin \frac{3\pi}{6}$, $\cos \frac{5\pi}{6} + i \sin \frac{5\pi}{6}$, $\cos \frac{7\pi}{6} + i \sin \frac{7\pi}{6}$, $\cos \frac{9\pi}{6} + i \sin \frac{9\pi}{6}$, or $\cos \frac{11\pi}{6} + i \sin \frac{11\pi}{6}$.

30. Solve $z^3 - i = 0$.

Answer: medium

$z^3 - i = 0 \quad \Leftrightarrow \quad z^3 = i \quad \Leftrightarrow \quad z^3 = \cos \frac{\pi}{2} + i \sin \frac{\pi}{2} \quad \Leftrightarrow \quad z = \cos \frac{1}{3}\left(\frac{\pi}{2} + 2k\pi\right) + i \sin \frac{1}{3}\left(\frac{\pi}{2} + 2k\pi\right) \quad \Rightarrow$ $z = \cos \frac{\pi}{6} + i \sin \frac{\pi}{6}$, $\cos \frac{5\pi}{6} + i \sin \frac{5\pi}{6}$, or $\cos \frac{9\pi}{6} + i \sin \frac{9\pi}{6}$.

31. Solve $z^2 - \sqrt{2}i = \sqrt{2}$.

Answer: medium

$z^2 - \sqrt{2}i = \sqrt{2} \quad \Leftrightarrow \quad z^2 = \sqrt{2} + \sqrt{2}i \quad \Leftrightarrow \quad z^2 = 2\left(\cos \frac{\pi}{4} + i \sin \frac{\pi}{4}\right) \quad \Leftrightarrow$ $z = \sqrt{2}\left(\cos\left(\frac{\pi}{4} + 2k\pi\right) + i \sin\left(\frac{\pi}{4} + 2k\pi\right)\right) \quad \Rightarrow \quad z = \sqrt{2}\left(\cos \frac{\pi}{4} + i \sin \frac{\pi}{4}\right)$ or $\sqrt{2}\left(\cos \frac{9\pi}{4} + i \sin \frac{9\pi}{4}\right)$.

SECTION 9.7

1. Given **u** and **v** in the figure sketch **u** + **v**.

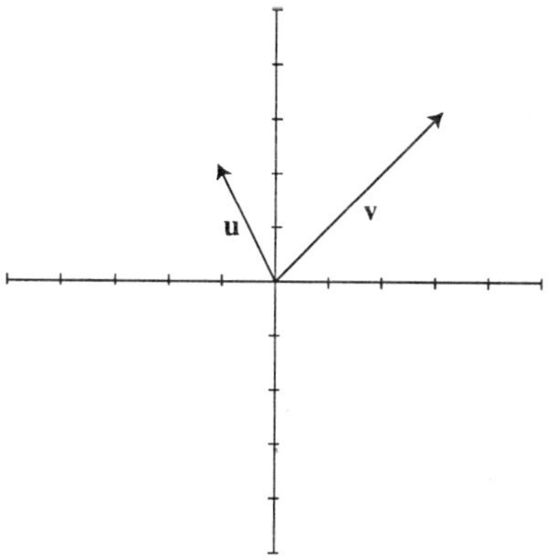

Answer: easy

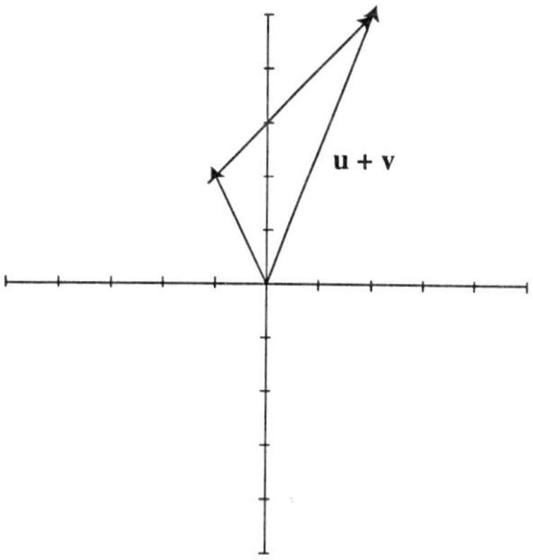

2. Given **u** and **v** in the figure sketch **v** − **u**.

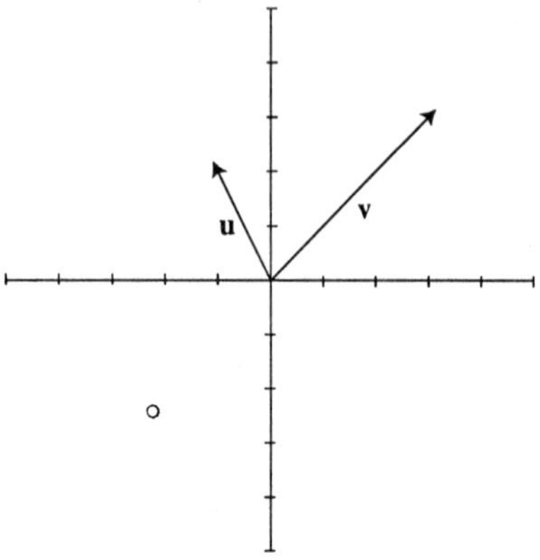

Answer: easy

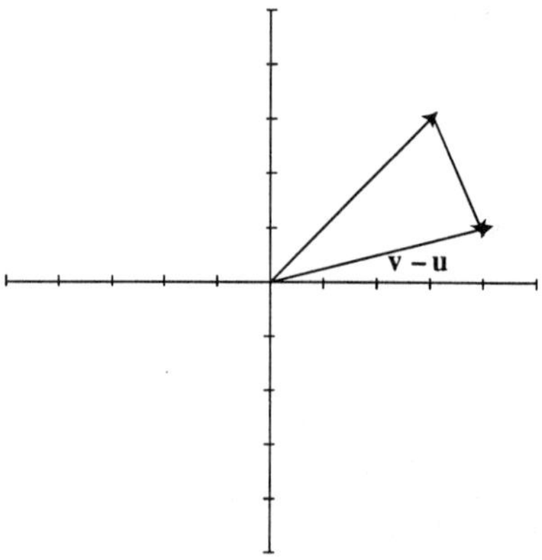

3. Given **u** and **v** in the figure sketch 2**u** − **v**.

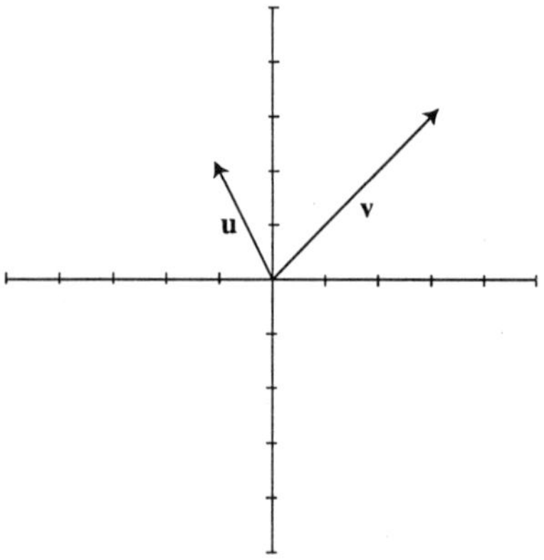

Answer: easy

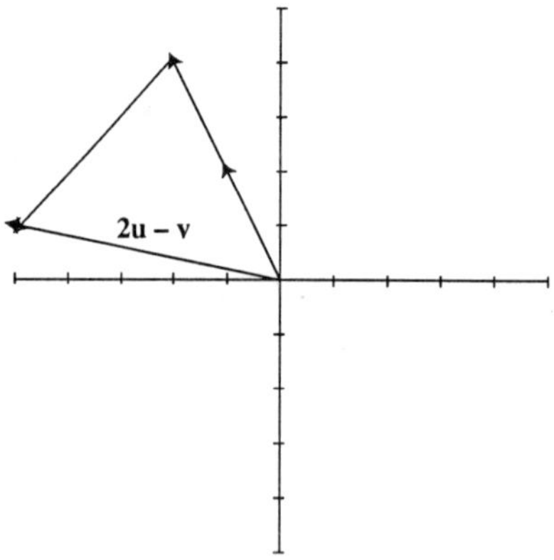

4. Find the vector with initial point $P(4, 2)$ and terminal point $Q(3, 5)$.
Answer: easy
The vector is given by $\langle 3 - 4, 5 - 2 \rangle = \langle -1, 3 \rangle$.

5. Find the vector with initial point $P(5, 1)$ and terminal point $Q(2, 2)$.
Answer: easy
The vector is given by $\langle 2 - 5, 2 - 1 \rangle = \langle -3, 1 \rangle$.

6. Find the vector with initial point $P(7, 3)$ and terminal point $Q(1, 4)$.
Answer: easy
The vector is given by $\langle 1 - 7, 4 - 3 \rangle = \langle -6, 1 \rangle$.

7. Find the vector with initial point $P(-2, 1)$ and terminal point $Q(3, 3)$.
Answer: easy
The vector is given by $\langle 3 - (-2), 3 - 1 \rangle = \langle 5, 2 \rangle$.

8. Find the vector with initial point $P(2, 4)$ and terminal point $Q(-3, 1)$.
Answer: easy
The vector is given by $\langle -3 - 2, 1 - 4 \rangle = \langle -5, -3 \rangle$.

9. Find the vector with initial point $P(6,3)$ and terminal point $Q(-6,-3)$.
Answer: easy
The vector is given by $\langle -6-6, -3-3\rangle = \langle -12, -6\rangle$.

10. Find $2\mathbf{u}$ and $\mathbf{u}+\mathbf{v}$, given that $\mathbf{u} = \langle 3, 2\rangle$ and $\mathbf{v} = \langle -1, 2\rangle$.
Answer: easy
$2\mathbf{u} = \langle 2\cdot 3, 2\cdot 2\rangle = \langle 6, 4\rangle$ and $\mathbf{u}+\mathbf{v} = \langle 3+(-1), 2+2\rangle = \langle 2, 4\rangle$.

11. Find $2\mathbf{u}$ and $\mathbf{u}+\mathbf{v}$, given that $\mathbf{u} = \langle -2, 3\rangle$ and $\mathbf{v} = \langle 4, 5\rangle$.
Answer: easy
$2\mathbf{u} = \langle 2\cdot(-2), 2\cdot 3\rangle = \langle -4, 6\rangle$ and $\mathbf{u}+\mathbf{v} = \langle -2+4, 3+5\rangle = \langle 2, 8\rangle$.

12. Find $2\mathbf{u}$ and $\mathbf{u}+\mathbf{v}$, given that $\mathbf{u} = \langle 1, -1\rangle$ and $\mathbf{v} = \langle -3, 7\rangle$.
Answer: easy
$2\mathbf{u} = \langle 2\cdot 1, 2\cdot(-1)\rangle = \langle 2, -2\rangle$ and $\mathbf{u}+\mathbf{v} = \langle 1+(-3), -1+7\rangle = \langle -2, 6\rangle$.

13. Find $-\mathbf{v}$ and $2\mathbf{u}+3\mathbf{v}$, given that $\mathbf{u} = 2\mathbf{i}$ and $\mathbf{v} = \mathbf{i}+\mathbf{j}$.
Answer: easy
$-\mathbf{v} = -\mathbf{i}-\mathbf{j}$, $2\mathbf{u}+3\mathbf{v} = 4\mathbf{i}+3\mathbf{i}+3\mathbf{j} = 7\mathbf{i}+3\mathbf{j}$.

14. Find $-\mathbf{v}$ and $2\mathbf{u}+3\mathbf{v}$, given that $\mathbf{u} = -2\mathbf{i}+\mathbf{j}$ and $\mathbf{v} = \mathbf{i}-\mathbf{j}$.
Answer: easy
$-\mathbf{v} = -\mathbf{i}+\mathbf{j}$, $2\mathbf{u}+3\mathbf{v} = -4\mathbf{i}+2\mathbf{j}+3\mathbf{i}-3\mathbf{j} = -\mathbf{i}-\mathbf{j}$.

15. Find $\mathbf{u}-\mathbf{v}$ and $-2\mathbf{u}+\mathbf{v}$, given that $\mathbf{u} = 2\mathbf{i}-\mathbf{j}$ and $\mathbf{v} = \mathbf{i}-2\mathbf{j}$.
Answer: easy
$\mathbf{u}-\mathbf{v} = (2\mathbf{i}-\mathbf{j})-(\mathbf{i}-2\mathbf{j}) = \mathbf{i}+\mathbf{j}$, $-2\mathbf{u}+\mathbf{v} = -2(2\mathbf{i}-\mathbf{j})+\mathbf{i}-2\mathbf{j} = -3\mathbf{i}$.

16. Find $|\mathbf{v}|$ and $|\mathbf{u}+\mathbf{v}|$, given that $\mathbf{u} = 2\mathbf{i}-\mathbf{j}$ and $\mathbf{v} = \mathbf{i}$.
Answer: easy
$|\mathbf{v}| = |\mathbf{i}| = 1$, $|\mathbf{u}+\mathbf{v}| = |2\mathbf{i}-\mathbf{j}+\mathbf{i}| = |3\mathbf{i}-\mathbf{j}| = \sqrt{3^2+(-1)^2} = \sqrt{10}$.

17. Find $\left|\frac{1}{2}\mathbf{v}\right|$, $|\mathbf{u}-\mathbf{v}|$ and $|\mathbf{u}|-|\mathbf{v}|$, given that $\mathbf{u} = -2\mathbf{i}+\mathbf{j}$ and $\mathbf{v} = \mathbf{i}-\mathbf{j}$.
Answer: easy
$\left|\frac{1}{2}\mathbf{v}\right| = \left|\frac{1}{2}\mathbf{i}-\frac{1}{2}\mathbf{j}\right| = \sqrt{\left(\frac{1}{2}\right)^2+\left(\frac{1}{2}\right)^2} = \sqrt{\frac{1}{2}} = \frac{\sqrt{2}}{2}$, $|\mathbf{u}-\mathbf{v}| = |-3\mathbf{i}| = 3$ and
$|\mathbf{u}|-|\mathbf{v}| = |-2\mathbf{i}+\mathbf{j}|-|\mathbf{i}-\mathbf{j}| = \sqrt{2^2+1}-\sqrt{1+1} = \sqrt{5}-\sqrt{2}$.

18. Find $|\mathbf{v}|$ and $|\mathbf{u}+\mathbf{v}|$, given that $\mathbf{u} = \langle 2, 3\rangle$ and $\mathbf{v} = \langle -1, 2\rangle$.
Answer: easy
$|\mathbf{v}| = |\langle -1, 2\rangle| = \sqrt{1^2+2^2} = \sqrt{5}$, $|\mathbf{u}+\mathbf{v}| = |\langle 2-1, 3+2\rangle| = |\langle -1, 5\rangle| = \sqrt{1^2+5^2} = \sqrt{26}$.

19. Find $\left|\frac{1}{2}\mathbf{v}\right|$, $|\mathbf{u}-\mathbf{v}|$ and $|\mathbf{u}|-|\mathbf{v}|$, given that $\mathbf{u} = \langle 1,3\rangle$ and $\mathbf{v} = \langle -2, 1\rangle$.
Answer: easy
$\left|\frac{1}{2}\mathbf{v}\right| = \left|\langle -1, \frac{1}{2}\rangle\right| = \sqrt{1^2+\left(\frac{1}{2}\right)^2} = \sqrt{\frac{5}{4}} = \frac{\sqrt{5}}{2}$, $|\mathbf{u}-\mathbf{v}| = |\langle 3, 2\rangle| = \sqrt{3^2+2^2} = \sqrt{13}$ and
$|\mathbf{u}|-|\mathbf{v}| = |\langle 1,3\rangle|-|\langle -2, 1\rangle| = \sqrt{1^2+3^2}-\sqrt{2^2+1} = \sqrt{10}-\sqrt{5}$.

20. Find the horizontal and vertical components of the vector with length $|\mathbf{v}| = 12$, and direction $\theta = 60°$.
Answer: easy
$\mathbf{v} = \langle a, b\rangle = \langle |\mathbf{v}|\cos\theta, |\mathbf{v}|\sin\theta\rangle = \langle 12\cos 60°, 12\sin 60°\rangle = \langle 6, 6\sqrt{3}\rangle$.

21. Find the horizontal and vertical components of the vector with length $|\mathbf{v}| = 2$, and direction $\theta = 135°$.
Answer: easy
$\mathbf{v} = \langle a, b\rangle = \langle |\mathbf{v}|\cos\theta, |\mathbf{v}|\sin\theta\rangle = \langle 2\cos 135°, 2\sin 135°\rangle = \langle -\sqrt{2}, \sqrt{2}\rangle$.

22. Find the horizontal and vertical components of the vector with length $|\mathbf{v}| = 240$, and direction $\theta = 300°$.
Answer: easy
$\mathbf{v} = \langle a, b\rangle = \langle |\mathbf{v}| \cos\theta, |\mathbf{v}| \sin\theta\rangle = \langle 240 \cos 300°, 240 \sin 300°\rangle = \langle 120, -120\sqrt{3}\rangle$.

23. A jet is flying in a direction N45°E with a speed of 430 mi/h, find the north and east components of the velocity..
Answer: easy
The north component is $430 \cos 45° = 215\sqrt{2}$ mi/h, the east component is $430 \sin 45° = 215\sqrt{2}$ mi/h.

24. Find the magnitude and direction (in degrees) of the vector $\mathbf{v} = \langle\sqrt{3}, 1\rangle$.
Answer: easy
The magnitude is $|\mathbf{v}| = \left|\langle\sqrt{3}, 1\rangle\right| = \sqrt{3+1} = 2$, direction is $\tan^{-1}\frac{1}{\sqrt{3}} = \frac{\pi}{6} = 30°$.

25. Find the magnitude and direction (in degrees) of the vector $\mathbf{v} = \mathbf{i} - \mathbf{j}$.
Answer: easy
The magnitude is $|\mathbf{v}| = |\mathbf{i} - \mathbf{j}| = \sqrt{1+1} = \sqrt{2}$, direction is $\tan^{-1}(-1) = -\frac{\pi}{4} = \frac{7\pi}{4} = 315°$.

26. Find the magnitude and direction (in degrees) of the vector $\mathbf{v} = -3\mathbf{i} + 2\mathbf{j}$.
Answer: easy
The magnitude is $|\mathbf{v}| = |-3\mathbf{i} + 2\mathbf{j}| = \sqrt{3^2 + 2^2} = \sqrt{13}$, direction is $180° + \tan^{-1}\left(-\frac{2}{3}\right) \approx 146.3°$.

27. Find the magnitude and direction (in degrees) of the vector $\mathbf{v} = \langle 5, 12\rangle$.
Answer: easy
The magnitude is $|\mathbf{v}| = |\langle 5, 12\rangle| = \sqrt{25 + 144} = 13$, direction is $\tan^{-1}\left(\frac{5}{12}\right) \approx 22.62°$.

28. A pilot flies a jet due north with speed 400 mi/h, the wind is blowing due west with a speed of 60 mi/h, find the true velocity of the jet as a vector, and the true speed and direction of the jet.
Answer: easy
The velocity of the jet is $\mathbf{v} = 400\mathbf{i} - 60\mathbf{j}$, the speed of the jet is $|\mathbf{v}| = \sqrt{400^2 + 60^2} = \sqrt{163600} \approx 404.5$ mi/h, and the direction of the jet is $180° + \tan^{-1} - \frac{60}{400} \approx 171.5°$.

29. Given that the forces $\mathbf{F}_1 = \langle 3, 4\rangle$, $\mathbf{F}_2 = \langle -1, 1\rangle$ and $\mathbf{F}_3 = \langle -1, -5\rangle$ are acting on a point P, find the resultant force and the additional force required in order for the forces to be in equilibrium.
Answer: medium
The resultant force is $\mathbf{F}_1 + \mathbf{F}_2 + \mathbf{F}_3 = \langle 3, 4\rangle + \langle -1, 1\rangle + \langle -1, -5\rangle = \langle 1, 0\rangle$, and so the force required to put the forces into equilibrium is $\mathbf{F}_4 = \langle -1, 0\rangle$.

30. Given that the forces $\mathbf{F}_1 = \mathbf{i} - \mathbf{j}$, $\mathbf{F}_2 = 3\mathbf{i} + \mathbf{j}$ and $\mathbf{F}_3 = -2\mathbf{i} - \mathbf{j}$ are acting on a point P, find the resultant force and the additional force required in order for the forces to be in equilibrium.
Answer: medium
The resultant force is $\mathbf{F}_1 + \mathbf{F}_2 + \mathbf{F}_3 = \mathbf{i} - \mathbf{j} + 3\mathbf{i} + \mathbf{j} + (-2\mathbf{i}) - \mathbf{j} = 2\mathbf{i} - \mathbf{j}$, and so the force required to put the forces into equilibrium is $\mathbf{F}_4 = -2\mathbf{i} + \mathbf{j}$.

SECTION 9.8

1. Find the dot product of the vectors $\langle 1, 2\rangle$ and $\langle 4, 5\rangle$.

(a) 14 (b) 15 (c) 17 (d) 19 (e) 21

Answer: easy (a)

2. What value of x will cause the two vectors $\langle x, 3\rangle$ and $\langle 4, 5\rangle$ to be perpendicular?

(a) $-\frac{15}{4}$ (b) $-\frac{3}{20}$ (c) $-\frac{20}{3}$ (d) $\frac{3}{20}$ (e) $\frac{20}{3}$

Answer: medium (a)

3. Find the cosine of the angle between the two vectors $\langle 1, 2\rangle$ and $\langle -2, 1\rangle$.

(a) $\frac{1}{\sqrt{2}}$ (b) 0 (c) $\frac{1}{\sqrt{3}}$ (d) $\frac{1}{2}$ (e) $\frac{3}{4}$

Answer: medium (b)

4. Find the cosine of the angle between the two vectors $\langle 1, 2\rangle$ and $\langle 2, 1\rangle$.

(a) $\frac{1}{\sqrt{2}}$ (b) 0 (c) $\frac{1}{\sqrt{3}}$ (d) $\frac{1}{\sqrt{5}}$ (e) $\frac{4}{5}$

Answer: medium (e)

5. Find the cosine of the angle between the two vectors $\langle 1, 2\rangle$ and $\langle 3, 4\rangle$.

(a) $13/\sqrt{117}$ (b) $13/\sqrt{125}$ (c) $11/\sqrt{117}$ (d) $11/\sqrt{125}$ (e) $11/\sqrt{132}$

Answer: medium (d)

6. A constant force with vector representation $\mathbf{F} = \mathbf{i} + 2\mathbf{j}$ moves an object along a straight line from the point $(2, 4)$ to the point $(5, 7)$. Find the work done in foot-pounds if force is measured in pounds and distance is measured in feet.

(a) 7 (b) 8 (c) 9 (d) 10 (e) 11

Answer: medium (c)

7. A person pushing a lawnmower exerts a force of 30 pounds in the direction of the handle, which makes an angle of 30° with the ground. How much work is done in moving the lawnmower 20 feet?

(a) $300\sqrt{3}$ (b) $\frac{300}{\sqrt{3}}$ (c) $\frac{200}{\sqrt{3}}$ (d) 200 (e) 400

Answer: medium (a)

8. Determine whether $\langle 3, 6\rangle$ and $\langle 12, -18\rangle$ are orthogonal.

Answer: easy

No

9. Determine whether $\langle 3, 6\rangle$ and $\langle 6, -12\rangle$ are orthogonal.

Answer: easy

No

10. Determine whether $\langle 1, 1\rangle$ and $\langle 4, -4\rangle$ are orthogonal.

Answer: easy

Yes

11. Determine whether $\langle 3, 4\rangle$ and $\langle 16, -12\rangle$ are orthogonal.

Answer: easy

Yes

12. Find the component of $\langle 3, 12\rangle$ along $\langle 2, 4\rangle$.
Answer: easy
$\frac{27\sqrt{5}}{5}$

13. Find the component of $12\mathbf{i} - 16\mathbf{j}$ along $60\mathbf{i} + 72\mathbf{j}$.
Answer: easy
$-\frac{36\sqrt{61}}{61}$

14. Find the work done by a force $\mathbf{F} = 20\mathbf{i} + 15\mathbf{j}$ in moving an object from $P(-4, 6)$ to $Q(4, 8)$.
Answer: medium
$W = 190$

15. Find the work done by a force $\mathbf{F} = 180\mathbf{i} - 120\mathbf{j}$ in moving an object from $P(55, 0)$ to $Q(200, -20)$.
Answer: medium
$W = 28{,}500$

16. What does the fact that $\mathbf{a} \cdot \mathbf{a} = 0$ imply about the vector $\mathbf{a}$?
Answer: hard
$\mathbf{a} = \mathbf{0}$

17. A package that weighs 340 lb is placed on an inclined plane. If a force of 95 lb is just sufficient to keep the package from moving, find the angle of inclination of the plane. (Assume that there are rollers on the plane, so you can ignore the effects of friction.)
Answer: medium
About 0.283 radians or 16.2°

18. Prove the fact that $(a\mathbf{u}) \cdot (b\mathbf{v}) = ab(\mathbf{u} \cdot \mathbf{v})$.
Answer: hard
If $\mathbf{u} = \langle u_1, u_2\rangle$ and $\mathbf{v} = \langle v_1, v_2\rangle$, then
$(a\mathbf{u}) \cdot (b\mathbf{v}) = \langle au_1, au_2\rangle \cdot \langle bv_1, bv_2\rangle = abu_1v_1 + abu_2v_2 = ab(u_1v_1 + u_2v_2) = ab(\mathbf{u} \cdot \mathbf{v})$.

CHAPTER 9 REVIEW

1. Write $\tan x \cos x \csc x$ in terms of $\sin x$ and $\cos x$ and simplify.
   Answer: easy
   $\tan x \cos x \csc x = \dfrac{\sin x \cos x}{\cos x \sin x} = 1.$

2. Simplify $\dfrac{\tan(-x)}{\sec x}$.
   Answer: easy
   $\dfrac{\tan(-x)}{\sec x} = -\dfrac{\sin x}{\cos x}\cos x = -\sin x.$

3. Substitute $2\tan\theta$ for x in $\dfrac{-x}{\sqrt{4+x^2}}$ and simplify. Assume that $0 \le \theta < \dfrac{\pi}{2}$.
   Answer: easy
   $\dfrac{-x}{\sqrt{4+x^2}} = \dfrac{-\tan\theta}{\sqrt{4+4\tan^2\theta}} = -\dfrac{\tan\theta}{2\sec\theta} = -\dfrac{1}{2}\sin\theta.$

4. Show that $\sin x^2 = 2\sin x$ is not an identity.
   Answer: medium
   $\sin\sqrt{\pi}^2 = 0$, but $2\sin\sqrt{\pi} > 0$.

5. Use addition or subtraction formulas to find the exact value of $\cos 75°$.
   Answer: medium
   $\cos 75° = \cos(30+45) = \cos 30\cos 45 - \sin 30 \sin 45 = \dfrac{\sqrt{3}}{2}\dfrac{\sqrt{2}}{2} - \dfrac{1}{2}\dfrac{\sqrt{2}}{2} = \dfrac{\sqrt{2}\left(\sqrt{3}-1\right)}{4}.$

6. Use an addition or subtraction formula to write $\cos 99\cos 9 + \sin 99 \sin 9$ and find its exact value.
   Answer: medium
   $\cos 99\cos 9 + \sin 99\sin 9 = \cos(99-9) = \cos 90 = 0.$

7. Use an addition or subtraction formula to write $\dfrac{\tan 10 + \tan 35}{1 - \tan 10 \tan 35}$ and find its exact value.
   Answer: medium
   $\dfrac{\tan 10 + \tan 35}{1 - \tan 10\tan 35} = \tan(10+35) = \tan 45 = 1.$

8. Write $\sin x\,\frac{\sqrt{2}}{2} - \cos x\,\frac{\sqrt{2}}{2}$ in terms of sine only.
   Answer: medium
   $\sin x\,\frac{\sqrt{2}}{2} - \cos x\,\frac{\sqrt{2}}{2} = \sin x\cos\frac{\pi}{4} - \cos x \sin\frac{\pi}{4} = \sin\left(x - \frac{\pi}{4}\right).$

9. Find $\sin 2x$, $\cos 2x$ and $\tan 2x$, given that $\sin x = \frac{3}{5}$ and x is in quadrant I.
   Answer: medium
   $\cos x = \frac{4}{5}$, so $\sin 2x = 2\sin x\cos x = 2\cdot\frac{3}{5}\cdot\frac{4}{5} = \frac{24}{25}$, $\cos 2x = \cos^2 x - \sin^2 x = \frac{4}{5}^2 - \frac{3}{5}^2 = \frac{7}{25}$, and $\tan 2x = \dfrac{\sin 2x}{\cos 2x} = \dfrac{24}{7}$.

10. Use a half angle formula to find the exact value of $\sin -22.5°$.
    Answer: medium
    $\sin(-22.5°) = -\sqrt{\dfrac{1-\cos(-45)}{2}} = -\sqrt{\dfrac{1-\frac{\sqrt{2}}{2}}{2}} = -\sqrt{\dfrac{2-\sqrt{2}}{4}} = \dfrac{\sqrt{\sqrt{2}-2}}{2}.$

11. Find $\sin\frac{x}{2}$, $\cos\frac{x}{2}$ and $\tan\frac{x}{2}$, given that $\cos x = -\frac{4}{5}$ and $90° < x < 180°$.
    Answer: medium
    $\sin x = \dfrac{3}{5}$, so $\sin\dfrac{x}{2} = \sqrt{\dfrac{1+\frac{4}{5}}{2}} = \dfrac{3\sqrt{10}}{10}$, $\cos\dfrac{x}{2} = \sqrt{\dfrac{1-\frac{4}{5}}{2}} = \dfrac{\sqrt{10}}{10}$ and $\tan\dfrac{x}{2} = \dfrac{1+\frac{4}{5}}{\frac{3}{5}} = 3.$

12. Write $\sin 7x + \sin 3x$ as a product.
Answer: medium
$\sin 7x + \sin 3x = 2\sin 5x \cos 2x$.

13. Find the value of $\tan^{-1}\left(-\sqrt{3}\right)$ if it is defined.
Answer: easy
$\tan^{-1}\left(-\sqrt{3}\right) = -\frac{\pi}{3}$.

14. Find an approximate value of $\tan^{-1}8$ correct to five decimal places.
Answer: easy
$\tan^{-1}8 \approx 1.44644$.

15. Find the value of $\sin^{-1}\left(\sin \frac{8\pi}{11}\right)$ if it is defined.
Answer: easy
$\sin^{-1}\left(\sin \frac{8\pi}{11}\right) = \frac{8\pi}{11}$.

16. Rewrite $\tan\left(2\cos^{-1}x\right)$ as an algebraic expression in x.
Answer: medium
$\tan\left(2\cos^{-1}x\right) = \dfrac{2\tan\left(\cos^{-1}x\right)}{1-\tan^2\left(\cos^{-1}x\right)} = 2\dfrac{\sin\left(\cos^{-1}x\right)}{\cos\left(\cos^{-1}x\right)}\dfrac{\cos^2\left(\cos^{-1}x\right)}{\cos^2\left(\cos^{-1}x\right)-\sin^2\left(\cos^{-1}x\right)} = 2\dfrac{\sqrt{1-x^2}}{x}\dfrac{x^2}{2x^2-1} = \dfrac{2x\sqrt{1-x^2}}{2x^2-1}$.

17. Find all solutions of $2\cos x + \sqrt{3} = 0$.
Answer: medium
$2\cos x + \sqrt{3} = 0 \quad\Leftrightarrow\quad \cos x = -\frac{\sqrt{3}}{2}$, solutions in $[0, 2\pi)$ are $x = \frac{5\pi}{6}$ and $x = 2\pi - \frac{5\pi}{6} = \frac{7\pi}{6}$, so the general solutions are $x = \frac{5\pi}{6} + 2k\pi$ and $x = \frac{7\pi}{6} + 2k\pi$.

18. Find all solutions of $\tan x \cos x - \sqrt{3}\cos x = 0$.
Answer: medium
$\tan x \cos x - \sqrt{3}\cos x = 0 \quad\Leftrightarrow\quad \left(\tan x - \sqrt{3}\right)\cos x = 0 \quad\Leftrightarrow\quad \tan x = \sqrt{3}$ or $\cos x = 0$. $\cos x = 0$ has solution $x = \frac{\pi}{2} + k\pi$, and $\tan x = \sqrt{3}$ has solution $x = \frac{\pi}{3} + k\pi$, so the general solutions are $x = \frac{\pi}{2} + k\pi$ and $x = \frac{\pi}{3} + k\pi$.

19. Use a calculator to solve $\sin x = 0.7$ to five decimal places on the interval $[0, 2\pi]$, and find all solutions.
Answer: medium
$\sin x = 0.3 \quad\Leftrightarrow\quad x = \sin^{-1}0.3 \quad\Rightarrow\quad x \approx 0.77540$ or $x \approx \pi - 0.77540 \approx 2.36620$. Thus the general solutions are $x \approx 0.77540 + 2k\pi$ and $x \approx 2.36620 + 2k\pi$.

20. Use an addition or subtraction formula to simplify $\sin x \cos 3x + \cos x \sin 3x = 0$ and then find all solutions in the interval $[0, 2\pi]$.
Answer: medium
$\sin x \cos 3x + \cos x \sin 3x = 0 \quad\Leftrightarrow\quad \sin(x + 3x) = 0 \quad\Leftrightarrow\quad \sin 4x = 0 \quad\Leftrightarrow\quad 4x = k\pi \quad\Leftrightarrow\quad x = \frac{k\pi}{4}$, so all the solutions in the interval $[0, 2\pi]$ are $0, \frac{\pi}{4}, \frac{\pi}{2}, \frac{3\pi}{4}, \pi, \frac{5\pi}{4}, \frac{3\pi}{2}, \frac{7\pi}{4}$ and 2π.

21. Write $-2 - 2i$ in trigonometric form, with θ between 0 and 2π.
Answer: easy
$-2 - 2i = \sqrt{2^2 + 2^2}\left(\cos \frac{5\pi}{4} + i \sin \frac{5\pi}{4}\right) = 2\sqrt{2}\left(\cos \frac{5\pi}{4} + i \sin \frac{5\pi}{4}\right)$.

22. Write $-4 + 3i$ in trigonometric form, with θ between 0 and 2π.
Answer: easy
$-4 + 3i = \sqrt{4^2 + 3^2}\left(\cos\left(\pi + \tan^{-1}\left(-\frac{3}{4}\right)\right) + i \sin\left(\pi + \tan^{-1}\left(-\frac{3}{4}\right)\right)\right) = 5\left(-\frac{4}{5} + \frac{3}{5}\right)$.

23. Write the product $z_1 z_2$, and quotient z_1/z_2, of $z_1 = \cos \frac{5\pi}{6} + i \sin \frac{5\pi}{6}$ and $z_2 = \cos\left(-\frac{\pi}{2}\right) + i \sin\left(-\frac{\pi}{2}\right)$ in trigonometric form.
Answer: medium
$z_1 z_2 = 1 \cdot 1\left(\cos\left(\frac{5\pi}{6} - \frac{\pi}{2}\right) + i \sin\left(\frac{5\pi}{6} - \frac{\pi}{2}\right)\right) = \cos \frac{\pi}{3} + i \sin \frac{\pi}{3}$;
$z_1/z_2 = 1/1\left(\cos\left(\frac{5\pi}{6} + \frac{\pi}{2}\right) + i \sin\left(\frac{5\pi}{6} + \frac{\pi}{2}\right)\right) = \cos \frac{4\pi}{3} + i \sin \frac{4\pi}{3}$.

24. Find $\left(4\sqrt{3}-4i\right)^6$ using DeMoivre's Theorem.

Answer: medium

$4\sqrt{3}-4i=\sqrt{\left(4\sqrt{3}\right)^2+4^2}\left(\cos\frac{11\pi}{6}+i\sin\frac{11\pi}{6}\right)=8\left(\cos\frac{11\pi}{6}+i\sin\frac{11\pi}{6}\right)$, so $\left(4\sqrt{3}-4i\right)^6=$ $8^6\left(\cos 6\frac{11\pi}{6}+i\sin 6\frac{11\pi}{6}\right)=8^6(\cos 11\pi+i\sin 11\pi)=8^6(\cos\pi+i\sin\pi)$.

25. Solve $z^7-i=0$.

Answer: medium

$z^7-i=0 \quad\Leftrightarrow\quad z^7=i \quad\Leftrightarrow\quad z^7=\cos\frac{\pi}{2}+i\sin\frac{\pi}{2} \quad\Leftrightarrow\quad z=\cos\frac{1}{7}\left(\frac{\pi}{2}+2k\pi\right)+i\sin\frac{1}{7}\left(\frac{\pi}{2}+2k\pi\right) \quad\Rightarrow$ $z=\cos\frac{\pi}{14}+i\sin\frac{\pi}{14}$, $\cos\frac{5\pi}{14}+i\sin\frac{5\pi}{14}$, $\cos\frac{9\pi}{14}+i\sin\frac{9\pi}{14}$, $\cos\frac{13\pi}{14}+i\sin\frac{13\pi}{14}$, $\cos\frac{17\pi}{14}+i\sin\frac{17\pi}{14}$, $\cos\frac{21\pi}{14}+i\sin\frac{21\pi}{14}$, or $\cos\frac{25\pi}{14}+i\sin\frac{25\pi}{14}$.

26. Find the vector with initial point $P(-3,5)$ and terminal point $Q(0,4)$.

Answer: easy

The vector is given by $\langle 0-(-3),4-5\rangle=\langle 3,-1\rangle$.

27. Find $2\mathbf{u}$ and $\mathbf{u}+\mathbf{v}$, given that $\mathbf{u}=\langle 2,1\rangle$ and $\mathbf{v}=\langle -1,4\rangle$.

Answer: easy

$2\mathbf{u}=\langle 2\cdot 2,2\cdot 1\rangle=\langle 4,2\rangle$ and $\mathbf{u}+\mathbf{v}=\langle 2+(-1),1+4\rangle=\langle -1,5\rangle$.

28. Find $|\mathbf{v}|$ and $|\mathbf{u}+\mathbf{v}|$, given that $\mathbf{u}=3\mathbf{i}-2\mathbf{j}$ and $\mathbf{v}=2\mathbf{j}$.

Answer: easy

$|\mathbf{v}|=|2\mathbf{j}|=2$, $|\mathbf{u}+\mathbf{v}|=|3\mathbf{i}-2\mathbf{j}+2\mathbf{j}|=|3\mathbf{i}|=3$.

29. Find the horizontal and vertical components of the vector with length $|\mathbf{v}|=100$, and direction $\theta=225°$.

Answer: easy

$\mathbf{v}=\langle a,b\rangle=\langle|\mathbf{v}|\cos\theta,|\mathbf{v}|\sin\theta\rangle=\langle 100\cos 225°,100\sin 225°\rangle=\langle -50\sqrt{2},-50\sqrt{2}\rangle$.

30. Find the magnitude and direction (in degrees) of the vector $\mathbf{v}=3\mathbf{i}-\mathbf{j}$.

Answer: easy

The magnitude is $|\mathbf{v}|=|3\mathbf{i}-\mathbf{j}|=\sqrt{3^2+1^2}=\sqrt{10}$, direction is $360°+\tan^{-1}\left(-\frac{1}{3}\right)\approx 341.6°$.

31. Find the dot product of the vectors $\langle 3,12\rangle$ and $\langle 11,-4\rangle$.

Answer: easy

-15

32. What value of x will cause the two vectors $\langle x,6\rangle$ and $\langle 2,7\rangle$ to be perpendicular?

Answer: medium

$x=-21$

33. Find the work done by a force $\mathbf{F}=15\mathbf{i}-12\mathbf{j}$ in moving an object from $P(3,20)$ to $Q(4,12)$.

Answer: medium

$W=111$

CHAPTER TEN

SECTION 10.1

1. Use the graphical method to find all the solutions of the system of equations $\begin{cases} y = 3 - 2x \\ y - 4 = 5x \end{cases}$ correct to two decimal places.

   Answer: easy
   $x \approx -0.14$, $y \approx 3.28$

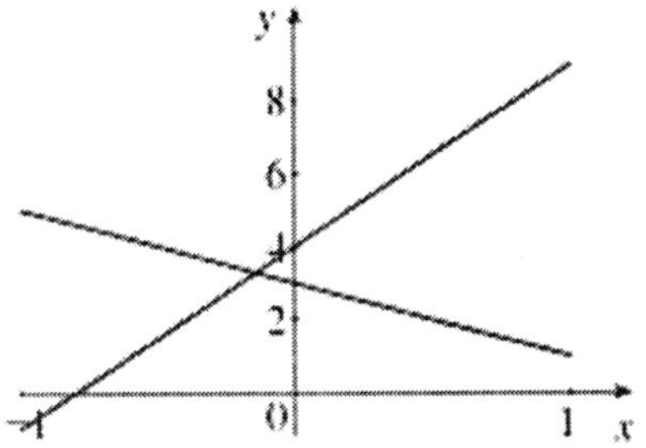

2. Use the graphical method to find all the solutions of the system of equations $\begin{cases} y = 1 - 8x^3 \\ y = 4x^3 - 2x^2 \end{cases}$ correct to two decimal places.

   Answer: easy
   $x = \frac{1}{2}$, $y = 0$

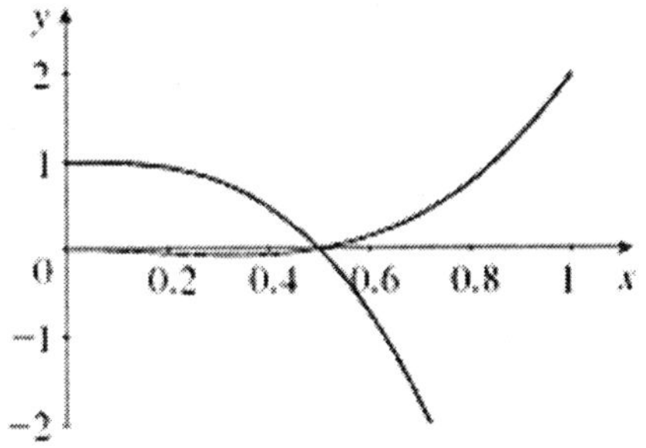

3. Use the graphical method to find all the solutions of the system of equations $\begin{cases} y + 4x = x^2 \\ y = -x^2 + 8x - 16 \end{cases}$ correct to two decimal places.

   Answer: $(2, -4)$, $(4, 0)$

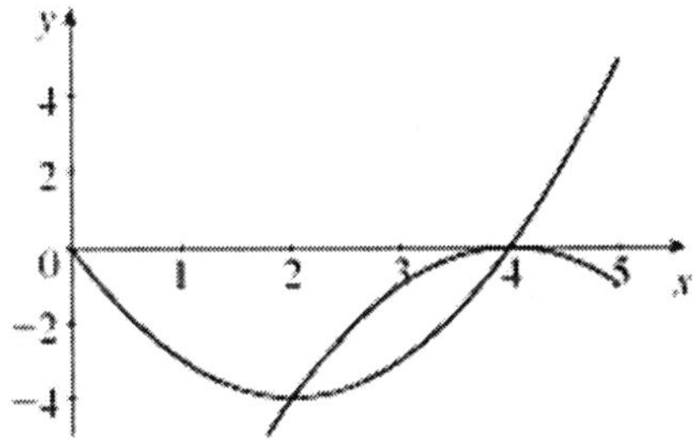

4. Use the graphical method to find all the solutions of the system of equations $\begin{cases} \frac{1}{4}x^2 + \frac{1}{25}y^2 = 1 \\ y = 4x^2 - 4 \end{cases}$ correct to two decimal places.

Answer: easy
$x \approx \pm 1.38$, $y \approx 3.61$

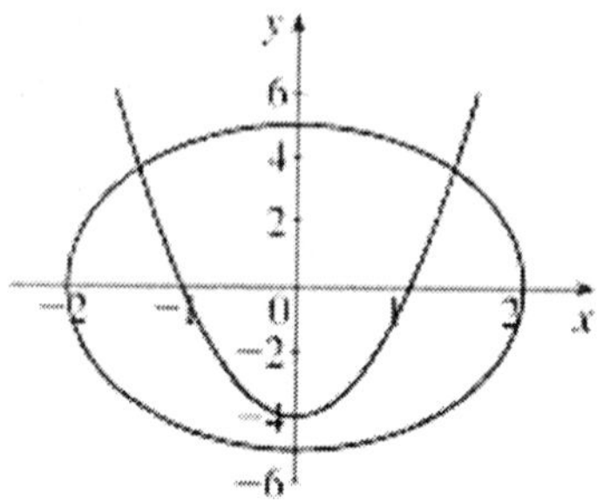

5. The volume of a right circular cone is 105 cm^3, and its radius is one-third its height. Find the cone's radius and its height.
Answer: medium
We have $\begin{cases} 105 = \dfrac{\pi r^2 h}{3} \\ r = \dfrac{h}{3} \end{cases}$ So $105 = \dfrac{\pi\left(\frac{1}{3}h\right)^2 h}{3} \Rightarrow 105 = \frac{1}{27}\pi h^3 \Rightarrow h^3 = \dfrac{27 \cdot 105}{\pi} \Rightarrow h \approx 9.66$ cm,
$r \approx \frac{9.66}{3} = 3.22$ cm.

6. The perimeter of a rectangle is 140 in and its diagonal is 50 in. Find its length and width.
Answer: medium
We have $\begin{cases} 2w + 2l = 140 \\ \sqrt{w^2 + l^2} = 50 \end{cases} \Rightarrow w = -l + 70 \Rightarrow \sqrt{(-l+70)^2 + l^2} = 50 \Rightarrow (-l+70)^2 + l^2 = 2500 \Rightarrow$
$(-l+70)^2 + l^2 = 2500 \Rightarrow 2l^2 - 140l + 2400 = 0 \Rightarrow l = 30$ in $\Rightarrow w = -30 + 70 = 40$ in.

7. Find all the solutions for the system $\begin{cases} x - y = 5 \\ x^2 - y^2 = 41 \end{cases}$

Answer: medium
$\begin{cases} x - y = 5 \\ x^2 - y^2 = 41 \end{cases} \Rightarrow x^2 - y^2 = 41 \Rightarrow (x-y)(x+y) = 41 \Rightarrow 5(x+y) = 41 \Rightarrow x = -y + \frac{41}{5} \Rightarrow$
$-y + \frac{41}{5} - y = 5$, so $y = \frac{8}{5}$, $x = \frac{33}{5}$.

8. Find all solutions for the system $\begin{cases} 5^x + 5^y = 10 \\ 25^x + 25^y = 75 \end{cases}$ Note that $25^x = 5^{2x} = (5^x)^2$.

Answer: medium
$\begin{cases} 5^x + 5^y = 10 \\ 25^x + 25^y = 75 \end{cases} \Rightarrow \begin{cases} 5^x + 5^y = 10 \\ (5^x)^2 + (5^y)^2 = 75 \end{cases} \Rightarrow 5^x + 5^y = 10 \Rightarrow 5^x = 10 - 5^y \Rightarrow (10 - 5^y)^2 - (5^y)^2 = 75$
$\Rightarrow 100 - 20(5^y) = 75 \Rightarrow 5^y = \frac{5}{4} \Rightarrow y = \dfrac{\ln\frac{5}{4}}{\ln 5}$. $5^x = 10 - 5^{(\ln\frac{5}{4})/(\ln 5)} \Rightarrow x = \dfrac{\ln\frac{35}{4}}{\ln 5}$. Note that $x = \dfrac{\ln\frac{5}{4}}{\ln 5}$,
$y = \dfrac{\ln\frac{35}{4}}{\ln 5}$ is also a solution.

SECTION 10.2

1. Graph the pair of lines $\begin{cases} 3x + y = 4 \\ -x + 3y = 12 \end{cases}$ on a single set of axes. Determine whether the lines are parallel or not, and if they are not parallel, estimate the coordinates of their point of intersection from your graph.

Answer: easy

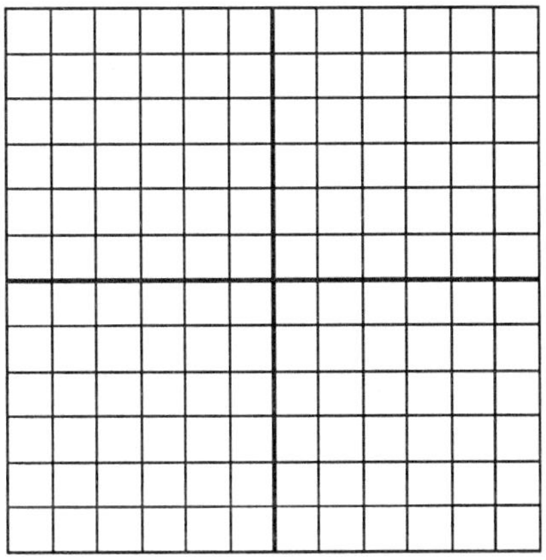

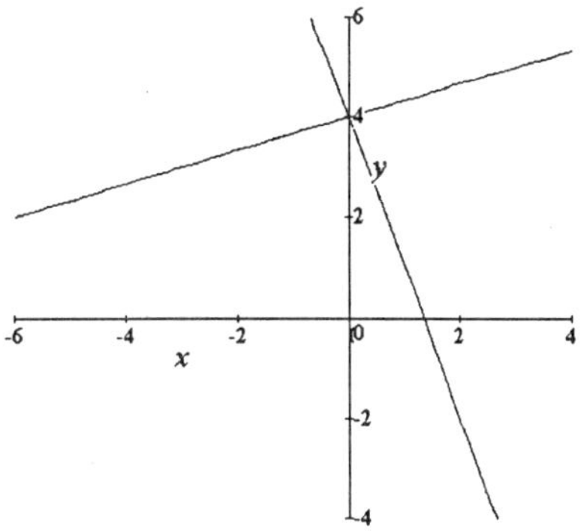

$\begin{cases} 3x + y = 4 \\ -x + 3y = 12 \end{cases}$ Not parallel. Intersect at $(0, 4)$.

2. Graph the pair of lines $\begin{cases} 2x + 4y = 14 \\ x + 2y = 7 \end{cases}$ on a single set of axes. Determine whether the lines are parallel or not, and if they are not parallel, estimate the coordinates of their point of intersection from your graph.

Answer: easy

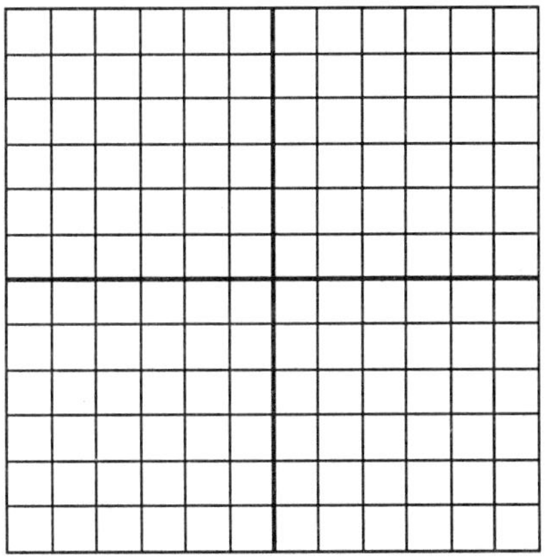

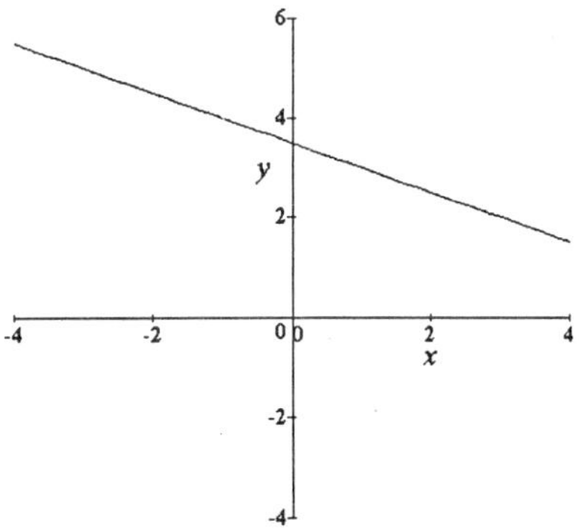

$\begin{cases} 2x + 4y = 14 \\ x + 2y - 7 \end{cases}$ The lines are identical. All points on the lines are points of intersection.

3. Graph the pair of lines $\begin{cases} -6x + 8y = 11 \\ 15x - 20y = 70 \end{cases}$ on a single set of axes. Determine whether the lines are parallel or not, and if they are not parallel, estimate the coordinates of their point of intersection from your graph.

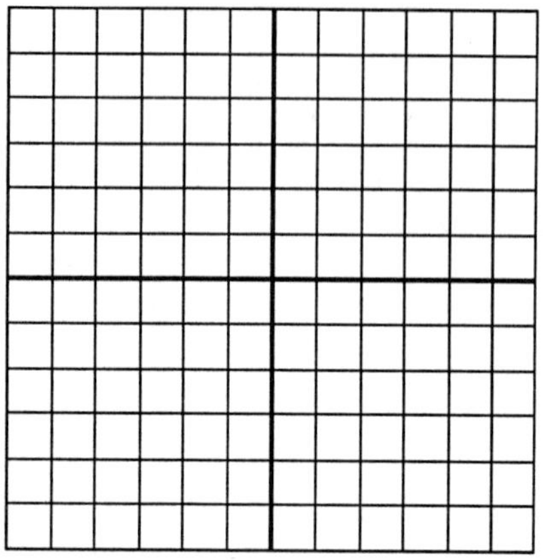

Answer: easy

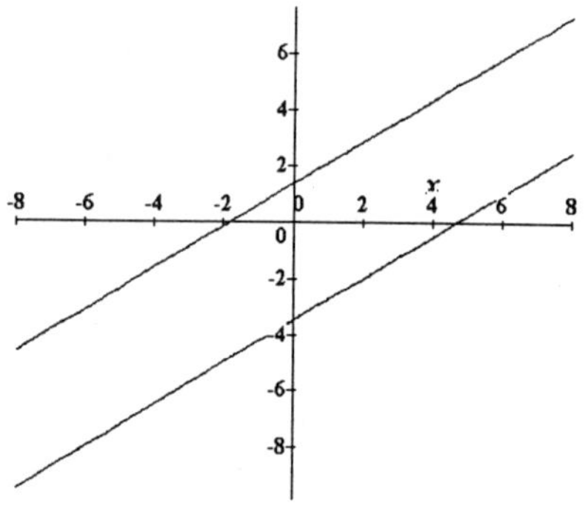

$\begin{cases} -6x + 8y = 11 \\ 15x - 20y = 70 \end{cases}$ Parallel. No intersection.

4. Solve the system $\begin{cases} 2x - y = 6 \\ 9x - 2y = -4 \end{cases}$ using the substitution method.

(a) $(-6, -5)$ (b) $\left(-\frac{16}{5}, -\frac{62}{5}\right)$ (c) $\left(-\frac{13}{5}, -\frac{52}{5}\right)$ (d) $\left(-\frac{17}{10}, -\frac{103}{10}\right)$ (e) $(-1, 1)$

Answer: easy (b)

$2x - y = 6 \quad \Leftrightarrow \quad y = 2x - 6$. Substituting for y into $9x - 2y = -4$ gives $9x - 2(2x - 6) = -4 \quad \Leftrightarrow \quad 5x = -16 \quad \Leftrightarrow$ $x = -\frac{16}{5}$, and so $y = 2\left(-\frac{16}{5}\right) - 6 = -\frac{62}{5}$. Thus the solution is $\left(-\frac{16}{5}, -\frac{62}{5}\right)$.

5. Solve the system $\begin{cases} -3x + 9y = 0 \\ 16x + 8y = 112 \end{cases}$ using the substitution method.

(a) $(-3, -2)$ (b) $(-2, 1)$ (c) $(-2, 2)$ (d) $(6, 2)$ (e) $(2, 8)$

Answer: easy (d)

$-3x + 9y = 0 \quad \Leftrightarrow \quad x = 3y$. Substituting for x into $16x + 8y = 112$ gives $16(3y) + 8y = 112 \quad \Leftrightarrow \quad 56y = 112 \quad \Leftrightarrow$ $y = 2$, and so $x = 3 \cdot 2 = 6$. Therefore, the solution is $(6, 2)$.

6. Solve the system $\begin{cases} 0.2x - 0.2y = -1.6 \\ -0.5x + 0.4y = 3 \end{cases}$ using the substitution method.

(a) $(2, 4)$ (b) $(3, 5)$ (c) $(2, 10)$ (d) $(3, 6)$ (e) $(4, 12)$

Answer: medium (c)

$0.2x - 0.2y = -1.6 \quad \Leftrightarrow \quad x = y - 8$. Substituting for x into $-0.5x + 0.4y = 3$ gives $-0.5(y - 8) + 0.4y = 3 \quad \Leftrightarrow$ $-0.1y = -1 \quad \Leftrightarrow \quad y = 10$, and so $x = 10 - 8 = 2$. Hence, the solution is $(2, 10)$.

7. Solve the system $\begin{cases} 5x + 3y = 12 \\ 4x - 2y = 14 \end{cases}$ using the elimination method. If the system has infinitely many solutions, write the general form for all the solutions.

(a) $(3, -4)$ (b) $(2, 2)$ (c) $(1, -2)$ (d) $(2, -3)$ (e) $(3, -1)$

Answer: easy (e)

$\begin{cases} 5x + 3y = 12 \\ 4x - 2y = 14 \end{cases} \quad \Leftrightarrow \quad \begin{cases} 10x + 6y = 24 \\ 12x - 6y = 42 \end{cases}$. Adding gives $22x = 66 \quad \Leftrightarrow \quad x = 3$, and so $4(3) - 2y = 14 \quad \Leftrightarrow \quad y = -1$.

So the solution is $(3, -1)$.

8. Solve the system $\begin{cases} 9x + 2y = 30 \\ x - y = -4 \end{cases}$ using the elimination method. If the system has infinitely many solutions, write the general form for all the solutions.

(a) $(2, 6)$ (b) $(1, 4)$ (c) $(2, 4)$ (d) $(3, -2)$ (e) $(5, -4)$

Answer: easy (a)

$\begin{cases} 9x + 2y = 30 \\ x - y = -4 \end{cases} \Leftrightarrow \begin{cases} 9x + 2y = 30 \\ 2x - 2y = -8 \end{cases}$. Thus $11x = 22 \Leftrightarrow x = 2$, and so $2 - y = -4 \Leftrightarrow y = 6$. Thus the solution is $(2, 6)$.

9. Solve the system $\begin{cases} 3x - y = -8 \\ 21x - 7y = 3 \end{cases}$ using the elimination method. If the system has infinitely many solutions, write the general form for all the solutions.

(a) $(2, 5)$ (b) $(-2, -5)$ (c) $(1, -2)$ (d) $(2, -3)$ (e) No solution

Answer: easy (e)

$\begin{cases} 3x - y = -8 \\ 21x - 7y = 3 \end{cases} \Leftrightarrow \begin{cases} 21x - 7y = -56 \\ 21x - 7y = 3 \end{cases}$. Subtracting gives $0 = -59$ which is false. Therefore there is no solution to this system.

10. Solve the system $\begin{cases} u - 4v = -16 \\ -3u + 10v = 20 \end{cases}$ using the elimination method. If the system has infinitely many solutions, write the general form for all the solutions.

(a) $(6, 8)$ (b) $(3, 12)$ (c) $(4, 14)$ (d) $(40, 14)$ (e) $(60, 124)$

Answer: easy (d)

$\begin{cases} u - 4v = -16 \\ -3u + 10v = 20 \end{cases} \Leftrightarrow \begin{cases} 3u - 12v = -48 \\ -3u + 10v = 20 \end{cases}$. So $-2v = -28 \Leftrightarrow v = 14$, and so $u - 4(14) = -16 \Leftrightarrow u = 40$. Thus the solution is $(u, v) = (40, 14)$.

11. Solve the system $\begin{cases} 2x - \frac{1}{3}y = 4 \\ 3x - \frac{2}{3}y = 6 \end{cases}$ using the elimination method. If the system has infinitely many solutions, write the general form for all the solutions.

(a) $(0, 0)$ (b) $(1, 3)$ (c) $(2, 0)$ (d) $(x, 3x + 5)$ (e) $(x, 2x - 2)$

Answer: medium (c)

$\begin{cases} 2x - \frac{1}{3}y = 4 \\ 3x - \frac{2}{3}y = 6 \end{cases} \Leftrightarrow \begin{cases} 6x - y = 12 \\ 6x - \frac{4}{3}y = 12 \end{cases}$. Hence $y = 0$, and so $x = 2$. Therefore, the solution is $(2, 0)$.

12. Solve the system $\begin{cases} 3.6x - 1.8y = 6 \\ 2.4x - 1.2y = 4 \end{cases}$ using the elimination method. If the system has infinitely many solutions, write the general form for all the solutions.

(a) $(2, 4)$ (b) $(3, 6)$ (c) $(1, -4)$ (d) $(x, 4x + 2)$ (e) $\left(x, 2x - \frac{10}{3}\right)$

Answer: medium (e)

$\begin{cases} 3.6x - 1.8y = 6 \\ 2.4x - 1.2y = 4 \end{cases} \Leftrightarrow \begin{cases} 1.2x - 0.6y = 2 \\ 1.2x - 0.6y = 2 \end{cases}$. Since the two equations represent the same line, there are infinitely many solutions which satisfy $1.2x - 0.6y = 2 \Leftrightarrow y = 2x - \frac{10}{3}$. Thus the solutions are $\left(x, 2x - \frac{10}{3}\right)$, where x is any real number.

13. Solve the system $\begin{cases} 2\sqrt{5}w - 2z = 10 - 2\sqrt{2} \\ 2w + 2\sqrt{10}z = 6\sqrt{5} \end{cases}$ using the elimination method. If the system has infinitely many solutions, write the general form for all the solutions.

(a) $\left(\sqrt{5}, \sqrt{2}\right)$ (b) $\left(\sqrt{3}, \sqrt{5}\right)$ (c) $(2, 5)$ (d) $\left(3, \sqrt{5}\right)$ (e) $\left(-\sqrt{2}, \sqrt{2}\right)$

Answer: hard (a)

$\begin{cases} 2\sqrt{5}w - 2z = 10 - 2\sqrt{2} \\ 2w + 2\sqrt{10}z = 6\sqrt{5} \end{cases} \Leftrightarrow \begin{cases} 2\sqrt{5}w - 2z = 10 - 2\sqrt{2} \\ 2\sqrt{5}w + 10\sqrt{2}z = 30 \end{cases}$. Subtracting gives

$-2z - 10\sqrt{2}z = 10 - 2\sqrt{2} - 30 \Leftrightarrow z = \dfrac{-2\sqrt{2} - 20}{-2 - 10\sqrt{2}} = \sqrt{2}$, and so $2\sqrt{5}w - 2\sqrt{2} = 10 - 2\sqrt{2} \Leftrightarrow w = \sqrt{5}$.

Therefore, the solution is $(w, z) = \left(\sqrt{5}, \sqrt{2}\right)$.

14. Solve the system $\begin{cases} 3r^2+4s^2=13 \\ 9r^2-3s^2=9 \end{cases}$ by first making a substitution that will turn the equations into equations of lines and then using the methods studied in the text. The appropriate substitution is made by letting $x=r^2$ and $y=s^2$.

(a) $\left(\pm\frac{\sqrt{15}}{3},\pm\sqrt{2}\right)$ (b) $(\pm 2,\pm 4)$ (c) $\left(\frac{5}{3},2\right)$ (d) $\left(\pm 1,\pm\sqrt{2}\right)$ (e) $\left(0,\pm\sqrt{3}\right)$

Answer: medium (a)

Let $x=r^2$, $y=s^2$. Then $\begin{cases} 3r^2+4s^2=13 \\ 9r^2-3s^2=9 \end{cases} \Leftrightarrow \begin{cases} 3x+4y=13 \\ 9x-3y=9 \end{cases} \Leftrightarrow \begin{cases} 9x+12y=39 \\ 9x-3y=9 \end{cases}$. So $15y=30 \Leftrightarrow y=2$, and so $9x-3\cdot 2=9 \Leftrightarrow x=\frac{5}{3}$. Thus the solution is $(r,s)=\left(\pm\sqrt{x},\pm\sqrt{y}\right)=\left(\pm\frac{\sqrt{15}}{3},\pm\sqrt{2}\right)$.

15. Solve the system $\begin{cases} \dfrac{1}{x}-\dfrac{2}{y^2}=4 \\ \dfrac{2}{x}-\dfrac{6}{y^2}=12 \end{cases}$ by first making a substitution that will turn the equations into equations of lines and then using the method studied in the text.

(a) $(0,-2)$ (b) $(-2,0)$ (c) $\left(\sqrt{2},0\right)$ (d) $\left(0,\sqrt{2}\right)$ (e) No solution

Answer: hard (e)

Let $u=\dfrac{1}{x}$, $v=\dfrac{1}{y^2}$. Then $\begin{cases} \dfrac{1}{x}-\dfrac{2}{y^2}=4 \\ \dfrac{2}{x}-\dfrac{6}{y^2}=12 \end{cases} \Leftrightarrow \begin{cases} u-2v=4 \\ 2u-6v=12 \end{cases} \Leftrightarrow \begin{cases} u-2v=4 \\ u-3v=6 \end{cases}$. So $v=-2$, and so $u=0$. Since there is no x such that $\dfrac{1}{x}=0$ and no y such that $-2=\dfrac{1}{y^2}$, there is no solution.

16. Solve the system $\begin{cases} 2x-4y=5x-7y-10 \\ x=6y+5 \end{cases}$

(a) $(2,-2)$ (b) $\left(1,-\frac{2}{3}\right)$ (c) $\left(3,-\frac{1}{3}\right)$ (d) $\left(-\frac{1}{9},6\right)$ (e) $(2,5)$

Answer: medium (c)

$\begin{cases} 2x-4y=5x-7y-10 \\ x=6y+5 \end{cases} \Leftrightarrow \begin{cases} 3x-3y=10 \\ 2x-12y=10 \end{cases} \Leftrightarrow \begin{cases} 12x-12y=40 \\ 2x-12y=10 \end{cases}$. Subtracting gives $10x=30 \Leftrightarrow x=3$, and so $2\cdot 3=12y+10 \Leftrightarrow y=-\frac{1}{3}$. Therefore, the solution is $\left(3,-\frac{1}{3}\right)$.

17. Solve the system $x=4x+y=3y+1$.

(a) $\left(-\frac{1}{3},\frac{2}{3}\right)$ (b) $(-10,10)$ (c) $\left(\frac{10}{3},12\right)$ (d) $\left(-\frac{1}{10},-\frac{3}{10}\right)$ (e) $\left(-\frac{3}{2},\frac{2}{3}\right)$

Answer: medium (d)

$\begin{cases} x=4x+y \\ x=3y+1 \end{cases} \Leftrightarrow \begin{cases} 3x+y=0 \\ x-3y=1 \end{cases} \Leftrightarrow \begin{cases} 3x+y=0 \\ 3x-9y=3 \end{cases}$. So $10y=-3 \Leftrightarrow y=-\frac{3}{10}$, and so $3x-\frac{3}{10}=0 \Leftrightarrow x=\frac{1}{10}$. Thus the solution is $\left(\frac{1}{10},-\frac{3}{10}\right)$.

18. $\begin{cases} ax+by=0 \\ x+y=1 \end{cases}$ $(a\neq b)$ Find x and y in terms of a and b.

(a) $\left(\dfrac{b}{b-a},\dfrac{a}{a-b}\right)$ (b) $\left(\dfrac{a}{a-b},\dfrac{b}{a-b}\right)$ (c) $\left(\dfrac{a}{a+b},\dfrac{-b}{a+b}\right)$ (d) $\left(\dfrac{-b}{a+b},\dfrac{a}{a+b}\right)$ (e) $(a+b,a-b)$

Answer: medium (a)

$\begin{cases} ax+by=0 \\ x+y=1 \end{cases} \Leftrightarrow \begin{cases} ax+by=0 \\ ax+ay=a \end{cases}$. So $(a-b)y=a \Leftrightarrow y=\dfrac{a}{a-b}$, and so $x+\dfrac{a}{a-b}=1 \Leftrightarrow x=\dfrac{b}{b-a}$.

Hence the solution is $\left(\dfrac{b}{b-a},\dfrac{a}{a-b}\right)$.

19. For the system $\begin{cases} ax+by=0 \\ a^2x+b^2y=1 \end{cases}$ with $a\neq 0$, $b\neq 0$ and $a\neq b$, find x and y in terms of a and b.

Answer: hard

$\begin{cases} ax+by=0 \\ a^2x+b^2y=1 \end{cases} \Leftrightarrow \begin{cases} a^2x+aby=0 \\ a^2x+b^2y=1 \end{cases}$. So $b(a-b)y=-1 \Leftrightarrow y=\dfrac{1}{b(b-a)}$, and so $ax+\dfrac{b}{b(b-a)}=0 \Leftrightarrow x=-\dfrac{1}{a(b-a)}$. Hence, the solution is $\left(-\dfrac{1}{a(b-a)},\dfrac{1}{b(b-a)}\right)=\left(\dfrac{1}{a^2-ab},\dfrac{1}{b^2-ab}\right)$.

20. The sum of two numbers is twice their difference. The larger number is 10 more than twice times the smaller. Find the numbers.

Answer: easy

Let the two numbers be x and y, and let the larger number be x. $\begin{cases} x+y=2(x-y) \\ x=10+2y \end{cases} \Leftrightarrow \begin{cases} -x+3y=0 \\ x-2y=10 \end{cases} \Rightarrow y=10$, and so $x=10+2\cdot 10=30$. Therefore, the two numbers are 30 and 10.

21. The admission fee at an amusement park is \$1.50 for children and \$3.50 for adults. On a certain day, 2400 people entered the park and the admission fees collected totaled \$5200. How many children and how many adults were admitted?

Answer: medium

Let c be the number of children and a be the number of adults. $\begin{cases} c+a=2400 \\ 1.50c+3.50a=5200 \end{cases} \Leftrightarrow \begin{cases} 3c+3a=7200 \\ 3c+7a=10400 \end{cases} \Rightarrow$ $4a=3200 \Leftrightarrow a=800$, and so $c+800=2400 \Leftrightarrow c=1600$. Therefore, the number of children admitted was 1600 and the number of adults admitted was 800.

22. A boat travels downstream between two points on a river 20 mi apart in 2 h. The return trip against the current takes $3\frac{1}{2}$ h. What is the boat's speed, and how fast does the current in the river flow?

Answer: medium

Let x be the speed of the boat in still water and y be the speed of the river flow.

$\begin{cases} 2x+2y=20 \\ 3.5x-3.5y=20 \end{cases} \Leftrightarrow \begin{cases} 7x+7y=70 \\ 7x-7y=40 \end{cases} \Rightarrow 14x=110 \Leftrightarrow x=\frac{55}{7}$, and so $y=\frac{15}{7}$. Therefore, the boat's speed is approximately 7.9 mph and the current in the river flows at approximately 2.1 mph.

23. A biologist has two brine solutions, one containing 5% salt and another containing 10% salt. How many milliliters of each solution should he mix to obtain 1 L of a solution that contains 7% salt?

Answer: medium

Let x and y be the number of milliliters of the two brine solutions. $\begin{cases} x+y=1000 \\ 0.05x+0.10y=0.07(x+y)=70 \end{cases} \Leftrightarrow$ $\begin{cases} x+y=1000 \\ x+2y=1400 \end{cases}$. Thus $y=400$, and so $x=600$. Therefore, 400 milliliters of the 10% solution and 600 milliliters of the 5% solution should be mixed.

24. A customer in a coffee shop wishes to purchase a blend of two coffees: Kenyan, costing \$3.20 a pound, and Sri Lankan, costing \$5.90 a pound. He ends up buying 3 lb of such a blend, which costs him \$10.95. How many pounds of each kind went into the mixture?

Answer: medium

Let x be the pounds of Kenyan coffee and y be the pounds of Sri Lankan coffee.

$\begin{cases} 3.20x+5.90y=10.95 \\ x+y=3 \end{cases} \Leftrightarrow \begin{cases} 32x+59y=109.5 \\ 32x+32y=96 \end{cases} \Rightarrow 27y=13.5 \Leftrightarrow y=0.5$, and so $x=2.5$. Thus the mixture contains 2.5 pounds of Kenyan coffee and 0.5 pounds of Sri Lankan coffee.

25. John and Mary leave their house at the same time and drive off in opposite directions. John drives at 65 mi/h and travels 40 mi farther than Mary, who drives at 45 mi/h. Mary's trip takes 20 min longer than John's. For what length of time does each of them drive?

Answer: hard

Let x be the length of time John drives and y be the length of time Mary drives.

$\begin{cases} y=x+\frac{1}{3} \\ 65x=45y+40 \end{cases} \Leftrightarrow \begin{cases} 45x-45y=-15 \\ 65x-45y=40 \end{cases} \Rightarrow 20x=55 \Leftrightarrow x=\frac{11}{4}$, and so $y=\frac{11}{4}+\frac{1}{3}=\frac{37}{12}$. Thus, John travels for 2:45 and Mary travels for 3:05.

26. The sum of the digits of a two-digit number is 6. When the digits are reversed, the number is increased by 36. Find the number.

Answer: hard

Let x be the tens digit and y be the ones digit of the number. $\begin{cases} x+y=6 \\ 10y+x=36+10x+y \end{cases} \Leftrightarrow \begin{cases} 9x+9y=54 \\ -9x+9y=36 \end{cases}$. So $18y=90 \Leftrightarrow y=5$, and so $x=1$. Thus the number is 15.

27. Find the area of the triangle that lies in the first quadrant (with its base on the x-axis) bounded by the lines $y = 4x - 2$ and $y = -4x + 24$.

Answer: hard

First, find the intersection point of the two lines. The y-coordinate of the intersection point is the height of the triangle. $\begin{cases} y = 4x - 2 \\ y = -4x + 24 \end{cases} \Rightarrow 2y = 22 \Leftrightarrow y = 11$ and so the triangle has height 11. Further, $y = 4x - 2$ intersects the x-axis at $x = \frac{1}{2}$, $y = -4x + 24$ intersects the x-axis at $x = 6$, and thus the base has length $6 - \frac{1}{2} = \frac{11}{2}$. Therefore, the area of the triangle is $A = \frac{1}{2} \cdot b \cdot h = \frac{1}{2} \cdot \frac{11}{2} \cdot 11 = \frac{121}{4}$ square units.

28. Find the equation of the parabola that passes through the point $(-2, 7)$, that has y-intercept 15, and whose vertex has x-coordinate -3. [Hint: Recall that the equation of a parabola with vertex (h, k) is of the form $y = a(x - h)^2 + k$.]

Answer: hard

$y = a(x - h)^2 + k$. Since the parabola passes through the point $(-2, 7)$, this implies that $7 = a(-2 - h)^2 + k \Leftrightarrow 7 = a(2 + h)^2 + k$. And since the parabola has y-intercept 15, we also have $15 = a(0 - h)^2 + k \Leftrightarrow 15 = ah^2 + k$. From the fact that the vertex of the parabola has x-coordinate -3, we have $\begin{cases} 7 = a + k \\ 15 = 9a + k \end{cases} \Rightarrow 8a = 8 \Leftrightarrow a = 1$ and so $k = 6$. Thus the equation of the parabola is $y = (x + 3)^2 + 6$.

29. Use a graphing device to graph the lines $\begin{cases} 19.45x - 13.62y = 14.20 \\ 21.90x - 15.33y = 13.55 \end{cases}$ in the same viewing rectangle. (Note that you must solve for y in terms of x before graphing if you are using a graphing calculator.) Solve the system by zooming in to the point of intersection and then using the cursor to find its coordinates, correct to two decimals.

Answer: hard

$$\begin{cases} 19.45x - 13.62y = 14.20 \\ 21.90x - 15.33y = 13.55 \end{cases} \Leftrightarrow \begin{cases} y = \frac{19.45}{13.62}x - \frac{14.20}{13.62} \\ y = \frac{21.90}{15.33}x - \frac{13.55}{15.33} \end{cases}.$$

The slope of the first line is $\frac{19.45}{13.62} \approx 1.4280$ and the slope of the second is $\frac{21.90}{15.33} \approx 1.4286$. Since the slopes are almost the same, the lines will look parallel on the graphing device, so the solution cannot be found with the graphing device.

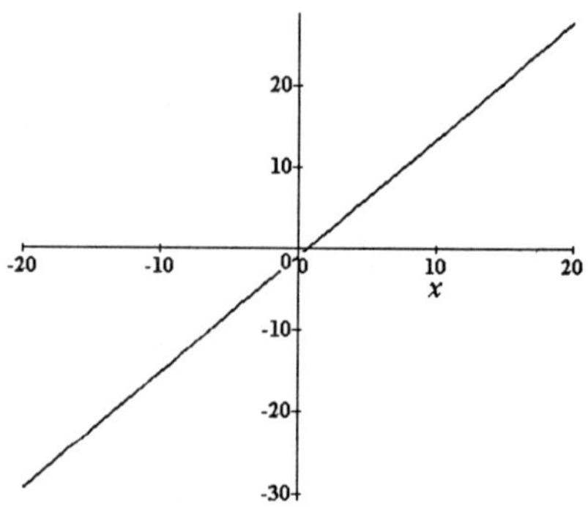

30. Use a graphing device to graph the lines $\begin{cases} -264x + 311y = 0 \\ 295x + 613y = 724 \end{cases}$ in the same viewing rectangle. (Note that you must solve for y in terms of x before graphing if you are using a graphing calculator.) Solve the system by zooming in to the point of intersection and then using the cursor to find its coordinates, correct to two decimals.

Answer: medium

$$\begin{cases} -264x + 311y = 0 \\ 295x + 613y = 724 \end{cases} \Leftrightarrow \begin{cases} y = \frac{264}{311}x \\ y = -\frac{295}{613}x + \frac{724}{613} \end{cases}$$

The solution is approximately $(0.89, 0.75)$.

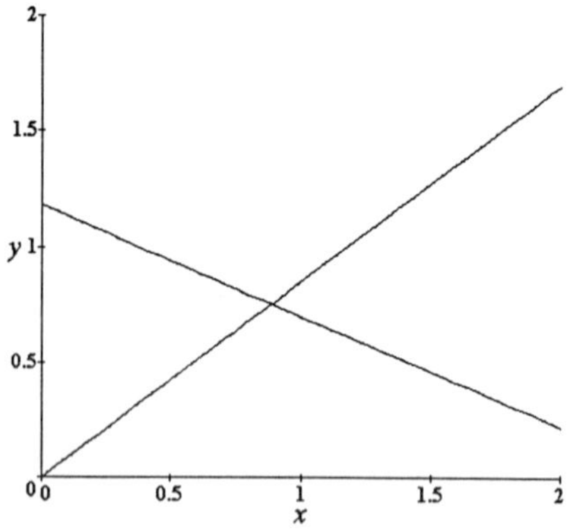

31. Use a graphing device to graph the lines $\begin{cases} \frac{1}{23}x - \frac{1}{37}y = 7 \\ \pi^2 x + \pi^3 y = 13.97 \end{cases}$ in the same viewing rectangle. (Note that you must solve for y in terms of x before graphing if you are using a graphing calculator.) Solve the system by zooming in to the point of intersection and then using the cursor to find its coordinates, correct to two decimals.
Answer: medium

$$\begin{cases} \frac{1}{23}x - \frac{1}{37}y = 7 \\ \pi^2 x + \pi^3 y = 13.97 \end{cases} \Leftrightarrow \begin{cases} y = \frac{37}{23}x - 259 \\ y = -\frac{1}{\pi}x + \frac{13.97}{\pi^3} \end{cases}$$

The solution is approximately $(134.64, -42.41)$.

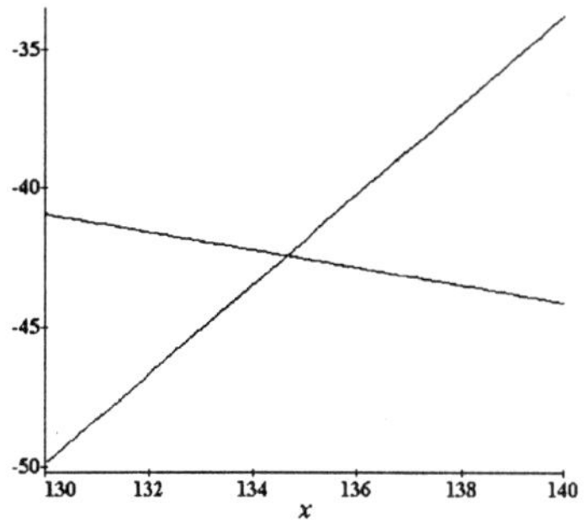

32. A man invests a total of \$21,000 in two accounts, one paying 5.5% simple interest per year and the other paying 8.5% simple interest per year. His annual interest from both accounts is \$1280. How much did he invest at each rate?
Answer: medium

$$\begin{cases} x + y = 21{,}000 \\ 0.055x + 0.085y = 1280 \end{cases} \Rightarrow x + y = 21{,}000 \Rightarrow y = -x + 21{,}000 \Rightarrow 0.055x + 0.085(-x + 21{,}000) = 1280$$

$\Rightarrow x = \$16{,}833.00 \Rightarrow y = -16{,}833.00 + 21{,}000 = \$4167$

33. A money manager invests a client's savings into two accounts, one paying 2% and the other 10% simple interest per year. He puts three times as much in the lower yielding account because it is less risky that the other one. His client's annual interest is \$4625. How much did he invest at each rate?
Answer: medium

Let x be the amount invested at the lower yield. $\begin{cases} 0.02x + 0.10y = 4625 \\ 3y = x \end{cases} \Rightarrow 3y = x \Rightarrow 0.02(3y) + 0.10y = 4625$

$\Rightarrow 0.16y = 4625 \Rightarrow y = \$28{,}906,\ x = 3(28{,}906) = \$86{,}718$

SECTION 10.3

1. State whether or not the equation $8xy - 4yz + 13zx = 0$ is linear.

Answer: easy
$8xy - 4yz + 13zx = 0$ is not linear, since it contains products of variables.

2. State whether or not the system of equations $e^3 x_1 - \pi x_2 - \sqrt{3} = x_3 - \frac{1}{3}x_4$ is linear.
Answer: easy
$e^3 x_1 - \pi x_2 - \sqrt{3} = x_3 - \frac{1}{3}x_4$ is linear.

3. State whether or not the system $\begin{cases} x - 2y = 14z + \frac{1}{\sqrt{2}} \\ x = 3ez \\ y - 2z = \frac{1}{\sqrt{37}}x \end{cases}$ is linear.

Answer: easy
$\begin{cases} x - 2y = 14z + \frac{1}{\sqrt{2}} \\ x = 3ez \\ y - 2z = \frac{1}{\sqrt{37}}x \end{cases}$ is a linear system of equations.

4. Write a system of equations that corresponds to the matrix $\begin{bmatrix} 2 & 3 & 5 & 6 \\ -1 & -1 & 5 & 2 \\ -2 & 3 & 0 & 11 \end{bmatrix}$.

Answer: easy

$$\begin{bmatrix} 2 & 3 & 5 & 6 \\ -1 & -1 & 5 & 2 \\ -2 & 3 & 0 & 11 \end{bmatrix} \quad \Leftrightarrow \quad \begin{cases} 2x + 3y + 5z = 6 \\ -x - y + 5z = 2 \\ -2x + 3y = 11 \end{cases}$$

5. Write a system of equations that corresponds to the matrix $\begin{bmatrix} 5 & 2 & 1 & 4 & 0 \\ -3 & 1 & 1 & 0 & 5 \\ -2 & 3 & 4 & 0 & 1 \\ 0 & 1 & 1 & 0 & -3 \end{bmatrix}$.

Answer: easy

$$\begin{bmatrix} 5 & 2 & 1 & 4 & 0 \\ -3 & 1 & 1 & 0 & 5 \\ -2 & 3 & 4 & 0 & 1 \\ 0 & 1 & 1 & 0 & -3 \end{bmatrix} \quad \Leftrightarrow \quad \begin{cases} 5x + 2y + z + 4w = 0 \\ -3x + y + z = 5 \\ -2x + 3y + 4z = 1 \\ y + z = -3 \end{cases}$$

6. Use Gaussian elimination to solve the system $\begin{cases} 3x - y = -9 \\ -x + 2y + 3z = 17 \\ x + y + z = 4 \end{cases}$.

(a) $(1, 2, -3)$ (b) $(-2, 3, 3)$ (c) $(4, 2, -2)$ (d) $(5, 1, 6)$ (e) $(-2, -2, 7)$

Answer: medium (b)

$$\begin{bmatrix} 3 & -1 & 0 & -9 \\ -1 & 2 & 3 & 17 \\ 1 & 1 & 1 & 4 \end{bmatrix} \quad \mathrm{R}_1 \leftrightarrow \mathrm{R}_3 \quad \begin{bmatrix} 1 & 1 & 1 & 4 \\ -1 & 2 & 3 & 17 \\ 3 & -1 & 0 & -9 \end{bmatrix} \quad \begin{matrix} \mathrm{R}_2 + \mathrm{R}_1 \to \mathrm{R}_2 \\ \mathrm{R}_3 - 3\mathrm{R}_1 \to \mathrm{R}_3 \end{matrix} \quad \begin{bmatrix} 1 & 1 & 1 & 4 \\ 0 & 3 & 4 & 21 \\ 0 & -4 & -3 & -21 \end{bmatrix} \quad 3\mathrm{R}_3 + 4\mathrm{R}_2 \to \mathrm{R}_3$$

$$\begin{bmatrix} 1 & 1 & 1 & 4 \\ 0 & 3 & 4 & 21 \\ 0 & 0 & 7 & 21 \end{bmatrix}$$

Then $7z = 21 \Leftrightarrow z = 3$; $3y + 4 \cdot 3 = 21 \Leftrightarrow y = 3$; $x + 3 + 3 = 4 \Leftrightarrow x = -2$. Therefore, the solution is $(-2, 3, 3)$.

7. Use Gaussian elimination to solve the system $\begin{cases} 2x - 3y + 5z = 15 \\ 4x + 2y - 3z = -6 \\ 6x + y + z = 9 \end{cases}$.

(a) $\left(\frac{1}{3}, -\frac{2}{3}, 4\right)$ (b) $(-2, 6, -1)$ (c) $(1, 0, -3)$ (d) $\left(\frac{1}{2}, 2, 4\right)$ (e) $(8, 2, 3)$

Answer: medium (d)

$$\begin{bmatrix} 2 & -3 & 5 & 15 \\ 4 & 2 & -3 & -6 \\ 6 & 1 & 1 & 9 \end{bmatrix} \quad \begin{matrix} \mathrm{R}_2 - 2\mathrm{R}_1 \to \mathrm{R}_2 \\ \mathrm{R}_3 - 3\mathrm{R}_1 \to \mathrm{R}_3 \end{matrix} \quad \begin{bmatrix} 2 & -3 & 5 & 15 \\ 0 & 8 & -13 & -36 \\ 0 & 10 & -14 & -36 \end{bmatrix} \quad 8\mathrm{R}_3 - 10\mathrm{R}_2 \to \mathrm{R}_3 \quad \begin{bmatrix} 2 & -3 & 5 & 15 \\ 0 & 8 & -13 & -36 \\ 0 & 0 & 18 & 72 \end{bmatrix}$$

Thus, $18z = 72 \Leftrightarrow z = 4$; $8y - 13 \cdot 4 = -36 \Leftrightarrow y = 2$; $2x - 3 \cdot 2 + 5 \cdot 4 = 15 \Leftrightarrow x = \frac{1}{2}$. The solution is $\left(\frac{1}{2}, 2, 4\right)$.

8. Use Gaussian elimination to solve the system $\begin{cases} 3x_1 + 2x_2 = 5 \\ x_1 - x_2 + x_3 = 6 \\ 4x_1 - 2x_2 + 3x_3 = 13 \end{cases}$.

(a) $(2, 3, 15)$ (b) $(17, -45, 20)$ (c) $\left(14, -2, \frac{2}{3}\right)$ (d) $\left(-\frac{1}{3}, \frac{5}{3}, 40\right)$ (e) $(15, -20, -29)$

Answer: medium (e)

$$\begin{bmatrix} 3 & 2 & 0 & 5 \\ 1 & -1 & 1 & 6 \\ 4 & -2 & 3 & 13 \end{bmatrix} \quad \begin{matrix} 3\mathrm{R}_2 - \mathrm{R}_1 \to \mathrm{R}_2 \\ \mathrm{R}_3 - 4\mathrm{R}_2 \to \mathrm{R}_3 \end{matrix} \quad \begin{bmatrix} 3 & 2 & 0 & 5 \\ 0 & -5 & 3 & 13 \\ 0 & 2 & -1 & -11 \end{bmatrix} \quad 5\mathrm{R}_3 + 2\mathrm{R}_2 \to \mathrm{R}_3 \quad \begin{bmatrix} 3 & 2 & 0 & 5 \\ 0 & -5 & 3 & 13 \\ 0 & 0 & 1 & -29 \end{bmatrix}$$

Then, $x_3 = -29$; $-5x_2 + 3(-29) = 13 \Leftrightarrow x_2 = -20$; $3x_1 + 2(-20) = 5 \Leftrightarrow x_1 = 15$. Therefore, the solution is $(x_1, x_2, x_3) = (15, -20, -29)$.

9. Use Gaussian elimination to solve the system $\begin{cases} 2x + 10y - 20z = 80 \\ 4x + 20y + 30z = -50 \\ -2x + 6y - 20z = 40 \end{cases}$.

(a) $(2, 1, 3)$ (b) $(10, 0, -3)$ (c) $(15, 7, -2)$ (d) $(14, 15, 5)$ (e) $(-2, -6, 3)$

Answer: medium (b)

$$\begin{bmatrix} 2 & 10 & -20 & 80 \\ 4 & 20 & 30 & -50 \\ -2 & 6 & -20 & 40 \end{bmatrix} \begin{array}{c} R_2 - 2R_1 \to R_2 \\ R_1 + R_3 \to R_3 \end{array} \begin{bmatrix} 2 & 10 & -20 & 80 \\ 0 & 0 & 70 & -210 \\ 0 & 16 & -40 & 120 \end{bmatrix} \; R_2 \leftrightarrow R_3 \; \begin{bmatrix} 2 & 10 & -20 & 80 \\ 0 & 16 & -40 & 120 \\ 0 & 0 & 70 & -210 \end{bmatrix}$$

Thus $70z = -210 \Leftrightarrow z = -3$; $16y - 40(-3) = 120 \Leftrightarrow y = 0$; $2x - 20(-3) = 80 \Leftrightarrow x = 10$. So the solution is $(10, 0, -3)$.

10. Use Gaussian elimination to solve the system $\begin{cases} \frac{1}{4}x + \frac{2}{3}y - z = 4 \\ \frac{1}{2}x - \frac{4}{3}y + 2z = -2 \\ -\frac{1}{6}x + y - 6z = 5 \end{cases}$.

(a) $\left(\frac{1}{3}, 2, \frac{4}{3}\right)$ (b) $\left(-2, 1, -\frac{1}{3}\right)$ (c) $\left(6, 3, -\frac{1}{2}\right)$ (d) $\left(2, 4, -\frac{1}{4}\right)$ (e) $\left(3, 6, \frac{19}{2}\right)$

Answer: medium (c)

$$\begin{bmatrix} \frac{1}{4} & \frac{2}{3} & -1 & 4 \\ \frac{1}{2} & -\frac{4}{3} & 2 & -2 \\ -\frac{1}{6} & 1 & -6 & 5 \end{bmatrix} \begin{array}{c} R_2 - 2R_1 \to R_2 \\ 6R_3 + 2R_2 \to R_3 \end{array} \begin{bmatrix} \frac{1}{4} & \frac{2}{3} & -1 & 4 \\ 0 & -\frac{8}{3} & 4 & -10 \\ 0 & \frac{10}{3} & -32 & 26 \end{bmatrix} \; 4R_3 + 5R_2 \to R_3 \; \begin{bmatrix} \frac{1}{4} & \frac{2}{3} & -1 & 4 \\ 0 & -\frac{8}{3} & 4 & -10 \\ 0 & 0 & -108 & 54 \end{bmatrix}$$

So, $-108z = 54 \Leftrightarrow z = -\frac{1}{2}$; $-\frac{8}{3}y + 4\left(-\frac{1}{2}\right) = -10 \Leftrightarrow y = 3$; $\frac{1}{4}x + \frac{2}{3}(3) - \left(-\frac{1}{2}\right) = 4 \Leftrightarrow x = 6$. Thus the solution is $\left(6, 3, -\frac{1}{2}\right)$.

11. Use Gaussian elimination to solve the system $\begin{cases} 2x - y + 3z = 1.9 \\ 4x - 8y + z = 6.4 \\ 4x + 5y - 3z = -2.2 \end{cases}$.

(a) $(0.35, -0.6, 0.2)$ (b) $(0.25, 0.4, -0.3)$ (c) $(0.3, -0.45, 0.6)$ (d) $(-0.45, 0.65, 0.3)$ (e) $(1.2, -0.2, 3.4)$

Answer: medium (a)

$$\begin{bmatrix} 2 & -1 & 3 & 1.9 \\ 4 & -8 & 1 & 6.4 \\ 4 & 5 & -3 & -2.2 \end{bmatrix} \begin{array}{c} R_2 - 2R_1 \to R_2 \\ R_3 - 2R_1 \to R_3 \end{array} \begin{bmatrix} 2 & -1 & 3 & 1.9 \\ 0 & -6 & -5 & 2.6 \\ 0 & 7 & -9 & -6.0 \end{bmatrix} \; 6R_3 + 7R_2 \to R_3 \; \begin{bmatrix} 2 & -1 & 3 & 1.9 \\ 0 & -6 & -5 & 2.6 \\ 0 & 0 & -89 & -17.8 \end{bmatrix}$$

Now $-89z = -17.8 \Leftrightarrow z = 0.2$; $-6y - 5(0.2) = 2.6 \Leftrightarrow y = -0.6$; $2x - (-0.6) + 3(0.2) = 1.9 \Leftrightarrow x = 0.35$. Thus the solution is $(0.35, -0.6, 0.2)$.

12. Use Gaussian elimination to solve the system $\begin{cases} -x + 2y + 3z - 3w = 3 \\ -3x - 4y + 2z + w = 6 \\ -x - y + 2z + w = 8 \\ 2x + y - 4z - 2w = -6 \end{cases}$.

(a) $(4, 3, -12, 6)$ (b) $(16, -10, 9, -4)$ (c) $(14, 5, 8, -2)$ (d) $(-2, -4, 8, 13)$ (e) $\left(2, \frac{4}{3}, 4, -11\right)$

Answer: hard (b)

$$\begin{bmatrix} -1 & 2 & 3 & -3 & 3 \\ -3 & -4 & 2 & 1 & 6 \\ -1 & -1 & 2 & 1 & 8 \\ 2 & 1 & -4 & -2 & -6 \end{bmatrix} \begin{array}{c} R_2 - 3R_1 \to R_2 \\ R_3 - R_1 \to R_3 \\ R_4 + 2R_1 \to R_4 \end{array} \begin{bmatrix} -1 & 2 & 3 & -3 & 3 \\ 0 & -10 & -7 & 10 & -3 \\ 0 & -3 & -1 & 4 & 5 \\ 0 & 5 & 2 & -8 & 0 \end{bmatrix} \begin{array}{c} 10R_3 - 3R_2 \to R_3 \\ 2R_4 + R_2 \to R_4 \end{array}$$

$$\begin{bmatrix} -1 & 2 & 3 & 3 & 3 \\ 0 & -10 & -7 & 10 & -3 \\ 0 & 0 & 11 & 10 & 59 \\ 0 & 0 & -3 & -6 & -3 \end{bmatrix} \; 11R_4 + 3R_3 \to R_4 \; \begin{bmatrix} -1 & 2 & 3 & -3 & 3 \\ 0 & -10 & -7 & 10 & -3 \\ 0 & 0 & 11 & 10 & 59 \\ 0 & 0 & 0 & -36 & 144 \end{bmatrix}$$

Therefore, $-36w = 144 \Leftrightarrow w = -4$, $11z + 10(-4) = 59 \Leftrightarrow z = 9$, $-10y - 7(9) + 10(-4) = -3 \Leftrightarrow y = -10$, and $-x + 2(-10) + 3(9) - 3(-4) = 3 \Leftrightarrow x = 16$. Hence, the solution is $(16, -10, 9, -4)$.

13. Use Gaussian elimination to solve the system $\left\{\begin{array}{l} 2x-y+3z-2w-u+v=24 \\ -y-3z-u-v=-2 \\ 4x-6z-8w-4v=-12 \\ -3y-3z+v=8 \\ 2x+y+3z+2w-u-v=0 \\ -2x+y+3z-2w=4 \end{array}\right.$.

(a) $\left(2,-1,\frac{1}{3},3,2,4\right)$ (b) $\left(-6,6,3,\frac{4}{3},5,-2\right)$ (c) $\left(\frac{1}{3},\frac{2}{3},4,-4,6,1\right)$ (d) $\left(1,-2,-4,3,6,\frac{1}{3}\right)$ (e) $\left(1,-2,\frac{4}{3},-2,-6,6\right)$

Answer: hard (e)

$$\begin{bmatrix} 2 & -1 & 3 & -2 & -1 & 1 & 24 \\ 0 & -1 & -3 & 0 & -1 & -1 & -2 \\ 4 & 0 & -6 & -8 & 0 & -4 & -12 \\ 0 & -3 & -3 & 0 & 0 & 1 & 8 \\ 2 & 1 & 3 & 2 & -1 & -1 & 0 \\ -2 & 1 & 3 & -2 & 0 & 0 & 4 \end{bmatrix} \begin{array}{c} R_3-2R_1 \to R_3 \\ R_4-3R_2 \to R_4 \\ R_5-R_1 \to R_5 \\ R_6+R_1 \to R_6 \end{array} \begin{bmatrix} 2 & -1 & 3 & -2 & -1 & 1 & 24 \\ 0 & -1 & -3 & 0 & -1 & -1 & -2 \\ 0 & 2 & -12 & -4 & 2 & -6 & -60 \\ 0 & 0 & 6 & 0 & 3 & 4 & 14 \\ 0 & 2 & 0 & 4 & 0 & -2 & -24 \\ 0 & 0 & 6 & -4 & -1 & 1 & 28 \end{bmatrix}$$

$$\begin{array}{c} R_3+2R_2 \to R_3 \\ R_5-R_3 \to R_5 \\ R_6-R_4 \to R_6 \end{array} \begin{bmatrix} 2 & -1 & 3 & -2 & -1 & 1 & 24 \\ 0 & -1 & -3 & 0 & -1 & -1 & -2 \\ 0 & 0 & -18 & -4 & 0 & -8 & -64 \\ 0 & 0 & 6 & 0 & 3 & 4 & 14 \\ 0 & 0 & 12 & 8 & -2 & 4 & 36 \\ 0 & 0 & 0 & -4 & -4 & -3 & 14 \end{bmatrix} \begin{array}{c} R_3+3R_4 \to R_4 \\ 2R_4-R_5 \to R_5 \end{array} \begin{bmatrix} 2 & -1 & 3 & -2 & -1 & 1 & 24 \\ 0 & -1 & -3 & 0 & -1 & -1 & -2 \\ 0 & 0 & -18 & -4 & 0 & -8 & -64 \\ 0 & 0 & 0 & -4 & 9 & 4 & -22 \\ 0 & 0 & 0 & -8 & 8 & 4 & -8 \\ 0 & 0 & 0 & -4 & -4 & -3 & 14 \end{bmatrix}$$

$$\begin{array}{c} 2R_4-R_5 \to R_5 \\ R_4-R_6 \to R_6 \end{array} \begin{bmatrix} 2 & -1 & 3 & -2 & -1 & 1 & 24 \\ 0 & -1 & -3 & 0 & -1 & -1 & -2 \\ 0 & 0 & -18 & -4 & 0 & -8 & -64 \\ 0 & 0 & 0 & -4 & 9 & 4 & -22 \\ 0 & 0 & 0 & 0 & 10 & 4 & -36 \\ 0 & 0 & 0 & 0 & 13 & 7 & -36 \end{bmatrix} \quad 13R_5-10R_6 \to R_6$$

$$\begin{bmatrix} 2 & -1 & 3 & -2 & -1 & 1 & 24 \\ 0 & -1 & -3 & 0 & -1 & -1 & -2 \\ 0 & 0 & -18 & -4 & 0 & -8 & -64 \\ 0 & 0 & 0 & -4 & 9 & 4 & -22 \\ 0 & 0 & 0 & 0 & 10 & 4 & -36 \\ 0 & 0 & 0 & 0 & 0 & -18 & -108 \end{bmatrix}$$

Thus $v=6$; $10u+4(6)=-36 \Leftrightarrow u=-6$; $-4w+9(-6)+4(6)=-22 \Leftrightarrow w=-2$; $-18z-4(-2)-8(6)=-64 \Leftrightarrow z=\frac{4}{3}$; $-y-3\left(\frac{4}{3}\right)-(-6)-6=-2 \Leftrightarrow y=-2$; $2x-(-2)+3\left(\frac{4}{3}\right)-2(-2)-(-6)+6=24 \Leftrightarrow x=1$. Thus the solution is $(x,y,z,w,u,v)=\left(1,-2,\frac{4}{3},-2,-6,6\right)$.

14. Use Gaussian elimination to solve the system $\left\{\begin{array}{l} 3y-5z=56 \\ y-3z=14 \\ 8x+2y=77 \end{array}\right.$.

(a) $\left(\frac{1}{7},\frac{2}{7},7\right)$ (b) $\left(\frac{49}{2},\frac{3}{2},\frac{7}{4}\right)$ (c) $\left(\frac{7}{2},\frac{49}{2},\frac{7}{2}\right)$ (d) $\left(\frac{2}{3},\frac{49}{3},\frac{3}{7}\right)$ (e) $\left(7,6,\frac{3}{2}\right)$

Answer: medium (c)

$$\begin{bmatrix} 0 & 3 & -5 & 56 \\ 0 & 1 & -3 & 14 \\ 8 & 2 & 0 & 77 \end{bmatrix} \begin{array}{c} R_1-3R_2 \to R_2 \\ R_3 \leftrightarrow R_1 \end{array} \begin{bmatrix} 8 & 2 & 0 & 77 \\ 0 & 0 & 4 & 14 \\ 0 & 3 & -5 & 56 \end{bmatrix} \quad R_2 \leftrightarrow R_3 \quad \begin{bmatrix} 8 & 2 & 0 & 77 \\ 0 & 3 & -5 & 56 \\ 0 & 0 & 4 & 14 \end{bmatrix}$$

Thus, $4z=14 \Leftrightarrow z=\frac{7}{2}$, $3y-5\left(\frac{7}{2}\right)=56 \Leftrightarrow y=\frac{49}{2}$, and $8x+2\left(\frac{49}{2}\right)=77 \Leftrightarrow x=\frac{7}{2}$. Thus the solution is $\left(\frac{7}{2},\frac{49}{2},\frac{7}{2}\right)$.

15. A chemist has three containers of acid solution at various concentrations. The first is 6% acid, the second is 12%, and the third is 32%. How many milliliters of each should he mix together to make 100 mL of acid at 15% concentration, if he has to use four times as much of the 12% solution as the 32% solution?

Answer: medium

Let x be the quantity, in mL, of 6% acid, y be the quantity of 12% acid, and z be the quantity of 32% acid. Then $\begin{cases} 0.06x + 0.12y + 0.32z = 15 \\ x + y + z = 100 \\ y - 4z = 0 \end{cases}$. Multiplying the first equation by 100 and writing the system as a matrix, we get:

$$\begin{bmatrix} 1 & 1 & 1 & 100 \\ 6 & 12 & 32 & 1500 \\ 0 & 1 & -4 & 0 \end{bmatrix} \quad \text{R}_2 - 6\text{R}_1 \rightarrow \text{R}_2 \quad \begin{bmatrix} 1 & 1 & 1 & 100 \\ 0 & 6 & 26 & 900 \\ 0 & 1 & -4 & 0 \end{bmatrix} \quad \text{R}_2 - 6\text{R}_3 \rightarrow \text{R}_3 \quad \begin{bmatrix} 1 & 1 & 1 & 100 \\ 0 & 6 & 26 & 900 \\ 0 & 0 & 50 & 900 \end{bmatrix}$$

Thus $50z = 900 \quad \Leftrightarrow \quad z = 18$; $6y + 26(18) = 900 \quad \Leftrightarrow \quad y = 72$; $x + 72 + 18 = 100 \quad \Leftrightarrow \quad x = 10$. So, he should mix together 10 mL of 6% acid, 72 mL of 12% acid, and 18 mL of 32% acid.

16. A small school has 120 students who occupy three classrooms: rooms A, B, and C. After the first class period of the school day, one half of the students in room A move to room B, one quarter of the students in room B move to room C, and one sixth of the students in room C move to room A. Nevertheless, the total number of students in each room is the same for each period. How many students occupy each room?

Answer: medium

Let a, b, and c be the number of students in classrooms A, B, and C, respectively where $a, b, c \geq 0$. Then, the system of equations is $\begin{cases} a + b + c = 120 \\ \frac{1}{2}a = \frac{1}{4}b \quad \Rightarrow \quad b = 2a \\ \frac{1}{4}b = \frac{1}{6}c \quad \Rightarrow \quad c = \frac{3}{2}b = 3a \end{cases}$

By substitution, $a + 2a + 3a = 120 \quad \Leftrightarrow \quad a = 20$, $b = 2 \cdot 20 = 40$, and $c = 3 \cdot 20 = 60$. So, there are 20 students in classroom A, 40 students in classroom B, and 60 students in classroom C.

17. Arnold, Beatrix, and Caliban enter a race in which they will have to run, swim, and cycle over a marked course. Arnold runs at an average of 10 mi/h, swims at 5 mi/h, and cycles at 18 mi/h during the race. Beatrix runs at 8 mi/h, swims at 6 mi/h, and cycles at 16 mi/h. Caliban runs at 14 mi/h, swims at 3 mi/h, and cycles at 42 mi/h. Caliban finishes first with a total time of 2 h. Arnold comes in second with a time of 2 h and 30 min. Beatrix finishes last with a time of 2 h and 45 min. How many miles long is each part of the race?

Answer: hard

Let x, y, and z represent the distance, in miles, of the run, swim, and cycle parts of the race respectively. Then, since $\text{time} = \dfrac{\text{distance}}{\text{speed}}$, we get the following equations from the three contestants' race times:

$$\begin{cases} \frac{1}{10}x + \frac{1}{5}y + \frac{1}{18}z = 2.5 \\ \frac{1}{8}x + \frac{1}{6}y + \frac{1}{16}z = 2.75 \\ \frac{1}{14}x + \frac{1}{3}y + \frac{1}{42}z = 2 \end{cases} \Leftrightarrow \begin{cases} 9x + 18y + 5 = 225 \\ 6x + 8y + 3z = 132 \\ 3x + 14y + z = 84 \end{cases}$$ which has the following matrix representation:

$$\begin{bmatrix} 9 & 18 & 5 & 225 \\ 6 & 8 & 3 & 132 \\ 3 & 14 & 1 & 84 \end{bmatrix} \quad \begin{matrix} 2\text{R}_1 - 3\text{R}_2 \rightarrow \text{R}_2 \\ 3\text{R}_3 - \text{R}_1 \rightarrow \text{R}_3 \end{matrix} \quad \begin{bmatrix} 9 & 18 & 5 & 225 \\ 0 & 12 & 1 & 54 \\ 0 & 24 & -2 & 27 \end{bmatrix} \quad 2\text{R}_2 - \text{R}_3 \rightarrow \text{R}_3 \quad \begin{bmatrix} 9 & 18 & 5 & 225 \\ 0 & 12 & 1 & 54 \\ 0 & 0 & 4 & 81 \end{bmatrix}$$

Thus, $4z = 81 \quad \Leftrightarrow \quad z = 20\frac{1}{4}$, $12y + 20\frac{1}{4} = 54 \quad \Leftrightarrow \quad y = 2\frac{13}{16}$, and $9x + 18(2\frac{13}{16}) + 5(20\frac{1}{4}) = 225 \quad \Leftrightarrow \quad x = 8\frac{1}{8}$. So, the race has an $8\frac{1}{8}$ mile run, a $2\frac{13}{16}$ mile swim, and a $20\frac{1}{4}$ mile cycle.

18. Determine a, b, c, and d so that the points $(-3, 203)$, $(-1, -11)$, $(2, 43)$, and $(4, 749)$ all lie on the graph of the function $f(x) = ax^4 + bx^2 + cx + d$.

Answer: hard

Since $f(-3) = 203$, $f(-1) = -11$, $f(2) = 43$, and $f(4) = 749$, we get the following equations:

$$\begin{cases} a(-3)^4 + b(-3)^2 + c(-3) + d = 203 \\ a(-1)^4 + b(-1)^2 + c(-1) + d = -11 \\ a(2)^4 + b(2)^2 + c(2) + d = 43 \\ a(4)^4 + b(4)^2 + c(4) + d = 749 \end{cases} \text{ or } \begin{cases} 81a + 9b - 3c + d = 203 \\ a + b - c + d = -11 \\ 16a + 4b + 2c + d = 43 \\ 256a + 16b + 4c + d = 749 \end{cases}$$ which has the following matrix representation:

$$\begin{bmatrix} 81 & 9 & -3 & 1 & 203 \\ 1 & 1 & -1 & 1 & -11 \\ 16 & 4 & 2 & 1 & 43 \\ 256 & 16 & 4 & 1 & 749 \end{bmatrix} \begin{array}{c} 81R_2 - R_1 \to R_2 \\ 16R_2 - R_3 \to R_3 \\ 16R_3 - R_4 \to R_4 \end{array} \begin{bmatrix} 81 & 9 & -3 & 1 & 203 \\ 0 & 72 & -78 & 80 & -1094 \\ 0 & 12 & -18 & 15 & -219 \\ 0 & 48 & 28 & 15 & -61 \end{bmatrix} \begin{array}{c} R_2 - 6R_3 \to R_3 \\ R_4 - 4R_3 \to R_4 \end{array}$$

$$\begin{bmatrix} 81 & 9 & -3 & 1 & 203 \\ 0 & 72 & -78 & 80 & -1094 \\ 0 & 0 & 30 & -10 & 220 \\ 0 & 0 & 100 & -45 & 815 \end{bmatrix} \begin{array}{c} 3R_4 - 10R_3 \to R_4 \end{array} \begin{bmatrix} 81 & 9 & -3 & 1 & 203 \\ 0 & 72 & -78 & 80 & -1094 \\ 0 & 0 & 30 & -10 & 220 \\ 0 & 0 & 0 & -35 & 245 \end{bmatrix}$$

Thus $-35d = 245 \Leftrightarrow d = -7$; $3c - (-7) = 22 \Leftrightarrow c = 5$; $72b - 78(5) + 80(-7) = -1094 \Leftrightarrow b = -2$; $81a + 9(-2) - 3(5) + (-7) = 203 \Leftrightarrow a = 3$. The function is $f(x) = 3x^4 - 2x^2 + 5x - 7$.

19. Solve the equation $6x + 3y = 2z - 3y = 6w - 6y = x + 2z = 2$.

Answer: hard

$$\begin{cases} 6x + 3y = 2 \\ -3y + 2z = 2 \\ -6y + 6w = 2 \\ x + 2z = 2 \end{cases}$$ has the following matrix representation:

$$\begin{bmatrix} 6 & 3 & 0 & 0 & 2 \\ 0 & -3 & 2 & 0 & 2 \\ 0 & -6 & 0 & 6 & 2 \\ 1 & 0 & 2 & 0 & 2 \end{bmatrix} \begin{array}{c} R_1 \leftrightarrow R_4 \end{array} \begin{bmatrix} 1 & 0 & 2 & 0 & 2 \\ 0 & -3 & 2 & 0 & 2 \\ 0 & -6 & 0 & 6 & 2 \\ 6 & 3 & 0 & 0 & 2 \end{bmatrix} \begin{array}{c} 2R_2 - R_3 \to R_3 \\ R_4 - 6R_1 \to R_4 \end{array} \begin{bmatrix} 1 & 0 & 2 & 0 & 2 \\ 0 & -3 & 2 & 0 & 2 \\ 0 & 0 & 4 & -6 & 2 \\ 0 & 3 & -12 & 0 & -10 \end{bmatrix}$$

$$\begin{array}{c} R_2 + R_4 \to R_4 \end{array} \begin{bmatrix} 1 & 0 & 2 & 0 & 2 \\ 0 & -3 & 2 & 0 & 2 \\ 0 & 0 & 4 & -6 & 2 \\ 0 & 0 & -10 & 0 & -8 \end{bmatrix} \begin{array}{c} 5R_3 + 2R_4 \to R_4 \end{array} \begin{bmatrix} 1 & 0 & 2 & 0 & 2 \\ 0 & -3 & 2 & 0 & 2 \\ 0 & 0 & 4 & -6 & 2 \\ 0 & 0 & 0 & -30 & -6 \end{bmatrix}$$

Therefore, $-30w = -6 \Leftrightarrow w = \frac{1}{5}$, $4z - 6\left(\frac{1}{5}\right) = 2 \Leftrightarrow z = \frac{4}{5}$, $-3y + 2\left(\frac{4}{5}\right) = 2 \Leftrightarrow y = -\frac{2}{15}$, and $x + 2\left(\frac{4}{5}\right) = 2 \Leftrightarrow x = \frac{2}{5}$. So, the solution is $(x, y, z, w) = \left(\frac{2}{5}, -\frac{2}{15}, \frac{4}{5}, \frac{1}{5}\right)$.

20. Solve the system of linear equations $\begin{cases} x + y + z = 10 \\ 2x + 3y + 2z = 12 \\ 8x - 4y - 2z = 16 \end{cases}$

Answer: medium

$$\begin{cases} x + y + z = 10 \\ 2x + 3y + 2z = 12 \\ 8x - 4y - 2z = 16 \end{cases} \Rightarrow \begin{bmatrix} 1 & 1 & 1 & 10 \\ 2 & 3 & 2 & 12 \\ 8 & -4 & -2 & 16 \end{bmatrix} \begin{array}{c} -2R_1 + R_2 \to R_2 \\ \Rightarrow \\ -8R_1 + R_3 \to R_3 \end{array} \begin{bmatrix} 1 & 1 & 1 & 10 \\ 0 & 1 & 0 & -8 \\ 0 & -12 & -10 & -64 \end{bmatrix} \begin{array}{c} 12R_2 + R_3 \to R_3 \\ \Rightarrow \\ \frac{R_3}{-10} \to R_3 \end{array}$$

$$\begin{bmatrix} 1 & 1 & 1 & 10 \\ 0 & 1 & 0 & -8 \\ 0 & 0 & 1 & 16 \end{bmatrix} \Rightarrow x = 2, y = -8, z = 16$$

21. Solve the system of linear equations $\begin{cases} 4x + 5y = 0 \\ 2x + 2y + 2z = -1 \\ x + 4y + 3z = 2 \end{cases}$

Answer: medium

$$\begin{bmatrix} 4 & 5 & 0 & 0 \\ 2 & 2 & 2 & -1 \\ 1 & 4 & 3 & 2 \end{bmatrix} \underset{R_1 \longleftrightarrow R_3}{\Rightarrow} \begin{bmatrix} 1 & 4 & 3 & 2 \\ 2 & 2 & 2 & -1 \\ 4 & 5 & 0 & 0 \end{bmatrix} \begin{matrix} -2R_1 \to R_2 \\ \Rightarrow \\ -4R_1 \to R_3 \end{matrix} \begin{bmatrix} 1 & 4 & 3 & 2 \\ 0 & -6 & -4 & -5 \\ 0 & -11 & -12 & -8 \end{bmatrix} \begin{matrix} -11R_2 + 6R_3 \to R_3 \\ \Rightarrow \end{matrix} \begin{bmatrix} 1 & 4 & 3 & 2 \\ 0 & -6 & -4 & -5 \\ 0 & 0 & 28 & -7 \end{bmatrix}$$

$\Rightarrow \quad x = -\frac{5}{4}, y = 1, z = -\frac{1}{4}$

22. Solve the system of linear equations $\begin{cases} x + 2y + w = 3 \\ 3x + 2y - z - w = 0 \\ y + 2w = 1 \\ y + z + w = 3 \end{cases}$

Answer: hard

$$\begin{bmatrix} 1 & 2 & 0 & 1 & 3 \\ 3 & 2 & -1 & -1 & 0 \\ 0 & 1 & 0 & 2 & 1 \\ 0 & 1 & 1 & 1 & 3 \end{bmatrix} \underset{-3R_1 + R_2 \to R_2}{\Rightarrow} \begin{bmatrix} 1 & 2 & 0 & 1 & 3 \\ 0 & -4 & -1 & -4 & -9 \\ 0 & 1 & 0 & 2 & 1 \\ 0 & 1 & 1 & 1 & 3 \end{bmatrix} \begin{matrix} 4R_3 + R_2 \to R_3 \\ \Rightarrow \\ 4R_4 + R_2 \to R_4 \end{matrix} \begin{bmatrix} 1 & 2 & 0 & 1 & 3 \\ 0 & -4 & -1 & -4 & -9 \\ 0 & 0 & -1 & 4 & -5 \\ 0 & 0 & 3 & 0 & 3 \end{bmatrix} \underset{3R_3 + R_4 \to R_4}{\Rightarrow}$$

$$\begin{bmatrix} 1 & 2 & 0 & 1 & 3 \\ 0 & -4 & -1 & -4 & -9 \\ 0 & 0 & -1 & 4 & -5 \\ 0 & 0 & 0 & 12 & -12 \end{bmatrix} \Rightarrow \quad x = -2, y = 3, z = 1, w = -1$$

23. Find an equation for the parabola $y = ax^2 + bx + c$ that passes through the points $(-3, 7)$, $(-2, 1)$, and $(1, 7)$.

Answer: medium

Passing through the point $(-3, 7) \Rightarrow 7 = 9a - 3b + c$. Through $(-2, 1) \Rightarrow 1 = 4a - 2b + c$. Through $(1, 7) \Rightarrow$ $7 = a + b + c$. Solve the system of linear equations: $\begin{cases} 9a - 3b + c = 7 \\ 4a - 2b + c = 1 \\ a + b + c = 7 \end{cases} \Rightarrow c = 1, a = 2, b = 4$. Thus, $y = 2x^2 + 4x + 1$.

SECTION 10.4

1. Let $B = \begin{bmatrix} 2 & \frac{1}{4} & 4 \\ -1 & 2 & 1 \end{bmatrix}$ and $F = \begin{bmatrix} 1 & 0 & 0 \\ 0 & 1 & 0 \\ 0 & 0 & 1 \end{bmatrix}$. Carry out the algebraic operation $B + F$ or explain why it cannot be performed.

Answer: easy

$B + F$ is undefined because $B(2 \times 3)$ and $F(3 \times 3)$ have incompatible dimensions.

2. Let $A = \begin{bmatrix} 1 & -4 \\ 0 & 3 \end{bmatrix}$. Carry out the algebraic operation $5A$ or explain why it cannot be performed.

(a) $\begin{bmatrix} -4 & 5 \\ 0 & 15 \end{bmatrix}$ (b) $\begin{bmatrix} 5 & -20 \\ 0 & 15 \end{bmatrix}$ (c) $\begin{bmatrix} -5 & 9 \\ 5 & -12 \end{bmatrix}$ (d) $\begin{bmatrix} 1 & -2 \\ 20 & 12 \end{bmatrix}$ (e) $\begin{bmatrix} 1 & -4 \\ 4 & -1 \end{bmatrix}$

Answer: easy (b)

$5A = 5\begin{bmatrix} 1 & -4 \\ 0 & 3 \end{bmatrix} = \begin{bmatrix} 5 & -20 \\ 0 & 15 \end{bmatrix}$

3. Let $A = \begin{bmatrix} 1 & -4 \\ 0 & 3 \end{bmatrix}$ and $C = \begin{bmatrix} 3 & -\frac{3}{2} & 0 \\ 0 & 1 & -2 \end{bmatrix}$. Carry out the algebraic operation $C - 5A$ or explain why it cannot be performed.

Answer: easy

$C - 5A$ is undefined because $C(2 \times 3)$ and $A(2 \times 2)$ have incompatible dimensions.

4. Let $A = \begin{bmatrix} 1 & -4 \\ 0 & 3 \end{bmatrix}$ and $D = [6 \quad 4]$. Carry out the algebraic operation DA or explain why it cannot be performed.

(a) $[3 \quad -14]$ (b) $[12 \quad -36]$ (c) $[1 \quad -2]$ (d) $[14 \quad -12]$ (e) $[6 \quad -12]$

Answer: easy (e)

$$DA = [6 \quad 4]\begin{bmatrix} 1 & -4 \\ 0 & 3 \end{bmatrix} = [6 \quad -12]$$

5. Let $B = \begin{bmatrix} 2 & \frac{1}{4} & 4 \\ -1 & 2 & 1 \end{bmatrix}$ and $C = \begin{bmatrix} 3 & -\frac{3}{2} & 0 \\ 0 & 1 & -2 \end{bmatrix}$. Carry out the algebraic operation BC or explain why it cannot be performed.

Answer: easy

BC is undefined because $B(2 \times 3)$ and $C(2 \times 3)$ have incompatible dimensions.

6. Let $F = \begin{bmatrix} 1 & 0 & 0 \\ 0 & 1 & 0 \\ 0 & 0 & 1 \end{bmatrix}$ and $G = \begin{bmatrix} 4 & -2 & 8 \\ 5 & 1 & 0 \\ -3 & 2 & 1 \\ 0 & 0 & 0 \end{bmatrix}$. Carry out the algebraic operation GF or explain why it cannot be performed.

(a) $\begin{bmatrix} 3 & -1 & 4 \\ 2 & 2 & 3 \\ 1 & 3 & 2 \\ 0 & 0 & 0 \end{bmatrix}$ (b) $\begin{bmatrix} 5 & 2 & 5 \\ 4 & 2 & 2 \\ -2 & 1 & 4 \\ 1 & 1 & 2 \end{bmatrix}$ (c) $\begin{bmatrix} 4 & -2 & 8 \\ 5 & 1 & 0 \\ -3 & 2 & 1 \\ 0 & 0 & 0 \end{bmatrix}$ (d) $\begin{bmatrix} 1 & 12 & 3 \\ 14 & -6 & 4 \\ 4 & 1 & 2 \\ 3 & 1 & 2 \end{bmatrix}$ (e) $\begin{bmatrix} -4 & 2 & 3 \\ -4 & 2 & 2 \\ 1 & -2 & 3 \\ 0 & 1 & 0 \end{bmatrix}$

Answer: easy (c)

$$GF = \begin{bmatrix} 4 & -2 & 8 \\ 5 & 1 & 0 \\ -3 & 2 & 1 \\ 0 & 0 & 0 \end{bmatrix}\begin{bmatrix} 1 & 0 & 0 \\ 0 & 1 & 0 \\ 0 & 0 & 1 \end{bmatrix} = \begin{bmatrix} 4 & -2 & 8 \\ 5 & 1 & 0 \\ -3 & 2 & 1 \\ 0 & 0 & 0 \end{bmatrix}$$

7. Let $A = \begin{bmatrix} 4 & -4 \\ 0 & 3 \end{bmatrix}$, $B = \begin{bmatrix} 2 & \frac{1}{4} & 4 \\ -1 & 2 & 1 \end{bmatrix}$, and $D = [6 \quad 4]$. Carry out the algebraic operation $D(AB)$ or explain why it cannot be performed.

(a) $[60 \quad -18 \quad 84]$ (b) $[12 \quad -4 \quad 14]$ (c) $[120 \quad 32 \quad 72]$ (d) $[15 \quad 24 \quad 18]$ (e) $[-42 \quad -9 \quad 12]$

Answer: medium (a)

$$D(AB) = [6 \quad 4] \cdot \begin{bmatrix} 4 & -4 \\ 0 & 3 \end{bmatrix}\begin{bmatrix} 2 & \frac{1}{4} & 4 \\ -1 & 2 & 1 \end{bmatrix} = [6 \quad 4]\begin{bmatrix} 12 & -7 & 12 \\ -3 & 6 & 3 \end{bmatrix} = [60 \quad -18 \quad 84]$$

8. Let $A = \begin{bmatrix} 1 & -4 \\ 0 & 3 \end{bmatrix}$. Carry out the algebraic operation A^2 or explain why it cannot be performed.

(a) $\begin{bmatrix} 2 & -12 \\ 2 & 4 \end{bmatrix}$ (b) $\begin{bmatrix} 4 & -12 \\ 3 & 4 \end{bmatrix}$ (c) $\begin{bmatrix} 1 & 16 \\ 0 & 9 \end{bmatrix}$ (d) $\begin{bmatrix} -1 & 16 \\ 0 & -9 \end{bmatrix}$ (e) $\begin{bmatrix} 1 & -16 \\ 0 & 9 \end{bmatrix}$

Answer: easy (e)

$$A^2 = \begin{bmatrix} 1 & -4 \\ 0 & 3 \end{bmatrix}\begin{bmatrix} 1 & -4 \\ 0 & 3 \end{bmatrix} = \begin{bmatrix} 1 & -16 \\ 0 & 9 \end{bmatrix}$$

9. Let $B = \begin{bmatrix} 2 & \frac{1}{2} & 4 \\ -1 & 2 & 1 \end{bmatrix}$, $C = \begin{bmatrix} 3 & -\frac{3}{2} & 0 \\ 0 & 1 & -2 \end{bmatrix}$, and $D = [6 \quad 4]$. Carry out the algebraic operation $DB + DC$ or explain why it cannot be performed.

(a) $[13 \quad 2 \quad -4]$ (b) $[26 \quad 6 \quad 20]$ (c) $[14 \quad -6 \quad 13]$ (d) $[-4 \quad 8 \quad 10]$ (e) $[-2 \quad 2 \quad 15]$

Answer: medium (b)

$$DB + DC = D(B + C) = [6 \quad 4]\left(\begin{bmatrix} 2 & \frac{1}{2} & 4 \\ -1 & 2 & 1 \end{bmatrix} + \begin{bmatrix} 3 & -\frac{3}{2} & 0 \\ 0 & 1 & -2 \end{bmatrix}\right) = [6 \quad 4]\begin{bmatrix} 5 & -1 & 4 \\ -1 & 3 & -1 \end{bmatrix} = [26 \quad 6 \quad 20]$$

10. Let $F = \begin{bmatrix} 1 & 0 & 0 \\ 0 & 1 & 0 \\ 0 & 0 & 1 \end{bmatrix}$. Carry out the algebraic operation F^2 or explain why it cannot be performed.

(a) $\begin{bmatrix} 0 & 0 & 1 \\ 0 & 1 & 0 \\ 1 & 0 & 0 \end{bmatrix}$ (b) $\begin{bmatrix} 1 & 1 & 1 \\ 1 & 1 & 1 \\ 1 & 1 & 1 \end{bmatrix}$ (c) $\begin{bmatrix} 1 & 0 & 0 \\ 0 & 1 & 0 \\ 0 & 0 & 1 \end{bmatrix}$ (d) $\begin{bmatrix} 0 & 0 & 0 \\ 0 & 0 & 0 \\ 0 & 0 & 0 \end{bmatrix}$ (e) $\begin{bmatrix} -1 & 0 & 0 \\ 0 & 1 & 0 \\ 0 & 0 & -1 \end{bmatrix}$

Answer: easy (c)

$$F^2 = \begin{bmatrix} 1 & 0 & 0 \\ 0 & 1 & 0 \\ 0 & 0 & 1 \end{bmatrix}\begin{bmatrix} 1 & 0 & 0 \\ 0 & 1 & 0 \\ 0 & 0 & 1 \end{bmatrix} = \begin{bmatrix} 1 & 0 & 0 \\ 0 & 1 & 0 \\ 0 & 0 & 1 \end{bmatrix}$$

11. What must be true about the dimensions of the matrices A and B if both products AB and BA are defined?

Answer: medium

Suppose A is $n \times m$ and B is $i \times j$. If the product AB is defined, then $m = i$. If the product BA is defined, then $j = n$. Thus if both products are defined and if A is $n \times m$, then B must be $m \times n$.

12. Write the system of equations $\begin{cases} 3x - y + 2z = 6 \\ x - 3z = 4 \\ 2y - 2z = 7 \end{cases}$ as a matrix equation.

Answer: easy

$\begin{cases} 3x - y + 2z = 6 \\ x - 3z = 4 \\ 2y - 2z = 7 \end{cases}$ as a matrix equation is $\begin{bmatrix} 3 & -1 & 2 \\ 1 & 0 & -3 \\ 0 & 2 & -2 \end{bmatrix}\begin{bmatrix} x \\ y \\ z \end{bmatrix} = \begin{bmatrix} 6 \\ 4 \\ 7 \end{bmatrix}$

13. Write the system of equations $\begin{cases} 2x - y + 3z = 1 \\ 3x - y - 5z = 1 \\ x + 2y + 3z = 1 \\ -x - y - z = 1 \end{cases}$ as a matrix equation.

Answer: easy

$\begin{cases} 2x - y + 3z = 1 \\ 3x - y - 5z = 1 \\ x + 2y + 3z = 1 \\ -x - y - z = 1 \end{cases}$ as a matrix equation is $\begin{bmatrix} 2 & -1 & 3 \\ 3 & -1 & -5 \\ 1 & 2 & 3 \\ -1 & -1 & -1 \end{bmatrix}\begin{bmatrix} x \\ y \\ z \end{bmatrix} = \begin{bmatrix} 1 \\ 1 \\ 1 \\ 1 \end{bmatrix}$

14. Let $C = \begin{bmatrix} 1 & 4 \\ 2 & 3 \\ 0 & 1 \end{bmatrix}$ and $D = \begin{bmatrix} 12 & 20 \\ 24 & 32 \\ 8 & 4 \end{bmatrix}$. Solve the matrix equation $4(X - C) = D$ for the unknown matrix X, or explain why there is no solution.

(a) $\begin{bmatrix} 4 & 9 \\ 8 & 11 \\ 2 & 2 \end{bmatrix}$ (b) $\begin{bmatrix} 2 & 3 \\ 4 & 6 \\ 7 & 1 \end{bmatrix}$ (c) $\begin{bmatrix} 3 & 2 \\ 11 & 14 \\ 6 & 6 \end{bmatrix}$ (d) $\begin{bmatrix} -4 & 3 \\ -1 & 6 \\ 3 & -2 \end{bmatrix}$ (e) $\begin{bmatrix} 0 & -1 \\ 3 & 2 \\ 1 & 2 \end{bmatrix}$

Answer: medium (a)

$4(X - C) = D \quad \Leftrightarrow \quad X - C = \frac{1}{4}D \quad \Leftrightarrow$

$$X = \tfrac{1}{4}D + C = \tfrac{1}{4}\begin{bmatrix} 12 & 20 \\ 24 & 32 \\ 8 & 4 \end{bmatrix} + \begin{bmatrix} 1 & 4 \\ 2 & 3 \\ 0 & 1 \end{bmatrix} = \begin{bmatrix} 3 & 5 \\ 6 & 8 \\ 2 & 1 \end{bmatrix} + \begin{bmatrix} 1 & 4 \\ 2 & 3 \\ 0 & 1 \end{bmatrix} = \begin{bmatrix} 4 & 9 \\ 8 & 11 \\ 2 & 2 \end{bmatrix}$$

15. Let $A = \begin{bmatrix} 4 & 6 \\ 1 & 3 \end{bmatrix}$ and $D = \begin{bmatrix} 12 & 20 \\ 24 & 32 \\ 8 & 4 \end{bmatrix}$. Solve the matrix equation $A + D = 3X$ for the unknown matrix X, or explain why there is no solution.

Answer: medium

$A + D = 3X \quad \Leftrightarrow \quad X = \frac{1}{3}(A + D)$ which is undefined because the dimensions, (2×2) and (3×2), are incompatible.

16. Prove that if A and B are 2×2 matrices, then $(A+B)^2 = A^2 + AB + BA + B^2$.

Answer: hard

Let $A = \begin{bmatrix} a & b \\ c & d \end{bmatrix}$ and $B = \begin{bmatrix} e & f \\ g & h \end{bmatrix}$. Then $A + B = \begin{bmatrix} a+e & b+f \\ c+g & d+h \end{bmatrix}$,

$$(A+B)^2 = \begin{bmatrix} a+e & b+f \\ c+g & d+h \end{bmatrix}\begin{bmatrix} a+e & b+f \\ c+g & d+h \end{bmatrix} = \begin{bmatrix} (a+e)^2 + (b+f)(c+g) & (a+e)(b+f) + (b+f)(d+h) \\ (c+g)(a+e) + (d+h)(c+g) & (c+g)(b+f) + (d+h)^2 \end{bmatrix},$$

$$A^2 = \begin{bmatrix} a & b \\ c & d \end{bmatrix}\begin{bmatrix} a & b \\ c & d \end{bmatrix} = \begin{bmatrix} a^2+bc & ab+bd \\ ac+cd & bc+d^2 \end{bmatrix}, \quad B^2 = \begin{bmatrix} e & f \\ g & h \end{bmatrix}\begin{bmatrix} e & f \\ g & h \end{bmatrix} = \begin{bmatrix} e^2+fg & ef+fh \\ eg+gh & fg+h^2 \end{bmatrix},$$

$$AB = \begin{bmatrix} a & b \\ c & d \end{bmatrix}\begin{bmatrix} e & f \\ g & h \end{bmatrix} = \begin{bmatrix} ae+bg & af+bh \\ ce+dg & cf+dh \end{bmatrix}, \text{ and } BA = \begin{bmatrix} e & f \\ g & h \end{bmatrix}\begin{bmatrix} a & b \\ c & d \end{bmatrix} = \begin{bmatrix} ae+cf & be+df \\ ag+ch & bg+dh \end{bmatrix}. \text{ Then}$$

$$\begin{aligned} A^2 + AB + BA + B^2 &= \begin{bmatrix} a^2+bc+ae+bg+ae+cf+e^2+fg & ab+bd+ef+fh+af+bh+be+df \\ ac+cd+eg+gh+ce+dg+ag+ch & bc+d^2+fg+h^2+cf+dh+bg+dh \end{bmatrix} \\ &= \begin{bmatrix} a^2+2ae+e^2+b(c+g)+f(c+g) & a(b+f)+e(b+f)+b(d+h)+f(d+h) \\ c(a+e)+g(a+e)+d(c+g)+h(c+g) & c(b+f)+g(b+f)+d^2+2dh+h^2 \end{bmatrix} \\ &= \begin{bmatrix} (a+e)^2+(b+f)(c+g) & (a+e)(b+f)+(b+f)(d+h) \\ (c+g)(a+e)+(d+h)(c+g) & (c+g)(b+f)+(d+h)^2 \end{bmatrix} \\ &= (A+B)^2 \end{aligned}$$

17. Let $A = \begin{bmatrix} 1 & 1 \\ 0 & 1 \end{bmatrix}$

(a) Calculate A^2, A^3, and A^4.

(b) Find a general formula for A^n.

Answer: medium

(a) $A^2 = \begin{bmatrix} 1 & 1 \\ 0 & 1 \end{bmatrix}\begin{bmatrix} 1 & 1 \\ 0 & 1 \end{bmatrix} = \begin{bmatrix} 1 & 2 \\ 0 & 1 \end{bmatrix}$, $A^3 = A \cdot A^2 = \begin{bmatrix} 1 & 1 \\ 0 & 1 \end{bmatrix}\begin{bmatrix} 1 & 2 \\ 0 & 1 \end{bmatrix} = \begin{bmatrix} 1 & 3 \\ 0 & 1 \end{bmatrix}$,

$A^4 = A \cdot A^3 = \begin{bmatrix} 1 & 1 \\ 0 & 1 \end{bmatrix}\begin{bmatrix} 1 & 3 \\ 0 & 1 \end{bmatrix} = \begin{bmatrix} 1 & 4 \\ 0 & 1 \end{bmatrix}$

(b) $A^n = \begin{bmatrix} 1 & n \\ 0 & 1 \end{bmatrix}$

18. A small fast-food chain has restaurants in Santa Monica, Long Beach, and Anaheim. Only burgers, fries and shakes are sold by this chain. On a certain day, sales were distributed according to the following matrix:

	Santa Monica	Long Beach	Anaheim
Burgers	5200	1400	4200
Fries	500	750	600
Shakes	600	400	800

$= A$

The price of each item is given by the matrix

Burger	Fries	Shake
\$1.10	\$0.75	\$1.25

$= B$

(a) Calculate the product BA.

(b) Interpret the entries in the product matrix BA.

Answer: medium

(a) $BA = [\$1.10 \quad \$0.75 \quad \$1.25]\begin{bmatrix} 5200 & 1400 & 4200 \\ 500 & 750 & 600 \\ 600 & 400 & 800 \end{bmatrix} = [\$6845 \quad \$2602.50 \quad \$6070]$

(b) The entries in the product matrix represent the total revenue for the restaurants in Santa Monica, Long Beach, and Anaheim respectively.

19. Let $A = \begin{bmatrix} 2 & 0 & 2 & -3 \\ 1 & \frac{1}{2} & 3 & 0 \end{bmatrix}$, $B = [2 \quad 5 \quad -4 \quad 3]$, and $C = \begin{bmatrix} -1 \\ 2 \\ 1 \\ 0 \end{bmatrix}$.

Determine which of the following products are defined, and calculate the ones that are: $ABC, ACB, BAC, BCA, CAB, CBA$.

Answer: medium

ABC is undefined because the dimension of A, (2×4), is not compatible with the dimension of B, (1×4).

$$ACB = \begin{bmatrix} 0 \\ 3 \end{bmatrix} [2 \quad 5 \quad -4 \quad 3] = \begin{bmatrix} 0 & 0 & 0 & 0 \\ 6 & 15 & -12 & 9 \end{bmatrix}$$

BAC is undefined because the dimension of B, (1×4), is not compatible with the dimension of A, (2×4).

BCA is undefined because the dimension of C, (4×1), is not compatible with the dimension of A, (2×4).

CAB is undefined because the dimension of C, (4×1), is not compatible with the dimension of A, (2×4).

CBA is undefined because the dimension of B, (1×4), is not compatible with the dimension of A, (2×4).

20. Let $A = \begin{bmatrix} 6\,5 \end{bmatrix}$, $B = \begin{bmatrix} -5 & 4 \\ 6 & 7 \end{bmatrix}$, $C = \begin{bmatrix} 0\,1\,2 \\ 6\,2\,3 \end{bmatrix}$, $D = \begin{bmatrix} 1 \\ 5 \\ 0 \end{bmatrix}$. Carry out the algebraic operation $AB \cdot CD$, or explain why it cannot be performed.

Answer: easy

$$AB = \begin{bmatrix} 6\,5 \end{bmatrix} \begin{bmatrix} -5 & 4 \\ 6 & 7 \end{bmatrix} = \begin{bmatrix} 0\,59 \end{bmatrix}$$

$$CD = \begin{bmatrix} 0\,1\,2 \\ 6\,2\,3 \end{bmatrix} \begin{matrix} 1 \\ 5 \\ 0 \end{matrix} = \begin{matrix} 5 \\ 16 \end{matrix}$$

$$AB \cdot CD = 0\,59 \begin{matrix} 5 \\ 16 \end{matrix} = 944$$

21. Solve $\begin{bmatrix} 2 & x \\ 6 & 0 \end{bmatrix} \begin{bmatrix} y & 3 \\ -2 & 2 \end{bmatrix} = \begin{bmatrix} 1 & 3 \\ -6 & 18 \end{bmatrix}$ for x and y.

Answer: medium

$$\begin{bmatrix} 2 & x \\ 6 & 0 \end{bmatrix} \begin{bmatrix} y & 3 \\ -2 & 2 \end{bmatrix} = \begin{bmatrix} 2y - 2x & 6 + 2x \\ 6y & 18 \end{bmatrix} = \begin{bmatrix} 1 & 3 \\ -6 & 18 \end{bmatrix} \Rightarrow \begin{cases} 2y - 2x = 1 \\ 6 + 2x = 3 \end{cases}$$

Therefore, the solution is $x = -\frac{3}{2}$, $y = -1$.

22. Solve $\begin{bmatrix} 1 & 7 \\ 2 & 3 \end{bmatrix} \begin{bmatrix} x \\ y \end{bmatrix} = \begin{bmatrix} -6 \\ -2 \end{bmatrix}$ for x and y.

Answer: medium

$$\begin{bmatrix} 1 & 7 \\ 2 & 3 \end{bmatrix} \begin{bmatrix} x \\ y \end{bmatrix} = \begin{bmatrix} x + 7y \\ 2x + 3y \end{bmatrix} = \begin{bmatrix} -6 \\ -2 \end{bmatrix} \Rightarrow \begin{cases} x + 7y = -6 \\ 2x + 3y = -2 \end{cases} \Rightarrow x = \frac{4}{11}, y = -\frac{10}{11}$$

SECTION 10.5

1. Find the inverse of the matrix $B = \begin{bmatrix} 2 & 0 & 3 \\ 0 & -2 & 1 \\ 2 & -1 & 0 \end{bmatrix}$, and verify that $B^{-1}B = BB^{-1} = I_3$.

(a) $\frac{1}{7}\begin{bmatrix} 2 & -4 & 5 \\ -3 & 5 & 1 \\ 0 & -2 & 4 \end{bmatrix}$ (b) $\begin{bmatrix} \frac{2}{5} & -\frac{3}{5} & 12 \\ -5 & 2 & -1 \\ -1 & \frac{3}{2} & -2 \end{bmatrix}$ (c) $\frac{1}{14}\begin{bmatrix} 1 & -3 & 6 \\ 2 & -6 & -2 \\ 4 & 2 & -4 \end{bmatrix}$ (d) $\frac{1}{21}\begin{bmatrix} 3 & -4 & -4 \\ 2 & 0 & -6 \\ 5 & 3 & 2 \end{bmatrix}$ (e) Does not exist

Answer: medium (c)

$B = \begin{bmatrix} 2 & 0 & 3 \\ 0 & -2 & 1 \\ 2 & -1 & 0 \end{bmatrix}$. Begin with a 3×6 matrix whose left half is B and whose right half is I_3.

$$\begin{bmatrix} 2 & 0 & 3 & 1 & 0 & 0 \\ 0 & -2 & 1 & 0 & 1 & 0 \\ 2 & -1 & 0 & 0 & 0 & 1 \end{bmatrix} \quad R_1 - R_3 \to R_3 \quad \begin{bmatrix} 2 & 0 & 3 & 1 & 0 & 0 \\ 0 & -2 & 1 & 0 & 1 & 0 \\ 0 & 1 & 3 & 1 & 0 & -1 \end{bmatrix} \quad \begin{matrix} \frac{1}{2}R_1 \to R_1 \\ -\frac{1}{2}R_2 \to R_2 \\ 2R_3 + R_2 \to R_3 \end{matrix} \quad \begin{bmatrix} 1 & 0 & \frac{3}{2} & \frac{1}{2} & 0 & 0 \\ 0 & 1 & -\frac{1}{2} & 0 & -\frac{1}{2} & 0 \\ 0 & 0 & 7 & 2 & 1 & -2 \end{bmatrix}$$

$$\begin{matrix} R_1 - \frac{3}{14}R_3 \to R_1 \\ R_2 + \frac{1}{14}R_3 \to R_2 \\ \frac{1}{7}R_3 \to R_3 \end{matrix} \quad \begin{bmatrix} 1 & 0 & 0 & \frac{1}{14} & -\frac{3}{14} & \frac{3}{7} \\ 0 & 1 & 0 & \frac{1}{7} & -\frac{3}{7} & -\frac{1}{7} \\ 0 & 0 & 1 & \frac{2}{7} & \frac{1}{7} & -\frac{2}{7} \end{bmatrix}$$

Then $B^{-1} = \frac{1}{14}\begin{bmatrix} 1 & -3 & 6 \\ 2 & -6 & -2 \\ 4 & 2 & -4 \end{bmatrix}$ and $B^{-1}B = \frac{1}{14}\begin{bmatrix} 1 & -3 & 6 \\ 2 & -6 & -2 \\ 4 & 2 & -4 \end{bmatrix}\begin{bmatrix} 2 & 0 & 3 \\ 0 & -2 & 1 \\ 2 & -1 & 0 \end{bmatrix} = \frac{1}{14}\begin{bmatrix} 14 & 0 & 0 \\ 0 & 14 & 0 \\ 0 & 0 & 14 \end{bmatrix} = I_3$ and

$$BB^{-1} = \begin{bmatrix} 2 & 0 & 3 \\ 0 & -2 & 1 \\ 2 & -1 & 0 \end{bmatrix}\frac{1}{14}\begin{bmatrix} 1 & -3 & 6 \\ 2 & -6 & -2 \\ 4 & 2 & -4 \end{bmatrix} = \frac{1}{14}\begin{bmatrix} 14 & 0 & 0 \\ 0 & 14 & 0 \\ 0 & 0 & 14 \end{bmatrix} = I_3.$$

2. Find the inverse of the matrix of $\begin{bmatrix} 12 & 5 \\ 7 & 3 \end{bmatrix}$, if it exists.

(a) $\begin{bmatrix} 3 & 7 \\ 5 & 10 \end{bmatrix}$ (b) $\begin{bmatrix} 3 & -5 \\ -7 & 12 \end{bmatrix}$ (c) $\begin{bmatrix} 12 & 2 \\ 4 & -13 \end{bmatrix}$ (d) $\begin{bmatrix} 2 & 5 \\ 1 & 4 \end{bmatrix}$ (e) Does not exist

Answer: easy (b)

$$\begin{bmatrix} 12 & 5 \\ 7 & 3 \end{bmatrix}^{-1} = \frac{1}{36 - 35}\begin{bmatrix} 3 & -5 \\ -7 & 12 \end{bmatrix} = \begin{bmatrix} 3 & -5 \\ -7 & 12 \end{bmatrix}$$

3. Find the inverse of the matrix of $\begin{bmatrix} -3 & 5 \\ 7 & -3 \end{bmatrix}$, if it exists.

(a) $\frac{1}{13}\begin{bmatrix} 2 & 1 \\ -7 & 3 \end{bmatrix}$ (b) $\frac{1}{6}\begin{bmatrix} 5 & 3 \\ 9 & 12 \end{bmatrix}$ (c) $\frac{1}{34}\begin{bmatrix} 2 & 5 \\ 17 & 5 \end{bmatrix}$ (d) $\frac{1}{26}\begin{bmatrix} 3 & 5 \\ 7 & 3 \end{bmatrix}$ (e) Does not exist

Answer: easy (d)

$$\begin{bmatrix} -3 & 5 \\ 7 & -3 \end{bmatrix}^{-1} = \frac{1}{9 - 35}\begin{bmatrix} -3 & -5 \\ -7 & -3 \end{bmatrix} = \frac{1}{26}\begin{bmatrix} 3 & 5 \\ 7 & 3 \end{bmatrix}$$

4. Find the inverse of the matrix of $\begin{bmatrix} \frac{1}{4} & \frac{1}{6} \\ 12 & 12 \end{bmatrix}$, if it exists.

(a) $\begin{bmatrix} -2 & \frac{1}{3} \\ 12 & -\frac{3}{8} \end{bmatrix}$ (b) $\begin{bmatrix} 12 & -\frac{2}{3} \\ \frac{15}{4} & \frac{1}{3} \end{bmatrix}$ (c) $\begin{bmatrix} 12 & -\frac{1}{6} \\ -12 & \frac{1}{4} \end{bmatrix}$ (d) $\begin{bmatrix} 3 & 4 \\ \frac{3}{8} & -\frac{2}{3} \end{bmatrix}$ (e) Does not exist

Answer: easy (c)

$$\begin{bmatrix} \frac{1}{4} & \frac{1}{6} \\ 12 & 12 \end{bmatrix}^{-1} = \frac{1}{3 - 2}\begin{bmatrix} 12 & -\frac{1}{6} \\ -12 & \frac{1}{4} \end{bmatrix} = \begin{bmatrix} 12 & -\frac{1}{6} \\ -12 & \frac{1}{4} \end{bmatrix}$$

5. Find the inverse of the matrix of $\begin{bmatrix} 3 & -2 & -5 \\ -1 & 1 & 1 \\ -3 & 2 & 6 \end{bmatrix}$, if it exists.

(a) $\begin{bmatrix} 1 & 3 & 4 \\ 5 & 5 & 4 \\ 0 & 2 & 1 \end{bmatrix}$ (b) $\begin{bmatrix} 1 & 5 & 4 \\ 4 & 4 & 3 \\ 0 & 6 & 2 \end{bmatrix}$ (c) $\begin{bmatrix} 5 & 3 & 4 \\ 3 & 2 & 5 \\ 7 & -\frac{1}{2} & 2 \end{bmatrix}$ (d) $\begin{bmatrix} 4 & 2 & 3 \\ 3 & 3 & 2 \\ 1 & 0 & 1 \end{bmatrix}$ (e) Does not exist

Answer: medium (d)

$\left[\begin{array}{ccc|ccc} 3 & -2 & -5 & 1 & 0 & 0 \\ -1 & 1 & 1 & 0 & 1 & 0 \\ -3 & 2 & 6 & 0 & 0 & 1 \end{array}\right]$ $\begin{array}{c} 3R_2 + R_1 \to R_2 \\ R_3 + R_1 \to R_3 \end{array}$ $\left[\begin{array}{ccc|ccc} 3 & -2 & -5 & 1 & 0 & 0 \\ 0 & 1 & -2 & 1 & 3 & 0 \\ 0 & 0 & 1 & 1 & 0 & 1 \end{array}\right]$ $R_1 + 2R_2 \to R_1$ $\left[\begin{array}{ccc|ccc} 3 & 0 & -9 & 3 & 6 & 0 \\ 0 & 1 & -2 & 1 & 3 & 0 \\ 0 & 0 & 1 & 1 & 0 & 1 \end{array}\right]$

$\begin{array}{c} \frac{1}{3}(R_1 + 9R_3) \to R_1 \\ R_2 + 2R_3 \to R_2 \end{array}$ $\left[\begin{array}{ccc|ccc} 1 & 0 & 0 & 4 & 2 & 3 \\ 0 & 1 & 0 & 3 & 3 & 2 \\ 0 & 0 & 1 & 1 & 0 & 1 \end{array}\right]$

Therefore, the inverse matrix is $\begin{bmatrix} 4 & 2 & 3 \\ 3 & 3 & 2 \\ 1 & 0 & 1 \end{bmatrix}$.

6. Find the inverse of the matrix of $\begin{bmatrix} 26 & 7 & -25 \\ -3 & -1 & 3 \\ -27 & -7 & 26 \end{bmatrix}$, if it exists.

(a) $\begin{bmatrix} 5 & 7 & 4 \\ 3 & -1 & 3 \\ 6 & 7 & 5 \end{bmatrix}$ (b) $\begin{bmatrix} 2 & 4 & -3 \\ -2 & 1 & 4 \\ 5 & 4 & 7 \end{bmatrix}$ (c) $\begin{bmatrix} 3 & 5 & -2 \\ -1 & 4 & 5 \\ 2 & 3 & 1 \end{bmatrix}$ (d) $\begin{bmatrix} 7 & 8 & 12 \\ -13 & 4 & 5 \\ -6 & -7 & 9 \end{bmatrix}$ (e) Does not exist

Answer: medium (a)

$\left[\begin{array}{ccc|ccc} 26 & 7 & -25 & 1 & 0 & 0 \\ -3 & -1 & 3 & 0 & 1 & 0 \\ -27 & -7 & 26 & 0 & 0 & 1 \end{array}\right]$ $\begin{array}{c} R_3 + R_1 \to R_1 \\ 9R_2 - R_3 \to R_2 \end{array}$ $\left[\begin{array}{ccc|ccc} -1 & 0 & 1 & 1 & 0 & 1 \\ 0 & -2 & 1 & 0 & 9 & -1 \\ -27 & -7 & 26 & 0 & 0 & 1 \end{array}\right]$ $\begin{array}{c} -R_1 \to R_1 \\ R_3 - 27R_1 \to R_3 \end{array}$

$\left[\begin{array}{ccc|ccc} 1 & 0 & -1 & -1 & 0 & -1 \\ 0 & -2 & 1 & 0 & 9 & -1 \\ 0 & -7 & -1 & -27 & 0 & -26 \end{array}\right]$ $2R_3 - 7R_2 \to R_3$ $\left[\begin{array}{ccc|ccc} 1 & 0 & -1 & -1 & 0 & -1 \\ 0 & -2 & 1 & 0 & 9 & -1 \\ 0 & 0 & -9 & -54 & -63 & -45 \end{array}\right]$ $\begin{array}{c} R_1 - \frac{1}{9}R_3 \to R_1 \\ -\frac{1}{2}R_2 \to R_2 \\ -\frac{1}{9}R_3 \to R_3 \end{array}$

$\left[\begin{array}{ccc|ccc} 1 & 0 & 0 & 5 & 7 & 4 \\ 0 & 1 & -\frac{1}{2} & 0 & -\frac{9}{2} & \frac{1}{2} \\ 0 & 0 & 1 & 6 & 7 & 5 \end{array}\right]$ $R_2 + \frac{1}{2}R_3 \to R_2$ $\left[\begin{array}{ccc|ccc} 1 & 0 & 0 & 5 & 7 & 4 \\ 0 & 1 & 0 & 3 & -1 & 3 \\ 0 & 0 & 1 & 6 & 7 & 5 \end{array}\right]$

Therefore, the inverse matrix is $\begin{bmatrix} 5 & 7 & 4 \\ 3 & -1 & 3 \\ 6 & 7 & 5 \end{bmatrix}$.

7. Find the inverse of the matrix of $\begin{bmatrix} -1 & -1 & 2 \\ 3 & 2 & -4 \\ -\frac{1}{2} & 0 & \frac{1}{2} \end{bmatrix}$, if it exists.

(a) $\begin{bmatrix} 1 & 2 & -2 \\ 3 & 4 & -4 \\ 1 & 2 & 1 \end{bmatrix}$ (b) $\begin{bmatrix} 1 & 5 & 2 \\ 2 & 3 & 0 \\ 0 & 2 & 0 \end{bmatrix}$ (c) $\begin{bmatrix} 2 & 1 & 0 \\ 1 & 1 & 4 \\ 2 & 1 & 2 \end{bmatrix}$ (d) $\begin{bmatrix} -3 & -1 & 0 \\ -2 & 4 & -1 \\ 2 & -1 & 3 \end{bmatrix}$ (e) Does not exist

Answer: medium (c)

$\left[\begin{array}{ccc|ccc} -1 & -1 & 2 & 1 & 0 & 0 \\ 3 & 2 & -4 & 0 & 1 & 0 \\ -\frac{1}{2} & 0 & \frac{1}{2} & 0 & 0 & 1 \end{array}\right]$ $\begin{array}{c} -R_1 \to R_1 \\ R_2 + 3R_1 \to R_2 \\ 2R_3 - R_1 \to R_3 \end{array}$ $\left[\begin{array}{ccc|ccc} 1 & 1 & -2 & -1 & 0 & 0 \\ 0 & -1 & 2 & 3 & 1 & 0 \\ 0 & 1 & -1 & -1 & 0 & 2 \end{array}\right]$ $\begin{array}{c} R_1 + R_2 \to R_1 \\ -R_2 \to R_2 \\ R_3 + R_2 \to R_3 \end{array}$

$\left[\begin{array}{ccc|ccc} 1 & 0 & 0 & 2 & 1 & 0 \\ 0 & 1 & -2 & -3 & -1 & 0 \\ 0 & 0 & 1 & 2 & 1 & 2 \end{array}\right]$ $R_2 + 2R_3 \to R_2$ $\left[\begin{array}{ccc|ccc} 1 & 0 & 0 & 2 & 1 & 0 \\ 0 & 1 & 0 & 1 & 1 & 4 \\ 0 & 0 & 1 & 2 & 1 & 2 \end{array}\right]$

Therefore, the inverse matrix is $\begin{bmatrix} 2 & 1 & 0 \\ 1 & 1 & 4 \\ 2 & 1 & 2 \end{bmatrix}$.

8. Find the inverse of the matrix of $\begin{bmatrix} 1 & 0 & -1 \\ 1 & 0 & -\frac{3}{2} \\ -6 & 1 & \frac{13}{2} \end{bmatrix}$, if it exists.

(a) $\begin{bmatrix} 3 & -2 & 0 \\ 5 & 1 & 1 \\ 2 & -2 & 0 \end{bmatrix}$ (b) $\begin{bmatrix} 1 & -3 & 1 \\ -2 & 5 & 4 \\ 3 & -3 & 2 \end{bmatrix}$ (c) $\begin{bmatrix} 2 & -1 & 3 \\ 1 & 2 & 4 \\ 2 & 1 & -3 \end{bmatrix}$ (d) $\begin{bmatrix} -2 & 1 & 0 \\ -2 & -4 & -1 \\ 2 & -1 & 3 \end{bmatrix}$ (e) Does not exist

Answer: medium (a)

$$\left[\begin{array}{ccc|ccc} 1 & 0 & -1 & 1 & 0 & 0 \\ 1 & 0 & -\frac{3}{2} & 0 & 1 & 0 \\ -6 & 1 & \frac{13}{2} & 0 & 0 & 1 \end{array}\right] \begin{array}{c} R_2 - R_1 \to R_2 \\ R_3 + 6R_1 \to R_3 \end{array} \left[\begin{array}{ccc|ccc} 1 & 0 & -1 & 1 & 0 & 0 \\ 0 & 0 & -\frac{1}{2} & -1 & 1 & 0 \\ 0 & 1 & \frac{1}{2} & 6 & 0 & 1 \end{array}\right] \begin{array}{c} R_1 - 2R_2 \to R_1 \\ R_2 \leftrightarrow R_3 \end{array}$$

$$\left[\begin{array}{ccc|ccc} 1 & 0 & 0 & 3 & -2 & 0 \\ 0 & 1 & \frac{1}{2} & 6 & 0 & 1 \\ 0 & 0 & -\frac{1}{2} & -1 & 1 & 0 \end{array}\right] \begin{array}{c} R_2 + R_3 \to R_2 \\ -2R_3 \to R_3 \end{array} \left[\begin{array}{ccc|ccc} 1 & 0 & 0 & 3 & -2 & 0 \\ 0 & 1 & 0 & 5 & 1 & 1 \\ 0 & 0 & 1 & 2 & -2 & 0 \end{array}\right]$$

Therefore, the inverse matrix is $\begin{bmatrix} 3 & -2 & 0 \\ 5 & 1 & 1 \\ 2 & -2 & 0 \end{bmatrix}$.

9. Find the inverse of the matrix of $\begin{bmatrix} 1 & 0 & 1 & 0 \\ 0 & 1 & 0 & 1 \\ 1 & 1 & 1 & 0 \\ 1 & 1 & 1 & 1 \end{bmatrix}$, if it exists.

(a) $\begin{bmatrix} 0 & 1 & -1 & 0 \\ 0 & -1 & 1 & 1 \\ 1 & 2 & 1 & 1 \\ 1 & 0 & 0 & 1 \end{bmatrix}$ (b) $\begin{bmatrix} -1 & 1 & 2 & 1 \\ 0 & 1 & -2 & 1 \\ -3 & 1 & 0 & 1 \\ 1 & 0 & 1 & 2 \end{bmatrix}$ (c) $\begin{bmatrix} 2 & 2 & 1 & 2 \\ 0 & 2 & 3 & 1 \\ 2 & 0 & 0 & 1 \\ 2 & 0 & 1 & 2 \end{bmatrix}$ (d) $\begin{bmatrix} 0 & 1 & 0 & 1 \\ 1 & 0 & 1 & 0 \\ 0 & 0 & 0 & 1 \\ 0 & 0 & 0 & 1 \end{bmatrix}$ (e) Does not exist

Answer: medium (e)

$$\left[\begin{array}{cccc|cccc} 1 & 0 & 1 & 0 & 1 & 0 & 0 & 0 \\ 0 & 1 & 0 & 1 & 0 & 1 & 0 & 0 \\ 1 & 1 & 1 & 0 & 0 & 0 & 1 & 0 \\ 1 & 1 & 1 & 1 & 0 & 0 & 0 & 1 \end{array}\right] \begin{array}{c} R_3 - R_1 \to R_3 \\ R_4 - R_1 \to R_4 \end{array} \left[\begin{array}{cccc|cccc} 1 & 0 & 1 & 0 & 1 & 0 & 0 & 0 \\ 0 & 1 & 0 & 1 & 0 & 1 & 0 & 0 \\ 0 & 1 & 0 & 0 & -1 & 0 & 1 & 0 \\ 0 & 1 & 0 & 1 & -1 & 0 & 0 & 1 \end{array}\right] \begin{array}{c} R_3 - R_2 \to R_3 \\ R_4 - R_2 \to R_4 \end{array} \left[\begin{array}{cccc|cccc} 1 & 0 & 1 & 0 & 1 & 0 & 0 & 0 \\ 0 & 1 & 0 & 1 & 0 & 1 & 0 & 0 \\ 0 & 0 & 0 & -1 & -1 & -1 & 1 & 0 \\ 0 & 0 & 0 & 0 & -1 & -1 & 0 & 1 \end{array}\right]$$

Therefore there is no inverse matrix.

10. Solve the system of equations $\begin{cases} 3x + 5y = 10 \\ 4x + 7y = 20 \end{cases}$ by converting to a matrix equation and using the inverse of the coefficient matrix.

Answer: medium

$\begin{cases} 3x + 5y = 10 \\ 4x + 7y = 20 \end{cases}$ is equivalent to the matrix equation $\begin{bmatrix} 3 & 5 \\ 4 & 7 \end{bmatrix}\begin{bmatrix} x \\ y \end{bmatrix} = \begin{bmatrix} 10 \\ 20 \end{bmatrix} \Leftrightarrow \begin{bmatrix} x \\ y \end{bmatrix} = \begin{bmatrix} 7 & -5 \\ -4 & 3 \end{bmatrix}\begin{bmatrix} 10 \\ 20 \end{bmatrix} = \begin{bmatrix} -30 \\ 20 \end{bmatrix}$

Therefore, $x = -30$ and $y = 20$.

11. Solve the system of equations $\begin{cases} -3x + 3y = 10 \\ 7x + 3y = 100 \end{cases}$ by converting to a matrix equation and using the inverse of the coefficient matrix.

Answer: medium

$\begin{cases} -3x + 3y = 10 \\ 7x + 3y = 100 \end{cases}$ is equivalent to the matrix equation $\begin{bmatrix} -3 & 3 \\ 7 & 3 \end{bmatrix}\begin{bmatrix} x \\ y \end{bmatrix} = \begin{bmatrix} 10 \\ 100 \end{bmatrix} \Leftrightarrow$

$\begin{bmatrix} x \\ y \end{bmatrix} = \begin{bmatrix} -\frac{1}{10} & \frac{1}{10} \\ \frac{7}{30} & \frac{1}{10} \end{bmatrix}\begin{bmatrix} 10 \\ 100 \end{bmatrix} = \begin{bmatrix} 9 \\ \frac{37}{3} \end{bmatrix}$ Therefore $x = 9$ and $y = \frac{37}{3}$.

12. Solve the system of equations $\begin{cases} 2x+y=1 \\ x+y+4z=1 \\ 2x+y+2z=1 \end{cases}$ by converting to a matrix equation and using the inverse of the coefficient matrix.

Answer: medium

$\begin{cases} 2x+y=1 \\ x+y+4z=1 \\ 2x+y+2z=1 \end{cases}$ is equivalent to $\begin{bmatrix} 2 & 1 & 0 \\ 1 & 1 & 4 \\ 2 & 1 & 2 \end{bmatrix}\begin{bmatrix} x \\ y \\ z \end{bmatrix} = \begin{bmatrix} 1 \\ 1 \\ 1 \end{bmatrix} \Leftrightarrow \begin{bmatrix} x \\ y \\ z \end{bmatrix} = \begin{bmatrix} -1 & -1 & 2 \\ 3 & 2 & -4 \\ -\frac{1}{2} & 0 & \frac{1}{2} \end{bmatrix}\begin{bmatrix} 1 \\ 1 \\ 1 \end{bmatrix} = \begin{bmatrix} 0 \\ 1 \\ 0 \end{bmatrix}$ Thus, $x=0$, $y=1$, and $z=0$.

13. Solve the system of equations $\begin{cases} x+2y+3w=0 \\ y+z+w=1 \\ y+w=2 \\ x+2y+2w=3 \end{cases}$ by converting to a matrix equation and using the inverse of the coefficient matrix.

Answer: hard

$\begin{cases} x+2y+3w=0 \\ y+z+w=1 \\ y+w=2 \\ x+2y+2w=3 \end{cases}$ is equivalent to $\begin{bmatrix} 1 & 2 & 0 & 3 \\ 0 & 1 & 1 & 1 \\ 0 & 1 & 0 & 1 \\ 1 & 2 & 0 & 2 \end{bmatrix}\begin{bmatrix} x \\ y \\ z \\ w \end{bmatrix} = \begin{bmatrix} 0 \\ 1 \\ 2 \\ 3 \end{bmatrix} \Leftrightarrow \begin{bmatrix} x \\ y \\ z \\ w \end{bmatrix} = \begin{bmatrix} 0 & 0 & -2 & 1 \\ -1 & 0 & 1 & 1 \\ 0 & 1 & -1 & 0 \\ 1 & 0 & 0 & -1 \end{bmatrix}\begin{bmatrix} 0 \\ 1 \\ 2 \\ 3 \end{bmatrix} = \begin{bmatrix} -1 \\ 5 \\ -1 \\ -3 \end{bmatrix}$

Therefore, $x=-1$, $y=5$, $z=-1$, and $w=-3$.

14. Solve the matrix equation $\begin{bmatrix} 4 & 2 & 3 \\ 3 & 3 & 2 \\ 1 & 0 & 1 \end{bmatrix}\begin{bmatrix} x & u \\ y & v \\ z & w \end{bmatrix} = \begin{bmatrix} 3 & 6 \\ 6 & 12 \\ 0 & 0 \end{bmatrix}$ by multiplying both sides by the appropriate inverse matrix.

Answer: medium

Using the inverse matrix $\begin{bmatrix} 4 & 2 & 3 \\ 3 & 3 & 2 \\ 1 & 0 & 1 \end{bmatrix}^{-1} = \begin{bmatrix} 3 & -2 & -5 \\ -1 & 1 & 1 \\ -3 & 2 & 6 \end{bmatrix}$, we find that

$$\begin{bmatrix} x & u \\ y & v \\ z & w \end{bmatrix} = \begin{bmatrix} 3 & -2 & -5 \\ -1 & 1 & 1 \\ -3 & 2 & 6 \end{bmatrix}\begin{bmatrix} 3 & 6 \\ 6 & 12 \\ 0 & 0 \end{bmatrix} = \begin{bmatrix} -3 & -6 \\ 3 & 6 \\ 3 & 6 \end{bmatrix}.$$

15. Find the inverse of the matrix $\begin{bmatrix} 1/x & -1 \\ 1 & x \end{bmatrix}$, $x \neq 0$.

Answer: hard

$$\begin{bmatrix} 1/x & -1 \\ 1 & x \end{bmatrix}^{-1} = \frac{1}{1+1}\begin{bmatrix} x & 1 \\ -1 & 1/x \end{bmatrix} = \begin{bmatrix} \frac{x}{2} & \frac{1}{2} \\ -\frac{1}{2} & \frac{1}{2x} \end{bmatrix}$$

16. Find the inverse matrix of $\begin{bmatrix} 2 & e^x & 0 \\ e^x & -e^{2x} & 0 \\ 0 & 0 & 1 \end{bmatrix}$.

Answer: hard

$\begin{bmatrix} 2 & e^x & 0 & 1 & 0 & 0 \\ e^x & -e^{2x} & 0 & 0 & 1 & 0 \\ 0 & 0 & 1 & 0 & 0 & 1 \end{bmatrix}$ $2R_2 - e^xR_1 \to R_2$ $\begin{bmatrix} 2 & c^x & 0 & 1 & 0 & 0 \\ 0 & -3e^{2x} & 0 & -e^x & 2 & 0 \\ 0 & 0 & 1 & 0 & 0 & 1 \end{bmatrix}$ $R_2 + 3e^xR_1 \to R_1$, $-\frac{1}{3}e^{-2x}R_2 \to R_2$

$\begin{bmatrix} 6e^x & 0 & 0 & 2e^x & 2 & 0 \\ 0 & 1 & 0 & \frac{1}{3}e^{-x} & -\frac{2}{3}e^{-2x} & 0 \\ 0 & 0 & 1 & 0 & 0 & 1 \end{bmatrix}$ $\frac{1}{6}e^xR_1 \to R_1$ $\begin{bmatrix} 1 & 0 & 0 & \frac{1}{3} & \frac{1}{3}e^x & 0 \\ 0 & 1 & 0 & \frac{1}{3}e^{-x} & -\frac{2}{3}e^{-2x} & 0 \\ 0 & 0 & 1 & 0 & 0 & 1 \end{bmatrix}$

Therefore, the inverse matrix is $\begin{bmatrix} \frac{1}{3} & \frac{1}{3}e^x & 0 \\ \frac{1}{3}e^{-x} & -\frac{2}{3}e^{-2x} & 0 \\ 0 & 0 & 1 \end{bmatrix}$.

17. Find the inverse matrix of $\begin{bmatrix} a & 0 & 0 & 0 \\ 0 & b & 0 & 0 \\ 0 & 0 & c & 0 \\ 0 & 0 & 0 & d \end{bmatrix}$ where $a, b, c, d \neq 0$.

Answer: hard

The matrix $\begin{bmatrix} a & 0 & 0 & 0 \\ 0 & b & 0 & 0 \\ 0 & 0 & c & 0 \\ 0 & 0 & 0 & d \end{bmatrix}$ has inverse $\begin{bmatrix} 1/a & 0 & 0 & 0 \\ 0 & 1/b & 0 & 0 \\ 0 & 0 & 1/c & 0 \\ 0 & 0 & 0 & 1/d \end{bmatrix}$.

18. Find the inverse of the matrix $\begin{bmatrix} \frac{1}{2} & 4 \\ 8 & -1 \end{bmatrix}$, if it exists.

(a) $\begin{bmatrix} \frac{2}{65} & \frac{8}{65} \\ \frac{16}{65} & -\frac{1}{65} \end{bmatrix}$ (b) $\begin{bmatrix} \frac{16}{65} & \frac{8}{65} \\ -\frac{2}{65} & -\frac{1}{65} \end{bmatrix}$ (c) $\begin{bmatrix} \frac{2}{65} & \frac{4}{65} \\ -\frac{16}{65} & -\frac{5}{65} \end{bmatrix}$

(d) $\begin{bmatrix} \frac{4}{65} & \frac{8}{65} \\ -\frac{16}{65} & -\frac{5}{65} \end{bmatrix}$ (e) $\begin{bmatrix} \frac{5}{65} & \frac{8}{65} \\ -\frac{16}{65} & -\frac{3}{65} \end{bmatrix}$

Answer: easy

(a). $\begin{bmatrix} \frac{1}{2} & 4 \\ 8 & -1 \end{bmatrix}^{-1} = \dfrac{1}{\frac{1}{2}(-1) - 4(8)} \begin{bmatrix} -1 & -4 \\ -8 & \frac{1}{2} \end{bmatrix} = \begin{bmatrix} \frac{2}{65} & \frac{8}{65} \\ \frac{16}{65} & -\frac{1}{65} \end{bmatrix}$.

19. Solve the system of equations $\begin{cases} 5x + 3y = -3 \\ 4x + 7y = 5 \end{cases}$ by converting to a matrix equation and using its inverse coefficient matrix $\begin{bmatrix} \frac{7}{23} & -\frac{3}{23} \\ -\frac{4}{23} & \frac{5}{23} \end{bmatrix}$.

Answer: easy

$\begin{cases} 5x + 3y = -3 \\ 4x + 7y = 5 \end{cases} \Rightarrow \begin{bmatrix} 5 & 3 \\ 4 & 7 \end{bmatrix} \begin{bmatrix} x \\ y \end{bmatrix} = \begin{bmatrix} -3 \\ 5 \end{bmatrix} \Leftrightarrow \begin{bmatrix} x \\ y \end{bmatrix} = \begin{bmatrix} \frac{7}{23} & -\frac{3}{23} \\ -\frac{4}{23} & \frac{5}{23} \end{bmatrix} \begin{bmatrix} -3 \\ 5 \end{bmatrix} = \begin{bmatrix} -\frac{36}{23} \\ \frac{37}{23} \end{bmatrix}$. Therefore, $x = -\frac{36}{23}, y = \frac{37}{23}$.

20. Solve the system of equations $\begin{cases} x + 3y - 2z = 1 \\ 2x + 4y + z = -4 \\ 2x + 6y - z = 2 \end{cases}$ by converting to a matrix equation and using its inverse coefficient matrix $\begin{bmatrix} \frac{5}{3} & \frac{3}{2} & -\frac{11}{6} \\ -\frac{2}{3} & -\frac{1}{2} & \frac{5}{6} \\ -\frac{2}{3} & 0 & \frac{1}{3} \end{bmatrix}$.

Answer: medium

$\begin{cases} x + 3y - 2z = 1 \\ 2x + 4y + z = -4 \\ 2x + 6y - z = 2 \end{cases} \Rightarrow \begin{bmatrix} 1 & 3 & -2 \\ 2 & 4 & 1 \\ 2 & 6 & -1 \end{bmatrix} \begin{bmatrix} x \\ y \\ z \end{bmatrix} = \begin{bmatrix} 1 \\ -4 \\ 2 \end{bmatrix} \Leftrightarrow \begin{bmatrix} x \\ y \\ z \end{bmatrix} = \begin{bmatrix} \frac{5}{3} & \frac{3}{2} & -\frac{11}{6} \\ -\frac{2}{3} & -\frac{1}{2} & \frac{5}{6} \\ -\frac{2}{3} & 0 & \frac{1}{3} \end{bmatrix} \begin{bmatrix} 1 \\ -4 \\ 2 \end{bmatrix} = \begin{bmatrix} -8 \\ 3 \\ 0 \end{bmatrix}$. Therefore, $x = -8, y = 3, z = 0$.

21. Solve the system of equations $\begin{cases} 3x+3y-2z=1 \\ 3x+4y-z=-4 \\ 3x+y-5z=0 \end{cases}$ by converting to a matrix equation and using its inverse coefficient matrix $\begin{bmatrix} \frac{19}{3} & -\frac{13}{3} & -\frac{5}{3} \\ -4 & 3 & 1 \\ 3 & -2 & -1 \end{bmatrix}$.

Answer: medium

$\begin{cases} 3x+3y-2z=1 \\ 3x+4y-z=-4 \\ 3x+y-5z=0 \end{cases} \Rightarrow \begin{bmatrix} 3 & 3 & -2 \\ 3 & 4 & -1 \\ 3 & 1 & -5 \end{bmatrix} \begin{bmatrix} x \\ y \\ z \end{bmatrix} = \begin{bmatrix} 1 \\ -4 \\ 0 \end{bmatrix} \Leftrightarrow \begin{bmatrix} x \\ y \\ z \end{bmatrix} = \begin{bmatrix} \frac{19}{3} & -\frac{13}{3} & -\frac{5}{3} \\ -4 & 3 & 1 \\ 3 & -2 & -1 \end{bmatrix} \begin{bmatrix} 1 \\ -4 \\ 0 \end{bmatrix} = \begin{bmatrix} \frac{71}{3} \\ -16 \\ 11 \end{bmatrix}$. Therefore, $x = \frac{71}{3}$, $y = -16$, $z = 11$.

SECTION 10.6

1. Find the determinant of the matrix $\begin{bmatrix} -4 & 6 \\ 2 & -5 \end{bmatrix}$, if it exists.

(a) -3 (b) 8 (c) 4 (d) -2 (e) Does not exist

Answer: easy (b)

$\begin{bmatrix} -4 & 6 \\ 2 & -5 \end{bmatrix}$ has determinant $|D| = (-4)(-5) - (2)(6) = 8$.

2. Find the determinant of the matrix $\begin{bmatrix} 4 \\ 2 \end{bmatrix}$, if it exists.

(a) 0 (b) 2 (c) 4 (d) -8 (e) Does not exist

Answer: easy (e)

$\begin{bmatrix} 4 \\ 2 \end{bmatrix}$ does not have a determinant because the matrix is not square.

3. Find the determinant of the matrix $[0]$, if it exists.

(a) 0 (b) 1 (c) -1 (d) 2 (e) Does not exist

Answer: easy (a)

$[0]$ has determinant $|D| = 0$.

4. Find the determinant of the matrix $\begin{bmatrix} \frac{3}{5} & -0.6 \\ \frac{1}{3} & 4 \end{bmatrix}$, if it exists.

(a) $\frac{3}{7}$ (b) $\frac{13}{5}$ (c) $\frac{12}{7}$ (d) $\frac{14}{3}$ (e) Does not exist

Answer: easy (b)

$\begin{bmatrix} \frac{3}{5} & -0.6 \\ \frac{1}{3} & 4 \end{bmatrix}$ has determinant $|D| = \frac{3}{5} \cdot 4 + \frac{1}{3} \cdot 0.6 = \frac{3}{5}\left(4 + \frac{1}{3}\right) = \frac{3}{5} \cdot \frac{13}{3} = \frac{13}{5}$.

5. Evaluate the minor M_{33} and cofactor A_{33} using the matrix $A = \begin{bmatrix} 2 & 1 & \frac{1}{3} \\ -1 & 4 & 5 \\ 2 & 0 & 0 \end{bmatrix}$.

Answer: easy

$M_{33} = 2 \cdot 4 + 1 \cdot 1 = 9$, $A_{33} = (-1)^6 M_{33} = 9$

6. Evaluate the minor M_{13} and cofactor A_{13} using the matrix $A = \begin{bmatrix} 2 & 1 & \frac{1}{3} \\ -1 & 4 & 5 \\ 2 & 0 & 0 \end{bmatrix}$.

Answer: easy
$M_{13} = -1 \cdot 0 - 4 \cdot 2 = -8,\ A_{13} = (-1)^4 M_{13} = -8$

7. Evaluate the minor M_{23} and cofactor A_{23} using the matrix $A = \begin{bmatrix} 2 & 1 & \frac{1}{3} \\ -1 & 4 & 5 \\ 2 & 0 & 0 \end{bmatrix}$.

Answer: easy
$M_{23} = 2 \cdot 0 - 1 \cdot 2 = -2,\ A_{23} = (-1)^5 M_{23} = 2$

8. Find the determinant of the matrix $\begin{bmatrix} -1 & -2 & 3 \\ 1 & 3 & 0 \\ 2 & \frac{1}{2} & 1 \end{bmatrix}$. Determine whether the matrix has an inverse or not, but do not calculate the inverse.

Answer: easy
$M = \begin{bmatrix} -1 & -2 & 3 \\ 1 & 3 & 0 \\ 2 & \frac{1}{2} & 1 \end{bmatrix}$. Therefore, $|M| = -\begin{vmatrix} -2 & 3 \\ \frac{1}{2} & 1 \end{vmatrix} + 3\begin{vmatrix} -1 & 3 \\ 2 & 1 \end{vmatrix} = -\left(-\frac{7}{2}\right) + 3(-7) = -\frac{35}{2}$, and the matrix has an inverse.

9. Find the determinant of the matrix $\begin{bmatrix} 2 & 1 & 7 \\ -1 & -2 & 3 \\ 4 & 2 & 1 \end{bmatrix}$. Determine whether the matrix has an inverse or not, but do not calculate the inverse.

Answer: easy
$M = \begin{bmatrix} 2 & 1 & 7 \\ -1 & -2 & 3 \\ 4 & 2 & 1 \end{bmatrix}$. Therefore, $|M| = 2\begin{vmatrix} -2 & 3 \\ 2 & 1 \end{vmatrix} - \begin{vmatrix} -1 & 3 \\ 4 & 1 \end{vmatrix} + 7\begin{vmatrix} -1 & -2 \\ 4 & 2 \end{vmatrix} = 2(-8) - (-13) + 7(6) = 39$, and so the matrix has an inverse.

10. Find the determinant of the matrix $\begin{bmatrix} 1 & 3 & 0 & 1 \\ 2 & -3 & 1 & 2 \\ 0 & 4 & 2 & 0 \\ 1 & 0 & -1 & 0 \end{bmatrix}$. Determine whether the matrix has an inverse or not, but do not calculate the inverse.

Answer: medium
$M = \begin{bmatrix} 1 & 3 & 0 & 1 \\ 2 & -3 & 1 & 2 \\ 0 & 4 & 2 & 0 \\ 1 & 0 & -1 & 0 \end{bmatrix}$. So $|M| = -\begin{vmatrix} 3 & 0 & 1 \\ -3 & 1 & 2 \\ 4 & 2 & 0 \end{vmatrix} + \begin{vmatrix} 1 & 3 & 1 \\ 2 & -3 & 2 \\ 0 & 4 & 0 \end{vmatrix} = -\left(3\begin{vmatrix} 1 & 2 \\ 2 & 0 \end{vmatrix} + \begin{vmatrix} -3 & 1 \\ 4 & 2 \end{vmatrix}\right) - 4\begin{vmatrix} 1 & 1 \\ 2 & 2 \end{vmatrix} = 22$, so M^{-1} exists.

11. Evaluate the determinant $\begin{vmatrix} -1 & 4 & 2 & 3 \\ 1 & 6 & 3 & 4 \\ -7 & -3 & 0 & 2 \\ 2 & -4 & -2 & 0 \end{vmatrix}$. Use row or column operations whenever possible to simplify your work.

(a) 12 (b) -32 (c) 36 (d) -48 (e) 64

Answer: medium (d)
$M = \begin{bmatrix} -1 & 4 & 2 & 3 \\ 1 & 6 & 3 & 4 \\ -7 & -3 & 0 & 2 \\ 2 & -4 & -2 & 0 \end{bmatrix}$ Then

$$|M| = \begin{vmatrix} -1 & 4 & 2 & 3 \\ 1 & 6 & 3 & 4 \\ -7 & -3 & 0 & 2 \\ 2 & -4 & -2 & 0 \end{vmatrix} \underset{=}{\text{C}_2 - 2\text{C}_3 \to \text{C}_2} \begin{vmatrix} -1 & 0 & 2 & 3 \\ 1 & 0 & 3 & 4 \\ -7 & -3 & 0 & 2 \\ 2 & 0 & -2 & 0 \end{vmatrix} = 3\begin{vmatrix} -1 & 2 & 3 \\ 1 & 3 & 4 \\ 2 & -2 & 0 \end{vmatrix} = 3\left(2\begin{vmatrix} 2 & 3 \\ 3 & 4 \end{vmatrix} + 2\begin{vmatrix} -1 & 3 \\ 1 & 4 \end{vmatrix}\right) = 3(-2 - 14)$$
$= -48$

12. Evaluate the determinant $\begin{bmatrix} 3 & -1 & 4 & 2 \\ 5 & 3 & -1 & 4 \\ 6 & -2 & 4 & 3 \\ 2 & 2 & 6 & 2 \end{bmatrix}$. Use row or column operations whenever possible to simplify your work.

(a) 92 (b) 102 (c) 104 (d) 112 (e) 116

Answer: medium (a)

$M = \begin{bmatrix} 3 & -1 & 4 & 2 \\ 5 & 3 & -1 & 4 \\ 6 & -2 & 4 & 3 \\ 2 & 2 & 6 & 2 \end{bmatrix}$. Then

$$|M| = \begin{vmatrix} 3 & -1 & 4 & 2 \\ 5 & 3 & -1 & 4 \\ 6 & -2 & 4 & 3 \\ 2 & 2 & 6 & 2 \end{vmatrix} \overset{C_1+3C_2 \to C_1}{=} \begin{vmatrix} 0 & -1 & 4 & 2 \\ 14 & 3 & -1 & 4 \\ 0 & -2 & 4 & 3 \\ 8 & 2 & 6 & 2 \end{vmatrix} = -14 \begin{vmatrix} -1 & 4 & 2 \\ -2 & 4 & 3 \\ 2 & 6 & 2 \end{vmatrix} - 8 \begin{vmatrix} -1 & 4 & 2 \\ 3 & -1 & 4 \\ -2 & 4 & 3 \end{vmatrix}$$

$$= -14 \begin{vmatrix} 1 & 0 & -1 \\ -2 & 4 & 3 \\ 2 & 6 & 2 \end{vmatrix} - 8 \begin{vmatrix} 1 & 0 & -1 \\ 3 & -1 & 4 \\ -2 & 4 & 3 \end{vmatrix} = -14\left(\begin{vmatrix} 4 & 3 \\ 6 & 2 \end{vmatrix} - \begin{vmatrix} -2 & 4 \\ 2 & 6 \end{vmatrix}\right) - 8\left(\begin{vmatrix} -1 & 4 \\ 4 & 3 \end{vmatrix} - \begin{vmatrix} 3 & -1 \\ -2 & 4 \end{vmatrix}\right)$$

$$= -14(-10+20) - 8(-19-10) = 92$$

13. Consider the system $\begin{cases} 2x + y + 3z = 4 \\ -2x - 3y + 4z = 7 \\ 3x + y + 6z = 6 \end{cases}$

(a) Verify that $x = -31$, $y = 33$, $z = 11$ is a solution of the system.

(b) Find the determinant of the coefficient matrix.

(c) Without solving the system, determine whether or not there are any other solutions.

(d) Can Cramer's Rule be used to solve this system? Why or why not?

Answer: hard

$$\begin{cases} 2x + y + 3z = 4 \\ -2x - 3y + 4z = 7 \\ 3x + y + 6z = 6 \end{cases}$$

(a) If $x = -31$, $y = 33$, and $z = 11$, then $2x + y + 3z = -62 + 33 + 3 \cdot 11 = 4$, $-2x - 3y + 4z = 62 - 99 + 44 = 7$, and $3x + y + 6z = -93 + 33 + 6 \cdot 11 = 6$. Therefore, $x = -31$, $y = 33$, $z = 11$ is a solution of the system.

(b) $M = \begin{bmatrix} 2 & 1 & 3 \\ -2 & -3 & 4 \\ 3 & 1 & 6 \end{bmatrix}$. Then $|M| = \begin{vmatrix} 2 & 1 & 3 \\ -2 & -3 & 4 \\ 3 & 1 & 6 \end{vmatrix} \overset{\substack{R_1+R_2 \to R_1 \\ R_3-R_1 \to R_3}}{=} \begin{vmatrix} 0 & -2 & 7 \\ -2 & -3 & 4 \\ 1 & 0 & 3 \end{vmatrix} = 2\begin{vmatrix} -2 & 7 \\ 0 & 3 \end{vmatrix} + \begin{vmatrix} -2 & 7 \\ -3 & 4 \end{vmatrix} = 1.$

(c) We can write the system as a matrix equation:
$\begin{bmatrix} 2 & 1 & 3 \\ -2 & -3 & 4 \\ 3 & 1 & 6 \end{bmatrix} \begin{bmatrix} x \\ y \\ z \end{bmatrix} = \begin{bmatrix} 4 \\ 7 \\ 6 \end{bmatrix}$, or $MX = B$. Since $|M| \neq 0$, M has an inverse. If we multiply both sides of the matrix equation by M^{-1}, then we get a unique solution for X, namely $X = M^{-1}B$. Thus the equation has no other solutions.

(d) Yes, since $|M| \neq 0$.

14. Use Cramer's Rule to solve the system $\begin{cases} 3x + 6y = 14 \\ 5x + 2y = 8 \end{cases}$.

(a) $\left(\frac{2}{3}, \frac{4}{5}\right)$ (b) $\left(\frac{2}{5}, \frac{6}{5}\right)$ (c) $\left(\frac{5}{6}, \frac{23}{12}\right)$ (d) $\left(\frac{1}{3}, \frac{2}{5}\right)$ (e) $\left(0, -\frac{11}{64}\right)$

Answer: medium (c)

$\begin{cases} 3x + 6y = 14 \\ 5x + 2y = 8 \end{cases}$ Then $|D| = \begin{vmatrix} 3 & 6 \\ 5 & 2 \end{vmatrix} = -24$, $|D_x| = \begin{vmatrix} 14 & 6 \\ 8 & 2 \end{vmatrix} = -20$, and $|D_y| = \begin{vmatrix} 3 & 14 \\ 5 & 8 \end{vmatrix} = -46$. Hence

$x = \dfrac{|D_x|}{|D|} = \dfrac{-20}{-24} = \dfrac{5}{6}$, $y = \dfrac{|D_y|}{|D|} = \dfrac{-46}{-24} = \dfrac{23}{12}$, and so the solution is $\left(\dfrac{5}{6}, \dfrac{23}{12}\right)$.

15. Use Cramer's Rule to solve the system $\begin{cases} x+3y=4 \\ \frac{1}{3}x-2y=-3 \end{cases}$.

(a) $\left(\frac{1}{4},\frac{5}{9}\right)$ (b) $\left(-\frac{1}{12},\frac{5}{12}\right)$ (c) $\left(\frac{2}{3},-\frac{4}{3}\right)$ (d) $\left(\frac{1}{9},\frac{2}{9}\right)$ (e) $\left(-\frac{1}{3},\frac{13}{9}\right)$

Answer: medium (e)

$\begin{cases} x+3y=4 \\ \frac{1}{3}x-2y=-3 \end{cases}$. Then, $|D|=\begin{vmatrix} 1 & 3 \\ \frac{1}{3} & -2 \end{vmatrix}=-3$, $|D_x|=\begin{vmatrix} 4 & 3 \\ -3 & -2 \end{vmatrix}=1$, and $|D_y|=\begin{vmatrix} 1 & 4 \\ \frac{1}{3} & -3 \end{vmatrix}=-\frac{13}{3}$. Hence $x=\dfrac{|D_x|}{|D|}=\dfrac{1}{-3}=-\dfrac{1}{3}$, $y=\dfrac{|D_y|}{|D|}=\dfrac{-13/3}{-3}=\dfrac{13}{9}$, and so the solution is $\left(-\dfrac{1}{3},\dfrac{13}{9}\right)$.

16. Use Cramer's Rule to solve the system $\begin{cases} 8x-4y=3 \\ 2x-3y=-2 \end{cases}$.

(a) $\left(\frac{2}{3},3\right)$ (b) $\left(\frac{17}{16},\frac{11}{8}\right)$ (c) $\left(\frac{11}{4},\frac{3}{16}\right)$ (d) $\left(\frac{5}{4},\frac{3}{4}\right)$ (e) $\left(0,\frac{2}{3}\right)$

Answer: medium (b)

$\begin{cases} 8x-4y=3 \\ 2x-3y=-2 \end{cases}$. Then $|D|=\begin{vmatrix} 8 & -4 \\ 2 & -3 \end{vmatrix}=-16$, $|D_x|=\begin{vmatrix} 3 & -4 \\ -2 & -3 \end{vmatrix}=-17$, and $|D_y|=\begin{vmatrix} 8 & 3 \\ 2 & -2 \end{vmatrix}=-22$. Hence $x=\dfrac{|D_x|}{|D|}=\dfrac{-17}{-16}=\dfrac{17}{16}$, $y=\dfrac{|D_y|}{|D|}=\dfrac{-22}{-16}=\dfrac{11}{8}$, and so the solution is $\left(\dfrac{17}{16},\dfrac{11}{8}\right)$.

17. Use Cramer's Rule to solve the system $\begin{cases} 2x-y+2z=5 \\ 3y-4z=10 \\ 3x+5y=-6 \end{cases}$.

(a) $\left(\frac{106}{17},-\frac{84}{17},-\frac{211}{34}\right)$ (b) $\left(\frac{106}{13},-\frac{84}{13},-\frac{13}{26}\right)$ (c) $\left(-\frac{13}{17},-\frac{12}{17},\frac{3}{34}\right)$ (d) $\left(\frac{12}{13},\frac{5}{13},1\right)$ (e) $\left(\frac{3}{16},\frac{5}{16},-\frac{1}{2}\right)$

Answer: medium (a)

$\begin{cases} 2x-y+2z=5 \\ 3y-4z=10 \\ 3x+5y=-6 \end{cases}$ Then $|D|=\begin{vmatrix} 2 & -1 & 2 \\ 0 & 3 & -4 \\ 3 & 5 & 0 \end{vmatrix}=2\begin{vmatrix} 3 & -4 \\ 5 & 0 \end{vmatrix}+3\begin{vmatrix} -1 & 2 \\ 3 & -4 \end{vmatrix}=34$,

$$|D_x|=\begin{vmatrix} 5 & -1 & 2 \\ 10 & 3 & -4 \\ -6 & 5 & 0 \end{vmatrix}=2\begin{vmatrix} 10 & 3 \\ -6 & 5 \end{vmatrix}+4\begin{vmatrix} 5 & -1 \\ -6 & 5 \end{vmatrix}=136+76=212,$$

$$|D_y|=\begin{vmatrix} 2 & 5 & 2 \\ 0 & 10 & -4 \\ 3 & -6 & 0 \end{vmatrix}=2\begin{vmatrix} 10 & -4 \\ -6 & 0 \end{vmatrix}+3\begin{vmatrix} 5 & 2 \\ 10 & -4 \end{vmatrix}=-48-120=-168, \text{ and}$$

$$|D_z|=\begin{vmatrix} 2 & -1 & 5 \\ 0 & 3 & 10 \\ 3 & 5 & -6 \end{vmatrix}=2\begin{vmatrix} 3 & 10 \\ 5 & -6 \end{vmatrix}+3\begin{vmatrix} -1 & 5 \\ 3 & 10 \end{vmatrix}=-136-75=-211.$$

Therefore, the solution is $\left(\frac{106}{17},-\frac{84}{17},-\frac{211}{34}\right)$.

18. Use Cramer's Rule to solve the system $\begin{cases} -a+2c=3 \\ 2a+3b-c=4 \\ 3a+4b+c=11 \end{cases}$.

(a) $\left(\frac{1}{3},\frac{2}{3},\frac{4}{9}\right)$ (b) $\left(\frac{5}{9},\frac{2}{3},-\frac{2}{9}\right)$ (c) $\left(\frac{13}{9},\frac{10}{9},\frac{20}{9}\right)$ (d) $\left(\frac{17}{16},\frac{3}{16},\frac{3}{8}\right)$ (e) $\left(\frac{1}{14},\frac{3}{7},\frac{6}{7}\right)$

Answer: medium (c)

$\begin{cases} -a+2c=3 \\ 2a+3b-c=4 \\ 3a+4b+c=11 \end{cases} \Rightarrow |D|=\begin{vmatrix} -1 & 0 & 2 \\ 2 & 3 & -1 \\ 3 & 4 & 1 \end{vmatrix}=-1\begin{vmatrix} 3 & -1 \\ 4 & 1 \end{vmatrix}+2\begin{vmatrix} 2 & 3 \\ 3 & 4 \end{vmatrix}=-7-2=-9,$

$$|D_a|=\begin{vmatrix} 3 & 0 & 2 \\ 4 & 3 & -1 \\ 11 & 4 & 1 \end{vmatrix}=3\begin{vmatrix} 3 & -1 \\ 4 & 1 \end{vmatrix}+2\begin{vmatrix} 4 & 3 \\ 11 & 4 \end{vmatrix}=21-34=-13,$$

$$|D_b|=\begin{vmatrix} -1 & 3 & 2 \\ 2 & 4 & -1 \\ 3 & 11 & 1 \end{vmatrix}=-\begin{vmatrix} 4 & -1 \\ 11 & 1 \end{vmatrix}-3\begin{vmatrix} 2 & -1 \\ 3 & 1 \end{vmatrix}+2\begin{vmatrix} 2 & 4 \\ 3 & 11 \end{vmatrix}=-15-15+20=-10, \text{ and}$$

$$|D_c|=\begin{vmatrix} -1 & 0 & 3 \\ 2 & 3 & 4 \\ 3 & 4 & 11 \end{vmatrix}=-\begin{vmatrix} 3 & 4 \\ 4 & 11 \end{vmatrix}+3\begin{vmatrix} 2 & 3 \\ 3 & 4 \end{vmatrix}=-17-3=-20.$$ Hence, the solution is $\left(\frac{13}{9},\frac{10}{9},\frac{20}{9}\right)$.

19. Use Cramer's Rule to solve the system $\begin{cases} 3x - y = 2 \\ 4x + 2z = 12 \\ 4y + 5z = 15 \end{cases}$.

(a) $\left(\frac{2}{3}, -\frac{5}{3}, 6\right)$ (b) $\left(-\frac{1}{2}, -\frac{3}{2}, \frac{13}{2}\right)$ (c) $(-2, -3, 5)$ (d) $\left(-\frac{1}{3}, -\frac{7}{3}, 9\right)$ (e) $\left(-\frac{7}{2}, -\frac{25}{2}, 13\right)$

Answer: medium (e)

$\begin{cases} 3x - y = 2 \\ 4x + 2z = 12 \\ 4y + 5z = 15 \end{cases} \Rightarrow |D| = \begin{vmatrix} 3 & -1 & 0 \\ 4 & 0 & 2 \\ 0 & 4 & 5 \end{vmatrix} = 3\begin{vmatrix} 0 & 2 \\ 4 & 5 \end{vmatrix} + \begin{vmatrix} 4 & 2 \\ 0 & 5 \end{vmatrix} = -24 + 20 = -4,$

$|D_x| = \begin{vmatrix} 2 & -1 & 0 \\ 12 & 0 & 2 \\ 15 & 4 & 5 \end{vmatrix} = 2\begin{vmatrix} 0 & 2 \\ 4 & 5 \end{vmatrix} + 1\begin{vmatrix} 12 & 2 \\ 15 & 5 \end{vmatrix} = -16 + 30 = 14,$

$|D_y| = \begin{vmatrix} 3 & 2 & 0 \\ 4 & 12 & 2 \\ 0 & 15 & 5 \end{vmatrix} = 3\begin{vmatrix} 12 & 2 \\ 15 & 5 \end{vmatrix} - 2\begin{vmatrix} 4 & 2 \\ 0 & 5 \end{vmatrix} = 90 - 40 = 50$, and

$|D_z| = \begin{vmatrix} 3 & -1 & 2 \\ 4 & 0 & 12 \\ 0 & 4 & 15 \end{vmatrix} = 3\begin{vmatrix} 0 & 12 \\ 4 & 15 \end{vmatrix} - 4\begin{vmatrix} -1 & 2 \\ 4 & 15 \end{vmatrix} = -144 + 92 = -52$. Thus the solution is $\left(-\frac{7}{2}, -\frac{25}{2}, 13\right)$.

20. Use Cramer's Rule to solve the system $\begin{cases} 3x - 4y = 2 \\ x + 2y - z = 5 \\ 2x + 3z = 0 \end{cases}$.

Answer: medium

$\begin{cases} 3x - 4y = 2 \\ x + 2y - z = 5 \\ 2x + 3z = 0 \end{cases} \Rightarrow |D| = \begin{vmatrix} 3 & -4 & 0 \\ 1 & 2 & -1 \\ 2 & 0 & 3 \end{vmatrix} = 3\begin{vmatrix} 2 & -1 \\ 0 & 3 \end{vmatrix} + 4\begin{vmatrix} 1 & -1 \\ 2 & 3 \end{vmatrix} = 18 + 20 = 38,$

$|D_x| = \begin{vmatrix} 2 & -4 & 0 \\ 5 & 2 & -1 \\ 0 & 0 & 3 \end{vmatrix} = 3\begin{vmatrix} 2 & -4 \\ 5 & 2 \end{vmatrix} = 72,$

$|D_y| = \begin{vmatrix} 3 & 2 & 0 \\ 1 & 5 & -1 \\ 2 & 0 & 3 \end{vmatrix} = 3\begin{vmatrix} 5 & -1 \\ 0 & 3 \end{vmatrix} - 2\begin{vmatrix} 1 & -1 \\ 2 & 3 \end{vmatrix} = 45 - 10 = 35$, and $|D_z| = \begin{vmatrix} 3 & -4 & 2 \\ 1 & 2 & 5 \\ 2 & 0 & 0 \end{vmatrix} = 2\begin{vmatrix} -4 & 2 \\ 2 & 5 \end{vmatrix} = -48$. Thus the solution is $x = \frac{36}{19}$, $y = \frac{35}{38}$, and $z = -\frac{24}{19}$.

21. Use Cramer's Rule to solve the system $\begin{cases} \theta + 2\phi + 3\psi = 6 \\ 2\theta - \phi + 5\psi = 4 \\ \theta - 3\phi + \psi = 0 \end{cases}$.

Answer: medium

$\begin{cases} \theta + 2\phi + 3\psi = 6 \\ 2\theta - \phi + 5\psi = 4 \\ \theta - 3\phi + \psi = 0 \end{cases} \Rightarrow |D| = \begin{vmatrix} 1 & 2 & 3 \\ 2 & -1 & 5 \\ 1 & -3 & 1 \end{vmatrix} = \begin{vmatrix} -1 & 5 \\ -3 & 1 \end{vmatrix} - 2\begin{vmatrix} 2 & 5 \\ 1 & 1 \end{vmatrix} + 3\begin{vmatrix} 2 & -1 \\ 1 & -3 \end{vmatrix} = 14 + 6 - 15 = 5,$

$|D_\theta| = \begin{vmatrix} 6 & 2 & 3 \\ 4 & -1 & 5 \\ 0 & -3 & 1 \end{vmatrix} = 6\begin{vmatrix} -1 & 5 \\ -3 & 1 \end{vmatrix} - 4\begin{vmatrix} 2 & 3 \\ -3 & 1 \end{vmatrix} = 84 - 44 = 40$, $|D_\phi| = \begin{vmatrix} 1 & 6 & 3 \\ 2 & 4 & 5 \\ 1 & 0 & 1 \end{vmatrix} = \begin{vmatrix} 6 & 3 \\ 4 & 5 \end{vmatrix} + \begin{vmatrix} 1 & 6 \\ 2 & 4 \end{vmatrix} = 18 - 8 = 10,$

and $|D_\psi| = \begin{vmatrix} 1 & 2 & 6 \\ 2 & -1 & 4 \\ 1 & -3 & 0 \end{vmatrix} = \begin{vmatrix} 2 & 6 \\ -1 & 4 \end{vmatrix} + 3\begin{vmatrix} 1 & 6 \\ 2 & 4 \end{vmatrix} = 14 - 24 = -10$. Hence, the solution is $\theta = 8$, $\phi = 2$, and $\psi = -2$.

22. Evaluate the determinant $\begin{vmatrix} a & a & a & a \\ 0 & a & a & a \\ 0 & 0 & a & a \\ 0 & 0 & 0 & a \end{vmatrix}$.

Answer: medium

$$\begin{vmatrix} a & a & a & a \\ 0 & a & a & a \\ 0 & 0 & a & a \\ 0 & 0 & 0 & a \end{vmatrix} = a\begin{vmatrix} a & a & a \\ 0 & a & a \\ 0 & 0 & a \end{vmatrix} = a^2\begin{vmatrix} a & a \\ 0 & a \end{vmatrix} = a^4.$$

23. Solve $\begin{vmatrix} 1 & 1 & x \\ x & 1 & 1 \\ x & 1 & x \end{vmatrix} = 0$ for x.

Answer: medium

$$\begin{vmatrix} 1 & 1 & x \\ x & 1 & 1 \\ x & 1 & x \end{vmatrix} = \begin{vmatrix} 1 & 1 \\ 1 & x \end{vmatrix} - \begin{vmatrix} x & 1 \\ x & x \end{vmatrix} + x\begin{vmatrix} x & 1 \\ x & 1 \end{vmatrix} = x - 1 - (x^2 - x) = -(x^2 - 2x + 1) = 0 \quad \Leftrightarrow \quad -(x-1)^2 = 0 \quad \Leftrightarrow$$

$x = 1$.

24. Solve $\begin{vmatrix} a & b & x-a \\ x & x+b & x \\ 0 & 1 & 1 \end{vmatrix} = 0$ for x.

Answer: hard

$$\begin{vmatrix} a & b & x-a \\ x & x+b & x \\ 0 & 1 & 1 \end{vmatrix} = -1\begin{vmatrix} a & x-a \\ x & x \end{vmatrix} + \begin{vmatrix} a & b \\ x & x+b \end{vmatrix} = -1[ax - x(x-a)] + a(x+b) - bx$$

$$= -ax + x^2 - ax + ax + ab - bx = x^2 - ax - bx + ab = (x-a)(x-b) = 0 \quad \Leftrightarrow \quad x = a \text{ or } x = b.$$

25. Find the determinant of the matrix $\begin{bmatrix} \frac{1}{8} & \frac{1}{4} \\ \frac{3}{2} & 1 \end{bmatrix}$, if it exists.

(a) $\frac{1}{4}$ (b) $-\frac{1}{4}$ (c) $\frac{1}{32}$ (d) $-\frac{1}{8}$ (e) $\frac{3}{8}$

Answer: easy (b)

$$\begin{bmatrix} \frac{1}{8} & \frac{1}{4} \\ \frac{3}{2} & 1 \end{bmatrix} = \left(\tfrac{1}{8}\right)(1) - \left(\tfrac{3}{2}\right)\left(\tfrac{1}{4}\right) = -\tfrac{1}{4}$$

26. Evaluate the minor M_{32} and the cofactor A_{32} for the matrix $A = \begin{bmatrix} 7 & 0 & 9 \\ 1 & -8 & -3 \\ 5 & 4 & 6 \end{bmatrix}$.

(a) $M_{32} = -12$, $A_{32} = 12$

(b) $M_{32} = -30$, $A_{32} = 30$

(c) $M_{32} = 30$, $A_{32} = 30$

(d) $M_{32} = 12$, $A_{32} = 30$

(e) $M_{32} = 30$, $A_{32} = 12$

Answer: easy (b)

$$A = \begin{bmatrix} 7 & 0 & 9 \\ 1 & -8 & -3 \\ 5 & 4 & 6 \end{bmatrix} \quad \Rightarrow \quad M_{32} = \begin{vmatrix} 7 & 9 \\ 1 & -3 \end{vmatrix} = -30,\ A_{32} = (-1)^{3+2} M_{32} = -1(-30) = 30$$

27. Evaluate the minor M_{33} and the cofactor A_{33} for the matrix $A = \begin{bmatrix} 0 & \frac{1}{4} & 8 \\ \frac{-1}{2} & -7 & -3 \\ 5 & 4 & 22 \end{bmatrix}$.

(a) $M_{33} = \frac{1}{8}$, $A_{33} = \frac{1}{8}$

(b) $M_{33} = 0$, $A_{33} = \frac{1}{8}$

(c) $M_{33} = \frac{1}{8}$, $A_{33} = -\frac{1}{8}$

(d) $M_{33} = 0$, $A_{33} = \frac{1}{6}$

(e) $M_{33} = \frac{1}{6}$, $A_{33} = \frac{1}{6}$

Answer: medium (a)

$$A = \begin{bmatrix} 0 & \frac{1}{4} & 8 \\ \frac{-1}{2} & -7 & -3 \\ 5 & 4 & 22 \end{bmatrix} \quad \Rightarrow \quad M_{33} = \begin{vmatrix} 0 & \frac{1}{4} \\ -\frac{1}{2} & -7 \end{vmatrix} = \tfrac{1}{8},\ A_{33} = (-1)^{3+3} M_{33} = 1\left(\tfrac{1}{8}\right) = \tfrac{1}{8}$$

28. Determine whether the matrix $A = \begin{bmatrix} 2 & 1 & 6 \\ -2 & 5 & -3 \\ 8 & 1 & 5 \end{bmatrix}$ has an inverse or not, but do not calculate the inverse.

Answer: easy

$$|A| = \begin{vmatrix} 2 & 1 & 6 \\ -2 & 5 & -3 \\ 8 & 1 & 5 \end{vmatrix} = 2 \begin{vmatrix} 5 & -3 \\ 1 & 5 \end{vmatrix} - 1 \begin{vmatrix} -2 & -3 \\ 8 & 5 \end{vmatrix} + 6 \begin{vmatrix} -2 & 5 \\ 8 & 1 \end{vmatrix}$$

$$= 2(25+3) - 1(-10+24) + 6(-2-40) = -210$$

and so the matrix has an inverse.

29. Solve $\begin{vmatrix} x & -4 & 10 \\ 0 & x-1 & 5 \\ 0 & 0 & x+2 \end{vmatrix} = 0$ for x.

Answer: medium

$$\begin{vmatrix} x & -4 & 10 \\ 0 & x-1 & 5 \\ 0 & 0 & x+2 \end{vmatrix} = 0 \quad \Leftrightarrow \quad x \begin{vmatrix} x-1 & 5 \\ 0 & x+2 \end{vmatrix} = 0 \quad \Leftrightarrow \quad x(x-1)(x+2) = 0 \quad \Leftrightarrow \quad x = 0, x = 1, x = -2$$

30. Use Cramer's Rule to solve the system $\begin{cases} x + 2y - 4z = 0 \\ 2x + 5y = 4 \\ y - z = 3 \end{cases}$.

(a) $\left(\frac{-52}{9}, \frac{-28}{9}, \frac{-1}{9}\right)$ (b) $\left(\frac{52}{9}, \frac{28}{9}, 1\right)$ (c) $(52, 28, 9)$ (d) $\left(\frac{52}{9}, \frac{28}{9}, 9\right)$ (e) $\left(-\frac{52}{9}, \frac{28}{9}, \frac{1}{9}\right)$

Answer: medium (e)

$$\begin{cases} x + 2y - 4z = 0 \\ 2x + 5y = 4 \\ y - z = 3 \end{cases} \Leftrightarrow \quad |D| = \begin{vmatrix} 1 & 2 & -4 \\ 2 & 5 & 0 \\ 0 & 1 & -1 \end{vmatrix} = -9. \ |D_x| = \begin{vmatrix} 0 & 2 & -4 \\ 4 & 5 & 0 \\ 3 & 1 & -1 \end{vmatrix} = 52, \ |D_y| = \begin{vmatrix} 1 & 0 & -4 \\ 2 & 4 & 0 \\ 0 & 3 & -1 \end{vmatrix} = -28, \ |D_z| = \begin{vmatrix} 1 & 2 & 0 \\ 2 & 5 & 4 \\ 0 & 1 & 3 \end{vmatrix} = -1.$$

Thus, the solution is $x = -\frac{52}{9}$, $y = \frac{28}{9}$, $z = \frac{1}{9}$.

31. Use Cramer's Rule to solve the system $\begin{cases} 5x + 2y - 3z = 1 \\ 2x + 5y + z = -4 \\ -3x + 9y - z = 5 \end{cases}$.

(a) $(1, 0, 2)$ (b) $(-171, 0, -2)$ (c) $(-1, 0, -2)$ (d) $(1, 342, -171)$ (e) $\left(-\frac{1}{2}, 0, -1\right)$

Answer: medium (c)

$$\begin{cases} 5x + 2y - 3z = 1 \\ 2x + 5y + z = -4 \\ -3x + 9y - z = 5 \end{cases} \Leftrightarrow \quad |D| = \begin{vmatrix} 5 & 2 & -3 \\ 2 & 5 & 1 \\ -3 & 9 & -1 \end{vmatrix} = -171. \ |D_x| = \begin{vmatrix} 1 & 2 & -3 \\ -4 & 5 & 1 \\ 5 & 9 & -1 \end{vmatrix} = 171, \ |D_y| = \begin{vmatrix} 5 & 1 & -3 \\ 2 & -4 & 1 \\ -3 & 5 & -1 \end{vmatrix} = 0,$$

$|D_z| = \begin{vmatrix} 5 & 2 & 1 \\ 2 & 5 & -4 \\ -3 & 9 & 5 \end{vmatrix} = 342$. Thus, the solution is $x = -1$, $y = 0$, $z = -2$.

32. Solve $\begin{vmatrix} x & 1 & 1 \\ 1 & x & x \\ x & 1 & x+1 \end{vmatrix} = 0$ for x.

Answer: $\begin{vmatrix} x & 1 & 1 \\ 1 & x & x \\ x & 1 & x+1 \end{vmatrix} = 0 \Leftrightarrow x\begin{vmatrix} x & x \\ 1 & x+1 \end{vmatrix} - \begin{vmatrix} 1 & x \\ x & x+1 \end{vmatrix} + \begin{vmatrix} 1 & x \\ x & 1 \end{vmatrix} = 0 \Leftrightarrow x^3 - x = 0 \Leftrightarrow x = 0, x = 1, x = -1$

SECTION 10.7

1. Graph the inequality $x \leq 4$.

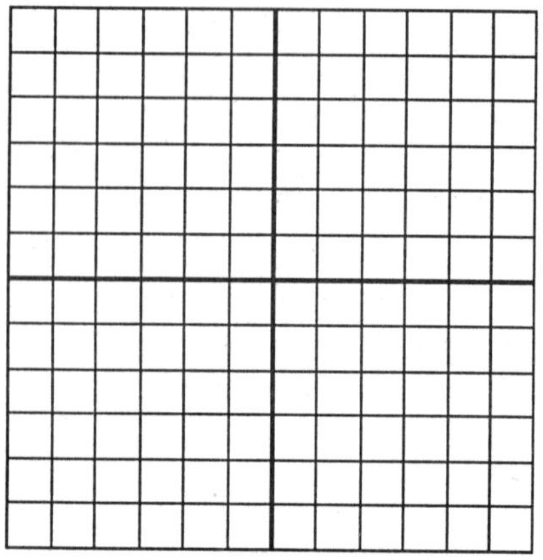

Answer: easy

$x \leq 4$

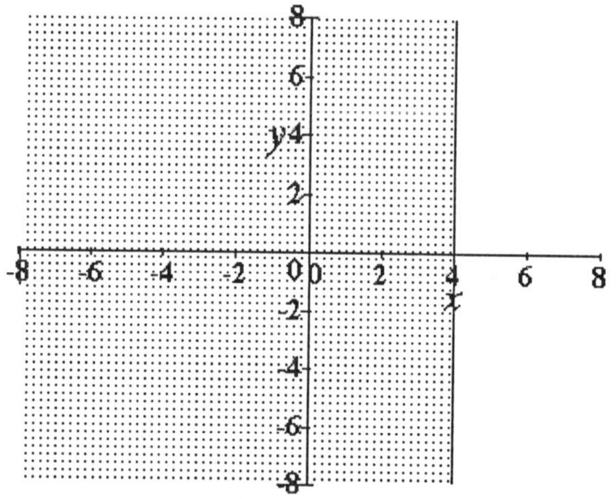

2. Graph the inequality $y < -x + 3$.

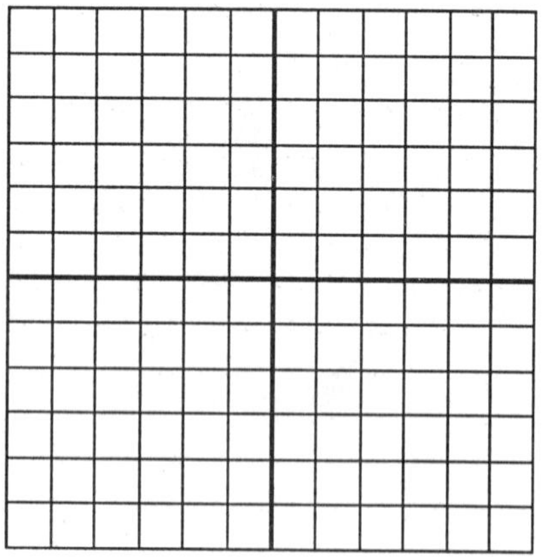

Answer: easy

$y < -x + 3$

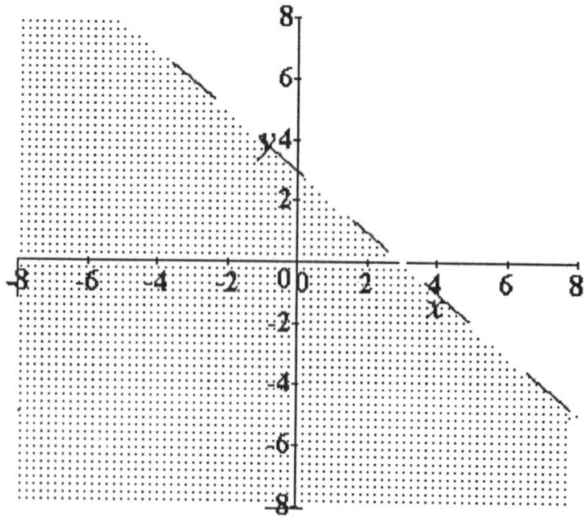

3. Graph the inequality $4x + 3y + 12 > 0$.

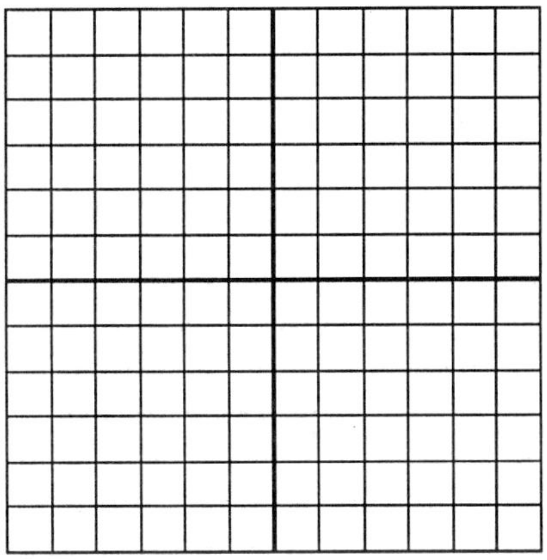

Answer: easy

$4x + 3y + 12 > 0$

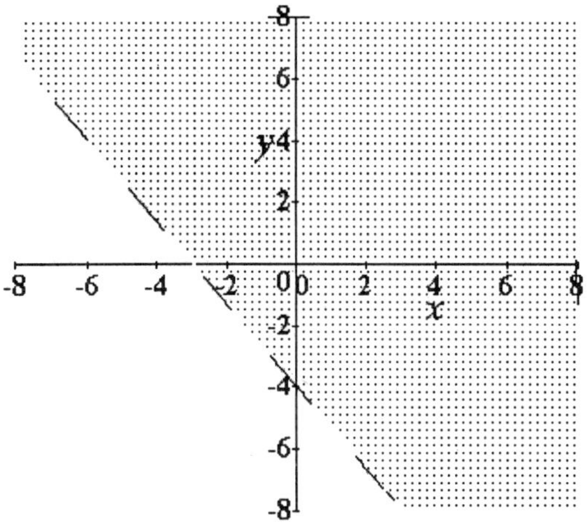

4. Graph the inequality $-x^2 + y \geq 16$.

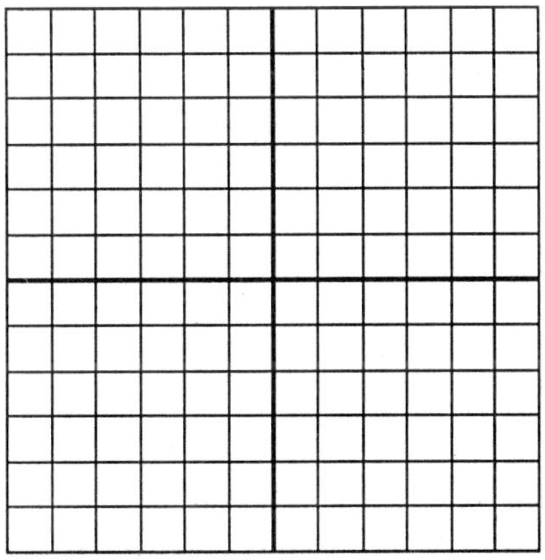

Answer: easy

$-x^2 + y \geq 16$

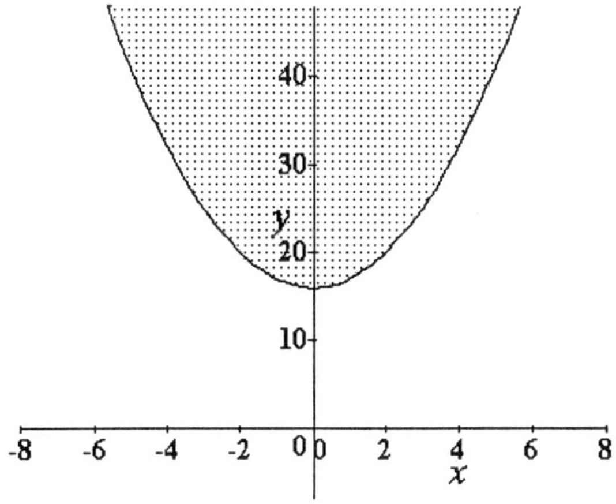

5. Graph the inequality $x - 2y^2 \le 0$.

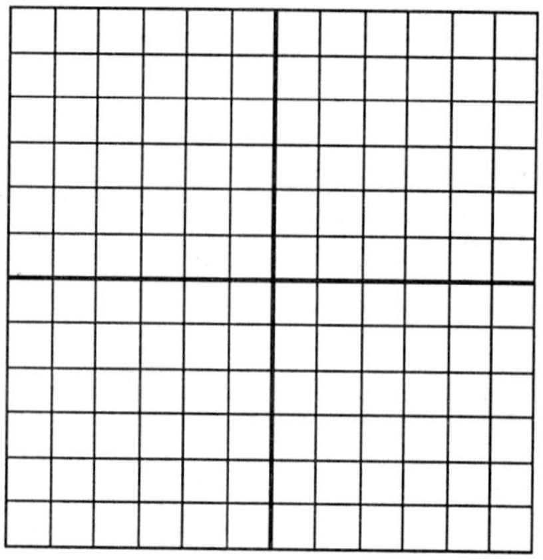

Answer: easy

$x - 2y^2 \le 0$

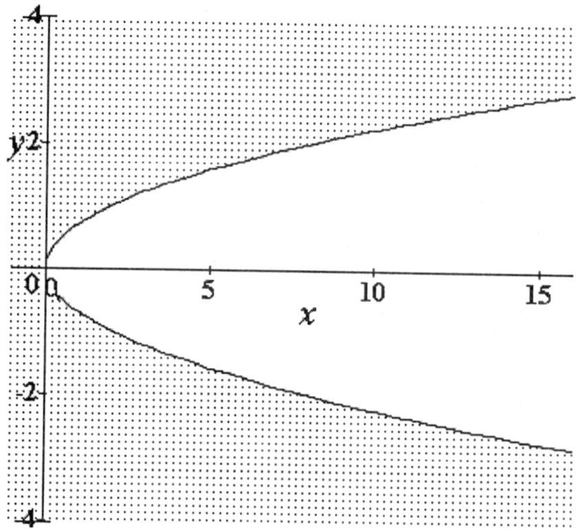

6. Graph the inequality $xy < 9$.

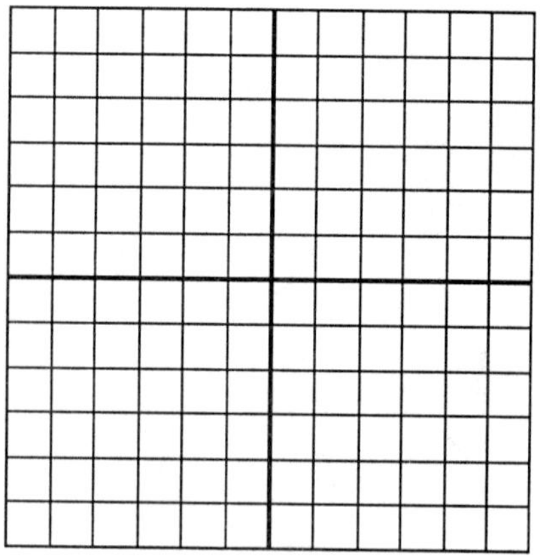

Answer: medium

$xy < 9$

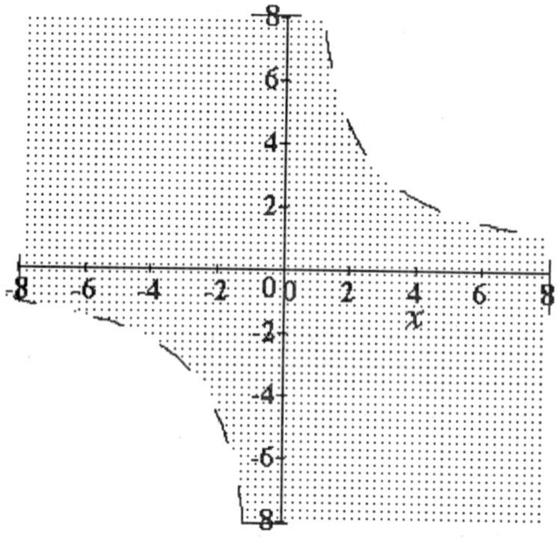

7. Graph the solution of the system of inequalities $\begin{cases} 3x + 2y > 9 \\ 2x - y < 20 \end{cases}$. Find the coordinates of all vertices and determine whether or not the solution set is bounded.

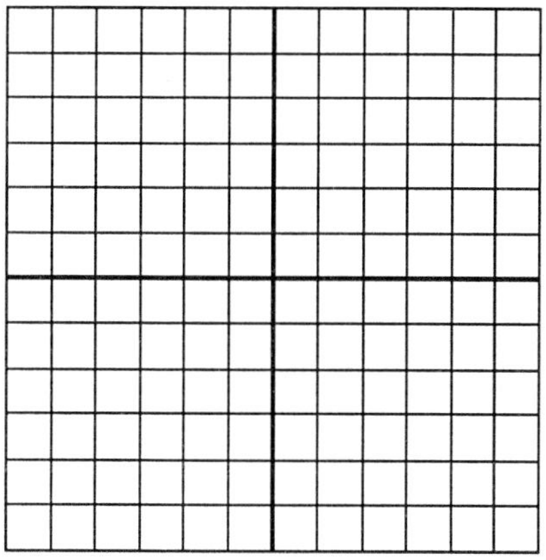

Answer: medium

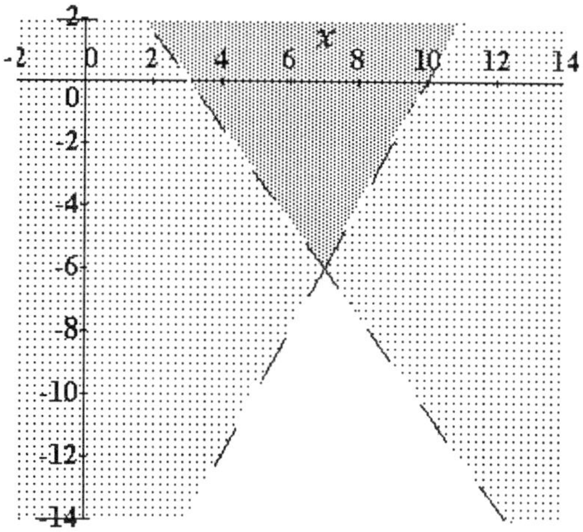

$\begin{cases} 3x + 2y > 9 \\ 2x - y < 20 \end{cases}$. The vertices occur where $\begin{cases} 3x + 2y = 9 \\ 2x - y = 20 \end{cases}$ $\Leftrightarrow$ $\begin{cases} 3x + 2y = 9 \\ 4x - 2y = 40 \end{cases}$ and adding the two equations gives $7x = 49$ $\Leftrightarrow$ $x = 7$, and so $y = -6$. Therefore, the vertex is at $(7, -6)$ and the solution set is not bounded.

8. Graph the solution of the system of inequalities $\begin{cases} x - y \leq 0 \\ 6 + y \leq 2x \end{cases}$. Find the coordinates of all vertices and determine whether or not the solution set is bounded.

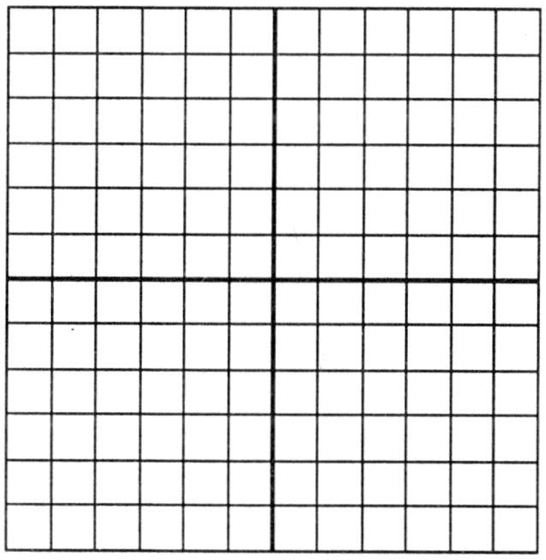

Answer: medium

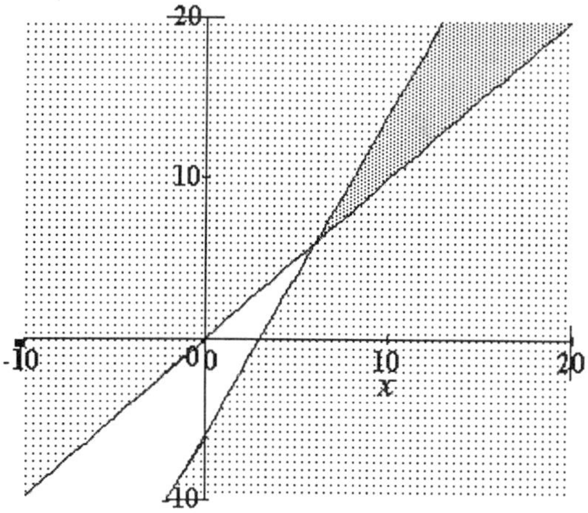

$\begin{cases} x - y \leq 0 \\ 6 + y \leq 2x \end{cases}$ The vertices occur where $\begin{cases} x - y = 0 \\ 6 + y = 2x \end{cases}$ $\Leftrightarrow$ $\begin{cases} x = y \\ x = 6 \end{cases}$. So the vertex is $(6, 6)$ and the solution is not bounded.

9. Graph the solution of the system of inequalities $\begin{cases} x < 8 \\ y \geq 4 \\ 2x - y > -3 \end{cases}$. Find the coordinates of all vertices and determine whether or not the solution set is bounded.

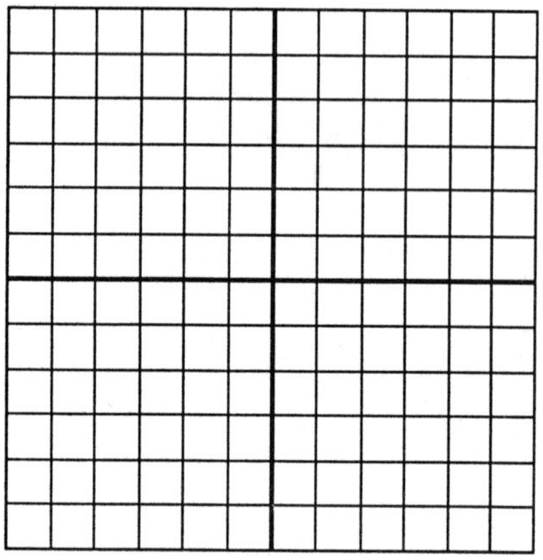

Answer: medium

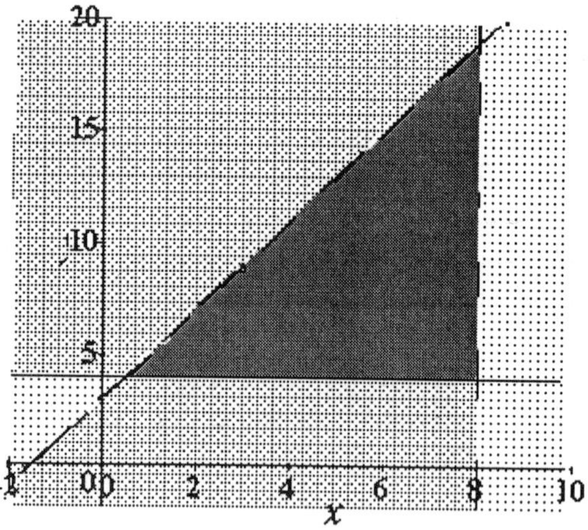

$\begin{cases} x < 8 \\ y \geq 4 \\ 2x - y > -3 \end{cases}$. From the graph, the vertices occur at $\left(\frac{1}{2}, 4\right)$, $(8, 4)$, and $(8, 19)$. The solution set is bounded.

10. Graph the solution of the system of inequalities $\begin{cases} x < y^2 \\ x + y \leq 4 \end{cases}$. Find the coordinates of all vertices and determine whether or not the solution set is bounded.

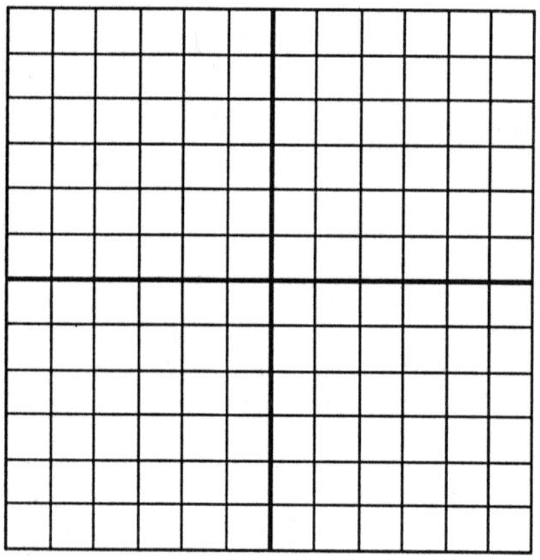

Answer: medium

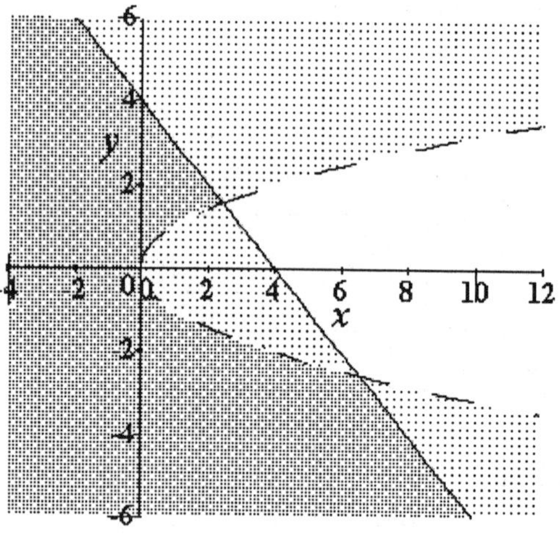

$\begin{cases} x < y^2 \\ x + y \leq 4 \end{cases}$. The vertices occur where $\begin{cases} x = y^2 \\ x + y = 4 \end{cases}$. Substituting for x gives $y^2 + y = 4 \quad \Leftrightarrow \quad y^2 + y - 4 = 0 \quad \Leftrightarrow$

$y = \dfrac{-1 \pm \sqrt{1 + 4 \cdot 4}}{2} = \dfrac{-1 \pm \sqrt{17}}{2}$. Therefore, the vertices occur at $\left(\dfrac{9 + \sqrt{17}}{2}, \dfrac{-1 - \sqrt{17}}{2}\right)$ and $\left(\frac{9-\sqrt{17}}{2}, \frac{-1+\sqrt{17}}{2}\right)$ and the solution set is not bounded.

11. Graph the solution of the system of inequalities $\begin{cases} x > 0, y > 0 \\ x + y < 9 \\ x^2 + y^2 > 49 \end{cases}$. Find the coordinates of all vertices and determine whether or not the solution set is bounded.

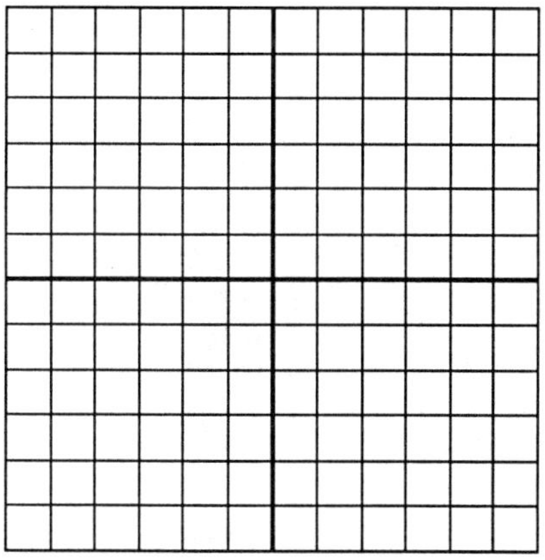

Answer: medium

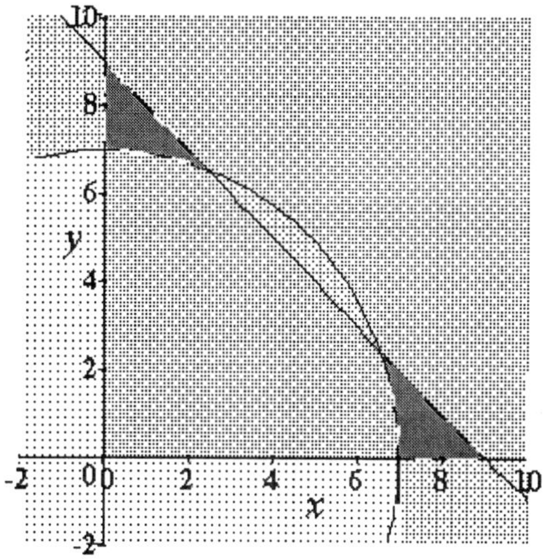

$\begin{cases} x > 0, y > 0 \\ x + y < 9 \\ x^2 + y^2 > 49 \end{cases}$. The vertices occur at $(7,0)$, $(9,0)$, $\left(\frac{9}{2} + \frac{\sqrt{17}}{2}, \frac{9}{2} - \frac{\sqrt{17}}{2}\right)$, $(0,7)$, $(0,9)$, and $\left(\frac{9}{2} - \frac{\sqrt{17}}{2}, \frac{9}{2} + \frac{\sqrt{17}}{2}\right)$. The solution set is bounded.

12. Graph the solution of the system of inequalities $\begin{cases} x^2 + y^2 \le 4 \\ 2x^2 + y \ge 1 \end{cases}$. Find the coordinates of all vertices and determine whether or not the solution set is bounded.

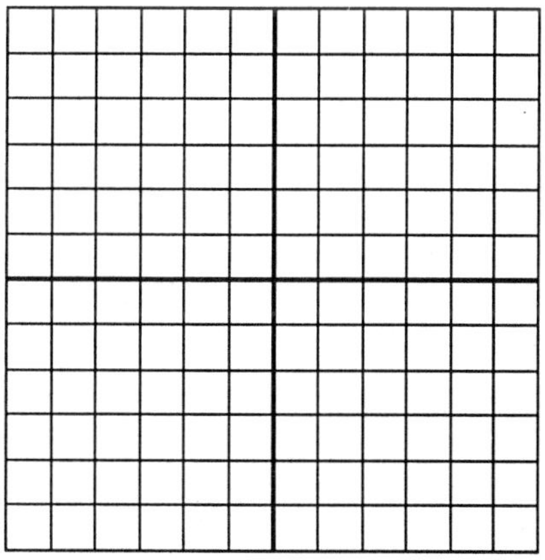

Answer: medium

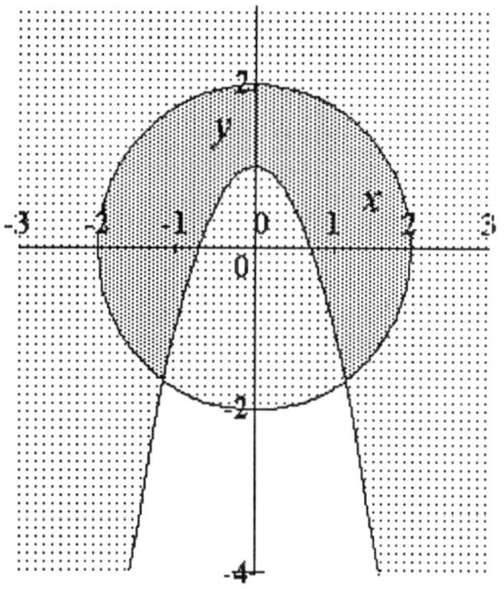

$\begin{cases} x^2 + y^2 \le 4 \\ 2x^2 + y \ge 1 \end{cases}$. From the graph, the vertices occur at approximately $(\pm 1.15, -1.64)$. The solution set is bounded.

13. Graph the solution of the system of inequalities $\begin{cases} y \leq x+2 \\ 4x+3y \leq 12 \\ x-2y > 3 \end{cases}$. Find the coordinates of all vertices and determine whether or not the solution set is bounded.

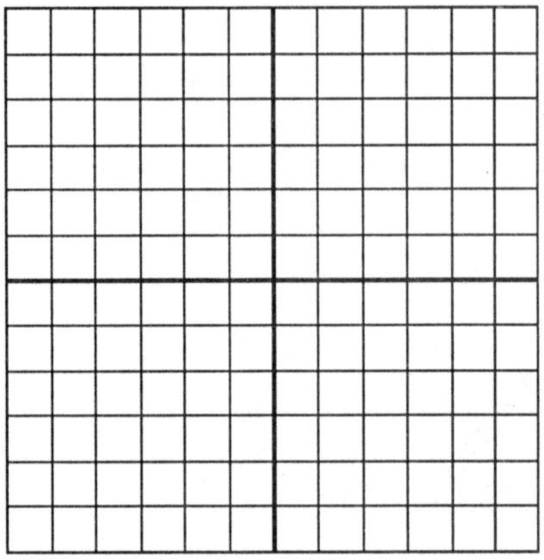

Answer: medium

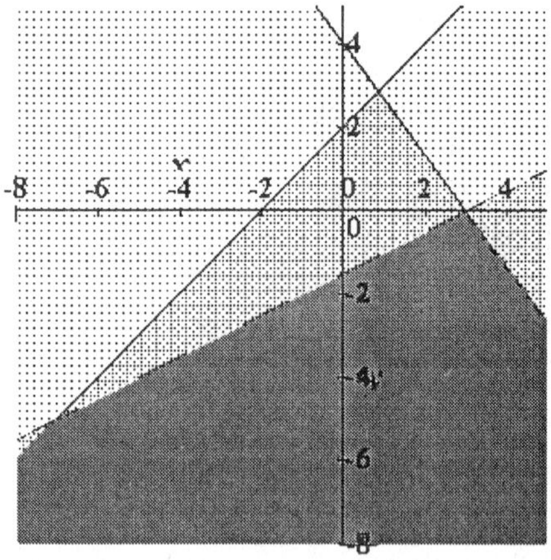

$\begin{cases} y \leq x+2 \\ 4x+3y \leq 12 \\ x-2y > 3 \end{cases}$. The vertices occur where $\begin{cases} y = x+2 \\ 4x+3y = 12 \\ x-2y = 3 \end{cases}$. By solving these equations, we find that vertices occur at $(-7,-5)$, $\left(\frac{6}{7}, \frac{20}{7}\right)$, and $(3,0)$. The solution set is not bounded.

14. Graph the solution of the system of inequalities $\begin{cases} x \geq 0, y \geq 0 \\ y \leq 6 \\ 2x+y \leq 20 \\ -3x+4y \leq 12 \end{cases}$. Find the coordinates of all vertices and determine whether or not the solution set is bounded.

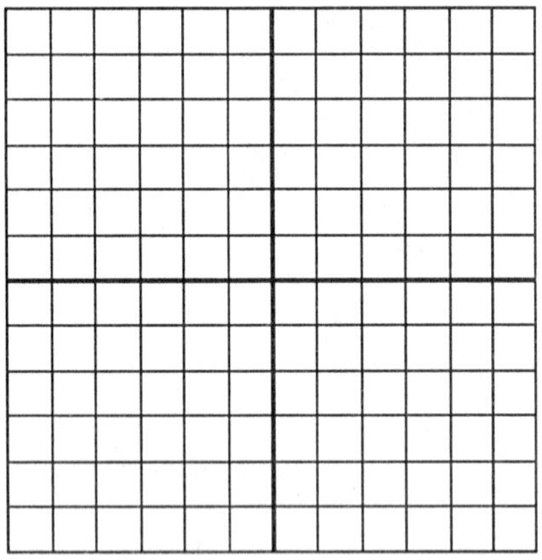

Answer: medium

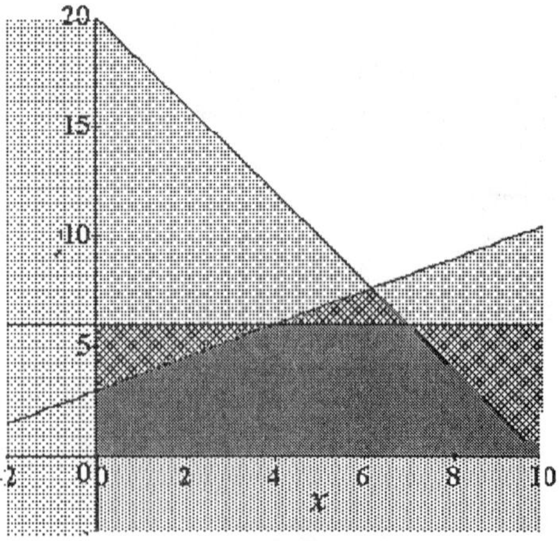

$\begin{cases} x \geq 0, y \geq 0 \\ y \leq 6 \\ 2x+y \leq 20 \\ -3x+4y \leq 12 \end{cases}$. The points of intersection are $(-4,0)$, $(0,0)$, $(0,3)$, $(0,20)$, $(4,6)$, $(6,8)$, $(7,6)$, and $(10,0)$. However, the points $(-4,0)$, $(0,20)$, and $(6,8)$ are not in the neighborhood of the solution set. Therefore, the vertices are $(0,0)$, $(0,3)$, $(4,6)$, $(7,6)$, and $(10,0)$. The solution set is bounded.

15. A man and his daughter manufacture unfinished tables and chairs. Each table requires 1 h of sawing and 2 h of assembly. Each chair requires 2 h of sawing and 2 h of assembly. The two of them can do a total of up to 8 h of sawing and 8 h of assembly work each day. Find a system of inequalities that describes all possible combinations of tables and chairs that they can make daily. Graph the solution.

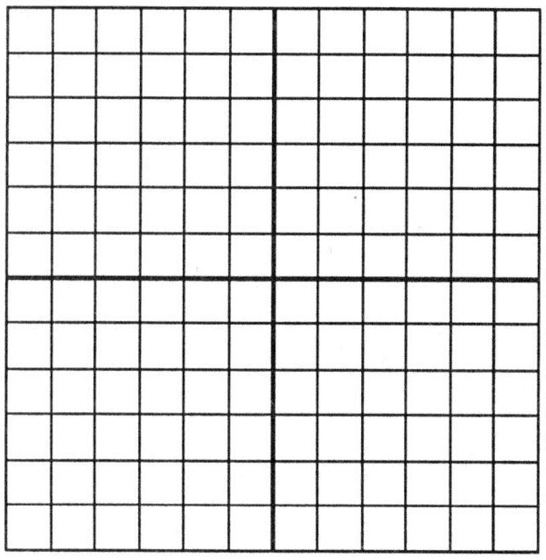

Answer: medium

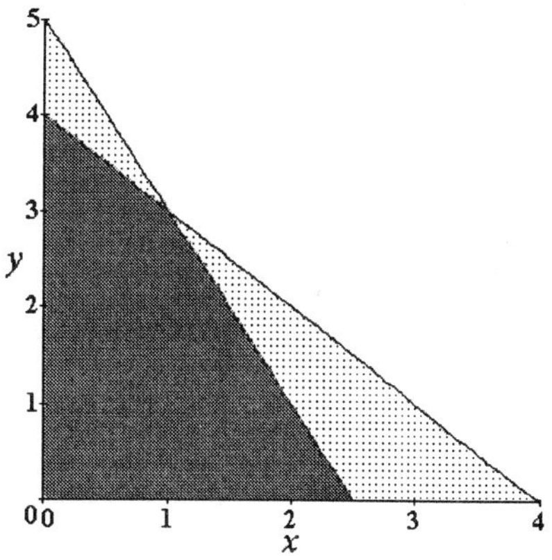

Let x be the number of tables made and y be the number of chairs made. Then, the following system of inequalities holds: $\begin{cases} x + 2y \le 8 \\ 2x + 2y \le 8 \\ x \ge 0, y \ge 0 \end{cases}$. The intersection points occur at $(0,0)$, $(0,4)$, $(0,5)$, $(1,3)$, $\left(\frac{5}{2},0\right)$, and $(4,0)$. Since the points $(0,5)$ and $(4,0)$ are not in the solution set, the vertices are $(0,0)$, $(0,4)$, $(1,3)$, and $\left(\frac{5}{2},0\right)$.

SECTION 10.8

1. Write out the form of the partial fraction decomposition of the rational function $\dfrac{2x+4}{x^2-3x+2}$. Do not determine the numerical values of the coefficients.

(a) $\dfrac{A}{x+2} + \dfrac{B}{x+1}$
(b) $\dfrac{Ax+B}{x+2} + \dfrac{Cx+D}{x+1}$
(c) $\dfrac{A}{x-2} + \dfrac{B}{x-1}$
(d) $\dfrac{Ax+B}{x-2} + \dfrac{Cx+D}{x-1}$
(e) $\dfrac{Ax^2+Bx+C}{x+2} + \dfrac{Dx^2+Ex+F}{x+1}$

Answer: easy (c)

$$\frac{2x+4}{x^2-3x+2} = \frac{2x+4}{(x-2)(x-1)} = \frac{A}{x-2} + \frac{B}{x-1}.$$

2. Write out the form of the partial fraction decomposition of the rational function $\dfrac{8x-5}{3x^2-x}$. Do not determine the numerical values of the coefficients.

(a) $\dfrac{A}{x+1} + \dfrac{B}{3x}$
(b) $\dfrac{A}{3x} + \dfrac{B}{x-1}$
(c) $\dfrac{A}{x^2} + \dfrac{B}{2}$
(d) $\dfrac{A}{x} + \dfrac{B}{3x-1}$
(e) $\dfrac{A}{x} + \dfrac{B}{x^2} + \dfrac{C}{3x-1}$

Answer: easy (d)

$$\frac{8x-5}{3x^2-x} = \frac{8x-5}{x(3x-1)} = \frac{A}{x} + \frac{B}{3x-1}.$$

3. Write out the form of the partial fraction decomposition of the rational function $\dfrac{5x-3}{x^3-2x^2-3x}$. Do not determine the numerical values of the coefficients.

(a) $\dfrac{A}{x}+\dfrac{B}{x^2}+\dfrac{C}{x+3}+\dfrac{D}{x-1}$

(b) $\dfrac{A}{x}+\dfrac{B}{x-3}+\dfrac{C}{x+1}$

(c) $\dfrac{A}{x}+\dfrac{B}{x^2}+\dfrac{C}{x-3}+\dfrac{D}{x+1}$

(d) $\dfrac{A}{x}+\dfrac{B}{x+3}+\dfrac{C}{x-1}$

(e) $\dfrac{A}{x}+\dfrac{B}{x^2}+\dfrac{C}{x^3}+\dfrac{D}{x+3}+\dfrac{E}{x-1}$

Answer: easy (b)

$$\frac{5x-3}{x^3-2x^2-3x}=\frac{5x-3}{x(x-3)(x+1)}=\frac{A}{x}+\frac{B}{x-3}+\frac{C}{x+1}.$$

4. Write out the form of the partial fraction decomposition of the rational function $\dfrac{-3x^2-3x+27}{(x+3)(2x^2+3x-2)}$. Do not determine the numerical values of the coefficients.

(a) $\dfrac{A}{x+3}+\dfrac{B}{2x-1}+\dfrac{C}{x+2}$

(b) $\dfrac{A}{x+2}+\dfrac{B}{2x-3}+\dfrac{C}{x+3}$

(c) $\dfrac{A}{x+3}+\dfrac{Bx+C}{2x^2+3x-2}$

(d) $\dfrac{A}{x}+\dfrac{B}{x^2-1}+\dfrac{C}{x+2}$

(e) $\dfrac{A}{x+3}+\dfrac{B}{(x+3)^2}+\dfrac{C}{2x^2+3x-2}$

Answer: easy (a)

$$\frac{-3x^2-3x+27}{(x+3)(2x^2+3x-2)}=\frac{-3x^2-3x+27}{(x+3)(2x-1)(x+2)}=\frac{A}{x+3}+\frac{B}{2x-1}+\frac{C}{x+2}.$$

5. Write out the form of the partial fraction decomposition of the rational function $\dfrac{3x^2+2x-10}{(3x+4)(x^2-6x+9)}$. Do not determine the numerical values of the coefficients.

(a) $\dfrac{A}{3x+4}+\dfrac{Bx+C}{(x-3)^2}$

(b) $\dfrac{A}{3x+4}+\dfrac{B}{(x-3)^2}$

(c) $\dfrac{A}{3x+4}+\dfrac{B}{(3x+4)^2}+\dfrac{C}{x-3}$

(d) $\dfrac{A}{3x}+\dfrac{B}{4}+\dfrac{C}{x-3}+\dfrac{D}{(x-3)^2}$

(e) $\dfrac{A}{3x+4}+\dfrac{B}{x-3}+\dfrac{C}{(x-3)^2}$

Answer: easy (e)

$$\frac{3x^2+2x-10}{(3x+4)(x^2-6x+9)}=\frac{3x^2+2x-10}{(3x+4)(x-3)^2}=\frac{A}{3x+4}+\frac{B}{x-3}+\frac{C}{(x-3)^2}.$$

6. Find the partial fraction decomposition of the rational function $\dfrac{x-3}{(2x-3)^2}$.

(a) $\dfrac{3/2}{(2x)^2}-\dfrac{2/3}{(2x-3)^2}$

(b) $\dfrac{1}{2x-3}-\dfrac{2}{(2x-3)^2}$

(c) $\dfrac{3/2}{2x-3}-\dfrac{1/2}{(2x-3)^2}$

(d) $\dfrac{1/2}{2x-3}-\dfrac{3/2}{(2x-3)^2}$

(e) $\dfrac{1}{2x-3}-\dfrac{3}{(2x-3)^2}$

Answer: medium (d)

$\dfrac{x-3}{(2x-3)^2}=\dfrac{A}{2x-3}+\dfrac{B}{(2x-3)^2}$. Hence, $x-3=A(2x-3)+B=2Ax+(-3A+B)$, and so $\begin{cases}2A=1\\-3A+B=-3\end{cases}$ $\Rightarrow$ $A=\frac{1}{2}$, and $-3\cdot\frac{1}{2}+B=-3$ $\Leftrightarrow$ $B=-\frac{3}{2}$. Therefore, $\dfrac{x-3}{(2x-3)^2}=\dfrac{1/2}{2x-3}-\dfrac{3/2}{(2x-3)^2}$.

7. Find the partial fraction decomposition of the rational function $\dfrac{x^3-3x^2-3x+4}{x^4}$.

(a) $\dfrac{1}{x}-\dfrac{3}{x^2}-\dfrac{3}{x^3}+\dfrac{4}{x^4}$
(b) $\dfrac{3}{x}-\dfrac{1/3}{x^2}-\dfrac{1/3}{x^3}+\dfrac{2}{x^4}$
(c) $\dfrac{4}{x}-\dfrac{3}{x^2}-\dfrac{3}{x^3}+\dfrac{1}{x^4}$
(d) $\dfrac{1/4}{x}-\dfrac{3/4}{x^2}-\dfrac{3/4}{x^3}+\dfrac{1}{x^4}$
(e) $\dfrac{1/3}{x}-\dfrac{1}{x^2}-\dfrac{1}{x^3}+\dfrac{4/3}{x^4}$

Answer: medium (a)

$\dfrac{x^3-3x^2-3x+4}{x^4}=\dfrac{A}{x}+\dfrac{B}{x^2}+\dfrac{C}{x^3}+\dfrac{D}{x^4}$. Hence, $x^3-3x^2-3x+4=Ax^3+Bx^2+Cx+D$, and so $A=1$, $B=-3$, $C=-3$, and $D=4$. Therefore $\dfrac{x^3-3x^2-3x+4}{x^4}=\dfrac{1}{x}-\dfrac{3}{x^2}-\dfrac{3}{x^3}+\dfrac{4}{x^4}$.

8. Find the partial fraction decomposition of the rational function $\dfrac{-3x^2+4x-2}{x^4-x^3-3x^2+5x-2}$.

Answer: medium

$\dfrac{-3x^2+4x-2}{x^4-x^3-3x^2+5x-2}=\dfrac{-3x^2+4x-2}{(x-1)(x^3-3x+2)}=\dfrac{-3x^2+4x-2}{(x-1)^3(x+2)}=\dfrac{A}{x+2}+\dfrac{B}{x-1}+\dfrac{C}{(x-1)^2}+\dfrac{D}{(x-1)^3}$. Thus

$$\begin{aligned}-3x^2+4x-2&=A(x-1)^3+B(x+2)(x-1)^2+C(x+2)(x-1)+D(x+2)\\&=A\left(x^3-3x^2+3x-1\right)+B(x+2)\left(x^2-2x+1\right)+C\left(x^2+x-2\right)+D(x+2)\\&=A\left(x^3-3x^2+3x-1\right)+B\left(x^3-3x+2\right)+C\left(x^2+x-2\right)+D(x+2)\\&=(A+B)x^3+(-3A+C)x^2+(3A-3B+C+D)x+(-A+2B-2C+2D)\end{aligned}$$

Therefore we must solve the system $\begin{bmatrix}1&1&0&0&0\\-3&0&1&0&-3\\3&-3&1&1&4\\-1&2&-2&2&-2\end{bmatrix}$ $\begin{array}{l}3R_1+R_2\to R_2\\R_2+R_3\to R_3\\R_1+R_4\to R_4\end{array}$ $\begin{bmatrix}1&1&0&0&0\\0&3&1&0&-3\\0&-3&2&1&1\\0&3&-2&2&-2\end{bmatrix}$

$\begin{array}{l}R_2+R_3\to R_3\\R_3+R_4\to R_4\end{array}$ $\begin{bmatrix}1&1&0&0&0\\0&3&1&0&-3\\0&0&3&1&-2\\0&0&0&3&-1\end{bmatrix}$. Hence $3D=-1$ $\Leftrightarrow$ $D=-\frac{1}{3}$, $3C-\frac{1}{3}=-2$ $\Leftrightarrow$ $C=-\frac{5}{9}$, $3B-\frac{5}{9}=-3$ $\Leftrightarrow$ $B=-\frac{22}{27}$, and $A-\frac{22}{27}=0$ $\Leftrightarrow$ $A=\frac{22}{27}$. Therefore

$\dfrac{-3x^2+4x-2}{x^4-x^3-3x^2+5x-2}=\dfrac{22/27}{x+2}-\dfrac{22/27}{x-1}-\dfrac{5/9}{(x-1)^2}-\dfrac{1/3}{(x-1)^3}$.

9. Find the partial fraction decomposition of the rational function $\dfrac{x^2+3x-6}{x^4-2x^2+1}$.

Answer: medium

$\dfrac{x^2+3x-6}{x^4-2x^2+1}=\dfrac{x^2+3x-6}{(x^2-1)^2}=\dfrac{x^2+3x-6}{(x-1)^2(x+1)^2}=\dfrac{A}{x+1}+\dfrac{B}{(x+1)^2}+\dfrac{C}{x-1}+\dfrac{D}{(x-1)^2}$. Thus

$$\begin{aligned}x^2+3x-6&=A(x+1)(x-1)^2+B(x-1)^2+C(x+1)^2(x-1)+D(x+1)^2\\&=A\left(x^3-x^2-x+1\right)+B\left(x^2-2x+1\right)+C\left(x^3+x^2-x-1\right)+D\left(x^2+2x+1\right)\\&=(A+C)x^3+(-A+B+C+D)x^2+(-A-2B-C+2D)x+(A+B-C+D)\end{aligned}$$

which leads to the system $\begin{bmatrix}1&0&1&0&0\\-1&1&1&1&1\\-1&-2&-1&2&3\\1&1&-1&1&-6\end{bmatrix}$ $\begin{array}{l}R_2+R_1\to R_2\\R_3+R_1\to R_3\\R_4-R_1\to R_4\end{array}$ $\begin{bmatrix}1&0&1&0&0\\0&1&2&1&1\\0&-2&0&2&3\\0&1&-2&1&-6\end{bmatrix}$ $\begin{array}{l}R_3+2R_2\to R_3\\R_2-R_4\to R_4\end{array}$

$\begin{bmatrix}1&0&1&0&0\\0&1&2&1&1\\0&0&4&4&5\\0&0&4&0&7\end{bmatrix}$. Hence $4C=7$ $\Leftrightarrow$ $C=\frac{7}{4}$, $7+4D=5$ $\Leftrightarrow$ $D=-\frac{1}{2}$, $B+\frac{7}{2}-\frac{1}{2}=1$ $\Leftrightarrow$ $B=-2$, and

$A+\frac{7}{4}=0$ $\Leftrightarrow$ $A=-\frac{7}{4}$. Therefore, $\dfrac{x^2+3x-6}{x^4-2x^2+1}=-\dfrac{7/4}{x+1}-\dfrac{2}{(x+1)^2}+\dfrac{7/4}{x-1}-\dfrac{1/2}{(x-1)^2}$.

10. Find the partial fraction decomposition of the rational function $\dfrac{4x^2-3x+5}{x^3-2x^2+x-2}$.

Answer: medium

$\dfrac{4x^2-3x+5}{x^3-2x^2+x-2}=\dfrac{4x^2-3x+5}{(x^2+1)(x-2)}=\dfrac{Ax+B}{x^2+1}+\dfrac{C}{x-2}$. Thus

$4x^2-3x+5=(Ax+B)(x-2)+C(x^2+1)=(A+C)x^2+(-2A+B)x+(-2B+C)$, which leads to the system

$$\begin{bmatrix}1&0&1&4\\-2&1&0&-3\\0&-2&1&5\end{bmatrix}\quad \mathrm{R}_2+2\mathrm{R}_1\to\mathrm{R}_2\quad \begin{bmatrix}1&0&1&4\\0&1&2&5\\0&-2&1&5\end{bmatrix}\quad 2\mathrm{R}_2+\mathrm{R}_3\to\mathrm{R}_3\quad \begin{bmatrix}1&0&1&4\\0&1&2&5\\0&0&5&15\end{bmatrix}.$$

Hence, $5C=15 \quad\Leftrightarrow\quad C=3,\ B+6=5 \quad\Leftrightarrow\quad B=-1$, and $A+3=4 \quad\Leftrightarrow\quad A=1$. Therefore,

$\dfrac{4x^2-3x+5}{x^3-2x^2+x-2}=\dfrac{x-1}{x^2+1}+\dfrac{3}{x-2}$.

11. Find the partial fraction decomposition of the rational function $\dfrac{x^2+2x+2}{2x^4+5x^2+2}$.

Answer: medium

$\dfrac{x^2+2x+2}{2x^4+5x^2+2}=\dfrac{x^2+2x+2}{(2x^2+1)(x^2+2)}=\dfrac{Ax+B}{2x^2+1}+\dfrac{Cx+D}{x^2+2}$. Thus

$x^2+2x+2=(Ax+B)(x^2+2)+(Cx+D)(2x^2+1)=Ax^3+2Ax+Bx^2+2B+2Cx^3+2Dx^2+Cx+D$

$=(A+2C)x^3+(B+2D)x^2+(2A+C)x+(2B+D)$, which leads to the system

$$\begin{bmatrix}1&0&2&0&0\\0&1&0&2&1\\2&0&1&0&2\\0&2&0&1&2\end{bmatrix}\quad \begin{matrix}2\mathrm{R}_1-\mathrm{R}_3\to\mathrm{R}_3\\2\mathrm{R}_2-\mathrm{R}_4\to\mathrm{R}_4\end{matrix}\quad \begin{bmatrix}1&0&2&0&0\\0&1&0&2&1\\0&0&3&0&-2\\0&0&0&3&0\end{bmatrix}.$$

Hence $D=0$, $C=-\frac{2}{3}$, $B=1$, and $A=\frac{4}{3}$. Therefore

$\dfrac{x^2+2x+2}{2x^4+5x^2+2}=\dfrac{\frac{4}{3}x+1}{2x^2+1}-\dfrac{\frac{2}{3}x}{x^2+2}$.

12. Find the partial fraction decomposition of the rational function $\dfrac{3x^2-2x+5}{(x^2+2)^2}$.

Answer: medium

$\dfrac{3x^2-2x+5}{(x^2+2)^2}=\dfrac{Ax+B}{x^2+2}+\dfrac{Cx+D}{(x^2+2)^2}$. Thus,

$3x^2-2x+5=(Ax+B)(x^2+2)+Cx+D=Ax^3+Bx^2+2Ax+2B+Cx+D$

$=Ax^3+Bx^2+(2A+C)x+(2B+D)$, and so $A=0$, $B=3$, $C=-2$, and $D=-1$. Therefore

$\dfrac{3x^2-2x+5}{(x^2+2)^2}=\dfrac{3}{x^2+2}-\dfrac{2x+1}{(x^2+2)^2}$.

13. Find the partial fraction decomposition of the rational function $\dfrac{x^5+x^3+3x^2-4x+2}{(x-1)^2(x^2+2)}$.

Answer: hard

$$\frac{x^5+x^3+3x^2-4x+2}{(x-1)^2(x^2+2)} = \frac{x^5+x^3+3x^2-4x+2}{x^4-2x^3+3x^2-4x+2} = r(x)$$

$$\begin{array}{r|rrrrrr}
 & x & +2 & & & & \\
\hline
x^4-2x^3+3x^2-4x+2 & x^5 & & +x^3 & +3x^2 & -4x & +2 \\
 & x^5 & -2x^4 & +3x^3 & -4x^2 & +2x & \\
\hline
 & & 2x^4 & -2x^3 & +7x^2 & -6x & +2 \\
 & & 2x^4 & -4x^3 & +6x^2 & -8x & +4 \\
\hline
 & & & 2x^3 & +x^2 & +2x & -2
\end{array}$$

Thus $r(x) = x+2+\dfrac{2x^3+x^2+2x-2}{(x-1)^2(x^2+2)}$. Now $\dfrac{2x^3+x^2+2x-2}{(x-1)^2(x^2+2)} = \dfrac{A}{x-1}+\dfrac{B}{(x-1)^2}+\dfrac{Cx+D}{x^2+2}$ and so

$$\begin{aligned}
2x^3+x^2+2x-2 &= A(x-1)(x^2+2)+B(x^2+2)+(Cx+D)(x-1)^2 \\
&= A(x^3-x^2+2x-2)+B(x^2+2)+(Cx+D)(x^2-2x+1) \\
&= (A+C)x^3+(-A+B-2C+D)x^2+(2A+C-2D)x+(-2A+2B+D)
\end{aligned}$$

Equating coefficients, we get the following system of equations:

$$\begin{bmatrix} 1 & 0 & 1 & 0 & 2 \\ -1 & 1 & -2 & 1 & 1 \\ 2 & 0 & 1 & -2 & 2 \\ -2 & 2 & 0 & 1 & -2 \end{bmatrix}
\quad \begin{array}{c} R_2+R_1 \to R_2 \\ 2R_1-R_3 \to R_3 \\ R_4+2R_1 \to R_4 \end{array} \quad
\begin{bmatrix} 1 & 0 & 1 & 0 & 2 \\ 0 & 1 & -1 & 1 & 3 \\ 0 & 0 & 1 & 2 & 2 \\ 0 & 2 & 2 & 1 & 2 \end{bmatrix}
\quad \begin{array}{c} R_2+R_3 \to R_2 \\ 2R_2-R_4 \to R_4 \end{array} \quad
\begin{bmatrix} 1 & 0 & 1 & 0 & 2 \\ 0 & 1 & 0 & 3 & 5 \\ 0 & 0 & 1 & 2 & 2 \\ 0 & 0 & -4 & 1 & 4 \end{bmatrix}$$

$4R_3+R_4 \to R_4$ $\begin{bmatrix} 1 & 0 & 1 & 0 & 2 \\ 0 & 1 & 0 & 3 & 5 \\ 0 & 0 & 1 & 2 & 2 \\ 0 & 0 & 0 & 9 & 12 \end{bmatrix}$. Then $D=\frac{4}{3}$, $C=-\frac{2}{3}$, $B=1$, and $A=\frac{8}{3}$. Therefore

$$\frac{x^5+x^3+3x^2-4x+2}{(x-1)^2(x^2+2)} = x+2+\frac{8/3}{x-1}+\frac{1}{(x-1)^2}-\frac{\frac{2}{3}x-\frac{4}{3}}{x^2+2}.$$

14. Give the form of the partial fraction decomposition of the rational function $\dfrac{1}{(x^6-1)(x^4-1)}$.

Answer: hard

$$\begin{aligned}
\frac{1}{(x^6-1)(x^4-1)} &= \frac{1}{(x^3-1)(x^3+1)(x^2-1)(x^2+1)} = \frac{1}{(x-1)(x^2+x+1)(x+1)(x^2-x+1)(x-1)(x+1)(x^2+1)} \\
&= \frac{1}{(x-1)^2(x+1)^2(x^2+x+1)(x^2-x+1)(x^2+1)} \\
&= \frac{A}{x-1}+\frac{B}{(x-1)^2}+\frac{C}{x+1}+\frac{D}{(x+1)^2}+\frac{Ex+F}{x^2+x+1}+\frac{Gx+H}{x^2-x+1}+\frac{Ix+J}{x^2+1}
\end{aligned}$$

15. Determine A, B, C, and D in terms of a and b for $\dfrac{ax^3+bx^2}{(x^2+1)^2} = \dfrac{Ax+B}{x^2+1}+\dfrac{Cx+D}{(x^2+1)^2}$.

Answer: hard

$\dfrac{ax^3+bx^2}{(x^2+1)^2} = \dfrac{Ax+B}{x^2+1}+\dfrac{Cx+D}{(x^2+1)^2}$. Hence

$ax^3+bx^2 = (Ax+B)(x^2+1)+Cx+D = Ax^3+Ax+Bx^2+B+Cx+D = Ax^3+Bx^2+(A+C)x+(B+D)$, and so $A=a$, $B=b$, $a+C=0 \Leftrightarrow C=-a$, and $b+D=0 \Leftrightarrow D=-b$. Therefore, $A=a$, $B=b$, $C=-a$, and $D=-b$.

16. Find the partial fraction decomposition of $\dfrac{2x^2+3x+6}{x^3+3x^2-4x-12}$.

(a) $\dfrac{2}{x-2}-\dfrac{3}{x+3}-\dfrac{2}{x+2}$ (b) $\dfrac{1}{x-2}-\dfrac{4}{x+3}-\dfrac{5}{x+3}$

(c) $\dfrac{1}{x-2}+\dfrac{3}{x+3}-\dfrac{2}{x+2}$ (d) $\dfrac{1}{x-2}-\dfrac{3}{x+2}-\dfrac{5}{x+2}$

(e) $\dfrac{5}{x-2}+\dfrac{6}{x+3}-\dfrac{3}{x+2}$

Answer: medium (c)

$\dfrac{2x^2+3x+6}{x^3+3x^2-4x-12}=\dfrac{2x^2+3x+6}{(x-2)(x+3)(x+2)}=\dfrac{A}{x-2}+\dfrac{B}{x+2}+\dfrac{C}{x+3}$. Thus

$$2x^2+3x+6=(x-2)(x+3)(x+2)\left(\frac{A}{x-2}+\frac{B}{x+2}+\frac{C}{x+3}\right)=Ax^2+5Ax+6A+Bx^2+Bx-6B+Cx^2-4C$$
$$=(A+B+C)x^2+(5A+B)x+(6A-6B-4C)$$

which leads to the following system: $\begin{cases}A+B+C=2\\5A+B=3\\6A-6B-4C=6\end{cases}\Rightarrow\begin{bmatrix}1&1&1&2\\5&1&0&3\\6&-6&-4&6\end{bmatrix}\;\underset{-6R_1+R_3\to R_3}{\overset{-5R_1+R_2\to R_2}{\Rightarrow}}\;\begin{bmatrix}1&1&1&2\\0&-4&-5&-7\\0&-12&-10&-6\end{bmatrix}$

$\underset{-3R_2+R_3\to R_3}{\Rightarrow}\begin{bmatrix}1&1&1&2\\0&-4&-5&-7\\0&0&-20&-60\end{bmatrix}\Rightarrow -20C=-60\Rightarrow C=3\Leftrightarrow -4B-5(3)=-7\Rightarrow B=-2\Leftrightarrow$

$A-2+3=2\Rightarrow A=1$. Therefore, $\dfrac{2x^2+3x+6}{x^3+3x^2-4x-12}=\dfrac{1}{x-2}+\dfrac{3}{x+3}-\dfrac{2}{x+2}$.

17. Find the partial fraction decomposition of $\dfrac{3x^2-6x+8}{x^3-4x^2+4x}$.

(a) $\dfrac{2}{x}+\dfrac{4}{(x-2)^2}+\dfrac{1}{x-2}$ (b) $\dfrac{1}{x}-\dfrac{4}{(x-2)^2}+\dfrac{2}{x-2}$

(c) $\dfrac{2}{x}+\dfrac{4}{(x-2)^2}+\dfrac{16}{x-2}$ (d) $\dfrac{8}{x}+\dfrac{4}{(x-2)^2}+\dfrac{4}{x-2}$

(e) $\dfrac{3}{x}-\dfrac{6}{(x-2)^2}+\dfrac{8}{x-2}$

Answer: medium (a)

$\dfrac{3x^2-6x+8}{x^3-4x^2+4x}=\dfrac{3x^2-6x+8}{x(x-2)^2}=\dfrac{A}{x}+\dfrac{B}{x-2}+\dfrac{C}{(x-2)^2}$. Thus

$$3x^2-6x+8=x(x-2)^2\left[\frac{A}{x}+\frac{B}{x-2}+\frac{C}{(x-2)^2}\right]=x^2A-4xA+4A+Bx^2-2Bx+Cx$$
$$=(A+B)x^2+x(-4A-2B+C)+4A$$

which leads to the following system: $\begin{cases}A+B=3\\-4A-2B+C=-6\\4A=8\end{cases}\Rightarrow B=1,\ A=2,\ C=4$. Therefore,

$\dfrac{3x^2-6x+8}{x^3-4x^2+4x}=\dfrac{2}{x}+\dfrac{4}{(x-2)^2}+\dfrac{1}{x-2}$.

18. Find the partial fraction decomposition of $\dfrac{x^4+9x^2+25+2x^3+10x}{x^5+10x^3+25x}$.

(a) $\dfrac{1}{x}-\dfrac{2}{x^2+5}+\dfrac{3x+1}{(x^2+5)^2}$
(b) $\dfrac{1}{x}+\dfrac{2}{x^2+5}+\dfrac{x}{(x^2+5)^2}$
(c) $\dfrac{1}{x}+\dfrac{5x+2}{x^2+5}-\dfrac{2x+1}{(x^2+5)^2}$
(d) $\dfrac{2}{x}+\dfrac{2x+1}{x^2+5}-\dfrac{x}{(x^2+5)^2}$
(e) $\dfrac{1}{x}+\dfrac{2}{x^2+5}-\dfrac{x}{(x^2+5)^2}$

Answer: medium (e)

$$\frac{x^4+9x^2+25+2x^3+10x}{x^5+10x^3+25x}=\frac{x^4+9x^2+25+2x^3+10x}{x\left(x^2+5\right)^2}=\frac{A}{x}+\frac{Bx+C}{x^2+5}+\frac{Dx+E}{\left(x^2+5\right)^2}.$$

$$\begin{aligned}x^4+9x^2+25+2x^3+10x&=x\left(x^2+5\right)^2\left[\frac{A}{x}+\frac{Bx+C}{x^2+5}+\frac{Dx+E}{\left(x^2+5\right)^2}\right]\\&=Ax^4+10x^2A+25A+Bx^4+5Bx^2+Cx^3+5Cx+Dx^2+xE\\&=(A+B)\,x^4+Cx^3+x^2\,(10A+5B+D)+x\,(5C+E)+25A\end{aligned}$$

which leads to the following system: $\begin{cases}A+B=1\\10A+5B+D=9\\5C+E=10\\25A=25\\C=2\end{cases}$ $\Rightarrow$ $C=2$, $D=-1$, $B=0$, $A=1$, $E=0$. Therefore,

$$\frac{x^4+9x^2+25+2x^3+10x}{x^5+10x^3+25x}=\frac{1}{x}+\frac{2}{x^2+5}-\frac{x}{\left(x^2+5\right)^2}.$$

19. Find the partial fraction decomposition of $\dfrac{x^4-4x^3-13x^2+49x-9}{x^3-4x^2-9x+36}$.

(a) $x-4-\dfrac{2}{x+3}+\dfrac{1}{x-3}-\dfrac{3}{x-4}$
(b) $\dfrac{2}{x+3}+\dfrac{1}{x-3}-\dfrac{3}{x-4}$
(c) $x+1-\dfrac{2}{x+3}+\dfrac{1}{x-3}-\dfrac{3}{x-4}$
(d) $2x-\dfrac{5}{x+3}+\dfrac{1}{x-3}-\dfrac{3}{x-4}$
(e) $x-\dfrac{2}{x+3}+\dfrac{1}{x-3}-\dfrac{3}{x-4}$

Answer: hard (e)

Using long division we have $\dfrac{x^4-4x^3-13x^2+49x-9}{x^3-4x^2-9x+36}=x-\dfrac{4x^2-13x+9}{(x-3)\,(x+3)\,(x-4)}$ $\Leftrightarrow$

$\dfrac{4x^2-13x+9}{(x-3)\,(x+3)\,(x-4)}=\dfrac{A}{x-3}+\dfrac{B}{x+3}+\dfrac{C}{x-4}$ $\Leftrightarrow$

$$\begin{aligned}4x^2-13x+9&=(x-3)\,(x+3)\,(x-4)\left(\frac{A}{x-3}+\frac{B}{x+3}+\frac{C}{x-4}\right)\\&=x^2A-xA-12A+Bx^2-7Bx+12B+Cx^2-9C\\&=(A+B+C)\,x^2+(-A-7B)\,x+(-12A+12B-9C)\end{aligned}$$

(continued over)

which leads to the system of equations: $\begin{cases} A+B+C=4 \\ -A-7B=-13 \\ -12A+12B-9C=9 \end{cases} \Rightarrow \begin{bmatrix} 1 & 1 & 1 & 4 \\ -1 & -7 & 0 & -13 \\ -12 & 12 & -9 & 9 \end{bmatrix} \xrightarrow[12R_1+R_3 \to R_3]{R_1+R_2 \to R_2}$

$\begin{bmatrix} 1 & 1 & 1 & 4 \\ 0 & -6 & 1 & -9 \\ 0 & 24 & 3 & 57 \end{bmatrix} \xrightarrow{4R_2+R_3 \to R_3} \begin{bmatrix} 1 & 1 & 1 & 4 \\ 0 & -6 & 1 & -9 \\ 0 & 0 & 7 & 21 \end{bmatrix}$. So $7C=21 \Rightarrow C=3 \Leftrightarrow -6B+3=-9 \Rightarrow B=-2 \Leftrightarrow$ $A-2+3=4 \Rightarrow A=3$. Therefore, $\dfrac{x^4-4x^3-13x^2+49x-9}{x^3-4x^2-9x+36} = x - \dfrac{2}{x+3} + \dfrac{1}{x-3} - \dfrac{3}{x-4}$.

CHAPTER 10 REVIEW

1. Solve the system of equations $\begin{cases} y=4x+4 \\ x=3y+10 \end{cases}$ and graph the lines.

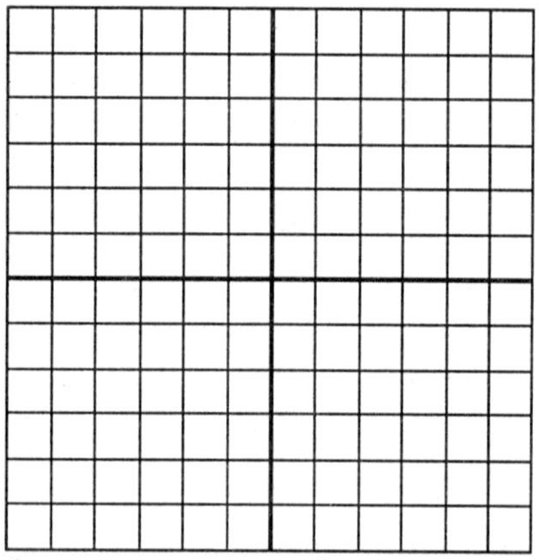

Answer: easy

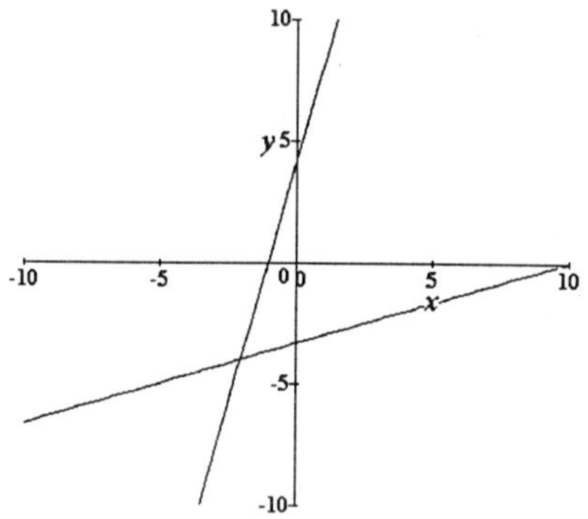

$\begin{cases} y=4x+4 \\ x=3y+10 \end{cases} \Leftrightarrow \begin{cases} 4x-y=-4 \\ x-3y=10 \end{cases} \Leftrightarrow \begin{cases} 4x-y=-4 \\ 4x-12y=40 \end{cases}$. Thus, $11y=-44 \Leftrightarrow y=-4$, and so $x=-2$. Hence the solution is $(-2,-4)$.

2. Solve the system of equations $\begin{cases} 4x-5y=12 \\ -\frac{8}{3}x+\frac{10}{3}y=-2 \end{cases}$ and graph the lines.

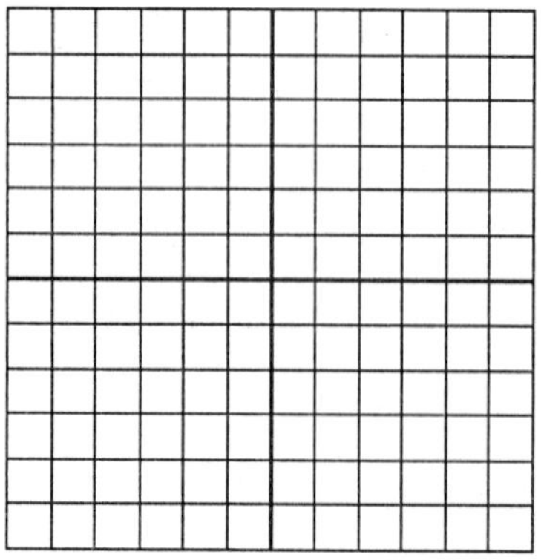

Answer: easy

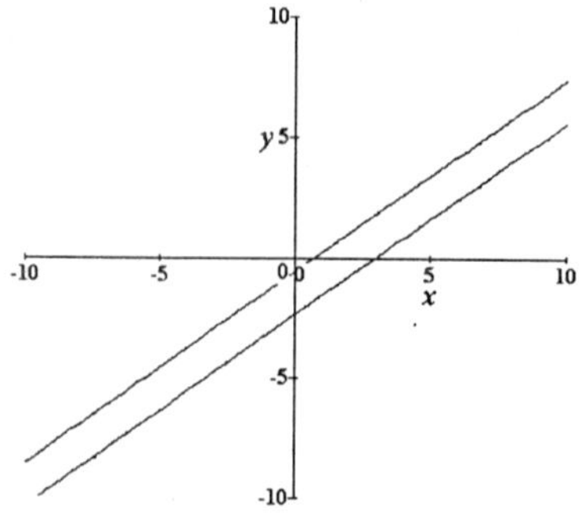

$\begin{cases} 4x-5y=12 \\ -\frac{8}{3}x+\frac{10}{3}y=-2 \end{cases} \Leftrightarrow \begin{cases} 4x-5y=12 \\ 4x-5y=3 \end{cases}$. Since $3 \neq 12$, there is no solution, and the lines are parallel.

3. Solve the system of equations $\begin{cases} 3x + 4y = 4 \\ -x + 3y = 1 \\ 14x - 3y = 12 \end{cases}$ and graph the lines.

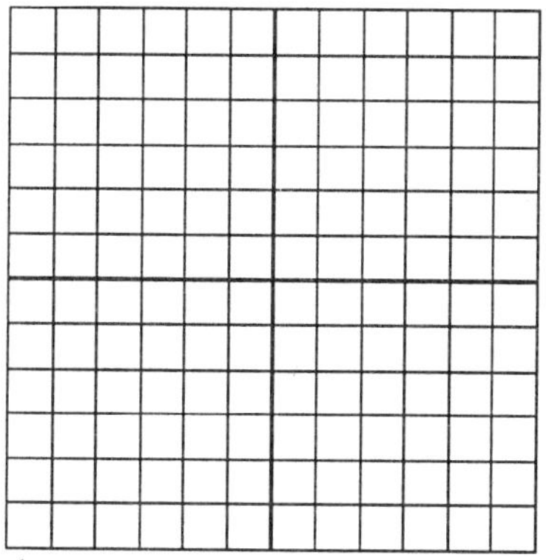

Answer: easy

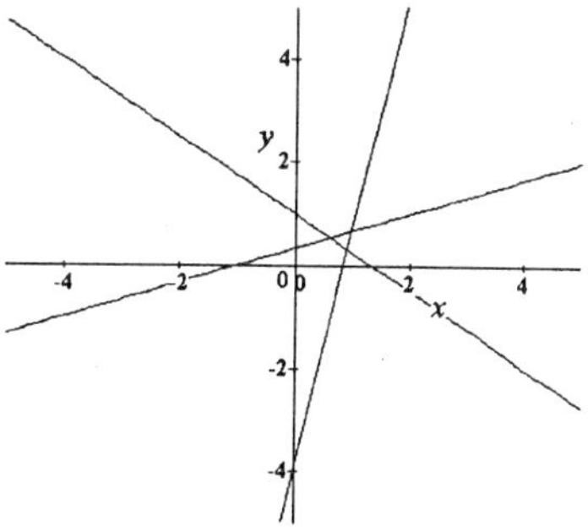

$\begin{cases} 3x + 4y = 4 \\ -x + 3y = 1 \\ 14x - 3y = 12 \end{cases} \Leftrightarrow \begin{cases} 13x = 18 \\ -x + 3y = 1 \\ 13x = 13 \end{cases} \Leftrightarrow \begin{cases} x = \frac{8}{13} \\ -x + 3y = 1 \\ x = 1 \end{cases}$ Thus there is no solution, and so the lines do not all intersect at one point.

4. Find the complete solution of the system $\begin{cases} 2x - y + 3z = 2 \\ 4x - 9z = 2 \\ 3x + 2y + 6z = 18 \end{cases}$ using Gaussian elimination, or show that there is no solution.

(a) $\left(1, 3, \frac{1}{3}\right)$ (b) $\left(2, 4, \frac{2}{3}\right)$ (c) $(3, 6, 2)$ (d) $\left(\frac{4}{3}, \frac{4}{3}, 3\right)$ (e) No solution

Answer: medium (b)

$\begin{bmatrix} 2 & -1 & 3 & 2 \\ 4 & 0 & -9 & 2 \\ 3 & 2 & 6 & 18 \end{bmatrix} \quad \begin{matrix} R_2 - 2R_1 \to R_2 \\ 2R_3 - 3R_1 \to R_3 \end{matrix} \quad \begin{bmatrix} 2 & -1 & 3 & 2 \\ 0 & 2 & -15 & -2 \\ 0 & 7 & 3 & 30 \end{bmatrix} \quad 2R_3 - 7R_2 \to R_3 \quad \begin{bmatrix} 2 & -1 & 3 & 2 \\ 0 & 2 & -15 & -2 \\ 0 & 0 & 111 & 74 \end{bmatrix}$

Thus, $111z = 74 \Leftrightarrow z = \frac{74}{111} = \frac{2}{3}$, $2y - 15 \cdot \frac{2}{3} = -2 \Leftrightarrow y = 4$, and $2x - 4 + 3 \cdot \frac{2}{3} = 2 \Leftrightarrow x = 2$, and so the solution is $\left(2, 4, \frac{2}{3}\right)$.

5. Find the complete solution of the system $\begin{cases} x - 2y + 2z - w = 2 \\ 2x - 6z = 5 \\ x + 4y - 4w = 9 \\ x - 2y + 4z - 3w = 5 \end{cases}$ using Gaussian elimination, or show that there is no solution.

(a) $\left(2, 1, 0, -\frac{1}{3}\right)$ (b) $\left(\frac{3}{2}, \frac{5}{2}, 2, 1\right)$ (c) $\left(-\frac{3}{2}, 2, 1, 0\right)$ (d) $\left(1, 0, -\frac{1}{2}, -2\right)$ (e) No solution

Answer: medium (d)

$\begin{bmatrix} 1 & -2 & 2 & -1 & 2 \\ 2 & 0 & -6 & 0 & 5 \\ 1 & 4 & 0 & -4 & 9 \\ 1 & -2 & 4 & -3 & 5 \end{bmatrix} \quad \begin{matrix} R_2 - 2R_1 \to R_2 \\ R_3 - R_1 \to R_3 \\ R_4 - R_1 \to R_4 \end{matrix} \quad \begin{bmatrix} 1 & -2 & 2 & -1 & 2 \\ 0 & 4 & -10 & 2 & 1 \\ 0 & 6 & -2 & -3 & 7 \\ 0 & 0 & 2 & -2 & 3 \end{bmatrix} \quad 2R_3 - 3R_2 \to R_3 \quad \begin{bmatrix} 1 & -2 & 2 & -1 & 2 \\ 0 & 4 & -10 & 2 & 1 \\ 0 & 0 & 26 & -12 & 11 \\ 0 & 0 & 2 & -2 & 3 \end{bmatrix}$

$R_3 - 13R_4 \to R_4 \quad \begin{bmatrix} 1 & -2 & 2 & -1 & 2 \\ 0 & 4 & -10 & 2 & 1 \\ 0 & 0 & 26 & -12 & 11 \\ 0 & 0 & 0 & 14 & -28 \end{bmatrix}$

Therefore, $14w = -28 \Leftrightarrow w = -2$, $26z + 24 = 11 \Leftrightarrow z = -\frac{1}{2}$, $4y + 5 - 4 = 1 \Leftrightarrow y = 0$, and $x - 1 + 2 = 2 \Leftrightarrow x = 1$. So, the solution is $\left(1, 0, -\frac{1}{2}, -2\right)$.

6. Find the complete solution of the system $\begin{cases} 3x+6y+4z=3 \\ 6x+10y+9z=13 \\ 3x+7z=0 \end{cases}$ using Gaussian elimination, or show that there is no solution.

(a) $(1,2,4)$ (b) $\left(3,2,\frac{1}{3}\right)$ (c) $\left(4,2,-\frac{1}{3}\right)$ (d) $\left(-\frac{1}{3},-\frac{2}{3},2\right)$ (e) No solution

Answer: easy (e)

$$\begin{bmatrix} 3 & 6 & 4 & 3 \\ 6 & 10 & 9 & 13 \\ 3 & 0 & 7 & 0 \end{bmatrix} \begin{array}{c} R_2-2R_1 \to R_2 \\ R_3-R_1 \to R_3 \end{array} \begin{bmatrix} 3 & 6 & 4 & 3 \\ 0 & -2 & 1 & 7 \\ 0 & -6 & 3 & -3 \end{bmatrix} \quad R_3-3R_2 \to R_3 \quad \begin{bmatrix} 3 & 6 & 4 & 3 \\ 0 & -2 & 1 & 7 \\ 0 & 0 & 0 & -24 \end{bmatrix}$$

Therefore the system is inconsistent and has no solution.

7. Find the complete solution of the system $\begin{cases} 2x+y-z+3w=2 \\ 2x+y-w=12 \\ 3y+2z+w=4 \\ x+2y-z=10 \end{cases}$ using Gaussian elimination, or show that there is no solution.

(a) $\left(\frac{19}{6},\frac{35}{12},-1,-\frac{11}{4}\right)$ (b) $\left(\frac{5}{6},\frac{7}{6},-2,-\frac{1}{2}\right)$ (c) $\left(-\frac{3}{2},-\frac{1}{2},\frac{1}{3},\frac{4}{3}\right)$ (d) $\left(2,1,-\frac{1}{3},\frac{4}{3}\right)$ (e) No solution

Answer: medium (a)

$$\begin{bmatrix} 2 & 1 & -1 & 3 & 2 \\ 2 & 1 & 0 & -1 & 12 \\ 0 & 3 & 2 & 1 & 4 \\ 1 & 2 & -1 & 0 & 10 \end{bmatrix} \begin{array}{c} R_2-R_1 \to R_2 \\ 2R_4-R_1 \to R_4 \end{array} \begin{bmatrix} 2 & 1 & -1 & 3 & 2 \\ 0 & 0 & 1 & -4 & 10 \\ 0 & 3 & 2 & 1 & 4 \\ 0 & 3 & -1 & -3 & 18 \end{bmatrix} \begin{array}{c} R_2 \leftrightarrow R_3 \\ R_3-R_4 \to R_4 \end{array} \begin{bmatrix} 2 & 1 & -1 & 3 & 2 \\ 0 & 3 & 2 & 1 & 4 \\ 0 & 0 & 1 & -4 & 10 \\ 0 & 0 & 3 & 4 & -14 \end{bmatrix}$$

$$R_4-3R_3 \to R_4 \quad \begin{bmatrix} 2 & 1 & -1 & 3 & 2 \\ 0 & 3 & 2 & 1 & 4 \\ 0 & 0 & 1 & -4 & 10 \\ 0 & 0 & 0 & 16 & -44 \end{bmatrix}$$

Therefore $16w=-44 \Leftrightarrow w=-\frac{11}{4}$, $z+\frac{44}{4}=10 \Leftrightarrow z=-1$, $3y-2-\frac{11}{4}=4 \Leftrightarrow y=\frac{35}{12}$, and $2x+\frac{35}{12}+1-\frac{33}{4}=2 \Leftrightarrow x=\frac{19}{6}$. Hence, the solution is $\left(\frac{19}{6},\frac{35}{12},-1,-\frac{11}{4}\right)$.

8. Find the values of a, b, and c if the parabola $y=ax^2+bx+c$ passes through the points $(5,-2)$, $(3,-6)$, and $(6,9)$.

(a) $a=1, b=4, c=-12$ (b) $a=2, b=14, c=-12$ (c) $a=3, b=-22, c=33$
(d) $a=-2, b=24, c=-12$ (e) $a=1, b=12, c=-12$

Answer: medium (c)

Since the parabola $y=ax^2+bx+c$ passes through the points $(5,-2)$, $(3,-6)$, and $(6,9)$, we substitute these points into the equation of the parabola and get the system:

$\begin{cases} 25a+5b+c=-2 \\ 9a+3b+c=-6 \\ 36a+6b+c=9 \end{cases}$. This has the matrix representation $\begin{bmatrix} 25 & 5 & 1 & -2 \\ 9 & 3 & 1 & -6 \\ 36 & 6 & 1 & 9 \end{bmatrix}$ $\begin{array}{c} 25R_2-9R_1 \to R_2 \\ 25R_3-36R_1 \to R_3 \end{array}$

$$\begin{bmatrix} 25 & 5 & 1 & -2 \\ 0 & 30 & 16 & -132 \\ 0 & -30 & -11 & 297 \end{bmatrix} \quad R_3+R_2 \to R_3 \quad \begin{bmatrix} 25 & 5 & 1 & -2 \\ 0 & 30 & 16 & -132 \\ 0 & 0 & 5 & 165 \end{bmatrix}$$

Therefore, $5c=165 \Leftrightarrow c=33$, $30b+16\cdot 33=-132 \Leftrightarrow b=-22$, and $25a-5\cdot 22+33=-2 \Leftrightarrow a=3$. So the solution is $(a,b,c)=(3,-22,33)$ and the equation of the parabola is $y=3x^2-22x+33$.

9. Carry out the operation $C-D$, or explain why it cannot be performed.

$C=\begin{bmatrix} 1 & 2 \\ \frac{3}{2} & 1 \\ -1 & 2 \end{bmatrix}$ and $D=\begin{bmatrix} 1 & 0 \\ -2 & 3 \\ 1 & -1 \end{bmatrix}$.

(a) $\begin{bmatrix} 1 & 3 \\ \frac{5}{2} & -1 \\ 3 & -2 \end{bmatrix}$ (b) $\begin{bmatrix} 0 & 1 \\ 3 & -\frac{2}{3} \\ -4 & 2 \end{bmatrix}$ (c) $\begin{bmatrix} 0 & 2 \\ \frac{7}{2} & -2 \\ -2 & 3 \end{bmatrix}$ (d) $\begin{bmatrix} 0 & 1 \\ 3 & -1 \\ -1 & 2 \end{bmatrix}$ (e) Incompatible dimensions

Answer: easy (c)

$$C-D=\begin{bmatrix} 1 & 2 \\ \frac{3}{2} & 1 \\ -1 & 2 \end{bmatrix}-\begin{bmatrix} 1 & 0 \\ -2 & 3 \\ 1 & -1 \end{bmatrix}=\begin{bmatrix} 0 & 2 \\ \frac{7}{2} & -2 \\ -2 & 3 \end{bmatrix}$$

10. Carry out the operation $5B - 2C$, or explain why it cannot be performed.

$$B = \begin{bmatrix} 0 & 1 & 3 \\ -1 & 2 & 1 \end{bmatrix} \text{ and } C = \begin{bmatrix} 1 & 2 \\ \frac{3}{2} & 1 \\ -1 & 2 \end{bmatrix}.$$

(a) $\begin{bmatrix} -1 & 1 & 2 \\ -2 & 1 & 0 \end{bmatrix}$ (b) $\begin{bmatrix} 0 & 1 \\ -1 & 1 \end{bmatrix}$ (c) $\begin{bmatrix} 2 & -1 \\ \frac{5}{2} & 2 \\ -2 & 1 \end{bmatrix}$ (d) $\begin{bmatrix} 1 & -2 & -3 \\ -\frac{3}{2} & \frac{1}{2} & -1 \\ -2 & 1 & 0 \end{bmatrix}$ (e) Incompatible dimensions

Answer: easy (e)

$5B - 2C$ cannot be performed because the matrix dimensions (2×3 and 3×2) are not compatible.

11. Carry out the operation AG, or explain why it cannot be performed.

$A = [1 \quad 0 \quad -2]$ and $G = [3]$.

(a) $[3 \quad 0 \quad -3]$ (b) $[-3]$ (c) $[1 \quad 0 \quad -3]$ (d) $[3 \quad 0 \quad -6]$ (e) Incompatible dimensions

Answer: easy (e)

AG cannot be performed because the matrix dimensions (1×3 and 1×1) are not compatible.

12. Carry out the operation CB, or explain why it cannot be performed.

$$B = \begin{bmatrix} 1 & 2 & -3 \\ 0 & 2 & 1 \end{bmatrix} \text{ and } C = \begin{bmatrix} 1 & 2 \\ \frac{3}{2} & 1 \\ -1 & 2 \end{bmatrix}.$$

(a) $\begin{bmatrix} 1 & 6 & -1 \\ \frac{3}{2} & 5 & -\frac{7}{2} \\ -1 & 2 & 5 \end{bmatrix}$ (b) $\begin{bmatrix} 2 & 1 & 3 \\ -\frac{1}{2} & 3 & -\frac{5}{2} \\ 1 & -3 & 4 \end{bmatrix}$ (c) $\begin{bmatrix} 0 & 3 & -2 \\ -1 & \frac{5}{2} & -\frac{3}{2} \\ -2 & -1 & 3 \end{bmatrix}$ (d) $\begin{bmatrix} 1 & 0 & 3 \\ 2 & 0 & -3 \\ -\frac{1}{2} & -\frac{3}{2} & -\frac{5}{2} \end{bmatrix}$ (e) Incompatible dimensions

Answer: easy (a)

$$CB = \begin{bmatrix} 1 & 2 \\ \frac{3}{2} & 1 \\ -1 & 2 \end{bmatrix} \begin{bmatrix} 1 & 2 & -3 \\ 0 & 2 & 1 \end{bmatrix} = \begin{bmatrix} 1 & 6 & -1 \\ \frac{3}{2} & 5 & -\frac{7}{2} \\ -1 & 2 & 5 \end{bmatrix}$$

13. Carry out the operation FC, or explain why it cannot be performed.

$$C = \begin{bmatrix} 1 & 2 \\ \frac{3}{2} & 1 \\ -1 & 2 \end{bmatrix} \text{ and } F = \begin{bmatrix} -2 & 1 & 0 \\ -1 & 1 & 3 \\ 2 & 3 & 1 \end{bmatrix}.$$

(a) $\begin{bmatrix} -2 & -1 \\ 2 & \frac{5}{2} \\ \frac{3}{2} & 4 \end{bmatrix}$ (b) $\begin{bmatrix} \frac{1}{2} & -1 \\ \frac{3}{2} & \frac{1}{2} \\ -2 & -3 \end{bmatrix}$ (c) $\begin{bmatrix} 1 & 2 \\ -1 & -3 \\ \frac{7}{2} & 2 \end{bmatrix}$ (d) $\begin{bmatrix} -\frac{1}{2} & -3 \\ -\frac{5}{2} & 5 \\ \frac{11}{2} & 9 \end{bmatrix}$ (e) Incompatible dimensions

Answer: easy (d)

$$FC = \begin{bmatrix} -2 & 1 & 0 \\ -1 & 1 & 3 \\ 2 & 3 & 1 \end{bmatrix} \begin{bmatrix} 1 & 2 \\ \frac{3}{2} & 1 \\ -1 & 2 \end{bmatrix} = \begin{bmatrix} -\frac{1}{2} & -3 \\ -\frac{5}{2} & 5 \\ \frac{11}{2} & 9 \end{bmatrix}$$

14. Carry out the operation $F(2C - D)$, or explain why it cannot be performed.

$$C = \begin{bmatrix} 1 & 2 \\ \frac{3}{2} & 1 \\ -1 & 2 \end{bmatrix}, D = \begin{bmatrix} 1 & 0 \\ -2 & 3 \\ 1 & -1 \end{bmatrix} \text{ and } F = \begin{bmatrix} -2 & 1 & 0 \\ -1 & 1 & 3 \\ 2 & 3 & 1 \end{bmatrix}.$$

(a) $\begin{bmatrix} 2 & -4 \\ 3 & -5 \\ 2 & 11 \end{bmatrix}$ (b) $\begin{bmatrix} 3 & -9 \\ -5 & 10 \\ 14 & 10 \end{bmatrix}$ (c) $\begin{bmatrix} -2 & -4 \\ 3 & -2 \\ \frac{15}{2} & 13 \end{bmatrix}$ (d) $\begin{bmatrix} 1 & 0 \\ -1 & -2 \\ 2 & 14 \end{bmatrix}$ (e) Incompatible dimensions

Answer: medium (b)

$$F(2C - D) = \begin{bmatrix} -2 & 1 & 0 \\ -1 & 1 & 3 \\ 2 & 3 & 1 \end{bmatrix} \cdot \left(\begin{bmatrix} 2 & 4 \\ 3 & 2 \\ -2 & 4 \end{bmatrix} - \begin{bmatrix} 1 & 0 \\ -2 & 3 \\ 1 & -1 \end{bmatrix} \right) = \begin{bmatrix} -2 & 1 & 0 \\ -1 & 1 & 3 \\ 2 & 3 & 1 \end{bmatrix} \begin{bmatrix} 1 & 4 \\ 5 & -1 \\ -3 & 5 \end{bmatrix} = \begin{bmatrix} 3 & -9 \\ -5 & 10 \\ 14 & 10 \end{bmatrix}$$

15. Find the determinant and, if possible, the inverse matrix of $\begin{bmatrix} 1 & 2 \\ 3 & -4 \end{bmatrix}$.

(a) $\begin{bmatrix} \frac{2}{5} & \frac{1}{5} \\ \frac{3}{10} & -\frac{1}{10} \end{bmatrix}$ (b) $\begin{bmatrix} \frac{2}{3} & \frac{1}{9} \\ \frac{4}{5} & -\frac{2}{5} \end{bmatrix}$ (c) $\begin{bmatrix} 1 & -2 \\ -3 & 1 \end{bmatrix}$ (d) $\begin{bmatrix} \frac{1}{5} & \frac{3}{5} \\ \frac{1}{12} & -\frac{4}{5} \end{bmatrix}$ (e) $|D| = 0$, no inverse

Answer: easy (a)

$D = \begin{bmatrix} 1 & 2 \\ 3 & -4 \end{bmatrix}$. Then, $|D| = 1(-4) - 2 \cdot 3 = -10$, and so $D^{-1} = -\frac{1}{10}\begin{bmatrix} -4 & -2 \\ -3 & 1 \end{bmatrix} = \begin{bmatrix} \frac{2}{5} & \frac{1}{5} \\ \frac{3}{10} & -\frac{1}{10} \end{bmatrix}$.

16. Find the determinant and, if possible, the inverse matrix of $\begin{bmatrix} 1 & 2 & 0 \\ -1 & 1 & 2 \\ 0 & 3 & 2 \end{bmatrix}$.

(a) $\frac{1}{12}\begin{bmatrix} -1 & 1 & 3 \\ 0 & -2 & 1 \\ -1 & 3 & 4 \end{bmatrix}$ (b) $\frac{1}{6}\begin{bmatrix} -2 & 1 & 4 \\ 5 & -5 & 11 \\ 2 & 0 & 1 \end{bmatrix}$ (c) $\begin{bmatrix} 2 & 1 & 2 \\ 0 & -1 & 1 \\ -2 & 2 & 1 \end{bmatrix}$ (d) $\begin{bmatrix} 4 & 2 & 3 \\ 3 & -2 & 1 \\ 1 & -2 & 2 \end{bmatrix}$ (e) $|D| = 0$, no inverse

Answer: medium (e)

$D = \begin{bmatrix} 1 & 2 & 0 \\ -1 & 1 & 2 \\ 0 & 3 & 2 \end{bmatrix}$. Then, $|D| = \begin{vmatrix} 1 & 2 \\ 3 & 2 \end{vmatrix} - 2\begin{vmatrix} -1 & 2 \\ 0 & 2 \end{vmatrix} = (2-6) - 2(-2) = 0$, and so D has no inverse.

17. Find the determinant and, if possible, the inverse matrix of $\begin{bmatrix} 2 & 0 & 0 & 2 \\ 0 & 3 & 0 & 3 \\ 0 & 0 & 4 & 4 \\ 0 & 0 & 0 & 5 \end{bmatrix}$.

Answer: medium

$D = \begin{bmatrix} 2 & 0 & 0 & 2 \\ 0 & 3 & 0 & 3 \\ 0 & 0 & 4 & 4 \\ 0 & 0 & 0 & 5 \end{bmatrix}$. Thus, $|D| = 2\begin{vmatrix} 3 & 0 & 3 \\ 0 & 4 & 4 \\ 0 & 0 & 5 \end{vmatrix} = 6\begin{vmatrix} 4 & 4 \\ 0 & 5 \end{vmatrix} = 120$.

$\begin{bmatrix} 2 & 0 & 0 & 2 & 1 & 0 & 0 & 0 \\ 0 & 3 & 0 & 3 & 0 & 1 & 0 & 0 \\ 0 & 0 & 4 & 4 & 0 & 0 & 1 & 0 \\ 0 & 0 & 0 & 5 & 0 & 0 & 0 & 1 \end{bmatrix}$ $\begin{matrix} \frac{1}{2}R_1 \to R_1 \\ \frac{1}{3}R_2 \to R_2 \\ \frac{1}{4}R_3 \to R_3 \\ \frac{1}{5}R_4 \to R_4 \end{matrix}$ $\begin{bmatrix} 1 & 0 & 0 & 1 & \frac{1}{2} & 0 & 0 & 0 \\ 0 & 1 & 0 & 1 & 0 & \frac{1}{3} & 0 & 0 \\ 0 & 0 & 1 & 1 & 0 & 0 & \frac{1}{4} & 0 \\ 0 & 0 & 0 & 1 & 0 & 0 & 0 & \frac{1}{5} \end{bmatrix}$ $\begin{matrix} R_1 - R_4 \to R_1 \\ R_2 - R_4 \to R_2 \\ R_3 - R_4 \to R_3 \end{matrix}$

$\begin{bmatrix} 1 & 0 & 0 & 0 & \frac{1}{2} & 0 & 0 & -\frac{1}{5} \\ 0 & 1 & 0 & 0 & 0 & \frac{1}{3} & 0 & -\frac{1}{5} \\ 0 & 0 & 1 & 0 & 0 & 0 & \frac{1}{4} & -\frac{1}{5} \\ 0 & 0 & 0 & 1 & 0 & 0 & 0 & \frac{1}{5} \end{bmatrix}$. Therefore $D^{-1} = \begin{bmatrix} \frac{1}{2} & 0 & 0 & -\frac{1}{5} \\ 0 & \frac{1}{3} & 0 & -\frac{1}{5} \\ 0 & 0 & \frac{1}{4} & -\frac{1}{5} \\ 0 & 0 & 0 & \frac{1}{5} \end{bmatrix}$.

18. Express the system of linear equations $\begin{cases} x + 2y + 4z = \frac{1}{3} \\ 2x + y + 3z = 1 \\ x + 2z = \frac{1}{3} \end{cases}$ as a matrix equation. Then solve the matrix equation by multiplying both sides by the inverse of the coefficient matrix.

Answer: hard

$\begin{bmatrix} 1 & 2 & 4 \\ 2 & 1 & 3 \\ 1 & 0 & 2 \end{bmatrix}\begin{bmatrix} x \\ y \\ z \end{bmatrix} = \begin{bmatrix} \frac{1}{3} \\ 1 \\ \frac{1}{3} \end{bmatrix}$. Let $A = \begin{bmatrix} 1 & 2 & 4 \\ 2 & 1 & 3 \\ 1 & 0 & 2 \end{bmatrix}$. Then $\begin{bmatrix} 1 & 2 & 4 & 1 & 0 & 0 \\ 2 & 1 & 3 & 0 & 1 & 0 \\ 1 & 0 & 2 & 0 & 0 & 1 \end{bmatrix}$ $\begin{matrix} 2R_1 - R_2 \to R_2 \\ R_1 - R_3 \to R_3 \end{matrix}$

$\begin{bmatrix} 1 & 2 & 4 & 1 & 0 & 0 \\ 0 & 3 & 5 & 2 & -1 & 0 \\ 0 & 2 & 2 & 1 & 0 & -1 \end{bmatrix}$ $\begin{matrix} R_1 - R_3 \to R_1 \\ 2R_2 - 3R_3 \to R_3 \end{matrix}$ $\begin{bmatrix} 1 & 0 & 2 & 0 & 0 & 1 \\ 0 & 3 & 5 & 2 & -1 & 0 \\ 0 & 0 & 4 & 1 & -2 & 3 \end{bmatrix}$ $\begin{matrix} 4R_2 - 5R_3 \to R_2 \\ \frac{1}{4}R_3 \to R_3 \end{matrix}$

$\begin{bmatrix} 1 & 0 & 2 & 0 & 0 & 1 \\ 0 & 12 & 0 & 3 & 6 & -15 \\ 0 & 0 & 1 & \frac{1}{4} & -\frac{1}{2} & \frac{3}{4} \end{bmatrix}$ $\begin{matrix} R_1 - 2R_3 \to R_1 \\ \frac{1}{12}R_2 \to R_2 \end{matrix}$ $\begin{bmatrix} 1 & 0 & 0 & -\frac{1}{2} & 1 & -\frac{1}{2} \\ 0 & 1 & 0 & \frac{1}{4} & \frac{1}{2} & -\frac{5}{4} \\ 0 & 0 & 1 & \frac{1}{4} & -\frac{1}{2} & \frac{3}{4} \end{bmatrix}$

Hence, $A^{-1} = \frac{1}{4}\begin{bmatrix} -2 & 4 & -2 \\ 1 & 2 & -5 \\ 1 & -2 & 3 \end{bmatrix}$ and $\begin{bmatrix} x \\ y \\ z \end{bmatrix} = \frac{1}{4}\begin{bmatrix} -2 & 4 & -2 \\ 1 & 2 & -5 \\ 1 & -2 & 3 \end{bmatrix}\begin{bmatrix} \frac{1}{3} \\ 1 \\ \frac{1}{3} \end{bmatrix} = \frac{1}{4}\begin{bmatrix} \frac{8}{3} \\ \frac{2}{3} \\ -\frac{2}{3} \end{bmatrix} = \begin{bmatrix} \frac{2}{3} \\ \frac{1}{6} \\ -\frac{1}{6} \end{bmatrix}$,

and so the solution is $\left(\frac{2}{3}, \frac{1}{6}, -\frac{1}{6}\right)$.

19. Solve the system $\begin{cases} 3x - 4y = 12 \\ 6x + 3y = 14 \end{cases}$ using Cramer's Rule.

Answer: easy

$|D| = \begin{vmatrix} 3 & -4 \\ 6 & 3 \end{vmatrix} = 33$, $|D_x| = \begin{vmatrix} 12 & -4 \\ 14 & 3 \end{vmatrix} = 92$, and $|D_y| = \begin{vmatrix} 3 & 12 \\ 6 & 14 \end{vmatrix} = -30$. Therefore, $x = \frac{92}{33}$ and $y = \frac{-30}{33} = -\frac{10}{11}$, and so the solution is $\left(\frac{92}{33}, -\frac{10}{11}\right)$.

20. Solve the system $\begin{cases} 2x + 3y - z = 2 \\ 4x - 2z = 3 \\ 2x + y + 3z = 4 \end{cases}$ using Cramer's Rule.

Answer: medium

$$|D| = \begin{vmatrix} 2 & 3 & -1 \\ 4 & 0 & -2 \\ 2 & 1 & 3 \end{vmatrix} = 2\begin{vmatrix} 0 & -2 \\ 1 & 3 \end{vmatrix} - 3\begin{vmatrix} 4 & -2 \\ 2 & 3 \end{vmatrix} - \begin{vmatrix} 4 & 0 \\ 2 & 1 \end{vmatrix} = 4 - 48 - 4 = -48,$$

$$|D_x| = \begin{vmatrix} 2 & 3 & -1 \\ 3 & 0 & -2 \\ 4 & 1 & 3 \end{vmatrix} = 2\begin{vmatrix} 0 & -2 \\ 1 & 3 \end{vmatrix} - 3\begin{vmatrix} 3 & -2 \\ 4 & 3 \end{vmatrix} - \begin{vmatrix} 3 & 0 \\ 4 & 1 \end{vmatrix} = 4 - 51 - 3 = -50,$$

$$|D_y| = \begin{vmatrix} 2 & 2 & -1 \\ 4 & 3 & -2 \\ 2 & 4 & 3 \end{vmatrix} = 2\begin{vmatrix} 3 & -2 \\ 4 & 3 \end{vmatrix} - 2\begin{vmatrix} 4 & -2 \\ 2 & 3 \end{vmatrix} - \begin{vmatrix} 4 & 3 \\ 2 & 4 \end{vmatrix} = 34 - 32 - 10 = -8, \text{ and}$$

$$|D_z| = \begin{vmatrix} 2 & 3 & 2 \\ 4 & 0 & 3 \\ 2 & 1 & 4 \end{vmatrix} = 2\begin{vmatrix} 0 & 3 \\ 1 & 4 \end{vmatrix} - 3\begin{vmatrix} 4 & 3 \\ 2 & 4 \end{vmatrix} + 2\begin{vmatrix} 4 & 0 \\ 2 & 1 \end{vmatrix} = -6 - 30 + 8 = -28.$$

Therefore $x = \frac{-50}{-48} = \frac{25}{24}$, $y = \frac{-8}{-48} = \frac{1}{6}$, and $z = \frac{-28}{-48} = \frac{7}{12}$, and so the solution is $\left(\frac{25}{24}, \frac{1}{6}, \frac{7}{12}\right)$.

21. Find all solutions of the system of equations $\begin{cases} x^2 + y^2 = 4 \\ x^2 + 3y^2 - 6y = 0 \end{cases}$.

Answer: medium

$\begin{cases} x^2 + y^2 = 4 \\ x^2 + 3y^2 - 6y = 0 \end{cases} \Leftrightarrow \begin{cases} x^2 = 4 - y^2 \\ x^2 + 3y^2 - 6y = 0 \end{cases}$. By substitution, $4 - y^2 + 3y^2 - 6y = 0 \Leftrightarrow 2y^2 - 6y + 4 = 0 \Leftrightarrow y^2 - 3y + 2 = 0 \Leftrightarrow y = 1$ or $y = 2$. If $y = 1$, then $x^2 = 3 \Leftrightarrow x = \pm\sqrt{3}$. If $y = 2$, $x^2 = 0 \Leftrightarrow x = 0$. Hence the solutions are $\left(\pm\sqrt{3}, 1\right)$ and $(0, 2)$.

22. Find all solutions of the system of equations $\begin{cases} x^2 + yz = 0 \\ y^2 + xz = 0 \\ x^2 + xy + y^2 = 6 \end{cases}$.

Answer: medium

$\begin{cases} x^2 + yz = 0 \\ y^2 + xz = 0 \\ x^2 + xy + y^2 = 6 \end{cases} \Leftrightarrow \begin{cases} x^3 + xyz = 0 \\ y^3 + xyz = 0 \\ x^2 + xy + y^2 = 6 \end{cases}$. From the first two equations, $x^3 = y^3 \Leftrightarrow x = y$, and substituting for y into the third equation gives $x^2 + x^2 + x^2 = 6 \Rightarrow x = \pm\sqrt{2}$. If $x = y = \sqrt{2}$, then $2 + \sqrt{2}z = 0 \Rightarrow z = -\sqrt{2}$. If $x = y = -\sqrt{2}$, then $2 - \sqrt{2}z = 0 \Rightarrow z = \sqrt{2}$. Therefore, the solutions are $\left(\sqrt{2}, \sqrt{2}, -\sqrt{2}\right)$ and $\left(-\sqrt{2}, -\sqrt{2}, \sqrt{2}\right)$.

23. Graph the solution set of the system of inequalities $\begin{cases} y - x^2 \geq 2 \\ y \leq 18 \end{cases}$. Find the coordinates of all vertices, and determine whether the solution set is bounded or unbounded.

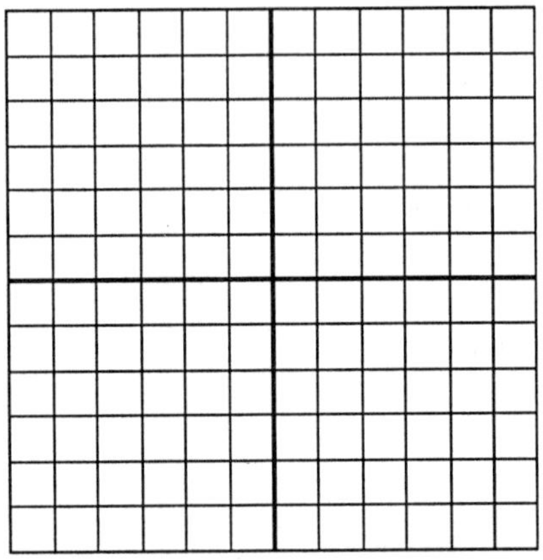

Answer: medium

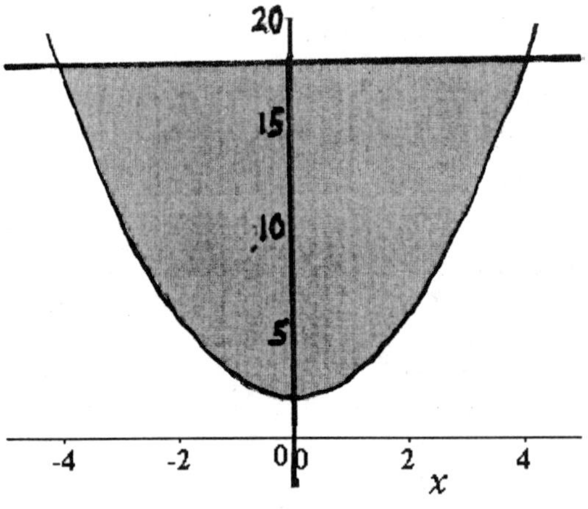

$\begin{cases} y - x^2 \geq 2 \\ y \leq 18 \end{cases}$. The vertices occur where $\begin{cases} y - x^2 = 2 \\ y = 18 \end{cases}$. By substitution, $x^2 + 2 = 18 \quad \Leftrightarrow \quad x^2 = 16 \quad \Leftrightarrow \quad x = \pm 4$, and $y = 18$. Thus the vertices occur at $(\pm 4, 18)$ and the solution set is bounded.

24. Graph the solution set of the system of inequalities $\begin{cases} x \geq 2 \\ x + y \geq 12 \\ x \leq y + 6 \end{cases}$. Find the coordinates of all vertices, and determine whether the solution set is bounded or unbounded.

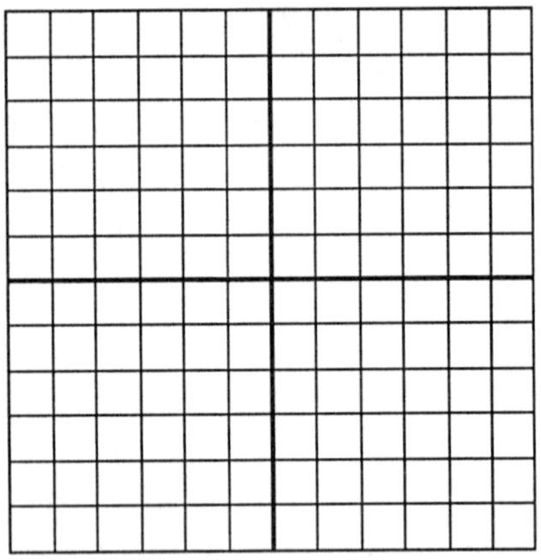

Answer: medium

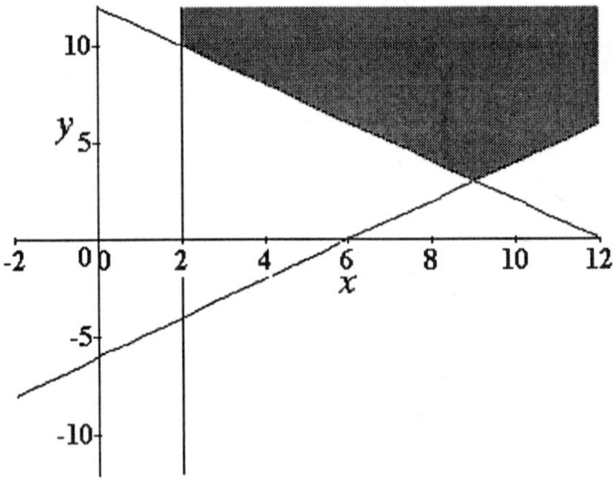

$\begin{cases} x \geq 2 \\ x + y \geq 12 \\ x \leq y + 6 \end{cases}$. The lines $x + y = 12$ and $x = y + 6$ intersect at the point $(9, 3)$. Hence, the vertices are $(9, 3)$ and $(2, 10)$.
[The point $(2, -4)$ is not in the solution set.] The solution set is not bounded.

25. For the system of equations $\begin{cases} ax + by + cz = a - b + c \\ bx + by + cz = c \\ cx + cy + cz = c \end{cases}$ with $a \neq b$, $b \neq c$, and $c \neq 0$ solve for x, y, and z in terms of a, b, and c.

Answer: medium

$\begin{cases} ax + by + cz = a - b + c \\ bx + by + cz = c \\ cx + cy + cz = c \end{cases}$ Subtracting the second equation from the first gives $(a - b)x = a - b \quad \Leftrightarrow \quad x = 1$.

Subtracting the third equation from the second, $(b - c)x + (b - c)y = 0 \quad \Leftrightarrow \quad y = -x = -1$. Thus $c - c + cz = c \quad \Leftrightarrow$ $z = 1$, and the solution is $(1, -1, 1)$.

26. For what value of k does the system $\begin{cases} kx - y + 2z = 0 \\ 2x + y + kz = 0 \\ -x + 4z = 0 \end{cases}$ have infinitely many solutions?

Answer: hard

$$\begin{bmatrix} k & -1 & 2 & 0 \\ 2 & 1 & k & 0 \\ -1 & 0 & 4 & 0 \end{bmatrix} \xrightarrow{R_1 \leftrightarrow R_2} \begin{bmatrix} 2 & 1 & k & 0 \\ k & -1 & 2 & 0 \\ -1 & 0 & 4 & 0 \end{bmatrix} \xrightarrow[2R_3 + R_1]{2R_2 - kR_1 \to R_2} \begin{bmatrix} 2 & 1 & k & 0 \\ 0 & -2-k & 4-k^2 & 0 \\ 0 & 1 & k+8 & 0 \end{bmatrix} \xrightarrow[R_2 \leftrightarrow R_3]{\frac{1}{2}R_1 \to R_1}$$

$$\begin{bmatrix} 1 & \frac{1}{2} & \frac{1}{2}k & 0 \\ 0 & 1 & k+8 & 0 \\ 0 & -2-k & 4-k^2 & 0 \end{bmatrix} \xrightarrow{R_3 + (k+2)R_2 \to R_3} \begin{bmatrix} 1 & \frac{1}{2} & \frac{1}{2}k & 0 \\ 0 & 1 & k+8 & 0 \\ 0 & 0 & 4-k^2+(k+2)(k+8) & 0 \end{bmatrix}$$

This system has infinitely many solutions when $4 - k^2 + (k+2)(k+8) = 0 \quad \Leftrightarrow \quad 4 - k^2 + k^2 + 10k + 16 = 0 \quad \Leftrightarrow$ $k = -2$.

GC 27. Use a graphing device to solve the system $\begin{cases} \sqrt{11}x - 2\sqrt{3}y = 420 \\ 563x + 3411y = 18{,}000 \end{cases}$ correct to the nearest hundredth.

Answer: medium

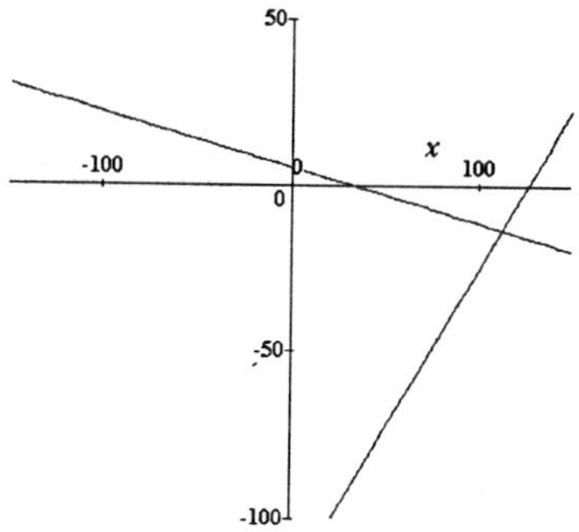

$\begin{cases} \sqrt{11}x - 2\sqrt{3}y = 420 \\ 563x + 3411y = 18{,}000 \end{cases} \Leftrightarrow \begin{cases} y = \frac{\sqrt{11}}{2\sqrt{3}}x - \frac{420}{2\sqrt{3}} \\ y = -\frac{563}{3411}x + \frac{18{,}000}{3411} \end{cases}$ has the approximate solution $(112.72, -13.33)$.

GC 28. Use a graphing device to solve the system $\begin{cases} y = 7^x + x \\ y = x^7 + 7 \end{cases}$ correct to the nearest hundredth.

Answer: medium

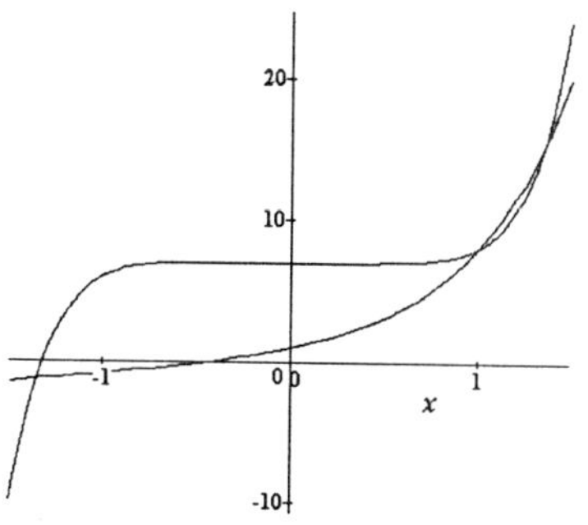

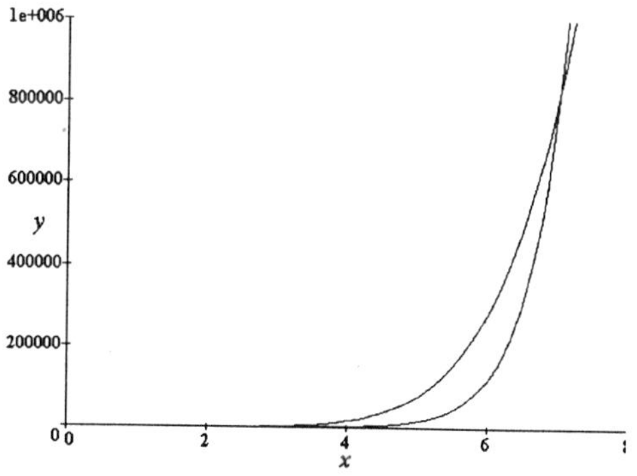

$\begin{cases} y = 7^x + x \\ y = x^7 + 7 \end{cases}$ has approximate solutions $(-1.35, -1.28)$, $(1, 8)$, $(1.35, 15.18)$, and $(7, 823550)$.

29. Find the partial fraction decomposition of the rational function $\frac{4}{x^3 - x}$.

Answer: medium

$\frac{4}{x^3 - x} = \frac{4}{x(x^2 - 1)} = \frac{4}{x(x + 1)(x - 1)} = \frac{A}{x} + \frac{B}{x - 1} + \frac{C}{x + 1}$. Then

$4 = A(x^2 - 1) + Bx(x + 1) + Cx(x - 1) = x^2(A + B + C) + x(B - C) - A$. So, $-A = 4 \quad \Leftrightarrow \quad A = -4, B - C = 0$

$\Leftrightarrow \quad B = C$, and $-4 + 2B = 0 \quad \Leftrightarrow \quad B = 2, C = 2$. Therefore $\frac{4}{x^3 - x} = -\frac{4}{x} + \frac{2}{x - 1} + \frac{2}{x + 1}$.

30. Find the partial fraction decomposition of the rational function $\frac{x + 1}{x^3 - x^2 + 2x - 2}$.

Answer: medium

$\frac{x + 1}{x^3 - x^2 + 2x - 2} = \frac{x + 1}{(x^2 + 2)(x - 1)} = \frac{Ax + B}{x^2 + 2} + \frac{C}{x - 1}$. Thus

$x + 1 = (Ax + B)(x - 1) + C(x^2 + 2) = Ax^2 - Ax + Bx - B + Cx^2 + 2C = x^2(A + C) + x(-A + B) + (-B + 2C)$,

and so $\begin{cases} A + C = 0 \\ -A + B = 1 \\ -B + 2C = 1 \end{cases}$ which has the matrix representation $\begin{bmatrix} 1 & 0 & 1 & 0 \\ -1 & 1 & 0 & 1 \\ 0 & -1 & 2 & 1 \end{bmatrix}$ $\xrightarrow{R_1 + R_2 \to R_2}$ $\begin{bmatrix} 1 & 0 & 1 & 0 \\ 0 & 1 & 1 & 1 \\ 0 & -1 & 2 & 1 \end{bmatrix}$

$\xrightarrow{R_2 + R_3 \to R_3}$ $\begin{bmatrix} 1 & 0 & 1 & 0 \\ 0 & 1 & 1 & 1 \\ 0 & 0 & 3 & 2 \end{bmatrix}$. Thus $3C = 2 \quad \Leftrightarrow \quad C = \frac{2}{3}, B + \frac{2}{3} = 1 \quad \Leftrightarrow \quad B = \frac{1}{3}$, and $A + \frac{2}{3} = 0 \quad \Leftrightarrow \quad A = -\frac{2}{3}$.

So $\frac{x + 1}{x^3 - x^2 + 2x - 2} = \frac{-\frac{2}{3}x + \frac{1}{3}}{x^2 + 2} + \frac{\frac{2}{3}}{x - 1}$.

CHAPTER ELEVEN

SECTION 11.1

1. Find the focus, directrix, and focal diameter of the parabola $x - 5y^2 = 0$. Then sketch the graph of the parabola.

Answer: medium

$x = 5y^2 \quad \Rightarrow \quad 4p = \frac{1}{5}$, so the focal diameter is $\frac{1}{5}$, the focus is $\left(\frac{1}{20}, 0\right)$ and the directrix is $x = -\frac{1}{20}$.

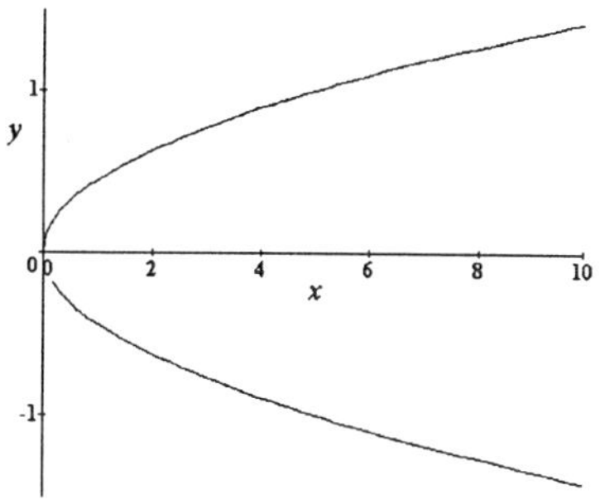

2. Find the focus, directrix, and focal diameter of the parabola $y = -4x^2$. Then sketch the graph of the parabola.

Answer: medium

$y = -4x^2 \quad \Rightarrow \quad 4p = -\frac{1}{4}$, so the focal diameter is $\frac{1}{4}$, the focus is $\left(0, -\frac{1}{16}\right)$ and the directrix is $y = \frac{1}{16}$.

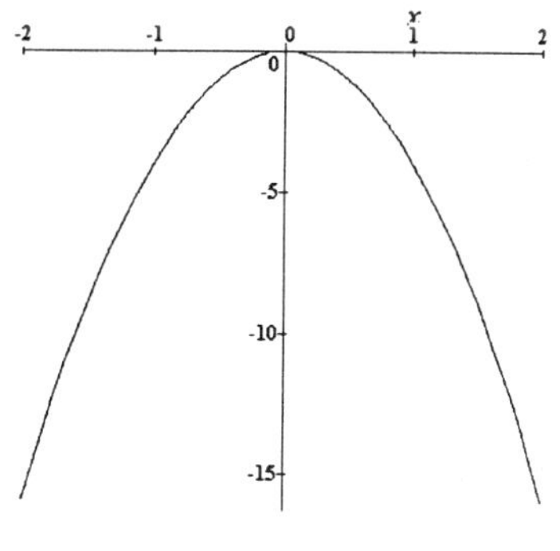

3. Find the focus, directrix, and focal diameter of the parabola $x + y^2 = 0$. Then sketch the graph of the parabola.

Answer: medium

$x + y^2 = 0 \quad \Rightarrow \quad 4p = -1$, so the focal diameter is 1, the focus is $\left(-\frac{1}{4}, 0\right)$ and the directrix is $x = \frac{1}{4}$.

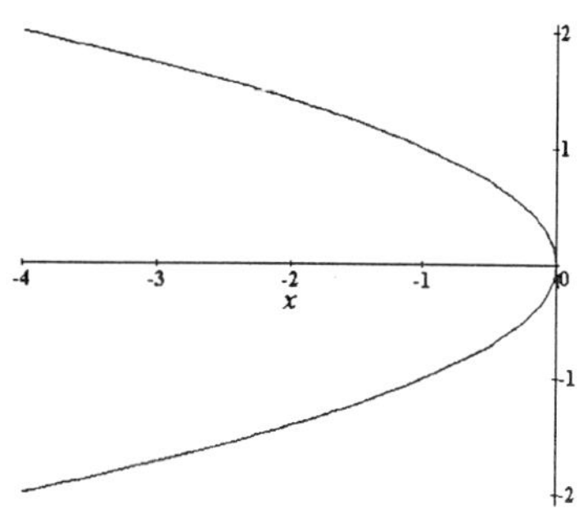

4. Find an equation for the parabola which has focus $F(0,4)$ and its vertex at the origin.
(a) $y = x^2$ (b) $y = \frac{1}{16}x^2$ (c) $y = \frac{1}{4}x^2$ (d) $x = -4y^2$ (e) $x = 2y^2$
Answer: medium (b)
$p = 4$, so the equation is $x^2 = 16y \quad \Leftrightarrow \quad y = \frac{1}{16}x^2$.

5. Find an equation for the parabola which has focus $F(7,0)$ and its vertex at the origin.
(a) $y = 4x^2$ (b) $y = 7x^2$ (c) $y = -\frac{7}{4}x^2$ (d) $x = 28y^2$ (e) $x = \frac{1}{28}y^2$
Answer: medium (e)
$p = 7$, so the equation is $y^2 = 28x \quad \Leftrightarrow \quad x = \frac{1}{28}y^2$.

6. Find an equation for the parabola which has directrix $x = 4$ and its vertex at the origin.
(a) $x = \frac{1}{12}y^2$ (b) $x = -\frac{1}{12}y^2$ (c) $x = -\frac{1}{16}y^2$ (d) $y = 12x^2$ (e) $y = -\frac{1}{12}x^2$
Answer: medium (c)
$p = -4$, so the equation is $y^2 = -16x \quad \Leftrightarrow \quad x = -\frac{1}{16}y^2$.

7. Find an equation for the parabola which has directrix $y = -2$ and its vertex at the origin.
(a) $x = 2y^2$ (b) $x = \frac{1}{2}y^2$ (c) $x = \frac{1}{4}y^2$ (d) $y = \frac{1}{8}x^2$ (e) $y = -\frac{1}{16}x^2$
Answer: medium (d)
$p = 2$, so the equation is $x^2 = 8y \quad \Leftrightarrow \quad y = \frac{1}{8}x^2$.

8. A reflector for a radiotelescope is parabolic in cross-section, with the receiver at the focus. The reflector is 20 ft wide from rim to rim, and is 4 ft deep. Find an equation for the parabola.
(a) $y = \frac{1}{25}x^2$ (b) $y = 2x^2$ (c) $y = \frac{1}{4}x^2$ (d) $y = \frac{2}{3}x^2$ (e) $y = \frac{5}{2}x^2$
Answer: medium (a)
Consider the reflector as a parabola with the form of $y = ax^2$. Hence the point $(10,4)$ is on $y = ax^2$. Thus $4 = a(10)^2 \quad \Leftrightarrow$ $a = \frac{1}{25}$. Therefore $y = \frac{1}{25}x^2$.

9. Find an equation of a parabola with horizontal axis and vertex at the origin which passes through the point $(-1,5)$.
(a) $y^2 = \frac{25}{4}x$ (b) $y^2 = -\frac{1}{25}x$ (c) $y^2 = -25x$ (d) $x^2 = 25y$ (e) $y^2 = -5x$
Answer: medium (c)
A parabola with horizontal axis has equation of the form $y^2 = 4px$. Passing through the point $(-1,5)$: $5^2 = 4p(-1)$ $\Leftrightarrow \quad 4p = -25$. Therefore an equation is $y^2 = -25x$.

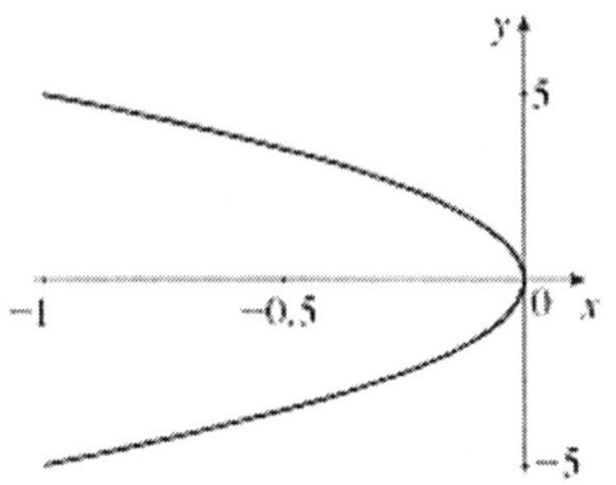

10. Find an equation of a parabola with vertex at the origin and focus on the negative x-axis, 7 units away from the directrix.
(a) $y^2 = -7x$ (b) $y^2 = -14x$ (c) $x^2 = \frac{7}{2}y$ (d) $y^2 = -28x$ (e) $x^2 = -\frac{14}{3}y$
Answer: medium (b)
The distance between the directrix and the focus is $2|p| = 7$, and the focus is on the negative x-axis, so $p < 0 \quad \Rightarrow \quad p = -\frac{7}{2}$, and the parabola has equation of the form $y^2 = 4px$, so an equation is $y^2 = 4\left(-\frac{7}{2}\right)x \quad \Rightarrow \quad y^2 = -14x$.

11. Linemen working for a local power company suspend an electric cable between two utility poles. The shape of the suspended cable is parabolic. If the poles are 324 ft apart, and the lowest point of the cable is 18 ft below the top of the poles, find the parabolic equation of the suspended cable, placing the origin of the coordinate system at the vertex.
Answer: Placing the vertex at the origin we use $x^2 = 4py$. The parabolic equation passes through the point $(162,18)$, therefore $(162)^2 = 4p(18) \quad \Leftrightarrow \quad 4p = 1458$. Therefore the parabolic equation of the suspended cable is $x^2 = 1458y$.

SECTION 11.2

1. Find the vertices, foci and eccentricity of the ellipse $\dfrac{x^2}{16}+\dfrac{y^2}{25}=1$. Determine the lengths of the major and minor axes, and sketch the graph.

Answer: medium

$\dfrac{x^2}{16}+\dfrac{y^2}{25}=1 \quad\Rightarrow\quad$ vertices $(0,\pm5)$, foci $(0,\pm3)$, eccentricity $\frac{3}{5}$, major axis has length $2a=10$, minor axis has length $2b=8$.

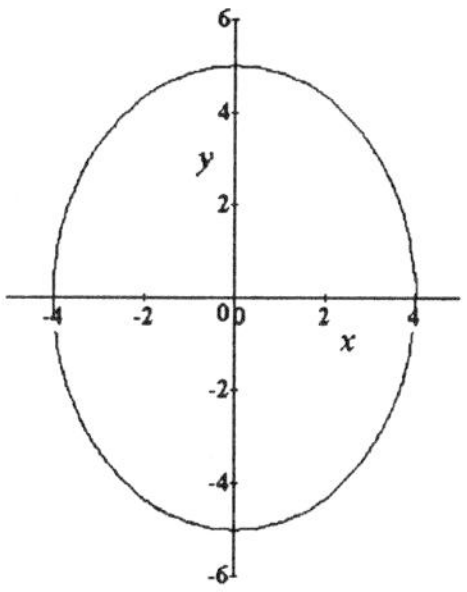

2. Find the vertices, foci and eccentricity of the ellipse $25x^2+4y^2=100$. Determine the lengths of the major and minor axes, and sketch the graph.

Answer: medium

$25x^2+4y^2=100 \quad\Rightarrow\quad \dfrac{x^2}{4}+\dfrac{y^2}{25}=1 \quad\Rightarrow$

vertices $(0,\pm5)$, foci $\left(0,\pm\sqrt{21}\right)$, eccentricity $\frac{\sqrt{21}}{5}$,

major axis has length $2a=10$, minor axis has length $2b=4$.

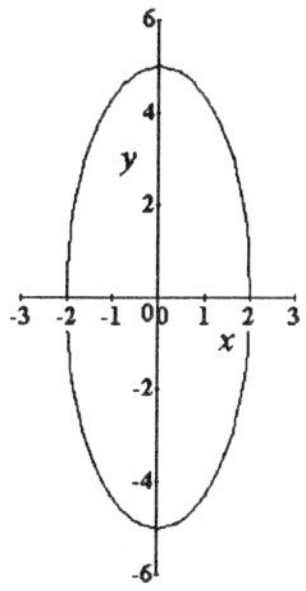

3. Find the vertices, foci and eccentricity of the ellipse $4x^2+y^2=12$. Determine the lengths of the major and minor axes, and sketch the graph.

Answer: medium

$4x^2+y^2=12 \quad\Rightarrow\quad \dfrac{x^2}{3}+\dfrac{y^2}{12}=1 \quad\Rightarrow\quad$ vertices $\left(0,\pm2\sqrt{3}\right)$,

foci $(0,\pm3)$, eccentricity $\frac{\sqrt{3}}{2}$, major axis has length $2a=4\sqrt{3}$,

minor axis has length $2b=2\sqrt{3}$.

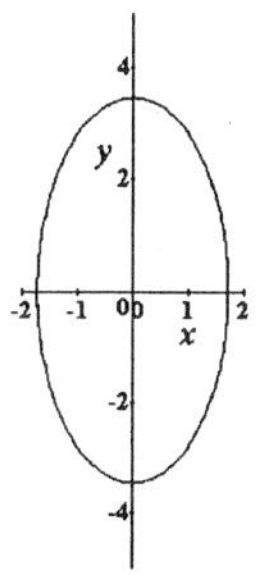

4. Find the vertices, foci and eccentricity of the ellipse $\frac{1}{8}x^2 + \frac{1}{4}y^2 = \frac{1}{4}$. Determine the lengths of the major and minor axes, and sketch the graph.

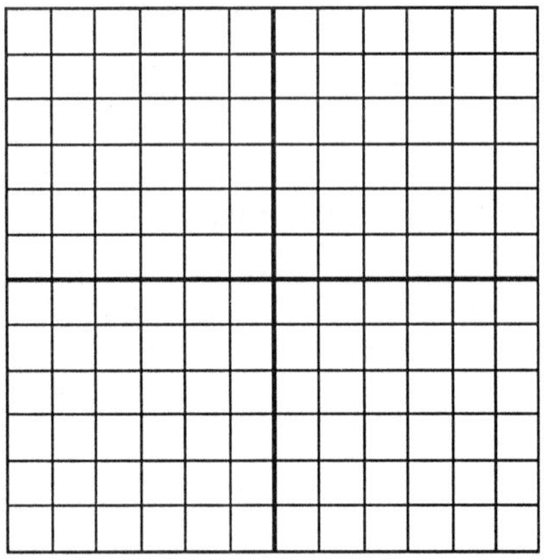

Answer: medium

$$\frac{x^2}{8} + \frac{y^2}{4} = \frac{1}{4} \quad \Rightarrow \quad \frac{x^2}{2} + y^2 = 1 \quad \Rightarrow \quad \text{vertices } \left(\pm\sqrt{2}, 0\right),$$

foci $(\pm 1, 0)$, eccentricity $\frac{\sqrt{2}}{2}$, major axis has length $2a = 2\sqrt{2}$, minor axis has length $2b = 2$.

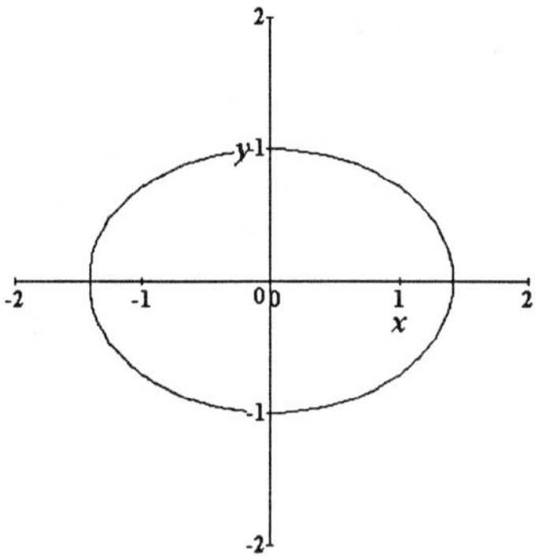

5. Find the vertices, foci and eccentricity of the ellipse $x^2 = 8 - 4y^2$. Determine the lengths of the major and minor axes, and sketch the graph.

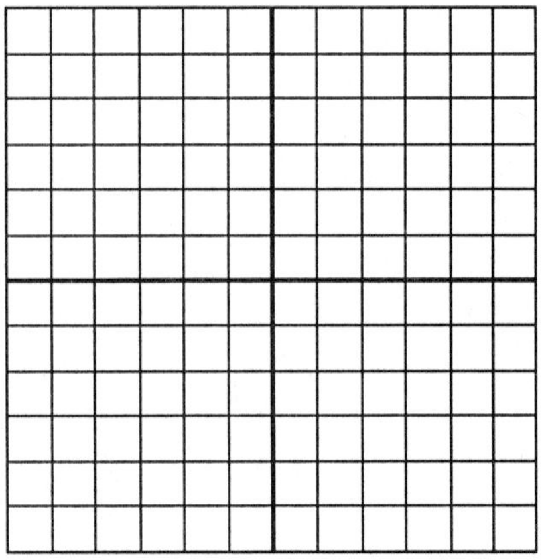

Answer: medium

$$x^2 = 8 - 4y^2 \quad \Rightarrow \quad \frac{x^2}{8} + \frac{y^2}{2} = 1 \quad \Rightarrow \quad \text{vertices } \left(\pm 2\sqrt{2}, 0\right),$$

foci $\left(\pm\sqrt{6}, 0\right)$, eccentricity $\frac{\sqrt{3}}{2}$, major axis has length $2a = 4\sqrt{2}$, minor axis has length $2b = 2\sqrt{2}$.

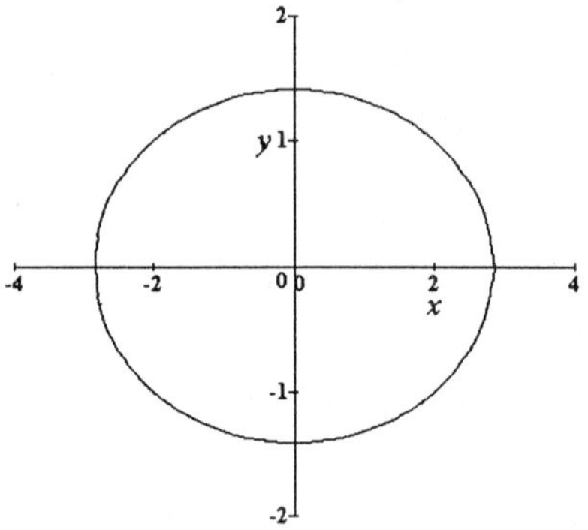

6. Find an equation for the ellipse which has foci $(\pm 2, 0)$ and vertices $(\pm 6, 0)$.

(a) $\frac{x^2}{36} + \frac{y^2}{32} = 1$ (b) $\frac{x^2}{6} + \frac{y^2}{16} = 1$ (c) $\frac{x^2}{32} + \frac{y^2}{6} = 1$ (d) $\frac{x^2}{2} + \frac{y^2}{4} = 1$

(e) $x^2 + 36y^2 = 1$

Answer: medium (a)

$c = 2$, $a = 6 \quad \Rightarrow \quad b = \sqrt{36 - 4} = 4\sqrt{2}$, so the equation is $\frac{x^2}{36} + \frac{y^2}{32} = 1$.

7. Find an equation for the ellipse which has foci $(0, \pm 4)$ and vertices $(0, \pm 8)$.

(a) $\dfrac{x^2}{12} + \dfrac{y^2}{16} = 1$ (b) $\dfrac{x^2}{12} + \dfrac{y^2}{8} = 1$ (c) $\dfrac{x^2}{48} + \dfrac{y^2}{64} = 1$ (d) $\dfrac{x^2}{2} + \dfrac{y^2}{6} = 1$

(e) $2x^2 + 12y^2 = 1$

Answer: medium (c)

$c = 4, a = 8 \quad \Rightarrow \quad b = \sqrt{64 - 16} = 4\sqrt{3}$, so the equation is $\dfrac{x^2}{48} + \dfrac{y^2}{64} = 1$.

8. Find an equation for the ellipse with major axis of length 8, minor axis of length 4, and whose foci lie on the y-axis.

(a) $\dfrac{x^2}{2} + \dfrac{y^2}{4} = 1$ (b) $\dfrac{x^2}{16} + \dfrac{y^2}{2} = 1$ (c) $x^2 + 4y^2 = 1$ (d) $\dfrac{x^2}{16} + \dfrac{y^2}{4} = 1$

(e) $\frac{x^2}{2} + \frac{y^2}{8} = 1$

Answer: medium (d)

$a = 4, b = 2 \quad \Rightarrow \quad$ the equation is $\dfrac{x^2}{16} + \dfrac{y^2}{4} = 1$.

9. Find an equation for the ellipse which has foci $(\pm 1, 0)$ and a minor axis with length 4.

(a) $x^2 + 2y^2 = 1$ (b) $\dfrac{x^2}{2} + \dfrac{y^2}{10} = 1$ (c) $\dfrac{x^2}{2} + \dfrac{y^2}{5} = 1$ (d) $2x^2 + 5y^2 = 1$

(e) $\frac{x^2}{5} + \frac{y^2}{4} = 1$

Answer: medium (e)

$c = 1, b = 2 \quad \Rightarrow \quad a = \sqrt{1 + 4} = \sqrt{5}$, so the equation is $\dfrac{x^2}{5} + \dfrac{y^2}{4} = 1$.

10. Find an equation for the ellipse which has foci $(0, \pm 7)$ and eccentricity 0.7.

(a) $\dfrac{x^2}{51} + \dfrac{y^2}{49} = 1$ (b) $\dfrac{x^2}{100} + \dfrac{y^2}{49} = 1$ (c) $\dfrac{x^2}{10} + \dfrac{y^2}{7} = 1$ (d) $\dfrac{x^2}{7} + \dfrac{y^2}{10} = 1$ (e) $\dfrac{x^2}{51} + \dfrac{y^2}{100} = 1$

Answer: medium (e)

Since the foci are on the y-axis, the major axis is vertical, and the equation is $\dfrac{x^2}{b^2} + \dfrac{y^2}{a^2} = 1$. The foci are $(0, \pm 7)$ so $c = 7$ and eccentricity $\frac{7}{10} \cdot \frac{7}{10} = \frac{7}{a} \quad \Leftrightarrow \quad a = 10, b^2 = a^2 - c^2 \quad \Rightarrow \quad b^2 = 100 - 49 = 51$. So the equation is $\dfrac{x^2}{51} + \dfrac{y^2}{100} = 1$.

11. Find an equation for the ellipse with center at the origin, major axis of length $2\sqrt{7}$, minor axis of length $2\sqrt{3}$, and whose foci lie on the y-axis.

(a) $\dfrac{x^2}{3} + \dfrac{y^2}{7} = 1$ (b) $\dfrac{x^2}{7} + \dfrac{y^2}{3} = 1$ (c) $\dfrac{x^2}{9} + \dfrac{y^2}{49} = 1$ (d) $\dfrac{x^2}{\sqrt{3}} + \dfrac{y^2}{7} = 1$ (e) $\dfrac{x^2}{\sqrt{3}} + \dfrac{y^2}{\sqrt{7}} = 1$

Answer: easy (a)

The major axis is $2a = 2\sqrt{7} \quad \Leftrightarrow \quad a = \sqrt{7}$, minor axis is $2b = 2\sqrt{3} \quad \Leftrightarrow \quad b = \sqrt{3}$. The equation is $\dfrac{x^2}{(\sqrt{3})^2} + \dfrac{y^2}{(\sqrt{7})^2} = 1$

$\Leftrightarrow \quad \dfrac{x^2}{3} + \dfrac{y^2}{7} = 1$.

12. Find an equation for the ellipse with minor axis of length $2\sqrt{3}$ and foci $\left(\pm\sqrt{29}, 0\right)$.

(a) $\dfrac{x^2}{3} + \dfrac{y^2}{32} = 1$ (b) $\dfrac{x^2}{32} + \dfrac{y^2}{3} = 1$ (c) $\dfrac{x^2}{9} + \dfrac{y^2}{32} = 1$ (d) $\dfrac{x^2}{32} + \dfrac{y^2}{9} = 1$ (e) $\dfrac{x^2}{\sqrt{3}} + \dfrac{y^2}{32} = 1$

Answer: medium (b)

Minor axis is $2b = 2\sqrt{3} \quad \Leftrightarrow \quad b = \sqrt{3}, c = \sqrt{29}$. $a^2 = c^2 + b^2 = \left(\sqrt{29}\right)^2 + \left(\sqrt{3}\right)^2 = 29 + 3 = 32$. Therefore the equation is $\dfrac{x^2}{32} + \dfrac{y^2}{3} = 1$.

13. By placing the origin at the center of Mercury's orbit and the Sun on the x-axis at one of the foci, find the equation of Mercury's orbit given that the length of its major axis is 1.159×10^{11} m and it has an eccentricity of 0.206.

Answer: hard The length of the major axis is $2a = 1.159 \times 10^{11} \quad \Leftrightarrow \quad a = 5.795 \times 10^{10}$ m.

Eccentricity : $\dfrac{c}{a} = 0.206 \quad \Leftrightarrow \quad c = 0.206a \quad \Leftrightarrow \quad c = 0.206\left(5.795 \times 10^{10}\right) = 1.1938 \times 10^{10}$.

$b^2 = a^2 - c^2 = \left(5.795 \times 10^{10}\right)^2 - \left(1.1938 \times 10^{10}\right)^2 = 3.2157 \times 10^{21}$, so the equation is

$$\frac{x^2}{\left(5.795 \times 10^{10}\right)^2} + \frac{y^2}{\left(5.6707 \times 10^{10}\right)^2} = 1 \quad \Leftrightarrow \quad \frac{x^2}{3.358 \times 10^{21}} + \frac{y^2}{3.215 \times 10^{21}} = 1.$$

SECTION 11.3

1. Find the vertices, foci and asymptotes of the hyperbola $\dfrac{x^2}{9} - \dfrac{y^2}{16} = 1$, and sketch its graph.

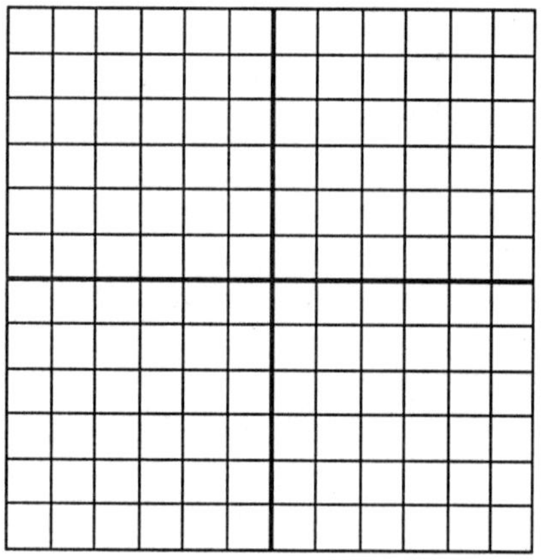

Answer: medium

$\dfrac{x^2}{9} - \dfrac{y^2}{16} = 1 \quad \Rightarrow \quad a = 3, b = 4$, so the vertices are $(\pm 3, 0)$, the foci are $(\pm 5, 0)$, and the asymptotes are $y = \pm\frac{4}{3}x$.

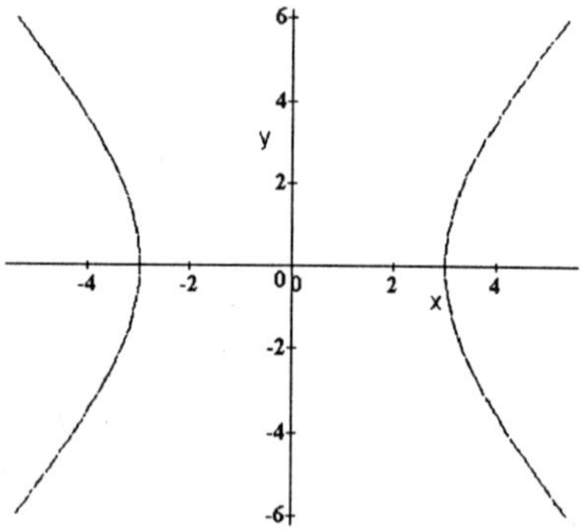

2. Find the vertices, foci and asymptotes of the hyperbola $\dfrac{y^2}{2} - x^2 = 1$, and sketch its graph.

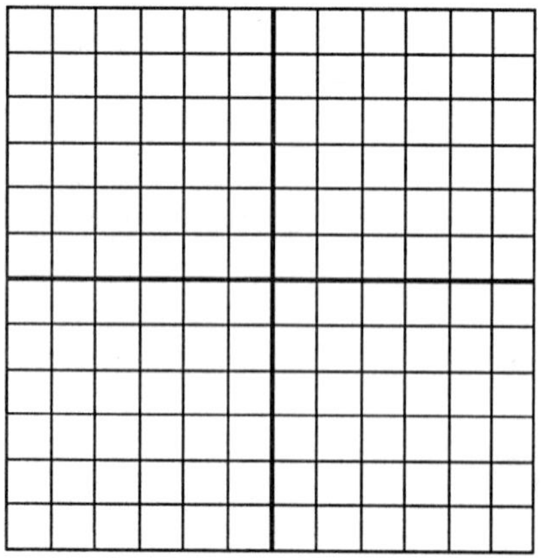

Answer: medium

$\dfrac{y^2}{2} - x^2 = 1 \quad \Rightarrow \quad a = \sqrt{2}, b = 1$, so the vertices are $\left(0, \pm\sqrt{2}\right)$, the foci are $\left(0, \pm\sqrt{3}\right)$, and the asymptotes are $y = \pm\sqrt{2}x$.

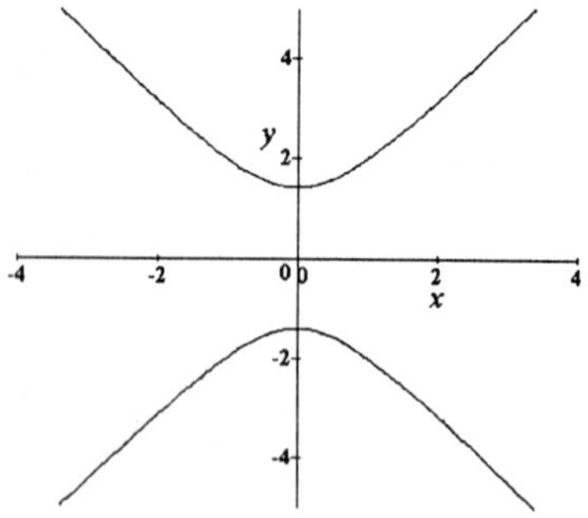

3. Find the vertices, foci and asymptotes of the hyperbola $x^2 - 6y^2 - 18 = 0$, and sketch its graph.

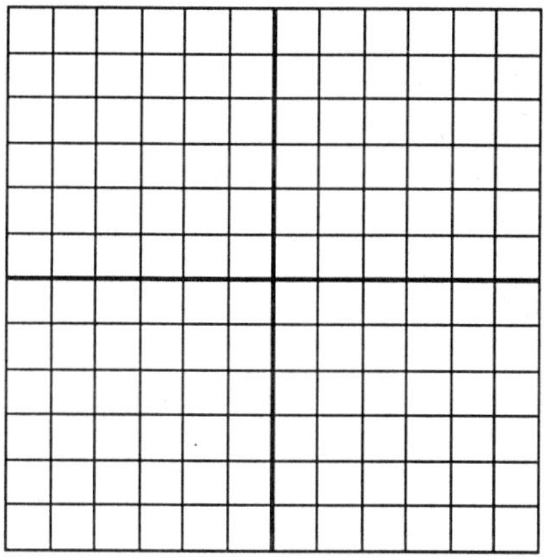

Answer: medium

$x^2 - 6y^2 - 18 = 0 \quad \Rightarrow \quad \dfrac{x^2}{18} - \dfrac{y^2}{3} = 1 \quad \Rightarrow \quad a = 3\sqrt{2}$, $b = \sqrt{3}$, so the vertices are $\left(\pm 3\sqrt{2}, 0\right)$, the foci are $\left(\pm\sqrt{21}, 0\right)$, and the asymptotes are $y = \pm\frac{\sqrt{6}}{6}x$.

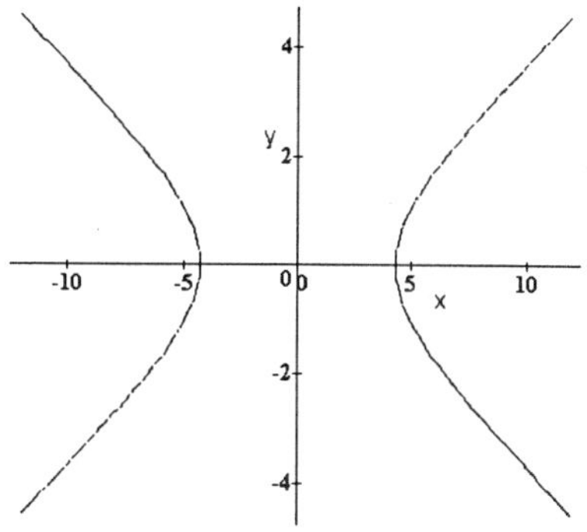

4. Find the vertices, foci and asymptotes of the hyperbola $y^2 - 16x^2 = 32$, and sketch its graph.

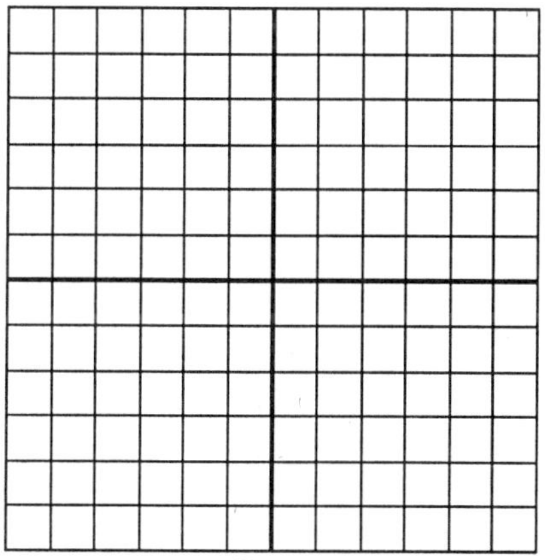

Answer: medium

$y^2 - 16x^2 = 32 \quad \Rightarrow \quad \dfrac{y^2}{32} - \dfrac{x^2}{2} = 1 \quad \Rightarrow \quad a = 4\sqrt{2}$, $b = \sqrt{2}$, so the vertices are $\left(0, \pm 4\sqrt{2}\right)$, the foci are $\left(0, \pm\sqrt{34}\right)$, and the asymptotes are $y = \pm 4x$.

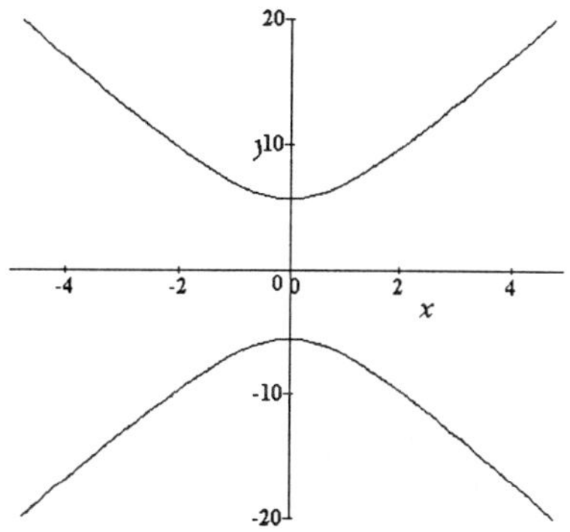

5. Find an equation for the hyperbola that has foci $(\pm 4, 0)$ and vertices $(\pm 2, 0)$.

(a) $\dfrac{x^2}{16} - \dfrac{y^2}{2} = 1$ (b) $\dfrac{x^2}{4} - \dfrac{y^2}{12} = 1$ (c) $\dfrac{x^2}{16} - \dfrac{y^2}{12} = 1$ (d) $\dfrac{x^2}{4} - \dfrac{y^2}{16} = 1$ (e) $x^2 - 12y^2 = 1$

Answer: easy (b)

$c = 4, a = 2 \quad \Rightarrow \quad b = \sqrt{16 - 4} = 2\sqrt{3}$, so the equation is $\dfrac{x^2}{4} - \dfrac{y^2}{12} = 1$.

6. Find an equation for the hyperbola that has foci $(0, \pm 12)$ and vertices $(0, \pm 6)$.

(a) $\dfrac{y^2}{36} - \dfrac{x^2}{108} = 1$ (b) $\dfrac{y^2}{36} - \dfrac{x^2}{12} = 1$ (c) $\dfrac{y^2}{24} - \dfrac{x^2}{18} = 1$ (d) $\dfrac{y^2}{4} - \dfrac{x^2}{12} = 1$ (e) $\dfrac{y^2}{42} - \dfrac{x^2}{36} = 1$

Answer: easy (a)

$a = 6, c = 12 \quad \Rightarrow \quad b^2 = 144 - 36 = 108$, so the equation is $\dfrac{y^2}{36} - \dfrac{x^2}{108} = 1$.

7. Find an equation for the hyperbola that has vertices $(\pm 2, 0)$ and asymptotes $y = \pm\frac{3}{2}x$.

(a) $\dfrac{x^2}{12} - \dfrac{y^2}{4} = 1$ (b) $\dfrac{x^2}{64} - \dfrac{y^2}{241} = 1$ (c) $\dfrac{x^2}{16} - \dfrac{y^2}{12} = 1$ (d) $\dfrac{x^2}{4} - \dfrac{y^2}{9} = 1$ (e) $x^2 - 9y^2 = 1$

Answer: easy (d)

$a = 2$, $\dfrac{b}{a} = \dfrac{3}{2} \quad\Rightarrow\quad b = 3$, so the equation is $\dfrac{x^2}{4} - \dfrac{y^2}{9} = 1$.

8. Find an equation for the hyperbola with foci $(0, \pm 2)$ and a transverse axis of length 2.

(a) $y^2 - \dfrac{x^2}{3} = 1$ (b) $\dfrac{y^2}{2} - \dfrac{x^2}{8} = 1$ (c) $\dfrac{x^2}{24} - \dfrac{y^2}{2} = 1$ (d) $\dfrac{x^2}{6} - \dfrac{y^2}{8} = 1$ (e) $\dfrac{x^2}{16} - \dfrac{y^2}{8} = 1$

Answer: easy (a)

$c = 2$, $2a = 2 \quad\Rightarrow\quad a = 1 \quad\Rightarrow\quad b^2 = 4 - 1 = 3$, so the equation is $y^2 - \dfrac{x^2}{3} = 1$.

9. Find an equation for the hyperbola with foci $\left(\pm\sqrt{7}, 0\right)$ and transverse axis of length 4.

(a) $\dfrac{x^2}{4} - \dfrac{y^2}{3} = 1$ (b) $\dfrac{x^2}{2} - \dfrac{y^2}{3} = 1$ (c) $\dfrac{x^2}{3} - \dfrac{y^2}{4} = 1$ (d) $\dfrac{x^2}{4} - \dfrac{y^2}{9} = 1$ (e) $\dfrac{x^2}{16} - \dfrac{y^2}{9} = 1$

Answer: medium (a)

Since the foci lie on the x-axis, the transverse axis is horizontal. $2a = 4 \quad\Leftrightarrow\quad a = 2$. $b^2 = c^2 - a^2 = 7 - 4 = 3$. Therefore the equation is $\dfrac{x^2}{4} - \dfrac{y^2}{3} = 1$.

10. Find the equation for the hyperbola that has asymptotes $y = \pm 3x$ and passes through the point $(3, -6)$.

(a) $\dfrac{x^2}{25} - \dfrac{y^2}{45} = 1$ (b) $\dfrac{y^2}{5} - \dfrac{x^2}{45} = 1$ (c) $\dfrac{x^2}{3} - \dfrac{y^2}{9} = 1$ (d) $\dfrac{x^2}{5} - \dfrac{y^2}{45} = 1$ (e) $\dfrac{x^2}{45} - \dfrac{y^2}{3} = 1$

Answer: medium (d)

Since $(3, -6)$ lies between the asymptotes and x-axis, the transverse axis is horizontal. Since $\dfrac{b}{a} = 3$, the equation of the hyperbola is $\dfrac{x^2}{a^2} - \dfrac{y^2}{b^2} = 1 \quad\Leftrightarrow\quad \dfrac{x^2}{a^2} - \dfrac{y^2}{(3a)^2} = 1$. Since the hyperbola passes through the point $(3, -6)$, $\dfrac{(3)^2}{a^2} - \dfrac{(-6)^2}{9a^2} = 1$ $\quad\Leftrightarrow\quad a^2 = 5$, and $b = 3a \quad\Leftrightarrow\quad b^2 = 9a^2 = 9(5) = 45$. So the equation is $\dfrac{x^2}{5} - \dfrac{y^2}{45} = 1$.

SECTION 11.4

1. Find the center, foci, and vertices of the ellipse $\dfrac{(x-1)^2}{4} + \dfrac{(y+2)^2}{8} = 1$, and determine the lengths of the major and minor axes. Then sketch the graph.

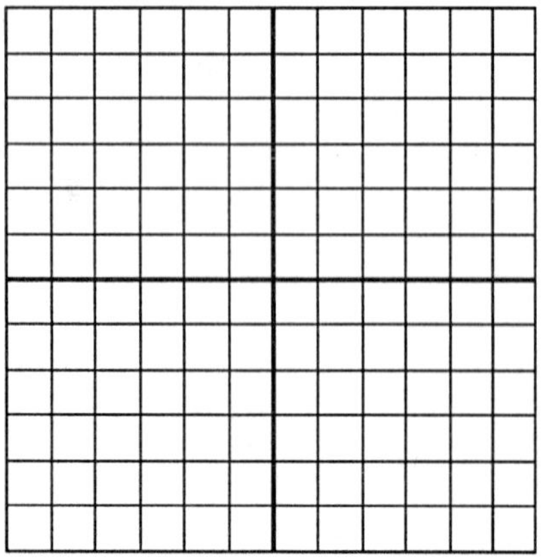

Answer: medium

$\dfrac{(x-1)^2}{4} + \dfrac{(y+2)^2}{8} = 1 \quad\Rightarrow\quad$ the center is $(1, -2)$, the foci are $(1, -4)$ and $(1, 0)$, the vertices are $\left(1, -2 \pm 2\sqrt{2}\right)$, $a = 2\sqrt{2}$ and $b = 2$, so the lengths of the axes are $4\sqrt{2}$ and 4.

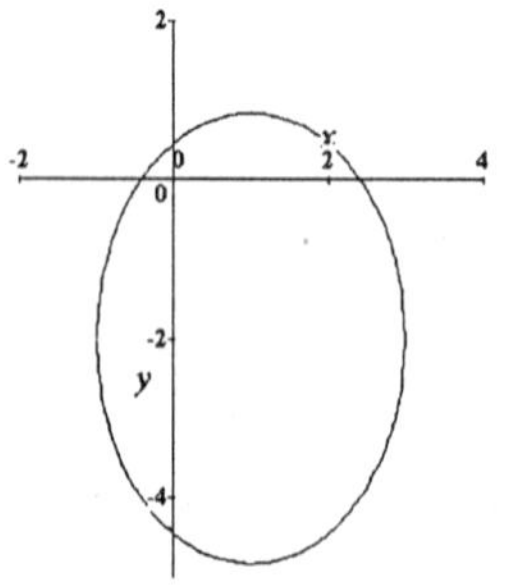

2. Find the vertex, focus, and directrix of the parabola $(y+2)^2 = -4x + 18$, and sketch the graph.

Answer: medium

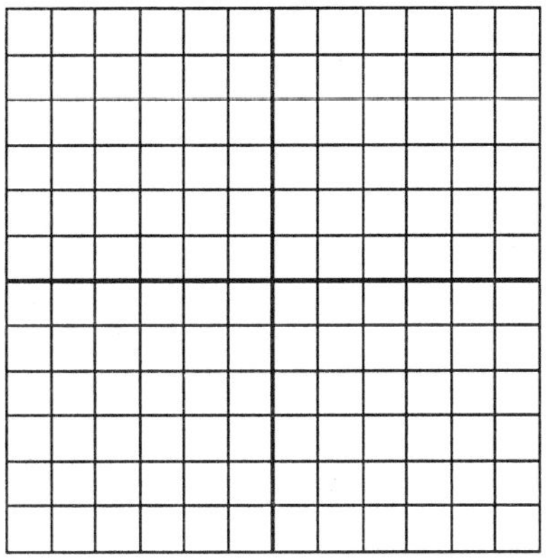

$(y+2)^2 = -4x+18 \quad \Rightarrow \quad (y+2)^2 = -4\left(x - \frac{9}{2}\right)$, so the vertex is $\left(\frac{9}{2}, -2\right)$, the focus is $\left(\frac{7}{2}, -2\right)$, and the directrix is $x = \frac{11}{2}$.

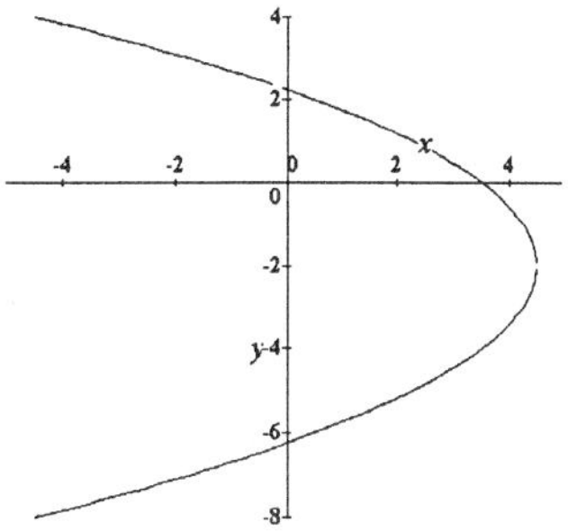

3. Find the center, foci, vertices, and asymptotes of the hyperbola $(x-4)^2 - (y+3)^2 = 1$. Then sketch the graph.

Answer: medium

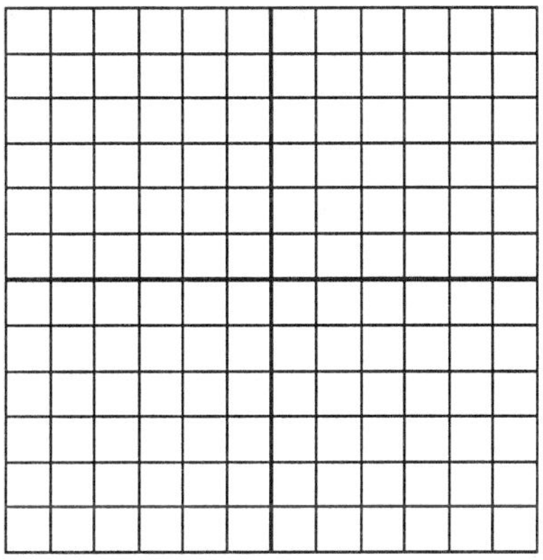

$(x-4)^2 - (y+3)^2 = 1 \quad \Rightarrow \quad a = b = 1 \quad \Rightarrow \quad c = \sqrt{2}$. The center is $(4, -3)$, the foci are $\left(4 \pm \sqrt{2}, -3\right)$, the vertices are $(3, -3)$ and $(5, -3)$, and the asymptotes are $(y+3) = \pm(x-4) \quad \Leftrightarrow \quad y = x - 7$ and $y = -x + 1$.

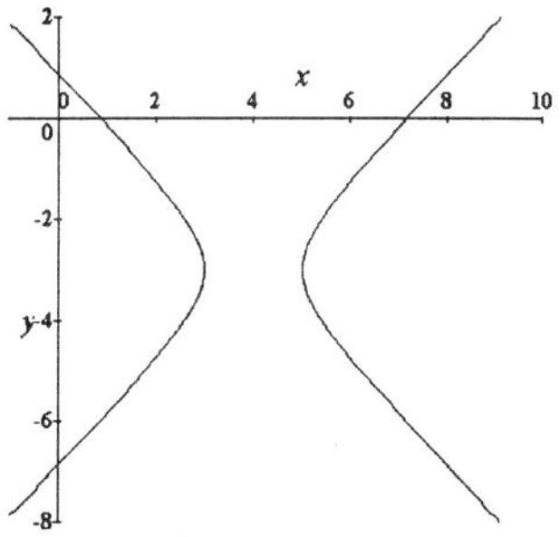

4. Complete the square in the equation $x^2 - 2x - 4y^2 - 16y = -13$ to determine the shape of its graph. Then sketch the graph. If it is an ellipse, find the center, foci, vertices, and lengths of the major and minor axes. If it is a parabola, find the vertex, focus, and directrix. If it is a hyperbola, find the center, foci, vertices, and asymptotes. If the equation has no graph, explain why.

Answer: hard

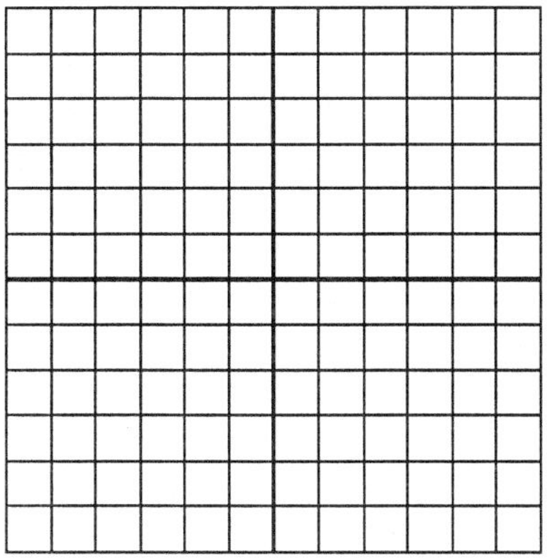

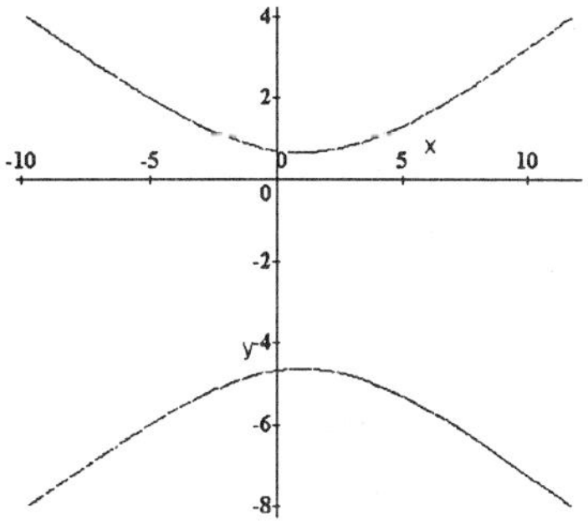

$x^2 - 2x - 4y^2 - 16y = -13 \quad \Leftrightarrow \quad x^2 - 2x + 1 - 4(y^2 + 4y + 4) = -13 + 1 - 16 \quad \Leftrightarrow \quad (x-1)^2 - 4(y+2)^2 = -28$
$\Leftrightarrow \quad \dfrac{(y+2)^2}{7} - \dfrac{(x-1)^2}{28} = 1$, so the graph is a hyperbola with center $(1, -2)$, foci $\left(1, -2 \pm \sqrt{35}\right)$, vertices $\left(1, -2 \pm \sqrt{7}\right)$, and asymptotes $y = \frac{1}{2}x - \frac{5}{2}$ and $y = -\frac{1}{2}x - \frac{1}{2}$.

5. Complete the square in the equation $9x^2 - 4y^2 - 18x + 16y = 43$ to determine the shape of its graph. If it is an ellipse, find the center, foci, vertices, and lengths of the major and minor axes. If it is a parabola, find the vertex, focus, and directrix. If it is a hyperbola, find the center, foci, vertices, and asymptotes. If the equation has no graph, explain why.

Answer: hard

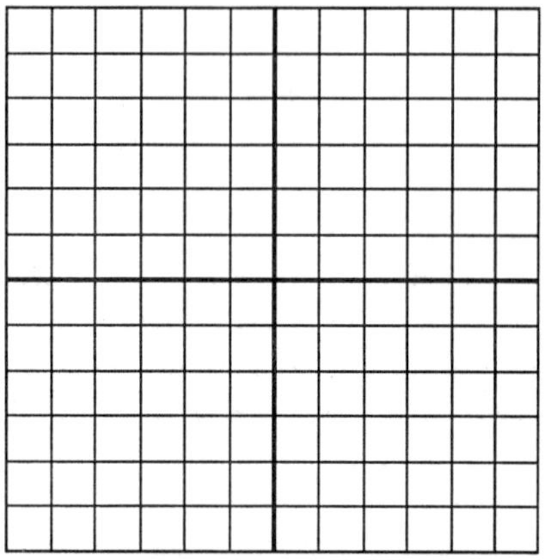

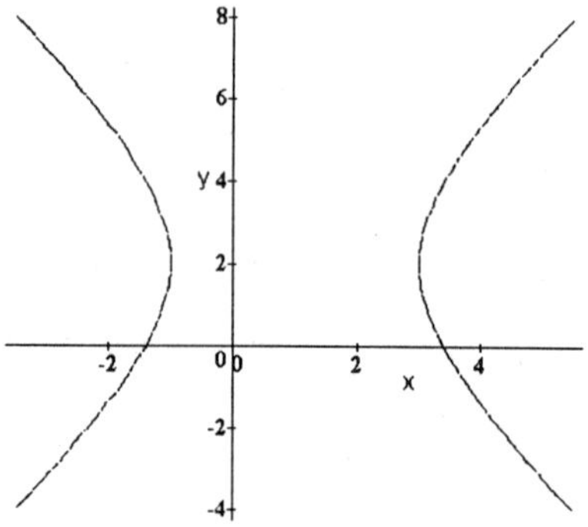

$9x^2 - 4y^2 - 18x + 16y = 43 \quad \Leftrightarrow \quad 9x^2 - 18x + 9 - 4(y^2 - 4y + 4) = 43 + 9 - 16 \quad \Leftrightarrow \quad 9(x-1)^2 - 4(y-2)^2 = 36$
$\Leftrightarrow \quad \dfrac{(x-1)^2}{4} - \dfrac{(y-2)^2}{9} = 1$, so the graph is a hyperbola with center $(1, 2)$, foci $\left(1 \pm \sqrt{13}, 2\right)$, vertices $(-1, 2)$ and $(3, 2)$, and asymptotes $(y - 2) = \pm\frac{3}{2}(x - 1) \quad \Leftrightarrow \quad y = \frac{3}{2}x + \frac{1}{2}$ or $y = -\frac{3}{2}x + \frac{7}{2}$.

6. Complete the square in the equation $16x^2 - 9y^2 + 96x + 288 = 0$ to determine the shape of its graph. If it is an ellipse, find the center, foci, vertices, and lengths of the major and minor axes. If it is a parabola, find the vertex, focus, and directrix. If it is a hyperbola, find the center, foci, vertices, and asymptotes. If the equation has no graph, explain why.

Answer: hard

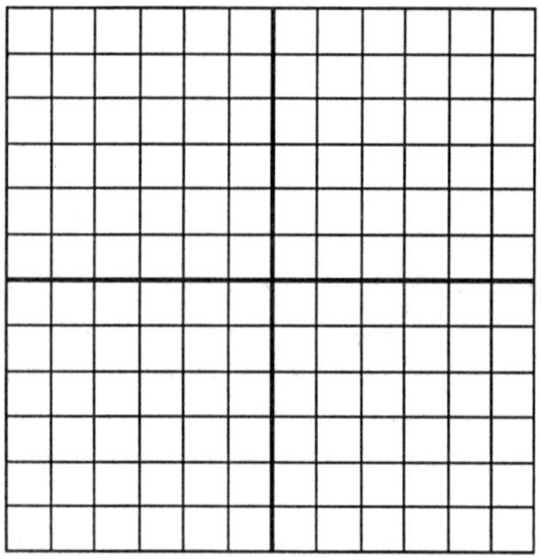

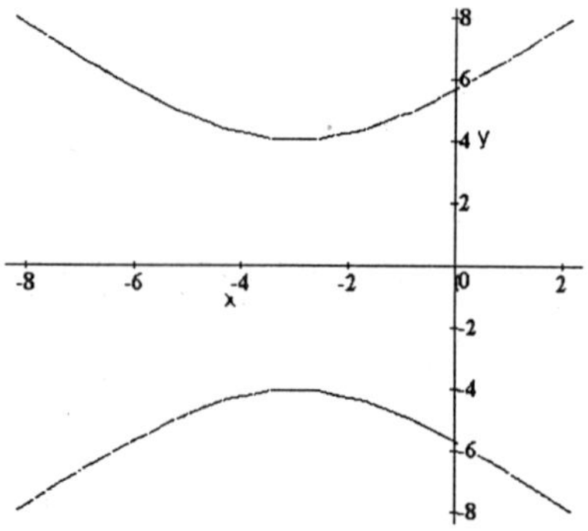

$16x^2 - 9y^2 + 96x + 288 = 0 \quad \Leftrightarrow \quad 16(x^2 + 6x + 9) - 9y^2 = -144 \quad \Leftrightarrow \quad 16(x+3)^2 - 9y^2 = -144 \quad \Leftrightarrow$
$\dfrac{y^2}{16} - \dfrac{(x+3)^2}{9} = 1$, so the graph is a hyperbola with center $(-3, 0)$, foci $(-3, \pm 5)$, vertices $(-3, \pm 4)$, and asymptotes $y = \pm\frac{4}{3}(x + 3)$.

7. Complete the square in the equation $24x^2 + 16y^2 = 160\left(\frac{3}{2}x + y\right) - 616$ to determine the shape of its graph. If it is an ellipse, find the center, foci, vertices, and lengths of the major and minor axes. If it is a parabola, find the vertex, focus, and directrix. If it is a hyperbola, find the center, foci, vertices, and asymptotes. If the equation has no graph, explain why.

Answer: hard

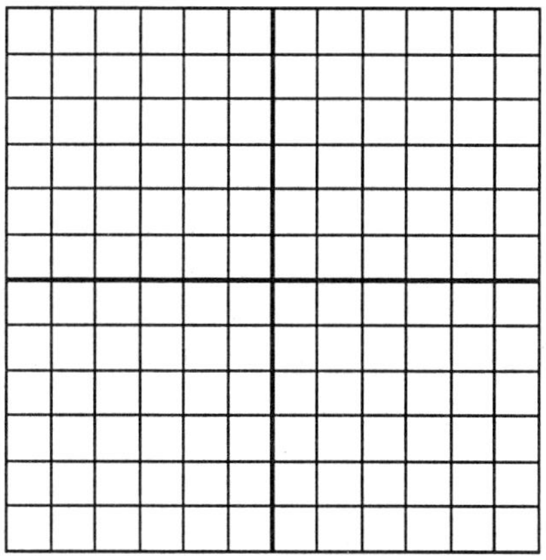

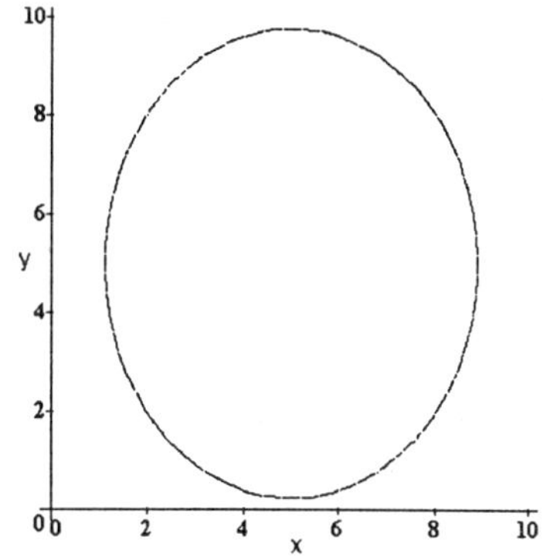

$24x^2 + 16y^2 = 160\left(\frac{3}{2}x + y\right) - 616 \quad \Leftrightarrow \quad 24(x^2 - 10x + 25) + 16(y^2 - 10y + 25) = 384 \quad \Leftrightarrow$ $\dfrac{(x-5)^2}{16} + \dfrac{(y-5)^2}{24} = 1$, so the graph is an ellipse with center $(5, 5)$, foci $\left(5, 5 \pm 2\sqrt{2}\right)$, vertices $\left(5, 5 \pm 2\sqrt{6}\right)$, major axis of length $4\sqrt{6}$, and minor axis of length 8.

8. Complete the square to determine whether the equation $y^2 + 25 = 14y - 16x$, represents an ellipse, parabola, hyperbola or degenerate conic. Then sketch the graph of the equation. If the graph is an ellipse, find the center, foci, vertices, and lengths of the major and minor axis. If it is a parabola, find the vertex, focus and directrix. If it is a hyperbola, find the center, foci, vertices, and asymptotes. If the equation has no graph, explain why.

Answer: hard

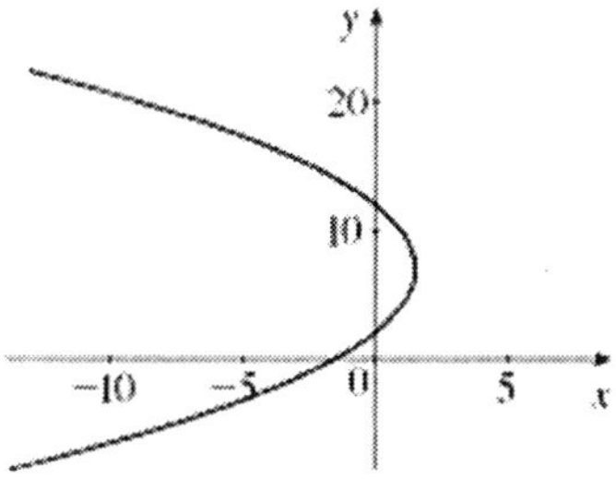

$y^2 + 25 = 14y - 16x \quad \Leftrightarrow \quad y^2 - 14y = -16x - 25 \quad \Leftrightarrow$
$y^2 - 14y + 49 = -16x - 25 + 49 \quad \Leftrightarrow$
$(y-7)^2 = -16x + 24 \quad \Leftrightarrow \quad (y-7)^2 = -16\left(x - \frac{3}{2}\right)$, so the graph is a parabola with vertex $\left(\frac{3}{2}, 7\right)$, focus $\left(-\frac{5}{2}, 7\right)$, directrix: $x = \frac{11}{2}$.

9. Complete the square to determine whether the equation $\frac{1}{36}x^2 - \frac{1}{2}x + \frac{153}{50} + \frac{1}{25}y^2 - \frac{9}{25}y = 1$ represents an ellipse, parabola, hyperbola or degenerate conic. Then sketch the graph of the equation. If the graph is an ellipse, find the center, foci, vertices, and lengths of the major and minor axis. If it is a parabola, find the vertex, focus and directrix. If it is a hyperbola, find the center, foci, vertices, and asymptotes. If the equation has no graph, explain why.

Answer: hard

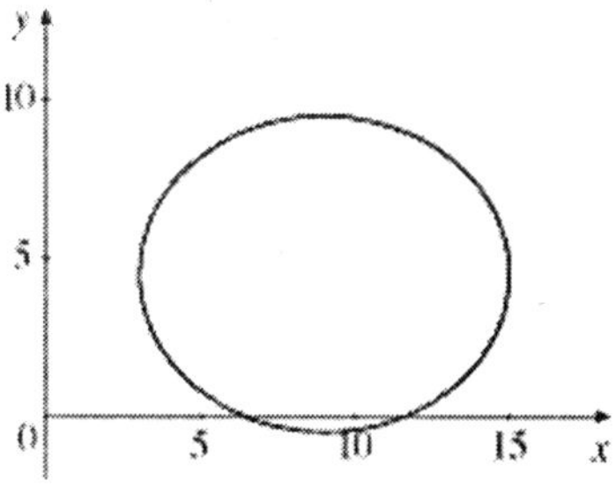

To help simplify the equation $\frac{1}{36}x^2 - \frac{1}{2}x + \frac{153}{50} + \frac{1}{25}y^2 - \frac{9}{25}y = 1$, multiply both sides of the equation by 900. $900\left(\frac{1}{36}x^2 - \frac{1}{2}x + \frac{153}{50} + \frac{1}{25}y^2 - \frac{9}{25}y\right) = 900 \quad \Leftrightarrow$
$25x^2 - 450x + 2754 + 36y^2 - 324y = 900 \quad \Leftrightarrow$
$25x^2 - 450x + 36y^2 - 324y = -1854 \quad \Leftrightarrow$
$25\left(x^2 - 18x + 81\right) + 36\left(y^2 - 9x + \frac{81}{4}\right) = -1854 + 2025 + 729 \quad \Leftrightarrow$
$25(x-9)^2 + 36\left(y - \frac{9}{2}\right)^2 = 900 \quad \Leftrightarrow \quad \dfrac{(x-9)^2}{36} + \dfrac{\left(y - \frac{9}{2}\right)^2}{25} = 1$. The ellipse has center $\left(9, \frac{9}{2}\right)$, vertices $\left(15, \frac{9}{2}\right)$, $\left(3, \frac{9}{2}\right)$, foci $\left(9 \pm \sqrt{11}\right)$, and length of major axis 12, and minor axis 10.

SECTION 11.5

1. Determine the XY-coordinates of $(1,1)$ if the axes are rotated through an angle $\phi = 30°$.
   Answer: easy
   $X = x\cos\phi + y\sin\phi = \frac{\sqrt{3}}{2} + \frac{1}{2} = \frac{\sqrt{3}+1}{2}$. $Y = -x\sin\phi + y\cos\phi = \frac{\sqrt{3}-1}{2}$

2. Determine the XY-coordinates of $(4,2)$ if the axes are rotated through an angle $\phi = 45°$.
   Answer: easy
   $X = x\cos\phi + y\sin\phi = 4\frac{\sqrt{2}}{2} + 2\frac{\sqrt{2}}{2} = 3\sqrt{2}$. $Y = -x\sin\phi + y\cos\phi = -4\frac{\sqrt{2}}{2} + 2\frac{\sqrt{2}}{2} = -\sqrt{2}$.

3. Determine the XY-coordinates of $(6,0)$ if the axes are rotated through an angle $\phi = 60°$.
   Answer: easy
   $X = x\cos\phi + y\sin\phi = 6\cdot\frac{1}{2} = 3$. $Y = -x\sin\phi + y\cos\phi = -6\frac{\sqrt{3}}{2} = -3\sqrt{3}$.

4. Determine the XY-coordinates of $(-2,5)$ if the axes are rotated through an angle $\phi = 30°$.
   Answer: easy
   $X = x\cos\phi + y\sin\phi = -2\frac{\sqrt{3}}{2} + 5\frac{1}{2} = \frac{5-2\sqrt{3}}{2}$. $Y = -x\sin\phi + y\cos\phi = 2\frac{1}{2} + 5\frac{\sqrt{3}}{2} = 1 + 5\frac{\sqrt{3}}{2}$

5. Determine the equation of the conic $y = (x+2)^2$ in XY-coordinates if the axes are rotated through an angle $\phi = 30°$.
   Answer: medium
   $x = X\cos 30 - Y\sin 30 = \frac{\sqrt{3}}{2}X - \frac{1}{2}Y$, $y = X\sin 30 + Y\cos 30 = \frac{1}{2}X + \frac{\sqrt{3}}{2}Y$. So $y = (x+2)^2 \quad\Leftrightarrow$
   $\frac{1}{2}X + \frac{\sqrt{3}}{2}Y = \left(\frac{\sqrt{3}}{2}X - \frac{1}{2}Y + 2\right)^2 \quad\Leftrightarrow\quad \frac{1}{2}X + \frac{\sqrt{3}}{2}Y = \frac{3}{4}X^2 - \frac{\sqrt{3}}{2}XY + 2\sqrt{3}X + \frac{1}{4}Y^2 - 2Y + 4 \quad\Leftrightarrow$
   $\frac{3}{4}X^2 - \frac{\sqrt{3}}{2}XY + \frac{1}{4}Y^2 + \left(2\sqrt{3} - \frac{1}{2}\right)X - \left(2 + \frac{\sqrt{3}}{2}\right)Y + 4 = 0.$

6. Determine the equation of the conic $x^2 - 2x = y^2$ in XY-coordinates if the axes are rotated through an angle $\phi = 45°$.
   Answer: medium
   $x = X\cos 45 - Y\sin 45 = \frac{1}{\sqrt{2}}X - \frac{1}{\sqrt{2}}Y$, $y = X\sin 45 + Y\cos 45 = \frac{1}{\sqrt{2}}X + \frac{1}{\sqrt{2}}Y$. So $x^2 - 2x = y^2 \quad\Leftrightarrow$
   $\left(\frac{1}{\sqrt{2}}X - \frac{1}{\sqrt{2}}Y\right)^2 - 2\left(\frac{1}{\sqrt{2}}X - \frac{1}{\sqrt{2}}Y\right) = \left(\frac{1}{\sqrt{2}}X + \frac{1}{\sqrt{2}}Y\right)^2 \quad\Leftrightarrow\quad \frac{1}{2}X^2 - XY + \frac{1}{2}Y^2 - \sqrt{2}X + \sqrt{2}Y = \frac{1}{2}X^2 + XY + \frac{1}{2}Y^2$
   $\Leftrightarrow\quad 2XY + \sqrt{2}X - \sqrt{2}Y = 0.$

7. Determine the equation of the conic $xy = x^2 - y^2$ in XY-coordinates if the axes are rotated through an angle $\phi = 60°$.
   Answer: medium
   $x = X\cos 60 - Y\sin 60 = \frac{1}{2}X - \frac{\sqrt{3}}{2}Y$, $y = X\sin 60 + Y\cos 60 = \frac{\sqrt{3}}{2}X + \frac{1}{2}Y$. $xy = x^2 - y^2 \quad\Leftrightarrow$
   $\left(\frac{1}{2}X - \frac{\sqrt{3}}{2}Y\right)\left(\frac{\sqrt{3}}{2}X + \frac{1}{2}Y\right) = \left(\frac{1}{2}X - \frac{\sqrt{3}}{2}Y\right)^2 - \left(\frac{\sqrt{3}}{2}X + \frac{1}{2}Y\right)^2 \quad\Leftrightarrow$
   $\frac{\sqrt{3}}{4}X^2 + \frac{1}{4}XY - \frac{3}{4}XY - \frac{\sqrt{3}}{4}Y^2 = \frac{1}{4}X^2 - \frac{\sqrt{3}}{2}XY + \frac{3}{4}Y^2 - \frac{3}{4}X^2 - \frac{\sqrt{3}}{2}XY - \frac{1}{4}Y^2 \quad\Leftrightarrow$
   $\left(\frac{\sqrt{3}}{4} + \frac{1}{2}\right)X^2 + \left(\sqrt{3} - \frac{1}{2}\right)XY - \left(\frac{\sqrt{3}}{4} - \frac{1}{2}\right)Y^2 = 0.$

8. Use the discriminant to determine if the graph of the equation, $3xy = -12$ is a parabola, an ellipse or a hyperbola, then use a rotation of axes to eliminate the xy-term, and sketch the graph.

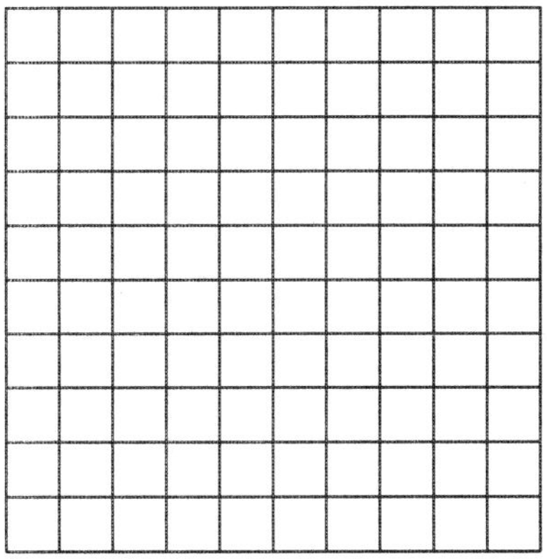

Answer:

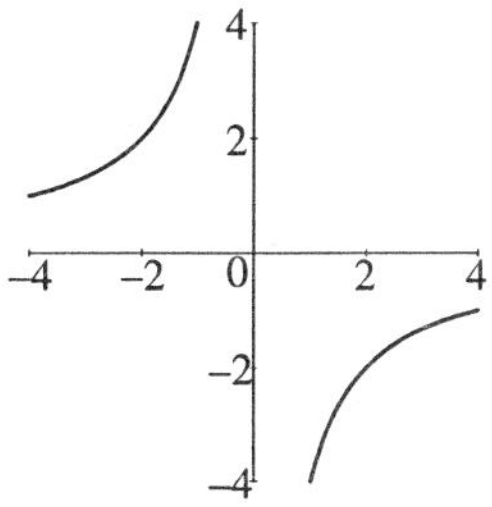

$3xy = -12 \quad \Leftrightarrow \quad xy + 4 = 0 \quad \Rightarrow \quad B^2 - 4AC = 1$, so the graph is a hyperbola. To eliminate the xy-term, rotate through ϕ, such that $\cot 2\phi = \frac{A-C}{B} = 0 \quad \Leftrightarrow \quad \cos 2\phi = 0$, so $\phi = \frac{\pi}{4}$. $x = X\cos\phi - Y\sin\phi = \frac{1}{\sqrt{2}}X - \frac{1}{\sqrt{2}}Y$,
$y = X\sin\phi + Y\cos\phi = \frac{1}{\sqrt{2}}X + \frac{1}{\sqrt{2}}Y$. So $xy + 4 = 0 \quad \Leftrightarrow \quad \left(\frac{1}{\sqrt{2}}X - \frac{1}{\sqrt{2}}Y\right)\left(\frac{1}{\sqrt{2}}X + \frac{1}{\sqrt{2}}Y\right) + 4 = 0 \quad \Leftrightarrow$
$\frac{1}{2}X^2 - \frac{1}{2}Y^2 = -4 \quad \Leftrightarrow \quad Y^2 - X^2 = 8$.

9. Use the discriminant to determine if the graph of the equation, $\frac{3}{4}x^2 + \frac{\sqrt{3}}{2}xy + \frac{1}{4}y^2 + 4x - 4\sqrt{3}y = 0$ is a parabola, an ellipse or a hyperbola, then use a rotation of axes to eliminate the xy-term, and sketch the graph.

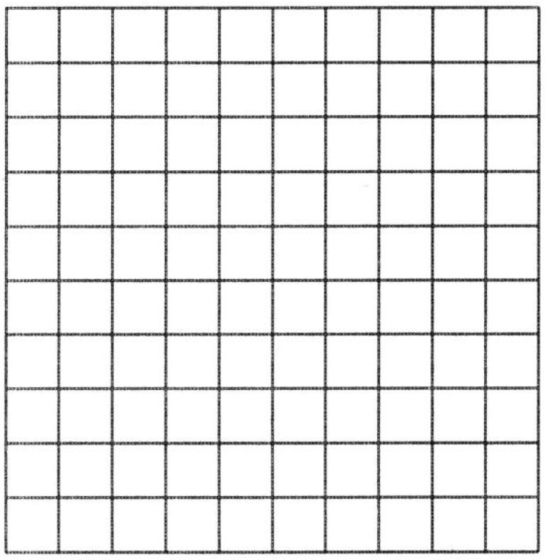

Answer:

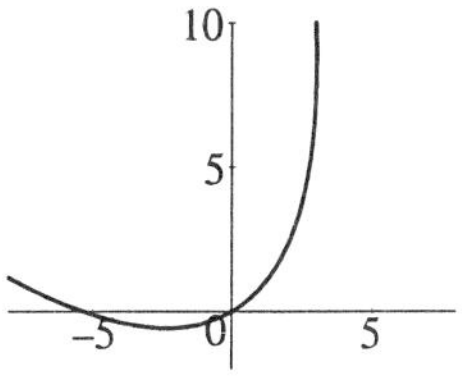

$B^2 - 4AC = \frac{3}{4} - 4\frac{3}{16} = 0$, so the graph is a parabola. $\cot 2\phi = \frac{A-C}{B} = \frac{1}{\sqrt{3}}$, so $\phi = 30°$. $x = X\frac{\sqrt{3}}{2} - Y\frac{1}{2}$, $y = X\frac{1}{2} + Y\frac{\sqrt{3}}{2}$
$\Rightarrow \quad \frac{3}{4}x^2 + \frac{\sqrt{3}}{2}xy + \frac{1}{4}y^2 + 4x - 4\sqrt{3}y = 0 \quad \Leftrightarrow$
$\frac{3}{4}\left(X\frac{\sqrt{3}}{2} - Y\frac{1}{2}\right)^2 + \frac{\sqrt{3}}{2}\left(X\frac{\sqrt{3}}{2} - Y\frac{1}{2}\right)\left(X\frac{1}{2} + Y\frac{\sqrt{3}}{2}\right) + \frac{1}{4}\left(X\frac{1}{2} + Y\frac{\sqrt{3}}{2}\right)^2 = 4\sqrt{3}\left(X\frac{1}{2} + Y\frac{\sqrt{3}}{2}\right) - 4\left(X\frac{\sqrt{3}}{2} - Y\frac{1}{2}\right) \quad \Leftrightarrow$
$\frac{3}{4}\left(\frac{3}{4}X^2 - \frac{\sqrt{3}}{2}XY + \frac{1}{4}Y^2\right) + \frac{\sqrt{3}}{2}\left(\frac{\sqrt{3}}{4}X^2 + \frac{3}{4}XY - \frac{1}{4}XY - \frac{\sqrt{3}}{4}Y^2\right) +$
$\frac{1}{4}\left(\frac{1}{4}X^2 + \frac{\sqrt{3}}{2}XY + \frac{3}{4}Y^2\right) = 2\sqrt{3}X + 6Y - 2\sqrt{3}X + 2Y \quad \Leftrightarrow$
$X^2\left(\frac{9}{16} + \frac{3}{8} + \frac{1}{16}\right) - XY\left(\frac{3}{4}\frac{\sqrt{3}}{2} - \frac{\sqrt{3}}{2}\frac{1}{2} - \frac{1}{4}\frac{\sqrt{3}}{2}\right) + Y^2\left(\frac{3}{16} - \frac{3}{8} + \frac{3}{16}\right) = 8Y \quad \Leftrightarrow \quad X^2 = 8Y$.

10. Use the discriminant to determine if the graph of the equation, $\frac{3}{4}x^2 - \frac{\sqrt{3}}{2}xy + \frac{1}{4}y^2 + 8x + 8\sqrt{3}y = 0$ is a parabola, an ellipse or a hyperbola, then use a rotation of axes to eliminate the xy-term, and sketch the graph.

Answer:

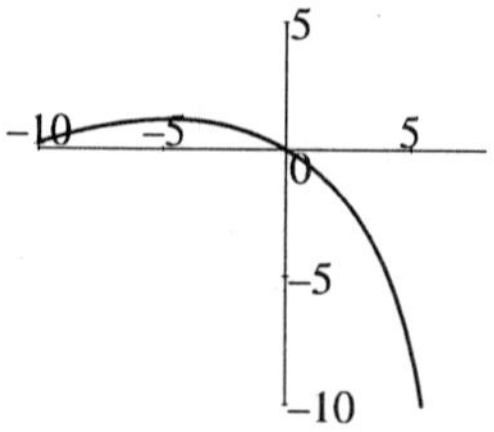

$B^2 - 4AC = \frac{3}{4} - \frac{3}{4} = 0$, so the graph is a parabola. $\cot 2\phi = \frac{A-C}{B} = -\frac{1}{\sqrt{3}}$, so $\phi = 60°$. $x = X\frac{1}{2} - Y\frac{\sqrt{3}}{2}$, $y = X\frac{\sqrt{3}}{2} + Y\frac{1}{2}$

$\Rightarrow \quad \frac{3}{4}x^2 - \frac{\sqrt{3}}{2}xy + \frac{1}{4}y^2 + 8x + 8\sqrt{3}y = 0 \quad \Leftrightarrow$

$\frac{3}{4}\left(X\frac{1}{2} - Y\frac{\sqrt{3}}{2}\right)^2 - \frac{\sqrt{3}}{2}\left(X\frac{1}{2} - Y\frac{\sqrt{3}}{2}\right)\left(X\frac{\sqrt{3}}{2} + Y\frac{1}{2}\right) + \frac{1}{4}\left(X\frac{\sqrt{3}}{2} + Y\frac{1}{2}\right)^2 = -8\left(X\frac{1}{2} - Y\frac{\sqrt{3}}{2}\right) - 8\sqrt{3}\left(X\frac{\sqrt{3}}{2} + Y\frac{1}{2}\right) \quad \Leftrightarrow$

$\frac{3}{16}X^2 - \frac{3\sqrt{3}}{8}XY + \frac{9}{16}Y^2 - \frac{3}{8}X^2 - \frac{\sqrt{3}}{8}XY + \frac{3\sqrt{3}}{8}XY + \frac{3}{8}Y^2 + \frac{3}{16}X^2 + \frac{\sqrt{3}}{8}XY + \frac{1}{16}Y^2 = -4X + 4\sqrt{3}Y - 12X - 4\sqrt{3}Y$

$\Leftrightarrow \quad X^2\left(\frac{3}{16} - \frac{3}{8} + \frac{3}{16}\right) - XY\left(\frac{3\sqrt{3}}{8} + \frac{\sqrt{3}}{8} - \frac{3\sqrt{3}}{8} - \frac{\sqrt{3}}{8}\right) + Y^2\left(\frac{9}{16} + \frac{3}{8} + \frac{1}{16}\right) = -16X \quad \Leftrightarrow \quad Y^2 = -16X.$

11. Use the discriminant to determine if the graph of the equation, $13x^2 - 10xy + 13y^2 - 72 = 0$ is a parabola, an ellipse or a hyperbola, then use a rotation of axes to eliminate the xy-term, and sketch the graph.

Answer:

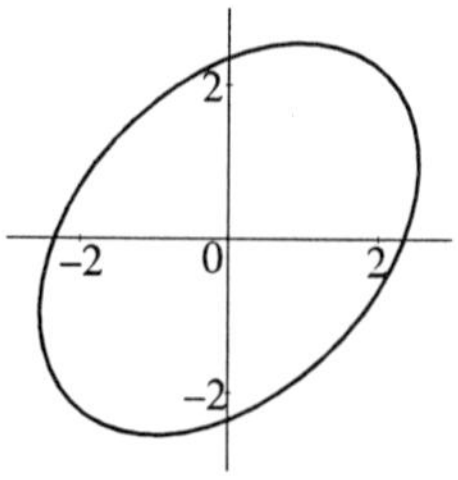

$B^2 - 4AC = 100 - 676 < 0$, so the graph is an ellipse. $\cot 2\phi = \frac{A-C}{B} = 0$, so $\phi = 45°$. $x = X\frac{1}{\sqrt{2}} - Y\frac{1}{\sqrt{2}}$, $y = X\frac{1}{\sqrt{2}} + Y\frac{1}{\sqrt{2}}$

$\Rightarrow \quad 13x^2 - 10xy + 13y^2 - 72 = 0 \quad \Leftrightarrow$

$13\left(X\frac{1}{\sqrt{2}} - Y\frac{1}{\sqrt{2}}\right)^2 - 10\left(X\frac{1}{\sqrt{2}} - Y\frac{1}{\sqrt{2}}\right)\left(X\frac{1}{\sqrt{2}} + Y\frac{1}{\sqrt{2}}\right) + 13\left(X\frac{1}{\sqrt{2}} + Y\frac{1}{\sqrt{2}}\right)^2 - 72 = 0 \quad \Leftrightarrow$

$\frac{13}{2}(X^2 - 2XY + Y^2) - 5(X^2 - Y^2) + \frac{13}{2}(X^2 + 2XY + Y^2) = 72 \quad \Leftrightarrow$

$X^2(13 - 10 + 13) - XY(26 - 26) + Y^2(13 + 10 + 13) = 144 \quad \Leftrightarrow \quad \frac{X^2}{9} + \frac{Y^2}{4} = 1.$

12. Use the discriminant to determine if the graph of the equation, $13x^2 + 6\sqrt{3}xy + 7y^2 = 64$ is a parabola, an ellipse or a hyperbola, then use a rotation of axes to eliminate the xy-term, and sketch the graph.

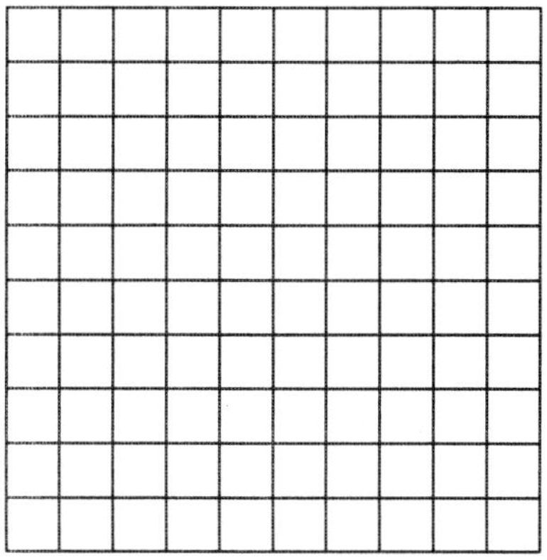

Answer:

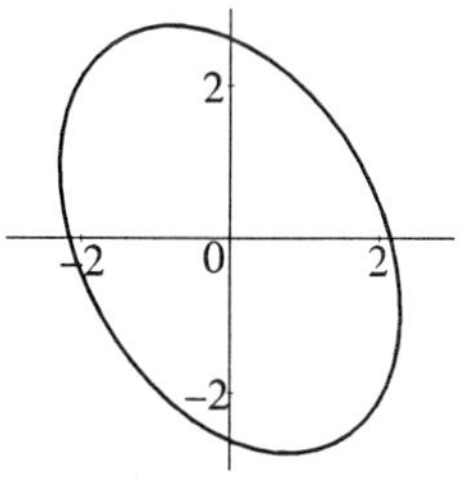

$B^2 - 4AC = 324 - 364 < 0$, so the graph is an ellipse. $\cot 2\phi = \frac{A-C}{B} = \frac{1}{\sqrt{3}}$, so $\phi = 30°$. $x = X\frac{\sqrt{3}}{2} - Y\frac{1}{2}$, $y = X\frac{1}{2} + Y\frac{\sqrt{3}}{2}$

$\Rightarrow \quad 13x^2 + 6\sqrt{3}xy + 7y^2 = 64 \quad \Leftrightarrow \quad 13\left(X\frac{\sqrt{3}}{2} - Y\frac{1}{2}\right)^2 + 6\sqrt{3}\left(X\frac{\sqrt{3}}{2} - Y\frac{1}{2}\right)\left(X\frac{1}{2} + Y\frac{\sqrt{3}}{2}\right) + 7\left(X\frac{1}{2} + Y\frac{\sqrt{3}}{2}\right)^2 = 64$

$\Leftrightarrow \quad 13\left(\frac{3}{4}X^2 - \frac{\sqrt{3}}{2}XY + \frac{1}{4}Y^2\right) + 6\sqrt{3}\left(\frac{\sqrt{3}}{4}X^2 + \frac{3}{4}XY - \frac{1}{4}XY - \frac{\sqrt{3}}{4}Y^2\right) + 7\left(\frac{1}{4}X^2 + \frac{\sqrt{3}}{2}XY + \frac{3}{4}Y^2\right) = 64 \quad \Leftrightarrow$

$X^2(39 + 18 + 7) - XY\left(26\sqrt{3} - 12\sqrt{3} - 14\sqrt{3}\right) + Y^2(13 - 18 + 21) = 256 \quad \Leftrightarrow \quad 64X^2 + 16Y^2 = 256 \quad \Leftrightarrow$

$\frac{X^2}{4} + \frac{Y^2}{16} = 1.$

13. Use the discriminant to determine if the graph of the equation, $-x^2 + 2\sqrt{3}xy + y^2 - 18 = 0$ is a parabola, an ellipse or a hyperbola, then use a rotation of axes to eliminate the xy-term, and sketch the graph.

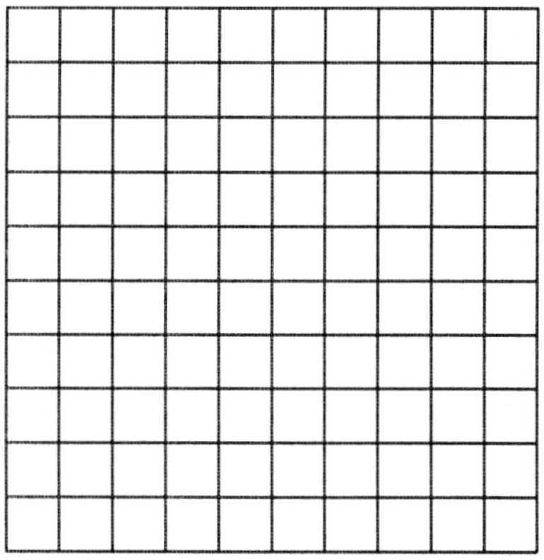

Answer:

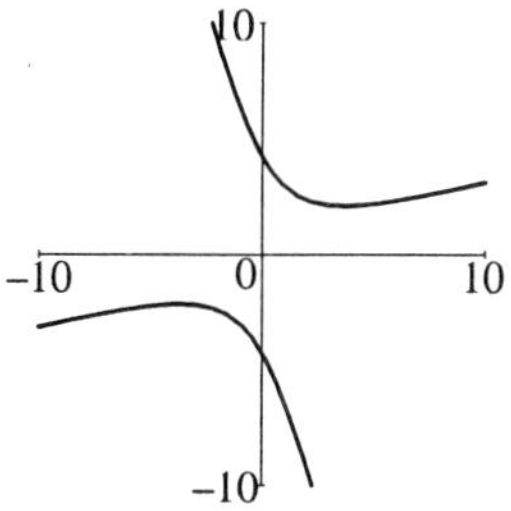

$B^2 - 4AC = 12 + 4 > 0$, so the graph is hyperbola. $\cot 2\phi = \frac{A-C}{B} = -\frac{1}{\sqrt{3}}$, so $\phi = 60°$. $x = X\frac{1}{2} - Y\frac{\sqrt{3}}{2}$, $y = X\frac{\sqrt{3}}{2} + Y\frac{1}{2}$

$\Rightarrow \quad -x^2 + 2\sqrt{3}xy + y^2 - 18 = 0 \quad \Leftrightarrow \quad -\left(X\frac{1}{2} - Y\frac{\sqrt{3}}{2}\right)^2 + 2\sqrt{3}\left(X\frac{1}{2} - Y\frac{\sqrt{3}}{2}\right)\left(X\frac{\sqrt{3}}{2} + Y\frac{1}{2}\right) + \left(X\frac{\sqrt{3}}{2} + Y\frac{1}{2}\right)^2 = 18$

$\Leftrightarrow \quad -\frac{1}{4}X^2 + \frac{\sqrt{3}}{2}XY - \frac{3}{4}Y^2 + \frac{3}{2}X^2 + \frac{\sqrt{3}}{2}XY - \frac{3\sqrt{3}}{2}XY - \frac{3}{2}Y^2 + \frac{3}{4}X^2 + \frac{\sqrt{3}}{2}XY + \frac{1}{4}Y^2 = 18 \quad \Leftrightarrow$

$X^2(-\frac{1}{4} + \frac{3}{2} + \frac{3}{4}) + XY\left(\frac{\sqrt{3}}{2} + \frac{\sqrt{3}}{2} - \frac{3\sqrt{3}}{2} + \frac{\sqrt{3}}{2}\right) + Y^2(-\frac{3}{4} - \frac{3}{2} + \frac{1}{4}) = 18 \quad \Leftrightarrow \quad X^2 - Y^2 = 9.$

14. Use the discriminant to determine if the graph of the equation, $3x^2 + 10xy + 3y^2 + 8 = 0$ is a parabola, an ellipse or a hyperbola, then use a rotation of axes to eliminate the xy-term, and sketch the graph.

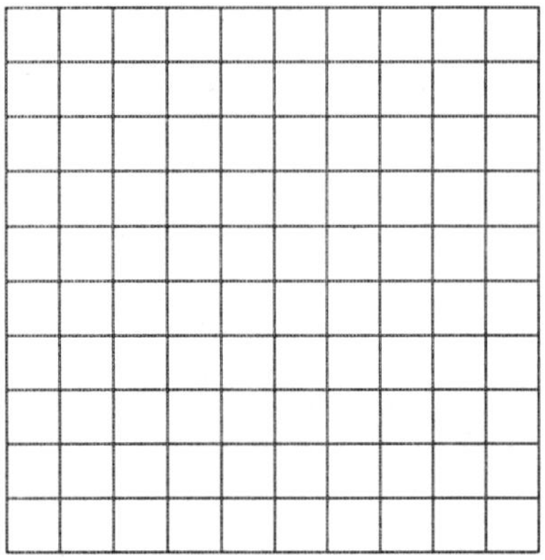

Answer:

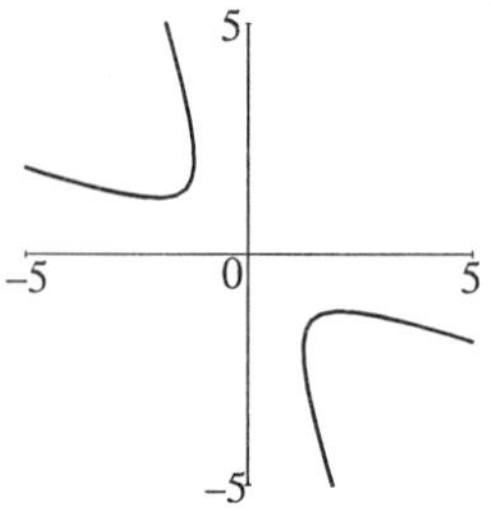

$B^2 - 4AC = 100 - 36 > 0$, so the graph is hyperbola. $\cot 2\phi = \frac{A-C}{B} = 0$, so $\phi = 45°$. $x = X\frac{1}{\sqrt{2}} - Y\frac{1}{\sqrt{2}}$, $y = X\frac{1}{\sqrt{2}} + Y\frac{1}{\sqrt{2}}$

$\Rightarrow \quad 3x^2 + 10xy + 3y^2 + 8 = 0 \quad \Leftrightarrow$

$3\left(X\frac{1}{\sqrt{2}} - Y\frac{1}{\sqrt{2}}\right)^2 + 10\left(X\frac{1}{\sqrt{2}} - Y\frac{1}{\sqrt{2}}\right)\left(X\frac{1}{\sqrt{2}} + Y\frac{1}{\sqrt{2}}\right) + 3\left(X\frac{1}{\sqrt{2}} + Y\frac{1}{\sqrt{2}}\right)^2 + 8 = 0 \quad \Leftrightarrow$

$3(X^2 - 2XY + Y^2) + 10(X^2 - Y^2) + 3(X^2 + 2XY + Y^2) = -16 \quad \Leftrightarrow$

$X^2(3 + 10 + 3) - XY(6 - 6) + Y^2(3 - 10 + 3) = -16 \quad \Leftrightarrow \quad \frac{Y^2}{4} - X^2 = 1.$

15. Use the discriminant to determine if the graph of the equation, $16x^2 - 24xy + 9y^2 - 60x - 80y = 0$ is a parabola, an ellipse or a hyperbola, then use a rotation of axes to eliminate the xy-term, and sketch the graph.

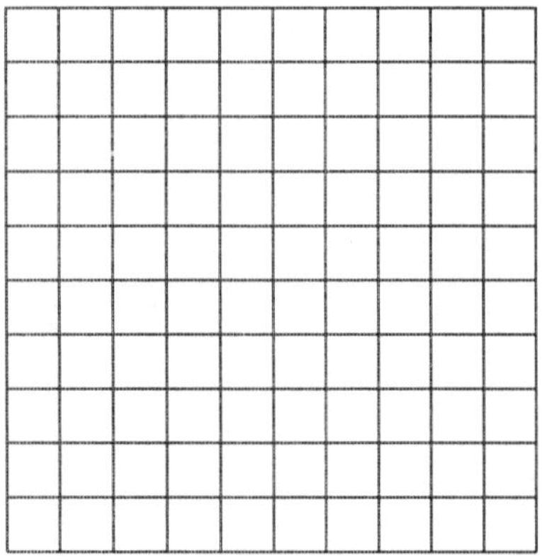

Answer:

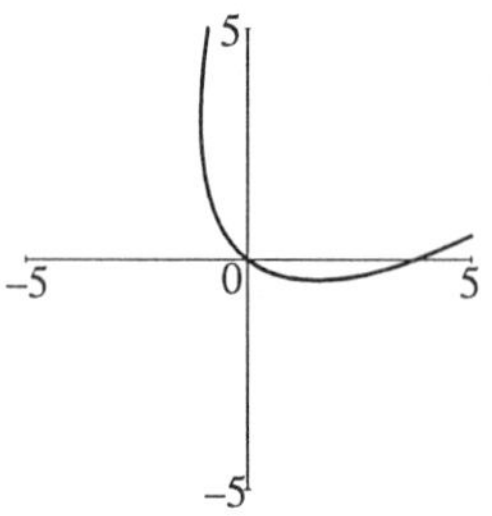

$B^2 - 4AC = 576 - 576 = 0$, so the graph is a parabola. $\cot 2\phi = \frac{A-C}{B} = -\frac{7}{24}$, by sketching a triangle, $\cos\phi = \frac{3}{5}$, $\sin\phi = \frac{4}{5}$.

$x = X\frac{3}{5} - Y\frac{4}{5}$, $y = X\frac{4}{5} + Y\frac{3}{5} \quad \Rightarrow \quad 16x^2 - 24xy + 9y^2 - 60x - 80y = 0 \quad \Leftrightarrow$

$16\left(X\frac{3}{5} - Y\frac{4}{5}\right)^2 - 24\left(X\frac{3}{5} - Y\frac{4}{5}\right)\left(X\frac{4}{5} + Y\frac{3}{5}\right) + 9\left(X\frac{4}{5} + Y\frac{3}{5}\right)^2 - 60\left(X\frac{3}{5} - Y\frac{4}{5}\right) - 80\left(X\frac{4}{5} + Y\frac{3}{5}\right) = 0 \quad \Leftrightarrow$

$\frac{16}{25}(9X^2 - 24XY + 16Y^2) - \frac{24}{25}(12X^2 + 9XY - 16XY - 12Y^2) + \frac{9}{25}(16X^2 + 24XY + 9Y^2) = 36X - 48Y + 64X + 48Y$

$\Leftrightarrow \quad X^2(144 - 288 + 144) - XY(24 \cdot 16 + 9 \cdot 24 - 16 \cdot 24 - 9 \cdot 24) + Y^2(256 + 288 + 81) = 2500X \quad \Leftrightarrow \quad Y^2 = 4X.$

SECTION 11.6

1. Plot the point with polar coordinates $(1, \pi/4)$ and give two other polar coordinate representations, one with $r < 0$, and the other with $r > 0$.

Answer:

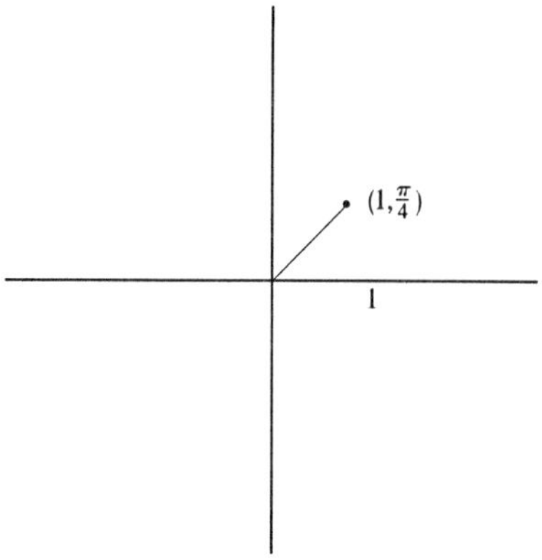

$(-1, 5\pi/4)$ and $(1, 9\pi/4)$ are two other polar coordinate representations.

2. Plot the point with polar coordinates $(-2, 8\pi/9)$ and give two other polar coordinate representations, one with $r < 0$, and the other with $r > 0$.

Answer:

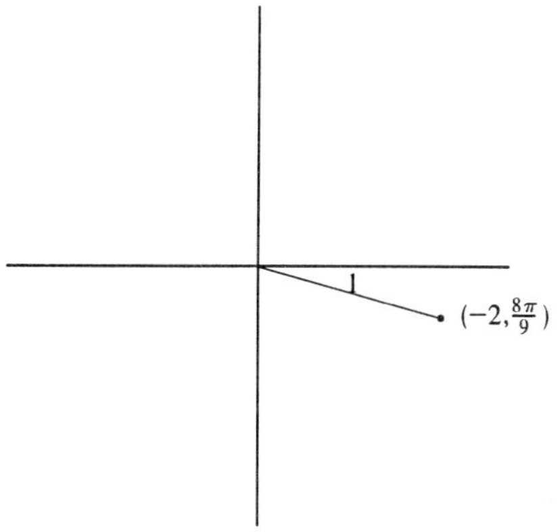

$(-2, 26\pi/9)$ and $(2, 17\pi/9)$ are two other polar coordinate representations.

3. Plot the point with polar coordinates $(3, 0)$ and give two other polar coordinate representations, one with $r < 0$, and the other with $r > 0$.

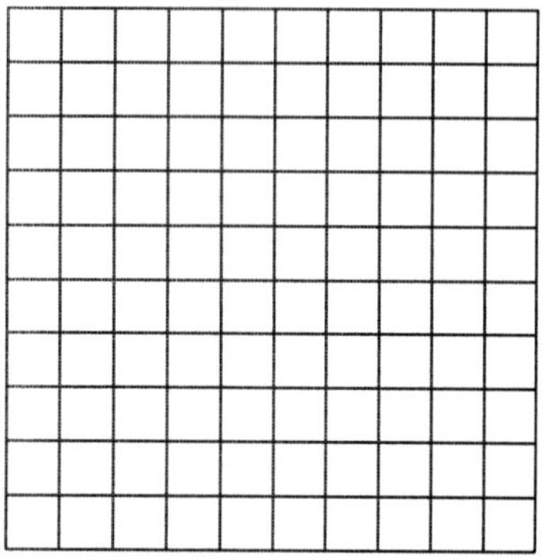

Answer: medium

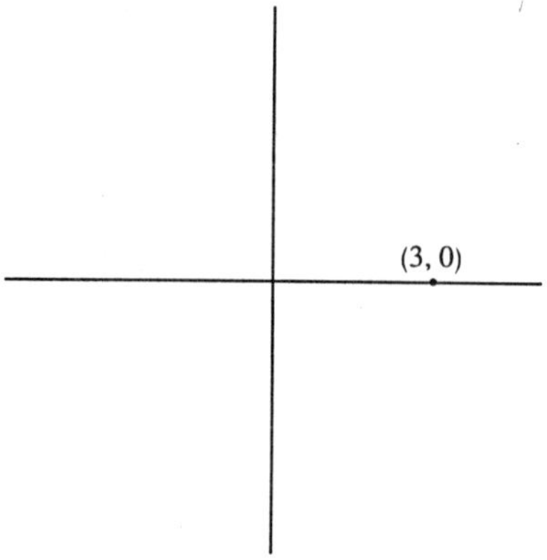

$(-3, \pi)$ and $(3, 2\pi)$ are two other polar coordinate representations.

4. Find the rectangular coordinates of the point with polar coordinates $(2, \pi/3)$.
Answer: easy
$x = r\cos\theta = 2\frac{1}{2} = 1$, $y = r\sin\theta = 2\frac{\sqrt{3}}{2} = \sqrt{3}$, so $(x, y) = \left(1, \sqrt{3}\right)$.

5. Find the rectangular coordinates of the point with polar coordinates $\left(\sqrt{3}, \pi/6\right)$.
Answer: easy
$x = r\cos\theta = \sqrt{3}\frac{\sqrt{3}}{2} = \frac{3}{2}$, $y = r\sin\theta = \sqrt{3}\frac{1}{2} = \frac{\sqrt{3}}{2}$, so $(x, y) = \left(\frac{3}{2}, \frac{\sqrt{3}}{2}\right)$.

6. Find the rectangular coordinates of the point with polar coordinates $\left(1/\sqrt{2}, 3\pi/4\right)$.
Answer: easy
$x = r\cos\theta = \frac{1}{\sqrt{2}} \cdot -\frac{1}{\sqrt{2}} = -\frac{1}{2}$, $y = r\sin\theta = \frac{1}{\sqrt{2}}\frac{1}{\sqrt{2}} = \frac{1}{2}$, so $(x, y) = \left(-\frac{1}{2}, \frac{1}{2}\right)$.

7. Find the rectangular coordinates of the point with polar coordinates $(0, \pi/27)$.
Answer: easy
$x = r\cos\theta = 0$, $y = r\sin\theta = 0$, so $(x, y) = (0, 0)$.

8. Find the polar coordinates of the point with rectangular coordinates $(-6, 6)$.
Answer: easy
$\tan\theta = -1 \quad\Rightarrow\quad \theta = 3\pi/4$, and $r = \sqrt{36 + 36} = 6\sqrt{2}$, so $(r, \theta) = \left(6\sqrt{2}, 3\pi/4\right)$.

9. Find the polar coordinates of the point with rectangular coordinates $\left(\sqrt{3}, -3\right)$.
Answer: easy
$\tan\theta = -\sqrt{3} \quad\Rightarrow\quad \theta = 5\pi/3$, and $r = \sqrt{3 + 9} = 2\sqrt{3}$, so $(r, \theta) = \left(2\sqrt{3}, 5\pi/3\right)$.

10. Find the polar coordinates of the point with rectangular coordinates $\left(-\sqrt{2}, -\sqrt{2}/\sqrt{3}\right)$.
Answer: easy
$\tan\theta = 1/\sqrt{3} \quad\Rightarrow\quad \theta = 7\pi/6$, and $r = \sqrt{2 + 2/3} = 2\sqrt{2}/\sqrt{3}$, so $(r, \theta) = \left(2\sqrt{2}/\sqrt{3}, 7\pi/6\right)$.

11. Convert the equation $xy = 1$ to polar form.
Answer: easy
$xy = 1 \quad\Leftrightarrow\quad (r\cos\theta)(r\sin\theta) = 1 \quad\Leftrightarrow\quad r^2\cos\theta\sin\theta = 1 \quad\Leftrightarrow\quad \frac{1}{2}r^2\sin 2\theta.$

12. Convert the equation $x + y = 2$ to polar form.

Answer: easy

$x + y = 2 \quad \Leftrightarrow \quad r\cos\theta + r\sin\theta = 2 \quad \Leftrightarrow \quad \cos\theta + \sin\theta = 2/r.$

13. Convert the equation $x^2 + xy + y^2 = 4$ to polar form.

Answer: medium

$x^2 + xy + y^2 = 4 \quad \Leftrightarrow \quad (r\cos\theta)^2 + (r\cos\theta)(r\sin\theta) + (r\sin\theta)^2 = 4 \quad \Leftrightarrow \quad r^2(\cos\theta\sin\theta + 1) = 4 \quad \Leftrightarrow$
$r^2\left(\frac{1}{2}\sin 2\theta + 1\right) = 4.$

14. Convert the polar equation $r = 3$ to rectangular coordinates.

Answer: easy

$r = 3 \quad \Leftrightarrow \quad \sqrt{x^2 + y^2} = 3 \quad \Leftrightarrow \quad x^2 + y^2 = 9.$

15. Convert the polar equation $\theta = \frac{\pi}{2}$ to rectangular coordinates.

Answer: easy

$\theta = \frac{\pi}{2} \quad \Leftrightarrow \quad \frac{y}{x} = \tan\theta \quad \Leftrightarrow \quad \frac{x}{y} = \cot\theta \quad \Leftrightarrow \quad \frac{x}{y} = 0 \quad \Leftrightarrow \quad x = 0.$

16. Convert the polar equation $r\sin\theta = 2$ to rectangular coordinates.

Answer: easy

$r\sin\theta = 2 \quad \Leftrightarrow \quad \sqrt{x^2 + y^2} \cdot y = 2 \quad \Leftrightarrow \quad (x^2 + y^2)y^2 = 4.$

17. Convert the polar equation $r^2 = \cot\theta$ to rectangular coordinates.

Answer: easy

$r^2 = \cot\theta \quad \Leftrightarrow \quad x^2 + y^2 = \frac{x}{y} \quad \Leftrightarrow \quad x^2y + y^3 = x.$

18. Convert the polar equation $1 = \dfrac{r}{1 + \cos\theta}$ to rectangular coordinates.

Answer: easy

$1 = \dfrac{r}{1 + \cos\theta} \quad \Leftrightarrow \quad r = 1 + \cos\theta \quad \Leftrightarrow \quad r^2 = r + r\cos\theta \quad \Leftrightarrow \quad x^2 + y^2 = \sqrt{x^2 + y^2} + x.$

19. Convert the polar equation $r = 3\csc\theta$ to rectangular coordinates.

Answer: easy

$r = 3\csc\theta \quad \Leftrightarrow \quad r\sin\theta = 3 \quad \Leftrightarrow \quad y = 3.$

20. Convert the polar equation $\sin 2\theta = 4$ to rectangular coordinates.

Answer: easy

$\sin 2\theta = 4 \quad \Leftrightarrow \quad 2\sin\theta\cos\theta = 4 \quad \Leftrightarrow \quad \dfrac{r\sin\theta\, r\cos\theta}{r^2} = 2 \quad \Leftrightarrow \quad \dfrac{xy}{x^2 + y^2} = 2.$

21. Convert the polar equation $r^3\cos\theta = 1$ to rectangular coordinates.

Answer: easy

$r^3\cos\theta = 1 \quad \Leftrightarrow \quad (x^2 + y^2)x = 1.$

22. Sketch the curve with polar equation $r = 2$.

Answer: easy

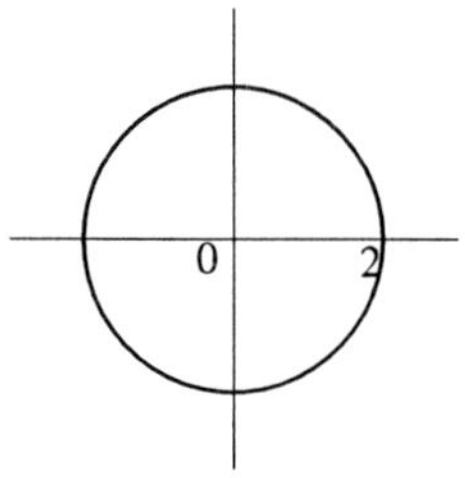

23. Sketch the curve with polar equation $\theta = -\frac{\pi}{4}$.

Answer: easy

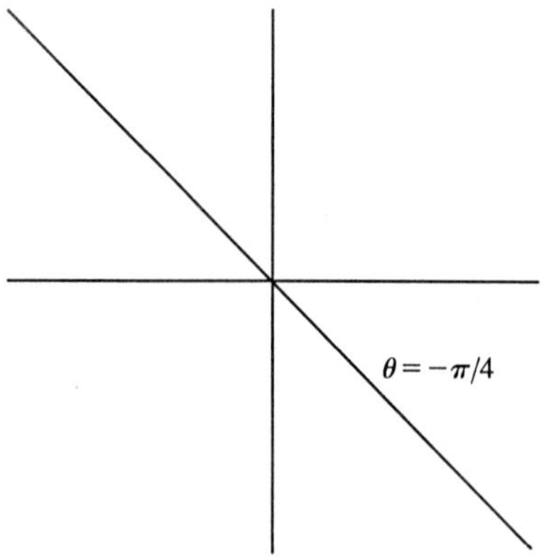

24. Sketch the curve with polar equation $r = -4\cos\theta$.

Answer: medium

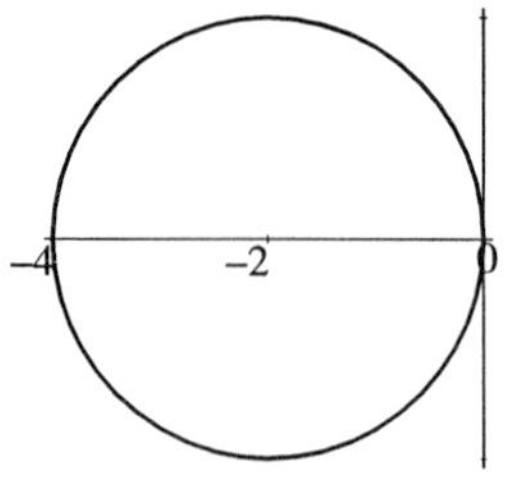

25. Sketch the curve with polar equation $r = \sin\theta - 2$.

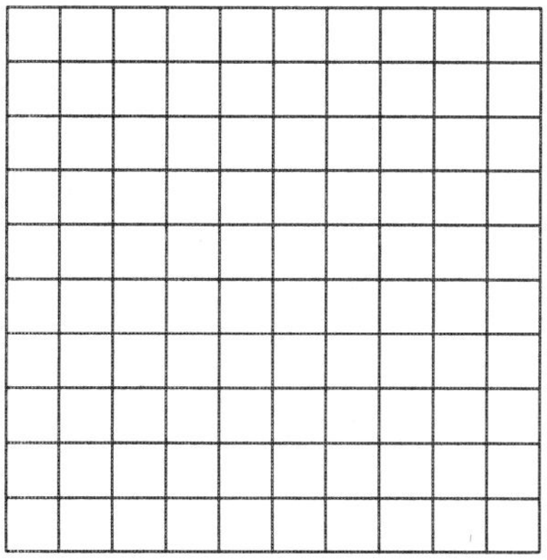

Answer: medium

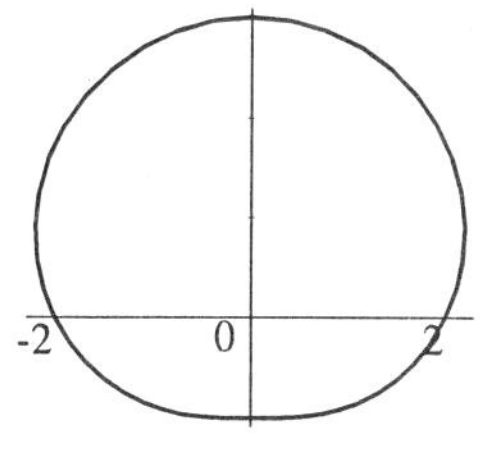

26. Sketch the curve with polar equation $r = 6 - 6\cos\theta$.

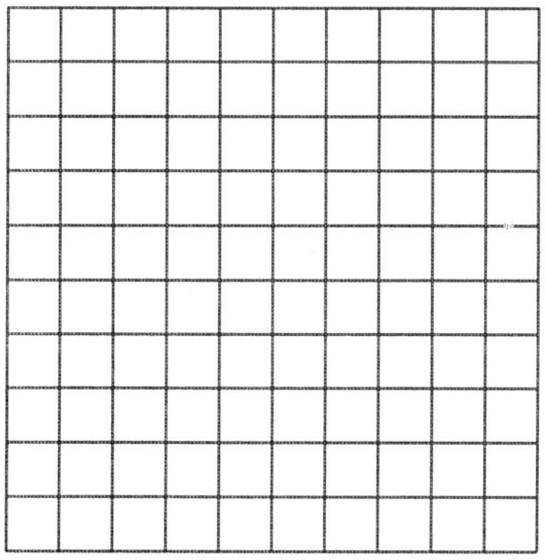

Answer: medium

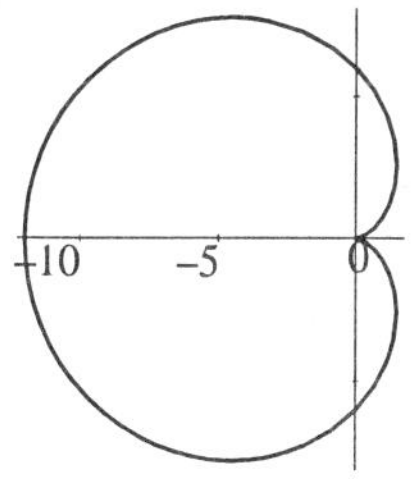

27. Sketch the curve with polar equation $r = 2\cos\theta - 2\sin\theta$.

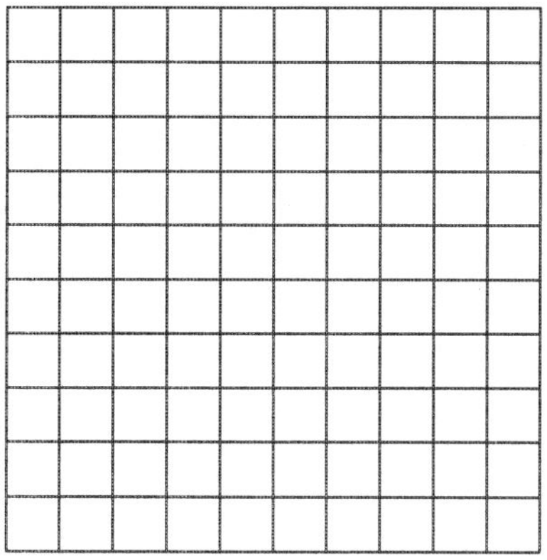

Answer: hard

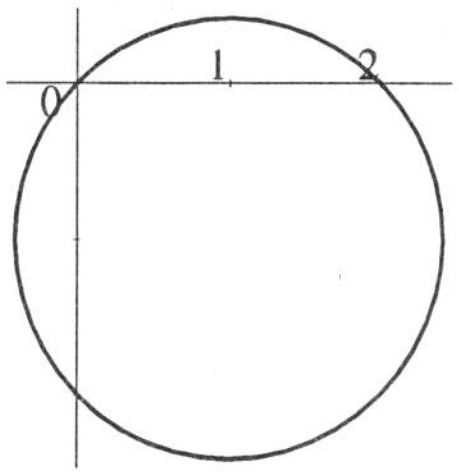

28. Sketch the curve with polar equation $r = \sin 3\theta$.

Answer: hard

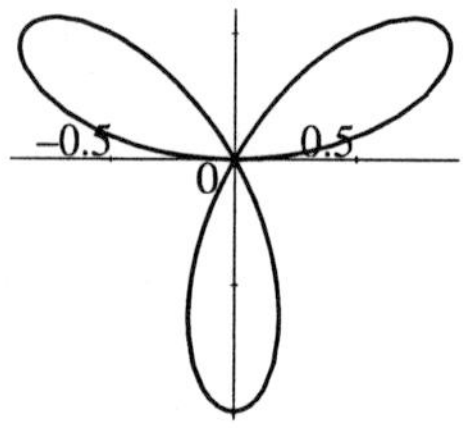

29. Sketch the curve with polar equation $r = 2 - \sec\theta$.

Answer: hard

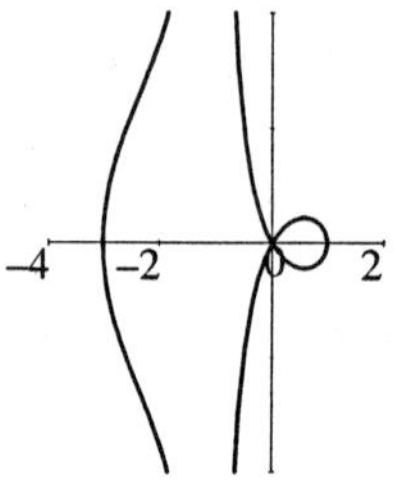

30. Sketch the curve with polar equation $r = \sin^2\theta$.

Answer: hard

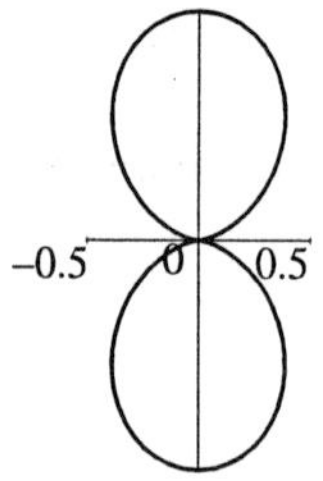

SECTION 11.7

1. Write a polar equation of an ellipse with eccentricity $\frac{3}{4}$ and directrix $x = 2$.

   Answer: easy

   $$r = \frac{\frac{3}{4} \cdot 2}{1 + \frac{3}{4}\cos\theta} \quad \Leftrightarrow \quad r = \frac{6}{4 + 3\cos\theta}.$$

2. Write a polar equation of a hyperbola with eccentricity 6 and directrix $x = -3$.

   Answer: easy

   $$r = \frac{6 \cdot 3}{1 - 6\cos\theta} \quad \Leftrightarrow \quad r = \frac{18}{1 - 6\cos\theta}.$$

3. Write a polar equation of a parabola with directrix $y = 5$.

   Answer: easy

   $$r = \frac{1 \cdot 5}{1 + \sin\theta} \quad \Leftrightarrow \quad r = \frac{5}{1 + \sin\theta}.$$

4. Write a polar equation of a parabola with vertex at $(2, 3\pi/2)$.

   Answer: easy

   Since the vertex is at $(2, 3\pi/2)$, the directrix is $y = -4$, and the equation is $r = \dfrac{1 \cdot 4}{1 - \sin\theta}$.

5. Write a polar equation of an ellipse with eccentricity 0.3 and directrix $y = -7$.

   Answer: easy

   $$r = \frac{0.3 \cdot 7}{1 + 0.3\sin\theta} \quad \Leftrightarrow \quad r = \frac{21}{10 - 3\sin\theta}.$$

6. Write a polar equation of a hyperbola with eccentricity 3 and directrix $r = 2\csc\theta$.

   Answer: easy

   $r = 2\csc\theta \quad \Leftrightarrow \quad r\sin\theta = 2 \quad \Leftrightarrow \quad y = 2$, so the equation is $r = \dfrac{3 \cdot 2}{1 + 3\sin\theta}$.

7. Write a polar equation of an ellipse with eccentricity $\frac{1}{6}$ and vertex at $(3, \pi)$.

   Answer: easy

   Since the vertex is at $(3, \pi)$, the directrix is $x = -d$ and the equation is $r = \dfrac{\frac{1}{6} \cdot d}{1 - \frac{1}{6}\cos\theta} \quad \Leftrightarrow \quad r = \dfrac{d}{6 - \cos\theta}$ and $d = r(6 - \cos\theta) = 3(6 + 1) = 21$. So, $r = \frac{21}{6 - \cos\theta}$.

8. Find the eccentricity and identify the conic given by $r = \dfrac{2}{1 + 2\cos\theta}$, sketch it and label its vertices.

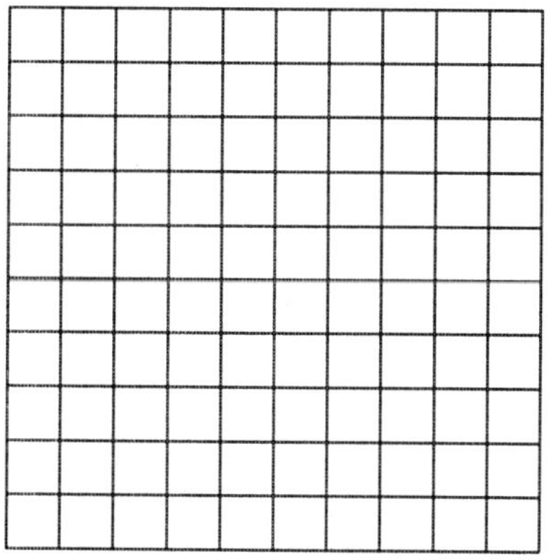

Answer: medium

$e = 2$, so the conic is a hyperbola.

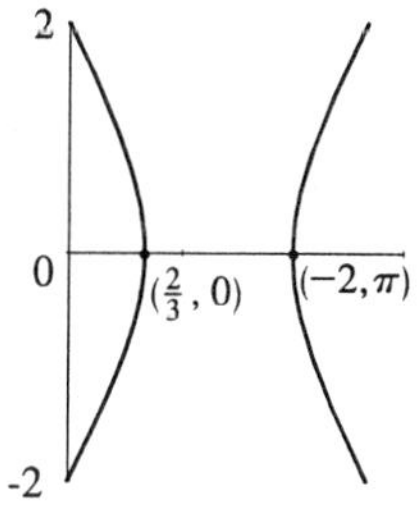

9. Find the eccentricity and identify the conic given by $r = \dfrac{2}{1 - \cos\theta}$, sketch it and label its vertices.

Answer: medium

$e = 1$, so the conic is a parabola.

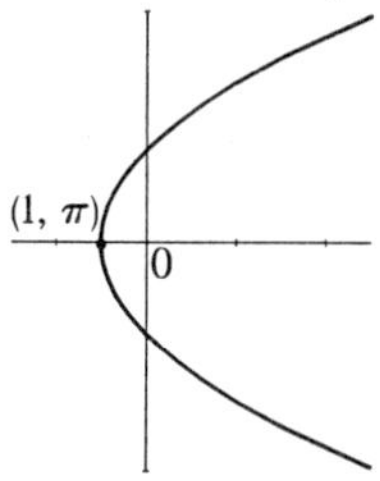

10. Find the eccentricity and identify the conic given by $r = \dfrac{7}{1 + 3\sin\theta}$, sketch it and label its vertices.

Answer: medium

$e = 3$, so the conic is a hyperbola.

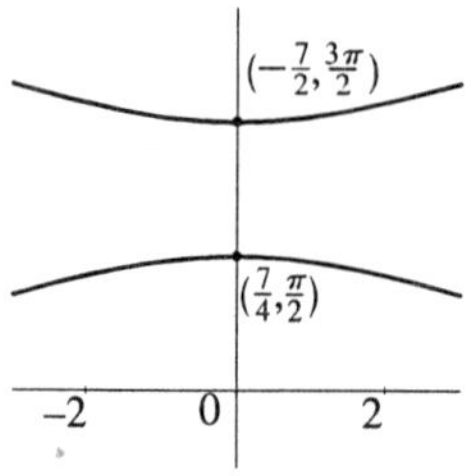

11. Find the eccentricity and identify the conic given by $r = \dfrac{6}{5 - 2\sin\theta}$, sketch it and label its vertices.

Answer: medium

$e = \frac{2}{5}$, so the conic is an ellipse.

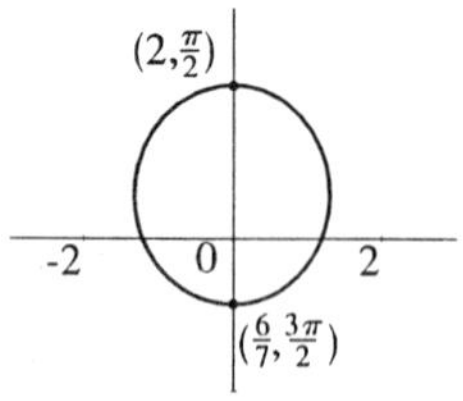

12. Find the eccentricity and identify the conic given by $r = \dfrac{8}{3 + 3\cos\theta}$, sketch it and label its vertices.

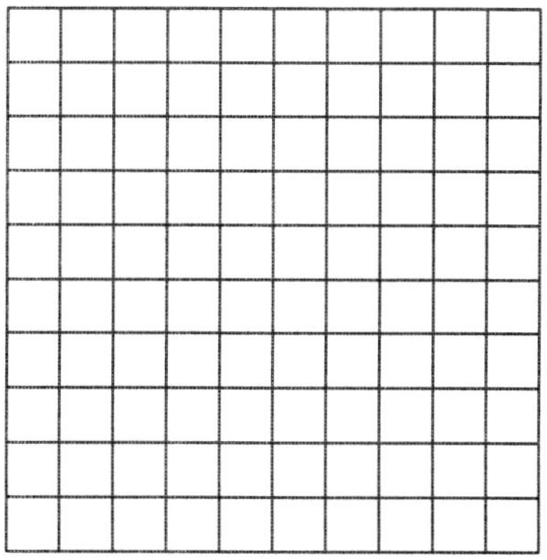

Answer: medium
$e = 1$, so the conic is a parabola.

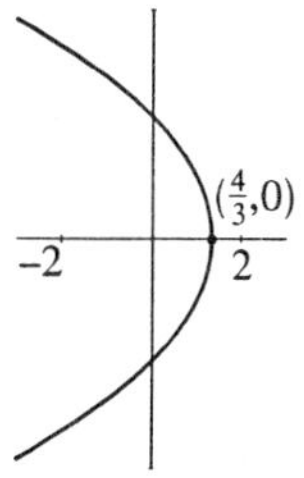

13. Find the eccentricity and identify the conic given by $r = \dfrac{5}{1 - \frac{1}{3}\cos\theta}$, sketch it and label its vertices.

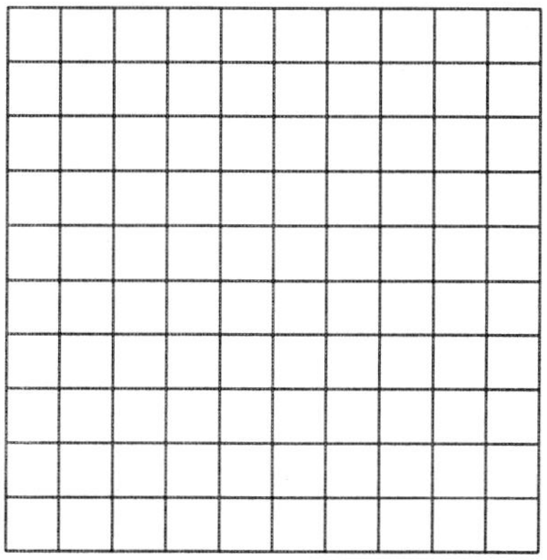

Answer: medium
$e = \frac{1}{5}$, so the conic is an ellipse.

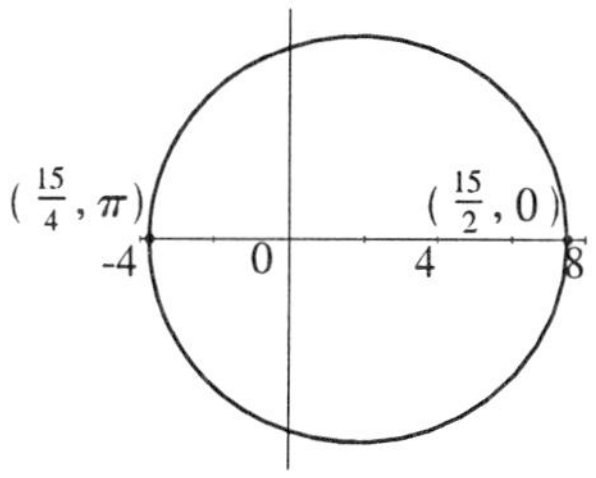

14. Find the eccentricity and directrix of the conic $r = \dfrac{3}{1 - \frac{1}{2}\cos\theta}$ and graph the conic. Find an equation for the conic rotated counterclockwise about the origin through an angle $\pi/6$, and graph it.

Answer: hard

$e = \frac{1}{2}$, $d = 6$, so the directrix is $x = -6$. The rotated conic is given by $r = \dfrac{3}{1 - \frac{1}{2}\cos\left(\theta - \frac{\pi}{6}\right)}$.

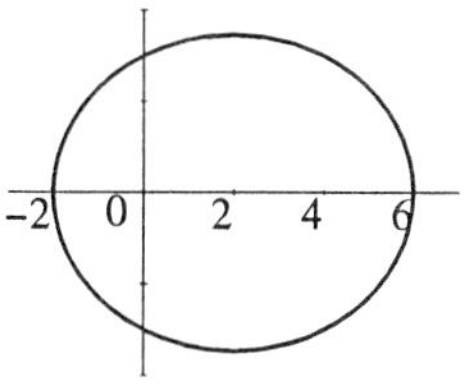

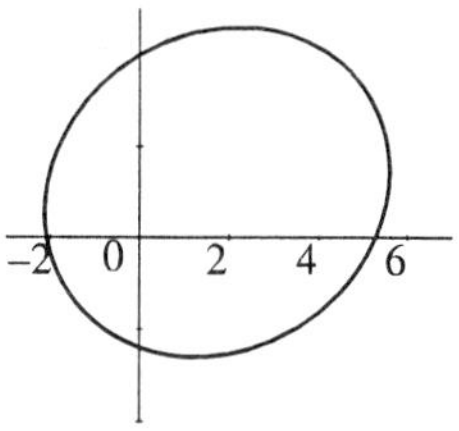

15. Find the eccentricity and directrix of the conic $r = \dfrac{6}{2 + 2\sin\theta}$ and graph the conic. Find an equation for the conic rotated clockwise about its vertex through an angle $3\pi/2$, and graph it.

Answer: hard

$e = 1$, $d = 3$, so the directrix is $y = 3$. The rotated conic is given by $r = \dfrac{6}{2 + 2\sin(\theta + 3\pi/2)}$.

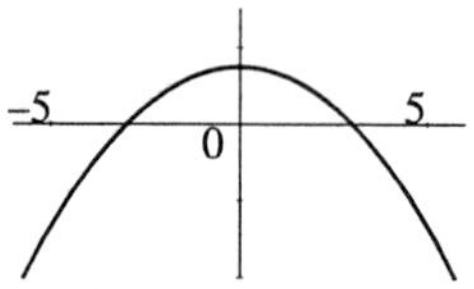

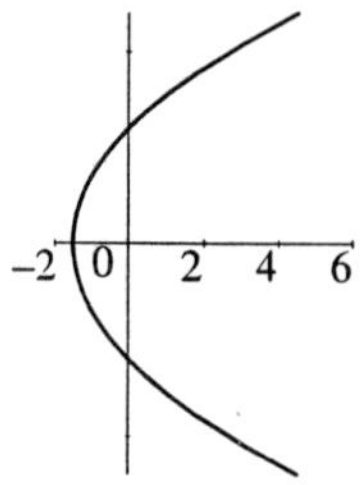

SECTION 11.8

1. Sketch the curve represented by $x = 3t$, $y = t - 2$ and find a rectangular-coordinate equation for the curve.

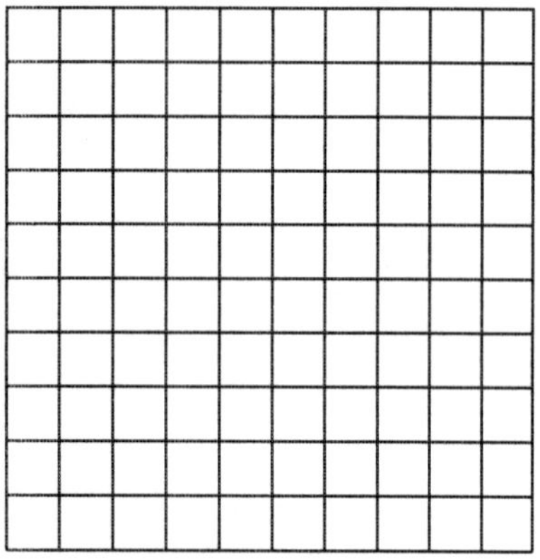

Answer: medium

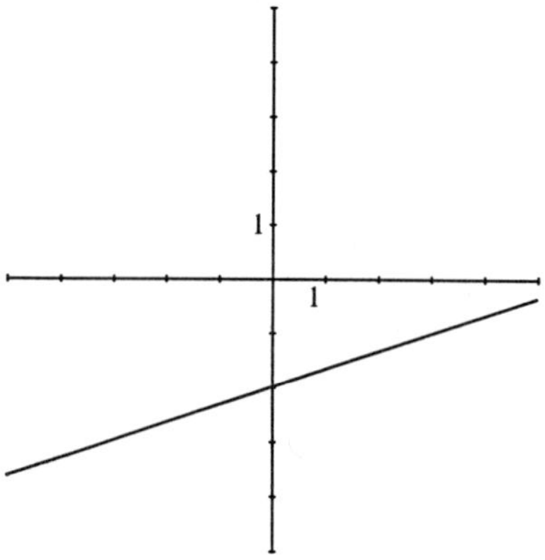

$y = t - 2 \quad \Leftrightarrow \quad t = y + 2$, so $x = 3t \quad \Rightarrow \quad x = 3y + 6$, which is a rectangular-coordinate equation.

2. Sketch the curve represented by $x = \frac{1}{2}t$, $y = t^2 + 3$ and find a rectangular-coordinate equation for the curve.

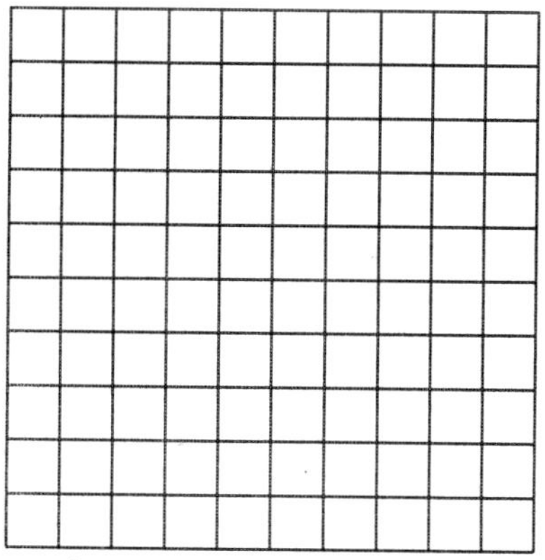

Answer: medium

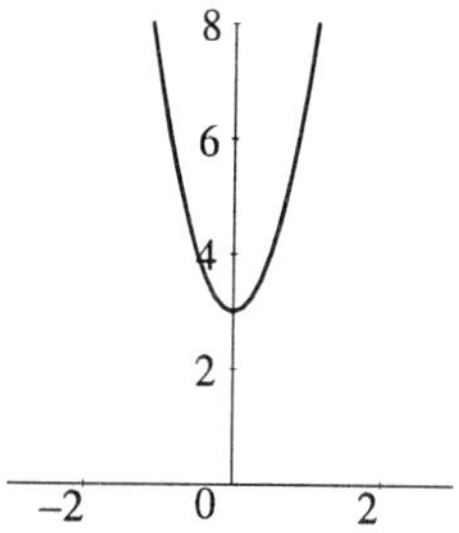

$x = \frac{1}{2}t \quad \Rightarrow \quad t = 2x$, so $y = t^2 + 3 \quad \Rightarrow \quad y = 4x^2 + 3$, which is a rectangular-coordinate equation.

3. Sketch the curve represented by $x = \sqrt{t}$, $y = t + 2$ and find a rectangular-coordinate equation for the curve.

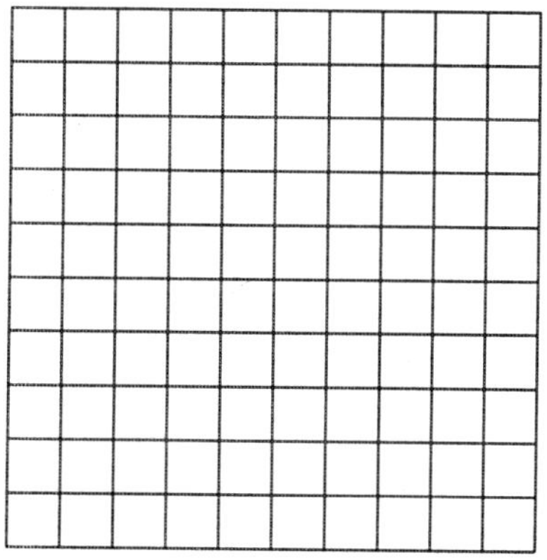

Answer: medium

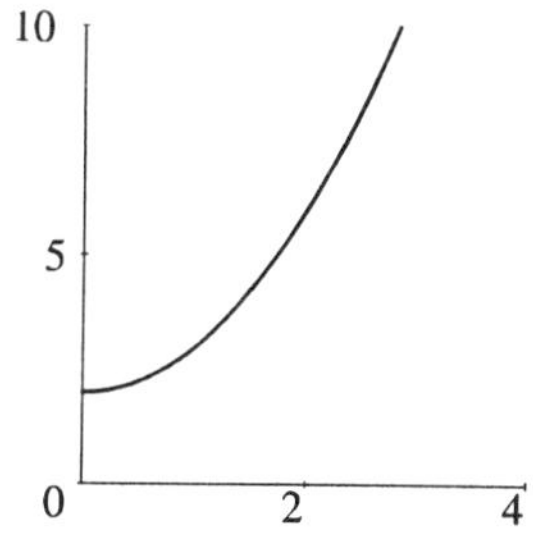

$y = t + 2 \quad \Leftrightarrow \quad t = y - 2$, so $x = \sqrt{y - 2}$, which is a rectangular-coordinate equation.

4. Sketch the curve represented by $x = 1 - t$, $y = \dfrac{2}{t}$ and find a rectangular-coordinate equation for the curve.

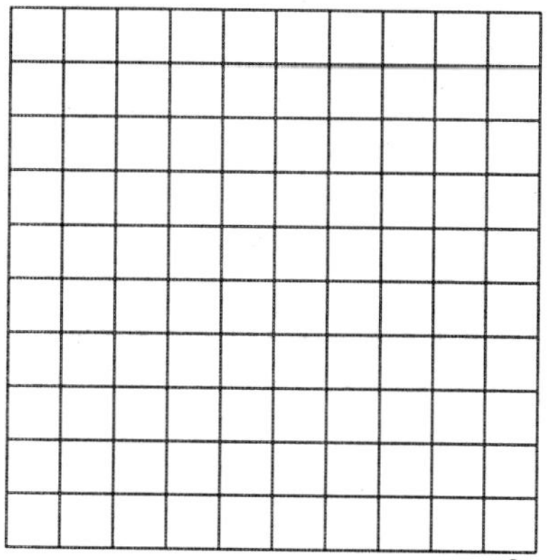

Answer: medium

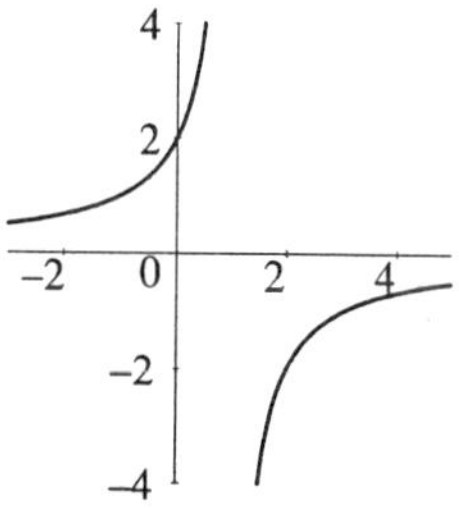

$x = 1 - t \quad \Leftrightarrow \quad t = 1 - x$, so $y = \dfrac{2}{t} \quad \Rightarrow \quad y = \dfrac{2}{1 - x}$, which is a rectangular-coordinate equation.

5. Sketch the curve represented by $x = 9t^2$, $y = 64t^3$ and find a rectangular-coordinate equation for the curve.

Answer: medium

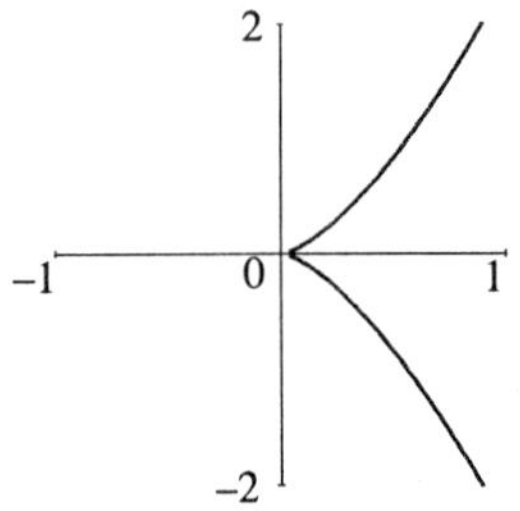

$x = 9t^2 \quad \Leftrightarrow \quad \pm\sqrt{x} = 3t \quad \Leftrightarrow \quad \pm x^{3/2} = 27t^3 \quad \Leftrightarrow \quad t^3 = \pm\frac{1}{27}x^{3/2}$, so $y = \pm\frac{64}{27}x^{3/2}$, which is a rectangular-coordinate equation.

6. Sketch the curve represented by $x = 3\sin t$, $y = \cos t$, $0 \le t \le 2\pi$ and find a rectangular-coordinate equation for the curve.

Answer: medium

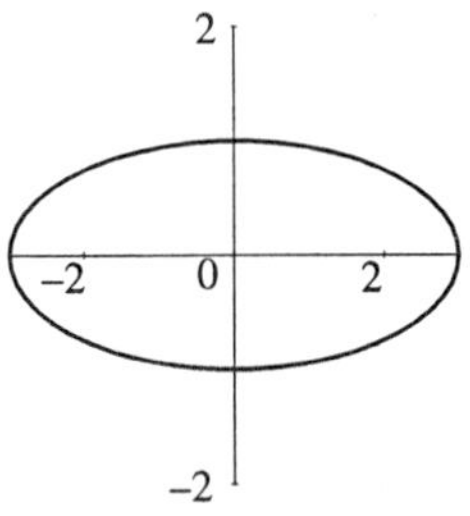

$x^2 + 9y^2 = 9\sin^2 t + 9\cos^2 t = 9 \quad \Rightarrow \quad x^2 + 9y^2 = 9$, which is a rectangular-coordinate equation.

7. Sketch the curve represented by $x = \cos^2 t$, $y = \cos^6 t$ and find a rectangular-coordinate equation for the curve.

Answer: medium

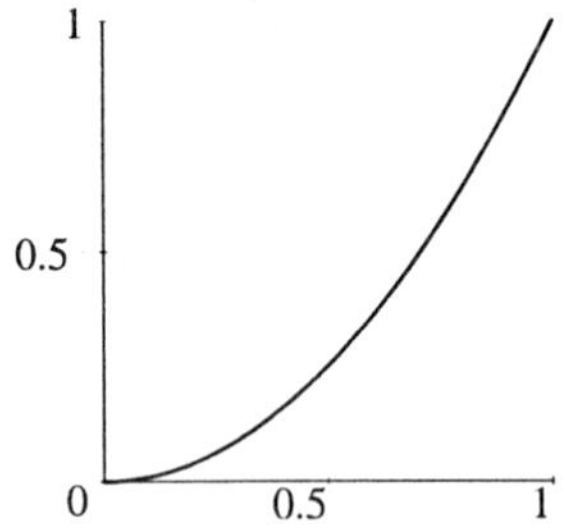

$x^3 = \cos^6 t = y$, so $y = x^3$, $0 \le x \le 1$, which is a rectangular-coordinate equation.

8. Sketch the curve represented by $x = \sin 2t$, $y = \sin t$ and find a rectangular-coordinate equation for the curve.

Answer: medium

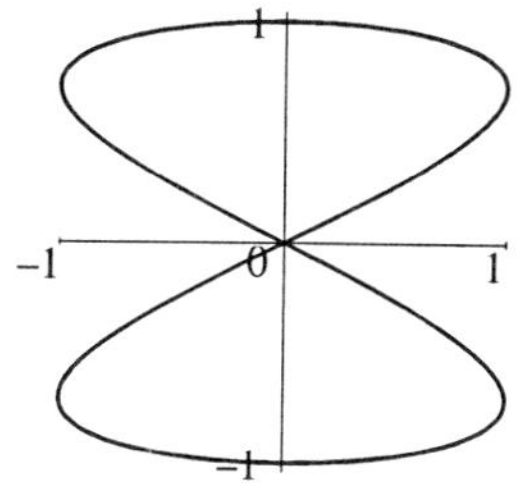

$x = \sin 2t = 2\sin t \cos t \quad \Rightarrow \quad x^2 = 4\sin^2 t \cos^2 t = 4y^2(1 - y^2)$, $-1 \le x \le 1$, $-1 \le y \le 1$, which is a rectangular-coordinate equation.

9. Sketch the curve represented by $x = \sec t$, $y = \cot t$ and find a rectangular-coordinate equation for the curve.

Answer: medium

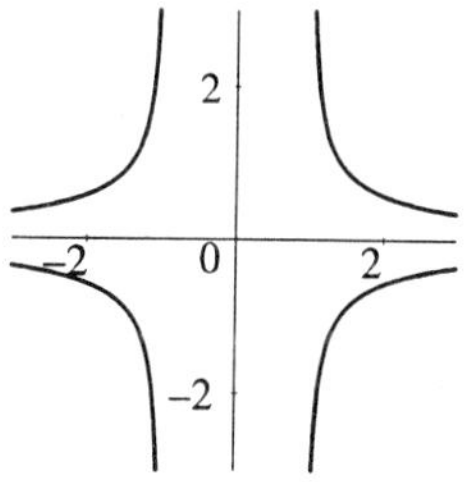

$y = \cot t \quad \Rightarrow \quad \dfrac{1}{y} = \tan t \quad \Rightarrow \quad 1 + \dfrac{1}{y^2} = 1 + \tan^2 t = \sec^2 t = x^2$, which is a rectangular-coordinate equation.

10. Sketch the curve represented by $x = -e^{2t}$, $y = e^t$ and find a rectangular-coordinate equation for the curve.

Answer: medium

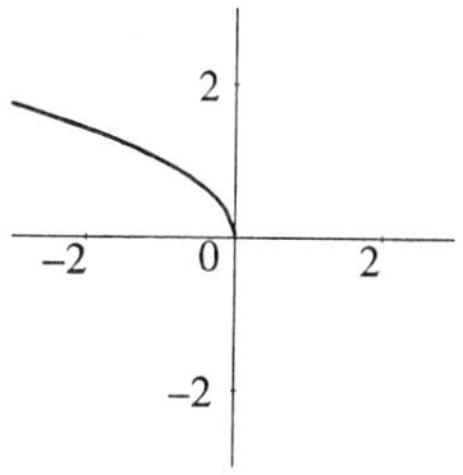

$x = -e^{2t} = -\left(e^t\right)^2 = -y^2$, $y > 0$, which is a rectangular-coordinate equation.

11. Sketch the curve represented by $x = e^t$, $y = e^{-2t}$ and find a rectangular-coordinate equation for the curve.

Answer: medium

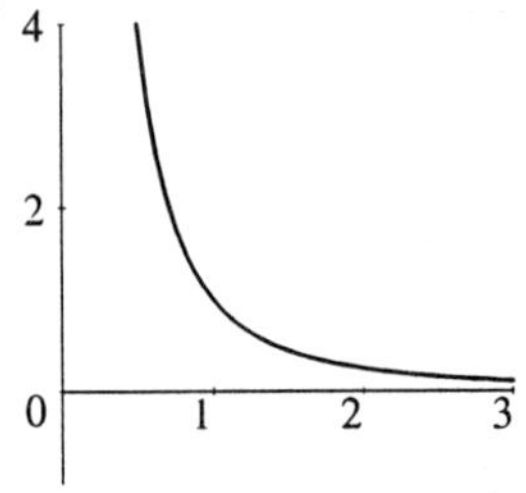

$y = (e^t)^{-2} = x^{-2} = \dfrac{1}{x^2}$, $x > 0$, which is a rectangular-coordinate equation.

12. Find parametric equations for the line of slope 2 passing through the point $(-2, 4)$.

Answer: easy

$x = -2 + t \quad y = 4 + 2t.$

13. Find parametric equations for the line of slope $-\frac{4}{5}$ passing through the point $(7, -1)$.

Answer: easy

$x = 7 - 5t \quad y = -1 + 4t.$

14. Find parametric equations for the line passing through the points $(3, 5)$ and $(-5, 7)$.

Answer: easy

The slope of the line is $\dfrac{7-5}{-5-3} = -\dfrac{1}{4}$, so $x = 3 - 4t \quad y = 5 + t$ are parametric equations for the line.

15. Find parametric equations for the parabola $y = x^2 - 3$.

Answer: easy

Let $x = t$, then the parametric equations are $x = t \quad y = t^2 - 3$.

16. Sketch the curve represented by $x = t \sin t$, $y = \cos t$.

Answer: hard

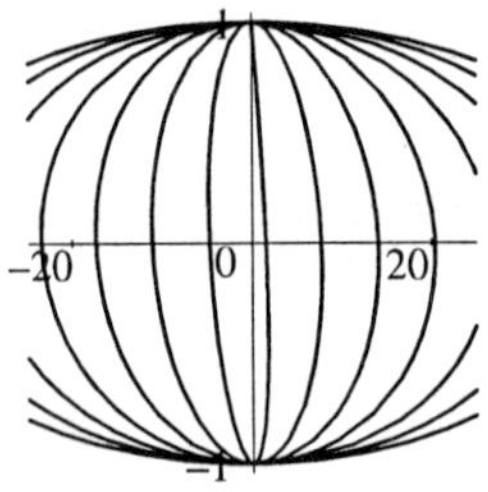

17. Sketch the curve represented by $x = \sin t$, $y = e^{t/2}$.

Answer: hard

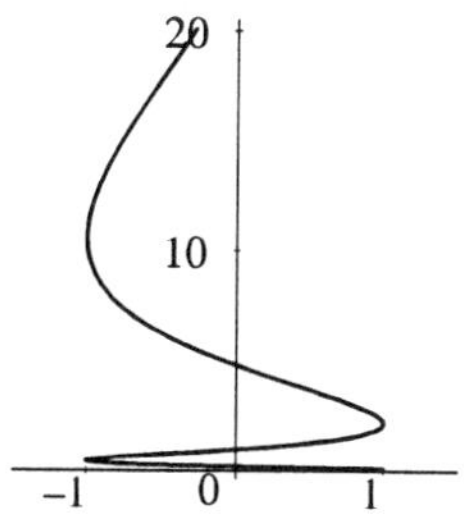

18. Use a graphing device to sketch the curve represented by $x = 2 \sin t \quad y = \cos 2t$.

Answer: easy

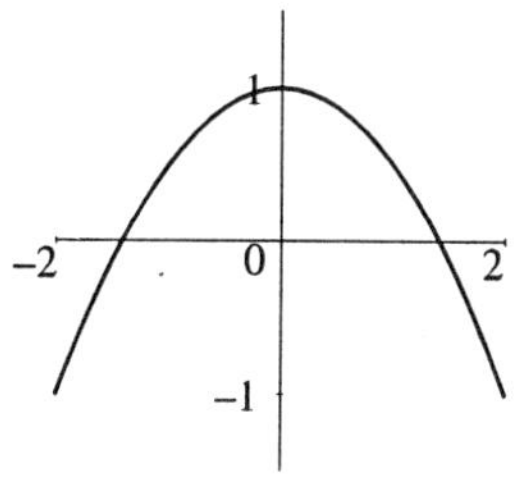

19. Use a graphing device to sketch the curve represented by $x = 2 \cos 7t \quad y = 7 \sin 2t$.

Answer: easy

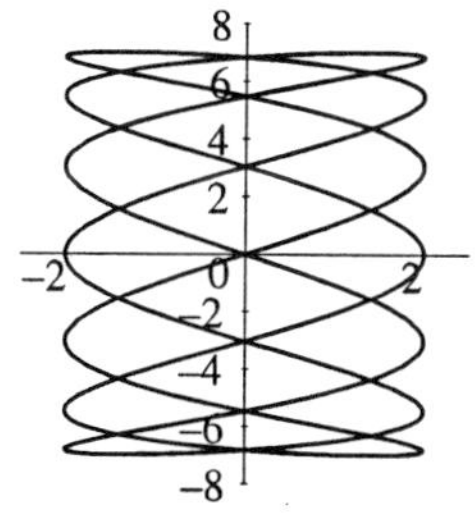

20. Usc a graphing dcvicc to skctch thc curvc rcprcscntcd by x — $\sin 5t \quad y$ — $\cos 2t$.

Answer: easy

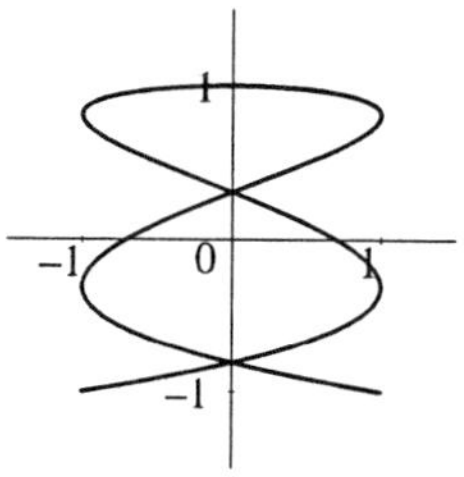

CHAPTER 11 REVIEW

1. Find the focus, directrix, and focal diameter of the parabola $y^2 = 24x$, and sketch its graph.

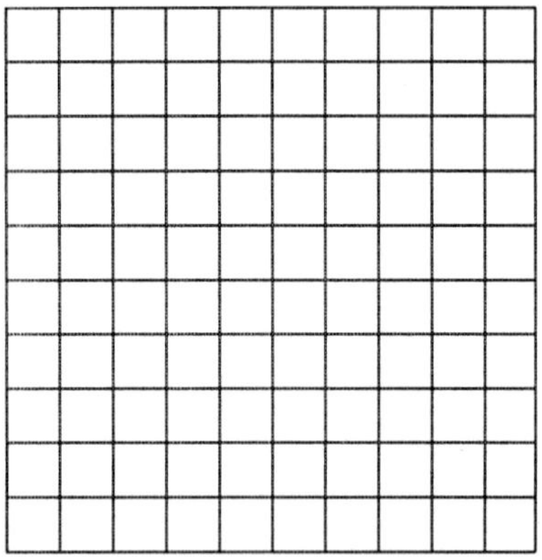

Answer: medium

$4p = 24 \quad \Rightarrow \quad p = 6$. So the focus is $F(6, 0)$, the directrix is $x = -6$, and the focal diameter is $4p = 24$.

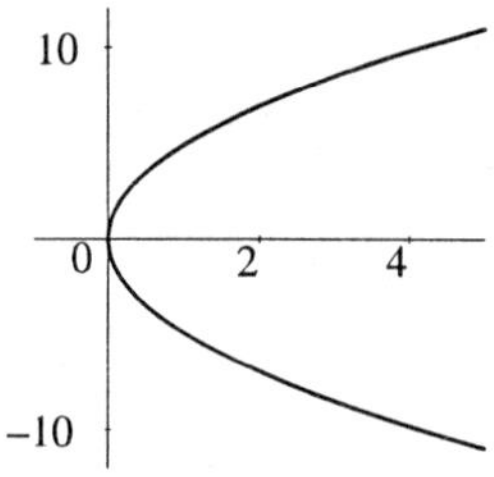

2. Find the focus, directrix, and focal diameter of the parabola $12x + 7y^2 = 0$, and sketch its graph.

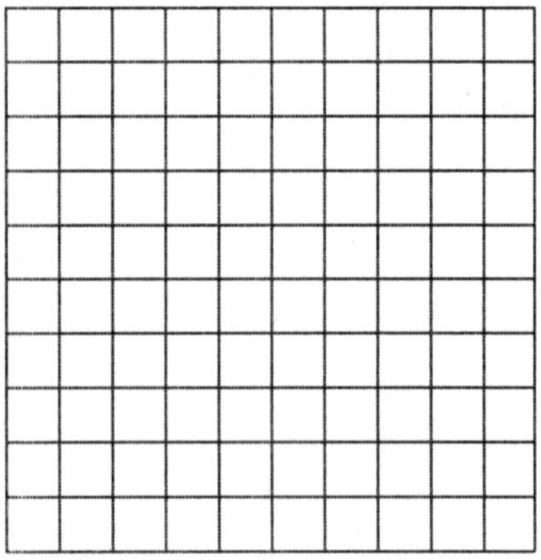

Answer: medium

$12x + 7y^2 = 0 \quad \Leftrightarrow \quad y^2 = -\frac{12}{7}x \quad \Rightarrow \quad 4p = -\frac{12}{7}$
$\Rightarrow \quad p = -\frac{3}{7}$. So the focus is $F\left(-\frac{3}{7}, 0\right)$, the directrix is $x = \frac{3}{7}$, and the focal diameter is $|4p| = \frac{12}{7}$.

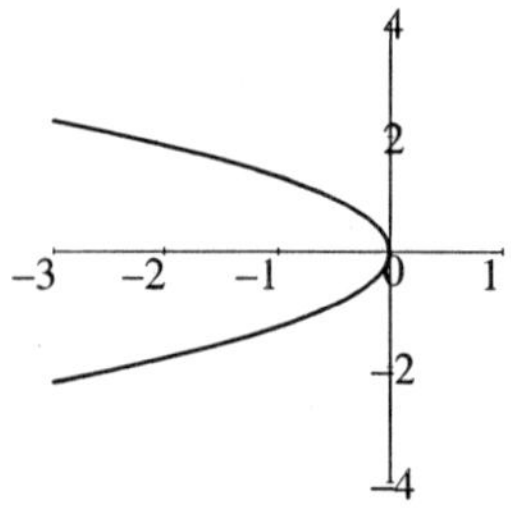

3. Find an equation for a parabola with vertex at the origin, and with focus $F(0, -5)$.
Answer: medium
$p = -5$, and the parabola is of the form $x^2 = 4py$, so its equation is $x^2 = -20y$.

4. Find an equation for a parabola with vertex at the origin, focal diameter 6, and focus on the positive y-axis.
Answer: medium
The length of the focal diameter is $4|p| = 6$. Focus on the positive y-axis means that $p > 0$, and the parabola is of the form $x^2 = 4py$, so its equation is $x^2 = 6y$.

5. Find the vertices, foci, and eccentricity of the ellipse given by $\frac{x^2}{144} + \frac{y^2}{9} = 1$. Determine the lengths of the major and minor axes, and sketch the graph.

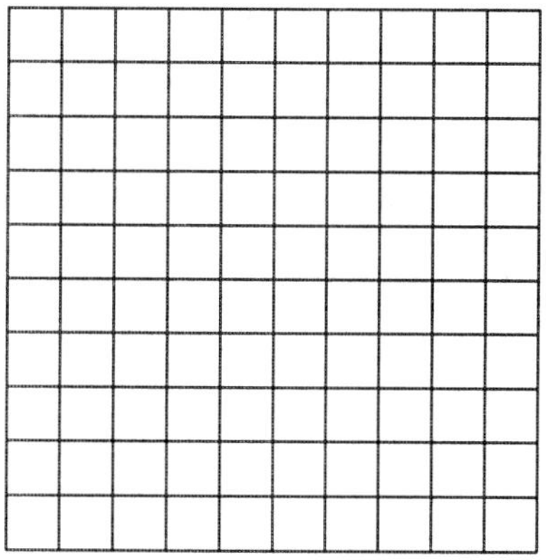

Answer: medium

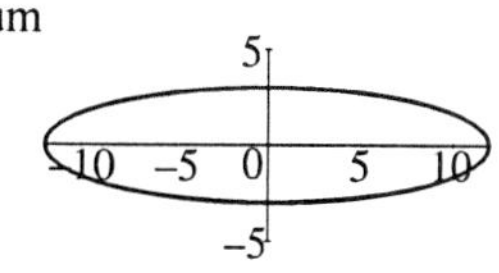

The denominator of x^2 is larger, so the ellipse has horizontal major axis. $a^2 = 144 \quad \Rightarrow \quad a = 12, b^2 = 9 \quad \Rightarrow \quad b = 3$. $c^2 = a^2 - b^2 = 135 \quad \Rightarrow \quad c = 3\sqrt{15}$. So the vertices are $(\pm 12, 0)$, foci are $\left(\pm 3\sqrt{15}, 0\right)$, eccentricity is $\frac{3\sqrt{15}}{12} = \frac{\sqrt{15}}{4}$. Length of major axis is 24, and minor is 6.

6. Find an equation for the ellipse whose graph is shown.

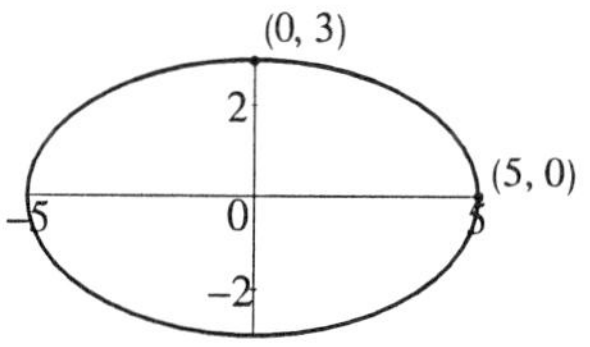

Answer: medium
From the graph, $a = 5$, $b = 3$, and the major axis is horizontal. Thus the equation of the ellipse is $\frac{x^2}{25} + \frac{y^2}{9} = 1$.

7. Find an equation for the ellipse whose foci are $(0, \pm 7)$, and whose vertices are $(0, \pm 3)$.
Answer: medium
$c = 3$, since the foci are $(0, \pm 3)$, and $a = 7$ since the vertices are $(0, \pm 7)$, so $b^2 = a^2 - c^2 = 40$. The major axis is vertical, so the equation of the ellipse is $\frac{x^2}{b^2} + \frac{y^2}{a^2} = 1 \quad \Leftrightarrow \quad \frac{x^2}{40} + \frac{y^2}{49} = 1$.

8. Find an equation for the ellipse with eccentricity $\frac{1}{2}$, and vertical major axis of length $4\sqrt{3}$.
Answer: medium
From the length of major axis, $a = 2\sqrt{3}$. $e = \frac{c}{a} \quad \Leftrightarrow \quad c = ae = 2\sqrt{3} \cdot \frac{1}{2} = \sqrt{3}$. $b^2 = a^2 - c^2 = 9$. The major axis is vertical, so the equation is $\frac{x^2}{b^2} + \frac{y^2}{a^2} = 1 \quad \Leftrightarrow \quad \frac{x^2}{9} + \frac{y^2}{12} = 1$

9. Find the vertices, foci, and asymptotes of the hyperbola $\dfrac{x^2}{49} - \dfrac{y^2}{100} = 1$, and sketch its graph.

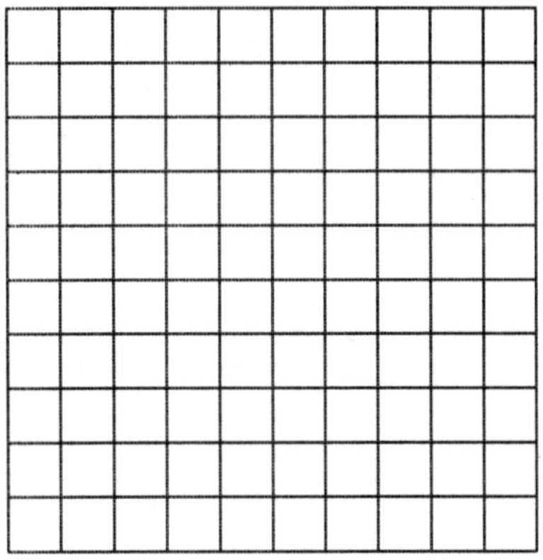

Answer: medium

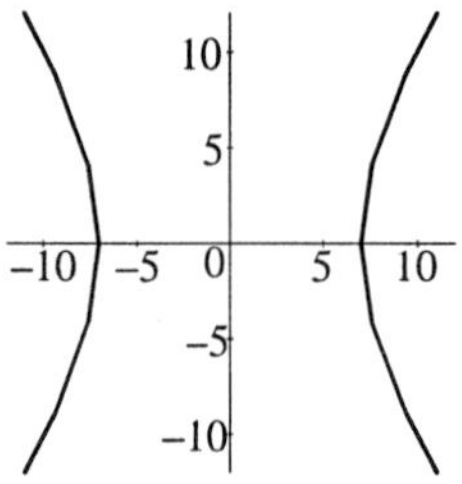

The transverse axis is horizontal of length $2a = 14$, so the vertices are $(\pm 7, 0)$. $c^2 = a^2 + b^2 = 149$, so the foci are $\left(\pm\sqrt{149}, 0\right)$. The asymptotes are $y = \pm\frac{10}{7}x$.

10. Find an equation for the hyperbola whose graph is shown.

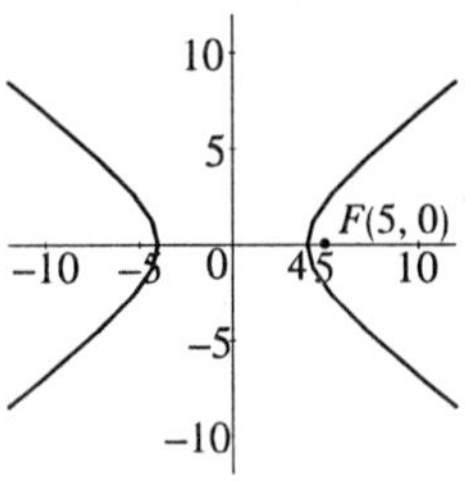

Answer: medium

From the graph, $a = 4$, $c = 5$, and the transverse axis is horizontal. $b^2 = c^2 - a^2 = 9$, so the equation of the hyperbola is $\dfrac{x^2}{16} - \dfrac{y^2}{9} = 1$.

11. Find an equation for the hyperbola that has foci $(0, \pm 7)$ and vertices $(0, \pm 4)$.

Answer: $a = 4$, and $c = 7$ $\Rightarrow$ $b^2 = c^2 - a^2 = 33$. Since the vertices lie on the y-axis, the equation of the hyperbola is $\dfrac{y^2}{a^2} - \dfrac{x^2}{b^2} = 1 \quad \Leftrightarrow \quad \dfrac{y^2}{16} - \dfrac{x^2}{33} = 1$.

12. Find an equation for the hyperbola that has foci $(\pm 4, 0)$ and passes through the point $(4, 6)$.

Answer: medium

Since the foci lie on the x-axis, the transverse axis is horizontal. $c = 4$. $c^2 = a^2 + b^2$, so an equation for the hyperbola is $\dfrac{x^2}{a^2} - \dfrac{y^2}{c^2 - a^2} = 1 \quad \Leftrightarrow \quad \dfrac{x^2}{a^2} - \dfrac{y^2}{16 - a^2} = 1$. Since $(4, 6)$ lies on the hyperbola, $\dfrac{16}{a^2} - \dfrac{36}{16 - a^2} = 1 \quad \Leftrightarrow$ $16(16 - a^2) - 36a^2 = a^2(16 - a^2) \quad \Leftrightarrow \quad 256 - 16a^2 - 36a^2 = 16a^2 - a^4 \quad \Leftrightarrow \quad a^4 - 68a^2 + 256 = 0 \quad \Leftrightarrow$ $a^2 = \dfrac{68 \pm \sqrt{68^2 - 4 \cdot 256}}{2} = \dfrac{68 \pm 60}{2} = 4, 64$. But $a^2 \neq 64$, since $b^2 = 16 - a^2$, so $a^2 = 4$, and $b^2 = 16 - 4 = 12$, and an equation for the hyperbola is $\dfrac{x^2}{4} - \dfrac{y^2}{12} = 1$.

13. Find the center, foci, and vertices of the ellipse given by $\dfrac{(x-3)^2}{16} + (y-2)^2 = 1$. Determine the lengths of the major and minor axes, and sketch the graph.

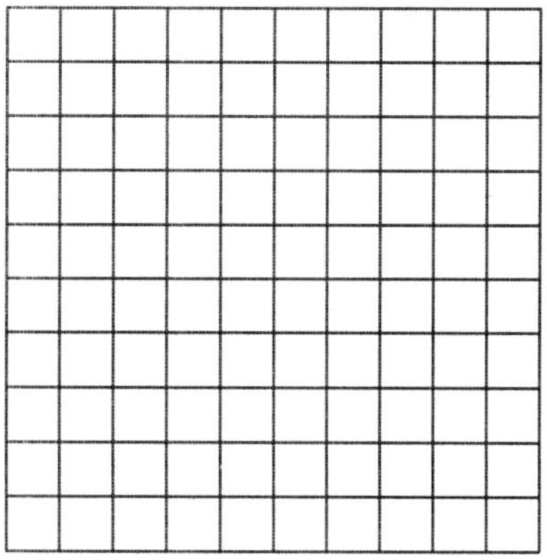

Answer: medium

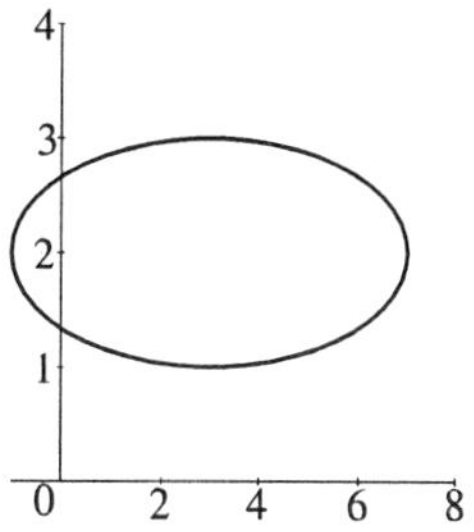

The center of the ellipse is $(3, 2)$. The denominator of $(x-3)^2$ is larger, so the ellipse has horizontal major axis. $a^2 = 16 \Rightarrow a = 4$, $b^2 = 1 \Rightarrow b = 1$. $c^2 = a^2 - b^2 = 15$. The foci of the shifted ellipse are $\left(3 \pm \sqrt{15}, 2\right)$. The vertices of the major axis are $(3 \pm 4, 2)$, and the vertices of the minor axis are $(3, 2 \pm 1)$. The lengths of the major and minor axes are 8 and 2.

14. Find an equation for the conic whose graph is shown.

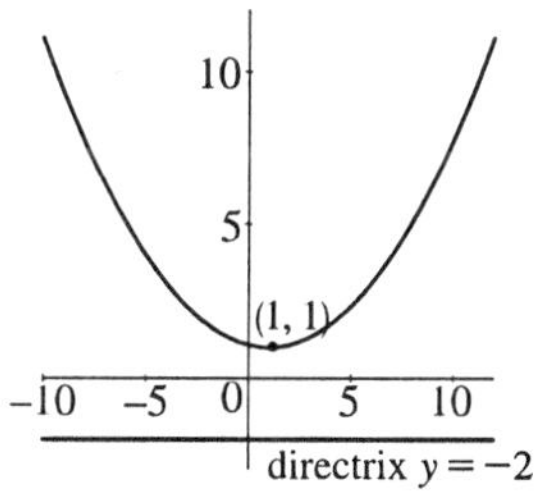

Answer: medium
$p = 3$, so the equation of the unshifted parabola is $x^2 = 12y$. The vertex of the parabola is $(1, 1)$, so the shifted parabola is $(x-1)^2 = 12(y-1)$

15. Complete the square to determine whether the equation, $x^2 + 9y^2 - 2x + 18y + 1 = 0$, represents an ellipse, a parabola, a hyperbola, or a degenerate conic. Then sketch the graph of the equation. If the graph is an ellipse, find the center, foci, vertices, and lengths of the major and minor axes. If it is a parabola, find the vertex, focus and directrix. If it is a hyperbola, find the center, foci, vertices, and asymptotes. If the equation has no graph, explain why.

Answer: hard

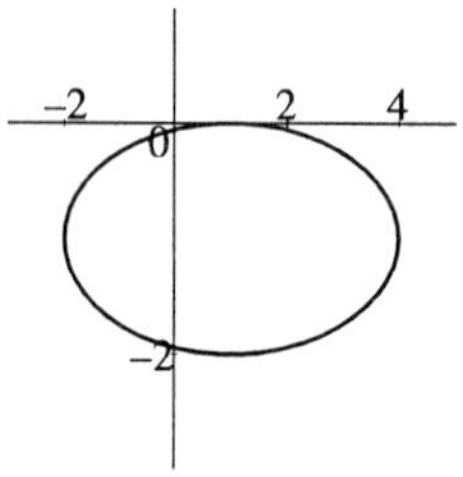

$x^2 + 9y^2 - 2x + 18y + 1 = 0 \quad \Leftrightarrow \quad x^2 - 2x + 1 + 9y^2 + 18y + 9 = 9 \quad \Leftrightarrow \quad (x-1)^2 + 9(y+1)^2 = 9 \quad \Leftrightarrow$ $\dfrac{(x-1)^2}{9} + (y+1)^2 = 1$, which is an equation for an ellipse. The center of the ellipse is $(1, -1)$, the major axis is horizontal of length 6, minor is of length 2. The vertices are $(-2, 0)$, $(4, 0)$, $(1, -2)$, $(1, 0)$. $c^2 = a^2 - b^2 = 8$, so the foci are $\left(1 \pm 2\sqrt{2}, -1\right)$.

16. Complete the square to determine whether the equation, $y^2 - 10y + 2x^2 + 8x = -33$, represents an ellipse, a parabola, a hyperbola, or a degenerate conic. Then sketch the graph of the equation. If the graph is an ellipse, find the center, foci, vertices, and lengths of the major and minor axes. If it is a parabola, find the vertex, focus and directrix. If it is a hyperbola, find the center, foci, vertices, and asymptotes. If the equation has no graph, explain why.

Answer: hard

$y^2 - 10y + 2x^2 + 8x = -33 \quad \Leftrightarrow$

$y^2 - 10y + 25 + 2(x^2 + 4x + 4) = 0 \quad \Leftrightarrow$

$(y-5)^2 + 2(x+2)^2 = -7$, which has no solution since the sum of two squares cannot be negative.

17. Determine the XY-coordinates of $(3, 7)$ if the axes are rotated through an angle $\phi = 30°$.

Answer: $X = x\cos\phi + y\sin\phi = 3\frac{\sqrt{3}}{2} + 7\frac{1}{2} = \frac{3\sqrt{3}+7}{2}$. $Y = -x\sin\phi + y\cos\phi = \frac{7\sqrt{3}-3}{2}$

18. Determine the equation of the conic $3xy = 2x^2 + y^2$ in XY-coordinates if the axes are rotated through an angle $\phi = 60°$.
Answer: medium
$x = X\cos 60 - Y\sin 60 = \frac{1}{2}X - \frac{\sqrt{3}}{2}Y$, $y = X\sin 60 + Y\cos 60 = \frac{\sqrt{3}}{2}X + \frac{1}{2}Y$. $3xy = 2x^2 + y^2 \quad \Leftrightarrow$
$3\left(\frac{1}{2}X - \frac{\sqrt{3}}{2}Y\right)\left(\frac{\sqrt{3}}{2}X + \frac{1}{2}Y\right) = 2\left(\frac{1}{2}X - \frac{\sqrt{3}}{2}Y\right)^2 + \left(\frac{\sqrt{3}}{2}X + \frac{1}{2}Y\right)^2 \quad \Leftrightarrow$
$3\left(\frac{\sqrt{3}}{4}X^2 + \frac{1}{4}XY - \frac{3}{4}XY - \frac{\sqrt{3}}{4}Y^2\right) = 2\left(\frac{1}{4}X^2 - \frac{\sqrt{3}}{2}XY + \frac{3}{4}Y^2\right) + \left(\frac{3}{4}X^2 + \frac{\sqrt{3}}{2}XY + \frac{1}{4}Y^2\right) \quad \Leftrightarrow$
$X^2\left(3\frac{\sqrt{3}}{4} - \frac{1}{2} - \frac{3}{4}\right) + XY\left(-\frac{3}{2} + \sqrt{3} - \frac{\sqrt{3}}{2}\right) + Y^2\left(-\frac{\sqrt{3}}{2} - \frac{3}{2} - \frac{1}{4}\right) = 0 \quad \Leftrightarrow \quad \frac{3\sqrt{3}-5}{4}X^2 + \frac{\sqrt{3}-3}{2}XY - \frac{2\sqrt{3}+7}{4}Y^2 = 0.$

19. Use the discriminant to determine if the graph of the equation, $3x^2 - 2xy + 3y^2 = 24$ is a parabola, an ellipse or a hyperbola, then use a rotation of axes to eliminate the xy-term, and sketch the graph.

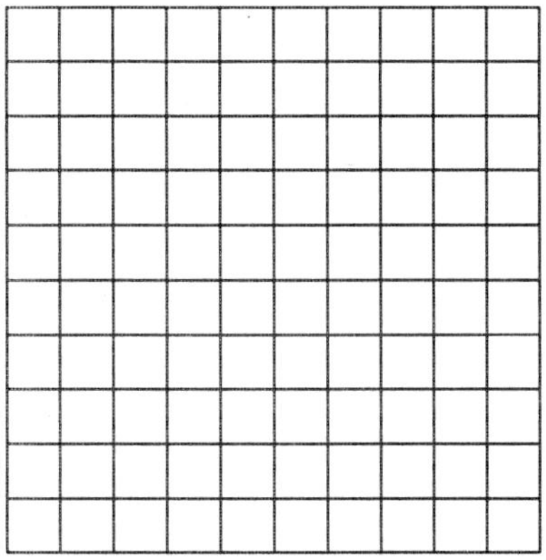

Answer: hard

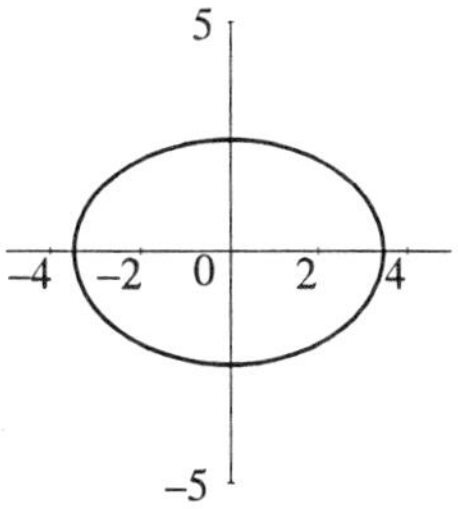

$B^2 - 4AC = 4 - 36 < 0$, so the graph is an ellipse. $\cot 2\phi = \frac{A-C}{B} = 0$, so $\phi = 45°$. $x = X\frac{1}{\sqrt{2}} - Y\frac{1}{\sqrt{2}}$, $y = X\frac{1}{\sqrt{2}} + Y\frac{1}{\sqrt{2}}$
$\Rightarrow \quad 3x^2 - 2xy + 3y^2 = 24 \quad \Leftrightarrow \quad 3\left(X\frac{1}{\sqrt{2}} - Y\frac{1}{\sqrt{2}}\right)^2 - 2\left(X\frac{1}{\sqrt{2}} - Y\frac{1}{\sqrt{2}}\right)\left(X\frac{1}{\sqrt{2}} + Y\frac{1}{\sqrt{2}}\right) + 3\left(X\frac{1}{\sqrt{2}} + Y\frac{1}{\sqrt{2}}\right)^2 = 24$
$\Leftrightarrow \quad \frac{3}{2}(X - Y)^2 - (X - Y)(X + Y) + \frac{3}{2}(X + Y)^2 = 24 \quad \Leftrightarrow$
$3(X^2 - 2XY + Y^2) - 2(X^2 - Y^2) + 3(X^2 + 2XY + Y^2) = 48 \quad \Leftrightarrow \quad 4X^2 + 8Y^2 = 48 \quad \Leftrightarrow \quad \dfrac{X^2}{12} + \dfrac{Y^2}{6} = 1.$

20. Find the rectangular coordinates of the point with polar coordinates $(7, \pi/6)$.
Answer: easy
$x = r\cos\theta = \frac{7\sqrt{3}}{2}$, $y = r\sin\theta = \frac{7}{2}$, so $(x, y) = \left(\frac{7\sqrt{3}}{2}, \frac{7}{2}\right)$.

21. Find the polar coordinates of the point with rectangular coordinates $\left(-6, 6/\sqrt{3}\right)$.
Answer: easy
$\tan\theta = -1/\sqrt{3} \quad \Rightarrow \quad \theta = 5\pi/6$, and $r = \sqrt{36 + 36/3} = 4\sqrt{3}$, so $(r, \theta) = \left(4\sqrt{3}, 5\pi/6\right)$.

22. Convert the equation $2x - y = 2$ to polar form.
Answer: easy
$2x - y = 2 \quad \Leftrightarrow \quad 2r\cos\theta - r\sin\theta = 2 \quad \Leftrightarrow \quad 2\cos\theta - \sin\theta = 2/r.$

23. Sketch the curve with polar equation $r = \sin^3\theta$.

Answer: medium

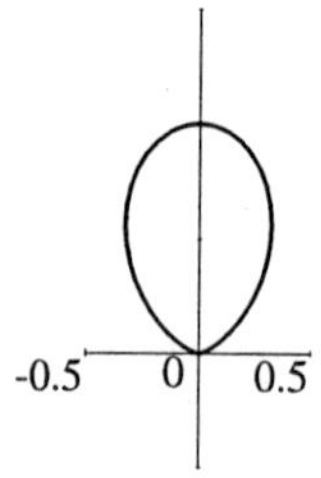

24. Write a polar equation of an ellipse with eccentricity $\frac{1}{4}$ and directrix $x = 3$.

Answer: easy

$$r = \frac{\frac{1}{4} \cdot 3}{1 + \frac{1}{4}\cos\theta} \quad \Leftrightarrow \quad r = \frac{3}{4 + \cos\theta}.$$

25. Write a polar equation of a hyperbola with eccentricity 1.2 and directrix $y = -4$.

Answer: easy

$$r = \frac{1.2 \cdot 4}{1 + 1.2\sin\theta} \quad \Leftrightarrow \quad r = \frac{24}{5 - 6\sin\theta}.$$

26. Find the eccentricity and identify the conic given by $r = \dfrac{8}{1 + \sin\theta}$, sketch it and label its vertices.

Answer: medium

$e = 1$, so the conic is a parabola.

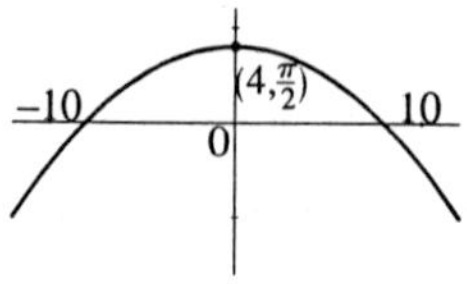

27. Find the eccentricity and directrix of the conic $r = \dfrac{4}{1 - 2\sin\theta}$ and graph the conic and its directrix. Find an equation for the conic rotated clockwise about its vertex through an angle $\pi/3$, and graph it.

Answer: medium

$e = 2$, $d = 2$, so the directrix is $y = -2$. The rotated conic is given by $r = \dfrac{4}{1 - 2\sin(\theta + \pi/3)}$.

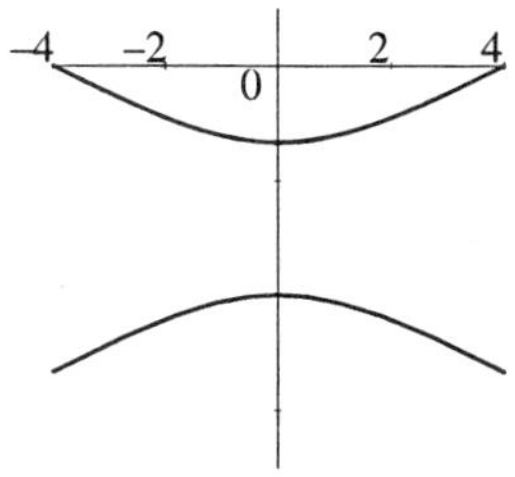

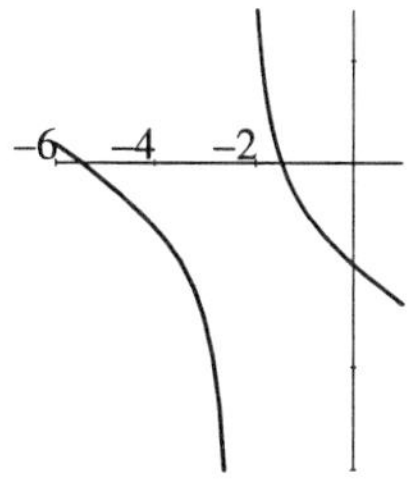

28. Sketch the curve represented by $x = 7t$, $y = 2t - 3$ and find a rectangular-coordinate equation for the curve.

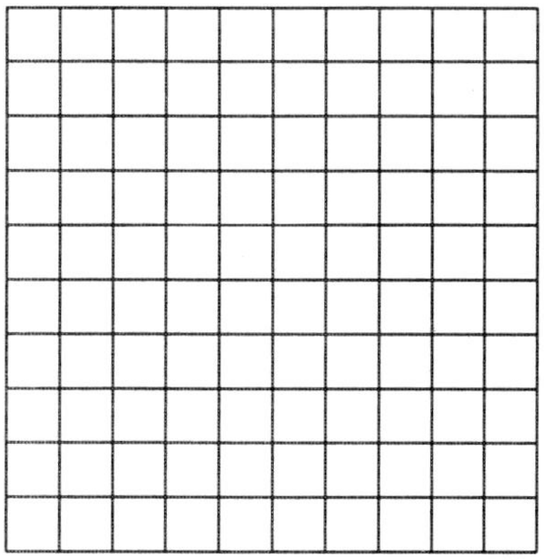

Answer: medium

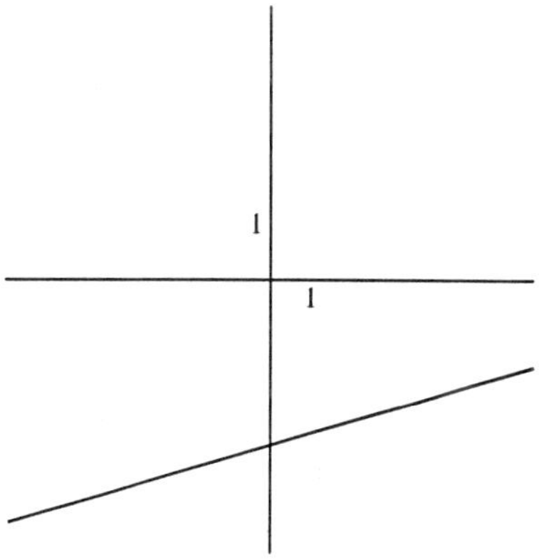

$x = 7t \quad \Leftrightarrow \quad t = \frac{1}{7}x$, so $y = 2t - 3 \quad \Rightarrow \quad y = \frac{2}{7}x - 3$, which is a rectangular-coordinate equation.

29. Find parametric equations for the line passing through the points $(1, -4)$ and $(-2, 3)$.

Answer: The slope of the line is $\dfrac{3 - -4}{-2 - 1} = -\dfrac{7}{3}$, so $x = 1 - 3t \quad y = -4 + 7t$ are parametric equations for the line.

30. Use a graphing device to sketch the curve represented by $x = 3\sin 4t \quad y = 2\cos 6t$.

Answer: easy

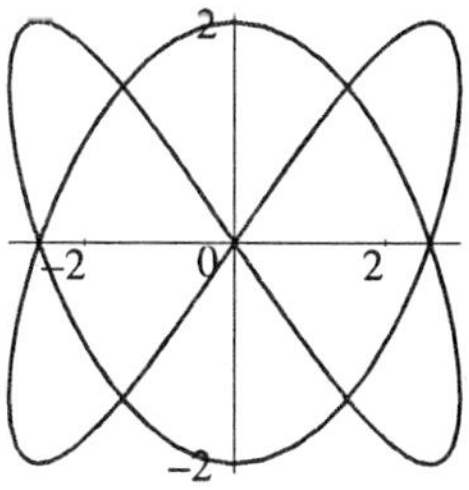

CHAPTER TWELVE

SECTION 12.1

1. Find the first four terms as well as the 1000th term of the sequence $a_n = n^3 + n^2 + n + 1$.

   Answer: easy

   $a_n = n^3 + n^2 + n + 1$. Then $a_1 = 1^3 + 1^2 + 1 + 1 = 4$, $a_2 = 2^3 + 2^2 + 2 + 1 = 15$, $a_3 = 3^3 + 3^2 + 3 + 1 = 40$, $a_4 = 4^3 + 4^2 + 4 + 1 = 85$, $a_{1000} = 1000^3 + 1000^2 + 1000 + 1 = 1{,}001{,}001{,}001$.

2. Find the first four terms as well as the 1000th term of the sequence $a_n = \dfrac{1}{2n^2}$.

   Answer: easy

   $a_n = \dfrac{1}{2n^2}$. Then $a_1 = \dfrac{1}{2 \cdot 1^2} = \dfrac{1}{2}$, $a_2 = \dfrac{1}{2 \cdot 2^2} = \dfrac{1}{8}$, $a_3 = \dfrac{1}{2 \cdot 3^2} = \dfrac{1}{18}$, $a_4 = \dfrac{1}{2 \cdot 4^2} = \dfrac{1}{32}$, $a_{1000} = \dfrac{1}{2 \cdot 1000^2} = \dfrac{1}{2{,}000{,}000}$

3. Find the first four terms as well as the 1000th term of the sequence $a_n = 2 - \dfrac{1}{2^n}$.

   Answer: easy

   $a_n = 2 - \dfrac{1}{2^n}$. Then $a_1 = 2 - \dfrac{1}{2^1} = \dfrac{3}{2}$, $a_2 = 2 - \dfrac{1}{2^2} = \dfrac{7}{4}$, $a_3 = 2 - \dfrac{1}{2^3} = \dfrac{15}{8}$, $a_4 = 2 - \dfrac{1}{2^4} = \dfrac{31}{16}$, $a_{1000} = 2 - \dfrac{1}{2^{1000}}$.

4. Find the first four terms as well as the 1000th term of the sequence $a_n = \dfrac{n^2}{n^3 + 1}$.

   Answer: easy

   $a_n = \dfrac{n^2}{n^3 + 1}$. Then $a_1 = \dfrac{1^2}{1^3 + 1} = \dfrac{1}{2}$, $a_2 = \dfrac{2^2}{2^3 + 1} = \dfrac{4}{9}$, $a_3 = \dfrac{3^2}{3^3 + 1} = \dfrac{9}{28}$, $a_4 = \dfrac{4^2}{4^3 + 1} = \dfrac{16}{65}$, $a_{1000} = \dfrac{1000^2}{1000^3 + 1}$.

5. Find the first four terms as well as the 1000th term of the sequence $a_n = 5n - 2$.

   Answer: easy

   $a_n = 5n - 2$. Then $a_1 = 5 \cdot 1 - 2 = 3$, $a_2 = 5 \cdot 2 - 2 = 8$, $a_3 = 5 \cdot 3 - 2 = 13$, $a_4 = 5 \cdot 4 - 2 = 18$, $a_{1000} = 5 \cdot 1000 - 2 = 4998$.

6. Find the first four terms as well as the 1000th term of the sequence $a_n = 6$.

   Answer: easy

   $a_n = 6$. Then $a_1 = a_2 = a_3 = a_4 = a_{1000} = 6$.

7. Find the first five terms of the recursively defined sequence $a_n = \dfrac{a_{n-1}}{3}$ and $a_1 = -81$.

   Answer: easy

   $a_n = \dfrac{a_{n-1}}{3}$. $a_1 = -81$, $a_2 = \dfrac{-81}{3} = -27$, $a_3 = \dfrac{-27}{3} = -9$, $a_4 = \dfrac{-9}{3} = -3$, $a_5 = \dfrac{-3}{3} = -1$.

8. Find the first five terms of the recursively defined sequence $a_n = (a_{n-1})^2$ and $a_1 = -2$.

   Answer: easy

   $a_n = (a_{n-1})^2$. $a_1 = -2$, $a_2 = (-2)^2 = 4$, $a_3 = 4^2 = 16$, $a_4 = 16^2 = 256$, $a_5 = 256^2 = 65{,}536$.

9. Find the first five terms of the recursively defined sequence $a_n = \dfrac{2}{2 + a_{n-1}}$ and $a_1 = 2$.

   Answer: medium

   $a_n = \dfrac{2}{2 + a_{n-1}}$. $a_1 = 2$, $a_2 = \dfrac{2}{2+2} = \dfrac{1}{2}$, $a_3 = \dfrac{2}{2 + 1/2} = \dfrac{4}{5}$, $a_4 = \dfrac{2}{2 + 4/5} = \dfrac{5}{7}$, $a_5 = \dfrac{2}{2 + 5/7} = \dfrac{14}{19}$.

10. Find the first five terms of the recursively defined sequence $a_n = a_{n-1} - a_{n-2} + a_{n-3}$, $a_1 = a_3 = 1$, and $a_2 = -1$.

    Answer: medium

    $a_n = a_{n-1} - a_{n-2} + a_{n-3}$. $a_1 = 1$, $a_2 = -1$, $a_3 = 1$, $a_4 = 1 - (-1) + 1 = 3$, $a_5 = 3 - 1 + (-1) = 1$.

11. Find the nth term of the sequence $-\frac{1}{2}, \frac{1}{4}, -\frac{1}{8}, \frac{1}{16}, \ldots$.

(a) $a_n = (-1)^n$ (b) $a_n = \dfrac{(-1)^n}{2^n}$ (c) $a_n = \dfrac{(-1)^n}{3^n}$ (d) $a_n = \dfrac{-n}{2^n}$ (e) $a_n = (-2)^n$

Answer: medium (b)

$-\dfrac{1}{2}, \dfrac{1}{4}, -\dfrac{1}{8}, \dfrac{1}{16}, \ldots \quad \Rightarrow \quad \dfrac{(-1)^1}{2^1}, \dfrac{(-1)^2}{2^2}, \dfrac{(-1)^3}{2^3}, \dfrac{(-1)^4}{2^4}, \ldots \quad \Rightarrow \quad a_n = \dfrac{(-1)^n}{2^n}.$

12. Find the nth term of the sequence $4, -16, 64, -256, \ldots$.

(a) $a_n = (-4)^n$ (b) $a_n = (-4)^{n+1}$ (c) $a_n = -4^n$ (d) $a_n = (-1)^{n+1} \cdot 4^n$ (e) $a_n = (-2)^{2n+1}$

Answer: medium (d)

$4, -16, 64, -256, \ldots \quad \Rightarrow \quad (-1)^2 \cdot 4^1, (-1)^3 \cdot 4^2, (-1)^4 \cdot 4^3, (-1)^5 \cdot 4^4, \ldots \quad \Rightarrow \quad a_n = (-1)^{n+1} \cdot 4^n.$

13. Find the nth term of the sequence $\frac{4}{5}, \frac{5}{6}, \frac{6}{7}, \frac{7}{8} \ldots$.

(a) $a_n = \dfrac{n+3}{n+4}$ (b) $a_n = \dfrac{n+1}{n+2}$ (c) $a_n = \dfrac{2n+1}{2n+2}$ (d) $a_n = \dfrac{n-2}{n-1}$ (e) $a_n = \dfrac{2n}{2n+1}$

Answer: easy (a)

$\dfrac{4}{5}, \dfrac{5}{6}, \dfrac{6}{7}, \dfrac{7}{8} \ldots \quad \Rightarrow \quad \dfrac{1+3}{1+4}, \dfrac{2+3}{2+4}, \dfrac{3+3}{3+4}, \dfrac{4+3}{4+4}, \ldots \quad \Rightarrow \quad a_n = \dfrac{n+3}{n+4}.$

14. Find the nth term of the sequence $a, a+2d, a+4d, a+6d, \ldots$.

(a) $a_n = a + 2nd$ (b) $a_n = a + 2(n+1)d$ (c) $a_n = a + 2(n-1)d$
(d) $a_n = 2a + nd$ (e) $a_n = 2a + (n-1)d$

Answer: medium (c)

$a, a+2d, a+4d, a+6d, \ldots \quad \Rightarrow \quad a_n = a + 2(n-1)d.$

15. Find the nth term of the sequence $1, -\frac{1}{2}, 3, -\frac{1}{4}, 5, -\frac{1}{6}, \ldots$.

(a) $a_n = (-1)^{n+1} n^{(-1)^{n+1}}$ (b) $a_n = (-n)^{-(n+1)}$ (c) $a_n = \dfrac{1}{(-n)^{-1/n}}$

(d) $a_n = (-1)^n n^{1-n}$ (e) $a_n = \left(\dfrac{1}{n}\right)^{(1/n)-1}$

Answer: hard (a)

$1, -\dfrac{1}{2}, 3, -\dfrac{1}{4}, 5, -\dfrac{1}{6}, \ldots \quad \Rightarrow \quad (1)(1)^1, (-1)(2)^{-1}, (1)(3)^1, (-1)(4)^{-1}, (1)(5)^1, (-1)(6)^{-1}, \ldots \quad \Rightarrow$
$a_n = (-1)^{n+1} n^{(-1)^{n+1}}.$

16. Find two different sequences that begin $2, 4, 8, 16, \ldots$

Answer: hard

$2, 4, 8, 16$. Since $2, 4, 8, 16 \Rightarrow 2^1, 2^2, 2^3, 2^4$, one sequence that begins this way is $a_n = 2^n$. Another sequence might be $a_n = 2^n + (n-1)(n-2)(n-3)(n-4)$.

17. Find the first 100 terms of the sequence defined by $a_{n+1} = \begin{cases} \frac{1}{2}a_n \text{ if } a_n \text{ is an even number} \\ 3a_n + 1 \text{ if } a_n \text{ is an odd number} \end{cases}$ and $a_1 = 13$.

Answer: hard

$a_{n+1} = \begin{cases} \frac{1}{2}a_n \text{ if } a_n \text{ is an even number} \\ 3a_n + 1 \text{ if } a_n \text{ is an odd number} \end{cases}$

If $a_1 = 13$, then

$a_2 = 40, a_3 = 20, a_4 = 10, a_5 = 5, a_6 = 16, a_7 = 8, a_8 = 4, a_9 = 2, a_{10} = 1, a_{11} = 4, a_{12} = 2, a_{13} = 1, a_{14} = 4, a_{15} = 2, a_{16} = 1, a_{17} = 4, a_{18} = 2, \ldots$ (with 4, 2, 1 repeating).

18. Find the first ten terms of the sequence defined recursively by $a_n = a_{n-a_{n-1}} + a_{n-a_{n-2}}$ and having the following first two terms:

(a) $a_1 = 1$ and $a_2 = 1$.
(b) $a_1 = 1$ and $a_2 = 2$.

Answer: hard

$a_n = a_{n-a_{n-1}} + a_{n-a_{n-2}}$

(a) $a_1 = 1, a_2 = 1$. Then

$a_3 = a_2 + a_2 = 1 + 1 = 2, a_4 = a_2 + a_3 = 1 + 2 = 3, a_5 = a_2 + a_3 = 1 + 2 = 3, a_6 = a_3 + a_3 = 2 + 2 = 4,$
$a_7 = a_3 + a_4 = 2 + 3 = 5, a_8 = a_3 + a_4 = 2 + 3 = 5, a_9 = a_4 + a_4 = 3 + 3 = 6, a_{10} = a_4 + a_5 = 3 + 3 = 6.$

(b) $a_1 = 1, a_2 = 2$. Then

$a_3 = a_1 + a_2 = 1 + 2 = 3, a_4 = a_1 + a_2 = 1 + 2 = 3, a_5 = a_2 + a_2 = 2 + 2 = 4, a_6 = a_2 + a_3 = 2 + 3 = 5,$
$a_7 = a_2 + a_3 = 2 + 3 = 5, a_8 = a_3 + a_3 = 3 + 3 = 6, a_9 = a_3 + a_4 = 3 + 3 = 6, a_{10} = a_4 + a_4 = 3 + 3 = 6.$

19. Find the sum $\sum_{n=1}^{3} \frac{1}{n}$.

(a) $\frac{11}{6}$ (b) $\frac{14}{3}$ (c) $\frac{16}{3}$ (d) $\frac{7}{4}$ (e) $\frac{11}{4}$

Answer: easy (a)

$$\sum_{n=1}^{3} \frac{1}{n} = \frac{1}{1} + \frac{1}{2} + \frac{1}{3} = \frac{6+3+2}{6} = \frac{11}{6}$$

20. Find the sum $\sum_{j=1}^{50} (-1)^j$.

(a) -1 (b) 0 (c) 1 (d) 50 (e) 100

Answer: medium (b)

$$\sum_{j=1}^{50} (-1)^j = (-1)^1 + (-1)^2 + (-1)^3 + (-1)^4 + \cdots + (-1)^{49} + (-1)^{50} = 0$$

21. Find the sum $\sum_{n=3}^{12} 10$.

(a) 10 (b) 20 (c) 80 (d) 100 (e) 200

Answer: easy (d)

$$\sum_{n=3}^{12} 10 = 10 + 10 + 10 + 10 + 10 + 10 + 10 + 10 + 10 + 10 = 100$$

22. Find the sum $\sum_{j=1}^{10} \frac{2}{j+3}$.

(a) $\frac{235{,}361}{335{,}152}$ (b) $\frac{485{,}333}{180{,}180}$ (c) $\frac{733{,}104}{356{,}360}$ (d) $\frac{891{,}362}{1{,}351{,}330}$ (e) 5

Answer: medium (b)

$$\sum_{j=1}^{10} \frac{2}{j+3} = \frac{2}{1+3} + \frac{2}{2+3} + \frac{2}{3+3} + \frac{2}{4+3} + \frac{2}{5+3} + \frac{2}{6+3} + \frac{2}{7+3} + \frac{2}{8+3} + \frac{2}{9+3} + \frac{2}{10+3} = \frac{485,333}{180,180}$$

23. Find the sum $\sum_{n=1}^{4} n \cdot 2^n$.

(a) 44 (b) 66 (c) 98 (d) 116 (e) 172

Answer: easy (c)

$$\sum_{n=1}^{4} n \cdot 2^n = 1 \cdot 2^1 + 2 \cdot 2^2 + 3 \cdot 2^3 + 4 \cdot 2^4 = 2 + 8 + 24 + 64 = 98$$

24. Find the sum $\sum_{k=1}^{1} k^{151}$.

(a) 0 (b) 1 (c) $\frac{151}{2}$ (d) 151 (e) 302

Answer: medium (b)

$$\sum_{k=1}^{1} k^{151} = 1^{151} = 1$$

25. Write the sum $\sum_{n=1}^{4} \frac{2n-1}{2n+1}$ without using sigma notation.

Answer: easy

$$\sum_{n=1}^{4} \frac{2n-1}{2n+1} = \frac{2 \cdot 1 - 1}{2 \cdot 1 + 1} + \frac{2 \cdot 2 - 1}{2 \cdot 2 + 1} + \frac{2 \cdot 3 - 1}{2 \cdot 3 + 1} + \frac{2 \cdot 4 - 1}{2 \cdot 4 + 1} = \frac{1}{3} + \frac{3}{5} + \frac{5}{7} + \frac{7}{9}$$

26. Write the sum $\sum_{k=4}^{50} x^k$ without using sigma notation.

Answer: easy

$$\sum_{k=4}^{50} x^k = x^4 + x^5 + x^6 + \cdots + x^{50}$$

27. Write the sum $\sum_{k=4}^{7} k(k+3)$ without using sigma notation.

Answer: easy

$$\sum_{k=4}^{7} k(k+3) = 4 \cdot 7 + 5 \cdot 8 + 6 \cdot 9 + 7 \cdot 10 = 28 + 40 + 54 + 70$$

28. Write the sum $\sum_{j=1}^{4} \frac{x^j}{j^2}$ without using sigma notation.

Answer: easy

$$\sum_{j=1}^{4} \frac{x^j}{j^2} = \frac{x}{1} + \frac{x^2}{2^2} + \frac{x^3}{3^2} + \frac{x^4}{4^2} = x + \frac{x^2}{4} + \frac{x^3}{9} + \frac{x^4}{16}$$

29. Write the sum $2 + 4 + 6 + \cdots + 4n$ in sigma notation.

(a) $\sum_{k=1}^{2n} 2k$ (b) $\sum_{k=1}^{4n} k$ (c) $\sum_{k=1}^{2n} 4k$ (d) $\sum_{k=1}^{n} 4k$ (e) $\sum_{k-1}^{n} k$

Answer: easy (a)

$$2 + 4 + 6 + \cdots + 4n = \sum_{k=1}^{2n} 2k$$

30. Write the sum $1 + x + x^2 + x^3 + \cdots + x^{200}$ in sigma notation.

(a) $\sum_{k=0}^{100} x^{2k}$ (b) $\sum_{k=0}^{200} x^k$ (c) $\sum_{k=100}^{200} x^k$ (d) $\sum_{k=1}^{199} x^k$ (e) $\sum_{k=0}^{200} k$

Answer: easy (b)

$$1 + x + x^2 + x^3 + \cdots + x^{200} = \sum_{k=0}^{200} x^k$$

31. Write the sum $\frac{10}{16} + \frac{11}{17} + \frac{12}{18} + \cdots + \frac{100}{106}$ in sigma notation.

(a) $\sum_{k=0}^{10} \frac{k}{k+6}$ (b) $\sum_{k=10}^{100} (k+6)$ (c) $10\sum_{k=10}^{100} \frac{1}{k+6}$ (d) $\sum_{k=10}^{100} \frac{k}{k+6}$ (e) $\sum_{k=0}^{100} \frac{k+6}{106}$

Answer: easy (d)

$$\frac{10}{16} + \frac{11}{17} + \frac{12}{18} + \cdots + \frac{100}{106} = \sum_{k=10}^{100} \frac{k}{k+6}$$

32. Write the sum $\frac{1}{1 \cdot 3} + \frac{1}{2 \cdot 4} + \frac{1}{3 \cdot 5} + \cdots + \frac{1}{999 \cdot 1001}$ in sigma notation.

(a) $\sum_{k=1}^{1000} \frac{1}{k(k-1)}$ (b) $\sum_{k=1}^{1000} \frac{1}{k(k-2)}$ (c) $\sum_{k=1}^{999} \frac{1}{k^2}$ (d) $\sum_{k=1}^{999} \frac{1}{k(k+2)}$ (e) $\sum_{k=1}^{1000} \frac{1}{k^2}$

Answer: easy (d)

$$\frac{1}{1 \cdot 3} + \frac{1}{2 \cdot 4} + \frac{1}{3 \cdot 5} + \cdots + \frac{1}{999 \cdot 1001} = \sum_{k=1}^{999} \frac{1}{k(k+2)}$$

33. Write the sum $\frac{\sqrt[3]{1}}{1^2} + \frac{\sqrt[3]{2}}{2^2} + \frac{\sqrt[3]{3}}{3^2} + \cdots + \frac{\sqrt[3]{n}}{n^2}$ using sigma notation.

(a) $\sum_{k=1}^{n} \frac{\sqrt[3]{k}}{k^2}$ (b) $\sum_{n=0}^{k} \frac{\sqrt[3]{n}}{n^2}$ (c) $\sum_{k=1}^{n} k^2$ (d) $\sum_{k=n}^{2n} k$ (e) $\sum_{k=1}^{n} \frac{3}{k}$

Answer: easy (a)

$$\frac{\sqrt[3]{1}}{1^2} + \frac{\sqrt[3]{2}}{2^2} + \frac{\sqrt[3]{3}}{3^2} + \cdots + \frac{\sqrt[3]{n}}{n^2} = \sum_{k=1}^{n} \frac{\sqrt[3]{k}}{k^2}$$

34. Find the first four terms and the 1000th term of the sequence $a_n = \dfrac{n^{2n}}{n^{n+1}}$.

Answer: easy

$a_n = \dfrac{n^{2n}}{n^{n+1}} = n^{2n-(n+1)} = n^{n-1}$. $a_1 = 1^0 = 1, a_2 = 2^1 = 2, a_3 = 3^2 = 9, a_4 = 4^3 = 64, a_{1000} = 1000^{999}$.

35. Find the first six terms of the given recursively defined sequence $a_n = \dfrac{a_{n-1} + a_{n-2}}{2a_{n-2}}$ and $a_1 = 1, a_2 = 2$.

Answer: medium

$a_n = \dfrac{a_{n-1} + a_{n-2}}{2a_{n-2}}$. $a_1 = 1, a_2 = 2, a_3 = \dfrac{1}{2}\dfrac{a_2 + a_1}{a_1} = \dfrac{1}{2}\left(\dfrac{2+1}{1}\right) = \dfrac{3}{2}, a_4 = \dfrac{1}{2}\dfrac{a_3 + a_2}{a_2} = \dfrac{1}{2}\left(\dfrac{\frac{3}{2}+2}{2}\right) = \dfrac{7}{8},$
$a_5 = \dfrac{1}{2}\dfrac{a_4 + a_3}{a_3} = \dfrac{1}{2}\left(\dfrac{\frac{7}{8}+\frac{3}{2}}{\frac{3}{2}}\right) = \dfrac{19}{24}, a_6 = \dfrac{1}{2}\dfrac{a_5 + a_4}{a_4} = \dfrac{1}{2}\left(\dfrac{\frac{19}{24}+\frac{7}{8}}{\frac{7}{8}}\right) = \dfrac{20}{21}$

36. Find the nth term of the sequence $\frac{1}{4}, -\frac{1}{9}, \frac{1}{16}, -\frac{1}{25}, \ldots$.

Answer: medium

$\dfrac{1}{4}, -\dfrac{1}{9}, \dfrac{1}{16}, \dfrac{-1}{25}, \ldots \Rightarrow \dfrac{(-1)^{1+1}}{(1+1)^2}, \dfrac{(-1)^{2+1}}{(2+1)^2}, \dfrac{(-1)^{3+1}}{(3+1)^2}, \dfrac{(-1)^{4+1}}{(4+1)^2}, \ldots, \dfrac{(-1)^{n+1}}{(n+1)^2}$. $a_n = \dfrac{(-1)^{n+1}}{(n+1)^2}$

37. Find the first six partial sums $S_1, S_2, S_3, S_4, S_5, S_6$ of the sequence 1, 8, 27, 64, ...

Answer: easy

$$\begin{aligned}
S_1 &= 1 && = 1\\
S_2 &= 1 + 8 && = 9\\
S_3 &= 1 + 8 + 27 && = 36\\
S_4 &= 1 + 8 + 27 + 64 && = 100\\
S_5 &= 1 + 8 + 27 + 64 + 125 && = 225\\
S_6 &= 1 + 8 + 27 + 64 + 125 + 6^3 && = 441
\end{aligned}$$

38. Find the first six partial sums $S_1, S_2, S_3, S_4, S_5, S_6$ of the sequence $\frac{1}{2}, \frac{1}{2^2}, \frac{1}{2^3}, \ldots$.

Answer: easy

$$\begin{aligned}
S_1 &= \tfrac{1}{2} && = \tfrac{1}{2}\\
S_2 &= \tfrac{1}{2} + \tfrac{1}{2^2} && = \tfrac{3}{4}\\
S_3 &= \tfrac{1}{2} + \tfrac{1}{2^2} + \tfrac{1}{2^3} && = \tfrac{7}{8}\\
S_4 &= \tfrac{1}{2} + \tfrac{1}{2^2} + \tfrac{1}{2^3} + \tfrac{1}{2^4} && = \tfrac{15}{16}\\
S_5 &= \tfrac{1}{2} + \tfrac{1}{2^2} + \tfrac{1}{2^3} + \tfrac{1}{2^4} + \tfrac{1}{2^5} && = \tfrac{31}{32}\\
S_6 &= \tfrac{1}{2} + \tfrac{1}{2^2} + \tfrac{1}{2^3} + \tfrac{1}{2^4} + \tfrac{1}{2^5} + \tfrac{1}{2^6} && = \tfrac{63}{64}
\end{aligned}$$

39. Find the first four partial sums and the nth partial sum of the sequence $a_n = \sqrt{n-1} - \sqrt{n}$.

Answer: medium

$$\begin{aligned}
S_1 &= \sqrt{0} - \sqrt{1} && = -\sqrt{1}\\
S_2 &= (\sqrt{0} - \sqrt{1}) + (\sqrt{1} - \sqrt{2}) && = -\sqrt{2}\\
S_3 &= (\sqrt{0} - \sqrt{1}) + (\sqrt{1} - \sqrt{2}) + (\sqrt{2} - \sqrt{3}) && = -\sqrt{3}\\
S_4 &= (\sqrt{0} - \sqrt{1}) + (\sqrt{1} - \sqrt{2}) + (\sqrt{2} - \sqrt{3}) + (\sqrt{3} - \sqrt{4}) && = -\sqrt{4}\\
S_n & && = -\sqrt{n}
\end{aligned}$$

40. Find the first four partial sums and the nth partial sum of the sequence $a_n = \dfrac{1}{n+2} - \dfrac{1}{n+3}$.

Answer: medium

$$\begin{aligned}
S_1 &= \left(\tfrac{1}{3} - \tfrac{1}{4}\right) &&= \tfrac{1}{3} - \tfrac{1}{4}\\
S_2 &= \left(\tfrac{1}{3} - \tfrac{1}{4}\right) + \left(\tfrac{1}{4} - \tfrac{1}{5}\right) &&= \tfrac{1}{3} - \tfrac{1}{5}\\
S_3 &= \left(\tfrac{1}{3} - \tfrac{1}{4}\right) + \left(\tfrac{1}{4} - \tfrac{1}{5}\right) + \left(\tfrac{1}{5} - \tfrac{1}{6}\right) &&= \tfrac{1}{3} - \tfrac{1}{6}\\
S_4 &= \left(\tfrac{1}{3} - \tfrac{1}{4}\right) + \left(\tfrac{1}{4} - \tfrac{1}{5}\right) + \left(\tfrac{1}{5} - \tfrac{1}{6}\right) + \left(\tfrac{1}{6} - \tfrac{1}{7}\right) &&= \tfrac{1}{3} - \tfrac{1}{7}\\
S_n &&&= \tfrac{1}{3} - \tfrac{1}{n+3}
\end{aligned}$$

41. Find the first four partial sums and the nth partial sum of the sequence $a_n = \dfrac{3}{4^n}$.

Answer: medium

$$\begin{aligned}
S_1 &= \tfrac{3}{4^1} &&= \tfrac{3}{4}\\
S_2 &= \tfrac{3}{4^1} + \tfrac{3}{4^2} &&= \tfrac{15}{16}\\
S_3 &= \tfrac{3}{4^1} + \tfrac{3}{4^2} + \tfrac{3}{4^3} &&= \tfrac{63}{64}\\
S_4 &= \tfrac{3}{4^1} + \tfrac{3}{4^2} + \tfrac{3}{4^3} + \tfrac{3}{4^4} &&= \tfrac{255}{256}\\
S_n &&&= \tfrac{4^n-1}{4^n}
\end{aligned}$$

SECTION 12.2

1. Determine the common difference, the fifth term, the nth term, and the 100th term for the arithmetic sequence $12, 9, 6, 3, \ldots$.
Answer: easy
$12, 9, 6, 3, \ldots$. Then $d = 9 - 12 = -3$, $a_5 = a_4 - 3 = 3 - 3 = 0$, $a_n = 12 - 3(n-1)$, $a_{100} = 12 - 3 \cdot 99 = -285$.

2. Determine the common difference, the fifth term, the nth term, and the 100th term for the arithmetic sequence $\frac{5}{6}, \frac{4}{3}, \frac{11}{6}, \frac{7}{3}, \ldots$.
Answer: easy
$\frac{5}{6}, \frac{4}{3}, \frac{11}{6}, \frac{7}{3}, \ldots \Rightarrow \frac{5}{6}, \frac{8}{6}, \frac{11}{6}, \frac{14}{6}, \ldots$. Then $d = \frac{8}{6} - \frac{5}{6} = \frac{1}{2}$, $a_5 = \frac{7}{3} + \frac{1}{2} = \frac{17}{6}$, $a_n = \frac{5}{6} + \frac{1}{2}(n-1)$,
$a_{100} = \frac{5}{6} + \frac{1}{2} \cdot 99 = \frac{5+297}{6} = \frac{151}{3}$.

3. Determine the common difference, the fifth term, the nth term, and the 100th term for the arithmetic sequence $15.3, 12.6, 9.9, 7.2, \cdots$.
Answer: easy
$15.3, 12.6, 9.9, 7.2, \ldots$. Then $d = 12.6 - 15.3 = -2.7$, $a_5 = 7.2 - 2.7 = 4.5$, $a_n = 15.3 - 2.7(n-1)$,
$a_{100} = 15.3 - 2.7 \cdot 99 = -252$.

4. Determine the common difference, the fifth term, the nth term, and the 100th term for the arithmetic sequence $t, t-3, t-6, t-9, \ldots$.
Answer: medium
$t, t-3, t-6, t-9, \ldots$. Then $d = (t-3) - t = -3$, $a_5 = t - 9 - 3 = t - 12$, $a_n = t - 3(n-1)$,
$a_{100} = t - 3 \cdot 99 = t - 297$.

5. The 11th term of an arithmetic sequence is 32 and the fifth term is 4. Find the 20th term.
(a) $a_{20} = 13$ (b) $a_{20} = 48$ (c) $a_{20} = 52$ (d) $a_{20} = 74$ (e) $a_{20} = 103$
Answer: medium (d)
$a_{11} = 32$, $a_5 = 4$, and $a_n = a + d(n-1)$. Then $a_5 = a + 4d = 4 \quad\Leftrightarrow\quad a = 4 - 4d$. Substituting into $a_{11} = a + 10d = 32$ gives $4 - 4d + 10d = 32 \quad\Leftrightarrow\quad 6d = 28 \quad\Leftrightarrow\quad d = \frac{14}{3}$, and then $a = 4 - \frac{56}{3} = -\frac{44}{3}$. Hence, the 20th term is
$a_{20} = -\frac{44}{3} + 19 \cdot \frac{14}{3} = 74$.

6. Show that the sequence defined recursively by $a_{n+1} = \dfrac{4a_n + 1}{4}$ and $a_1 = c$ is an arithmetic sequence and find the common difference.

Answer: medium

$a_{n+1} = \dfrac{4a_n + 1}{4}$ and $a_1 = c$. Thus $a_{n+1} - a_n = \dfrac{4a_n + 1}{4} - a_n = a_n + \dfrac{1}{4} - a_n = \dfrac{1}{4}$, and so the sequence is arithmetic with common difference $d = \frac{1}{4}$.

7. Show that a right triangle whose sides are in arithmetic progression is similar to a triangle with sides of length 3, 4, and 5.

Answer: hard

Let x denote the side with neither the smallest nor the largest length. Thus the length of the three sides of the triangle are $x - a$, x, and $x + a$, for some $a > 0$. Since $x + a$ is the longest side, it is the hypotenuse. By the Pythagorean Theorem, $(x - a)^2 + x^2 = (x + a)^2 \Leftrightarrow x^2 - 2ax + a^2 + x^2 = x^2 + 2ax + a^2 \Leftrightarrow x^2 - 4ax = 0 \Leftrightarrow x(x - 4a) = 0 \Rightarrow x = 4a$. Thus, the lengths of the three sides are $x - a = 4a - a = 3a$, $x = 4a$, and $x + a = 4a + a = 5a$. The lengths $3a$, $4a$, $5a$ are proportional to 3, 4, 5 and so the triangle is similar to a 3, 4, 5 triangle.

8. The sum of five consecutive terms of an arithmetic sequence is 340. Find the middle term.

Answer: medium

Let x denote the middle term of the sequence and d the common difference. Then $(x - 2d) + (x - d) + x + (x + d) + (x + 2d) = 340 \Leftrightarrow 5x = 340 \Leftrightarrow x = 68$. Therefore, the middle term is 68.

9. If the numbers $a_1, a_2, \ldots, a_n$ form an arithmetic sequence, then $a_2, a_3, \ldots, a_{n-1}$ are called arithmetic means between a_1 and a_n. Insert three arithmetic means between 3 and 15. (Notice that if only one arithmetic mean is inserted between two numbers, then it is their average.)

Answer: hard

Since we have 5 terms, let us denote $a_1 = 3$ and $a_5 = 15$. Also, $a_5 - a_1 = 4d$ because the sequence is arithmetic and so $4d = 12 \Leftrightarrow d = 3$. Therefore, $a_2 = 6$, $a_3 = 9$, and $a_4 = 12$ are three arithmetic means between 3 and 15.

10. A sequence is called harmonic if the reciprocals of the terms of the sequence form an arithmetic sequence. Determine whether the sequence $1, \frac{3}{7}, \frac{3}{11}, \frac{1}{5}, \ldots$ is harmonic.

Answer: medium

$1, \frac{3}{7}, \frac{3}{11}, \frac{1}{5}, \ldots$ is harmonic if $1, \frac{7}{3}, \frac{11}{3}, 5, \ldots$ forms an arithmetic sequence. Since $\frac{7}{3} - 1 = \frac{11}{3} - \frac{7}{3} = 5 - \frac{11}{3} = \frac{4}{3}$, the sequence of reciprocals is arithmetic and thus the original sequence is harmonic.

11. Find the sum S_n of the arithmetic series that satisfies the conditions $a = 50$, $d = -2$, and $n = 6$.

(a) 240 (b) 270 (c) 310 (d) 330 (e) 380

Answer: easy (b)

$a = 50$, $d = -2$, $n = 6$. Then $S_n = \frac{1}{2}n[2a + (n - 1)d] = \frac{6}{2}[2 \cdot 50 - 5 \cdot 2] = 270$

12. Find the sum S_n of the arithmetic series that satisfies the conditions $a_2 = 4$, $a_5 = 8$, and $n = 12$.

(a) 80 (b) 104 (c) 120 (d) 166 (e) 850

Answer: medium (c)

$a_2 = 4$, $a_5 = 8$, $n = 12$. Thus $a + d = 4$ and $a + 4d = 8$. Subtracting the first equation from the second gives $3d = 4 \Leftrightarrow d = \frac{4}{3}$. Substituting into the first equation gives $a + \frac{4}{3} = 4 \Leftrightarrow a = \frac{8}{3}$. Thus $S_{12} = \frac{12}{2}[2 \cdot \frac{8}{3} + 11 \cdot \frac{4}{3}] = 120$

13. Find the sum S_n of the arithmetic series that satisfies the conditions $a_4 = 16$, $d = 4$, and $n = 8$.

(a) 84 (b) 90 (c) 102 (d) 118 (e) 144

Answer: easy (e)

$a_4 = 16$, $d = 4$, $n = 8$. So $a_4 = a + 3d = 16 \Rightarrow a + 12 = 16 \Leftrightarrow a = 4$. Thus $S_8 = \frac{8}{2}[2 \cdot 4 + 7 \cdot 4] = 144$

14. Find the sum of the arithmetic series $-3 + \left(-\frac{3}{2}\right) + 0 + \frac{3}{2} + 3 + \cdots + 33$.

(a) 177 (b) $\frac{361}{2}$ (c) 244 (d) 375 (e) 412

Answer: easy (d)

$-3 + \left(-\frac{3}{2}\right) + 0 + \frac{3}{2} + 3 + \cdots + 33$ is an arithmetic series. Then $a = -3$, $d = \frac{3}{2}$, and $a_n = 33 = -3 + \frac{3}{2}(n - 1) \Leftrightarrow n = 25$. So, $S_{25} = \frac{25}{2}\left[-6 + 24\left(\frac{3}{2}\right)\right] = 375$.

15. Find the sum of the arithmetic series $-100 - 99.9 - 99.8 - \cdots - 0.1$.

(a) $-50{,}050$ (b) $-20{,}045$ (c) $-15{,}500$ (d) -505 (e) -405

Answer: easy (a)

$-100 + (-99.9) + (-99.8) + \cdots + (-0.1)$ is an arithmetic series. Then $a = -100$, $d = 0.1$, and $a_n = -0.1 = -100 + 0.1(n-1) \quad \Leftrightarrow \quad n = 1000$. So, $S_{1000} = \frac{1000}{2}(-200 + 99.9) = -50{,}050$.

16. The seventh term of an arithmetic sequence is -12 and the seventeenth term is -62. Find the twenty-fourth term.

Answer: medium

$$a_7 = -12, a_{17} = -62 \quad \Rightarrow \quad \begin{cases} -12 = a + 6d \\ -62 = a + 16d \end{cases} \quad \Rightarrow \quad a = 18, d = -5.\ a_{24} = 18 + 23(-5) = -97$$

17. Which term of the arithmetic sequence $\frac{1}{2}, 2, \frac{7}{2}, \ldots$, is 23?

Answer: medium

$d = \frac{3}{2}, a = \frac{1}{2} \quad \Rightarrow \quad 23 = \frac{1}{2} + (n-1)\frac{3}{2} \quad \Leftrightarrow \quad n = 16, a_{16} = 23.$

18. The first term of the arithmetic sequence is $\frac{1}{3}$ and the common difference is $-\frac{2}{3}$. Which term of this sequence is $\frac{-23}{3}$?

Answer: medium

$a = \frac{1}{3}, d = \left(-\frac{2}{3}\right) \quad \Rightarrow \quad \left(\frac{-23}{3}\right) = \frac{1}{3} + (n-1)\left(-\frac{2}{3}\right) \quad \Leftrightarrow \quad n = 13, a_{13} = \frac{-23}{3}.$

19. Given the partial sum of an arithmetic sequence $2.1 + 2.7 + 3.3 + \cdots + 13.5$, find its sum.

Answer: medium

$$13.5 = 2.1 + (n-1)(0.6) \quad \Leftrightarrow \quad n = 20.\ S_n = n\left(\frac{a + a_n}{2}\right) \quad \Rightarrow \quad S_{20} = 20\left(\frac{2.1 + 13.5}{2}\right) = 156.$$

20. Find the sum of the first 50 even integers.

Answer: medium

$S_{50} = 2 + 4 + 6 + 8 + \ldots. + 100 = 50\left(\frac{2+100}{2}\right) = 2550$

21. A large outdoor amphitheater has 85 rows of seats with 20 seats in the first row, 24 seats in the second, 28 in the third and so on. How many seats are in row 72? How many seats are there in the theater? How many seats total are there in rows 32 through 64?

Answer: medium

(a) In row 72, there are $a_{72} = 20 + 71(4) = 304$ seats.

(b) $a_{85} = 20 + 84(4) = 356 \quad \Rightarrow \quad S_{85} = 85\left(\frac{20+356}{2}\right) = 15{,}980$ seats.

(c) $S_{64} = \frac{64}{2}[2(20) + 63(4)] = 9344$, $S_{32} = \frac{32}{2}[2(20) + 31(4)] = 2624$. $S_{64} - S_{32} = 9344 - 2624 = 6720$ seats.

SECTION 12.3

1. Determine the common ratio, the fifth term, and the nth term for the geometric sequence $8, \frac{16}{3}, \frac{32}{9}, \frac{64}{27}, \ldots$.

Answer: easy

$8, \frac{16}{3}, \frac{32}{9}, \frac{64}{27}, \ldots$. Then $r = \frac{16/3}{8} = \frac{2}{3}, a_5 = \frac{64}{27} \cdot \frac{2}{3} = \frac{128}{81}, a_n = 8 \cdot \left(\frac{2}{3}\right)^{n-1}$.

2. Determine the common ratio, the fifth term, and the nth term for the geometric sequence $1, \sqrt{3}, 3, 3\sqrt{3}, \cdots$.

Answer: easy

$1, \sqrt{3}, 3, 3\sqrt{3}, \ldots$. Then $r = \frac{\sqrt{3}}{1} = \sqrt{3}, a_5 = 3\sqrt{3} \cdot \sqrt{3} = 9, a_n = \left(\sqrt{3}\right)^{n-1}$.

3. Determine the common ratio, the fifth term, and the nth term for the geometric sequence $-16, -4, -1, -\frac{1}{4}, \ldots$.

Answer: easy

$-16, -4, -1, -\frac{1}{4}, \ldots$. Then $r = \frac{-4}{-16} = \frac{1}{4}, a_5 = -\frac{1}{4} \cdot \frac{1}{4} = -\frac{1}{16}, a_n = -16\left(\frac{1}{4}\right)^{n-1}$.

4. Determine the common ratio, the fifth term, and the nth term for the geometric sequence $t, \frac{t^3}{2}, \frac{t^5}{4}, \frac{t^7}{8}, \ldots$.

Answer: medium

$t, \frac{t^3}{2}, \frac{t^5}{4}, \frac{t^7}{8}, \ldots$. Then $r = \frac{t^3/2}{t} = \frac{t^2}{2}, a_5 = \frac{t^7}{8} \cdot \frac{t^2}{2} = \frac{t^9}{16}, a_n = t\left(\frac{t^2}{2}\right)^{n-1}$.

5. Determine the common ratio, the fifth term, and the nth term for the geometric sequence $5, 5^{\sqrt{2}+1}, 5^{2\sqrt{2}+1}, 5^{3\sqrt{2}+1}, \ldots$.

Answer: medium

$5, 5^{\sqrt{2}+1}, 5^{2\sqrt{2}+1}, 5^{3\sqrt{2}+1}, \ldots$. Then

$r = \frac{5^{\sqrt{2}+1}}{5} = 5^{\sqrt{2}}, a_5 = 5^{3\sqrt{2}+1} \cdot 5^{\sqrt{2}} = 5^{4\sqrt{2}+1}, a_n = 5\left(5^{\sqrt{2}}\right)^{n-1} = 5 \cdot 5^{\sqrt{2}n-\sqrt{2}} = 5^{\sqrt{2}n-\sqrt{2}+1}$.

6. The first four terms of a sequence are $\frac{5}{3}, \frac{7}{3}, 3, \frac{11}{3}, \ldots$. Determine whether the terms can be the terms of an arithmetic sequence, a geometric sequence, or neither. Find the next term if the sequence is arithmetic or geometric.

(a) Arithmetic, $a_5 = 4$ (b) Arithmetic, $a_5 = \frac{13}{3}$ (c) Geometric, $a_5 = 11$ (d) Geometric, $a_5 = 4$ (e) Neither

Answer: medium (b)

d_{ij} is the difference between the j and i terms. $d_{12} = \frac{7}{3} - \frac{5}{3} = \frac{2}{3}$, $d_{23} = 3 - \frac{7}{3} = \frac{2}{3}$, $d_{34} = \frac{11}{3} - 3 = \frac{2}{3}$. Therefore, the sequence is arithmetic with $d = \frac{2}{3}$, and $a_5 = \frac{11}{3} + \frac{2}{3} = \frac{13}{3}$.

7. The first four terms of a sequence are $-1, 1, -1, 1, \ldots$. Determine whether the terms can be the terms of an arithmetic sequence, a geometric sequence, or neither. Find the next term if the sequence is arithmetic or geometric.

(a) Arithmetic, $a_5 = 2$ (b) Arithmetic, $a_5 = 1$ (c) Geometric, $a_5 = -1$ (d) Geometric, $a_5 = 1$ (e) Neither

Answer: medium (c)

r_{ij} is the ratio between the j and i terms. $r_{12} = \frac{1}{-1} = -1$, $r_{23} = \frac{-1}{1} = -1$, $r_{34} = \frac{1}{-1} = -1$. Thus, the sequence is geometric with $r = -1$, and $a_5 = 1(-1) = -1$.

8. The first four terms of a sequence are $-3, 2, 5, 9, \ldots$. Determine whether the terms can be the terms of an arithmetic sequence, a geometric sequence, or neither. Find the next term if the sequence is arithmetic or geometric.

(a) Arithmetic, $a_5 = 11$ (b) Arithmetic, $a_5 = 12$ (c) Geometric, $a_5 = \frac{17}{3}$ (d) Geometric, $a_5 = \frac{35}{3}$ (e) Neither

Answer: medium (e)

d_{ij} is the difference between the j and i terms while r_{ij} is the ratio. $d_{12} = 2 - (-3) = 5$, $d_{23} = 5 - 2 = 3$, $d_{34} = 9 - 5 = 4$, so the sequence is not arithmetic. Also $r_{12} = -\frac{2}{3}$ and $r_{23} = \frac{5}{2}$, so the sequence is not geometric. Therefore, the sequence is neither arithmetic nor geometric.

9. The first four terms of a sequence are $-\frac{5}{2}, -1, \frac{1}{2}, 2, \ldots$. Determine whether the terms can be the terms of an arithmetic sequence, a geometric sequence, or neither. Find the next term if the sequence is arithmetic or geometric.

(a) Arithmetic, $a_5 = \frac{9}{2}$ (b) Arithmetic, $a_5 = \frac{7}{2}$ (c) Geometric, $a_5 = 4$ (d) Geometric, $a_5 = 5$ (e) Neither

Answer: medium (b)

d_{ij} is the difference between the j and i terms. $d_{12} = -1 - \left(-\frac{5}{2}\right) = \frac{3}{2}$, $d_{23} = \frac{1}{2} - (-1) = \frac{3}{2}$, $d_{34} = 2 - \frac{1}{2} = \frac{3}{2}$. Thus, the sequence is arithmetic with $d = \frac{3}{2}$, and $a_5 = 2 + \frac{3}{2} = \frac{7}{2}$.

10. The first four terms of a sequence are $-\sqrt{5}, \sqrt[3]{5}, -\sqrt[6]{5}, 1, \ldots$. Determine whether the terms can be the terms of an arithmetic sequence, a geometric sequence, or neither. Find the next term if the sequence is arithmetic or geometric.

(a) Arithmetic, $a_5 = 2\sqrt{5}$ (b) Arithmetic, $a_5 = 3\sqrt{5}$ (c) Geometric, $a_5 = -\frac{1}{\sqrt[6]{5}}$ (d) Geometric, $a_5 = 2\sqrt{5}$ (e) Neither

Answer: medium (c)

d_{ij} is the difference between the j and i terms while r_{ij} is the ratio. $d_{12} = \sqrt[3]{5} + \sqrt{5}$, but $d_{23} = -\sqrt[6]{5} - \sqrt[3]{5}$. Thus the sequence is not arithmetic. Now $r_{12} = \frac{\sqrt[3]{5}}{-\sqrt{5}} = -5^{-1/6}$, $r_{23} = \frac{-\sqrt[6]{5}}{\sqrt[3]{5}} = -5^{-1/6}$, and $r_{34} = \frac{1}{-\sqrt[6]{5}} = -5^{-1/6}$ Hence, the sequence is geometric with $r = -5^{-1/6}$, and $a_5 = 1 \cdot \left(-5^{-1/6}\right) = -\frac{1}{\sqrt[6]{5}}$.

11. The first term of a geometric sequence is 8 and the second term is 2. Find the fifth term.

(a) $a_5 = \frac{1}{128}$ (b) $a_5 = \frac{1}{32}$ (c) $a_5 = \frac{1}{16}$ (d) $a_5 = \frac{1}{8}$ (e) $a_5 = 1$

Answer: medium (b)

$a_1 = 8$, $a_2 = 2$. Thus $r = \frac{2}{8} = \frac{1}{4}$, and $a_5 = 8\left(\frac{1}{4}\right)^4 = \frac{8}{256} = \frac{1}{32}$.

12. The first term of a geometric sequence is 1 and the sixth term is 7^5. Find the common ratio, assuming it is positive.

(a) $r = 5$ (b) $r = \sqrt{7}$ (c) $r = 7$ (d) $r = 7\sqrt{7}$ (e) $r = 40$

Answer: easy (c)

$a_1 = 1$, $a_6 = 7^5$. Thus $r^5 = 7^5 \quad \Leftrightarrow \quad r = 7$ (assuming r is positive).

13. The common ratio in a geometric sequence is $\frac{3}{2}$ and the fifth term is 2. Find the first three terms.

(a) $a_1 = 1, a_2 = \frac{2}{3}, a_3 = \frac{4}{9}$ (b) $a_1 = \frac{2}{3}, a_2 = 1, a_3 = \frac{4}{3}$ (c) $a_1 = \frac{3}{16}, a_2 = \frac{4}{3}, a_3 = \frac{4}{9}$

(d) $a_1 = \frac{32}{81}, a_2 = \frac{16}{27}, a_3 = \frac{8}{9}$ (e) $a_1 = \frac{14}{3}, a_2 = \frac{2}{7}$, and $a_3 = \frac{9}{64}$

Answer: medium (d)

$r = \frac{3}{2}$, $a_5 = 2$. Since $a_5 = ar^4 = a\left(\frac{3}{2}\right)^4 = 2 \quad \Leftrightarrow \quad a = 2 \cdot \left(\frac{2}{3}\right)^4 = \frac{32}{81}$. Therefore, $a_1 = \frac{32}{81}$, $a_2 = \frac{32}{81} \cdot \frac{3}{2} = \frac{16}{27}$, and $a_3 = \frac{32}{81} \cdot \left(\frac{3}{2}\right)^2 = \frac{8}{9}$.

14. Which term of the geometric sequence, 2, 6, 18, ... is 354,294?

(a) 5th (b) 7th (c) 9th (d) 11th (e) 12th

Answer: medium (e)

The geometric sequence is 2, 6, 18, ... Thus $a_n = 2 \cdot 3^{n-1} = 354{,}294 \quad \Leftrightarrow \quad 3^{n-1} = 177{,}147 \quad \Leftrightarrow$ $n - 1 = \log_3 177{,}147 = 11 \quad \Leftrightarrow \quad n = 12$. Therefore, 354,294 is the 12th term.

15. The second term and the fifth term of a geometric sequence are 10 and 1250, respectively. Is 34,650 a term of this sequence? If so which term is it?

Answer: hard

$a_2 = 10$, $a_5 = 1250$. Thus, $r^3 = \frac{1250}{10} = 125 \quad \Leftrightarrow \quad r = 5$, and $a_1 = \frac{1}{5}a_2 = \frac{10}{5} = 2$. So, $a_n = 2 \cdot 5^{n-1} = 34{,}650 \quad \Leftrightarrow$ $5^{n-1} = 17{,}325 \quad \Leftrightarrow \quad n - 1 = \log_5 17{,}325 \approx 6.06417$. Therefore, 34,650 is not a term of the sequence.

16. The size of a bacteria culture increases by 12% every hour and there are 4000 bacteria present initially. How many bacteria will be present at the end of 5 hours? Find a formula for the number of bacteria present after n hours.

Answer: medium

$a = 4000$, $r = 1.12$. After 5 hours, there are $4000(1.12)^5 \approx 7049$ bacteria. After n h, the number of bacteria is $a_{n+1} = 4000 \cdot 1.12^n$.

17. A certain radioactive substance decays so that at the end of each year there is only 0.65 as much as there was at the beginning of the year. If there was originally 1 kg of the substance, find the amount that remains after 5 yr. Find a formula for the amount that remains after n years.

Answer: medium

$a = 1$, $r = 0.65$. After n yr, the amount of substance remaining is $1 \cdot 0.65^n$. Hence after 5 yr, the amount of substance remaining is $1 \cdot 0.65^5 \approx 0.116$, or 11.6%.

18. If $a_1, a_2, a_3, \ldots$ is a geometric sequence with common ratio r, show that the sequence $\dfrac{1}{a_1}, \dfrac{1}{a_2}, \dfrac{1}{a_3}, \ldots$ is also a geometric sequence and find the common ratio.

Answer: hard

$a_1, a_2, a_3, \ldots$ is a geometric sequence with common ratio r. Thus $a_2 = a_1 \cdot r$, $a_3 = a_1 \cdot r^2, \ldots, a_n = a_1 \cdot r^{n-1}$. Hence $\dfrac{1}{a_2} = \dfrac{1}{a_1 \cdot r} = \dfrac{1}{a_1} \cdot \dfrac{1}{r}, \dfrac{1}{a_3} = \dfrac{1}{a_1 \cdot r^2} = \dfrac{1}{a_1} \cdot \dfrac{1}{r^2}, \ldots \dfrac{1}{a_n} = \dfrac{1}{a_1 \cdot r^{n-1}} = \dfrac{1}{a_1} \cdot \dfrac{1}{r^{n-1}} = \dfrac{1}{a_1} \cdot \left(\dfrac{1}{r}\right)^{n-1}$ and so $\dfrac{1}{a_1}, \dfrac{1}{a_2}, \dfrac{1}{a_3}, \ldots$ is a geometric sequence with common ratio $1/r$.

19. If $a_1, a_2, a_3, \ldots$ is a geometric sequence with common ratio $r > 0$ and $a_1 > 0$, show that the sequence $\log a_1, \log a_2, \log a_3, \ldots$ is an arithmetic sequence and find the common difference.

Answer: hard

$a_1, a_2, a_3, \ldots$ is a geometric sequence with common ratio r. Thus $a_2 = a_1 \cdot r$, $a_3 = a_1 \cdot r^2, \ldots, a_n = a_1 \cdot r^{n-1}$. Hence $\log a_2 = \log(a_1 \cdot r) = \log a_1 + \log r$,
$\log a_3 = \log(a_1 \cdot r^2) = \log a_1 + 2\log r, \ldots, \log a_n = \log(a_1 \cdot r^{n-1}) = \log a_1 + (n-1)\log r$ and so $\log a_1, \log a_2, \log a_3, \ldots$ is an arithmetic sequence with common difference $\log r$.

20. If the product of three consecutive terms in a geometric sequence is 343 and their sum is 21, find the three terms. (Hint: Let x denote the middle term.)

Answer: hard

Let x denote the middle term and r the common ratio. Then $\left(\frac{x}{r}\right) \cdot x \cdot xr = 343 \quad \Leftrightarrow \quad x^3 = 343 \quad \Leftrightarrow \quad x = 7$. Also, $\frac{x}{r} + x + xr = 21 \quad \Leftrightarrow \quad \frac{7}{r} + 7 + 7r = 21 \quad \Leftrightarrow \quad \frac{7}{r} + 7r = 14 \quad \Leftrightarrow \quad 7r^2 - 14r + 7 = 0 \quad \Leftrightarrow \quad r = 1$. For $x = 7$ and $r = 1$ the three terms are 7, 7, and 7.

21. Find the sum of the infinite geometric series $\frac{1}{3} + \frac{2}{9} + \frac{4}{27} + \frac{8}{81} + \cdots$.

Answer: easy

$a = \frac{1}{3}, r = \frac{2/9}{1/3} = \frac{2}{3} \quad \Rightarrow \quad S = \frac{1/3}{1-(2/3)} = 1$

22. Find the sum of the infinite geometric series $\frac{2}{\sqrt{2}} + 1 + \frac{1}{\sqrt{2}} + \frac{1}{2} + \cdots$.

Answer: easy

$a = \frac{2}{\sqrt{2}} = \sqrt{2}, r = \frac{\sqrt{2}}{2} \quad \Rightarrow \quad S = \frac{\sqrt{2}}{1-(\sqrt{2}/2)} = \frac{\sqrt{2}}{(2-\sqrt{2})/2} = \frac{2\sqrt{2}}{2-\sqrt{2}}$.

23. Express the repeating decimal $3.1\overline{68}$ as a fraction.

Answer: medium

$3.1\overline{68} = \frac{31}{10} + \frac{68}{1000} + \frac{68}{100{,}000} + \cdots$. $a = \frac{68}{1000}, r = \frac{1}{100} \quad \Rightarrow \quad S = \frac{\frac{68}{1000}}{1 - \frac{1}{100}} = \frac{34}{495}$. Thus $3.1\overline{68} = \frac{31}{10} + \frac{34}{495} = \frac{3137}{990}$.

24. Express the repeating decimal $0.345345345\ldots$ as a fraction.

Answer: medium

$0.\overline{345} = \frac{345}{1000} + \frac{345}{1{,}000{,}000} + \cdots$. $a = \frac{345}{1000}, r = \frac{1}{1000} \quad \Rightarrow \quad S = \frac{\frac{345}{1000}}{1 - \frac{1}{1000}} = \frac{115}{333}$. Thus $0.\overline{345} = \frac{115}{333}$.

25. A ball rebounds to one-quarter the height from which it was dropped. Approximate the total vertical distance the ball travels after being dropped from 3 ft above the ground, until it comes to rest.

Answer: hard

The distance traveled on each rebound is $2 \times 3\left(\frac{1}{4}\right)^n = 6\left(\frac{1}{4}\right)^n$ (once up and once down). Thus the total vertical distance is

$3 + 6\left(\frac{1}{4}\right) + 6\left(\frac{1}{4}\right)^2 + 6\left(\frac{1}{4}\right)^3 + \ldots = 3 + \frac{\frac{3}{2}}{1 - \frac{1}{4}} = 5$ ft.

SECTION 12.4

1. Find the amount of an annuity that consists of 40 semiannual payments of \$600 each into an account that pays 7% interest a year compounded semiannually.

(a) \$17,365.22 (b) \$19,336.12 (c) \$32,066.43 (d) \$50,730.17 (e) \$84,321.40

Answer: medium (d)

$R = 600$, $n = 40$, $i = \frac{0.07}{2} = 0.035$. Then $A = 600 \cdot \frac{1.035^{40} - 1}{0.035} \approx \$50{,}730.17$

2. How much money should be invested every quarter at 12% a year compounded quarterly in order to have \$5000 in 3 years?

(a) \$187.96 (b) \$352.31 (c) \$402.21 (d) \$435.16 (e) \$507.90

Answer: medium (b)

$A = 5000$, $n = 4 \cdot 3 = 12$, $i = \frac{0.12}{4} = 0.03 \quad \Rightarrow \quad 5000 = R \cdot \frac{1.03^{12} - 1}{0.03} \quad \Leftrightarrow \quad R = \frac{150}{1.03^{12} - 1} \approx \$352.31$

3. What is the present value of an annuity that consists of 12 semiannual payments of \$1000 at the interest rate of 10% compounded semiannually?

(a) \$8,863.25 (b) \$8,942.11 (c) \$9,066.67 (d) \$10,255.12 (e) \$10,275.52

Answer: medium (a)

$R = 1000$, $n = 12$, $i = \frac{0.1}{2} = 0.05 \quad \Rightarrow \quad A_p = 1000 \cdot \dfrac{1 - (1.05)^{-12}}{0.05} \approx \$8{,}863.25$

4. A 55-year old man deposits \$40, 000 to set up an annuity with an insurance company at 7% a year compounded semiannually until age 70. If he is to be paid twice a year, how large is each payment he receives?

(a) \$1457.11 (b) \$1933.26 (c) \$2033.80 (d) \$2039.95 (e) \$2174.85

Answer: medium (e)

Present Value $= 40{,}000$, $n = 15 \cdot 2 = 30$, $i = \frac{0.07}{2} = 0.035$. Then $40{,}000 = R \cdot \dfrac{1.035^{30} - 1}{0.035 \cdot 1.035^{30}} \quad \Leftrightarrow$

$R = \dfrac{40,000 \cdot 0.035 \cdot 1.035^{30}}{1.035^{30} - 1} \approx \$2174.85$

5. What is the monthly payment on a 30-yr mortgage of \$120, 000 at 12% interest?

(a) \$941.11 (b) \$967.13 (c) \$1033.42 (d) \$1234.34 (e) \$1440.20

Answer: medium (d)

$A = 120,000$, $n = 30 \cdot 12 = 360$, $i = \frac{0.12}{12} = 0.01$. Then $R = \dfrac{120,000 \cdot 0.01}{1 - 1.01^{-360}} = 1234.34$ and so the monthly payment is \$1234.34.

6. A couple can afford to pay \$750 per month toward buying a house. If the mortgage rates are 10% and they intend to secure a 30-year mortgage, how much can they borrow?

Answer: medium

$R = 750$, $n = 12 \cdot 30 = 360$, $i = \frac{0.1}{12} = \frac{1}{120}$. Then $A_p = 750 \cdot \dfrac{1 - \left(\frac{121}{120}\right)^{-360}}{\frac{1}{120}} \approx \$85{,}463.11$. Therefore, the couple can afford a loan of \$85,463.11.

7. Jane agrees to buy a car by putting a down payment of \$2400 and making payments of \$240 per month for three years. If the interest rate is 9% compounded monthly, what is the actual purchase price of the car?

Answer: medium

$R = 240$, $n = 12 \cdot 3 = 36$, $i = \frac{0.09}{12} = 0.0075$. The amount borrowed is A_p. But $A_p = 240 \cdot \dfrac{1 - (1.0075)^{-36}}{0.0075} \approx \$7547.23$ So she purchased the car for $\$7{,}547.23 + 2400 = \$9947.23$.

GC 8. Janet decides to buy a \$13,500 car by paying \$440 a month for three years. Assuming that interest is compounded monthly, what interest rate is she paying on the loan?

Answer: hard

$A_p = \$13,500$, $R = \$440$, $n = 36$. We want to solve for the interest rate using the equation $R = \dfrac{iA_P}{1 - (1+i)^{-n}}$. Let x be the interest rate.

Then $i = \frac{1}{12}x$. Then $R(x) = \dfrac{\frac{1}{12}x \cdot 13{,}500}{1 - \left(1 + \frac{1}{12}x\right)^{-36}}$.

We graph $R(x)$ in the rectangle $[0.10, 0.14] \times [430, 460]$, and find that if Janet's payments are \$440, then she is paying an interest rate of approximately 10.7%.

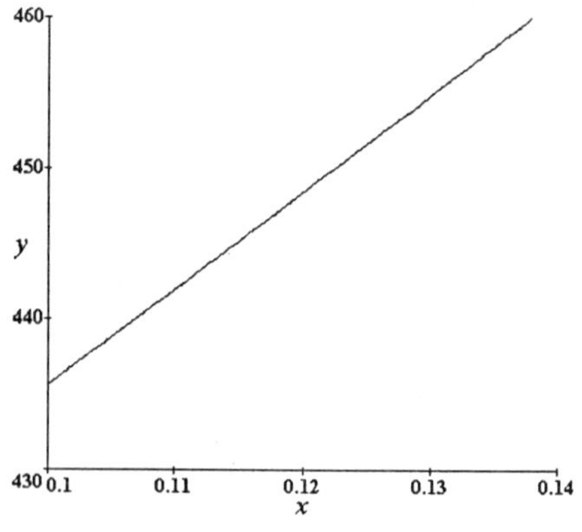

GC 9. A woman purchases an \$1800 diamond ring by making a down payment of \$100 and paying monthly installments of \$80 for two years. Assuming that interest is compounded monthly, what interest rate is she paying?
Answer: medium
$A_p = \$1800 - \$100 = \$1700$, $R = \$80$, $n = 24$. We want to solve for the interest rate using the equation $R = \dfrac{iA_P}{1-(1+i)^{-n}}$. Let x be the interest rate. Then $i = \frac{1}{12}x$. Then $R(x) = \dfrac{\frac{1}{12}x \cdot 1700}{1-\left(1+\frac{1}{12}x\right)^{-24}}$.
We graph $R(x)$ in the rectangle $[0.10, 0.14] \times [79, 81]$, and find that she is paying an interest rate of approximately 12%.

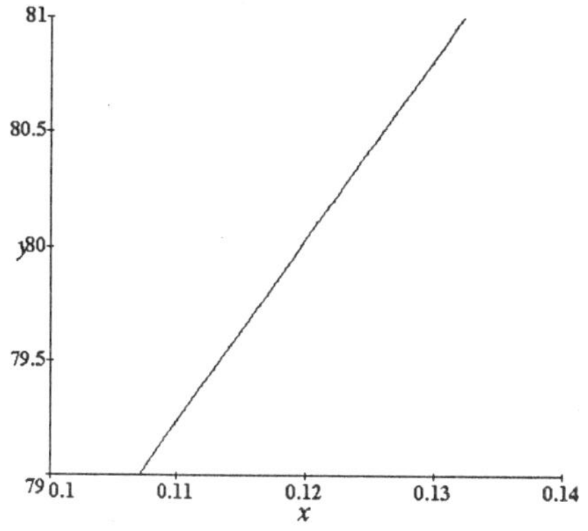

10. How much money must be invested now at 6.5% per year, compounded quarterly, to fund an annuity of 15 quarterly payments of \$500 each, the first payment being three months after the initial investment?
Answer: easy

$$A_p = 500\left[\frac{1-\left(1+\frac{0.065}{4}\right)^{-15}}{\frac{0.065}{4}}\right] \approx \$6608.60$$

11. A buyer decides to finance a new vehicle by paying \$416 a month for 8 years with an APR of 8.5%. What is the total interest paid for the vehicle?
Answer: medium

The amount of the loan is $416 = A_p\left[\dfrac{\frac{0.085}{12}}{1-\left(1+\frac{0.085}{12}\right)^{-96}}\right] \Leftrightarrow A_p = 416\left[\dfrac{1-\left(1+\frac{0.085}{12}\right)^{-96}}{\left(\frac{0.085}{12}\right)}\right] = \$28{,}905.$

Total interest paid $= 416\,(96) - 28{,}905 = \$11{,}031$.

SECTION 12.5

1. Use mathematical induction to prove that the formula $1^2 + 2^2 + 3^2 + \cdots + n^2 = \dfrac{n(n+1)(2n+1)}{6}$ is true for all natural numbers n.
Answer: medium
Let $P(n)$ denote the statement $1^2 + 2^2 + 3^2 + \cdots + n^2 = \dfrac{n(n+1)(2n+1)}{6}$.
Step 1: We need to show that $P(1)$ is true. But $P(1)$ says that $1^2 = \dfrac{1 \cdot 2 \cdot 3}{6}$ which is clearly true.
Step 2: Assume that $P(k)$ is true; that is, $1^2 + 2^2 + 3^2 + \cdots + k^2 = \dfrac{k(k+1)(2k+1)}{6}$. We want to show that $P(k+1)$ is true. Let us add $(k+1)^2$ to both sides of $P(k)$. Then

$$\begin{aligned}1^2 + 2^2 + 3^2 + \cdots + k^2 + (k+1)^2 &= \frac{k(k+1)(2k+1)}{6} + (k+1)^2 = \frac{k(k+1)(2k+1) + 6(k+1)(k+1)}{6}\\ &= \frac{(k+1)(2k^2+k+6k+6)}{6} = \frac{(k+1)(k+2)(2k+3)}{6}\\ &= \frac{(k+1)(k+1+1)[2(k+1)+1]}{6}\end{aligned}$$

which is exactly $P(k+1)$. Thus, $P(n)$ is true for all n.

2. Use mathematical induction to prove that the formula $1 \cdot 3 + 2 \cdot 4 + 3 \cdot 5 + \cdots + n(n+2) = \dfrac{n(n+1)(2n+7)}{6}$ is true for all natural numbers n.

Answer: medium

Let $P(n)$ denote the statement $1 \cdot 3 + 2 \cdot 4 + 3 \cdot 5 + \cdots + n(n+2) = \dfrac{n(n+1)(2n+7)}{6}$.

Step 1: We need to show that $P(1)$ is true. But $P(1)$ says that $1 \cdot 3 = \frac{1 \cdot 2 \cdot 9}{6}$ which is clearly true.

Step 2: Assume that $P(k)$ is true; that is, $1 \cdot 3 + 2 \cdot 4 + 3 \cdot 5 + \cdots + k(k+2) = \dfrac{k(k+1)(2k+7)}{6}$. We want to show that $P(k+1)$ is true. Let us add $(k+1)(k+3)$ to both sides of $P(k)$. Then

$$\begin{aligned}1 \cdot 3 + 2 \cdot 4 + \cdots + k(k+2) + (k+1)(k+3) &= \frac{k(k+1)(2k+7)}{6} + (k+1)(k+3) = \frac{k(k+1)(2k+7) + 6(k+1)(k+3)}{6} \\ &= \frac{(k+1)(2k^2+7k+6k+18)}{6} = \frac{(k+1)(k+2)(2k+9)}{6} \\ &= \frac{(k+1)(k+1+1)[2(k+1)+7]}{6}\end{aligned}$$

which is exactly $P(k+1)$. Thus, $P(n)$ is true for all n.

3. Use mathematical induction to prove that the formula $1^3 + 3^3 + 5^3 + \cdots + (2n-1)^3 = n^2(2n^2-1)$ is true for all natural numbers n.

Answer: hard

Let $P(n)$ denote the statement $1^3 + 3^3 + 5^3 + \cdots + (2n-1)^3 = n^2(2n^2-1)$.

Step 1: We need to show that $P(1)$ is true. But $P(1)$ says that $1^3 = 1^2(2 \cdot 1^2 - 1)$ which is clearly true.

Step 2: Assume that $P(k)$ is true; that is, $1^3 + 3^3 + 5^3 + \cdots + (2k-1)^3 = k^2(2k^2-1)$. We want to show that $P(k+1)$ is true. Let us add $[2(k+1)-1]^3 = (2k+1)^3$ to both sides of $P(k)$. Then

$$\begin{aligned}1^3 + 3^3 + 5^3 + \cdots + (2k-1)^3 + (2k+1)^3 &= k^2(2k^2-1) + (2k+1)^3 = k^2(2k^2-1) + (2k+1)^2(2k+1) \\ &= [(k+1)^2 - 2k - 1](2k^2-1) + [(k+1)^2 + 3k^2 + 2k](2k+1) \\ &= (k+1)^2(2k-1) - (2k+1)(2k^2-1) + (k+1)^2(2k+1) + (3k^2+2k)(2k+1) \\ &= (k+1)^2[(2k^2-1) + (2k+1)] + (2k+1)[-(2k^2-1) + 3k^2 + 2k] \\ &= (k+1)^2[2k^2+2k] + (2k+1)[k^2+2k+1] \\ &= (k+1)^2[2k^2+2k] + (2k+1)(k+1)^2 = (k+1)^2(2k^2+2k+2k+1) \\ &= (k+1)^2(2k^2+4k+1) = (k+1)^2[2(k+1)^2-1]\end{aligned}$$

which is exactly $P(k+1)$. Thus, $P(n)$ is true for all n.

4. Use mathematical induction to prove that the formula $\dfrac{1}{1 \cdot 2} + \dfrac{1}{2 \cdot 3} + \dfrac{1}{3 \cdot 4} + \cdots + \dfrac{1}{n(n+1)} = \dfrac{n}{n+1}$ is true for all natural numbers n.

Answer: medium

Let $P(n)$ denote the statement $\dfrac{1}{1 \cdot 2} + \dfrac{1}{2 \cdot 3} + \dfrac{1}{3 \cdot 4} + \cdots + \dfrac{1}{n(n+1)} = \dfrac{n}{n+1}$.

Step 1: We need to show that $P(1)$ is true. But $P(1)$ says that $\frac{1}{1 \cdot 2} = \frac{1}{2}$ which is clearly true.

Step 2: Assume that $P(k)$ is true; that is, $\dfrac{1}{1 \cdot 2} + \dfrac{1}{2 \cdot 3} + \dfrac{1}{3 \cdot 4} + \cdots + \dfrac{1}{k(k+1)} = \dfrac{k}{k+1}$. We want to show that $P(k+1)$ is true. Let us add $\dfrac{1}{(k+1)(k+2)}$ to both sides of $P(k)$. Then

$$\begin{aligned}\frac{1}{1 \cdot 2} + \frac{1}{2 \cdot 3} + \cdots + \frac{1}{k(k+1)} + \frac{1}{(k+1)(k+2)} &= \frac{k}{k+1} + \frac{1}{(k+1)(k+2)} = \frac{k(k+2)+1}{(k+1)(k+2)} = \frac{k^2+2k+1}{(k+1)(k+2)} \\ &= \frac{(k+1)^2}{(k+1)(k+2)} = \frac{k+1}{k+2} = \frac{k+1}{k+1+1}\end{aligned}$$

which is exactly $P(k+1)$. Therefore, $P(n)$ is true for all n.

5. Prove that $n < 2^n$, for all natural numbers n.

Answer: medium

Let $P(n)$ denote the statement $n < 2^n$.

Step 1: $P(1)$ is the statement that $1 < 2^1 = 2$ which is clearly true.

Step 2: Assume that $P(k)$ is true; that is, $k < 2^k$. We want to show that $P(k+1)$ is true. Let us add 1 to both sides of $P(k)$. Then $k + 1 < 2^k + 1 < 2 \cdot 2^k = 2^{k+1}$ which is exactly $P(k+1)$. Therefore, $P(n)$ is true for all n.

6. Prove that $(n+1)^2 < 2n^2$ for all natural numbers $n \geq 3$.

Answer: medium

Let $P(n)$ denote the statement $(n+1)^2 < 2n^2$.

Step 1: $P(3)$ is the statement that $(3+1)^2 < 2 \cdot 3^2$ or $16 < 18$ which is true.

Step 2: Assume that $P(k)$ is true; that is, $(k+1)^2 < 2k^2$.

We want to show that $P(k+1)$ is true. Now

$(k+2)^2 = k^2 + 4k + 4 = (k^2 + 2k + 1) + (2k + 3) = (k+1)^2 + 2k + 3 \leq 2k^2 + 2k + 3$ (by induction hypothesis)

$= 2\left(k^2 + k + \frac{3}{2}\right) \leq 2\left(k^2 + k + \frac{3}{2} + k - \frac{1}{2}\right)$ (because $k - \frac{1}{2} \geq 0$) $= 2(k^2 + 2k + 1) = 2(k+1)^2$

which is exactly $P(k+1)$. Hence, $P(n)$ is true for all $n \geq 3$.

7. Show that $n^3 - n + 3$ is divisible by 3 for all natural numbers n.

Answer: medium

Let $P(n)$ denote the statement that $n^3 - n + 3$ is divisible by 3.

Step 1: $P(1)$ is the statement that $1^3 - 1 + 3 = 3$ is divisible by 3, which is true.

Step 2: Assume that $P(k)$ is true; that is, $k^3 - k + 3$ is divisible by 3. We want to show that $P(k+1)$ is true. Now, $(k+1)^3 - (k+1) + 3 = k^3 + 3k^2 + 3k + 1 - k - 1 + 3 = k^3 - k + 3 + 3k^2 + 3k = (k^3 - k + 3) + 3(k^2 + k)$ which is divisible by 3 because $k^3 - k + 3$ is divisible by 3 by the induction hypothesis and clearly $3(k^2 + k)$ is divisible by 3. Thus $P(k+1)$ is true, and so is $P(n)$ for all n.

8. Show that $3^{2n} - 1$ is divisible by 8 for every natural number n.

Answer: medium

Let $P(n)$ denote the statement that $3^{2n} - 1$ is divisible for 8.

Step 1: $P(1)$ is the statement that $3^2 - 1 = 8$ is divisible by 8 which is clearly true.

Step 2: Assume that $P(k)$ is true; that is, $3^{2k} - 1$ is divisible by 8. We want to show that $P(k+1)$ is true. Now $3^{2(k+1)} - 1 = 9 \cdot 3^{2k} - 1 = 9(3^{2k} - 1) + 8$ which is divisible by 8 because $3^{2k} - 1$ is divisible by 8 by the induction hypothesis. So, $P(k+1)$ is true, and therefore, $P(n)$ is true for all n.

9. A sequence is defined recursively by $a_{n+1} = 3a_n - 8$ and $a_1 = 4$. Find an explicit formula for a_n and prove that the formula you found is true using mathematical induction.

Answer: medium

$a_{n+1} = 3a_n - 8$ and $a_1 = 4$. Then $a_2 = 3 \cdot 4 - 8 = 4$, $a_3 = 3 \cdot 4 - 8 = 4$, $a_4 = 3 \cdot 4 - 8 = 4, \ldots$ and the conjecture is that $a_n = 4$. Let $P(n)$ denote the statement that $a_n = 4$.

Step 1: $P(1)$ is the statement that $a_1 = 4$, which is true.

Step 2: Assume that $P(k)$ is true; that is, $a_k = 4$. We want to show that $P(k+1)$ is true. Now $a_{k+1} = 3 \cdot a_k - 8 = 3 \cdot 4 - 8 = 4$ which is exactly $P(k+1)$. Thus, $P(n)$ is true for all n.

10. Show that $x + y$ is a factor of $x^{2n-1} + y^{2n-1}$ for all natural numbers n.

Answer: medium

Let $P(n)$ be the statement that $x + y$ is a factor of $x^{2n-1} + y^{2n-1}$.

Step 1: $P(1)$ is the statement that $x + y$ is a factor of $x^1 + y^1$ which is clearly true.

Step 2: Assume that $P(k)$ is true; that is, $x + y$ is a factor of $x^{2k-1} + y^{2k-1}$. We want to show that $P(k+1)$ is true. Now $x^{2(k+1)-1} + y^{2(k+1)-1} = x^{2k+1} + y^{2k+1} = x^{2k+1} - x^{2k-1}y^2 + x^{2k-1}y^2 + y^{2k+1} = x^{2k-1}(x^2 - y^2) + (x^{2k-1} + y^{2k-1})y^2$ for which $x + y$ is a factor because $x + y$ is a factor of $x^2 - y^2 = (x+y)(x-y)$ and $(x+y)$ is a factor of $x^{2k-1} + y^{2k-1}$ by our induction hypothesis. So $P(k+1)$ is true, and therefore $P(n)$ is true for all n.

11. If F_n denotes the nth term of the Fibonacci sequence, use mathematical induction to prove the statement that, for all natural numbers n, F_{3n} is even.

Answer: medium

Let $P(n)$ denote the statement that F_{3n} is even for all natural numbers n.

Step 1: $P(1)$ is the statement that $F_3 = F_2 + F_1 = 1 + 1 = 2$ is even, which is true.

Step 2: Assume that $P(k)$ is true; that is, F_{3k} is even. We want to show that $P(k+1)$ is true. Now $F_{3(k+1)} = F_{3k+3} = F_{3k+2} + F_{3k+1} = F_{3k+1} + F_{3k} + F_{3k+1} = F_{3k} + 2 \cdot F_{3k+1}$ which is even because F_{3k} is even by the induction hypothesis and $2 \cdot F_{3k+1}$ is even for all n (twice any natural number is an even number). Since this is $P(k+1)$, $P(n)$ is true for all n.

12. If F_n denotes the nth term of the Fibonacci sequence, use mathematical induction to prove the statement $F_1^2 + F_2^2 + F_3^2 + \cdots + F_n^2 = F_n \cdot F_{n+1}$.

Answer: medium

Let $P(n)$ denote the statement that $F_1^2 + F_2^2 + F_3^2 + \cdots + F_n^2 = F_n \cdot F_{n+1}$.

Step 1: $P(1)$ is the statement that $F_1^2 = F_1 \cdot F_2$ or $1^2 = 1 \cdot 1$ which is true.

Step 2: Assume that $P(k)$ is true, then $F_1^2 + F_2^2 + F_3^2 + \cdots + F_k^2 = F_k \cdot F_{k+1}$. Then $F_1^2 + F_2^2 + F_3^2 + \cdots + F_k^2 + F_{k+1}^2 = F_k \cdot F_{k+1} + F_{k+1}^2$ (adding F_{k+1}^2 to both sides) $= F_{k+1}(F_k + F_{k+1}) = F_{k+1} \cdot F_{k+2}$ (by definition of the Fibonacci sequence). Thus $F_1^2 + F_2^2 + F_3^2 + \cdots + F_{k+1}^2 = F_{k+1} \cdot F_{k+2}$. But this is exactly $P(k+1)$. Therefore $P(n)$ is true for all n.

13. If F_n denotes the nth term of the Fibonacci sequence, use mathematical induction to prove the statement that, for all $n \geq 2$, $\begin{bmatrix} 1 & 1 \\ 1 & 0 \end{bmatrix}^n = \begin{bmatrix} F_{n+1} & F_n \\ F_n & F_{n-1} \end{bmatrix}$.

Answer: hard

Let $P(n)$ denote the statement $\begin{bmatrix} 1 & 1 \\ 1 & 0 \end{bmatrix}^n = \begin{bmatrix} F_{n+1} & F_n \\ F_n & F_{n-1} \end{bmatrix}$.

Step 1: Since $\begin{bmatrix} 1 & 1 \\ 1 & 0 \end{bmatrix}^2 = \begin{bmatrix} 1 & 1 \\ 1 & 0 \end{bmatrix}\begin{bmatrix} 1 & 1 \\ 1 & 0 \end{bmatrix} = \begin{bmatrix} 2 & 1 \\ 1 & 1 \end{bmatrix} = \begin{bmatrix} F_3 & F_2 \\ F_2 & F_1 \end{bmatrix}$, it follows that $P(2)$ is true.

Step 2: Assume that $P(k)$ is true; that is, $\begin{bmatrix} 1 & 1 \\ 1 & 0 \end{bmatrix}^k = \begin{bmatrix} F_{k+1} & F_k \\ F_k & F_{k-1} \end{bmatrix}$. We show that $P(k+1)$ follows from this. Now

$$\begin{bmatrix} 1 & 1 \\ 1 & 0 \end{bmatrix}^{k+1} = \begin{bmatrix} 1 & 1 \\ 1 & 0 \end{bmatrix}^k \begin{bmatrix} 1 & 1 \\ 1 & 0 \end{bmatrix} = \begin{bmatrix} F_{k+1} & F_k \\ F_k & F_{k-1} \end{bmatrix}\begin{bmatrix} 1 & 1 \\ 1 & 0 \end{bmatrix} = \begin{bmatrix} F_{k+1} + F_k & F_{k+1} \\ F_k + F_{k-1} & F_k \end{bmatrix} = \begin{bmatrix} F_{k+2} & F_{k+1} \\ F_{k+1} & F_k \end{bmatrix}$$

(by the properties of the Fibonacci sequence). This shows that $P(k+1)$is true. Therefore, $P(n)$ is true for all $n \geq 2$.

14. Use mathematical induction to prove that the sum of a partial geometric sequence is $\dfrac{a(1-r^n)}{1-r}$.

Answer: hard

Let $P(n)$ be the statement that $a + ar^1 + ar^2 + ar^3 + \ldots. + ar^{n-1} = \dfrac{a(1-r^n)}{1-r}$, $r \neq 1$. *Step 1:* Show that $P(1)$ is true. $a = a(1)$, which is clearly true. *Step 2:* Assume $P(k)$ is true; that is, $a + ar^1 + ar^2 + ar^3 + \ldots. + ar^{k-1} = \dfrac{a(1-r^k)}{1-r}$.

Show $P(k+1) = a + ar^1 + ar^2 + ar^3 + \ldots. + ar^{k-1} + ar^k = \dfrac{a(1-r^{k+1})}{1-r}$.

$P(k+1) = P(k) + ar^k = \dfrac{a(1-r^k)}{1-r} + ar^k = \dfrac{a}{1-r} - \dfrac{a}{1-r}r^k + \dfrac{ar^k - ar^k r}{1-r} = \dfrac{a - ar^{k+1}}{1-r} = \dfrac{a(1-r^{k+1})}{1-r}$. Thus $P(n)$ is true for all n.

15. Use mathematical induction to prove that the sum of a finite arithmetic sequence is $\frac{n}{2}[2a+(n-1)d]$.

Answer: hard

Let $P(n)$ be the statement that $a+(a+d)+(a+2d)+\ldots.+[a+(n-1)d]=\frac{n}{2}[2a+(n-1)d]$. *Step 1:* Show that $P(1)$ is true. $a=\frac{1}{2}[2a+(1-1)d]$, which is clearly true. *Step 2:* Assume $P(k)$ is true; that is, $a+(a+d)+(a+2d)+\ldots.+[a+(k-1)d]=\frac{k}{2}[2a+(k-1)d]$. Show

$P(k+1)=a+(a+d)+(a+2d)+\cdots+[a+(k-1)d]+[a+(k)d]=\frac{k+1}{2}[2a+(k)d]$:

$P(k+1)=P(k)+[a+(k)d]=\frac{k}{2}[2a+(k-1)d]+[a+(k)d]=\left[a+\frac{1}{2}kd+ak+\frac{1}{2}k^2d\right]=\frac{1}{2}(k+1)(2a+kd)=$ $\frac{k+1}{2}[2a+kd]$. Thus, $P(n)$ is true for all n.

SECTION 12.6

1. Use Pascal's Triangle to expand the expression $(3x+1)^4$.

Answer: easy

$(3x+1)^4=(3x)^4+4(3x)^3+6(3x)^2+4\cdot 3x+1=81x^4+108x^3+54x^2+12x+1$

2. Use Pascal's Triangle to expand the expression $(x+y)^5$.

Answer: easy

$(x+y)^5=x^5+5x^4y+10x^3y^2+10x^2y^3+5xy^4+y^5$

3. Use Pascal's Triangle to expand the expression $\left(\sqrt{a}-\sqrt{b}\right)^6$.

Answer: medium

$$\begin{aligned}\left(\sqrt{a}-\sqrt{b}\right)^6&=a^3-6a^2\sqrt{a}\sqrt{b}+15a^2b-20ab\sqrt{a}\sqrt{b}+15ab^2-6\sqrt{a}b^2\sqrt{b}+b^3\\&=a^3-6a^2\sqrt{ab}+15a^2b-20ab\sqrt{ab}+15ab^2-6b^2\sqrt{ab}+b^3\end{aligned}$$

4. Use Pascal's Triangle to expand the expression $\left(1+\sqrt{3}\right)^6$.

Answer: easy

$$\begin{aligned}\left(1+\sqrt{3}\right)^6&=1^6+6\cdot 1^5\cdot\sqrt{3}+15\cdot 1^4\cdot 3+20\cdot 1^3\cdot 3\sqrt{3}+15\cdot 1^2\cdot 9+6\cdot 1\cdot 9\sqrt{3}+3^3\\&=1+6\sqrt{3}+45+60\sqrt{3}+135+54\sqrt{3}+27=208+120\sqrt{3}\end{aligned}$$

5. Use Pascal's Triangle to expand the expression $(1+x^2)^3$.

Answer: easy

$(1+x^2)^3=1^3+3\cdot 1^2\cdot x^2+3\cdot 1(x^2)^2+(x^2)^3=1+3x^2+3x^4+x^6$

6. Use Pascal's Triangle to expand the expression $\left(3+\frac{1}{3}x\right)^5$.

Answer: medium

$$\begin{aligned}\left(3+\tfrac{1}{3}x\right)^5&=3^5+5\cdot 3^4\cdot\tfrac{1}{3}x+10\cdot 3^3\cdot\left(\tfrac{1}{3}x\right)^2+10\cdot 3^2\cdot\left(\tfrac{1}{3}x\right)^3+5\cdot 3\cdot\left(\tfrac{1}{3}x\right)^4+\left(\tfrac{1}{3}x\right)^5\\&=243+135x+30x^2+\tfrac{10}{3}x^3+\tfrac{5}{27}x^4+\tfrac{1}{243}x^5\end{aligned}$$

7. Evaluate the expression $\binom{10}{6}$.

(a) 210 (b) 240 (c) 256 (d) 320 (e) 420

Answer: easy (a)

$\binom{10}{6}=\frac{10!}{6!\,4!}=\frac{10\cdot 9\cdot 8\cdot 7\cdot 6!}{6!\cdot 4\cdot 3\cdot 2\cdot 1}=10\cdot 3\cdot 7=210$

8. Evaluate the expression $\binom{10}{4}$.

(a) 196 (b) 210 (c) 240 (d) 256 (e) 280

Answer: easy (b)

$$\binom{10}{4} = \frac{10!}{4!\,6!} = \frac{10\cdot 9\cdot 8\cdot 7\cdot 6!}{4\cdot 3\cdot 2\cdot 1\cdot 6!} = 10\cdot 3\cdot 7 = 210$$

9. Evaluate the expression $\binom{7}{0} - \binom{7}{1} + \binom{7}{2} - \binom{7}{3} + \binom{7}{4} - \binom{7}{5} + \binom{7}{6} - \binom{7}{7}$.

(a) -241 (b) -7 (c) 0 (d) 7 (e) 241

Answer: medium (c)

$$\binom{7}{0} - \binom{7}{1} + \binom{7}{2} - \binom{7}{3} + \binom{7}{4} - \binom{7}{5} + \binom{7}{6} - \binom{7}{7} = 1 - \frac{7!}{1!\,6!} + \frac{7!}{2!\,5!} - \frac{7!}{3!\,4!} + \frac{7!}{4!\,3!} - \frac{7!}{5!\,2!} + \frac{7!}{6!\,1!} - 1 = 0$$

10. Use the Binomial Theorem to expand the expression $(1+x)^5$.

Answer: easy

$$(1+x)^5 = \binom{5}{0}1^5 + \binom{5}{1}1^4\cdot x + \binom{5}{2}1^3\cdot x^2 + \binom{5}{3}1^2\cdot x^3 + \binom{5}{4}1\cdot x^4 + \binom{5}{5}x^5 = 1 + 5x + 10x^2 + 10x^3 + 5x^4 + x^5$$

11. Use the Binomial Theorem to expand the expression $(A^2+2B)^4$.

Answer: medium

$$(A^2+2B)^4 = \binom{4}{0}A^8 + \binom{4}{1}A^6\cdot 2B + \binom{4}{2}A^4\cdot (2B)^2 + \binom{4}{3}A^2\cdot (2B)^3 + \binom{4}{4}(2B)^4$$
$$= A^8 + 8A^6B + 24A^4B^2 + 32A^2B^3 + 16B^4$$

12. Find the first four terms in the expansion of $(x^{1/2}+2)^{30}$.

Answer: medium

The first four terms in the expansion of $(x^{1/2}+2)^{30}$ are $\binom{30}{0}x^{15} = x^{15}$, $\binom{30}{1}x^{29/2}\cdot 2 = 60x^{29/2}$, $\binom{30}{2}x^{14}\cdot 2^2 = 1740x^{14}$, and $\binom{30}{3}x^{27/2}\cdot 2^3 = 32{,}480x^{27/2}$.

13. Find the first three terms in the expansion of $\left(x+\frac{1}{x}\right)^{20}$.

Answer: medium

The first three terms in the expansion of $\left(x+\frac{1}{x}\right)^{20}$ are: $\binom{20}{0}x^{20} = x^{20}$, $\binom{20}{1}x^{19}\cdot\frac{1}{x} = 20x^{18}$, and $\binom{20}{2}x^{18}\cdot\frac{1}{x^2} = 190x^{16}$.

14. Find the fifth term in the expansion of $(ab-1)^{16}$.

(a) $900a^{12}$ (b) $900a^{12}b^{12}$ (c) $1340a^{16}b^{12}$ (d) $1820a^{12}b^{12}$ (e) $2400a^{12}b^{12}$

Answer: medium (d)

The fifth term in the expansion of $(ab-1)^{16}$ is $\binom{16}{4}(ab)^{12}\cdot 1^4 = 1820a^{12}b^{12}$.

15. Find the 24th term in the expansion of $(A-B)^{30}$.

(a) $324{,}620A^{23}B^7$ (b) $106{,}400A^7B^{23}$ (c) $-106{,}400A^7B^{23}$ (d) $-1{,}463{,}240A^{30}B^{23}$ (e) $-2{,}035{,}800A^7B^{23}$

Answer: medium (e)

The 24th term in the expansion of $(A-B)^{30}$ is $\binom{30}{23}A^7(-B)^{23} = -2{,}035{,}800A^7B^{23}$.

16. Find the third term in the expansion of $\left(x^2-\frac{1}{x}\right)^{25}$.

(a) $140x^{42}$ (b) $300x^{44}$ (c) $320x^{46}$ (d) $180x^{24}$ (e) $220x^{24}$

Answer: medium (b)

The third term in the expansion of $\left(x^2-\frac{1}{x}\right)^{25}$ is $\binom{25}{2}(x^2)^{23}\left(-\frac{1}{x}\right)^2 = 300x^{44}$.

17. Find the term containing y^3 in the expansion of $\left(\sqrt{3}+y\right)^{12}$.

(a) $17{,}820\sqrt{3}\,y^3$ (b) $24{,}660y^3$ (c) $12{,}330y^3$ (d) $14{,}121\sqrt{3}\,y^3$ (e) $16{,}230\sqrt{3}\,y^3$

Answer: hard (a)

The rth term in the expansion of $\left(\sqrt{3}+y\right)^{12}$ is $\binom{12}{r}\left(\sqrt{3}\right)^r y^{12-r}$. The term that contains y^3 occurs when $12-r=3$ $\Leftrightarrow$ $r=9$. Therefore, the term is $\binom{12}{9}\left(\sqrt{3}\right)^9 \cdot y^3 = 17{,}820\sqrt{3}\,y^3$.

18. Find the term containing a^2 in the expansion of $\left(\sqrt{a}+1/\sqrt{a}\right)^{10}$.

(a) $80a^2$ (b) $96a^2$ (c) $120a^2$ (d) $180a^2$ (e) $240a^2$

Answer: hard (c)

The rth term in the expansion of $\left(\sqrt{a}+\dfrac{1}{\sqrt{a}}\right)^{10}$ is

$\binom{10}{r}\left(\sqrt{a}\right)^r\left(\dfrac{1}{\sqrt{a}}\right)^{10-r} = \binom{10}{r}a^{r/2}a^{-\frac{1}{2}(10-r)} = \binom{10}{r}a^{r/2}a^{-5+r/2} = \binom{10}{r}a^{-5+r}$. The term that contains a^2 occurs when $-5+r=2$ $\Leftrightarrow$ $r=7$. Thus, the term is $\binom{10}{7}\left(\sqrt{a}\right)^7\left(\dfrac{1}{\sqrt{a}}\right)^3 = 120a^2$.

19. Find the term that does not contain m in the expansion of $\left(m^2n+m^{-4}\right)^6$.

(a) $5n^4$ (b) $5n^6$ (c) $8n^4$ (d) $15n^2$ (e) $15n^4$

Answer: hard (e)

The rth term in the expansion of $\left(m^2n+m^{-4}\right)^6$ is $\binom{6}{r}\left(m^2n\right)^r\left(m^{-4}\right)^{6-r} = \binom{6}{r}m^{2r}n^rm^{-24+4r} = \binom{6}{r}m^{-24+6r}n^r$. The term that does not contain m occurs when $-24+6r=0$ $\Leftrightarrow$ $r=4$. Thus this term is $\binom{6}{4}\left(m^2n\right)^4\left(m^{-4}\right)^2 = 15n^4$.

20. Find the coefficient of r^{-5} in the expansion of $\left(\dfrac{r^3}{4}-\dfrac{4}{r^2}\right)^5$.

(a) 140 (b) 240 (c) 320 (d) 420 (e) 560

Answer: hard (c)

The kth term in the expansion of $\left(\dfrac{r^3}{4}-\dfrac{4}{r^2}\right)^5$ is

$\binom{5}{k}\left(\dfrac{r^3}{4}\right)^k\left(-\dfrac{4}{r^2}\right)^{5-k} = \binom{5}{k}\left(\dfrac{1}{4}\right)^k(-4)^{5-k}\cdot\dfrac{r^{3k}}{r^{2(5-k)}} = \binom{5}{k}(-1)^{5-k}4^{5-2k}r^{5k-10}$. The term that contains r^{-5} has $5k-10=-5$ $\Leftrightarrow$ $k=1$. Thus the desired coefficient is $\binom{5}{1}(-1)^4 4^3 = 320$.

21. Find the number of distinct terms in the expansion of the expression $\left(x^2+\dfrac{1}{x}\right)^8$.

(a) 7 (b) 8 (c) 9 (d) 10 (e) 12

Answer: medium (c)

$$\left(x^2+\frac{1}{x}\right)^8 - \binom{8}{0}x^{16} + \binom{8}{1}x^{14}\cdot\frac{1}{x} + \binom{8}{2}x^{12}\cdot\frac{1}{x^2} + \binom{8}{3}x^{10}\cdot\frac{1}{x^3} + \binom{8}{4}x^8\cdot\frac{1}{x^4} + \binom{8}{5}x^6\cdot\frac{1}{x^5} + \binom{8}{6}x^4\cdot\frac{1}{x^6} + \binom{8}{7}x^2\cdot\frac{1}{x^7} + \binom{8}{8}\frac{1}{x^8}$$

which has 9 distinct terms.

22. Simplify the expression $(x-2)^4+8(x-2)^3+24(x-2)^2+32(x-2)+16$.

(a) x^4-2 (b) x^4-1 (c) x^4 (d) $2x^4$ (e) $4x^2$

Answer: medium (c)

$(x-2)^4+8(x-2)^3+24(x-2)^2+32(x-2)+16 = [(x-2)+2]^4 = x^4$

23. Simplify the expression $x^{12} + 4x^9y + 6x^6y^2 + 4x^3y^3 + y^4$.

(a) $(x^3+y)^4$ (b) $(x-y^3)^4$ (c) $(x+y^4)^3$ (d) $(x-y)^{12}$ (e) $(x^3-y)^4$

Answer: hard (a)

$$x^{12} + 4x^9y + 6x^6y^2 + 4x^3y^3 + y^4 = \binom{4}{0}(x^3)^4 + \binom{4}{1}(x^3)^3y + \binom{4}{2}(x^3)^2y^2 + \binom{4}{3}x^3y^3 + \binom{4}{4}y^4 = (x^3+y)^4$$

24. Show that $(1.01)^{100} > 2$. [Hint: Note that $(1.01)^{100} = (1+0.01)^{100}$ and use the Binomial Theorem.]

Answer: medium

$$1.01^{100} = (1+0.01)^{100} = 1^{100} + \binom{100}{1}1^{99}\cdot 0.01 + \binom{100}{2}1^{98}\cdot 0.01^2 + \cdots = 1 + 1 + 0.495 + \cdots > 2.$$

25. Show that $\binom{n}{0} = 1$ and $\binom{n}{n} = 1$.

Answer: easy

$$\binom{n}{0} = \frac{n!}{0!\,n!} = \frac{n!}{1\cdot n!} = 1, \text{ and } \binom{n}{n} = \frac{n!}{n!\,0!} = \frac{n!}{n!\cdot 1} = 1.$$

26. Show that $\binom{n}{r} = \binom{n}{n-r}$ for $0 \le r \le n$.

Answer: medium

$$\binom{n}{r} = \frac{n!}{r!(n-r)!} = \frac{n!}{(n-r)!r!} = \binom{n}{n-r} \text{ for } 0 \le r \le n.$$

27. Show that $\binom{n}{0} - \binom{n}{1} + \binom{n}{2} - \cdots + (-1)^k\binom{n}{k} + \cdots + (-1)^n\binom{n}{n} = 0$.

Answer: hard

$$\begin{aligned} 0 = 0^n = [1+(-1)]^n &= \binom{n}{0}(-1)^0\cdot 1^n + \binom{n}{1}(-1)^1\cdot 1^{n-1} + \binom{n}{2}(-1)^2\cdot 1^{n-2} + \cdots + \binom{n}{n}(-1)^n\cdot 1^0 \\ &= \binom{n}{0} - \binom{n}{1} + \binom{n}{2} - \cdots + (-1)^k\binom{n}{k} + \cdots + (-1)^n\binom{n}{n} \end{aligned}$$

28. Use Pascal's Triangle to expand the expression $(2x^2 - y^2)^5$.

Answer: medium

$$\begin{aligned} (2x^2-y^2)^5 &= 1(2x^2)^5(-y^2)^0 + 5(2x^2)^4(-y^2)^1 + 10(2x^2)^3(-y^2)^2 \\ &\quad + 10(2x^2)^2(-y^2)^3 + 5(2x^2)^1(-y^2)^4 + 1(2x^2)^0(-y^2)^5 \\ &= 32x^{10} - 80x^8y^2 + 80x^6y^4 - 40x^4y^6 + 10x^2y^8 - y^{10} \end{aligned}$$

29. Use Pascal's Triangle to expand the expression $\left(\frac{1}{y} - x\right)^4$.

Answer: medium

$$\begin{aligned} \left(\frac{1}{y}-x\right)^4 &= 1\left(\frac{1}{y}\right)^4(-x)^0 + 4\left(\frac{1}{y}\right)^3(-x)^1 + 6\left(\frac{1}{y}\right)^2(-x)^2 + 4\left(\frac{1}{y}\right)^1(-x)^3 + 1\left(\frac{1}{y}\right)^0(-x)^4 \\ &= \frac{1}{y^4} - \frac{4}{y^3}x + \frac{6}{y^2}x^2 - \frac{4}{y}x^3 + x^4. \end{aligned}$$

30. Evaluate the expression $\binom{8}{4}\binom{8}{5}$.

Answer: easy

$$\binom{8}{4}\binom{8}{5} = \left(\frac{8!}{4!\,(8-4)!}\right)\left(\frac{8!}{5!\,(8-5)!}\right) = 3920$$

31. Find the middle term in the expansion of $(2x-y)^{22}$.
Answer: medium
The middle term in the expansion of $(2x-y)^{22}$ is $\binom{22}{11}(2x)^{11}(-y)^{11} = -1{,}444{,}724{,}736x^{11}y^{11}$.

32. Simplify $\dfrac{(x-h)^4 - x^4}{h}$ using the Binomial Theorem.
Answer: easy
$$\frac{(x-h)^4 - x^4}{h} = \frac{x^4 - 4x^3h + 6x^2h^2 - 4xh^3 + h^4 - x^4}{h} = -4x^3 + 6x^2h - 4xh^2 + h^3$$

33. Simplify $\dfrac{(x+h)^5 - x^5}{5h}$ using the Binomial Theorem.
Answer: easy
$$\frac{(x+h)^5 - x^5}{5h} = \frac{x^5 + 5x^4h + 10x^3h^2 + 10x^2h^3 + 5xh^4 + h^5 - x^5}{5h} = x^4 + 2x^3h + 2x^2h^2 + xh^3 + \tfrac{1}{5}h^4$$

34. Factor $x^5 + 5x^4y + 10x^3y^2 + 10x^2y^3 + 5xy^4 + y^5$ using the Binomial Theorem.
Answer: medium
$x^5 + 5x^4y + 10x^3y^2 + 10x^2y^3 + 5xy^4 + y^5 = (x+y)^5$

35. Factor $x^4 - 12x^3y + 54x^2y^2 - 108xy^3 + 81y^4$ using the Binomial Theorem.
Answer: medium
$x^4 - 12x^3y + 54x^2y^2 - 108xy^3 + 81y^4 = (x-3y)^4$

CHAPTER 12 REVIEW

1. Find the first four terms as well as the tenth term of the sequence $a_n = (-1)^n \dfrac{n}{2^n}$.
Answer: easy
$a_n = (-1)^n \dfrac{n}{2^n}$. Then $a_1 = (-1)^1 \dfrac{1}{2^1} = -\dfrac{1}{2}$, $a_2 = (-1)^2 \dfrac{2}{2^2} = \dfrac{1}{2}$, $a_3 = (-1)^3 \dfrac{3}{2^3} = -\dfrac{3}{8}$, $a_4 = (-1)^4 \dfrac{4}{2^4} = \dfrac{1}{4}$, and $a_{10} = (-1)^{10} \dfrac{10}{2^{10}} = \dfrac{10}{1024} = \dfrac{5}{512}$.

2. Find the first four terms as well as the tenth term of the sequence $a_n = \binom{n+2}{3}$.
Answer: medium
$a_n = \binom{n+2}{3}$. Then $a_1 = \binom{1+2}{3} = 1$, $a_2 = \binom{2+2}{3} = \dfrac{4!}{3!\,1!} = 4$, $a_3 = \binom{3+2}{3} = \dfrac{5!}{3!\,2!} = 10$, $a_4 = \binom{4+2}{3} = \dfrac{6!}{3!\,3!} = 20$, and $a_{10} = \binom{10+2}{3} = \dfrac{12!}{3!\,9!} = 220$.

3. The sequence $a_n = \dfrac{a_{n-1}}{2n}$, $a_1 = 1$ is defined recursively. Find the first seven terms of the sequence.
Answer: medium
$a_n = \dfrac{a_{n-1}}{2n}$. $a_1 = 1$, $a_2 = \dfrac{a_1}{2 \cdot 2} = \dfrac{1}{4}$, $a_3 = \dfrac{a_2}{6} = \dfrac{1}{24}$, $a_4 = \dfrac{a_3}{8} = \dfrac{1}{192}$, $a_5 = \dfrac{a_4}{10} = \dfrac{1}{1920}$, $a_6 = \dfrac{a_5}{12} = \dfrac{1}{23{,}040}$, and $a_7 = \dfrac{a_6}{14} = \dfrac{1}{322{,}560}$.

4. The sequence $a_n = \binom{n+2}{a_{n-1}}$, $a_1 = 1$ is defined recursively. Find the first seven terms of the sequence.

Answer: medium

$a_n = \binom{n+2}{a_{n-1}}$. $a_1 = 1$, $a_2 = \binom{2+2}{a_1} = 4$, $a_3 = \binom{3+2}{a_2} = 5$, $a_4 = \binom{4+2}{a_3} = 6$, $a_5 = \binom{5+2}{a_4} = 7$, $a_6 = \binom{6+2}{a_5} = 8$, and $a_7 = \binom{7+2}{a_6} = 9$.

5. The first four terms of a sequence are $1, -\frac{5}{2}, 2$, and $-\frac{3}{2}$. Determine whether the given terms can be the terms of an arithmetic sequence, a geometric sequence, or neither. If the sequence is arithmetic or geometric, find the fifth term.

Answer: medium

$1, -\frac{5}{2}, 2, -\frac{3}{2}, \ldots$. Since $-\frac{5}{2} - 1 \neq 2 - \left(-\frac{5}{2}\right)$, and $\frac{-5/2}{1} \neq \frac{2}{-5/2}$, the series is neither arithmetic nor geometric.

6. The first four terms of a sequence are $\sqrt{3}, 3, 3\sqrt{3}$, and 9. Determine whether the given terms can be the terms of an arithmetic sequence, a geometric sequence, or neither. If the sequence is arithmetic or geometric find the fifth term.

Answer: easy

$\sqrt{3}, 3, 3\sqrt{3}, 9, \ldots$. Since $\frac{3}{\sqrt{3}} = \frac{3\sqrt{3}}{3} = \frac{9}{3\sqrt{3}} = \sqrt{3}$, this is a geometric sequence with $a_1 = \sqrt{3}$ and $r = \sqrt{3}$. Then $a_5 = a_4 \cdot r = 9\sqrt{3}$.

7. The first four terms of a sequence are $\left(\frac{a}{c}\right)^2, \frac{a}{c}, 1$, and $\frac{c}{a}$. Determine whether the given terms can be the terms of an arithmetic sequence, a geometric sequence, or neither. If the sequence is arithmetic or geometric find the fifth term.

Answer: easy

$\left(\frac{a}{c}\right)^2, \frac{a}{c}, 1, \frac{c}{a}, \ldots$. Since $\frac{a/c}{(a/c)^2} = \frac{1}{a/c} = \frac{c/a}{1} = \frac{c}{a}$, this is a geometric sequence with $a_1 = \left(\frac{a}{c}\right)^2$ and $r = \frac{c}{a}$. Then $a_5 = a_4 \cdot r = \frac{c}{a} \cdot \frac{c}{a} = \left(\frac{c}{a}\right)^2$.

8. The first four terms of a sequence are $a^2, a, 1, 1/a, \ldots$. Determine whether the given terms can be the terms of an arithmetic sequence, a geometric sequence, or neither. If the sequence is arithmetic or geometric find the fifth term.

Answer: easy

$a^2, a, 1, 1/a, \ldots$. Since $\frac{a}{a^2} = \frac{1}{a} = \frac{1/a}{1}$, this is a geometric sequence with $a_1 = a^2$ and $r = \frac{1}{a}$. Then $a_5 = a_4 \cdot r = \frac{1}{a} \cdot \frac{1}{a} = \frac{1}{a^2}$.

9. Find the nth term of the geometric sequence $3, 3+3i, 6i, -6+6i, -12, \ldots$ (where $i^2 = -1$).

(a) $a_n = (3+i)^{n-1}$ (b) $a_n = 3(1+i)^{n-1}$ (c) $a_n = 6(1+i)^{n-2}$ (d) $a_n = -i(1+i)^{n-2}$ (e) $a_n = 3^{n-1} + 2ni$

Answer: medium (b)

The sequence $3, 3+3i, 6i, -6+6i, -12, \ldots$ is geometric, since $\frac{a_2}{a_1} = \frac{a_3}{a_2} = \frac{a_4}{a_3} = \frac{a_5}{a_4} = 1+i$. Thus the common ratio is $1+i$ and the first term is 3. So, the nth term is $a_n = a_1 r^{n-1} = 3(1+i)^{n-1}$.

10. The 20th term of an arithmetic sequence is 44 and the common difference is 4. Find the nth term.

(a) $a_n = -32 + 6n$ (b) $a_n = -32 + 4n$ (c) $a_n = -36 + 4n$ (d) $a_n = 4n$ (e) $a_n = 44 + 4n$

Answer: easy (c)

$a_{20} = 44 = a + 19 \cdot 4 = a + 76 \quad \Leftrightarrow \quad a = 44 - 76 = -32$. Then $a_n = -32 + 4(n-1) = -36 + 4n$.

11. The second term of a geometric sequence is 10 and the fifth term is 7290. Find the nth term.

(a) $a_n = \frac{4}{3}(3)^{n-1}$ (b) $a_n = \frac{10}{9}(4)^{n+1}$ (c) $a_n = \frac{10}{9}(9)^{n-1}$ (d) $a_n = 729(3)^{n-3}$ (e) $a_n = \frac{2}{3}(9)^{n-2}$

Answer: medium (c)

$a_2 = 10$ and $a_5 = 7290$. Then $\frac{a_5}{a_2} = r^3 = \frac{7290}{10} = 729 \quad \Leftrightarrow \quad r = 9$ and $a = a_1 = \frac{a_2}{r} = \frac{10}{9}$. Therefore $a_n = ar^{n-1} = \frac{10}{9}(9)^{n-1}$.

12. A person has two parents, four grandparents, eight great-grandparents, and so on. How many ancestors does a person have 12 generations back?

(a) 256 (b) 512 (c) 1024 (d) 2048 (e) 4096

Answer: medium (e)

Let a_n denote the number of ancestors a person has n generations back. Then $a_1 = 2$, $a_2 = 4$, $a_3 = 8, \ldots$. Since $\frac{4}{2} = \frac{8}{4} = 2$, and so on, this is a geometric sequence with $r = 2$. Therefore, $a_{12} = 2 \cdot 2^{11} = 2^{12} = 4096$.

13. If $a_1, a_2, a_3, \ldots$ and $b_1, b_2, b_3, \ldots$ are arithmetic sequences, show that $a_1 + b_1, a_2 + b_2, a_3 + b_3, \ldots$ is also an arithmetic sequence.

Answer: medium

Let d be the common difference in the arithmetic sequence $a_1, a_2, a_3, \ldots$, so that $a_n = a_1 + (n-1)d$ for $n = 1, 2, 3, \ldots$, and let e be the common difference for $b_1, b_2, b_3, \ldots$, so that $b_n = b_1 + (n-1)e$. Then $a_n + b_n = [a_1 + (n-1)d] + [b_1 + (n-1)e] = (a_1 + b_1) + (n-1)(d+e)$ for $n = 1, 2, 3, \ldots$. Thus $a_1 + b_1, a_2 + b_2, a_3 + b_3, \ldots$ is an arithmetic sequence with common difference $d + e$.

14. (a) If $a_1, a_2, a_3, \ldots$ is an arithmetic sequence, is the sequence $a_1 + 2, a_2 + 2, a_3 + 2, \ldots$ arithmetic?

(b) If $a_1, a_2, a_3, \ldots$ is a geometric sequence is the sequence $5a_1, 5a_2, 5a_3, \ldots$ geometric?

Answer: hard

(a) Yes. If the common difference is a, then $a_n = a_1 + (n-1)d$, so $a_n + 2 = a_1 + 2 + (n-1)d$, and thus the sequence $a_1 + 2, a_2 + 2, a_3 + 2, \ldots$ is also arithmetic, with the same common difference d.

(b) Yes. If the common ratio is r, then $a_n = a_1 \cdot r^{n-1}$ and so $5a_n = (5a_1) \cdot r^{n-1}$, and the sequence $5a_1, 5a_2, 5a_3, \ldots$ is also geometric, with the same common ratio r.

15. Find the values of x and y for which the sequence $3, x, y, 14, \ldots$ is

(a) arithmetic.

(b) geometric.

Answer: medium

(a) $3, x, y, 14, \ldots$ is arithmetic. Therefore $14 - 3 = 11 = 3d \quad \Leftrightarrow \quad d = \frac{11}{3}$ and so $x = 3 + \frac{11}{3} = \frac{20}{3}$ and $y = \frac{20}{3} + \frac{11}{3} = \frac{31}{3}$.

(b) $3, x, y, 14, \ldots$ is geometric. Therefore $\frac{14}{3} = r^3 \quad \Leftrightarrow \quad r = \sqrt[3]{\frac{14}{3}}$ and so $x = 3\sqrt[3]{\frac{14}{3}} = 3\left(\frac{14}{3}\right)^{1/3}$ and $y = 3\left(\frac{14}{3}\right)^{2/3}$.

16. Find the sum $\displaystyle\sum_{i=1}^{4} \frac{2i}{2i+1}$.

(a) $\frac{4}{3}$ (b) $\frac{16}{3}$ (c) $\frac{981}{296}$ (d) $\frac{1012}{315}$ (e) $\frac{3241}{640}$

Answer: easy (d)

$$\sum_{i=1}^{4} \frac{2i}{2i+1} = \frac{2 \cdot 1}{2 \cdot 1 + 1} + \frac{2 \cdot 2}{2 \cdot 2 + 1} + \frac{2 \cdot 3}{2 \cdot 3 + 1} + \frac{2 \cdot 4}{2 \cdot 4 + 1} = \frac{2}{3} + \frac{4}{5} + \frac{6}{7} + \frac{8}{9} = \frac{1012}{315}.$$

17. Find the sum $\displaystyle\sum_{m=1}^{5} 3^{m-1}$.

(a) 101 (b) 111 (c) 121 (d) 143 (e) 243

Answer: easy (c)

$$\sum_{m=1}^{5} 3^{m-1} = 3^0 + 3^1 + 3^2 + 3^3 + 3^4 = 1 + 3 + 9 + 27 + 81 = 121.$$

18. Write the sum $\displaystyle\sum_{j=2}^{100} \frac{1}{j-1}$ without using sigma notation. Do not evaluate.

Answer: easy

$$\sum_{j=2}^{100} \frac{1}{j-1} = \frac{1}{1} + \frac{1}{2} + \frac{1}{3} + \frac{1}{4} + \frac{1}{5} + \cdots + \frac{1}{98} + \frac{1}{99}$$

19. Write the sum $\sum_{n=1}^{10} n^2 2^n$ without using sigma notation. Do not evaluate.

Answer: easy

$\sum_{n=1}^{10} n^2 2^n = 1^2 \cdot 2^1 + 2^2 \cdot 2^2 + 3^2 \cdot 2^3 + \cdots + 9^2 \cdot 2^9 + 10^2 \cdot 2^{10}$

20. Write the sum $1^3 + 2^3 + 3^3 + \cdots + 100^3$ using sigma notation. Do not evaluate.

Answer: easy

$1^3 + 2^3 + 3^3 + \cdots + 100^3 = \sum_{k=1}^{100} k^3$

21. Write the sum $\frac{1}{1 \cdot 2} + \frac{1}{2 \cdot 3} + \frac{1}{3 \cdot 4} + \cdots + \frac{1}{9999 \cdot 10000}$ using sigma notation. Do not evaluate.

Answer: easy

$\frac{1}{1 \cdot 2} + \frac{1}{2 \cdot 3} + \frac{1}{3 \cdot 4} + \cdots + \frac{1}{9999 \cdot 10000} = \sum_{k=1}^{9999} \frac{1}{k(k+1)}$

22. Determine whether the series $3 + 3.8 + 4.6 + \cdots + 7$ is arithmetic or geometric and find its sum.

(a) Arithmetic, 24 (b) Arithmetic, 30 (c) Geometric, 33 (d) Geometric, 35 (e) Geometric, 71

Answer: easy (b)

$3 + 3.8 + 4.6 + \cdots + 7$ is an arithmetic series with $a = 3$ and $d = 0.8$. Then $a_n = 7 = 3 + 0.8(n-1) \quad \Leftrightarrow \quad n = 6$, and so the sum of the series is $S_6 = \frac{6}{2}(3+7) = 30$.

23. Determine whether the series $\sqrt{3} + 2\sqrt{3} + 3\sqrt{3} + \cdots + 100\sqrt{3}$ is arithmetic or geometric and find its sum.

(a) Arithmetic, $5050\sqrt{3}$ (b) Arithmetic, $2400\sqrt{3}$ (c) Arithmetic, $200\sqrt{3}$ (d) Geometric, $9460\sqrt{3}$ (e) Geometric, $460\sqrt{3}$

Answer: medium (a)

$\sqrt{3} + 2\sqrt{3} + 3\sqrt{3} + \cdots + 100\sqrt{3}$ is an arithmetic series with $a = \sqrt{3}$ and $d = \sqrt{3}$. Then $a_n = 100\sqrt{3} = \sqrt{3} + \sqrt{3}(n-1) \quad \Leftrightarrow \quad n = 100$, and so the sum is $S_{100} = \frac{100}{2}\left(\sqrt{3} + 100\sqrt{3}\right) = 5050\sqrt{3}$.

24. Determine whether the series $a + ab^2c^3 + ab^4c^6 + ab^6c^9 + \cdots + ab^{16}c^{24}$ is arithmetic or geometric and find its sum.

(a) Arithmetic, $\frac{a^2}{2}\left(1 + b^{16}c^{24}\right)$ (b) Arithmetic, $a^2\left(b^2c^3 + b^{14}c^{22}\right)$ (c) Arithmetic, $a^9b^{18}c^{32}$

(d) Geometric, $\frac{a(1 - b^{18}c^{27})}{1 - b^2c^3}$ (e) Geometric, $\frac{a^2(1 - b^8c^9)}{1 - b^4c^6}$

Answer: medium (d)

$a + ab^2c^3 + ab^4c^6 + ab^6c^9 + \cdots + ab^{16}c^{24}$ is a geometric series with first term a and common ratio b^2c^3. Therefore, $n = 9$ and so the sum of the series is $S_9 = a \cdot \frac{1 - (b^2c^3)^9}{1 - b^2c^3} = \frac{a(1 - b^{18}c^{27})}{1 - b^2c^3}$.

25. A geometric sequence has first term $a = 16$ and common ratio $r = -\frac{3}{4}$. How many terms of this sequence must be added to get $\frac{181}{16}$?

(a) 3 (b) 4 (c) 5 (d) 6 (e) 7

Answer: medium (c)

Geometric sequence with $a = 16$ and $r = -\frac{3}{4}$. Then $S_n = \frac{181}{16} = 16 \cdot \frac{1 - \left(-\frac{3}{4}\right)^n}{1 - \left(-\frac{3}{4}\right)} = \frac{64}{7}\left[1 - \left(-\frac{3}{4}\right)^n\right] \quad \Leftrightarrow$

$1 - \left(-\frac{3}{4}\right)^n = \frac{1267}{1024} \quad \Leftrightarrow \quad \left(-\frac{3}{4}\right)^n = -\frac{243}{1024} \quad \Leftrightarrow \quad n = 5$. Therefore 5 terms must be added.

26. The sum of the first three terms of a geometric series is 26 and the common ratio is $r = 3$. Find the first term.

(a) -1 (b) 2 (c) 4 (d) 6 (e) 13

Answer: medium (b)

Geometric series with $S_3 = 26$ and $r = 3$. Then $S_3 = 26 = a + 3a + 9a = 13a \quad \Leftrightarrow \quad a = 2$, and so the first term is 2.

27. A person has two parents, four grandparents, eight great-grandparents, and so on. What is the total number of ancestors of a person in 12 generations?

(a) 8190 (b) 16,551 (c) 18,054 (d) 21,062 (e) 34,494

Answer: medium (a)

Geometric sequence with $a = 2$ and $r = 2$. Then $S_{12} = 2 \cdot \dfrac{1-2^{12}}{1-2} = 2(2^{12}-1) = 8190$, and so the total number of ancestors is 8190.

28. How much money should be invested every quarter at 10% a year compounded quarterly in order to have $12,000 in one year?

(a) $1034.27 (b) $1733.23 (c) $1942.11 (d) $2041.46 (e) $2889.81

Answer: medium (e)

$A = 12,000$, $i = 0.025$, and $n = 4$. Thus, $12,000 = R\dfrac{1.025^4-1}{0.025} \quad \Leftrightarrow \quad R = \dfrac{12{,}000 \cdot 0.025}{1.025^4-1} \approx \$2889.81$.

29. Find the sum of the infinite geometric series $1 - \frac{3}{5} + \frac{9}{25} - \frac{27}{125} + \cdots$.

(a) $\frac{4}{7}$ (b) $\frac{5}{7}$ (c) $\frac{5}{8}$ (d) $\frac{7}{8}$ (e) $\frac{17}{16}$

Answer: medium (c)

$1 - \frac{3}{5} + \frac{9}{25} - \frac{27}{125} + \cdots$ is a geometric series with $a = 1$ and $r = -\frac{3}{5}$. Therefore, the sum is $S = \dfrac{a}{1-r} = \dfrac{1}{1-\left(-\frac{3}{5}\right)} = \dfrac{5}{8}$.

30. Find the sum of the infinite geometric series $1 + \dfrac{1}{3^{1/4}} + \dfrac{1}{3^{1/2}} + \dfrac{1}{3^{3/4}} + \cdots$.

(a) $\dfrac{1}{3^{4/3}}$ (b) $\dfrac{2}{3-\sqrt{3}}$ (c) $\dfrac{\sqrt{3}}{3+2\sqrt{3}}$ (d) $\dfrac{3}{3-3^{3/4}}$ (e) $\dfrac{4^{2/3}}{3^{3/4}-1}$

Answer: medium (d)

$1 + \dfrac{1}{3^{1/4}} + \dfrac{1}{3^{1/2}} + \dfrac{1}{3^{3/4}} + \cdots$ is an infinite geometric series with $a = 1$ and $r = \dfrac{1}{\sqrt[4]{3}}$. Thus the sum is $S = \dfrac{1}{1-\left(1/\sqrt[4]{3}\right)} = \dfrac{3}{3-3^{3/4}}$.

31. Evaluate the expression $\dbinom{10}{2} + \dbinom{10}{4}$.

(a) 204 (b) 255 (c) 275 (d) 314 (e) 592

Answer: medium (b)

$$\binom{10}{2} + \binom{10}{4} = \frac{10!}{2!\,8!} + \frac{10!}{4!\,6!} = \frac{10 \cdot 9}{2} + \frac{10 \cdot 9 \cdot 8 \cdot 7}{4 \cdot 3 \cdot 2} = 45 + 210 = 255$$

32. Evaluate the expression $\displaystyle\sum_{k=0}^{6} \binom{6}{k}\binom{6}{6-k}$

(a) 714 (b) 924 (c) 944 (d) 960 (e) 1014

Answer: medium (b)

$$\sum_{k=0}^{6} \binom{6}{k}\binom{6}{6-k} = 2\binom{6}{0}\binom{6}{6} + 2\binom{6}{1}\binom{6}{5} + 2\binom{6}{2}\binom{6}{4} + \binom{6}{3}\binom{6}{3} = 2 + 2 \cdot 6^2 + 2 \cdot 15^2 + 20^2 = 924$$

33. Expand the expression $(x+2y)^4$.

Answer: medium

$$(x+2y)^4 = \binom{4}{0}x^4 + \binom{4}{1}x^3(2y) + \binom{4}{2}x^2(2y)^2 + \binom{4}{3}x(2y)^3 + \binom{4}{4}(2y)^4 = x^4 + 8x^3y + 24x^2y^2 + 32xy^3 + 16y^4$$

34. Find the first three terms in the expansion of $\left(b^{1/3} + b^{-2/3}\right)^{20}$.

Answer: medium

The first three terms in the expansion of $\left(b^{1/3} + b^{-2/3}\right)^{20}$ are $\dbinom{20}{0}\left(b^{1/3}\right)^{20} = b^{20/3}$, $\dbinom{20}{1}\left(b^{1/3}\right)^{19}\left(b^{-2/3}\right) = 20b^{17/3}$, and $\dbinom{20}{2}\left(b^{1/3}\right)^{18}\left(b^{-2/3}\right)^2 = 190b^{14/3}$.

35. Find the term containing A^6 in the expansion of $(A + 4B)^{10}$.

Answer: hard

The rth term in the expansion of $(A + 4B)^{10}$ is $\binom{10}{r} A^r (4B)^{10-r}$. The term that contains A^6 occurs when $r = 6$. Thus, the term is $\binom{10}{6} A^6 (4B)^4 = 210A^6 \cdot 256B^4 = 53{,}760A^6B^4$.

36. Simplify the expression $x^6 + 6x^5y + 15x^4y^2 + 20x^3y^3 + 15x^2y^4 + 6xy^5 + y^6$.

Answer: medium

$x^6 + 6x^5y + 15x^4y^2 + 20x^3y^3 + 15x^2y^4 + 6xy^5 + y^6 = (x + y)^6$

CHAPTER THIRTEEN

SECTION 13.1

1. Evaluate the expression $3!\,5!$.

   (a) 144 (b) 15 (c) 120 (d) 720 (e) 1540

   Answer: easy (d)

   $3!\,5! = 3 \cdot 2 \cdot 1 \cdot 5 \cdot 4 \cdot 3 \cdot 2 \cdot 1 = 720$

2. Evaluate the expression $\dfrac{5!}{2!\,3!}$.

   (a) 12 (b) $\dfrac{5}{6}$ (c) 10 (d) 20 (e) 30

   Answer: easy (d)

   $\dfrac{5!}{2!\,3!} = \dfrac{5 \cdot 4 \cdot 3!}{2 \cdot 1 \cdot 3!} = \dfrac{20}{2} = 10$

3. Evaluate the expression $\dfrac{13!\,15!}{12!\,14!}$.

   (a) 167 (b) 180 (c) 144 (d) 169 (e) 195

   Answer: medium (e)

   $\dfrac{13!15!}{12!14!} = \dfrac{13 \cdot 12! \cdot 15 \cdot 14!}{12!\,14!} = 195$

4. Evaluate the expression $\dfrac{2001!}{2000!}$.

   (a) 2001 (b) 2000 (c) 1001 (d) 1000 (e) 501

   Answer: easy (a)

   $\dfrac{2001!}{2000!} = \dfrac{2001 \cdot 2000!}{2000!} = 2001$

5. Evaluate the expression $\dfrac{50! - 49!}{48!}$.

   (a) 2425 (b) 2500 (c) 2401 (d) 2001 (e) 1402

   Answer: medium (c)

   $\dfrac{50! - 49!}{48!} = \dfrac{48!(50 \cdot 49 - 49)}{48!} = 49^2 = 2401$

6. Simplify the expression $\dfrac{n!(n+1)}{(n+1)!}$.

   (a) n (b) $n+1$ (c) 1 (d) n^2 (e) n^3

   Answer: medium (c)

   $\dfrac{n!(n+1)}{(n+1)!} = \dfrac{(n+1)!}{(n+1)!} = 1$

7. Simplify the expression $\dfrac{(n+1)!}{(n-2)!\,n}$.

   (a) n^2 (b) $(n-1)^2$ (c) $n-1$ (d) n^2-1 (e) n^3-8

   Answer: medium (d)

   $\dfrac{(n+1)!}{(n-2)!n} = \dfrac{(n+1) \cdot n \cdot (n-1)(n-2)!}{(n-2)!\,n} = (n+1)(n-1) = n^2 - 1$

8. In how many ways can four people be seated in a row of four seats?

   (a) 4 (b) 20 (c) 24 (d) 120 (e) 720

   Answer: easy (c)

   By the Fundamental Counting Principle, the possible number of ways in seating 4 people in a row of 4 seats is $4 \cdot 3 \cdot 2 \cdot 1 = 4! = 24$.

9. In how many ways can six different mathematics books be placed next to each other on a shelf?

(a) 6 (b) 120 (c) 720 (d) 1440 (e) 28,800

Answer: easy (c)

By the Fundamental Counting Principle, the possible number of ways in arranging 6 mathematics books is $6 \cdot 5 \cdot 4 \cdot 3 \cdot 2 \cdot 1 = 6! = 720$.

10. In a family with five children, how many different boy-girl combinations are possible? (The birth order of the children matters, so that BBBBG is different from BBBGB.)

(a) 5 (b) 10 (c) 24 (d) 32 (e) 36

Answer: easy (d)

The possible number of different boy-girl combinations of 5 children $= 2^5 = 32$.

11. A red and a blue die are tossed and the numbers showing are recorded. How many different outcomes are possible?

(a) 36 (b) 72 (c) 6 (d) 12 (e) 128

Answer: easy (a)

The number of different outcomes of a red and a blue die $= 6^2 = 36$.

12. Two cards are chosen in order from a deck. In how many ways can this be done if the first card must be a spade and the second must be a heart?

(a) 156 (b) 161 (c) 169 (d) 172 (e) 144

Answer: easy (c)

The number of ways that the first card is a spade and the second is a heart $= (13)(13) = 169$.

13. Two cards are chosen in order from a deck. In how many ways can this be done if both cards must be spades?

(a) 156 (b) 161 (c) 169 (d) 172 (e) 144

Answer: easy (a)

The number of ways that the first card is a spade and the second is a spade $= (13)(12) = 156$.

14. A company decides to give ID numbers to its employees that consist of one letter followed by four digits. How many different such ID numbers are there?

Answer: easy

The number of possible ID numbers consisting of one letter followed by 4 digits $= (26)(10)(10)(10)(10) = 260{,}000$.

15. An all-star baseball team has a roster of six pitchers and three catchers. How many pairs of pitchers and catchers can the manager choose from the roster?

Answer: easy

The number of possible pairs of a pitcher and a catcher $= (6)(3) = 18$.

16. A combination lock has 50 different positions. To open the lock its dial is turned to a certain number in the clockwise direction, then to a number in the counterclockwise direction, and finally to a third number in the clockwise direction. If consecutive numbers in the combination can not be the same, how many different combinations are there?

Answer: medium

The number of possible combinations consisting of a number in the clockwise direction, then a number in the counterclockwise direction, and a number in the clockwise direction $= (50)(49)(49) = 120{,}050$.

17. A new car dealer offers four different models. Each car comes in a choice of five colors, three types of stereo equipment, with or without air conditioning, and with or without sunroof. In how many different ways can a customer order a car from this dealer?

Answer: easy

The possible number of different cars consisting of a model, a color, a type of stereo, air conditioning, and sunroof $= (4)(5)(3)(2)(2) = 240$.

18. Explain why in any group of 677 people at least two must have the same first and last initials.

Answer: medium

Since there are 26 letters, the possible number of combinations of the first and the last initials $= (26)(26) = 676$, then there must be at least 2 people that have the same first and the last initials in any group of 677 people.

19. A state has 17 million registered cars and the design of their license plates allows for six places on each plate to be used for a fixed number of letters followed by a fixed number of digits.

(For example, one letter followed by five digits, or two numbers followed by four digits, etc.).

(a) If the state decides on a system consisting of one letter followed by five digits, will there be enough different plates to go around?

(b) Find such a system that will be sufficient if the smallest possible number of letters is to be used.

Answer: medium

(a) The number of different plates consisting of 1 letter followed by 5 digits $= (26)(10^5) = 2{,}600{,}000 = 2.6$ million, hence there are not enough different plates to go around for 17 million registered cars.

(b) The system with two letters followed by 4 digits gives $(26^2)(10^4) = 6{,}760{,}000 < 17$ million which is not enough. The system consisting of 3 letters followed by 3 digits gives $(26^3)(10^3) = 17{,}576{,}000 > 17$ million. Thus the fewest number of letters that will provide enough different plates is 3.

20. In how many ways can a president, vice-president, and secretary be chosen from a class of 20 girls and 30 boys if the president must be a girl and the vice-president a boy?

Answer: medium

The number of ways that a president, a vice-president and a secretary can be chosen $= (20)(30)(48) = 28{,}800$.

21. Telephone area codes have three digits. The first digit must not be 0 or 1, the middle digit must be 0 or 1, and the last digit must not be 0. How many different area codes are possible?

Answer: easy

The possible number of telephone area codes consisting of 3 digits $= (8)(2)(9) = 144$.

22. How many seven-letter palindromes are there?

(A palindrome is a sequence of letters that reads the same backward and forward, such as XCVZVCX.)

Answer: medium

The possible number of 7-letter palindromes $= (26)(26)(26)(26)(1)(1)(1) = 456{,}976$.

23. How many different three-character code words consisting of letters or digits are there if

(a) the first entry must be a letter?

(b) the first entry cannot be zero?

Answer: medium

(a) The number of possible different 3-character code words with a letter as the first entry $= 26 \cdot 36 \cdot 36 = 33{,}696$.

(b) The number of possible different 3-character code words in which zero is not the first entry $= 35 \cdot 36 \cdot 36 = 45{,}360$.

24. In how many ways can five different mathematics books be placed on a shelf if the three algebra books are to be placed next to each other?

Answer: medium

Since the three algebra books must be next to each other, then by treating the three algebra books as one object, the number of ways to arrange the 5 books is 3!. But there are $3! = 6$ possible ways to arrange three algebra books. Thus the number of ways that 5 mathematics books may be placed on a shelf if the three algebra books are to be next to each other $= 6 \cdot 3! = 36$.

25. Three-digit numbers are formed using the digits 2, 4, 5, and 8, with repetition of digits allowed. How many such numbers are there that are

(a) less than 800?

(b) even?

(c) divisible by 5?

Answer: medium

(a) Since the numbers to be formed are less than 800, the first digit must be either 2, 4 or 5. This gives 3 possible ways to choose the first digit. The remaining 2 digits can be selected form 2, 4, 5 and 8. Therefore the number of 3-digit numbers that can be formed using the digits 2, 4, 5 and 8 if the numbers are less than 800 is $3 \cdot 4 \cdot 4 = 48$.

(b) Since the numbers to be formed must be even, the last digit must be either 2, 4, or 8. The first 2 digits can be chosen from 2, 4, 5, and 8. Therefore the number of 3-digit numbers that can be formed using the digits 2, 4, 5 and 8 if the numbers are even is $(4)(4)(3) = 48$.

(c) Since the numbers to be formed must be divisible by 5, the last digit must be 5. The first 2 digits can be chosen from 2, 4, 5, and 8. Therefore the number of 3-digit numbers that can be formed using the digits 2, 4, 5 and 8 if the numbers are divisible by 5 is $(4)(4)(1) = 16$.

26. A true-false test contains 8 questions. In how many different ways can the test be completed?

Answer: medium

The true-false test can be completed $(2)(2)(2)(2)(2)(2)(2)(2) = 2^8 = 256$ ways.

27. Part one of an exam has 10 multiple choice questions with four choices for each question. Part two consists of 8 true-false questions. In how many different ways can the exam be completed?

Answer: hard

Part one can be completed in $(4)(4)(4)(4)(4)(4)(4)(4)(4)(4) = 4^{10}$ ways. Part two can be completed in $(2)(2)(2)(2)(2)(2)(2)(2) = 2^8$ ways. The exam can be completed in $4^{10}2^8 = 268{,}435{,}456$ ways.

28. A particular university issues ID numbers to its students that consist of 3 digits and 2 letters. How many different ID numbers are possible if repetition of the digits and letters are not allowed?

Answer: hard

The number of possible ID numbers is $10 \cdot 9 \cdot 8 \cdot 26 \cdot 25 = 468{,}000$.

SECTION 13.2

1. Evaluate the expression $P(10, 3)$.

(a) 120 (b) 720 (c) 764 (d) 864 (e) 944

Answer: easy (b)

$$P(10,3) = \frac{10!}{(10-3)!} = \frac{10!}{7!} = 10 \times 9 \times 8 = 720$$

2. Evaluate the expression $P(7, 3)$.

(a) 210 (b) 221 (c) 224 (d) 243 (e) 333

Answer: easy (a)

$$P(7,3) = \frac{7!}{4!} = 210$$

3. Evaluate the expression $P(99, 4)$.

(a) 40,210,116 (b) 82,202,168 (c) 87,266,480 (d) 90,345,024 (e) 103,454,545

Answer: easy (d)

$$P(99,4) = \frac{99!}{95!} = 90{,}345{,}024$$

4. Evaluate the expression $P(n+1, 2)$.
(a) $n^2 + n$ (b) $n^2 - 1$ (c) $n^2 + n + 1$ (d) $n^2 + 2n + 1$ (e) $2n^2 - n - 1$
Answer: medium (a)
$$P(n+1, 2) = \frac{(n+1)!}{[(n+1)-2]!} = \frac{(n+1)!}{(n-1)!} = (n+1)n = n^2 + n$$

5. Evaluate the expression $P(n+1, n)$.
(a) $(n+1)!$ (b) $n! + 1$ (c) $n^2 + 1$ (d) $n^2 - n$ (e) $n^3 - 3n$
Answer: medium (a)
$$P(n+1, n) = \frac{(n+1)!}{[(n+1)-n]!} = (n+1)!$$

6. In how many different ways can five of ten people be seated in a row of five chairs?
(a) 16,656 (b) 30,240 (c) 33,144 (d) 36,663 (e) 49,044
Answer: easy (b)
The number of ways of seating 5 people from 10 people in a row of 5 chairs $= P(10, 5) = 10 \times 9 \times 8 \times 7 \times 6 = 30{,}240$.

7. How many permutations are there of the letters of the word VOLE?
Answer: easy
The number of ways of ordering 4 letters from the word VOLE is $P(4, 4) = 4! = 24$.

8. A pianist plans to play six pieces at a recital. In how many ways can she arrange these pieces in the program?
Answer: easy
The number of ways of ordering 6 pieces in order $= P(6, 6) = 6! = 720$.

9. A ship carries four signal flags of different colors. How many different signals can be sent by hoisting these flags on the ship's flagpole in different orders?
Answer: easy
The number of ways of ordering 4 distinct color flags $= P(4, 4) = 4! = 24$.

10. In how many ways can a president, secretary, and treasurer be chosen from a class of 20 students?
Answer: easy
The number of ways of choosing 3 officers from 20 students $= P(20, 3) = 6840$.

11. In how many ways can nine students be seated in a row if Jack insists on sitting in the middle chair?
Answer: medium
The number of ways of seating the remaining 8 students $= P(8, 8) = 8! = 40{,}320$.

12. There are twelve vacant seats on an airplane. In how many ways can four passengers be assigned seats?
Answer: easy
The number of ways of seating 4 passengers in 12 seats $= P(12, 4) = 11{,}880$.

13. In how many different ways can four red balls, three white balls, and five blue balls be arranged in a row?
Answer: medium
The number of distinguishable permutations of 12 objects (the 12 balls) which can be partitioned into like groups of size 4 (the red balls), of size 3 (the white balls) and of size 5 (the blue balls) $= \dfrac{12!}{4!3!5!} = 27{,}720$.

14. In how many different ways can the letters of the word AVOCADO be arranged?
Answer: medium
The number of distinguishable permutations of 7 objects (the 7 letters) which can be partitioned into two like groups of size 2 (the A's and the O's), and into 3 like groups of size 1 (the V, the C, and the D) $= \dfrac{7!}{2!\,2!\,1!\,1!\,1!} = 1260$.

15. In how many ways can the algebraic expression a^4b^3 be written without using exponents? (For example, $ababbaa$ is one way.)

Answer: medium

The number of distinguishable permutations of 7 objects (the 7 factors) which can be partitioned into like groups of size 4 (the a's) and of size 3 (the b's) $= \dfrac{7!}{4!\,3!} = 35$.

16. A man bought five vanilla ice cream cones, five chocolate cones, two strawberry cones, and three butterscotch cones for his fifteen children. In how many ways can he distribute the cones among his children?

Answer: medium

The number of distinguishable permutations of 15 objects (the 15 ice cream cones) which can be partitioned into like groups of sizes 5, 5, 2, and 3 is $\dfrac{15!}{5!\,5!\,3!\,2!} = 7{,}567{,}560$.

17. Nine workers are cleaning a large house. Five are needed to clean windows, three to clean the carpets, and one to clean the rest of the house. In how many different ways can these tasks be assigned to the nine workers?

Answer: medium

The number of distinguishable permutations of 9 objects (the 9 workers) which can be partitioned into three groups of size 5, 3, and 1 workers is $\dfrac{9!}{5!\,3!\,1!} = 504$.

18. A computer company is giving away a million dollars to whoever guesses the order in which they have rearranged the letters of the name of their computer, which is DEEPTHOUGHT. Each entry must be mailed in a separate envelope. How many postage stamps would have to be used to submit all possible entries?

Answer: medium

The number of distinguishable permutations of 11 objects (the 11 letters) which can be partitioned into like groups of sizes 2, 2, 2, 1, 1, 1, 1, and 1 is $\dfrac{11!}{2!\,2!\,2!\,1!\,1!\,1!\,1!\,1!} = 6{,}652{,}800$. Therefore, postage would have to be paid for 6,652,800 separate envelopes.

19. Evaluate the expression $C(12, 3)$.

(a) 110 (b) 220 (c) 240 (d) 250 (e) 650

Answer: easy (b)

$$C(12,3) = \frac{12!}{3!(12-3)!} = \frac{12!}{3!\,9!} = \frac{12 \times 11 \times 10}{3 \times 2 \times 1} = 220$$

20. Evaluate the expression $C(9, 4)$.

(a) 84 (b) 106 (c) 126 (d) 204 (e) 240

Answer: easy (c)

$$C(9,4) = \frac{9!}{4!\,5!} = \frac{9 \times 8 \times 7 \times 6}{4 \times 3 \times 2 \times 1} = 126$$

21. Evaluate the expression $C(101, 6)$.

(a) 492,261,458 (b) 894,204,920 (c) 914,911,956 (d) 1,132,127,130 (e) 1,267,339,920

Answer: easy (e)

$$C(101,6) = \frac{101!}{6!\,95!} = \frac{101 \times 100 \times 99 \times 98 \times 97 \times 96}{6 \times 5 \times 4 \times 3 \times 2 \times 1} = 1{,}267{,}339{,}920$$

22. Evaluate the expression $C(n, 3)$.

(a) $\dfrac{n(n-1)(n-2)}{6}$ (b) $\dfrac{(n+1)n(n-1)}{4}$ (c) $\dfrac{n(n-1)(n-2)}{4}$ (d) $\dfrac{(n+1)^2}{6}$ (e) $\dfrac{n(n-1)^2}{12}$

Answer: medium (a)

$$C(n,3) = \frac{n!}{3!(n-3)!} = \frac{n(n-1)(n-2)}{6}$$

23. Evaluate the expression $C(n + 1, n)$.

(a) $n + 1$ (b) n (c) $n - 1$ (d) $n(n + 1)$ (e) $\dfrac{(n+1)!}{12}$

Answer: medium (a)

$$C(n+1,n) = \frac{(n+1)!}{n!\,1!} = n+1$$

24. In how many different ways can a committee of five members be chosen from a club with 22 members?

(a) 14,060 (b) 18,368 (c) 21,056 (d) 24,012 (e) 26,334

Answer: easy (e)

The number of combinations of 22 objects (the 22 members) taken 5 at a time $= C(22,5) = \dfrac{22!}{5!\,17!} = 26{,}334$.

25. How many six-card hands can be picked from a deck of 52 cards?

(a) 14,265,440 (b) 20,358,520 (c) 24,642,960 (d) 32,857,404 (e) 528,165,160

Answer: easy (b)

The number of combinations of 52 objects (the 52 cards) taken 6 at a time $= C(52,6) = \dfrac{52!}{6!\,46!} = 20{,}358{,}520$.

26. A pizza parlor offers a choice of 20 different toppings. How many different four-topping pizzas are possible?

(a) 3864 (b) 4000 (c) 4012 (d) 4845 (e) 5484

Answer: easy (d)

The number of combinations of 20 different toppings taken 4 at a time $= C(20,4) = \dfrac{20!}{4!\,16!} = 4845$.

27. If a student has nine skirts, in how many different ways can she choose four of these to take on a weekend trip?

Answer: easy

The number of combinations of 9 skirts taken 4 at a time $= C(9,4) = \dfrac{9!}{4!\,5!} = 126$.

28. In how many ways can six students from a class of 25 students be chosen for a field trip if Jack must go on the field trip?

Answer: medium

The number of combinations of 24 students taken 5 at a time $= C(24,5) = \dfrac{24!}{5!\,19!} = 42{,}504$.

29. In the 7/49 lottery game a player picks seven numbers from 1 to 49. How many different choices does the player have?

Answer: easy

The number of combinations of 49 numbers taken 7 at a time $= C(49,7) = \dfrac{49!}{7!\,42!} = 85{,}900{,}584$.

30. In the Alabama Super-7 lottery a player chooses seven different numbers from 1 to 75. The Lottery Commission chooses 10 numbers randomly. A player is a winner if his seven numbers are among the 10 chosen by the commission.

(a) In how many ways can a player choose seven numbers for his lottery ticket?

(b) How many winning combinations are possible on each draw?

Answer: medium

(a) The number of combinations of 75 numbers taken 7 at a time $= C(75,7) = \dfrac{75!}{7!\,68!} = 1{,}984{,}829{,}850$.

(b) The number of combinations of 10 numbers taken 7 at a time $= C(10,7) = 120$.

31. A student is required to answer eight of 12 questions on a test. In how many ways can she do this if

(a) there are no restrictions on which questions she must answer?

(b) she definitely cannot do the first or last question?

(c) she must do four of the first six questions?

Answer: medium

(a) The number of combinations of 12 questions taken 8 at a time $= C(12,8) = \frac{12!}{8!\,4!} = 495$.

(b) The number of combinations of 10 questions taken 8 at a time $= C(10,8) = \frac{10!}{8!\,2!} = 45$.

(c) By the Fundamental Counting Principle, the possible number of combinations of questions to be answered is

$$\begin{pmatrix}\text{number of ways that 4 questions}\\ \text{are chosen from the first 6}\end{pmatrix}\begin{pmatrix}\text{number of ways that 4 questions}\\ \text{are chosen from the remaining 8}\end{pmatrix} = C(6,4)\cdot C(8,4) = 15\cdot 70 = 1050.$$

32. Maria has four pennies, three nickels, two dimes, and five quarters. In how many different ways can she give some of these coins to her brother?

Answer: medium

The number of distinguishable combinations of 14 objects (the 14 coins) consisting of like groups of sizes 2, 3, 4 and 5, is $(2+1)(3+1)(4+1)(5+1) = 3 \cdot 4 \cdot 5 \cdot 6 = 360$.

33. Jane's bag of marbles has five red, two white, and six blue marbles. If she reaches in to pick some without looking, how many different selections might she make?

Answer: medium

The number of distinguishable combinations of 13 marbles consisting of like groups of size 5, 2, and 6 is $(5+1)(2+1)(6+1) = (6)(3)(7) = 126$.

34. Find the number of divisors of 1080 that are:

(a) Odd numbers (b) Even numbers (c) Divisible by 5 (d) Divisible by 3

Answer: hard

The prime factorization of the number 1080 is $2 \times 2 \times 2 \times 3 \times 3 \times 3 \times 5$.

(a) Odd numbers do not have 2 as a factor. So the number of odd factors is the number of distinguishable combinations of four factors consisting of like groups of size 3 (the 3's), and of size 1 (the 5). Thus there are $(3+1)(1+1) = (4)(2) = 8$ odd factors.

(b) An even factor must have 2 as a factor. This leaves only two 2's, three 3's, and one 5 to choose from to make factors of 1080. Thus the number of even factors is the number of distinguishable combinations of 6 objects of like groups of sizes 2, 3, and 1. Thus there are $(2+1)(3+1)(1+1) = (3)(4)(2) = 24$ even factors.

(c) The desired factors must have 5 as a factor. This leaves only three 2's and three 3's to choose from to make factors of 1080. Thus the number of factors divisible by five is the number of distinguishable combinations of 6 objects of like groups of sizes 3 and 3. Thus there are $(3+1)(3+1) = (4)(4) = 16$ factors divisible by five.

(d) The desired factors must have 3 as a factor. This leaves only three 2's, two 3's, and one 5 to choose from to make factors of 1080. Thus the number of factors divisible by three is the number of distinguishable combinations of 6 objects of like groups of sizes 3, 2, and 1. Thus there are $(3+1)(2+1)(1+1) = (4)(3)(2) = 24$ factors divisible by three.

35. A hamburger chain gives their customers a choice of eight different toppings for their hamburgers. In how many different ways can a customer order a hamburger?

Answer: easy

The number of possible subsets of the 8 toppings $= 2^8 = 256$.

36. (a) Calculate $C(1000, 0)$ and $C(1000, 1000)$.

(b) Show that in general $C(n, 0) = C(n, n) = 1$.

(c) Give a counting argument that explains why this should be true.

Answer: hard

(a) $C(1000, 0) = \dfrac{1000!}{0!\,1000!} = 1$ and $C(1000, 1000) = 1$

(b) $C(n, 0) = \dfrac{n!}{0!\,n!} = \dfrac{n!}{1 \cdot n!} = 1$ and $C(n, n) = \dfrac{n!}{n!\,0!} = \dfrac{n!}{n! \cdot 1} = 1$

(c) $C(n, 0)$ is the number of ways of choosing 0 objects at a time, that is no objects, which can only be done in one way. $C(n, n)$ is the number of ways of choosing all of the n objects at a time which can only be done in one way.

37. A chess player has a trophy for first place, one for second, one for third and one for fourth place. In how many ways can he arrange his trophies on his trophy case?
Answer: medium
$4 \cdot 3 \cdot 2 \cdot 1 = 24$

38. A little league baseball coach has 13 players at tryouts. In how many ways can the coach select 9 starters (one at each unique position)?
Answer: hard
We need to choose 9 from 13. The number of permutations is $P(13, 9) = \dfrac{13!}{(13-9)!} = 259,459{,}200$.

39. Seven volunteers are needed to operate a car wash fundraiser. Three are needed to wash, two are needed to rinse, and two are needed to dry. In how many different ways can the volunteers be assigned to perform these tasks?
Answer: medium
We need groups of 3, 2 and 2. Thus, the answer is $\dfrac{7!}{3!\,2!\,2!} = 210$.

40. There are 9 campers at a campsite. Four are to collect firewood, while the other five are to set up the campsite. In how many ways can the group to collect firewood be selected?
Answer: hard
$C(9, 4) = \dfrac{9!}{4!\,(9-4)!} = 126$ ways

SECTION 13.3

1. A fair die is rolled. Find the probability that the number showing is

(a) a four or a five.

(b) an even number.

(c) a number divisible by 2.

Answer: easy

(a) Let E be the event of rolling a four or a five. Then $P(E) = \dfrac{n(E)}{n(S)} = \dfrac{2}{6} = \dfrac{1}{3}$.

(b) Let E be the event of rolling an even number. Then $P(E) = \dfrac{n(E)}{n(S)} = \dfrac{3}{6} = \dfrac{1}{2}$.

(c) Let E be the event of rolling a number divisible by 2. Then $P(E) = \dfrac{n(E)}{n(S)} = \dfrac{1}{2}$.

2. A card is randomly picked from a 52-card deck. Find the probability that the card selected is
(a) a heart.
(b) either a heart or a spade.
(c) a heart, a diamond, or a spade.
Answer: easy
(a) Let E be the event of choosing a heart. Then $P(E) = \dfrac{n(E)}{n(S)} = \dfrac{13}{52} = \dfrac{1}{4}$.
(b) Let E be the event of choosing a heart or a spade. Then $P(E) = \dfrac{n(E)}{n(S)} = \dfrac{26}{52} = \dfrac{1}{2}$.
(c) Let E be the event of choosing a heart, a diamond or a spade. Then $P(E) = \dfrac{n(E)}{n(S)} = \dfrac{39}{52} = \dfrac{3}{4}$.

3. A ball is randomly selected from an urn that contains four red balls, two white balls, and one yellow ball. Find the probability that the ball selected is

(a) neither white nor yellow. (b) red, white, or yellow. (c) not white.

Answer: medium

(a) Let E be the event of selecting a white or a yellow ball. Then $P(E') = 1 - P(E) = 1 - \frac{n(E)}{n(S)} = 1 - \frac{3}{7} = \frac{4}{7}$.

(b) Let E be the event of selecting a red, a white or a yellow ball. Then $P(E) = \frac{n(E)}{n(S)} = \frac{7}{7} = 1$.

(c) Let E be the event of selecting a white ball. Then $P(E') = 1 - P(E) = 1 - \frac{n(E)}{n(S)} = 1 - \frac{2}{7} = \frac{5}{7}$.

4. A drawer contains an unorganized collection of socks. Of these, three pairs are red, two pairs are white, and five pairs are black.

(a) If one sock is drawn at random from the drawer, what is the probability it is a red sock?

(b) Once a sock is drawn and discovered to be red, what is the probability of drawing another red sock to make a matching pair?

Answer: medium

(a) Let E be the event of drawing a red sock. Then $P(E) = \frac{n(E)}{n(S)} = \frac{6}{20} = \frac{3}{10}$.

(b) Let E be the event of drawing another red sock. Since there are 19 socks left of which 5 are red, $P(E) = \frac{n(E)}{n(S)} = \frac{5}{19}$.

5. A child's game has a spinner which consists of a circle divided into eleven equal wedges, with the wedges numbered 1 through 11, and an arrow which spins around the center of the circle. Find the probability that the spinner stops at

(a) an even number.

(b) an odd number or a number greater than 4.

Answer: medium

(a) Let E be the event that the spinner stops at an even number. Then $P(E) = \frac{5}{11}$.

(b) Let E be the event that the spinner stops at an odd number or a number greater than 4. Since there are 7 numbers greater than 4 and 6 odd numbers of which 2 are less than or equal to 4, then there are 9 numbers which either are greater than 4 or odd. Therefore $P(E) = \frac{n(E)}{n(S)} = \frac{9}{11}$.

6. A poker hand consisting of five cards is dealt from a deck of 52 cards. Find, to significant digit, the probability of all five cards being spades.

(a) 0.00002 (b) 0.0003 (c) 0.0005 (d) 0.0008 (e) 0.02

Answer: medium (c)

Let E be the event of dealing 5 spades. Then $P(E) = \frac{C(13,5)}{C(52,5)} \approx 0.0005$.

7. A poker hand consisting of five cards is dealt from a deck of 52 cards. Find, to two significant digits, the probability that the hand contains no face cards.

(a) 0.0065 (b) 0.034 (c) 0.11 (d) 0.25 (e) 0.36

Answer: medium (d)

Let E be the event of dealing no face cards among the 5 cards. Then $P(E) = \frac{C(40,5)}{C(52,5)} \approx 0.25$.

8. A poker hand consisting of five cards is dealt from a deck of 52 cards. Find, to three decimal places, the probability of a royal flush (ace, king, queen, jack, and 10 of the same suit).

(a) 2.257×10^{-7} (b) 3.166×10^{-7} (c) 5.225×10^{-7} (d) 9.007×10^{-7} (e) 1.539×10^{-6}

Answer: medium (e)

Let E be the event of dealing a royal flush. Since there are 4 such hands, $P(E) = \dfrac{4}{C(52,5)} \approx 1.539 \times 10^{-6}$.

9. A poker hand consisting of five cards is dealt from a deck of 52 cards. Find, to two decimal places, the probability that the hand contains at least one king.

(a) 0.22 (b) 0.34 (c) 0.44 (d) 0.46 (e) 0.54

Answer: medium (b)

Let E be the event of dealing at least one king. Then E' is the event of dealing no kings. Then $P(E) = 1 - P(E') = 1 - \dfrac{C(48,5)}{C(52,5)} \approx 1 - 0.65884 = 0.34116$.

10. A pair of dice is rolled and the numbers showing are observed.

(a) List the sample space of this experiment.

(b) Find the probability of getting a sum of 8.

(c) Find the probability of getting a sum of 10.

(d) Find the probability that the two dice show the same number.

(e) Find the probability that the two dice show different numbers.

(f) Find the probability of getting a sum of 10 or higher.

Answer: medium

(a) The sample space is

$$\begin{aligned}&\{(1,1), (1,2), (1,3), (1,4), (1,5), (1,6),\\ &(2,1), (2,2), (2,3), (2,4), (2,5), (2,6),\\ &(3,1), (3,2), (3,3), (3,4), (3,5), (3,6),\\ &(4,1), (4,2), (4,3), (4,4), (4,5), (4,6),\\ &(5,1), (5,2), (5,3), (5,4), (5,5), (5,6),\\ &(6,1), (6,2), (6,3), (6,4), (6,5), (6,6)\}.\end{aligned}$$

(b) Let E be the event of getting a sum of 8. Then $P(E) = \frac{5}{36}$.

(c) Let E be the event of getting a sum of 10. Then $P(E) = \frac{3}{36} = \frac{1}{12}$.

(d) Let E be the event that the two dice show the same number. Then $P(E) = \frac{6}{36} = \frac{1}{6}$.

(e) Let E be the event that the two dice show different numbers. Then E' is the event that the two dice show the same number. Thus $P(E) = 1 - P(E') = 1 - \frac{1}{6} = \frac{5}{6}$.

(f) Let E be the event of getting a sum of 10 or higher. Then $P(E) = \frac{6}{36} = \frac{1}{6}$.

11. What is the probability that a 13-card bridge hand consists of all cards from the same suit?

(a) 6.29908×10^{-12} (b) 8.41884×10^{-12} (c) 2.38657×10^{-11} (d) 5.28765×10^{-11} (e) 0.22

Answer: medium (a)

Let E be the event that a 13-card bridge hand consists of all cards from the same suit. Since there are exactly 4 such hands, $P(E) = \dfrac{4}{C(52,13)} \approx 6.29908 \times 10^{-12}$.

12. A toddler has wooden blocks with the following letters on them: A, E, L, P, T. What is the probability that the child

(a) arranges the letters to spell the word PLATE?

(b) arranges the letters in alphabetical order?

Answer: medium

(a) Let E be the event that the toddler arranges the word PLATE. Since there are $P(5,5)$ ways of arranging the blocks of which only one spells the word plate, $P(E) = \dfrac{1}{P(5,5)} = \dfrac{1}{120}$.

(b) Let E be the event that the toddler arranges the letters in alphabetical order. Since there are $P(5,5)$ ways of arranging the blocks of which only one is in alphabetical order, $P(E) = \dfrac{1}{P(5,5)} = \dfrac{1}{120}$.

13. The California Lotto game used to be a $6/49$ lottery game where a player picks six numbers from 1 to 49. In order to have fewer winners and so increase the size of the jackpots from draw to draw (and hence also increase interest in the game) the lottery officials decided to change the lottery to a $6/59$ game. What is the probability of winning this new game?

(a) 1.41×10^{-9} (b) 6.23×10^{-9} (c) 1.04×10^{-8} (d) 2.22×10^{-8} (e) 3.02×10^{-8}

Answer: medium (d)

Let E be the event of picking the 6 winning numbers. Then $P(E) = \dfrac{1}{C(59,6)} \approx 2.22 \times 10^{-8}$.

14. The president of a large company selects a group of six of his employees for a special bonus. He claims that the group is randomly chosen from among the 40 employees of which 24 are women and 16 are men. What is the probability that no women are chosen?

(a) 0.000340 (b) 0.00209 (c) 0.00558 (d) 0.011 (e) 0.028

Answer: medium (b)

Let E be the event that no women are chosen. Since the number of ways that no women are chosen is the same as the number of ways that all men are chosen which is $C(16,6)$. Thus $P(E) = \dfrac{C(16,6)}{C(40,6)} \approx 0.00209$.

15. To control the quality of their product, the Bright-Light company inspects four light bulbs out of each batch of ten bulbs. If a defective bulb is found, the batch is discarded. Suppose that a batch actually contains two defective bulbs. What is the probability that it will be discarded?

(a) $\frac{2}{9}$ (b) $\frac{4}{9}$ (c) $\frac{2}{3}$ (d) $\frac{7}{9}$ (e) $\frac{8}{9}$

Answer: hard (c)

Let E be the event that the batch will be discarded. Thus E is the event that at least one defective bulb is found. It is easier to find E', the event that no defective bulbs are found. Since there 10 bulbs in the batch of which 8 are nondefective bulbs, $P(E') = \dfrac{C(8,4)}{C(10,4)}$. Thus $P(E) = 1 - P(E') = 1 - \dfrac{C(8,4)}{C(10,4)} = 1 - \frac{1}{3} = \frac{2}{3}$.

16. Thirty students are arranged randomly in a row for a class picture. The class has 18 boys and 12 girls. What is the probability that all the boys will be standing together and all the girls together?

Answer: hard

Let E be the event that all the boys are standing together and all the girls together. If we consider all 18 boys as one object and all 12 girls as one object, then there are two ways to arrange these 2 objects. Since there are 18! possible ways to arrange 18 boys and 12! possible ways to arrange 12 girls, thus there are $2 \cdot 18! \cdot 12!$ possible ways to arrange 18 boys and 12 girls in which all the boys are standing together and all the girls are standing together. Therefore $P(E) = \dfrac{2 \cdot 18! \cdot 12!}{30!} \approx 2.31 \times 10^{-8}$.

17. A monkey is trained to arrange wooden blocks in a straight line. He is then given blocks with the letters I, O, L, R, S, T, and U. What is the probability that he will arrange them to spell the word TROILUS?

Answer: medium

Let E be the event that the monkey will arrange the 7 blocks to spell "TROILUS". Then $P(E) = \dfrac{1}{7!} = \dfrac{1}{5040}$.

18. An experiment consists of rolling a die.
    (a) Find the sample space.
    (b) Find the probability of getting an even number.
    (c) Find the probability of getting a number greater than 4.
    Answer: medium
    (a) $\{1, 2, 3, 4, 5, 6\}$
    (b) E: Number is even $\Rightarrow$ $P(E) = \frac{3}{6} = \frac{1}{2}$
    (c) G: Number is greater than 4 $\Rightarrow$ $P(G) = \frac{2}{6} = \frac{1}{3}$.

19. In a class of nine students, the final grades were as follows: Two boys earned A's, three girls earned A's, three boys earned C's and one girl earned an F. If a student is chosen at random, what is the probability that the student is a boy who earned an A or the girl that earned an F?
    Answer: medium
    Let E be the boy that earned an A, and F be the girl that earned an F. $P(E \cup F) = P(E) + P(F) = \frac{2}{9} + \frac{1}{9} = \frac{3}{9} = \frac{1}{3}$.

20. If a card is chosen at random from a standard deck. What is the probability that the card is either less than 5 or a spade?
    Answer: hard
    E: The card is less than 5; F: The card is a spade. $P(E) = \frac{16}{52} = \frac{4}{13}$ and $P(F) = \frac{13}{52} = \frac{1}{4}$. There are 4 cards that are both less than 5 and spades. Thus $P(E \cup F) = \frac{4}{13} + \frac{1}{4} - \frac{4}{52} = \frac{25}{52} \approx 0.481$.

21. A couple decides to have three children. If having a boy is equally likely to having a girl, what is the probability of the couple having a boy, then a girl, then another boy?
    Answer: hard
    B: Child is a boy; G: Child is a girl. $P(BGB) = \frac{1}{2} \cdot \frac{1}{4} \cdot \frac{1}{8} = \frac{1}{64}$

SECTION 13.4

1. Jane gets \$8 if a die shows a 6 and loses \$1 otherwise. What is her expectation?
   (a) \$0.33 (b) \$0.50 (c) \$0.62 (d) \$0.80 (e) \$1.00
   Answer: easy (b)
   Jane gets \$8 with probability $\frac{1}{6}$ and loses \$1 with probability $\frac{5}{6}$. Thus $E = (8)\left(\frac{1}{6}\right) + (-1)\left(\frac{5}{6}\right) = \frac{1}{2}$, and so her expectation is \$0.50.

2. Tim gets \$2 if a coin shows up heads and \$1 if it shows up tails. What is his expected value?
   (a) \$1.00 (b) \$1.05 (c) \$1.25 (d) \$1.32 (e) \$1.50
   Answer: easy (e)
   The expected value of this game is $E = (2)\left(\frac{1}{2}\right) + (1)\left(\frac{1}{2}\right) = \frac{3}{2} = 1.5$. So Tim's expected value is \$1.50 per game.

3. A card is drawn from a deck. You get \$52 if the card is an ace, \$13 if it is a face card, and \$26 if it is the 8 of clubs. What is your expectation?
   (a) \$7.50 (b) \$8.25 (c) \$9.00 (d) \$11.45 (e) \$22.50
   Answer: easy (a)
   The expected value of this game is $E = (52)\left(\frac{4}{52}\right) + (13)\left(\frac{12}{52}\right) + (26)\left(\frac{1}{52}\right) = \$7.50$.

4. In the game of roulette as played in Monaco, the wheel has 33 slots. One slots is numbered 0, and the rest are numbered from 1 to 32. A \$1 bet on any number other than 0 wins \$32 (\$31 plus the \$1 bet). Find the expected value of this game.
   (a) $\$\left(-\frac{1}{31}\right)$ (b) $\$\left(-\frac{1}{32}\right)$ (c) $\$\left(-\frac{1}{33}\right)$ (d) \$0 (e) $\$\,\frac{1}{32}$
   Answer: easy (c)
   The expected value of this game is $E = (31)\left(\frac{1}{33}\right) + (-1)\left(\frac{32}{33}\right) = \$\left(-\frac{1}{33}\right)$.

5. A sweepstakes contest offers a first prize of one million dollars, a second prize of $200,000, and a third prize of $40,000. Suppose that three million people enter the contest and three names are selected randomly for the three prizes.
(a) What are the expected winnings of a person participating in this contest?
(b) Is it worth paying $0.50 to enter this contest?
Answer: medium

(a) $P(\text{winning the first prize}) = \dfrac{1}{3\times 10^6}$. After the first prize winner is selected, then $P(\text{winning the second prize}) = \dfrac{1}{3\times 10^6 - 1}$. Similarly, $P(\text{winning the third prize}) = \dfrac{1}{3\times 10^6 - 2}$. So the expected value of this game is $E = (10^6)\left(\dfrac{1}{3\times 10^6}\right) + (2\times 10^5)\left(\dfrac{1}{3\times 10^6 - 1}\right) + (4\times 10^4)\left(\dfrac{1}{3\times 10^6 - 2}\right) \approx \$0.41333$.

(b) Since we expect to win $0.41333 on average per game, if we pay $0.50, then our net outcome is a loss of $0.08667 per game. Hence it is not worth playing.

6. A safe contains one billion dollars. You pay one dollar to guess once at the 9-digit combination of the lock. If you open the lock, you get to keep the billion dollars. What is your expectation?
(a) $(−0.05) (b) 0 (c) $0.50 (d) $1.00 (e) $1,000,000
Answer: medium (b)
The expected value of this game is $E = (10^9 - 1)\left(\dfrac{1}{10^9}\right) + (-1)\left(\dfrac{10^9 - 1}{10^9}\right) = 0$.

7. A slot machine has three wheels. Each wheel has the ten digits 0, 1, 2, … , 9 and the picture of a watermelon. When a quarter is placed in the machine and the handle is pulled, the three wheels spin independently and come to rest. Getting three watermelons pays $10; otherwise, nothing is paid. What is the expected value of this game?
Answer: medium
The expected value of this game is $E = (9.75)\left(\dfrac{1}{11^3}\right) + (-0.25)\left(1 - \dfrac{1}{11^3}\right) \approx \$(-0.2425)$.

8. In the 7/49 lottery game a player pays $5 and picks seven numbers from 1 to 49. Any player who picks the seven winning numbers wins ten million dollars. Assuming this is the only way to win, what is the expected value of this game?
Answer: medium
The expected value of this game is $E = (10^7 - 5)\left[\dfrac{1}{C(49,7)}\right] + (-5)\left(1 - \dfrac{1}{C(49,7)}\right) \approx \$(-4.88)$.

9. A game consists of picking a card from a deck. You get $6.50 if you pick an ace. What is a "fair price" to pay to play this game? A fair price means a price at which the player can be expected to break even if he plays the game many times; in other words, the price at which his expectation is 0.
Answer: medium
Let x be the fair price to pay to play this game. Then $E = 0 \quad\Leftrightarrow\quad (6.50 - x)\left(\frac{4}{52}\right) + (-x)\left(\frac{48}{52}\right) = 0 \quad\Leftrightarrow\quad 26 - 52x = 0 \quad\Leftrightarrow\quad x = 0.5$ Thus, a fair price to play this game is $0.50.

10. A bag contains six white balls and four black balls. If three balls are drawn at random from the bag, how many white balls should we expect to get?
Answer: medium
We get one white ball with probability $\dfrac{C(6,1)\cdot C(4,2)}{C(10,3)}$. We get two white balls with probability $\dfrac{C(6,2)\cdot C(4,1)}{C(10,3)}$. We get three white balls with probability $\dfrac{C(6,3)}{C(10,3)}$. Thus the expected number of white balls is
$E = 1\cdot\dfrac{C(6,1)\cdot C(4,2)}{C(10,3)} + 2\cdot\dfrac{C(6,2)\cdot C(4,1)}{C(10,3)} + 3\cdot\dfrac{C(6,3)}{C(10,3)} = \dfrac{9}{5}$. Thus we expect to get $\frac{9}{5}$ white balls, on average.

11. A fair coin is tossed, and Lucy receives \$1 if it is heads, and must pay \$1.25 if it is tails. What is Lucy's expectation?
Answer: medium
$E = \frac{1}{2} \cdot 1 - \frac{1}{2}(1.25) = \frac{1}{2} - 0.625 = -0.125$. If Lucy plays this game many times, she would expect to lose about 13 cents per game.

12. A die is rolled, and Steve wins \$1.50 if the number is 3 or less and must pay \$1.25 if the number is 4 or greater. Find Steve's expectation.
Answer: medium
$E = \frac{1}{2} \cdot (1.50) - \frac{1}{2} \cdot (1.25) = 0.125$. If Steve plays this game many times, Steve could expect to win \$0.13 on average.

13. A game consists of drawing a card from a standard 52-card deck. You win \$24 if you draw an ace. What is a "fair price" to pay to play this game?
Answer: medium
$0 = 24 \cdot \frac{4}{52} - x \cdot \frac{48}{52} \quad \Leftrightarrow \quad -x \cdot \frac{48}{52} = -24 \cdot \frac{4}{52} \quad \Leftrightarrow \quad x = 2$. If the game is to be played many times, the player can pay \$2 per game in order to break even.

14. A bag contains 12 objects, 4 fifty-cent coins and 8 slugs. By paying 25 cents you can reach into the bag and take one object out, which you get to keep. Is 25 cents a fair price to pay for this game?
Answer: medium
Yes. The fair price is given by $0 = 0.50\left(\frac{4}{12}\right) - x\left(\frac{8}{12}\right) \quad \Rightarrow \quad -0.50\left(\frac{4}{12}\right) = -x\left(\frac{8}{12}\right) \quad \Leftrightarrow \quad x = \$0.25$.

CHAPTER 13 REVIEW

1. How many four-digit numbers can be formed using the digits 1, 2, 3, 4, 5, 6, and 7 if a given digit can be used
(a) only once?
(b) any number of times?
Answer: easy

(a) The number of 4-digit numbers that can be formed using the digits 1-7 if a digit can be used only once is $(7)(6)(5)(4) = 840$.

(b) The number of 4-digit numbers that can be formed using the digits 1-7 if a digit can be used any number of times is $(7)(7)(7)(7) = 2401$.

2. A pizza parlor advertises that they prepare 1024 different types of pizza. How many toppings does this parlor offer?
(a) 10 (b) 12 (c) 15 (d) 64 (e) 1024
Answer: medium (a)
Since $2^{10} = 1024$, the parlor offers 10 toppings.

3. A group of students determines that there are 720 different ways that they can stand in a row for their class picture. How many students are there in this class?
(a) 4 (b) 6 (c) 8 (d) 10 (e) 24
Answer: easy (b)
Since $6! = 720$, there are 6 students in this class.

4. An airline company overbooks a particular flight and six passengers are "bumped" from the flight. If there are 120 passengers booked on this flight, in how many ways can the six passengers to be bumped be chosen?
(a) 1,035,335,480 (b) 2,247,406,336 (c) 2,462,486,840 (d) 2,965,300,020 (e) 3,652,745,460
Answer: easy (e)
The number of ways that 6 passengers can be bumped is $C(120, 6) = 3{,}652{,}745{,}460$

5. A test has twelve true-false questions and four multiple choice questions with five choices each. In how many ways can this test be completed?

(a) $1,894,560$ (b) $2,560,000$ (c) $2,808,330$ (d) $6,225,480$ (e) $12,147,420$

Answer: easy (b)

There are 2^{12} ways to answer 12 true-false questions and 5^4 ways to answer 4 multiple choice questions with 5 choices for each. So the number of ways that this test can be completed is $(2^{12})(5^4) = 2,560,000$.

6. An ice-cream store offers 14 flavors of ice cream. The store specialty is a banana split with four scoops of ice cream. If each scoop must be a different flavor, how many such banana splits are possible?

(a) 840 (b) 901 (c) 912 (d) 1001 (e) 1420

Answer: easy (d)

The number of ways to have a banana split is $C(14, 4) = 1001$.

7. The Yukon Territory in Canada uses a license plate system for its cars that consists of two letters followed by three numbers. Explain how we can know that there are fewer than $700,000$ licensed cars in the Yukon.

Answer: medium

The number of ways to form a license plate consisting of 2 letters followed by 3 numbers $= (26)(26)(10)(10)(10) = 676,000$. Since there are fewer possible license plates than $700,000$, there must be fewer than $700,000$ licensed cars in the Yukon.

8. (a) How many four-digit numbers can be formed using the digits $0, 1, \ldots, 9$? (Remember, a three-digit number cannot have a 0 as the leftmost digit.)

(b) If a number is chosen randomly from the set $\{0, 1, 2, \ldots, 10000\}$, what is the probability that the number chosen is a four-digit number?

Answer: medium

(a) Since the leftmost digit of a 4-digit number cannot be zero, the number of ways to form a 4-digit number from 0-9 is $(9)(10)(10)(10) = 9000$.

(b) There are 10001 numbers from 0-10000. From part (a), there are 9000 4-digit numbers. Therefore the probability that the number chosen is a 4-digit number is $P(E) = \frac{9000}{10001} \approx 0.8999$.

9. An anagram of a word is a permutation of the letters of that word. For example, anagrams of the word ARGUE include GUEAR, RUAGE, and GRUAE. How many anagrams of the word ARGUE are there in which

(a) the vowels are in alphabetical order?

(b) all the vowels are together?

(c) all the letters are in alphabetical order?

Answer: medium

(a) The number of anagrams of the word ARGUE is 5!. The vowels in each anagram can be permuted within their same three positions in 3! ways. Only one of these ways has the vowels in alphabetical order. Thus the number of anagrams of the word ARGUE in which the vowels are in alphabetical order is $\frac{5!}{3!} = 20$.

(b) The number of anagrams of the word ARGUE in which all the vowels are together is $P(3,3) \cdot P(3,3) = 36$.

(c) The number of anagrams of the word ARGUE in which all the letters are in alphabetical order is 1.

10. An anagram of a word is a permutation of the letters of that word. For example, anagrams of the word REGENT include GERENT, GREENT, and TERGEN. How many anagrams are there of the word MISSOURI in which all the vowels are together?

(a) 12 (b) 108 (c) 144 (d) 288 (e) 341

Answer: medium (c)

Imagine the four vowels of the word MISSOURI tied together as one object, but rearrangeable in $\frac{4!}{2!} = 12$ distinguishable ways. Each of these groups of vowels can be arranged with the other 4 letters, two of which are indistinguishable (the S's). Thus the number of anagrams in which all of the vowels are together is $12 \cdot \frac{4!}{2!} = 144$.

11. The genetic code is based on the four nucleotides adenine (abbreviated A), cytosine (C), guanine (G), and thymine (T). These are connected together in long strings to form DNA molecules. For example, a sequence in the DNA may look like CAGTGGTACC.... The code uses "words", all of the same length, that are composed of these nucleotides ("letters"). It is known that there are at least 20 different words. What is the minimum length of word needed?

(a) 2 (b) 3 (c) 4 (d) 5 (e) 8

Answer: medium (b)

There are $4^2 = 16$ different words of length 2 which is less than 20. Since 3 is the next number and $4^3 = 64$, 3 is the minimum length of word needed.

12. An urn contains eight red balls labeled $0, 1, 2, \ldots, 7$ and five white balls labeled $0, 1, 2, 3, 4$. What is the probability that a ball drawn from the urn is

(a) red? (b) even-numbered?
(c) white and odd-numbered? (d) red or odd-numbered?

Answer: easy

(a) The probability that the ball is red is $\frac{8}{13}$.
(b) The probability that the ball is even-numbered is $\frac{7}{13}$.
(c) The probability that the ball is white and odd-numbered is $\frac{2}{13}$.
(d) The probability that the ball is red or odd-numbered is $\frac{10}{13}$.

13. Three balls are drawn one at a time and without replacement from an urn that contains seven white and five black balls. What is the probability that

(a) the first is white, the second black, and the third black?
(b) the first is black, the second white, and the third black?
(c) the first is black, the second black, and the third white?
(d) exactly one is white?

Answer: medium

(a) The probability the first is white, the second is black, and the third is black is $\frac{7}{12} \cdot \frac{5}{11} \cdot \frac{4}{10} = \frac{7}{66}$.

(b) The probability the first is black, the second is white, and the third is black is $\frac{5}{12} \cdot \frac{7}{11} \cdot \frac{4}{10} = \frac{7}{66}$.

(c) The probability the first is black, the second is black, and the third is white is $\frac{5}{12} \cdot \frac{4}{11} \cdot \frac{7}{10} = \frac{7}{66}$.

(d) We give two solutions.

Solution 1: By parts (a), (b), and (c),

$$\begin{aligned} P(\text{exactly one white ball}) &= P(\text{first white, others black } \textit{or} \text{ second white, others black } \textit{or} \text{ third white, others black}) \\ &= P(\text{first white, others black}) + P(\text{second white, others black}) + P(\text{third white, others black}) \\ &= \frac{7}{66} + \frac{7}{66} + \frac{7}{66} = \frac{21}{66} = \frac{7}{22} \end{aligned}$$

Solution 2: We can think of drawing all three balls at once. The number of ways of doing this is $C(12, 3)$. To get exactly one white ball we must get exactly two black balls and the number of ways of doing this is $C(7, 1)C(5, 2)$. Thus

$$P(\text{exactly one white ball}) = \frac{C(7,1)C(5,2)}{C(12,3)} = \frac{7}{22}.$$

14. A card is drawn from a deck, a die is rolled, and a coin is tossed. What is the probability of getting

(a) the ace of spades, a 1, and a tail?
(b) a spade, a 1, and a tail?
(c) a face card, a number greater than 2, and a tail?

Answer: medium

(a) The probability of getting the ace of spades, a 1, and a tail is $\frac{1}{52} \cdot \frac{1}{6} \cdot \frac{1}{2} = \frac{1}{624}$.

(b) The probability of getting a spade, a 1, and a tail is $\frac{13}{52} \cdot \frac{1}{6} \cdot \frac{1}{2} = \frac{1}{48}$.

(c) The probability of getting a face card, a number greater than 2, and a tail is $\frac{12}{52} \cdot \frac{4}{6} \cdot \frac{1}{2} = \frac{1}{13}$

15. Two dice are rolled. John gets \$6 if they all show the same number and pays \$1 if they show different numbers. What is his expectation?

(a) $\frac{1}{36}$ (b) $\frac{1}{18}$ (c) $\frac{3}{18}$ (d) $\frac{1}{6}$ (e) $\frac{1}{2}$

Answer: medium (d)

There are 36 possible outcomes in rolling two dice and 6 ways in which two dice show the same numbers, namely, $(1,1)$, $(2,2)$, $(3,3)$, $(4,4)$, $(5,5)$, and $(6,6)$. So the expected value of this game is $E = (6)\left(\frac{6}{36}\right) + (-1)\left(\frac{30}{36}\right) = \frac{1}{6}$.

16. Three dice are rolled. What is the probability of getting

(a) a total of 5?

(b) a total of less than 6?

Answer: medium

(a) There are $6^3 = 216$ possible outcomes in rolling 3 dice. Since there are 6 possible ways for 3 dice to show a total of 5, that is, $(1,1,3)$, $(1,2,2)$, $(1,3,1)$, $(2,1,2)$, $(2,2,1)$, and $(3,1,1)$, the probability of getting a total of 5 is $\frac{6}{216} = \frac{1}{36}$.

(b) Since there are 10 possible ways for 3 dice to show a total of less than 6, namely, the combinations listed in part (a) as well as $(1,1,1)$, $(1,1,2)$, $(1,2,1)$, and $(2,1,1)$, the probability of getting a total of less than 6 is $\frac{10}{216} \approx 0.046296$.

17. A pizza parlor offers 10 different toppings, one of which is anchovies. If a pizza is ordered at random, what is the probability that it has anchovies as one of the toppings?

(a) $\frac{1}{10}$ (b) $\frac{1}{8}$ (c) $\frac{1}{3}$ (d) $\frac{2}{5}$ (e) $\frac{1}{2}$

Answer: medium (e)

The number of ways a pizza can have anchovies is the number of subsets of the other 9 items, that is 2^9. (You can think of this as follows: you first put anchovies on the pizza then you choose one subset of the other 9 items.) The total number of different pizzas is the number of subsets of the 10 items, that is 2^{10}. Thus $P(\text{anchovies}) = \dfrac{2^9}{2^{10}} = \dfrac{1}{2}$.

Note: A probability of $\frac{1}{2}$ makes intuitive sense — you either get anchovies or you don't.

18. (a) A volleyball team has nine players. In how many ways can a starting lineup be chosen consisting of two forward players and three defense players?

(b) Use a counting argument to explain why $C(n,2) \cdot C(n-2,3) = C(n,3) \cdot C(n-3,2)$.

Answer: medium

(a) We give two solutions.

Solution 1: First choose the two forwards; there are $C(9,2)$ ways of doing this. Then choose the 3 defenders from the remaining 7 players; there are $C(7,3)$ ways of doing this. Thus the number of ways of choosing a starting lineup is $C(9,2) \cdot C(7,3) = 1260$.

Solution 2: First choose the 3 defensive players; there are $C(9,3)$ ways of doing this. Then choose the 2 forwards from the remaining 6 players; there are $C(6,2)$ ways of doing this. Thus the number of ways of choosing a starting lineup is $C(9,3) \cdot C(6,2) = 1260$

(b) The number of ways of choosing 2 objects from n objects, then choosing 3 objects from the remaining objects is the same as first choosing 3 objects followed by choosing 2 objects from the remaining ones. We can see this by examining part (a). Thus $C(n,2) \cdot C(n-2,3) = C(n,3) \cdot C(n-3,2)$.

19. (a) How many factors are there of the expression $a^7b^3c^4$?

(b) How many different ways can the expression $a^7b^3c^4$ be written without using exponents?

Answer: medium

(a) Using the rule for the number of distinguishable combinations, the number of factors of the expression $a^7b^3c^4$ is $(7+1)(3+1)(4+1) = 160$.

(b) The number of ways that the expression $a^7b^3c^4$ is written without using exponents is the number of distinguishable permutations of 14 digits of which there are groups of 7, 3, and 4 indistinguishable digits, which is $\dfrac{14!}{7!\,3!\,4!} = 120{,}120$.

20. In the U.S. Senate there are two senators from each of the 50 states. In how many ways can a committee of six senators be chosen if no state is to have two members on the committee?

Answer: medium

$$\begin{pmatrix}\text{number}\\ \text{of}\\ \text{committees}\end{pmatrix} = \begin{pmatrix}\text{number of}\\ \text{ways to choose}\\ \text{6 states}\end{pmatrix} \cdot \begin{pmatrix}\text{number of ways}\\ \text{to choose the senators}\\ \text{from these states}\end{pmatrix} = C(50,6) \times 2^6 = 1{,}017{,}004{,}800$$